TECHNOLOGY AND VALUES

T0206148

EDITED BY CRAIG HANKS

TECHNOLOGY
AND VALUES
ESSENTIAL READINGS

⟨W⟩WILEY-BLACKWELL

A John Wiley & Sons, Ltd., Publication

This edition first published 2010
Editorial material and organization © 2010 Blackwell Publishing Ltd

Blackwell Publishing was acquired by John Wiley & Sons in February 2007. Blackwell's publishing program has been merged with Wiley's global Scientific, Technical, and Medical business to form Wiley-Blackwell.

Registered Office
John Wiley & Sons Ltd, The Atrium, Southern Gate, Chichester, West Sussex, PO19 8SQ, United Kingdom

Editorial Offices
350 Main Street, Malden, MA 02148-5020, USA
9600 Garsington Road, Oxford, OX4 2DQ, UK
The Atrium, Southern Gate, Chichester, West Sussex, PO19 8SQ, UK

For details of our global editorial offices, for customer services, and for information about how to apply for permission to reuse the copyright material in this book please see our website at www.wiley.com/wiley-blackwell.

The right of Craig Hanks to be identified as the author of the editorial material in this work has been asserted in accordance with the Copyright, Designs and Patents Act 1988.

Library of Congress Cataloging-in-Publication Data

Technology and values / edited by Craig Hanks.
    p. cm. — (Essential readings)
Includes bibliographical references and index.
ISBN 978-1-4051-4900-6 (hardcover : alk. paper) — ISBN 978-1-4051-4901-3 (pbk. : alk. paper)
1. Technology—Moral and ethical aspects. I. Hanks, Craig.

BJ59.T44 2010
601—dc22

2008054060

A catalogue record for this book is available from the British Library.

Set in 9.5/11.5pt Minion by Graphicraft Limited, Hong Kong

Dedicated to Larry, John, and Rick
Teachers, Mentors, Friends

Contents

Figures

Acknowledgments

First thanks go to David Kaplan and Andrew Light for encouraging and advising this project from the beginning. My departmental colleagues at Texas State University-San Marcos and Stevens Institute of Technology provided supportive and intellectually lively work environments and insightful conversations about teaching. Texas State University provided support through a Research Enhancement Grant and Faculty Development Leave, along with travel funds. The staff of the Alkek Library at Texas State helped me track down materials, and a generous Library Research Grant allowed me to select many of the works excerpted here for our campus collection. My editor, Jeff Dean, provided advice throughout. He, along with Lindsay Pullen, Jamie Harlan, and Danielle Descoteaux at Blackwell, patiently guided this book throughout, and special thanks are due to Tiffany Mok for assistance in the final stages. Amiable work atmosphere and good coffee were found at Zebu in Hoboken, NJ, and Arlo and Esme in New York City. Friends and colleagues in the Society for Philosophy and Technology and the Society for the Advancement of American Philosophy have long embodied the best sorts of philosophical practice and human community, as well as sympathetic and critical ears for my thoughts on technology and values. Over nearly 20 years, I have had the privilege to teach courses in Engineering Ethics, Ethics of Science and Technology, Bioethics, Environmental Ethics, Computer Ethics, and Philosophy of Technology to wonderful students at the University of Alabama in Huntsville, the Stevens Institute of Technology, and Texas State University. The selection of readings here and the impetus to pursue this project owe much to them. Most importantly, gratitude is due to María Inmaculada de Melo-Martín, for endless conversations about philosophy and life, and from whom I have learned more than I can say.

Craig Hanks

Source Acknowledgments

The editor and publisher gratefully acknowledge the permission granted to reproduce the copyright material in this book:

Chapter 1: From *The Hastings Center Report* 9, no. 1 (1979): 40–59. Reprinted by permission of The Hastings Center. Reprinted by permission of Eleanore Jonas.

Chapter 2: From *Philosophy Today* (Summer 1982): 103–17.

Chapter 3: From W. W. Lowrance, *Modern Science and Human Values* (New York: Oxford University Press, 1986), pp. 145–50. By permission of Oxford University Press, Inc.

Chapter 4: From Bruno Latour, *Pandora's Hope: Essays on the Reality of Science Studies* (Cambridge, MA: Harvard University Press, 1999), pp. 198–215. Reprinted by permission of the publisher. Copyright © 1999 by the President and Fellows of Harvard College.

Chapter 5: Originally: "Technology," in Lawrence C. Becker and Charlotte B. Becker, eds., *Encyclopedia of Ethics*, vol. 2 (New York: Garland Publishing, 1992), pp. 1231–4. © 1992. Reproduced by permission of Routledge, Inc., a division of Informa plc.

Chapter 6: In *The Technological Society*, trans. John Wilkinson (New York: Alfred A. Knopf, 1964), pp. 333–43. Translation copyright © 1964 by Alfred A. Knopf, a division of Random House, Inc. Used by permission of Alfred A. Knopf, a division of Random House, Inc.

Chapter 7: From Langdon Winner, *Autonomous Technology: Technics-out-of-control as a Theme in Political Thought* (Cambridge, MA: MIT Press, 1977), pp. 191–209. © 1977 Massachusetts Institute of Technology, by permission of MIT Press.

Chapter 8: From Joseph Pitt, *Thinking about Technology: Foundations of Philosophy of Technology* (New York: Seven Bridges Press, 2000), pp. 87–99.

Chapter 9: From Martin Heidegger, *The Question Concerning Technology and Other Essays*, trans. William Lovitt (New York: Harper & Row, 1977), pp. 287–317. English language translation © 1977 by Harper & Row, Publishers, Inc. Reprinted by permission of HarperCollins Publishers.

Chapter 10: From José Ortega y Gasset, *History as a System and Other Essays Toward a Philosophy of History* (New York: W. W. Norton & Company), pp. 109–55. © 1941, 1961, by W. W. Norton & Company, Inc. Used by permission of W. W. Norton & Company Inc.

Chapter 11: From Albert Borgmann, *Technology and the Character of Contemporary Life* (Chicago: University of Chicago Press, 1984), pp. 196–210. Reprinted by permission of the publisher, The University of Chicago Press, and by Albert Borgmann.

Chapter 12: From Don Ihde, *Technology and the Lifeworld* (Bloomington: Indiana University Press, 1990), pp. 72–100, 105–12. Reprinted by permission of the publisher, Indiana University Press.

Chapter 13: From Herbert Marcuse, *One Dimensional Man* (Boston: Beacon Press, 1964), pp. 1–18.

Chapter 14: From Jürgen Habermas, *Toward a Rational Society*, trans. Jeremy Shapiro (Boston: Beacon Press, 1970), pp. 50–61.

Chapter 15: From Andrew Feenberg, *Transforming Technology: A Critical Theory Revisited* (Oxford: Oxford University Press, 2002), pp. 162–90. By permission of Oxford University Press, Inc.

Chapter 16: From *John Dewey: The Later Works, 1925–1953*, vol. 6: *1931–1932*, ed. Jo Ann Boydston (Carbondale, IL: Southern Illinois University Press, 1985), pp. 53–63.

Chapter 17: From Larry Hickman, *Philosophical Tools for Technological Culture* (Bloomington, IN: Indiana University Press, 2001), pp. 44–64. Reprinted by permission of the publisher, Indiana University Press.

Chapter 18: From Donna Haraway, *Simians, Cyborgs and Women: The Reinvention of Nature* (New York: Routledge, 1991), pp. 149–81.

Chapter 19: From Eric Higgs, Andrew Light, and David Strong, eds., *Technology and the Good Life* (Chicago: University of Chicago Press, 2000), pp. 219–33. Reprinted by permission of the publisher, The University of Chicago Press, and by Diane Michelfelder.

Chapter 20: From John McDermott, *Streams of Experience: Reflections on History and Philosophy in the American Grain* (Amherst, MA: University of Massachusetts Press, 1986), pp. 129–40. © 1986 by The University of Massachusetts Press.

Chapter 21: From Judy Wajcman, *Feminism Confronts Technology* (University Park, PA: The Pennsylvania State University Press), pp. 81–109.

Chapter 22: In Larry A. Hickman, ed., *Technology as a Human Affair* (New York: McGraw Hill Publishing, 1990), pp. 172–7. Reprinted by permission of Douglas Browning.

Chapter 23: From *The Hastings Center Report*, vol. 33, no. 2 (March–April 2003): 19–22. Reprinted by permission of The Hastings Center. Reprinted by permission of Dan Callahan.

Chapter 24: From Laura M. Purdy, *Reproducing Persons: Issues in Feminist Bioethics* (Ithaca, NY: Cornell University Press, 1996), pp. 39–49. © 1996 by Cornell University. Used by permission of the publisher, Cornell University Press.

Chapter 25: From *The New Republic* (May 21, 2001), pp. 265–76. ©2001 by Leon R. Kass. Reprinted by permission of the author.

Chapter 27: From Nina V. Federoff and Nancy Marie Brown, eds., *Mendel in the Kitchen* (Washington DC: Joseph Henry Press, 2004), pp. 295–316.

Chapter 28: From *Plant Physiology* 132 (May 2003): 10–16. © American Society of Plant Biologists.

Chapter 29: From Lewis Mumford, *The Urban Prospect* (New York: Harcourt Brace Jovanovich, 1968), pp. 92–107.

Chapter 30: From David Schmidtz and Elizabeth Willott, eds., *Environmental Ethics: What Really Matters, What Really Works* (Oxford: Oxford University Press, 2002), pp. 426–30. By permission of Oxford University Press, Inc.

Chapter 31: From Steven Moore, *Technology and Place: Sustainable Architecture and the Blueprint Farm* (Austin, TX: The University of Texas Press, 2001), pp. 45–85.

Chapter 32: In Joseph Grange, *The City: An Urban Cosmology* (Albany, Suny Press: 1999), pp. 175–92.

Chapter 33: From *Journal of Social Philosophy*, 34 (1) (Spring 2003): 44–63. Reprinted by permission of Blackwell Publishing.

Chapter 34: From Michael Pollan, *Second Nature: A Gardener's Education* (New York: Dell Publishing, 1991), pp. 65–78.

Chapter 35: From Dale Jamieson, ed., *A Companion to Environmental Philosophy* (Malden, MA: Blackwell Publishing, 2003), pp. 339–448. Reprinted by permission of Blackwell Publishing.

Chapter 36: From J. R. Engel and J. G. Engel, eds., *Ethics of Environment and Development: Global Challenge, International Response* (London: Belhaven Press, 1990), pp. 228–37.

Chapter 37: From J. Baird Callicott, *Companion to a Sand County Almanac* (Madison: University of Wisconsin Press, 1987), pp. 186–217.

Chapter 38: From Bill Devall and George Sessions, *Deep Ecology* (Salt Lake, UT: Peregrine Smith, 1985), pp. 65–77.

Chapter 39: From *Environmental Ethics* 11 (1989): 71–83. Reprinted by permission of Ramachandra Guha.

Chapter 40: From Laura Westra and Bill E. Lawson, *Faces of Environmental Racism* (Lanham, MD: Rowman and Littlefield, 2001), pp. 57–71. Reprinted by permission of Rowman and Littlefield Publishers, Inc.

Chapter 41: From Luciano Floridi, ed., *The Blackwell Guide to Philosophy of Computing and Information Technology* (Oxford: Blackwell Publishing, 2003), pp. 327–36. Reprinted by permission of Blackwell Publishing.

Chapter 42: From Sven Birkerts, *The Gutenberg Elegies: The Fate of Reading in an Electronic Age* (New York: Fawcett Columbine: 1994), pp. 117–33.

Chapter 43: From Wendell Berry, *What are People For?* (New York: North Point Press, 2000), pp. 171–7.

Chapter 44: From Shoshana Zuboff, *In the Age of the Smart Machine* (New York: Basic Books, 1989), pp. 3–12. Reprinted by permission of Basic Books, a member of Perseus Books Group.

Chapter 45: From John Seely Brown and Paul Duguid, *The Social Life of Information* (Boston, MA: Harvard Business School Press, 2000), pp. 11–33. © President and Fellows of Harvard College. Reprinted by permission of Harvard Business School Press.

Chapter 46: From Ben Shneiderman, *Leonardo's Laptop: Human Needs and the New Computing Technologies* (Cambridge, MA: MIT Press, 2003), pp. 35–49. © 2002 Massachusetts Institute of Technology. By permission of MIT Press.

General Introduction

Craig Hanks

Technology, and the constantly new world it brings, is an abiding concern for human beings. How should we fight disease? How should we move ourselves and our goods around the planet? Can we have more technology and more democracy? How do our technologies affect our natural environment? What goods ought we produce, value, need, or desire? How do we protect ourselves? What is possible, and permissible, to do to ourselves? Do our tools set us free, or are we building our own prison? These are all questions about the relations between technology and values. The study of technology and values is a critical, reflective examination of the effects and transformation of technologies upon human activities and societies. The aim is to develop critical forms of thought that allow us to understand, evaluate, appreciate, and criticize the ways in which technologies reflect, as well as change, human life, individually, socially, and culturally.

In spite of the prevalence of technology in our society and its influence on all aspects of our lives, the systematic evaluation of technology as a bearer and transformer of values is relatively new. In "Philosophy of Technology: Retrospective and Prospective Views," Paul Durbin, one of the founders of the International Society for Philosophy and Technology (SPT) observed that "Philosophers have become interested in technology and technological problems only recently – though Karl Marx in the nineteenth century as well as Plato and Aristotle in the classical period had paid some attention either to technical work or to its social implications."

One reason for the lack of concern with technology is the traditional bias of philosophy against everyday affairs and applied issues.[1] The idea that philosophy must concern itself with universals has prevented many philosophers from critically engaging questions about technology and values. Another, perhaps more important, reason for the historical lack of critical analysis of technology has to do with the newness of technology as a clear and central dimension in human life. As José Ortega y Gassett argued, only at a certain recent point did technology as such stand out from the ongoing processes of human life, and call for self-conscious direction and evaluation. Jacques Ellul claimed that technology is the gamble of the twentieth century (and, we might add, the twenty-first). Ortega y Gassett further argued that just as technology begins to stand apart, to be sufficiently developed that it is possible to examine it in its own right, it has become a "second nature" and thus taken for granted as the background of our existence.

Fortunately, over the past century, and especially over the past 30 years, philosophers have turned their attention to examining technology as a central part of human existence. Joseph Pitt, one of the founders of the Science, Technology, and Society program at Virginia Tech, who has long argued for the philosophical importance of technology, noted that the directions we take in creating our visions of the good life, and of the future, have been a direct function of what we have thought is possible – that is, such visions are a function of our technological development. Thus, Pitt argues, if the set of technologies we command are central to our way of life and to our future, if they reflect our value system, or even if they merely affect the economic structure of our society, we need to know what this means and how it happens.

Trends in the Philosophy of Technology

This entire collection explores aspects of the philosophy of technology and what has come to be known as Science and Technology Studies (STS). There are some excellent general introductions to the field,[2] and thus my comments here will focus on some issues most relevant to our theme, the relations between technology and values.

One way to approach the field is to consider a series of important distinctions that can help us understand changes in the philosophical consideration of technology over the past 100 years. We will here consider five distinctions: between classical and contemporary philosophies of technology, between transcendental and empirical approaches, between humanities and engineering philosophies of technology, between dystopian and utopian views (between pessimists and optimists), and between the position that technology determines values and the view that technology is value-neutral. There are important overlaps between these typologies; for example, most representatives of the classical philosophy of technology take a transcendental view of technology, but a brief examination of each can help us understand and better appreciate the complexity of the debates about how best to understand ourselves and our technologized world.

Classical philosophy of technology refers to the early–mid-twentieth-century practitioners who responded in a systematic way to what Robert Hughes called "The Shock of the New" – the seemingly relentless pursuit of change through scientific and technological change. The twentieth century opened with significant hope that the powers unleashed by modern technology could be harnessed with methods of scientific rationality to

usher in a new era of peace, prosperity, and flourishing. This view was profoundly challenged by the machine-driven destruction of the First World War and by what appeared as an increasing dehumanization of work and social life.[3] Peter-Paul Verbeek has argued that classical philosophy of technology was largely transcendental in character, concerned with understanding the nature of technology as such.[4] As we shall see below, this contrasts with a contemporary turn toward a greater empirical emphasis on examining particular technologies in their social context.

Rapid technological change, and the hopes and fears it inspired lead to a bifurcated response into what Carl Mitcham has called "Engineering" and "Humanities" philosophy of technology.[5] Engineering philosophy of technology has been, in both classical and contemporary guise, largely optimistic about the prospects of a technological society. Mostly engineers and technologists, these thinkers (ranging from Ernst Kapp and Thorstein Veblen to Juan David García Bacca, Buckminster Fuller, and Mario Bunge) consider technology as central to human life, and argue that a consistent application of technology and technical principles presents our greatest hopes for improving human life. More recent examples of engineering philosophy of technology, found below in the work of Federoff and Brown, and Shneiderman, continue this emphasis on the promise of technologies through a focus on particular technologies.

Humanities philosophy of technology follows a similar trajectory from classical to contemporary, and from transcendental to empirical. Classical humanities philosophers of technology, including Martin Heidegger, Ellul, Hans Jonas, Herbert Marcuse, Ortega y Gassett, Lewis Mumford, and John Dewey, developed accounts of the nature of technology and its place in human existence. This approach produced many insights into the changes wrought by a transition to a technological culture. With the exception of Dewey, each of these thinkers develops an account of how social, historical, and cultural factors present the conditions of possibility for modern technology. Again, with the exception of Dewey, these thinkers are largely pessimistic about the changes technology brings. Drawing on the work of Marx, Sigmund Freud, and Max Weber, they find alienation an iron cage of our own making, and an increasingly autonomous sphere of technology. In this respect, many classical humanities philosophers of technology articulate some version of the thesis of technological determinism.[6] Determinists argue that technology is itself a force that transforms the nature of human thought and action, so that the interesting questions concern not what we do with technology, but what technology does with/to us.

Beginning in the 1980s, humanities philosophy of technology took an empirical turn, exemplified below in the work of Larry Hickman, Andrew Feenberg, Diane Michelfelder, and most of the authors in Section 2, "Applied Reflections on Technology and Value." These thinkers began to examine the specific processes of development and adoption of technologies, leading to philosophies of design and planning. Some, such as Pitt, argue that it makes no sense to have an account of technology as such, nor to view technology as determining the content of the rest of culture. Thus, many also focus more carefully on the roles played by specific technologies in culture and everyday life. While classical humanities philosophers of technology are rather dystopian in their outlook (again, with the exception of Dewey), contemporary empirically oriented humanities philosophers of technology present a more nuanced and complex picture, but tend to argue that technological development can be a good, but only when engaged critically throughout the process from design through disposal.[7] One aspect of more contemporary work is an attempt to understand how technology mediates experience. Because so much of popular culture presents technological advance in glowing terms, even to the point of revisiting earlier utopian tropes[8] (consider the hopes invested in human embryonic stem cell research), critical thinkers who reject dystopian views of technology can still appear as cynics about technological advance. One of our tasks in engaging these texts is to cultivate the possibilities of hopeful, but critical, engagement with our technologies and our values, or, in other words, a critical engagement with ourselves.

Technology and Values

Of course, this increasing attention to questions related to technology and values has not come accompanied by a unified response. Thus, in general the traditional answers to the question of the relations between technology and values can be grouped into three. First, some authors have argued that there is no connection whatsoever between our technologies and our values. Tools, and techniques, are value-free. The only interesting question about technologies is how people will decide to use them. Second, some scholars have seen technology as the means to obtain human freedom. Here we find optimism that, as technological capacity expands, human beings will lead lives of greater freedom and ease. Finally, some authors have portrayed technology as a growing monster of our own making that will increasingly determine the course and content of our lives and sap what remains of our humanity.

The position that technology (either technology as such, or a specific technology) is value-laden claims that embedded within the technology itself are values, that the technology carries values along with it. The position that technology is value-neutral claims that technology (and, in another version, science) is itself neutral with respect to values, and that it is only the use humans make of the technology that has any value implications. On this account, technology is purely instrumental, a mere tool.

One clear example of the value-neutral conception of science and technology can be found in the words of James Watson, one of the sources of the double-helix model of DNA and one of the leaders of the Human Genome Project: "Genetics per se can never be evil. It is only when we use or misuse it that morality comes in. That we want to find ways to lessen the impact of mental illness is inherently good. The killing by the Nazis of the German mental patients for reasons of supposed genetic inferiority, however, was barbarism at its worst."[9]

The claim here about the relation between genetic science and technology and values is that, at the very least, value questions arise only in use, and the stronger claim is that it is only in misuse that value questions arise.

Often we find that those who work creating technology adopt the view that technology is the path to improving human life, and many humanities scholars, especially within philosophy, have taken the position that technology is a threat to human values. As Carl Mitcham indicates in his *Thinking Through Technology*, technical people have often been optimists about technology, and humanists have been critics.[10] He notes that approaches to evaluations of technology from engineers, such as Samuel Florman, tend to take technological thought and action as the model for all human thought and action and attempt to explain or reformulate all apparently non-technological thought and action in technological terms.[11] On the other hand, humanities approaches to philosophy of technology tend to see technological thought and action as only one aspect or dimension of human experience, and thus seek to delimit or restrict the technological within a more expansive framework. Neither approach is wholly adequate, and a new approach that draws on both is needed.

This split in our ways of dealing with the relationships between technology and human values replicates the separation that C. P. Snow traced in *The Two Cultures* between the culture of science and technology and the culture of the arts and humanities.[12] As the twenty-first century opens, the still present disconnection between these two cultures appears more and more fruitless and the need to bring the realms of the humanities and science and technology to bear on each other becomes more crucial.

One of the goals of this book is precisely to help bridge this gap, to foster exactly this sort of dialogue, so as to gain insights into how technology influences our values and how our values shape our technologies. We will consider both optimistic and critical approaches to technology. In order to do so, we will consider the work of scholars who are known for their attempts to bring these two realms of human life, the technological and the ethical, into contact. Thus, Carl Mitcham's work bridges this distinction through a broad-ranging encounter with poetry, the artifacts of everyday life, the technological history of the twentieth century, and philosophically rich work from both the engineering and the humanistic traditions. Mitcham further argues that the insights of both traditions are necessary if we are to understand technology not merely as something we do, but as something of which we are a part.

Pessimism about technologies is found not only in what Mitcham calls "Humanities Philosophy of Technology," but such pessimism is common

fodder for popular culture and popular science. Just in the area of genetic technologies, we need only consider the horror stories of many little Hitlers (*The Boys from Brazil*), or of rigid social castes (*Gattaca*), or of policing crimes before they happen (*Minority Report*). Blogs and popular science publications raise the specter that (1) some mad-person might clone him/her self in an attempt at immortality, or attempt to create some super class of warriors; or (2) we will soon be cloning human beings to obtain their organs; or (3) people with any kind of genetic disease will be forced into sterilization, or denied health insurance or jobs because they happen to have the "wrong" kind of genes. These expressions of pessimism about what technology brings are forms of what Alan Drengson calls "technophobia."

This pessimism concerning technological change can blind us to the fact that society has made choices that have controlled technology. This is not to say that we are unaware of problematic uses of technology, but that one danger of technological pessimism is found in technological determinism and the idea that, while we must respond to the problems raised by modern technology, there is no possibility of directing technological development. However, in contrast to this view, in the second half of the twentieth century, public attitudes about technology have shifted from a positive expectation to a vocal concern about the negative or unanticipated effects of different technological achievements. The damaging effects of pesticides on humans and wildlife, the health problems and deaths caused by accidents in nuclear power plants and by radioactive waste management, the dangerous side effects of hormonal contraceptives, the widespread use of certain chemicals, and the threats of genetic research are some of the phenomena that have contributed to an increasing demand for public control over the kind of technology implemented in our society. Technology assessment, addressed by Kristin Shrader-Frechette below, is one of the more prominent responses to this demand.

Technological optimism does not seem to fare better than the catastrophist position on technology. According to this view, the advance of scientific knowledge is good because it allows us to do more things. It is also good because knowing more is directly related to our ability to understand the world and our place in it. This aids our capacity to be self-determining beings and to lead morally responsible lives. Technological advance is worthy then, because it gives us more options, make our lives easier, happier, even expands the realm of freedom. One form of technological optimism is what Drengson calls "technophilia," in which individual users come to have a relation to particular technologies that is similar to a love-affair. As with technological pessimism, technological optimism, including technophilia and even technological utopianism, is a common trope in popular culture, from celebratory publications such as WIRED magazine to the marketing of everything from automobiles to computer systems to everything Apple (iPods, iPhones, Macs).

Because technological optimism is a triumphalist viewpoint, this view requires a certain willful ignorance about the downsides of technological advance. It seems to ignore, for example, some of the legitimate fears expressed by the catastrophist view. Evidently, in neglecting to consider them (i.e., discrimination based on genetic records, global warming, threats to privacy), they may bypass the opportunity to offer control or regulations of technologies.

Of special interest to discussions about the relations between technology and values is the concern about the interactions between technology and democracy. Not surprisingly, the founding texts in philosophy of technology are particularly concerned with this issue. And not unexpectedly, either, these texts are either celebratory of the democratic possibilities of technology (Marshall McLuhan, Alvin Toffler, Karl Marx) or critical and dismissive of such a potential (Max Horkheimer, Theodor Adorno, Jacques Ellul). Of late, however, alternative positions have been developed that see technology as neither determining democratic values nor preventing them altogether. Two of these positions are American Pragmatism, as exemplified by Dewey and Hickman, and Critical Theory, as exemplified by Marcuse, Jürgen Habermas, and Feenberg. These two philosophical traditions examine the assumptions and consequences of scientific and technological change, and explore the nature and possibilities of democracy in light of science and technology. In this middle ground, we find work that is empirically rich and thinkers who are actively engaged in public dialogue on the place and meaning of technology. We also find a cautious optimism that we can have both more technology and more democracy. Hickman argues that there are good grounds for rejecting what he regards as the Luddism, pessimism, and romanticism that seem to infect much thinking about technology. He maintains that his grounds for rejecting these tendencies are neither optimistic nor pessimistic, but hopeful. He believes that the various forms of technological determinism that have been a central feature of many philosophical and sociological treatments of technology are only debilitating and counterproductive.[13]

Among the most pressing concerns facing our culture, and among those issues least often discussed by philosophers of technology, are those related to biotechnology and environmental philosophy. When these issues have been addressed, all too often the ethicists have approached technology as a pure outside critic, while the technologists have seen value questions as either interfering with the important issues or as simply foreign concerns, thereby replicating the two cultures problem. By including these topics and by including work that deals with both aspects of the problem, this text tries to address both gaps.

As Jonas argued, technology is more than mere tools.[14] It also includes modes of transforming how we live and who we are. The genetic revolution in biology promises to allow us to remake ourselves. For this reason, some, such as Leon Kass, have argued that the very nature of our humanness is at stake with the progress of human biotechnology. Others, such as Laura Purdy and Inmaculada de Melo-Martín and Marin Gillis, maintain not that we are moving into a "posthuman" future, but instead argue that the promise of genetics will affect people differentially, with greater burdens falling on women and traditionally disadvantaged groups. They reject both the value-neutral and the value-determining thesis.

We can also remake ourselves through remaking our foods. The biotechnology revolution in agriculture is far ahead of the biotechnology applied directly to humans. Paul B. Thompson has argued that most work on the ethical issues in food biotechnology and agriculture is written from the consumer perspective alone, and thus places food consumers (all of us) at odds with food producers (without whom we would not be food consumers). In *The Spirit of the Soil*,[15] he argues that environmentalists and food producers have common interests, and thus brings together often-competing voices with the goal of articulating a common view.

Similarly, the impacts of technology on the environment, evidence of our abilities to remake the world, are loci of conflict. Consideration of technology and the environment often returns to the debate over whether technology is good or not, and the debate tends to focus on rural and wilderness areas. In these debates, often the city is bad and the country is good, technology is bad and wilderness is good. Andrew Light rejects this view of technology and focuses as much on cities as on rural areas. He argues that neither can we do without cities nor without technology, and draws our attention again to the democratic potentials of our responses to the impacts of technology on our human and nonhuman environment.

Finally, our technologies remake our world in ways both intimate and global. The opening and closing sections of Section 2 in this volume address technology in everyday life, and technological networks and systems, specifically information technologies.

Notes

1 For an important discussion of this bias, see P.-P. Verbeek, *What Things Do: Philosophical Reflections on Technology, Agency, and Design*, trans. Robert P. Crease. University Park, PA: The Pennsylvania University Press, 2005.

2 C. Mitcham, *Thinking Through Technology*. Chicago: The University of Chicago Press, 1994;

V. Dusek, *Philosophy of Technology: An Introduction*. Malden, MA: Blackwell, 2006; J.-K. B. Olsen and E. Selinger, *Philosophy of Technology*. Automatic Press, 2007; F. Ferre, *The Philosophy of Technology*. Englewood Cliffs, NJ: Prentice Hall, 1988; D. Ihde, *Philosophy of Technology: An Introduction*. New York: Paragon, 1993; H. Achterhuis, *American Philosophy of Technology*. Bloomington: Indiana University Press, 2001.

3 R. Hughes, *The Shock of the New*, New York: Alfred A. Knopf, 1982.

4 Verbeek, *What Things Do*.

5 Mitcham, *Thinking Through Technology*.

6 M. R. Smith and L. Marx, *Does Technology Drive History: The Dilemma of Technological Determinism*. Cambridge, MA: MIT Press, 1994.

7 W. Mcdonough and M. Braungart, *Cradle to Cradle: Remaking the Way We Make Things*. New York: North Point Press, 2002.

8 H. P. See Segal, *Technological Utopianism in American Culture*. Chigaco: University of Chicago Press, 1985.

9 J. Watson, "Genes and Politics,": *Journal of Molecular Medicine* 75 (Sept. 1997): 632–6.

10 Mitcham, *Thinking Through Technology*.

11 See, for example, S. Florman, *Blaming Technology*. New York: St Martin's, 1982; *The Civilized Engineer*. New York: St Martin's, 1988; *The Existential Pleasures of Engineering*. New York: St Martin's, 1996; *The Introspective Engineer*. New York: St Martin's, 1997; and H. Petroski, *To Engineer is Human*. New York: Vintage, 1992.

12 C. P. Snow, *The Two Cultures*, Cambridge: Cambridge University Press, 1993.

13 L. A. Hickman, *Philosophical Tools for Technological Culture: Putting Pragmatism to Work*. Bloomington: Indiana University Press, 2001.

14 H. Jonas, *The Imperative of Responsibility: In Search of an Ethics for the Technological Age*. Chicago: University of Chicago Press, 1984.

15 P. B. Thompson, *The Spirit of the Soil*. New York: Routledge, 1994.

Section One

Theoretical Reflections on Technology

Part I

Introductory Considerations
of Technology

1

Toward a Philosophy of Technology

Hans Jonas

Hans Jonas (1903–93) was a German-born philosopher who fled Germany in 1933 to escape the Nazi regime. After many years in Palestine/Israel, and briefer periods in England and Canada, he spent 1955–76 as a professor at the New School for Social Research in New York City. Jonas was a student of Heidegger's, and he shares with Heidegger the view that technology is "the focal fact of modern life." In this piece, Jonas invites us to consider technology according to three dimensions: the "formal dynamics," the "material content," and the importance of valuation. The first is a consideration of technology according to its internal logic, a consideration of technology as a whole. The second is technology in use, and includes not only the history of particular artifacts (consider Henry Petroski's wonderful book *The Pencil* (1989)) but also the phenomenology of everyday interactions with technology (for example, see Douglas Browning's essay in chapter 22). The third dimension is where we take a critical distance on our tools and lives and ask where we are going and why. This is the aspect of understanding technology that most directly draws on the tradition of Western Philosophy that started with Socrates. In his book, *The Imperative of Responsibility*, Jonas argues that because of both the temporal reach and the power of contemporary technologies, traditional ethical theories are insufficient. Drawing on examples that range from nuclear war to human biotechnologies, he notes that we have the capacity to unleash broad ranging changes in our world, the full impacts of which will not be known for generations. This concern about the scope of our technological capabilities is a feature that Jonas' work shares with that of Lowrance (chapter 3). Because of this, we have obligations to know and to exercise caution that are new in human history.

From *The Hastings Center Report* 9, no. 1 (1979): 40–59. Reprinted by permission of The Hastings Center. Reprinted by permission of Eleanore Jonas.

Are there philosophical aspects to technology? Of course there are, as there are to all things of importance in human endeavor and destiny. Modern technology touches on almost everything vital to man's existence – material, mental, and spiritual. Indeed, what of man is *not* involved? The way he lives his life and looks at objects, his intercourse with the world and with his peers, his powers and modes of action, kinds of goals, states and changes of society, objectives and forms of politics (including warfare no less than welfare), the sense and quality of life, even man's fate and that of his environment: all these are involved in the technological enterprise as it extends in magnitude and depth. The mere enumeration suggests a staggering host of potentially philosophic themes.

To put it bluntly: if there is a philosophy of science, language, history, and art; if there is social, political, and moral philosophy; philosophy of thought and of action, of reason and passion, of decision and value – all facets of the inclusive philosophy of man – how then could there not be a philosophy of technology, the focal fact of modern life? And at that a philosophy so spacious that it can house portions from all the other branches of philosophy? It is almost a truism, but at the same time so immense a proposition that its challenge staggers the mind. Economy and modesty require that we select, for a beginning, the most obvious from the multitude of aspects that invite philosophical attention.

The old but useful distinction of "form" and "matter" allows us to distinguish between these two major themes: (1) the *formal dynamics* of technology as a continuing collective enterprise, which advances by its own "laws of motion"; and (2) the *substantive content* of technology in terms of the things it puts into human use, the powers it confers, the novel objectives it opens up or dictates, and the altered manner of human action by which these objectives are realized.

The first theme considers technology as an abstract whole of movement; the second considers its concrete uses and their impact on our world and our lives. The formal approach will try to grasp the pervasive "process properties" by which modern technology propels itself – through our agency, to be sure – into ever-succeeding and superseding novelty. The material approach will look at the species of novelties themselves, their

taxonomy, as it were, and try to make out how the world furnished with them looks. A third, overarching theme is the *moral* side of technology as a burden on human responsibility, especially its long-term effects on the global condition of man and environment. This – my own main preoccupation over the past years – will only be touched upon.

The Formal Dynamics of Technology

First some observations about technology's form as an abstract whole of movement. We are concerned with characteristics of *modern* technology and therefore ask first what distinguishes it *formally* from all previous technology. One major distinction is that modern technology is an enterprise and process, whereas earlier technology was a possession and a state. If we roughly describe technology as comprising the use of artificial implements for the business of life, together with their original invention, improvement, and occasional additions, such a tranquil description will do for most of technology through mankind's career (with which it is coeval), but not for modern technology. In the past, generally speaking, a given inventory of tools and procedures used to be fairly constant, tending toward a mutually adjusting, stable equilibrium of ends and means, which – once established – represented for lengthy periods an unchallenged optimum of technical competence.

To be sure, revolutions occurred, but more by accident than by design. The agricultural revolution, the metallurgical revolution that led from the neolithic to the iron age, the rise of cities, and such developments, *happened* rather than were consciously created. Their pace was so slow that only in the time-contraction of historical retrospect do they appear to be "revolutions" (with the misleading connotation that their contemporaries experienced them as such). Even where the change was sudden, as with the introduction first of the chariot, then of armed horsemen into warfare – a violent, if short-lived, revolution indeed – the innovation did not originate from within the military art of the advanced societies that it affected, but was thrust on it from outside by the (much less civilized) peoples of Central Asia. Instead of spreading through the technological universe of their time, other technical break-throughs, like Phoenician purple-dyeing, Byzantine "greek fire," Chinese porcelain and silk, and Damascene steel-tempering, remained jealously guarded monopolies of the inventor communities. Still others, like the hydraulic and steam playthings of Alexandrian mechanics, or compass and gunpowder of the Chinese, passed unnoticed in their serious technological potentials.[1]

On the whole (not counting rare upheavals), the great classical civilizations had comparatively early reached a point of technological saturation – the aforementioned "optimum" in equilibrium of means with acknowledged needs and goals – and had little cause later to go beyond it. From there on, convention reigned supreme. From pottery to monumental architecture, from food growing to shipbuilding, from textiles to engines of war, from time measuring to stargazing: tools, techniques, and objectives remained essentially the same over long times; improvements were sporadic and unplanned. Progress therefore – if it occurred at all[2] – was by inconspicuous increments to a universally high level that still excites our admiration and, in historical fact, was more liable to regression than to surpassing. The former at least was the more noted phenomenon, deplored by the epigones with a nostalgic remembrance of a better past (as in the declining Roman world). More important, there was, even in the best and most vigorous times, no proclaimed *idea* of a future of *constant progress* in the arts. Most important, there was never a deliberate method of going about it like "research," the willingness to undergo the risks of trying unorthodox paths, exchanging information widely about the experience, and so on. Least of all was there a "natural science" as a growing body of theory to guide such semitheoretical, prepractical activities, plus their social institutionalization. In routines as well as panoply of instruments, accomplished as they were for the purposes they served, the "arts" seemed as settled as those purposes themselves.[3]

Traits of modern technology

The exact opposite of this picture holds for modern technology, and this is its first philosophical aspect. Let us begin with some manifest traits.

1. Every new step in whatever direction of whatever technological field tends *not* to approach an equilibrium or saturation point in the process of fitting means to ends (nor is it meant to), but, on the contrary, to give rise, if successful, to further steps in all kinds of direction and with a fluidity of the ends themselves. "Tends to" becomes a compelling "is bound to" with any major or important step (this almost being its criterion); and the innovators themselves expect, beyond the accomplishment, each time, of their immediate task, the constant future repetition of their inventive activity.

2. Every technical innovation is sure to spread quickly through the technological world community, as also do theoretical discoveries in the sciences. The spreading is in terms of knowledge and of practical adoption, the first (and its speed) guaranteed by the universal intercommunication that is itself part of the technological complex, the second enforced by the pressure of competition.

3. The relation of means to ends is not unilinear but circular. Familiar ends of long standing may find better satisfaction by new technologies whose genesis they had inspired. But equally – and increasingly typical – new technologies may suggest, create, even impose new ends, never before conceived, simply by offering their feasibility. (Who had ever wished to have in his living room the Philharmonic orchestra, or open heart surgery, or a helicopter defoliating a Vietnam forest? or to drink his coffee from a disposable plastic cup? or to have artificial insemination, test-tube babies, and host pregnancies? or to see clones of himself and others walking about?) Technology thus adds to the very objectives of human desires, including objectives for technology itself. The last point indicates the dialectics or circularity of the case: once incorporated into the socioeconomic demand diet, ends first gratuitously (perhaps accidentally) generated by technological invention become necessities of life and set technology the task of further perfecting the means of realizing them.

4. Progress, therefore, is not just an ideological gloss on modern technology, and not at all a mere option offered by it, but an inherent drive which acts willynilly in the formal automatics of its *modus operandi* as it interacts with society. "Progress" is here not a value term but purely

descriptive. We may resent the fact and despise its fruits and yet must go along with it, for – short of a stop by the fiat of total political power, or by a sustained general strike of its clients or some internal collapse of their societies, or by self-destruction through its works (the last, alas, the least unlikely of these) – the juggernaut moves on relentlessly, spawning its always mutated progeny by coping with the challenges and lures of the now. But while not a value term, "progress" here is not a neutral term either, for which we could simply substitute "change." For it is in the nature of the case, or a law of the series, that a later stage is always, in terms of technology itself, *superior* to the preceding *stage*.[4] Thus we have here a case of the entropy-defying sort (organic evolution is another), where the internal motion of a system, left to itself and not interfered with, leads to ever "higher," not "lower" states of itself. Such at least is the present evidence.[5] If Napoleon once said, "Politics is destiny," we may well say today, "Technology is destiny."

These points go some way to explicate the initial statement that modern technology, unlike traditional, is an enterprise and not a possession, a process and not a state, a dynamic thrust and not a set of implements and skills. And they already adumbrate certain "laws of motion" for this restless phenomenon. What we have described, let us remember, were formal traits which as yet say little about the contents of the enterprise. We ask two questions of this descriptive picture: *why* is this so, that is, what *causes* the restlessness of modern technology; what is the nature of the thrust? And, what is the philosophical import of the facts so explained?

The nature of restless technology

As we would expect in such a complex phenomenon, the motive forces are many, and some causal hints appeared already in the descriptive account. We have mentioned *pressure of competition* – for profit, but also for power, security, and so forth – as one perpetual mover in the universal appropriation of technical improvements. It is equally operative in their origination, that is, in the process of invention itself, nowadays dependent on constant outside subsidy and even goal-setting: potent interests see to both. War, or

the threat of it, has proved an especially powerful agent. The less dramatic, but no less compelling, everyday agents are legion. To keep one's head above the water is their common principle (somewhat paradoxical, in view of an abundance already far surpassing what former ages would have lived with happily ever after). Of pressures other than the competitive ones, we must mention those of population growth and of impending exhaustion of natural resources. Since both phenomena are themselves already by-products of technology (the first by way of medical improvements, the second by the voracity of industry), they offer a good example of the more general truth that to a considerable extent technology itself begets the problems which it is then called upon to overcome by a new forward jump. (The Green Revolution and the development of synthetic substitute materials or of alternate sources of energy come under this heading.) These compulsive pressures for progress, then, would operate even for a technology in a noncompetitive, for example, a socialist setting.

A motive force more autonomous and spontaneous than these almost mechanical pushes with their "sink or swim" imperative would be the pull of the quasi-utopian *vision* of an ever better life, whether vulgarly conceived or nobly, once technology had proved the open-ended capacity for procuring the conditions for it: perceived possibility whetting the appetite ("the American dream," "the revolution of rising expectations"). This less palpable factor is more difficult to appraise, but its playing a role is undeniable. Its deliberate fostering and manipulation by the dream merchants of the industrial-mercantile complex is yet another matter and somewhat taints the spontaneity of the motive, as it also degrades the quality of the dream. It is also moot to what extent the vision itself is *post hoc* rather than *ante hoc*, that is, instilled by the dazzling feats of a technological progress already underway and thus more a response to than a motor of it.

Groping in these obscure regions of motivation, one may as well descend, for an explanation of the dynamism as such, into the Spenglerian mystery of a "Faustian soul" innate in Western culture, that drives it, nonrationally, to infinite novelty and unplumbed possibilities for their own sake; or into the Heideggerian depths of a fateful, metaphysical decision of the will for boundless power over the world of things – a decision equally peculiar to the Western mind: speculative intuitions which do strike a resonance in us, but are beyond proof and disproof.

Surfacing once more, we may also look at the very sober, functional facts of industrialism as such, of production and distribution, output maximization, managerial and labor aspects, which even apart from competitive pressure provide their own incentives for technical progress. Similar observations apply to the requirements of *rule* or control in the vast and populous states of our time, those giant territorial superorganisms which for their very cohesion depend on advanced technology (for example, in information, communication, and transportation, not to speak of weaponry) and thus have a stake in its promotion: the more so, the more centralized they are. This holds for socialist systems no less than for free-market societies. May we conclude from this that even a communist world state, freed from external rivals as well as from internal free-market competition, might still have to push technology ahead for purposes of control on this colossal scale? Marxism, in any case, has its own inbuilt commitment to technological progress beyond necessity. But even disregarding all dynamics of these conjectural kinds, the most monolithic case imaginable would, at any rate, still be exposed to those noncompetitive, natural pressures like population growth and dwindling resources that beset industrialism as such. Thus, it seems, the compulsive element of technological progress may not be bound to its original breeding ground, the capitalist system. Perhaps the odds for an eventual stabilization look somewhat better in a socialist system, provided it is worldwide – and possibly totalitarian in the bargain. As it is, the pluralism we are thankful for ensures the constancy of compulsive advance.

We could go on unravelling the causal skein and would be sure to find many more strands. But none nor all of them, much as they explain, would go to the heart of the matter. For all of them have one premise in common without which they could not operate for long: the premise that there *can* be indefinite progress because there *is* always something new and better to find. The, by no means obvious, givenness of this objective condition is also the pragmatic conviction of the performers in the technological drama; but

without its being true, the conviction would help as little as the dream of the alchemists. Unlike theirs, it is backed up by an impressive record of past successes, and for many this is sufficient ground for their belief. (Perhaps holding or not holding it does not even greatly matter.) What makes it more than a sanguine belief, however, is an underlying and well-grounded, theoretical view of the nature of things and of human cognition, according to which they do not set a limit to novelty of discovery and invention, indeed, that they of themselves will at each point offer another opening for the as yet unknown and undone. The corollary conviction, then, is that a technology tailored to a nature and to a knowledge of this indefinite potential ensures its indefinitely continued conversion into the practical powers, each step of it begetting the next, with never a cutoff from internal exhaustion of possibilities.

Only habituation dulls our wonder at this wholly unprecedented belief in virtual "infinity." And by all our present comprehension of reality, the belief is most likely true – at least enough of it to keep the road for innovative technology in the wake of advancing science open for a long time ahead. Unless we understand this ontologic-epistemological premise, we have not understood the inmost agent of technological dynamics, on which the working of all the adventitious causal factors is contingent in the long run.

Let us remember that the virtual infinitude of advance we here seek to explain is in essence different from the always avowed perfectibility of every human accomplishment. Even the undisputed master of his craft always had to admit as possible that he might be surpassed in skill or tools or materials; and no excellence of product ever foreclosed that it might still be bettered, just as today's champion runner must know that his time may one day be beaten. But these are improvements within a given genus, not different in kind from what went before, and they must accrue in diminishing fractions. Clearly, the phenomenon of an exponentially growing *generic* innovation is qualitatively different.

Science as a source of restlessness

The answer lies in the interaction of *science* and *technology* that is the hallmark of modern progress, and thus ultimately in the kind of nature which modern science progressively discloses. For it is here, in the movement of *knowledge*, where relevant novelty first and constantly occurs. This is itself a novelty. To Newtonian physics, nature appeared simple, almost crude, running its show with a few kinds of basic entities and forces by a few universal laws, and the application of those well-known laws to an ever greater variety of composite phenomena promised ever widening knowledge indeed, but no real surprises. Since the mid-nineteenth century, this minimalistic and somehow finished picture of nature has changed with breathtaking acceleration. In a reciprocal interplay with the growing subtlety of exploration (instrumental and conceptual), nature itself stands forth as ever more subtle. The progress of probing makes the object grow richer in modes of operation, not sparer as classical mechanics had expected. And instead of narrowing the margin of the still-undiscovered, science now surprises itself with unlocking dimension after dimension of new depths. The very essence of matter has turned from a blunt, irreducible ultimate to an always reopened challenge for further penetration. No one can say whether this will go on forever, but a suspicion of intrinsic infinity in the very being of things obtrudes itself and therewith an anticipation of unending inquiry of the sort where succeeding steps will not find the same old story again (Descartes's "matter in motion"), but always add new twists to it. If then the art of technology is correlative to the knowledge of nature, technology too acquires from this source that potential of infinity for its innovative advance.

But it is not just that indefinite scientific progress offers the *option* of indefinite technological progress, to be exercised or not as other interests see fit. Rather the cognitive process itself moves by interaction with the technological, and in the most internally vital sense: for its own *theoretical* purpose, science must generate an increasingly sophisticated and physically formidable technology as its tool. What it finds with this help initiates new departures in the practical sphere, and the latter as a whole, that is, technology at work provides with its experiences a large-scale laboratory for science again, a breeding ground for new questions, and so on in an unending cycle. In brief, a mutual feedback operates between science and technology; each requires and propels

the other; and as matters now stand, they can only live together or must die together. For the dynamics of technology, with which we are here concerned, this means that (all external promptings apart) an agent of restlessness is implanted in it by its functionally integral bond with science. As long, therefore, as the cognitive impulse lasts, technology is sure to move ahead with it. The cognitive impulse, in its turn, culturally vulnerable in itself, liable to lag or to grow conservative with a treasured canon – that theoretical eros itself no longer lives on the delicate appetite for truth alone, but is spurred on by its hardier offspring, technology, which communicates to it impulsions from the broadest arena of struggling, insistent life. Intellectual curiosity is seconded by interminably self-renewing practical aim.

I am conscious of the conjectural character of some of these thoughts. The revolutions in science over the last fifty years or so are a fact, and so are the revolutionary style they imparted to technology and the reciprocity between the two concurrent streams (nuclear physics is a good example). But whether those scientific revolutions, which hold primacy in the whole syndrome, will be typical for science henceforth – something like a law of motion for its future – or represent only a singular phase in its longer run, is unsure. To the extent, then, that our forecast of incessant novelty for technology was predicated on a guess concerning the future of science, even concerning the nature of things, it is hypothetical, as such extrapolations are bound to be. But even if the recent past did not usher in a state of permanent revolution for science, and the life of theory settles down again to a more sedate pace, the scope for technological innovation will not easily shrink; and what may no longer be a revolution in science, may still revolutionize our lives in its practical impact through technology. "Infinity" being too large a word anyway, let us say that present signs of potential and of incentives point to an indefinite perpetuation and fertility of the technological momentum.

The philosophical implications

It remains to draw philosophical conclusions from our findings, at least to pinpoint aspects of philosophical interest. Some preceding remarks have already been straying into philosophy of science in the technical sense. Of broader issues, two will be ample to provide food for further thought beyond the limitations of this paper. One concerns the status of knowledge in the human scheme, the other the status of technology itself as a human goal, or its tendency to become that from being a means, in a dialectical inversion of the means-end order itself.

Concerning knowledge, it is obvious that the time-honored division of theory and practice has vanished for both sides. The thirst for pure knowledge may persist undiminished, but the involvement of knowing at the heights with doing in the lowlands of life, mediated by technology, has become inextricable; and the aristocratic self-sufficiency of knowing for its own (and the knower's) sake has gone. Nobility has been exchanged for utility. With the possible exception of philosophy, which still can do with paper and pen and tossing thoughts around among peers, all knowledge has become thus tainted, or elevated if you will, whether utility is intended or not. The technological syndrome, in other words, has brought about a thorough *socializing* of the theoretical realm, enlisting it in the service of common need. What used to be the freest of human choices, an extravagance snatched from the pressure of the world – the esoteric life of thought – has become part of the great public play of necessities and a prime necessity in the action of the play.[6] Remotest abstraction has become enmeshed with nearest concreteness. What this pragmatic functionalization of the once highest indulgence in impractical pursuits portends for the image of man, for the restructuring of a hallowed hierarchy of values, for the idea of "wisdom," and so on, is surely a subject for philosophical pondering.

Concerning technology itself, its actual role in modern life (as distinct from the purely instrumental definition of technology as such) has made the relation of means and ends equivocal all the way up from the daily living to the very vocation of man. There could be no question in former technology that its role was that of humble servant – pride of workmanship and esthetic embellishment of the useful notwithstanding. The Promethean enterprise of modern technology speaks a different language. The word "enterprise" gives the clue, and its unendingness another. We have mentioned that the effect of its innovations is disequilibrating rather than equilibrating with respect to the balance of wants and supply, always breeding its own new wants. This in itself compels the constant attention of the best minds, engaging the full capital of human ingenuity for meeting challenge after challenge and seizing the new chances. It is psychologically natural for that degree of engagement to be invested with the dignity of dominant purpose. Not only does technology dominate our lives in fact, it nourishes also a belief in its being of predominant worth. The sheer grandeur of the enterprise and its seeming infinity inspire enthusiasm and fire ambition. Thus, in addition to spawning new ends (worthy or frivolous) from the mere invention of means, technology as a grand venture tends to establish *itself* as the transcendent end. At least the suggestion is there and casts its spell on the modern mind. At its most modest, it means elevating *homo faber* to the essential aspect of man; at its most extravagant, it means elevating *power* to the position of his dominant and interminable goal. To become ever more masters of the world, to advance from power to power, even if only collectively and perhaps no longer by choice, can now be seen to be the chief vocation of mankind. Surely, this again poses philosophical questions that may well lead unto the uncertain grounds of metaphysics or of faith.

I here break off, arbitrarily, the formal account of the technological movement in general, which as yet has told us little of what the enterprise is about. To this subject I now turn, that is, to the new kinds of powers and objectives that technology opens to modern man and the consequently altered quality of human action itself.

The Material Works of Technology

Technology is a species of power, and we can ask questions about how and on what object any power is exercised. Adopting Aristotle's rule in *De anima* that for understanding a faculty one should begin with its objects, we start from them too – "objects" meaning both the visible *things* technology generates and puts into human use, and the *objectives* they serve. The objects of modern technology are first everything that had always been an object of human artifice and labor: food, clothing, shelter, implements, transportation

– all the material necessities and comforts of life. The technological intervention changed at first not the product but its production, in speed, ease, and quantity. However, this is true only of the very first stage of the industrial revolution with which large-scale scientific technology began. For example, the cloth for the steam-driven looms of Lancashire remained the same. Even then, one significant new product was added to the traditional list – the machines themselves, which required an entire new industry with further subsidiary industries to build them. These novel entities, machines – at first capital goods only, not consumer goods – had from the beginning their own impact on man's symbiosis with nature by being consumers themselves. For example: steam-powered water pumps facilitated coal mining, required in turn extra coal for firing their boilers, more coal for the foundries and forges that made those boilers, more for the mining of the requisite iron ore, more for its transportation to the foundries, more – both coal and iron – for the rails and locomotives made in these same foundries, more for the conveyance of the foundries' product to the pitheads and return, and finally more for the distribution of the more abundant coal to the users outside this cycle, among which were increasingly still more machines spawned by the increased availability of coal. Lest it be forgotten over this long chain, we have been speaking of James Watt's modest steam engine for pumping water out of mine shafts. This syndrome of self-proliferation – by no means a linear chain but an intricate web of reciprocity – has been part of modern technology ever since. To generalize, technology exponentially increases man's drain on nature's resources (of substances and of energy), not only through the multiplication of the final goods for consumption, but also, and perhaps more so, through the production and operation of its own mechanical means. And with these means – machines – it introduced a new category of goods, not for consumption, added to the furniture of our world. That is, among the objects of technology a prominent class is that of technological apparatus itself.

Soon other features also changed the initial picture of a merely mechanized production of familiar commodities. The final products reaching the consumer ceased to be the same, even if still serving the same age-old needs; new needs, or desires, were added by commodities of entirely new kinds which changed the habits of life. Of such commodities, machines themselves became increasingly part of the consumer's daily life to be used directly by himself, as an article not of production but of consumption. My survey can be brief as the facts are familiar.

New kinds of commodities

When I said that the cloth of the mechanized looms of Lancashire remained the same, everyone will have thought of today's synthetic fibre textiles for which the statement surely no longer holds. This is fairly recent, but the general phenomenon starts much earlier, in the synthetic dyes and fertilizers with which the chemical industry – the first to be wholly a fruit of science – began. The original rationale of these technological feats was substitution of artificial for natural materials (for reasons of scarcity or cost), with as nearly as possible the same properties for effective use. But we need only think of plastics to realize that art progressed from substitutes to the creation of really new substances with properties not so found in any natural one, raw or processed, thereby also initiating uses not thought of before and giving rise to new classes of objects to serve them. In chemical (molecular) engineering, man does more than in mechanical (molar) engineering which constructs machinery from natural materials; his intervention is deeper, redesigning the infra-patterns of nature, making substances to specification by arbitrary disposition of molecules. And this, be it noted, is done deductively from the bottom, from the thoroughly analyzed last elements, that is, in a real *via compositiva* after the completed *via resolutiva*, very different from the long-known empirical practice of coaxing substances into new properties, as in metal alloys from the bronze age on. Artificiality or creative engineering with abstract construction invades the heart of matter. This, in molecular biology, points to further, awesome potentialities.

With the sophistication of molecular alchemy we are ahead of our story. Even in straightforward hardware engineering, right in the first blush of the mechanical revolution, the objects of use that came out of the factories did not really remain the same, even where the objectives did. Take the old objective of travel. Railroads and ocean liners are relevantly different from the stage coach and from the sailing ship, not merely in construction and efficiency but in the very feel of the user, making travel a different experience altogether, something one may do for its own sake. Airplanes, finally, leave behind any similarity with former conveyances, except the purpose of getting from here to there, with no experience of what lies in between. And these instrumental objects occupy a prominent, even obtrusive place in our world, far beyond anything wagons and boats ever did. Also they are constantly subject to improvement of design, with obsolescence rather than wear determining their life span.

Or take the oldest, most static of artifacts: human habitation. The multistoried office building of steel, concrete, and glass is a qualitatively different entity from the wood, brick, and stone structures of old. With all that goes into it besides the structures as such – the plumbing and wiring, the elevators, the lighting, heating, and cooling systems – it embodies the end products of a whole spectrum of technologies and far-flung industries, where only at the remote sources human hands still meet with primary materials, no longer recognizable in the final result. The ultimate customer inhabiting the product is ensconced in a shell of thoroughly derivative artifacts (perhaps relieved by a nice piece of driftwood). This transformation into utter artificiality is generally, and increasingly, the effect of technology on the human environment, down to the items of daily use. Only in agriculture has the product so far escaped this transformation by the changed modes of its production. We still eat the meat and rice of our ancestors.[7]

Then, speaking of the commodities that technology injects into private use, there are machines themselves, those very devices of its own running, originally confined to the economic sphere. This unprecedented novum in the records of individual living started late in the nineteenth century and has since grown to a pervading mass phenomenon in the Western world. The prime example, of course, is the automobile, but we must add to it the whole gamut of household appliances – refrigerators, washers, dryers, vacuum cleaners

– by now more common in the lifestyle of the general population than running water or central heating were one hundred years ago. Add lawn mowers and other power tools for home and garden: we are mechanized in our daily chores and recreations (including the toys of our children) with every expectation that new gadgets will continue to arrive.

These paraphernalia are machines in the precise sense that they perform work and consume energy, and their moving parts are of the familiar magnitudes of our perceptual world. But an additional and profoundly different category of technical apparatus was dropped into the lap of the private citizen, not labor-saving and work-performing, partly not even utilitarian, but – with minimal energy input – catering to the senses and the mind: telephone, radio, television, tape recorders, calculators, record players – all the domestic terminals of the electronics industry, the latest arrival on the technological scene. Not only by their insubstantial, mind-addressed output, also by the subvisible, not literally "mechanical" physics of their functioning do these devices differ in kind from all the macroscopic, bodily moving machinery of the classical type. Before inspecting this momentous turn from power engineering, the hallmark of the first industrial revolution, to communication engineering, which almost amounts to a second industrial-technological revolution, we must take a look at its natural base: electricity.

In the march of technology to ever greater artificiality, abstraction, and subtlety, the unlocking of electricity marks a decisive step. Here is a universal force of nature which yet does not naturally appear to man (except in lightning). It is not a datum of uncontrived experience. Its very "appearance" had to wait for science, which contrived the experience for it. Here, then, a technology depended on science for the mere providing of its "object," the entity itself it would deal with – the first case where theory alone, not ordinary experience, wholly preceded practice (repeated later in the case of nuclear energy). And what sort of entity! Heat and steam are familiar objects of sensuous experience, their force bodily displayed in nature; the matter of chemistry is still the concrete, corporeal stuff mankind had always known. But electricity is an abstract object, disembodied, immaterial, unseen;

in its usable form, it is entirely an artifact, generated in a subtle transformation from grosser forms of energy (ultimately from heat via motion). Its theory indeed had to be essentially complete before utilization could begin.

Revolutionary as electrical technology was in itself, its purpose was at first the by now conventional one of the industrial revolution in general: to supply motive power for the propulsion of machines. Its advantages lay in the unique versatility of the new force, the ease of its transmission, transformation, and distribution – an unsubstantial commodity, no bulk, no weight, instantaneously delivered at the point of consumption. Nothing like it had ever existed before in man's traffic with matter, space, and time. It made possible the spread of mechanization to every home; this alone was a tremendous boost to the technological tide, at the same time hooking private lives into centralized public networks and thus making them dependent on the functioning of a total system as never before, in fact, for every moment. Remember, you cannot hoard electricity as you can coal and oil, or flour and sugar for that matter.

But something much more unorthodox was to follow. As we all know, the discovery of the universe of electromagnetics caused a revolution in theoretical physics that is still underway. Without it, there would be no relativity theory, no quantum mechanics, no nuclear and subnuclear physics. It also caused a revolution in technology beyond what it contributed, as we noted, to its classical program. The revolution consisted in the passage from electrical to electronic technology which signifies a new level of abstraction in means and ends. It is the difference between power and communication engineering. Its object, the most impalpable of all, is information. Cognitive instruments had been known before – sextant, compass, clock, telescope, microscope, thermometer, all of them for information and not for work. At one time, they were called "philosophical" or "metaphysical" instruments. By the same general criterion, amusing as it may seem, the new electronic information devices, too, could be classed as "philosophical instruments." But those earlier cognitive devices, except the clock, were inert and passive, not generating information actively, as the new instrumentalities do.

Theoretically as well as practically, electronics signifies a genuinely new phase of the scientific-technological revolution. Compared with the sophistication of its theory as well as the delicacy of its apparatus, everything which came before seems crude, almost natural. To appreciate the point, take the man-made satellites now in orbit. In one sense, they are indeed an imitation of celestial mechanics – Newton's laws finally verified by cosmic experiment: astronomy, for millennia the most purely contemplative of the physical sciences, turned into a practical art! Yet, amazing as it is, the astronomic imitation, with all the unleashing of forces and the finesse of techniques that went into it, is the least interesting aspect of those entities. In that respect, they still fall within the terms and feats of classical mechanics (except for the remote-control course corrections).

Their true interest lies in the instruments they carry through the voids of space and in what these do, their measuring, recording, analyzing, computing, their receiving, processing, and transmitting abstract information and even images over cosmic distances. There is nothing in all nature which even remotely foreshadows the kind of things that now ride the heavenly spheres. Man's imitative practical astronomy merely provides the vehicle for something else with which he sovereignly passes beyond all the models and usages of known nature.[8] That the advent of man portended, in its inner secret of mind and will, a cosmic event was known to religion and philosophy: now it manifests itself as such by fact of things and acts in the visible universe. Electronics indeed creates a range of objects imitating nothing and progressively added to by pure invention.

And no less invented are the ends they serve. Power engineering and chemistry for the most part still answered to the natural needs of man: for food, clothing, shelter, locomotion, and so forth. Communication engineering answers to needs of information and control solely created by the civilization that made this technology possible and, once started, imperative. The novelty of the means continues to engender no less novel ends – both becoming as necessary to the functioning of the civilization that spawned them as they would have been pointless for any former one. The world they help to constitute

and which needs computers for its very running is no longer nature supplemented, imitated,
improved, transformed, the original habitat made
more habitable. In the pervasive mentalization
of physical relationships it is a *trans-nature* of
human making, but with this inherent paradox:
that it threatens the obsolescence of man himself,
as increasing automation ousts him from the
places of work where he formerly proved his
humanhood. And there is a further threat: its
strain on nature herself may reach a breaking
point.

The last stage of the revolution?

That sentence would make a good dramatic
ending. But it is not the end of the story. There
may be in the offing another, conceivably the last,
stage of the technological revolution, after the
mechanical, chemical, electrical, electronic stages
we have surveyed, and the nuclear we omitted. All
these were based on physics and had to do with
what man can put to his use. What about biology? And what about the user himself? Are we,
perhaps, on the verge of a technology, based on
biological knowledge and wielding an engineering art which, this time, has man himself for its
object? This has become a theoretical possibility
with the advent of molecular biology and its
understanding of genetic programming; and it
has been rendered morally possible by the metaphysical neutralizing of man. But the latter,
while giving us the license to do as we wish, at
the same time denies us the guidance for knowing what to wish. Since the same evolutionary
doctrine of which genetics is a cornerstone has
deprived us of a valid image of man, the actual
techniques, when they are ready, may find us
strangely unready for their responsible use. The
anti-essentialism of prevailing theory, which
knows only of *de facto* outcomes of evolutionary
accident and of no valid essences that would
give sanction to them, surrenders our being to a
freedom without norms. Thus the technological
call of the new microbiology is the twofold one
of physical feasibility and metaphysical admiss-
ibility. Assuming the genetic mechanism to be
completely analyzed and its script finally decoded,
we can set about rewriting the text. Biologists
vary in their estimates of how close we are to the
capability; few seem to doubt the right to use it.

Judging by the rhetoric of its prophets, the idea
of taking our evolution into our own hands is
intoxicating even to many scientists.

In any case, the idea of making over man is no
longer fantastic, nor interdicted by an inviolable
taboo. If and when *that* revolution occurs, if
technological power is really going to tinker with
the elemental keys on which life will have to
play its melody in generations of men to come
(perhaps the only such melody in the universe),
then a reflection on what is humanly desirable and
what should determine the choice – a reflection,
in short, on the image of man, becomes an
imperative more urgent than any ever inflicted on
the understanding of mortal man. Philosophy, it
must be confessed, is sadly unprepared for this,
its first cosmic task.

Toward an Ethics of Technology

The last topic has moved naturally from the
descriptive and analytic plane, on which the
objects of technology are displayed for inspection,
onto the evaluative plane where their ethical
challenge poses itself for decision. The particular
case forced the transition so directly because
there the (as yet hypothetical) technological object
was man directly. But once removed, man is
involved in all the other objects of technology,
as these singly and jointly remake the worldly
frame of his life, in both the narrower and the
wider of its senses: that of the artificial frame of
civilization in which social man leads his life
proximately, and that of the natural terrestrial
environment in which this artifact is embedded
and on which it ultimately depends.

Again, because of the magnitude of technological effects on both these vital environments in
their totality, both the quality of human life and
its very preservation in the future are at stake in
the rampage of technology. In short, certainly the
"image" of man, and possibly the survival of the
species (or of much of it), are in jeopardy. This
would summon man's duty to his cause even if
the jeopardy were not of his own making. But it
is, and, in addition to his ageless obligation to meet
the threat of things, he bears for the first time the
responsibility of prime agent in the threatening
disposition of things. Hence nothing is more
natural than the passage from the objects to the

ethics of technology, from the things made to the duties of their makers and users.

A similar experience of inevitable passage from analysis of fact to ethical significance, let us remember, befell us toward the end of the first section. As in the case of the matter, so also in the case of the form of the technological dynamics, the image of man appeared at stake. In view of the quasi-automatic compulsion of those dynamics, with their perspective of indefinite progression, every existential and moral question that the objects of technology raise assumes the curiously eschatological quality with which we are becoming familiar from the extrapolating guesses of futurology. But apart from thus raising all challenges of present particular matter to the higher powers of future exponential magnification, the despotic dynamics of the technological movement as such, sweeping its captive movers along in its breathless momentum, poses its own questions to man's axiological conception of himself. Thus, form and matter of technology alike enter into the dimension of ethics.

The questions raised for ethics by the objects of technology are defined by the major areas of their impact and thus fall into such fields of knowledge as ecology (with all its biospheric subdivisions of land, sea, and air), demography economics, biomedical and behavioral sciences (even the psychology of mind pollution by television), and so forth. Not even a sketch of the substantive problems, let alone of ethical policies for dealing with them, can here be attempted. Clearly, for a normative rationale of the latter, ethical theory must plumb the very foundations of value, obligation, and the human good.

The same holds of the different kind of questions raised for ethics by the sheer fact of the formal dynamics of technology. But here, a question of another order is added to the straightforward ethical questions of both kinds, subjecting any resolution of them to a pragmatic proviso of harrowing uncertainty. Given the mastery of the creation over its creators, which yet does not abrogate their responsibility nor silence their vital interest, what are the chances and what are the means of gaining *control* of the process, so that the results of any ethical (or even purely prudential) insights can be translated into effective action? How in short can man's freedom prevail against the determinism he has created

for himself? On this most clouded question, whereby hangs not only the effectuality or futility of the ethical search which the facts invite (assuming it to be blessed with *theoretical* success!), but perhaps the future of mankind itself, I will make a few concluding, but – alas – inconclusive, remarks. They are intended to touch on the whole ethical enterprise.

*Problematic preconditions
of an effective ethics*

First, a look at the novel state of determinism. Prima facie, it would seem that the greater and more varied powers bequeathed by technology have expanded the range of choices and hence increased human freedom. For economics, for example, the argument has been made[9] that the uniform compulsion which scarcity and subsistence previously imposed on economic behavior with a virtual denial of alternatives (and hence – conjoined with the universal "maximization" motive of capitalist market competition – gave classical economics at least the appearance of a deterministic "science") has given way to a latitude of indeterminacy. The plenty and powers provided by industrial technology allow a pluralism of choosable alternatives (hence disallow scientific prediction). We are not here concerned with the status of economics as a science. But as to the altered state of things alleged in the argument, I submit that the change means rather that one, relatively homogeneous determinism (thus relatively easy to formalize into a law) has been supplanted by another, more complex, multifarious determinism, namely, that exercised by the human artifact itself upon its creator and user. We, abstractly speaking the possessors of those powers, are concretely subject to their emancipated dynamics and the sheer momentum of our own multitude, the vehicle of those dynamics.

I have spoken elsewhere[10] of the "new realm of necessity" set up, like a second nature, by the feedbacks of our achievements. The almighty we, or Man personified is, alas, an abstraction. *Man* may have become more powerful; *men* very probably the opposite, enmeshed as they are in more dependencies than ever before. What ideal Man now can do is not the same as what real men permit or dictate to be done. And here I am thinking not only of the immanent dynamism, almost automatism, of the impersonal techno-logical complex I have invoked so far, but also of the pathology of its client society. Its compulsions, I fear, are at least as great as were those of unconquered nature. Talk of the blind forces of nature! Are those of the sorcerer's creation less blind? They differ indeed in the serial shape of their causality: the action of nature's forces is cyclical, with periodical recurrence of the same, while that of the technological forces is linear, pro-gressive, cumulative, thus replacing the curse of constant toil with the threat of maturing crisis and possible catastrophe. Apart from this significant vector difference, I seriously wonder whether the tyranny of fate has not become greater, the lat-itude of spontaneity smaller; and whether man has not actually been weakened in his decision-making capacity by his accretion of collective strength.

However, in speaking, as I have just done, of "his" decision-making capacity, I have been guilty of the same abstraction I had earlier criti-cized in the use of the term "man." Actually, the subject of the statement was no real or rep-resentative individual but Hobbes' "Artificiall Man," "that great Leviathan, called a Common-Wealth," or the "large horse" to which Socrates likened the city, "which because of its great size tends to be sluggish and needs stirring by a gadfly." Now, the chances of there being such gadflies among the numbers of the common-wealth are today no worse nor better than they have ever been, and in fact they are around and stinging in our field of concern. In that respect, the free spontaneity of personal insight, judg-ment, and responsible action by speech can be trusted as an ineradicable (if also incalculable) endowment of humanity, and smallness of num-ber is in itself no impediment to shaking public complacency. The problem, however, is not so much complacency or apathy as the counterforces of active, and anything but complacent, interests and the complicity with them of all of us in our daily consumer existence. These interests themselves are factors in the determinism which technology has set up in the space of its sway. The question, then, is that of the possible chances of unselfish insight in the arena of (by nature) selfish *power*, and more particularly: of one long-range, interloping insight against the short-range goals of many incumbent powers. Is there hope

that wisdom itself can become power? This renews the thorny old subject of Plato's philosopher-king and – with that inclusion of realism which the utopian Plato did not lack – of the role of myth, not knowledge, in the education of the guardians. Applied to our topic: the *knowledge* of objective dangers and of values endangered, as well as of the technical remedies, is beginning to be there and to be disseminated; but to make it prevail in the marketplace is a matter less of the rational dissemination of truth than of public relations techniques, persuasion, indoctrination, and manipulation, also of unholy alliances, per-haps even conspiracy. The philosopher's descent into the cave may well have to go all the way to "if you can't lick them, join them."

That is so not merely because of the active resistance of special interests but because of the optical illusion of the near and the far which condemns the long-range view to impotence against the enticement and threats of the nearby: it is this incurable shortsightedness of animal-human nature more than ill will that makes it difficult to move even those who have no special axe to grind, but still are in countless ways, as we all are, beneficiaries of the untamed system and so have something dear in the present to lose with the inevitable cost of its taming. The taskmaster, I fear, will have to be actual pain beginning to strike, when the far has moved close to the skin and has vulgar optics on its side. Even then, one may resort to palliatives of the hour. In any event, one should try as much as one can to forestall the advent of emergency with its high tax of suffering or, at the least, prepare for it. This is where the scientist can redeem his role in the technological estate.

The incipient knowledge about technological danger trends must be developed, coordinated, sys-tematized, and the full force of computer-aided projection techniques be deployed to determine priorities of action, so as to inform preventive efforts wherever they can be elicited, to minimize the necessary sacrifices, and at the worst to pre-plan the saving measures which the terror of beginning calamity will eventually make people willing to accept. Even now, hardly a decade after the first stirrings of "environmental" con-sciousness, much of the requisite knowledge, plus the rational persuasion, is available inside and outside academia for any well-meaning powerholder to draw upon. To this, we – the growing band of concerned intellectuals – ought persistently to contribute our bit of competence and passion.

But the real problem is to get the well-meaning into power and have that power as little as possible beholden to the interests which the technological colossus generates on its path. It is the problem of the philosopher-king compounded by the greater magnitude and complexity (also sophistication) of the forces to contend with. Ethically, it becomes a problem of playing the game by its impure rules. For the servant of truth to join in it means to sacrifice some of his time-honored role: he may have to turn apostle or agitator or political operator. This raises moral questions beyond those which technology itself poses, that of sanctioning immoral means for a surpassing end, of giving unto Caesar so as to promote what is not Caesar's. It is the grave question of moral casuistry, or of Dostoevsky's Grand Inquisitor, or of regarding cherished liberties as no longer affordable luxuries (which may well bring the anxious friend of mankind into odious political company) – questions one excusably hesitates to touch but in the further tide of things may not be permitted to evade.

What is, prior to joining the fray, the role of philosophy, that is, of a philosophically grounded ethical knowledge, in all this? The somber note of the last remarks responded to the quasi-apocalyptic prospects of the technological tide, where stark issues of planetary survival loom ahead. There, no philosophical ethics is needed to tell us that disaster must be averted. Mainly, this is the case of the ecological dangers. But there are other, noncatastrophic things afoot in technology where not the existence but the image of man is at stake. They are with us now and will accompany us and be joined by others at every new turn technology may take. Mainly, they are in the biomedical, behavioral, and social fields. They lack the stark simplicity of the survival issue, and there is none of the (at least declaratory) unanimity on them which the specter of extreme crisis commands. It is here where a philosophical ethics or theory of values has its task. Whether its voice will be listened to in the dispute on policies is not for it to ask; perhaps it cannot even muster an authoritative voice with which to speak – a house divided, as philosophy is. But the philosopher must try for normative

knowledge, and if his labors fall predictably short of producing a compelling axiomatics, at least his clarifications can counteract rashness and make people pause for a thoughtful view.

Where not existence but "quality" of life is in question, there is room for honest dissent on goals, time for theory to ponder them, and freedom from the tyranny of the lifeboat situation. Here, philosophy can have its try and its say. Not so on the extremity of the survival issue. The philosopher, to be sure, will also strive for a theoretical grounding of the very proposition that there ought to be men on earth, and that present generations are obligated to the existence of future ones. But such esoteric, ultimate validation of the perpetuity imperative for the species – whether obtainable or not to the satisfaction of reason – is happily not needed for consensus in the face of ultimate threat. Agreement in favor of life is pretheoretical, instinctive, and universal. Averting disaster takes precedence over everything else, including pursuit of the good, and suspends otherwise inviolable prohibitions and rules. All moral standards for individual or group behavior, even demands for individual sacrifice of life, are premised on the continued existence of human life. As I have said elsewhere,[11] "No rules can be devised for the waiving of rules in extremities. As with the famous shipwreck examples of ethical theory, the less said about it, the better."

Never before was there cause for considering the contingency that all mankind may find itself in a lifeboat, but this is exactly what we face when the viability of the planet is at stake. Once the situation becomes desperate, then what there is to do for salvaging it must be done, so that there be life – which "then," after the storm has been weathered, can again be adorned by ethical conduct. The moral inference to be drawn from this lurid eventuality of a moral pause is that we must never allow a lifeboat situation for humanity to arise.[12] One part of the ethics of technology is precisely to guard the space in which any ethics can operate. For the rest, it must grapple with the cross-currents of value in the complexity of life.

A final word on the question of determinism versus freedom which our presentation of the technological syndrome has raised. The best hope of man rests in his most troublesome gift: the spontaneity of human acting which confounds all prediction. As the late Hannah Arendt never tired of stressing: the continuing arrival of new-born individuals in the world assures ever-new beginnings. We should expect to be surprised and to see our predictions come to naught. But those predictions themselves, with their warning voice, can have a vital share in provoking and informing the spontaneity that is going to confound them.

Notes

1 But as serious an actuality as the Chinese plough "wandered" slowly westward with little traces of its route and finally caused a major, highly beneficial revolution in medieval European agriculture, which almost no one deemed worth recording when it happened (cf. Paul Leser, *Entstehung und Verbreitung des Pfluges*, Münster, 1931; reprint: The International Secretariate for Research on the History of Agricultural Implements, Brede-Lingby, Denmark, 1971).

2 Progress did, in fact, occur even at the heights of classical civilizations. The Roman arch and vault, for example, were distinct engineering advances over the horizontal entablature and flat ceiling of Greek (and Egyptian) architecture, permitting spanning feats and thereby construction objectives not contemplated before (stone bridges, aqueducts, the vast baths and other public halls of Imperial Rome). But materials, tools, and techniques were still the same, the role of human labor and crafts remained unaltered, stonecutting and brickbaking went on as before. An existing technology was enlarged in its scope of performance, but none of its means or even goals made obsolete.

3 One meaning of "classical" is that those civilizations had somehow implicitly "defined" themselves and neither encouraged nor even allowed to pass beyond their innate terms. The – more or less – achieved "equilibrium" was their very pride.

4 This only seems to be but is not a value statement, as the reflection on, for example, an ever more destructive atom bomb shows.

5 There may conceivably be internal degenerative factors – such as the overloading of finite information-processing capacity – that may bring the (exponential) movement to a halt or even make the system fall apart. We don't know yet.

6 There is a paradoxical side effect to this change of roles. That very science which forfeited its place in the domain of leisure to become a busy toiler in the field of common needs, creates by its toils a growing domain of leisure for the masses, who reap this with the other fruits of technology as an additional (and no less novel) article of forced consumption. Hence leisure, from a privilege of the few, has become a problem for the many to cope with. Science, not idle, provides for the needs of this idleness too: no small part of technology is spent on filling the leisure-time gap which technology itself has made a fact of life.

7 Not so, objects my colleague Robert Heilbroner in a letter to me; "I'm sorry to tell you that meat and rice are both *profoundly* influenced by technology. Not even they are left untouched." Correct, but they are at least generically the same (their really profound changes lie far back in the original breeding of domesticated strains from wild ones – as in the case of all cereal plants under cultivation). I am speaking here of an order of transformation in which the results bear no resemblance to the natural materials at their source, nor to any naturally occurring state of them.

8 Note also that in radio technology, the medium of action is nothing material, like wires conducting currents, but the entirely immaterial electromagnetic "field," i.e., space itself. The symbolic picture of "waves" is the last remaining link to the forms of our perceptual world.

9 I here loosely refer to Adolph Lowe, "The Normative Roots of Economic Values," in Sidney Hook, ed., *Human Values and Economic Policy* (New York: New York University Press, 1967) and, more perhaps, to the many discussions I had with Lowe over the years. For my side of the argument, see "Economic Knowledge and the Critique of Goals," in R. L. Heilbroner, ed., *Economic Means and Social Ends* (Englewood Cliffs, NJ: Prentice-Hall, 1969), reprinted in Hans Jonas, *Philosophical Essays* (Englewood Cliffs, NJ: Prentice-Hall, 1969), reprinted in Hans Jonas, *Philosophical Essays* (Englewood Cliffs, NJ: Prentice-Hall, 1974).

10 "The Practical Uses of Theory," *Social Research* 26 (1959), reprinted in Hans Jonas, *The Phenomenon of Life* (New York, 1966). The reference is to pp. 209–10 in the latter edition.

11 "Philosophical Reflections on Experimenting with Human Subjects," in Paul A. Freund, ed., *Experimentation with Human Subjects* (New York: George Braziller, 1970), reprinted in Hans Jonas, *Philosophical Essays*. The reference is to pp. 124–5 in the latter edition.

12 For a comprehensive view of the demands which such a situation or even its approach would make on our social and political values, see Geoffrey Vickers, *Freedom in a Rocking Boat* (London, 1970).

2

Four Philosophies of Technology

Alan R. Drengson

Alan Drengson is an Emeritus Professor of Philosophy at the University of Victoria, BC, Canada, where he was a Director of Environmental Studies and a member of the Philosophy Department. In this essay Drengson develops some typologies of technology and of philosophy. He first presents four possible understandings of "philosophy" and argues that philosophy is best understood as "a sort of jazz played with concepts." He further identifies four stages of technological development: technological anarchy, technophilia, technophobia, and appropriate technology. The central criterion of demarcation is the dominant human attitude toward technology in each stage. Technological anarchy is a playful, anything goes stage, when the possibilities of a technology are explored and when there is no dominant standard. Technophilia is love of, and in some cases identification with technology. In this stage, as in early stages of a love affair, one often will not notice the downsides, limits, and problems of technology. The "personation" discussed by Doug Browning is a good example of technophilia. Technophobia is a fear or hatred of technology. It goes beyond a reasoned awareness of negative affects, and tends toward rejection. Drengson argues that appropriate technology, as a self-critical stage and attitude, is the most mature and philosophically rich. Appropriate technology urges us to balance all costs, maintain biodiversity, promote benign interactions between humans, non-human animals, and technology, and to promote human development. This view captures many of the insights and goals of current programs of sustainable development and sustainable, or green, engineering and design.

From *Philosophy Today* (Summer 1982): 103–17.

Philosophy and Creative Inquiry

The aims of this essay are threefold: First, to describe four main philosophies of technology manifest in our culture; second, to engage in a process of creative inquiry that will make it progressively more obvious the extent to which an unwitting adherence to some of these philosophies can affect perceptions of technological possibilities; third, to outline the interconnection between conception, action, and social process with the aim of clarifying the role of conceptual design in intentional technological innovation.

In order to advance the aims of this essay, it is first necessary to explain what is meant by "philosophy" in this context. There are three levels to the term here: At the lowest level, a philosophy can be nonexplicit; at an intermediate level, it is an explicit elaboration of a particular position which spells out assumptions, axioms, etc., and argues for its conclusions; in the final and mature sense, philosophy is a creative activity of conceptual inquiry which frees us of attachment to specific models and doctrines in order to develop more appropriate cultural practices.

In the title of this essay, then, I speak of philosophy in the sense that one can express and live by a philosophy which is neither explicit nor clear, but which forms the structure and quality of one's experience. By "philosophy," then, is meant a way of life formed by attitudes and assumptions which, taken together, constitute a systematic way of conceptualizing actions and experiences by means of an implicit process of unquestioned judgments and conditioned emotional responses. In some dimensions these are cultural, in others they are familial or personal. Together these responses and judgments, constituted by both assumptions and evaluations, and an articulation of them in word and deed, make up one's philosophy of life. Most of the four philosophies of technology analyzed in this essay are culturally at the first level. The aim of this essay is to raise them to the second level, and then to move them to the third level by engaging in creative philosophizing about technological innovation and appropriate design. "Appropriate" here refers to right and artful fit between technique, tool, and human, moral, and environmental limits.

A caveat needs to be made at this point. The four philosophies of technology described here each occupies a given range on the continuum of responses to current technological development. The precise boundaries between each are difficult to mark. Moreover, each of these "philosophies" has certain specific adaptive and economic advantages. For example, a technophobic reaction to modern technology involves in part an attempt to revive and preserve simple, "primitive" technologies which, in the event of disaster, could serve survival and preservation of certain culture values. Technological change is highly dynamic in terms of its material manifestations, and the four philosophies described herein represent dominant views associated with technologically advanced societies. Nonetheless, the attitudes these philosophies represent tend to be primary human responses to change. A specific person may go through stages of development that pass through each of these philosophies. The creative philosopher recognizes the usefulness and limitations of each within this whole developmental process. He or she also recognizes the importance of a balance between each (as represented by different groups within a society) and within the dynamics of healthy social change.

Creative philosophy, as a form of inquiry, aims to free us of an attachment to doctrines and views, but enables us to use such doctrines and views to facilitate positive change and growth in understanding. In order to achieve this end, various metaphors and models are used as part of the activity of creative reflection on the four philosophies of technology. The use of such devices has certain risks. As has been observed by numerous sages, philosophers, insightful psychologists, novelists, Zen masters, and others, human thought tends to become fixated on various stereotypes, metaphors, models, paradigms and belief systems. Creative philosophizing recognizes their inherent limitations, but uses these various models, paradigms, etc., as a way of freeing understanding of their dominance. Initially one uses such models and the like as a way of conceptualizing the world in order to gain understanding and to serve practical aims. However, when these paradigms and their accompanying ideas, ideals, beliefs, and so on, become part of a belief system, it is easy to invest one's identity in them. When we invest our identities in beliefs we resist reflecting on them, and we resist their change, for this can seem a threat to one's self-identity and sense of reality. Thus belief systems tend to become static. Since life is a dynamic process, flexibility and creative adaptation suffer, when cultural processes involving dynamic factors such as science (as inquiry) and technology (as creative technique) get out of harmony with these more static belief systems.

The four philosophies sketched here are offered as provisional models to facilitate insight into the patterns of philosophy of technological development inherent in our culture. The creative philosopher recognizes the limitations in these patterns of thought and approaches them with a serious, but playful attitude so that distinctions can be recast through a continuous process of conceptual adjustment, readjustment, and improvisation. In creative philosophy, concepts become tools, paradigms heuristic devices, clarity and insight products of philosophical activity. In creative philosophy the aim is not *a* philosophy, but the activity of philosophizing as a way of continuously clarifying human intelligence by freeing it from its conceptual constraints. The fully sound human understanding is one that sees the world as it is, while it also realizes that cultural adaptation (of which technology is a part) is a creative affair and has a range of possible options, given the nature of the world.

A final word of caution. Creative philosophizing in its mature form is a nonposition and an activity. It is a sort of jazz played with concepts. It is a creative art that one acquires through long practice. It is classically illustrated in many of Plato's Socratic dialogues. As was observed in *The Republic*, ultimate reality lies beyond all of our forms of thought. The contemporary creative philosopher realizes that as long as we do not identify with these forms, they can be adjusted to better fit reality as revealed through fully aware immediate experience. By approaching philosophy creatively, as a process of dialogue and interaction, of give and take, playfully adopting a variety of perspectives, we free our capacity for creative thought and insight. Insight involves (in part) a direct grasp of networks of relationships and a seeing of the world that reveals its significance and value intensity, which are part of a common ground in the unity of being.

In contemporary Western industrial culture there is wide disagreement about how we should develop resources, whether or how to exploit animal species, whether and which new technologies to develop, and how to manage our collective activities in relation to individual rights and to the biosphere. Thus, the four philosophies discussed here represent the kind of broad, pluralistic mix that one would expect in modern Western democracy. This is particularly evident if we think of this matrix as a dynamic process that displays dialectical features. Within democratic society as a whole, complete consensus is not possible, especially since different people are at different stages of development. The four philosophies to be discussed could be said to represent the stages of maturation of an industrial society, and its gradual transformation into a mature, postindustrial culture characterized by human-scaled, ecologically sound, appropriate technologies, consciously designed to achieve compatibility with fundamental moral values. These matters will be explored now in greater detail.

Four Philosophies

There are four fundamental attitudes toward technology that can be discerned in current cultural processes in the industrial West. These attitudes form a continuum from an extreme faith in, to a complete distrust of, technology. The degree to which the various possibilities in between are held varies from person to person and between various subcultural groups. They do not readily correspond to any particular economic philosophy. These four philosophies can be conceived of as nodal points or as dense nexus of social attitudes which are centered on constellations of paradigms and beliefs. Within the whole continuum of social response their features can be described. Since the culture as a whole is in process, and since individuals within the culture are also changing at varying rates, depending on their particular circumstances, these nodal points are not static. They do not define all or nothing positions for the culture as a whole. If Western culture were to become either too static, or too dynamic, these views could become polarized, and then precipitate unresolvable conflicts and statements. As it is, they now appear to represent developmental stages of a continuous growth in which each successively becomes emphasized, as persons and the culture evolve.

For the purposes of this discussion I will designate the four philosophies under consideration as the following: (1) technological anarchy, (2) technophilia, (3) technophobia, and (4) technological appropriateness.[1] I shall now discuss the essential characteristics of each position and the interrelationships between them.

Technological anarchy was a dominant philosophy throughout much of the nineteenth-century industrial development of the West. In brief, technological anarchy is the philosophy that technology and technical knowledge are good as instruments and should be pursued in order to realize wealth, power, and the taming of nature. Whatever can be done to serve these ends should be done. The fewer government regulations over technology and the marketplace, the better. Ideally, there should be none, but this is impossible, since some basic order is necessary to further private ends. The market alone will determine which technologies will prevail. Technological anarchy is a philosophy of exuberant, youthful curiosity and self-centeredness. It is an expression of optimistic self-assertion and individual opportunism.

Technological anarchy helped to stimulate rapid technological development. It tends to encourage technological diversity. As industrial development matures, technological anarchy (within a given culture) tends to become less dominant. Technology becomes a more powerful directing force in the whole social process. Technology begins to take on certain autonomous features on a large scale. Technology, which was originally pursued as an instrument to satisfy desires and needs, tends in such a context to become an end in itself. As this process completes itself, technological anarchy loses its dominant position, even though it rarely completely disappears. It then gives way to technophilia, which in turn develops into a structure with technocratic features. (At the international level, technological anarchy still seems a dominant force.)

Technophilia, as the word implies, is the love of technology. It is like the love of adolescence. Humans become enamored with their own mechanical cleverness, with their techniques and tricks, their technical devices and processes. The products of our technology become not only productive instruments but also our toys. Technology becomes our life game. This is like the adolescent affair in which we identify with the objects of our love. As a result they tend to control us, for our unconscious identification with them invests these objects with our person. This identification becomes a form of control over us, since we are unable to disassociate ourselves from our technology. We cannot see it objectively. This can be illustrated by our love affair with the automobile. We can become so infatuated with automobiles that they become extensions of our selves. "Insults" to them become personal affronts, and can be felt as threats to self-esteem. This represents a loss of an objective understanding of the positive and the negative features of the technology of the auto – which includes the whole infrastructure of factories, gas stations, parking lots, roads, freeways, legal structures, supported, of course, by a whole complex of human routines and skills. Thus, although the automobile was first a means to an end, viz., transportation, it and its supporting infrastructure eventually became a dominant feature of the culture as a whole. Cities, land use, and even economic well-being have become entangled with the technology of the auto. What began as an instrumental value, as a means to the end which was transport, becomes an end in itself. Paradoxically this works to frustrate the original human values involved. Finally, the technology of the automobile can become a threat to life, health, economy, the environment, and even to our way of life.

Technophilia, as the love of technology, turns the pursuit of technology into the main end of life. It eventually aims to apply technology to everything: To education, government, trade, office work, health care, personal psychology, sex, etc. In this way it becomes technocracy, for technology is now a governing force. This represents the overwhelming of spontaneity by technique. In its most complete form, as technocratic, it represents the rule by and for technological processes. At this point humans are technologized by their own love of the technical and of techniques. Life becomes mere mechanism. However, this is only the implied logical terminus of technophilia. It is unlikely that it could achieve a complete technocracy because the social process is a stream with diverse elements. The application of technology to nearly everything stirs counterforces, and the imagined logical end of this pursuit is unacceptable to many. The love affair with technology cools as the process of maturation leads many people to realize that technology is becoming an autonomous force endangering human and nonhuman values. Even the biosphere as a whole becomes threatened by the products and processes of human technological activity. The initial reaction to these imagined and perceived threats is first to attempt to control technology and its hazards by means of technique and the technological fix. But these are both only extensions of the technophilia which furthers the development of technocracy.

Technophobia emerges when it is realized that only human and humane values can curb the threats of a technology running out of human control. As an extreme reaction technophobia attempts to detechnologize human life, for to many persons the idea of applying engineering techniques and technocratic control to all aspects of human culture is repugnant. It is seen as a mechanization of the human, leading to the loss of the sensitive, spontaneous and vital organism. There is a natural desire to return to human autonomy, which was originally one of the motives in pursuing technology, but it is now seen as frustrated by the techno-structure. This autonomy is perceived to reside in the revitalization of crafts and arts, of simpler, "neoprimitive" technologies. A do-it-yourself attitude characterizes it. The aim is self-sufficiency; a distrust of complex technologies is one of its features. Even while this reaction is developing the forces of technocracy are consolidating their control of extensive industrial technologies, which in turn, by their own inner dynamics, are evolving toward post-industrial maturity through smaller scaled, flexible systems of production. Ultimately, technophobia aims to bring the large-scale technologies to an end, and to bring technology once more under local human control. It helps prepare the ground for evolution to appropriate technological design.

Technophobia can be compared to the disenchantments of early adulthood. One learns that attachments which are centered in romantic and erotic identification can frustrate growth and can generate suffering, pain, grief, and fear of loss.

Such loss is felt initially as a severe threat to one's self-image. Unable to accept full responsibility for oneself (in every dimension), because one does not understand the exact nature of the situation, one inevitably suffers disappointment and may attempt to avoid such relationships in the future. This is usually not possible, although it is probably necessary to take this "pledge" as a step toward more mature relationships with others. In a similar way, perceiving the dangerous character of the technological panoply can at first be very disorienting, especially since it was originally thought that building such a technostructure would make life easier and safer. However, direct planning and innovation has often been done by persons who were not able to be fully responsible because they lacked sufficient understanding of the nature and implications of powerful technologies, or because they were caught in structures that made responsibility difficult. When human imagination is harnessed to technophilia in order to create and to proliferate technologies (as in the chemicals industry, e.g.), and when competition becomes an important force (whether national or international), it then becomes very difficult to control these technological forces. Fearing that this technological power will ultimately lead to total control of humans, or even to ecocide, finally brings disenchantment with the whole process. The romantic entanglement with technology (technophilia) is now perceived as threatening human integrity and survival.

Technophobia rejects technological autonomy and asserts human autonomy over it. This accomplishes two important things. First, it brings renewed commitment to humane values. Second, as already noted, it leads to the revitalization and preservation of arts, crafts, techniques, and skills that *emphasize* personal and interpersonal development as more important than technological supremacy over humans and nature. This not only preserves simpler technologies, but it insures that the process of maturation will continue, since it is necessary to psychologically distance ourselves from these activities, if we are to understand them. This understanding is necessary, if we are to perceive the possibilities for new forms of technology that are under our control and that are more appropriate to human and to natural values.

It is realized at this stage that it is the relationship between technology and ourselves that we must understand. This means understanding the relationship between nature and technology as well, for humans are born as nature and through techné and other cultural activities they modify themselves. The tendency is to see this cultural process as fixed, rather than the stochastic process that it is. In this case, jazz is a good paradigm for the art of self-creation as a stochastic process, for here there is the possibility of both control *and* spontaneity. The culture provides different roles for us to play. Thus there are patterns through which our activities can cohere and gain meaningful harmony. We could compare this process to the capacity of learning how to learn. Becoming aware of the possibility of knowing how to learn sets the stage for a continuous, consciously ordered transformation. If one becomes adept at learning, then one is adept at adjustment to ongoing changes in the world. One then becomes sensitively attuned to these changes, and can stay with them. When one learns how to bring one's full attention to a subject, and becomes capable of learning all there is to learn about it, then one becomes a master learner. From this vantage point technophobia can be seen as one of the stages of growth that involves becoming aware of the use of technology in a consciously reflective, critical way. We have the chance to see it from a meta-level.

Appropriate technology represents the fourth stage of technological development we have been describing in terms of the evolution of philosophy of technology and technological design. The fourth stage involves a maturing of the reciprocal relationships between technology, person, and world. Appropriate technology requires that we reflect on our ends and values, before we commit ourselves to the development of new technologies, or even to the continuation and use of certain older ones. As in mature love, one becomes capable of compassion and helping others to attain their ends (this is the very essence of the compassionate person), so in this stage we become capable of mastering our technology as instrumental to ends about which we become progressively more clear.

In the philosophy of appropriate technology, technologies should be designed so that they meet the following requirements. First, they should preserve diversity; second, they should promote benign interactions between humans, their machines, and the biosphere; third, they should be thermodynamically sound in the generation and use of energy; fourth, they should dynamically balance all costs; fifth, they should promote human development through their use. Let us reflect upon these points. Diversity is one of the features of both stable ecosystems and stable economies. Diverse technologies provide a large range of options to individuals and to further social development. Benign, symbiotic interactions between technology and the biosphere are necessary features of future technologies, if we are to develop sustainable economies. Compatibility with ecosystem principles is a minimal requirement. This is emphasized in sound thermodynamic design, for ecological compatibility and thermodynamic soundness work together to balance social, economic, and environmental costs. Finally, when technologies evolve to the level of appropriateness, they can be designed in such a way as to facilitate human development. Such technologies are designed to allow humans to master whole processes as arts, which stimulate the development of the complete human person. Thus the maturation of appropriate technology involves the transformation of the technological process into an art. The technological processes then become a life-enhancing part of a significant set of values. Labor thus becomes meaningful work. Comparing the stage of maturity of appropriateness to the capacity to love, we can say that it corresponds to the capacity for compassion. The compassionate person loves in order to enhance the other. Here technology is designed to enhance individual persons, ecological integrity, and cultural health.

From what has been said so far, we can see that technology cannot be separated from the selves that create and perpetuate it. If its creators and its perpetuators are immature selves, then the technical process will reflect their characteristics in various ways. Some of the bad consequences of technology are the result of design unduly influenced by immaturity, ignorance, confusion of ends, impatience, and too narrow values.

Appropriate technology is the most complete philosophy of the four outlined, since it addresses more of the relevant values, and since it also brings subject and object together in a responsible, reciprocal interaction. Furthermore, it recognizes the useful roles that the other philosophies can play. At this stage there is the possibility for continuing technological development in ways that resolve the negative consequences of the technological imperative of modern human history.

At present we seem to be moving toward the emergence of the philosophy of appropriate technology as a major force in our society. The period of the maximum influence of technophobia might be waning, but this is by no means certain, for there remain powerful forces of technocratic intent which are supported by vast resources with great institutional momentum. This tends to increase political and environmental opposition to technocratic policies. Technophobia could wax, particularly if there are large-scale failures of major technological projects which fully reveal all of their hazardous dimensions, such as pollution, debt, tyranny, and their displacement of human workers.

Appropriate Technology, Innovation and Mastery

Each of the philosophies outlined has played a role throughout the process of industrialization in the West. The technological anarchy that was dominant earlier was important in exploring and developing options that led to the industrial revolution. The forces of technophilia and the technocratic mind-set helped to create large-scale processes and infrastructures of continental and global extent that have importance and value. Without the function of these four philosophies expressed in individual lives and in collective social activities, we would lack many positive things we have today.

It would seem that the revolution in modern electronics, the miniaturization of technologies, the emerging solar technologies, improved organic agriculture, and various forms of personal and spiritual growth taken together point toward the possible emergence of the philosophy of appropriate technology as a major cultural force. It is a philosophy conducive to, and compatible with, these postindustrial technologies. From the perspective of appropriate technology, we have the opportunity to create new benign technologies with a clear intent of purpose.

An important feature of appropriate technology is that it forces us to ask central questions for the philosophy of technology: What shall be our relationship to technology? How should we define it? These and other fundamental questions receive our conscious attention. We are able to articulate assumptions of current policy and evaluate them in terms of the human context. Ultimately, appropriate technology aims to transform our relationship with technology in such a way that it becomes a means to the realization of abiding values we fully understand and freely choose. This means that the limits of technology are clearly perceived, and the values of simplicity realized in reduced dependence on heavy technologies.

Appropriate technology will also help to promote the re-creation of community vitality. Many of our current systems are too centralized. It is now necessary to shift to community revitalization through the development of decentralized, human-scaled technologies that preserve the values of the places in which communities have their being. Some of the large processes that are built into the system have now generated spin-off technologies that make such down-scaling and decentralization possible. The transition to such appropriate technologies can be aided by government policies, such as tax incentives and facilitating citizen participation in planning, but ultimately it can only be fully realized as the result of community and personal commitments which grow out of a mature understanding of the values at stake. As Plato saw so clearly, beyond all ideas, at the very center of existence is the Good. Putting this in twentieth-century terms, we can say that appropriate technology leads us to reflect deeply on life as a whole, and mature reflection leads us to realize that life has value at its center. The division between fact and value is only a logical division of concepts with limited usefulness. The practice of science and technology is a value-laden activity. Understanding life requires cognizance and appreciation for its many dimensions of value.

The philosophy of appropriate technology can be further illuminated by considering the four levels of innovation it recognizes. With respect to technological innovation, appropriate technology recognizes four fundamental forms: (1) technological modification, (2) technological hybridization, (3) technological mutation, and (4) technological mastery and creation. Technological modification involves improvement of a technology by means of gradual modification. This process relies heavily on trial and error. In the case of hybridization, we have the merging of two or more technologies to form a new technology or a new technological solution to an existing problem. An example of this would be the design of hybrid vehicles such as a propane-electric automobile. Technological mutation is the transformation of a technology to some other form, or for some radically different purpose. For example, the Chinese used gunpowder for fireworks entertainment, but not to do work or to fight battles. The Mongols and then the Europeans transformed this technology and applied it not only to armaments and warfare, but also for use in the construction of roads, tunnels, dams, and other things. In a reverse direction, atomic bomb technology has been transformed to nonwarfare applications in medicine, the generation of electricity, and the propulsion of ships. These forms of innovation are recognized by the other philosophies of technology discussed here. But appropriate technology emphasizes technological mastery and creation, which involves the capacity to transcend technology and much of human dependence on it. At the same time it opens endless possibilities for the creation of new appropriate technologies. One masters an art by transcending one's fascination with techniques; for the master there is fluency and freedom in the art.[2] Rules and a breakdown of techniques are useful for instructing learners. Mastery transcends these since it leads to spontaneous, creative activity. We often depend on rules and techniques because we have not achieved complete mastery or fluency in the art. Technological mastery in the context of appropriate technology leads to the possibility of transcending technology as a force in human life that lies beyond our control.[3]

The philosophy of appropriate technology encompasses the possibility of mastery and creativity. In its mature form this can be seen as the possibility for a self-mastery that transcends self-manipulation and the desire to control others. In short, the end of domination by technology is seen to lie beyond technology in the realization of human possibilities for mastery of technology in a way that emphasizes the value of persons, develops creative community, and promotes communion with nature. This is the ultimate raison d'être of a fully mature, appropriate technology.

Technophilia and Appropriate Technology Compared: Four Examples

In this section we shall explore more fully the contrast between the approach of appropriate technology (which involves the self-mastery necessary for the wise use of technology) and the approach of technophilia (which uses technology as a means to provide the power to control nature and other humans). The philosophy of appropriate technology applies technology to the natural world in a way respectful of its intrinsic values, whereas technophilia seeks to impose technology upon a nature seen only as resources having instrumental values. The appropriate technologist is respectful of the values *in* the world, whereas the technocratic mind of technophilia attempts to impose patterns of its own devising *on* the world. The aim of appropriate technology is to understand the world and appreciate it, so that humans can interact with it to realize a maximum of reciprocal benefits and also of such values as wonder, delight, and compassion. Technophilia (in contrast) does not seek to know the other, to experience the other, but only to manipulate and control the other, to possess the other. It sees the other as object, not as subject. For the appropriate technologist, however, the living world is filled with subjects. Its dynamic, untamed, organic processes are interdependent. It cannot be approached in a fragmentary way, as a collection of objects to be subdued. It must be approached as a subject-other.

Appropriate technology is a philosophy that includes the human self as part of nature's selves. Questions of ends are primary, and ends depend upon knowing the kinds of beings that we are and can be. This finally leads beyond all techniques and tools, beyond their limits to our own limits. These limits are known through self-knowledge and self-mastery. Self-mastery leads to a mastery of technology that is appropriate

to ends worthy of human pursuit. In order to illustrate this important aspect of the philosophy of
appropriate technology let us now consider four
examples. These examples will help to illustrate
the difference between appropriate technology
and technophilia in the use and design of technology. The four examples we will discuss are
interpersonal conflict, alpine hiking, exercise,
and energy generation and use.

Consider, then, some of the levels of technology available for resolving interpersonal
conflicts. We shall use examples of warfare and a
specific martial art to illustrate the practical difference in philosophy between technophilia and
appropriate technology. Suppose that two tribes,
two countries, or two treaty groups have a disagreement that seems unresolvable and tending
toward violence. Naturally, in these situations
tempers can become inflamed. Tension builds
while the conflict simmers on. Under these
conditions fears arise. These fears magnify the
perception of what are interpreted as threats. At
a certain stage one or both of the antagonists will
think of resorting to force in order to remove the
tension. If they apply the full range of modern
technology to this conflict, the forces involved
could destroy one or even both sides. If they
think in terms of winning and losing, then an
all-out technological response would seem irrational, given nuclear arms. Hence, they are
forced to consider other options. Negotiation and
willingness to compromise could be buttressed by
this powerful technology, but only if the parties
know that there is no armed technological solution to their conflict. It becomes clear at this
point that the total use of this vast technological
power negates its practicality. It is no longer
useful for its originally designed purpose, for
technological power has undermined the rationale
of war. The pursuit of a technological solution
to the conflict could then lead beyond a focus
on technology, as a result of the very logic of
technophilia's total technological response. At
this point it can be seen that the appropriate
response to human conflict is not technological
warfare. The application of technology leads us
to realize that ignorance, immaturity, and lack
of self-mastery underlie much interpersonal and
international conflict. This can be brought out
more clearly at the level of an interpersonal
conflict restricted to two persons.

Let us consider the range of options open to
two persons, assuming a high level of conflict
between them. At the technical level they could
resort to bombs, guns, swords, knives, clubs,
stones, fists, and feet. If they are martial artists
they might use karate, judo, or boxing. Now the
technocratic approach is to try to control the
other person through the use of technology
and techniques. The philosophy of appropriate
technology can be illustrated by the martial art
of aikido. The aikido martial artist practices
the martial way but uses the energy that would
be spent on fighting to transcend fighting. The
master aikidoist is the ultimate martial artist,
since there can be no aggression and competition.
Aikido is such a complete art that it resolves
conflicts before they can progress to fighting. It
is highly subtle, since it masters the impulse to
fight by transcending the small self that would
fight. It leads one to understand others and the
reasons for our impulses toward aggression. This
is an art that has its origin in the techniques
of fighting, but ultimately it transcends fighting
and techniques by means of a practice which
leads toward self-mastery. Instead of attempting
to manipulate and control others through techniques and fighting technology, aikido resolves
conflicts through self-mastery, self-correction,
and understanding.

Consider as our second example alpine hiking.
Let us compare two hikers: One is loaded with
every conceivable camping device modern technology has produced. He is also involved in
learning all available techniques. His weekend
pack weighs at least 100 pounds. When he camps
he employs these various techniques and technology to make a well-organized, "comfortable"
camp. He engages in lots of wood craft, lots of
"wild-river Jim," nailing, chopping, and building.
He loads up his gear in the morning, after
spending two hours flipping pancakes on a fancy
griddle.

In contrast, the appropriate technology hiker
travels light. She is not a "live off the wilderness"
wildperson, digging up roots, rooting out berries,
and eating the flowers. She is there to celebrate
the joy of being alive, and the joy of being able
to know nature in an intimate way. She is there
to listen to the softer voices of the world and to
the deeper voices within herself. Her equipment
is carefully designed to be simple, light, durable,

minimally polluting, and harmless to the world in its production and use. She is comfortable, but not isolated from the elements of nature she would know. The rain is not an enemy, nor is the sunshine the only pleasure. She eats simple food, such as a breakfast of homemade granola, that requires no or minimal cooking, but nonetheless is optimally nutritious and aesthetically satisfying.

These two hikers illustrate the differences in philosophy between the technophiliac and the appropriate technologist. For the former, the equipment becomes a burden that isolates him from the natural world. For the latter, the equipment is a minimal intrusion which is efficient and enhances her enjoyment of the natural world. It is not a burden, but a joy to use.

As our third example let us consider the range of possibilities open to us with respect to technology and exercise. Ideally, the aims of exercise are self-discipline, fun, and a strong, healthy, flexible, and aesthetically balanced body. Technology can be used to assist in this process. However, the ultimate end of applying technology to exercise undermines many of these aims, as is seen in the exercise machines that do all the moving for you. There is no interaction. You become the manipulated. The other contrasting attitude approaches exercise as a form of self-discipline to be enjoyed also for its own sake.

In jogging, one needs only running shoes, nothing else. Aikido can be done with soft clothes, a padded floor, and one other person. Isometrics and calisthenics require no equipment or helpers. For the philosophy of appropriate technology, the approach to exercise is an integrated and elegant one that uses technology minimally, and would emphasize self-mastery instead of some "easy" technological solution to overweight and lack of sound conditioning. In the technocratic approach, machines become a substitute for this self-discipline and tend to alienate one from one's own body.

Finally, for our fourth example consider the generation and use of energy to illustrate the contrast between the technocratic thrust of technophilia and the approach of appropriate technology. The epitome of the technocratic approach is represented by nuclear power. The use of nuclear fission to boil water to generate steam to power electric generators involves the use of highly capitalized and centralized technology. In the form of electricity this power is distributed through complex grids to distant end users. Electricity is applied to a variety of uses, such as cooling, cooking, and space heating. Nuclear power is highly complex and requires vast subsidies in the form of publicly financed insurance and storage of dangerous wastes. It presents difficult problems of security and increases the probability of the spread of nuclear weapons. In terms of energy use it employs high-temperature processes to accomplish many practical ends which are of low thermodynamic quality. It adds thermopollution to rivers. For these and many other reasons, nuclear power is environmentally, economically, and thermodynamically unsound. It raises serious moral questions. Nonetheless, to the technocrat it is a "logical" way to go.

In contrast, for the appropriate technologist the aim is to diversify and decentralize the use and production of energy. Instead of relying on vast power systems (although some may be developed), the aim is to develop a large variety of smaller scale technologies such as photovoltaic, hydroelectric, and solar. Such approaches as cogeneration and conservation within communities create local systems that use generated power and heat over several times. It gives to local communities greater control over their future, lower costs and debt, and broader public participation, in contrast to many of the large-scale projects which promote complex bureaucratic management structures, increased environmental hazards, and large debt. Appropriate technology emphasizes thermodynamic soundness, doing more with less, conservation, and keeping open a large variety of options. It is rich in understanding of natural processes and takes advantage of the rhythms of natural sources of energy that are readily available on site. It relies on a mastery of design that blends technology and ecological processes, rather than imposing powerful technologies upon nature. In contrast, technocratic forces strive to master nature by controlling and overwhelming rather than working with it.

We can see from these examples, and from earlier comments in this essay, that attempting to resolve the problems caused by technology without first appreciating the human elements involved leads nowhere. The problems of technology that have social and personal implications are not just problems of technology. If we do not appreciate the influence of the particular philosophy of technology that underlies our own individual approach, and see its contrast with other views within our culture, then we will lack a perspective that enables us to move beyond the search for technical solutions to nontechnical problems. In philosophizing about these philosophies of technology I have attempted to sketch how their conceptions of technology affect self, society, and nature. If through this activity we are better able to attend to these attitudes directly, then the chances for a flexible, creative adjustment of our interactions with one another and the world will be increased.

Conclusion

The problems of technology do not all have technical solutions, for the root of some problems of technology lies in the problems of human life itself. Our attitudes toward technology define us, and they bind us to the creation of processes that magnify our initial failure to understand life as the interrelated, holistic process that it is. Powerful modern technologies express in their material forms problems for human life precisely because these technologies reflect the nonresolution of underlying uncertainties about existence and value. Martin Heidegger was one twentieth-century thinker who realized this. He saw that much modern technology grows out of a confused metaphysics that manifests itself in our material and other cultural processes. This confused metaphysics, he observed, is essentially the result of a failure to understand Being and what it means to dwell in the world. Our failure is not that we have linked our industrial technology to profit; it is rather that our pursuits and their technology fail to understand what it is to be in the world in the full openness (the mystery) of Being. Modern industrial technology, as often applied, is an example of a lack of comprehension of Being, a lack of care for the world, and a failure to perceive the fundamental essence of things. It lacks an understanding of the sense of life and of values. With this failure goes the inability to let others be. It begins with confused, calculative thinking, but once this thinking is

expressed in the material of technology, that technology then carries it across political and economic boundaries. This is why in the contemporary world industrial technologies and their negative features are transpolitical. The philosophy of appropriate technology recognizes these failings and is open to new possibilities. Because of this it can help us to free our minds of narrower technological concerns, and the sense of being overwhelmed by the "inevitability" of the domination of humans by their own technology. Technology need not be an alien power that overrides responsible human choice. We are better able to solve problems because we better understand their source. The dialogue of creative philosophy frees our minds, the philosophy of appropriate technology frees our practical work of technical and technological tyranny. Together they blend science and art in creative adaptation to a natural world that embodies values to which humans contribute.

Notes

1 "Appropriate technology" is a term sometimes used for intermediate technologies (Dunn, 1978). Intermediate technologies are designed for application in developing economies. As we use the term, "appropriate technology" refers to the philosophy we have here described. It is capable of guiding technological designs for many levels of development. Dunn's definition of appropriate technology in his first chapter is not incompatible with the one used here. For a more detailed discussion of the philosophy of appropriate technology, see my article, "Toward a Philosophy of Appropriate Technology," *Humboldt Journal of Social Relations*, Spring/Summer, 1982, vol. 9, no. 2, pp. 161–76. This issue of the journal is devoted entirely to appropriate technology.

2 On the mastery of arts as a form of self-development and self-transcendence, see my paper, "Masters and Mastery," *Philosophy Today*, Fall, 1983, vol. 27, no. 3/4, pp. 230–46. On the relationship between art, imagination, and technology, see my paper, "Art and Imagination in Technological Society," *Research in Philosophy and Technology*, Fall, 1983, vol. 6, pp. 77–91.

3 One example of the creation of a completely new technology would be learning how to directly influence the informational forms that underlie matter, and which direct energy to create specific material forms. Gene splicing would be another

example (perhaps just a different application of the former). Such new technologies depend on a deep understanding of natural processes, which could work with them, rather than attempting to subdue or overwhelm them. Many earlier (and present) industrial technologies are less subtle, poorer in understanding, and are often crudely overpowerful. However, biotechnologies carry some profound risks. There are also inherent limits to the pursuit of a technological fix. For an exploration of some of these issues, see my paper, "The Sacred and the Limits of the Technological Fix," *Zygon*, September, 1984, vol. 19, no. 3, pp. 259–75. This issue of *Zygon* contains other articles relevant to new biotechnologies.

Bibliography

Barrett, William. *The Illusion of Technique: A Search for Meaning in a Technological Civilization.* New York: Anchor/Doubleday, 1978.

Boulding, Kenneth E. *The Meaning of the 20th Century: The Great Transition.* New York: Harper & Row, 1965.

Commoner, Barry. *The Closing Circle: Nature, Man and Technology.* New York: Bantam, 1972.

Commoner, Barry. *The Poverty of Power.* New York: Knopf, 1976.

Dunn, P. D. *Appropriate Technology: Technology with a Human Face.* New York, Schocken Books, 1978.

Durban, P. T., ed. *Research in Philosophy and Technology.* 6 vols. Greenwich, Conn.: JAI Press, 1978–1983.

Ellul, Jacques. *The Technological Society.* New York: Vintage, 1964.

Fromm, Eric. *The Revolution of Hope: Toward a Humanized Technology.* New York: Bantam, 1968.

Galbraith, John K. *The New Industrial State.* New York: Signet, 1967.

Heidegger, Martin. *The Question Concerning Technology and Other Essays.* New York: Harper & Row, 1977.

Illich, Ivan. *Tools for Conviviality.* New York: Harper & Row, 1973.

Jantsche, Erich. *Design for Evolution.* New York: George Braziller, 1975.

Lovins, Amory B. *Soft Energy Paths: Toward a Durable Peace.* Cambridge, Mass.: Ballinger, 1977.

Mitcham, C., and R. Mackey, eds. *Philosophy and Technology.* New York: Free Press, 1972.

Mumford, Lewis. *The Myth of the Machine.* Vol. 1, *Technics and Human Development.* New York: Harcourt Brace, 1967.

Mumford, Lewis. *The Myth of the Machine.* Vol. 2, *The Pentagon of Power.* New York: Harcourt Brace, 1970.

Odum, Eugene. *Fundamentals of Ecology.* Philadelphia: Saunders, 1971.

Papanek, Victor. *Design for the Real World.* New York: Bantam, 1973.

Roszak, Theodore. *Personal Planet.* New York: Anchor/Doubleday, 1978.

Schumcher, E. F. *Small Is Beautiful: Economics As If People Mattered.* New York: Harper & Row, 1973.

Shepard, Paul, and Daniel McKinley, eds. *The Subversive Science: Essays Towards an Ecology of Man.* New York: Houghton Mifflin, 1969.

Stavrianos, L. S. *The Promise of the Coming Dark Age.* San Francisco: W. H. Freeman, 1976.

Tawney, R. H. *The Acquisitive Society.* New York: Harvest Books, 1948.

Watt, James. *The Titanic Effect.* Stanford: Senaur and Associates, 1974.

Weizenbaum, Joseph. *Computer Power and Human Reason.* San Francisco: W. H. Freeman, 1976.

Wilber, Ken. *The Spectrum of Consciousness.* Wheaton, Ill.: Quest, 1977.

Young, Arthur M. *The Reflexive Universe: Evolution of Consciousness.* San Francisco: Delacorte, 1976.

3

The Relation of Science and Technology to Human Values

William W. Lowrance

William W. Lowrance, formerly at Harvard University and at Stanford University, is now a consultant in health research policy and ethics. In this selection from his 1985 book, *Modern Science and Human Values*, Lowrance argues that two traditional positions on the relations between technology and values are wrong. Both the claim that technology has no relation to values, and the claim that technology strictly determines values are mistaken. Holding either of these positions limits our ability to understand the value assumptions and implications of technological change. He brings to bear the classical notion of the tragic to illustrate that this failure to understand technology also amounts to a failure to understand ourselves and the world we are making. Technology is not value neutral, but neither does it sap human values. He writes that attending to the tragic dimension in technology can lead to a better understanding of both the limits and possibilities of human existence. The tragic dimensions comes to the fore in many of the applied issues considered later in the book, especially human reproduction, environmental concerns, energy production, and war. Like Jonas, Lowrance would urge on us something like the "precautionary principle," – when we are uncertain of an outcome we must weigh not only probability but also the scope of possible outcomes and err on the side of precaution.

Admiration of the extraordinary powers of science often tempts people to hope that the laboratory and clinic will hand down social *oughts*, "Thou shalts." Convictions on the issue range from the view that science and technology can and must be used to generate human values, to the view that science and technology are just as value-neutral as banking or playing

From W. W. Lowrance, *Modern Science and Human Value* (New York: Oxford University Press, 1986), pp. 145–50. By permission of Oxford University Press, Inc.

soccer are and have little moral or political character.

Neither extreme, I will argue, is correct. Although scientific knowledge, once attained, may be considered *ambi-potent* for good or evil, the work of pursuing new science and developing technologies is by no means value-neutral (as, in the context of international politics, banking and playing soccer may not be, either). And although they don't dictate values, technical analyses and accomplishments profoundly influence social philosophies and choices.

Values and Facts: the Basic Relations

Does science generate moral oughts? Almost, on occasion, but usually not, and never without reference to embedded social values. Observations of unfitness in offspring of incestuous human mating, reinforced by analogous observations in nonhuman species and long proscription by most societies as a "crime against nature," support the almost universal taboo against incest. Technical estimates of the physical, biological, and social consequences of nuclear war make us dread it. But even in these extreme cases our responses still depend on value judgments that lie outside science: abhorrence of giving birth to defective children, abhorrence of genocide.[1]

Toward desired ends, enabling oughts can be formulated in the light of knowledge from many sources, including – powerfully – science. As we become aware of environmental connections and consequences, we weave ecological sensibilities into our values fabric. As we accumulate evidence on how life-habits affect personal health, we reassign social responsibilities for health promotion.

Social values simply cannot be derived from science qua science alone. To speak, as some cavalierly do, of "the values of science" may mislead. To be sure, over the past several centuries scientists have developed tenets of method, evidence, and proof, and they have cultivated an ethos of intellectual openness, truthfulness, and international fraternity. Some writers have been so impressed with the ethics and etiquette through which scientific work proceeds that they have urged that these mores be adopted as the foundation of social ethics. In *Science and the Social Order* Bernard Barber suggested that the "rationality, universalism, individualism, 'communality,' and 'disinterestedness'" that serve science so effectively "could even some day become the dominant moral values for the whole society" (1952, 90). Anatol Rapoport went so far as to say that "the ethics of science must become *the* ethics of humanity" (1957, 798). Jacob Bronowski argued that since science flourishes in societies fostering such "values of science" as "independence and originality, dissent and freedom and tolerance," such norms should be adopted for other social endeavors as well (1965, 62). But science has no monopoly on creativity, truthtelling, or tolerance, nor is it uniquely the definer of these traits. Science's precedent is hardly a sufficient model for the redesign of social ethics.

Technical people contribute richly to the alleviation of suffering and the enhancement of culture. Like everyone else, scientists hold deeply cherished personal convictions, which they express often and articulately. Groups of scientists, very large groups even, vigorously pursue social goals. Coalitions may go so far as to engage in partisan politicking, as the Scientists, Engineers, and Physicians for Johnson/Humphrey did in the 1964 presidential campaign. But such actions derive no more from the methods or scientific knowledge of Pasteur or Bohr than the 1983 Artists Call Against U.S. Intervention in Central America derived from the aesthetic tenets or oeuvre of Turner or Cézanne.

On the other hand, any assertion that scientific activity is value-free or value-neutral is disingenuous. Disclaimers have been made at least since Robert Hooke's 1663 proposed charter for the Royal Society: "The business and design of the Royal Society is – To improve the knowledge of naturall things, and of all useful Arts, Manufactures, Mechanick Practises, Engynes and Inventions by Experiments – (not meddling with Divinity, Metaphysics, Moralls, Politicks, Grammar, Rhetorick or Logick)" (Lyons 1944, 41). But even as Hooke drafted that antiseptic mandate, interpretations of Divinity and Metaphysics were brought under severe challenge by science; Mechanick Practises, Engynes, and Inventions were pursued that would profoundly affect the Moralls of warfare and Politicks of labor; studies of the Grammar of the world's languages grew through philology toward modern semantics;

and lines of Rhetorick and Logick were explored that would lead straight to G. E. Moore and Bertrand Russell.

When, in London in 1838, William Whewell, Charles Babbage, and their colleagues founded the *Journal of the Statistical Society* they chose as its symbol the wheatsheaf, to stand for the facts the journal would gather that "alone can form the basis of correct conclusions with respect to social and political government." On a band around the sheaf these canny masters emblazoned the motto "*Aliis Exterendum,*" "It must be threshed by others," as though the facts threshed aren't conditioned by the gathering and sheaving. But even the titles of the journal's first papers gave them away: "Social and moral statistics of criminal offenders," "Vicious extent and heavy expense of advertisements in England," "On the accumulation of capital by the different classes of society." They kept the wheatsheaf but dropped the motto in 1857.[2]

Occasionally even today a neutral gray waistcoat is donned against suggestions that the work of science and technology is value-freighted. But, just as in Hooke's day, and in Whewell and Babbage's, no major creative activities in society – especially those that are pragmatically or symbolically powerful – should be allowed to claim valuative or moral immunity.

Jacob Bronowski's distinction is key: "Those who think that science is ethically neutral confuse the findings of science, which are, with the activity of science, which is not" (1965, 63). Research, once accomplished, must be considered in the long run ambi-potent, usable for either good or evil. The anticholinesterase chemicals developed as nerve gases between the World Wars later turned out to be elegant research weapons in the protein biochemistry revolution, and botanical research on how trees drop their leaves in autumn led to development of the military defoliant Agent Orange. In the short run, of course, facts and know-how can be kept secret, or applied under close control, but they are likely to be revealed or discovered independently elsewhere, eventually.

But: We must vigorously resist any notion that researchers are helpless to make choices among envisionable future lines of research and development, or among possible conditions of pursuit. Although it may not be useful to regard already published knowledge as having any particular moral or ethical cast, surely it is wrong to view not-yet-accomplished research, which cannot be undertaken without commitment of will and resources, as being anything other than value-laden. And, because timing and pacing always are important, it would be naïve not to recognize that new knowledge may at the moment of its emergence have maleficent or beneficent potency that demands attention. Questions of practical ethics always lie in *what to do next*.

Technical activity must be considered value-laden in two senses: technical people's social values and value perceptions affect their research and service; and that work, in turn, affects the value-situation of others in the public.

Thus I consider to be value-laden: the undertaking and supporting of research (for example, committal of funds to research on acquired immune deficiency syndrome (AIDS), or other disease); the choosing of conditions of experimentation (diplomatic auspices of hurricane-seeding experiments in the South Pacific Basin); the marshaling of science to analyze, assess, and help decide on socially important problems (agricultural policymaking); the investigating of people's values, using social-scientific methods (studying jail guards' attitudes); the applying of technology and medicine to practical problems (treatment of breast cancer); and the incorporating of scientific knowledge into the fabric of social philosophies and policies (taking the findings of child psychology and of income-maintenance economics into account when revising child welfare programs).

Technical experts make crucial decisions for, and in the name of, the public. Although geology students usually don't view themselves as moving into a value-charged realm – what could be less social than rocks, after all? – as their careers progress geologists find themselves making seismicity assessments for hydroelectric or nuclear power plant siting decisions, advising on beach protection, municipal building codes, transnational water resources, seabed mining, and strategic minerals supply, and leading projects on causes of acidification of lakes and on underground disposal of radioactive waste. Teams of automotive engineers design cars and sell the designs through corporate management to the public. Pharmaceutical experts develop, test, and push drugs toward the market. Nuclear power plant designers weight the ratio of instant to delayed (cancer) death risks they design into reactors. Nuclear managers decide, in cleaning up after an accident, between exposing a few workers to radiation for relatively long times and exposing more workers for shorter times. Much research by social scientists – on the effect of school busing on educational achievement, on the effect of incarceration on criminal recidivism, on the influence of wage incentives on acceptance of occupational hazard – is so integral to policy-making that analysis can hardly be distinguished from advocacy. And of course some scientists and physicians themselves become high official decisionmakers in industry, labor, and government.

Experts' overt value stances can be argued about. Much harder to dig out and deal with are inarticulate premises. Genetic counselors' counsel is bound to be tempered by their attitude toward contraception and abortion. Marine ecologists' advice on the dumping of wastes into the ocean hinges on whether they think of the oceanic environment as being fragile or resilient, and on whether they prefer a pristine ocean to a "working" ocean. In psychotherapy, as Anne Seiden has argued, "the assumption that dependency, masochism, and passivity are normal for women and the tendency to treat assertiveness and aggression differently for women than for men" leads to "different standards of health for women and men"; therapists' practice thus is conditioned by their intuited, schooled, and inferred interpretation of gender (1976, 1116).

Later chapters will address such questions that arise on this account as: How should advisors and advisory committees, in their procedures and reports, deal with their factual biases and value preferences? (As with bias in a textile, "bias" simply means inclination, and isn't necessarily pejorative.) Since most of the work individual scientists do is guided by personal motives and morals rather than by grand ethical schemes, and since each researcher contributes only small increments to the overall technical enterprise, how should individual scientists' actions be oriented to "society's" values? How should scientific research freedoms be balanced against societal constraints?

Serious trouble arises when the distinction between facts and values is blurred or not recognized, or when disputants engage in mislabeling. Nothing has illustrated this more dramatically than the congressional abortion battle of 1981. Senate Bill 158 was introduced which would extend to unborn fetuses the rights of due process guaranteed by the Constitution's Fourteenth Amendment, circumventing Supreme Court decisions preserving women's right to abortion. In hearings, five medical researchers and physicians were drafted into testifying that a human being is formed at the moment sperm fuses with egg, and that this is a "scientific fact." Professor Jérôme Lejeune of the Medical College of Paris asserted, "To accept the fact that after fertilization has taken place a new human has come into being is no longer a matter of taste or of opinion. The human nature of the human being from conception to old age is not a metaphysical contention, it is plain experimental evidence." Boston physician Micheline Mathews-Roth insisted that "one is being scientifically accurate if one says that an individual human life begins at fertilization or conception." Entering the fray, Yale geneticist Leon Rosenberg protested that "the notion embodied in the phrase 'actual human life' is not a scientific one, but rather a philosophic and religious one." He quoted geneticist Joshua Lederberg: "'Modern man knows too much to pretend that life is merely the beating of the heart or the tide of breathing. Nevertheless he would like to ask biology to draw an absolute line that might relieve his confusion. The plea is in vain. There is no single, simple answer to 'When does life begin?'" Rosenberg emphasized, "I have no quarrel with anyone's ideas on this matter, so long as it is clearly understood that they are personal beliefs... and not scientific truths" (U.S. Senate, Subcommittee on Separation of Powers, 1981). The transgression was important enough to move the National Academy of Sciences to pass one of its rare resolutions (April 28, 1981):

It is the view of the National Academy of Sciences that the statement in Chapter 101, Section 1, of U.S. Senate Bill S158, 1981, cannot stand up to the scrutiny of science. This section reads "The Congress finds that present-day scientific evidence indicates a significant likelihood that actual human life exists from conception." This statement purports to derive its conclusions from science, but it deals with a question to which science can provide no answer. The proposal in S158 that the term "person" shall include "all human life" has no basis within our scientific understanding. Defining the time at which the developing embryo becomes a "person" must remain a matter of moral or religious values.

[. . .]

Meaning of "Values" and "Value"

"Values" and "value" have been given so many connotations by the public and by philosophers, theologians, psychologists, economists, and other specialists that I cannot hope to refine them into neat definitions. What I will try to do is draw some commonsense perspective, illustrate how values are manifested, and show how values enter into analysis and decisionmaking.

In ordinary usage, *values* are taken to be abstract aspirations: freedom of speech, cohesive family life, national security. Such goals may be neither perfectly definable nor attainable, but, as with the Constitution of the United States, they can serve as ideals. It is in this spirit that the World Health Organization constitution declares that all people have a right to the "highest possible level of health," which it defines as "a state of complete physical, mental, and social well-being, not merely the absence of disease or infirmity."

Values can, in narrower usage, be taken as potentially attainable states of affairs, as objectives: assurance of infant survival, eradication of leprosy, achievement of energy independence, maintenance of forests for future generations. Values can govern means as well: protection of entrepreneurial access to seabed minerals, requirement that behavioral experiments not be carried out unless the subjects freely grant informed consent.

Value I take to be ascribed worth, as reflected in social preferences and transactions: market value of zinc, information value of a blueprint, political value of a senator's endorsement, social value of literacy, aesthetic value of a cityscape, symbolic value of a new medical clinic.

Value-laden (-*freighted*, -*charged*, -*oriented*) then connotes that an analysis, decision, or action is influenced by personal or institutional proclivities and prejudices, and that the analysis, decision, or action may affect people's value situation – their opportunities, status, wealth, happiness, or aspirations.

In 1752 David Hume made clear how valuation regresses to deep "sentiments":[3]

> Ask a man *why he uses exercise*; he will answer, *because he desires to keep his health.* If you then inquire *why he desires health*, he will readily reply, *because sickness is painful.* If you push your inquiries further and desire a reason *why he hates pain*, it is impossible he can ever give any. This is an ultimate end, and is never referred to any other object.
>
> Perhaps to your second question, *why he desires health*, he may also reply that *it is necessary for the exercise of his calling.* If you ask *why he is anxious on that head*, he will answer, *because he desires to get money.* If you demand, *Why? It is the instrument of pleasure*, says he. . . . Something must be desirable on its own account, and because of its immediate accord or agreement with human sentiment and affection.

Three fundamental questions have pervaded value-laden decisions and actions throughout history. How should choices be made when options cannot all be pursued or when they conflict? (If people are differently vulnerable to health hazards, and workplaces can never be perfectly risk-free, how should equal health protection be reconciled with equal employment opportunity?) How should collective societal goods be pursued with least erosion of the rights and goods of affected individuals? (What degree of vaccination efficacy for a population should be judged to outweigh side-effect risks to individuals?) And how should specific guidance on particular real actions be derived from abstract high precepts? (What does "right to personal privacy" mean in our present world of electronic financial transactions, international campaigns against terrorism and drug smuggling, and institutionalized health recordkeeping?)

People's strivings toward what they value – home and neighborhood lifestyles, opportunities for their children, assistance to citizens of less fortunate countries – hardly can be expressed precisely. Within any group, preferences will differ, and values will change over time. It is especially hard to resolve grand goals (national security) into the prosaic objectives (missile deployment plans, titanium stockpile policies, computer export restrictions), themselves value-laden, required for achieving those goals.

Public opinion polls can to some extent reveal value concerns. I remain a skeptic, and believe that polls that don't force respondents to choose among real options and confront trade-offs are worthless. Actions speak louder than answers to surveys. To me, actual manifestations of valuation – in political actions, budgets, laws, treaties, regulatory policies, military strategies; in court, corporation, and labor union decisions; in consumer purchasing behavior; in medical preferences; and in wage differentials and insurance schemes – are much more telling than casually expressed "opinions" are.

Influence of Science and Technology on Social Philosophies and Choices

Although they overlap and are not a formal taxonomy, the following modes of influence can be distinguished.

Science deeply informs our cultural outlook. Science has transmuted quite a few major cultural myths; negated many superstitions; left us living in a "disenchanted" world; imparted substance to a host of miasmas, humors, auras, scourges, and vital forces; recast the mind–body, nature–nurture, and other classic mysteries; and conspicuously revealed the hand of Man where none was seen before but Fortune's. Science reveals fundamentals about the occurrence and causes of mortality, genetic inheritance, and material wealth; gives us insights into where we have come from, and into our place in the universe; enables us to understand how we perceive what we see and mean what we say; and not only describes particular cultures, but helps us elaborate the very notions of "culture" and "society."

Scientific and technological advance can create options for public consideration. Choices from intimately personal to Malthusianly global have been opened up by the invention of the condom, diaphragm, spermicidal foam, Pill, and IUD, while other choices remain distant because of the practical unavailability of a male Pill, reversible vas deferens valve, and other contraceptive options. In addition to such options for *doing*, technology can create options for *knowing* (and then perhaps doing): until the recent development of amniocentesis and related techniques, never had it been possible to know with any certainty the gender, genetics, or pathology of a fetus in utero and thus be presented with informed choices over carrying to term, or seeking pre- or post-natal therapeutics, or terminating the pregnancy.

Technology can strongly alter the relative attractiveness of competing social alternatives and can induce value changes. The increasing practicality of solar energy and conservation methods will have implications, in the long run, for issues ranging from insulation installers' pulmonary health to diplomatic relations with Persian Gulf sheikdoms. Advance toward such dreams as rooftop solar electric cells, or vaccines against venereal diseases, depends not just on more efficient manufacture or adaptation, but on fundamental scientific discovery. Possibility changes tend to induce value changes, although they don't necessarily do so: the development of surgical anesthesias in the nineteenth century led to revision of the risks and costs patients were willing to bear in the surgical "calculus of suffering" (Pernick 1985).

Science can identify and analyze consequences of choices and events, and help raise issues to public attention. Science describes causal conditionals. Social attitudes and actions are altered by the knowledge that masturbation does not, contrary to earlier dogma, lead to madness; that pellagra, far from being an inherited inferiority, is the result of a dietary niacin deficiency that can easily be remedied; that some forms of schizophrenia can be attributed not to a mystically evil soul but to treatable physiological anomalies; and that soft, fluffy, seemingly harmless textile dust causes brown lung disease. Such knowledge allows causes to be assigned, opportunities and liabilities identified, and ethical issues altered.

Science can anticipate and analyze perturbations in society itself, including impacts of technological change. It can estimate the effects of entering alternative energy futures, of building a high-speed train system between Los Angeles and San Francisco, of mandating kindergarten attendance. It can project demographic changes, the relation of future populations to the resources they will have available, and the likely future interactions between people and technologies.

The social sciences can observe and analyze expressed and implicit social values and valuation processes. Reports on social attitudes and practices, such as surveys of sexual habits, can prime the way for changes in the way people evaluate their own attitudes. Although social scientists from Max Weber through Clyde Kluckhohn to the present have hoped to become able to analyze values and valuation psychologies, their methods are just now gaining enough acuity to warrant practical use. Social scientists are making progress in constructing value typologies, in surveying voter, medical client, and consumer attitudes, and in analyzing the preferences implicit in economic and legal actions. All of this information can be brought to bear on social decisions.

Tractatus

The argument so far has been intended to establish these fundamentals.

- Social values cannot be derived from science qua science alone.
- Although scientific knowledge, once attained, may be considered ambi-potent for good or evil, the work of pursuing new science and developing technologies, which requires commitment of will and resources to undertake, is by no means value-neutral.
- Besides, at the moment of its emergence new knowledge may well have maleficent or beneficent potency that demands attention.
- Technical activity must be considered value-laden in two senses: technical people's social values and value perceptions affect their research and service; and that work, in turn, affects the value-situation of others in the public.
- Technical experts make crucial decisions for, and in the name of, the public.
- Science and technology affect our philosophies and choices by: deeply informing our cultural outlook; creating options for public consideration; altering the relative attractiveness of competing alternatives, and inducing value changes; identifying consequences of choices and events, and helping raise issues to public attention; anticipating and analyzing perturbations in society itself, including impacts

of technological change; and observing and analyzing expressed and implicit social values and valuation processes.

Technical Progress as Directed Tragedy

Because it pervades this essay's outlook, I must now introduce my view of technical progress as tragedy. Not mere sadness or misfortune, but tragedy in a high sense. And not fatalistic tragedy, but, in what has become a characteristic of modern society, deliberately directed tragedy.

"The myths warn us that the wresting and exploitation of knowledge are perilous acts, but that man must and will know, and once knowing, will not forget," David Landes has reminded us (1969, 555).

> Adam and Eve lost Paradise for having eaten the fruit of the tree of knowledge; but they retained the knowledge. Prometheus was punished, and indeed all of mankind, for Zeus sent Pandora with her box of evils to compensate the advantages of fire; but Zeus never took back the fire. Daedalus lost his son, but he was the founder of a school of sculptors and craftsmen and passed much of his cunning on to posterity.

Man – *bestia cupidissima rerum novarum*, the "species cupidinous of new things" – must and will know. That ambitiousness has long been embodied in our Western tragic sense of ourselves. Alfred North Whitehead provocatively recast it (1925, 14):

> The pilgrim fathers of the scientific imagination as it exists today are the great tragedians of ancient Athens, Aeschylus, Sophocles, Euripides. Their vision of fate, remorseless and indifferent, urging a tragic incident to its inevitable issue, is the vision possessed by science. Fate in Greek Tragedy becomes the order of nature in modern thought.

Crucially, Whitehead continued: "Let me remind you that the essence of dramatic tragedy is not unhappiness. It resides in the solemnity of the remorseless working of things." And I would add the emphasis, "... especially as human agency intervenes."

For present purposes, I take tragedy to mean the deliberate confrontation of deeply important but nearly irresolvable life issues. Tragedy begins in our knowing of causalities, in our intervening in particular causes, and in our technical enlargement of interventional possibilities.

Robert Oppenheimer's confessio – "In some sort of crude sense which no vulgarity, no humor, no overstatement can quite extinguish, the physicists have known sin" – sounds only to be a starkly lame *mea culpa*, unless one realizes that the ages-old service of science and technology to war-making was too well known to Oppenheimer and his colleagues (1948, 66). No; surely the emphasis was on the verb: "Physicists have *known* sin." And not bad-boy sin, but Original sin. I think it not unlikely that the father of the A-bomb would approve of my transmutation: *scientists have known tragedy*. It is in that knowing that many of the issues of this book reside.

Nowhere has this "remorseless working of things" been more profoundly evident than in the development of nuclear weapons. In a *Discovery* editorial in September 1939, C. P. Snow, noting that the idea of explosive chain reaction had become accepted among leading physicists, grimly predicted that a project to make an atomic bomb would "certainly be carried out somewhere in the world." The Manhattan Project, of course, went forward, as did its Japanese and German counterparts. Looking back on the 1945 decision to drop the Hiroshima weapon, Robert Wilson recalled (1970, 32):

> Things and events were happening on a scale of weeks: the death of Roosevelt, the fall of Germany, the 100-ton TNT test of May 7, the bomb test of July 16, each seemed to follow on the heels of the other. A person cannot react that fast. Then too, there was an absolutely Faustian fascination about whether the bomb would really work.

Similarly Norbert Wiener's melancholia of 1948 over the development of cybernetics, although it strikes me as giving in too easily (1948, 28):

> Those of us who have contributed to the new science of cybernetics stand in a moral position which is, to say the least, not very comfortable.

We have contributed to the initiation of a new science which embraces technical developments with great possibilities for good and for evil. We can only hand it over into the world that exists about us, and this is the world of Belsen and Hiroshima. We do not even have the choice of suppressing these new technical developments. They belong to the age. . . .

Once arcane knowledge is generated somewhere, its transmission, or independent reconstruction, is almost, though not absolutely, inevitable. As Dürrenmatt had Möbius say in *The Physicists*, "What has once been thought can't be unthought." After basic nuclear information was released after World War II, the question of proliferation became not *whether* but *when* and *under what circumstances* other countries would pursue nuclear options – and, in Whitehead's phrase, "tragic incident moved to inevitable issue."

Three fundamental tragic motifs can be recognized.

First, by describing flatly the way things are, science raises tragic awarenesses: that certain things are happening, others may happen, others will happen, others cannot happen; that events are determined by causes; that causes may reflect willful human agency and decision.

Even as scientists achieve long-sought humanitarian breakthroughs, the timeless lament of Ecclesiastes (1:18) resonates: "He that increaseth knowledge increaseth sorrow" ("sorrow," or "mental anguish," is a standard translation of the Hebrew, *mak'ôbâh*).

In absence of knowledge, people may resign themselves to Fate, to "blind chance" (mongoloid birth just happens). With knowledge (the chance of Down's syndrome increases sharply with maternal age above thirty-five), intentionality issues arise; chances still have to be taken, but odds may be altered or stakes adjusted deliberately. Tragedy lies not in resigned fatalism, but in considered confrontation of the near-irreconcilables (wanting to have a child, but wanting to pursue other early-life goals first, but of course wishing that the child not be born infirm).

Biomedical science is forcing us to confront basic facts about differences among people, such as differences in allergic vulnerability, color perception, reflex quickness, and lower-back resiliency, that can bring occupational health protection squarely into conflict with equal employment opportunity. Differences we have pretended don't exist will have to be recognized.

Second, in their inventions and in the systems they weave our lives into, technology and medicine confront us with tragic choices among life-extending options whose consequences we have at least some foreknowledge of.[4] In a trend that can only lead, eventually, to anguishing decisions, improvements in neonatal care, coupled with a desire to save all infants no matter what, are preserving evermore-premature babies (down to 500 grams, or a little over a pound, birthweight, now), at ever-increasing costs, with higher incidences of permanent medical deficiencies.[5] As society invests in life-extending medical technologies, sorts out endangered-species protection priorities, and debates the future of fourth-world nations, exceedingly traumatic choices will have to be confronted – not just made (we do that already, often by defaulting), but *confronted*. Weighed. Debated. Faced.

The answers do not reside in knowledge by itself. Again we hear a classic voice, Tiresias moaning when Oedipus commanded him to consult "bird-flight or any art of divination" to guide Thebes from the plague: "How dreadful knowledge of the truth can be when there is no help in truth."

Third, technological advance challenges society with tragic commitments to consequences. Warning that commercial nuclear power comes as a "Faustian bargain," in 1972 Alvin Weinberg, the director of Oak Ridge National Laboratory, said "the price that we demand of society for this magical energy source is both a vigilance and a longevity of our social institutions that we are quite unaccustomed to" (1972a, 33). Regardless of future decisions about nuclear weapons or nuclear power, the high-level radioactive waste already accumulated in many countries will demand curatorship for thousands of years; the stuff will not go away. The great system of dikes that creates and protects one-third of the Netherlands requires similar massive perpetual commitment. Now that we have eradicated smallpox worldwide as a clinical entity, we shall forever have to monitor our increasingly unvaccinated populace, to watch against recrudescence of the virus from who-knows-what lurking source.

These tragic awarenesses, choices, and commitments caused or mediated by technical ventures may leave us happy or not. But of their solemnity there can be no question.

Think for a moment about our progress in dealing with public health risks, in which all of these themes are so evident. Health in the industrial West surely is, in general, more robust than ever before. Many of the most dangerous infectious diseases, such as tuberculosis, diphtheria, smallpox, cholera, typhus, and polio, have been conquered, and progress has been made against many others. Scurvy, pellagra, iron deficiency anemia, and other nutritional diseases have been mastered. Many illnesses that have not yet been eliminated, such as diabetes, have at least been brought under control. Exposure to mercury, lead, arsenic, chromium, and other heavy-metal poisons has been substantially reduced, as has exposure to asbestos, halocarbon solvents, and many other chemicals. Through prediction and protection much damage from hurricanes, floods, and earthquakes has been mitigated. All over the world infant mortality continues to decrease, and life expectancy to increase. More people are living longer, healthier, more vigorous lives.

Nonetheless, almost ruefully, we have progressed to an inherently discomfiting state, a state in which we must expect to remain from now on. Why inherently discomfiting? Because steadily we have broadened our apprehensions to include not only natural catastrophes, infectious diseases, everyday mechanical accidents, and acute poisons, but also large-scale technological accidents, chronic low-level hazards from chemicals, radiation, and noise, and lifestyle vices, such as addictions to tobacco, alcohol, barbiturates, narcotics, caffeine, and rich foods. To our struggle against cancer and other classical illnesses we have added concern about reproductive, genetic, immunological, behavioral, and other debilitations. And of course we continue to create new hazards, to identify risks that have existed without being recognized, and to resolve to reduce previously tolerated risks.

Now, about many hazards we know enough, scientifically, to "worry," but not enough to know *how much* to worry – or how much protective action to invest. Knowledge has grown enormously, and we even have the luxury of going around searching for possible trouble. But many scientific disciplines are still in their adolescence and are unable to evaluate risks precisely. We can detect minuscule traces of manmade pesticides in mothers' milk all over the world, which is vaguely disturbing; but, with rare exceptions, we don't have a clue as to whether the chemicals exert any effect on mother or infant (Jensen 1983; Wolff 1983). Toxicology, epidemiology, and medicine have taken us out to their borders, but it's unruly territory.

At the same time that we are learning more, we are heightening our societal aspirations beyond all previous limits. We intend to help all infants get a vigorous start in life. And we strive to afford first-rate, broadly defined health protection to all citizens and noncitizens, even immigrant aliens, through an enormous range of risks, throughout their lives. No civilization ever before has had these ambitions.

The crux: In our knowing so much more, though imperfectly, and aspiring to so much more, we have passed beyond the sheltering blissfulness of ignorance and risk-enduring resignation. This has given rise to considerable social apprehensiveness, which is affecting both the outlook of individuals and the functioning of institutions.

Similar phase-changes have occurred historically when people became aware of specific causes of disease and deformity, as when it became clear that moral turpitude alone was not the cause of syphilis, and when societal aspirations, such as commitment to worker protection, rose. We are going through both kinds of change at the same time.

Risks and aspirations will continue to evolve. In his 1803 revised *Essay on Population* Thomas Malthus observed of Jenner's new vaccine, "I have not the slightest doubt that if the introduction of cowpox should extirpate the smallpox, we shall find . . . increased mortality of some other disease" (1803, 522). Malthus was right. As any risk is reduced, others inevitably increase in the mortality and morbidity tables – though perhaps setting in at later ages. Risks in the industrial West are evolving now as rural and agrarian risks are succeeded by urban, industrial, and medical-care risks. A similar progression is occurring, displaced in time, in less developed countries: cholera and fatal infant diarrhea are being

succeeded by cancer, heart disease, and drug side effects. I can't imagine that our societal aspirations won't expand even further, both for the industrial West and for the rest of the world.

"We are in for a sequentiality of improbable possibles," *Finnegans Wake's* Shem knew to expect. Some possibles are, of course, more predictable than others. Recent years' debates over such matters as food additives, contraceptives, pesticides, and energy sources have brought broad public recognition that nothing can be risk-free, that there are no rewards without risks, and that risktaking for benefit is the essence of human striving.

Our analyses and decisions are making us face risks ever more explicitly and comparatively. The Occupational Safety and Health Administration's 1983 standard for worker exposure to airborne inorganic arsenic was a striking example. In tightening the standard from 500 to 10 micrograms arsenic per cubic meter of air, OSHA concluded "that inorganic arsenic is a carcinogen, that no safe level of exposure can be demonstrated, and that 10 micrograms per cubic meter is the lowest possible level to which employee exposure could be controlled." Further (U.S. Occupational Safety and Health Administration 1983, 1867):

> The level of risk from working a lifetime of exposure at 10 micrograms per cubic meter is estimated at approximately 8 excess lung cancer deaths per 1000 employees. OSHA believes that this level of risk does not appear to be insignificant. It is below risk levels in high risk occupations but it is above risk levels in occupations with average levels of risk.

Surely this rationale is commendable (regardless of whether the particular numbers adopted survive current dispute), but the explicitness is unsettling.

. . . I must emphasize that even while I stress the essentially tragic nature of our awarenesses, decisions, and commitments, I believe that we are in a great many ways better off than ever before. Just as surely as science and technology are part of the problem, they will be part of the solution. And I would affirm that tragic confrontations are an essence of humanness.

The difference from the past is that – discomfiting though it may be – the conjunction of greatly improved technical abilities with heightened societal aspirations now puts us in a position to *direct* aspects of tragic progress. Experts perform center stage and in the wings. All of us speak from the citizens' chorus.

Notes

1 Edel (1955) and Margenau (1964) are typical but disappointing earlier efforts to strengthen the conjunction between science and values. Neither book removes the contingency of oughts on what Margenau called "primary value postulates," such as "respect for human life."
2 For history of the Statistical Society, see Mouat (1885) and Cullen (1975).
3 First appendix to *Inquiry Concerning the Principles of Morals* (1752).
4 Calabresi and Bobbitt (1978) ably essays on some life-and-death decisions, but despite its title only lightly addresses their "tragic" nature.

5 Phibbs, Williams, and Phibbs (1981) and Murray and Caplan (1985).

References

Calabresi, G. and Bobbitt, P. (1978) *Tragic Choices* (W. W. Norton).

Cullen, M. J. (1975) *The Statistical Movement in Early Victorian Britain* (Harvester).

Edel, A. (1955) *Ethical Judgment: The Use of Science in Ethics* (Free Press).

Margenau, H. (1964) *Ethics and Science* (Van Nostrand).

Mouat, F. J. (1885) *History of the Statistical Society.* Jubilee Volume of the Statistical Society.

Murray, T. and Caplan, A. (1985) *Which Babies Shall Live?* (Humana Press).

Phibbs, C. S., Williams, R. L., and Phibbs, R. H. (1981) "Newborn Risk Factors and Costs of Neonatal Intensive Care." *Pediatrics* 68 (3): 313–21.

4

A Collective of Humans and Nonhumans

Bruno Latour

Bruno Latour is a French sociologist and philosopher of science best known for his books *We Have Never Been Modern, Laboratory Life*, and *Science in Action*. From 1982 to 2006, he was Professor at the Centre de sociologie de l'Innovation at the Ecole nationale supérieure des mines in Paris. He has been a key thinker in the development of actor-network theory. In this essay Latour challenges the modernist distinction between humans and technology. Latour's early work was associated with the social constructivist view in the social studies of science. According to this view, scientific facts are mere social constructs, products of politics, religion, gender, race, and other social forces. The social constructivist view is at odds with the predominant view within philosophy of science, which holds that science aspires to objectivity. Some advocates of these views were participants in the "science wars" of the 1980s and '90s. Here Latour rejects each view, and argues that each mistakenly assumes the modernist view that it is possible to make a clear differentiation between subject and object. He argues that neither humans nor technologies exist without the other, and that science, technology, and society are becoming ever more entwined. We would do better to adopt a pragmatogony, or a view that subjects and objects are mutually creating, that attempts to understand humans as socio-technical animals. He borrows the concept of "pragmatogony" from his teacher Michel Serres. According to Serres and Latour, and as illustrated in this essay, instead of separating humans and machines, the pragmatogony tells of successive layers of ever more complex entanglements among them. On Latour's account, there have been at least 11 crossovers, or

From Bruno Latour, *Pandora's Hope: Essays on the Reality of Science Studies* (Cambridge, MA: Harvard University Press, 1999), pp. 198–215. Reprinted by permission of the publisher. Copyright © 1999 by the President and Fellows of Harvard College.

significant changes in either humans or non-humans (including
non-human animals and machines) that have brought changes in
the other.

A detailed case study of sociotechnical networks
ought to follow at this juncture, but many such
studies have already been written, and most have
failed to make their new social theory felt, as
the science wars have made painfully clear to all.
Despite the heroic efforts of these studies, many
of their authors are all too often misunderstood
by readers as cataloguing examples of the "social
construction" of technology. Readers account
for the evidence mustered in them according to
the dualist paradigm that the studies themselves
frequently undermine. The obstinate devotion to
"social construction" as an explanatory device,
whether by careless readers or "critical" authors,
seems to derive from the difficulty of disentan-
gling the various meanings of the catchword
sociotechnical. What I want to do, then, is to peel
away, one by one, these layers of meaning and
attempt a genealogy of their associations.

Moreover, having disputed the dualist paradigm
for years, I have come to realize that no one is
prepared to abandon an arbitrary but useful
dichotomy, such as that between society and
technology, if it is not replaced by categories

that have at least a semblance of providing the
same discriminating power as the one jettisoned.
Of course, I will never be able to do the same
political job with the pair human-nonhuman as
the subject-object dichotomy has accomplished,
since it was in fact to free science from politics
that I embarked on this strange undertaking.
In the meantime we can toss around the phrase
"sociotechnical assemblages" forever without moving beyond the dualist paradigm that we wish to
leave behind. To move forward I must convince
the reader that, pending the resolution of the
political kidnapping of science, *there is an alternative to the myth of progress*. At the heart of the
science wars lies the powerful accusation that
those who undermine the objectivity of science
and the efficiency of technology are trying to
lead us backward into some primitive, barbaric
dark age – that, incredibly, the insights of science
studies are somehow "reactionary."

In spite of its long and complex history, the
myth of progress is based on a very rudimentary
mechanism (Figure 4.1). What gives the thrust to
the arrow of time is that modernity at last breaks

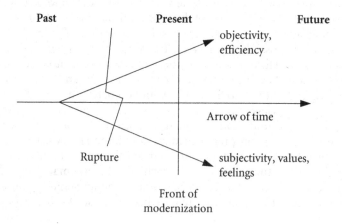

Figure 4.1   What makes the arrow of time thrust forward in the modernist narrative of progress is the
certainty that the past will differ from the future because what was confused will become distinct:
objectivity and subjectivity will no longer be mixed up. The result of this certainty is a front of
modernization that allows one to distinguish slips backward from steps forward.

out of a confusion, made in the past, between what objects really are in themselves and what subjectivity of humans believes them to be, projecting onto them passions, biases, and prejudices. What could be called a front of modernization – like the Western Frontier – thus clearly distinguishes the confused past from the future, which will be more and more radiant, no doubt about that, because it will distinguish even more clearly the efficiency and objectivity of the laws of nature from the values, rights, ethical requirements, subjectivity, and politics of the human realm. With this map in their hands, science warriors have no difficulty situating science studies: "Since they are always insisting that objectivity and subjectivity [the science warriors' terms for nonhumans and humans] are mixed up, science students are leading us in only one possible direction, into the obscure past out of which we must extract ourselves by a movement of radical conversion, the conversion through which a barbarian premodernity becomes a civilised modernity."

In an interesting case of cartographic incommensurability, however, science studies uses an entirely different map (Figure 4.2). The arrow of time is *still there*, it still has a powerful and maybe irresistible thrust, but an entirely different mechanism makes it tick. Instead of clarifying even further the relations between objectivity and subjectivity, time enmeshes, at an ever greater level of intimacy and on an ever greater scale, humans and nonhumans with each other. The feeling of time, the definition of where it leads, of what we should do, of what war we should wage, is entirely different in the two maps, since in the one I use, Figure 4.2, the confusion of humans and nonhumans is not only our past *but our future as well*. If there is one thing of which we may be as certain as we are of death and taxation, it is that we will live tomorrow in imbroglios of science, techniques, and society *even more tightly linked* than those of yesterday – as the mad cow affair has demonstrated so clearly to European beefeaters. The difference between the two maps is total, because what the modernist science warriors see as a horror to be avoided at all costs – the mixing up of objectivity and subjectivity – is for us, on the contrary, the hallmark of a civilized life, except that what time mixes up in the future even more than in the past *are not objects and subjects at all, but humans and nonhumans*, and that makes a world of difference. Of this difference the science warriors remain blissfully ignorant, convinced that we want to confuse objectivity and subjectivity.

I am now in the usual quandary. I have to offer an alternative picture of the world that can rely on none of the resources of common sense although, in the end, I aim at nothing but

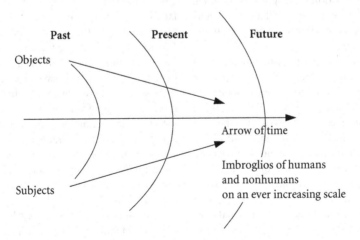

Figure 4.2 In the alternative "servant" narrative there is still an arrow of time, but it is registered very differently from Figure 4.1: the two lines of objects and subjects become more confused in the future than they were in the past, hence the feeling of instability. What is growing instead is the ever expanding scale at which humans and nonhumans are connected together.

common sense. The myth of progress has centuries of institutionalization behind it, and my little pragmatogony is helped by nothing but my miserable diagrams. And yet I have to go on, since the myth of progress is so powerful that it puts any discussion to an end.

Yes, I want to tell another tale. For my present pragmatogony* [mythic origin of technology], I have isolated eleven distinct layers. Of course I do not claim for these definitions, or for their sequence, any plausibility. I simply want to show that the tyranny of the dichotomy between objects and subjects is not inevitable, since it is possible to envision another myth in which it plays no role. If I succeed in opening some space for the imagination, then we are not forever stuck with the implausible myth of progress. If I could even begin to recite this pragmatogony – I use this word to insist on its fanciful character – I would have found an alternative to the myth of progress, that most powerful of all the modernist myths, the one that held my friend under its sway when he asked me, "Do we know more than we used to?" No, we don't know more, if by this expression we mean that every day we extract ourselves further from a confusion between facts, on the one hand, and society, on the other. But yes, we do know a good deal more, if by this we mean that our collectives are tying themselves ever more deeply, more intimately, into imbroglios of humans and nonhumans. Until we have an alternative to the notion of progress, provisional as it may be, science warriors will always be able to attach to science studies the infamous stigma of being "reactionary."

I will build this alternative with the strangest of means. I want to highlight the successive crossovers through which humans and nonhumans have exchanged their properties. Each of those crossovers results in a dramatic change in the scale of the collective, in its composition, and in the degree to which humans and nonhumans are enmeshed. To tell my tale I will open Pandora's box backward; that is, starting with the most recent types of folding. I will try to map the labyrinth until we find the earliest (mythical) folding. As we will see, contrary to the science warriors' fear, no dangerous regression is involved here, since all of the earlier steps are still with us today. Far from being a horrifying miscegenation between objects and subjects, they are simply the very hybridizations that make us humans and nonhumans.

Level 11: Political Ecology

Talk of a crossover between techniques and politics does not, in my pragmatogony, indicate belief in the distinction between a material realm and a social one. I am simply unpacking the eleventh layer of what is packed in the definitions of society and technique. The eleventh interpretation of the crossover – the swapping of properties – between humans and nonhumans is the simplest to define because it is the most *literal*. Lawyers, activists, ecologists, businessmen, political philosophers, are now seriously talking, in the context of our ecological crisis, of granting to nonhumans some sort of rights and even legal standing. Not so many years ago, contemplating the sky meant thinking of matter, or of nature. These days we look up at a sociopolitical imbroglio, since the depletion of the ozone layer brings together a scientific controversy, a political dispute between North and South, and immense strategic changes in industry. Political representation of nonhumans seems not only plausible now but necessary, when the notion would have seemed ludicrous or indecent not long ago. We used to deride primitive peoples who imagined that a disorder in society, a pollution, could threaten the natural order. We no longer laugh so heartily, as we abstain from using aerosols for fear the sky may fall on our heads. Like the "primitives," we fear the pollution caused by our negligence – which means of course that neither "they" nor "we" have ever been primitive.

As with all crossovers, all exchanges, this one mixes elements from both sides, the political with the scientific and technical, and this mixture is not a haphazard rearrangement. Technologies have taught us how to manage vast assemblies of nonhumans; our newest sociotechnical hybrid brings what we have learned to bear on the political system. The new hybrid remains a nonhuman, but not only has it lost its material and objective character, it has acquired properties of citizenship. It has, for instance, the right not to be enslaved. This first layer of meaning – the last in chronological sequence to arrive – is that of political ecology or, to use Michel Serres's term, "the natural contract." *Literally*, not symbolically as before, we have to manage the planet we inhabit, and must now define a politics of things.

Level 10: Technoscience

If I descend to the tenth layer, I see that our current definition of technology is itself due to the crossover between a previous definition of society and a particular version of what a nonhuman can be. To illustrate: some time ago, at the Institut Pasteur, a scientist introduced himself, "Hi, I am the coordinator of yeast chromosome 11." The hybrid whose hand I shook was, all at once, a person (he called himself "I"), a corporate body ("the coordinator"), and a natural phenomenon (the genome, the DNA sequence, of yeast). The dualist paradigm will not allow us to understand this hybrid. Place its social aspect on one side and yeast DNA on the other, and you will bungle not only the speaker's words but also the opportunity to grasp how a genome becomes known to an organization and how an organization is naturalized in a DNA sequence on a hard disk.

We again encounter a crossover here, but it is of a different sort and goes in a different direction, although it could also be called sociotechnical. For the scientist I interviewed there is no question of granting any sort of rights, of citizenship, to yeast. For him yeast is a strictly material entity. Still, the industrial laboratory where he works is a place in which new modes of organization of labor elicit completely new features in nonhumans. Yeast has been put to work for millennia, of course, for instance in the old brewing industry, but now it works for a network of thirty European laboratories where its genome is mapped, humanized, and socialized, as a code, a book, a program of action, compatible with our ways of coding, counting, and reading, retaining none of its material quality, the quality of an outsider. It is absorbed into the collective. Through technoscience – defined, for my purposes here, as a fusion of science, organization, and industry – the forms of coordination learned through "networks of power" (see Level 9) are extended to inarticulate entities. Nonhumans are endowed with speech, however primitive, with intelligence, fore-sight, self-control, and discipline, in a fashion both large-scale and intimate. Socialness is shared with nonhumans in an almost promiscuous way. While in this model, the tenth meaning of sociotechnical (see Figure 4.3), automata have no rights, they are much more than material entities; they are complex organizations.

Level 9: Networks of Power

Technoscientific organizations, however, are not purely social, because they themselves recapitulate, in my story, nine prior crossovers of humans and nonhumans. Alfred Chandler and Thomas

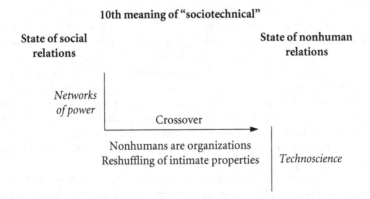

Figure 4.3 Each step in the mythical pragmatogony may be sketched as a crossover through which skills and properties learned in social relations are made relevant for establishing relations within nonhumans. By convention, the next step will be understood as going in the opposite direction.

Hughes have each traced the interpenetration of technical and social factors in what Chandler terms the "global corporation" and Hughes terms "networks of power." Here again the phrase "sociotechnical imbroglio" would be apt, and one could replace the dualist paradigm with the "seamless web" of technical and social factors so beautifully traced by Hughes. But the point of my little genealogy is also to identify, inside the seamless web, properties borrowed from the social world in order to socialize nonhumans and properties borrowed from nonhumans in order to naturalize and expand the social realm. For each layer of meaning, whatever happens happens as if we are learning, in our contacts with one side, ontological properties that are then reimported to the other side, generating new, completely unexpected effects.

The extension of networks of power in the electrical industry, in telecommunications, in transportation, is impossible to imagine without a massive mobilization of material entities. Hughes's book is exemplary for students of technology because it shows how a technical invention (electric lighting) led to the establishment (by Edison) of a corporation of unprecedented scale, its scope directly related to the physical properties of electrical networks. Not that Hughes in any way talks of the infrastructure triggering changes in the superstructure; on the contrary, his networks of power are complete hybrids, though hybrids of a peculiar sort – they lend their nonhuman qualities to what were until then weak, local, and scattered corporate bodies. The management of large masses of electrons, clients, power stations, subsidiaries, meters, and dispatching rooms acquires the formal and universal character of scientific laws.

This ninth layer of meaning resembles the eleventh, since in both cases the crossover goes roughly from nonhumans to corporate bodies. (What can be done with electrons can be done with electrors.) But the intimacy of human and nonhuman is less apparent in networks of power than in political ecology. Edison, Bell, and Ford mobilized entities that looked like matter, that seemed nonsocial, whereas political ecology involves the fate of nonhumans already socialized, so closely related to us that they have to be protected by delineation of their legal rights.

Level 8: Industry

Philosophers and sociologists of techniques tend to imagine that there is no difficulty in defining material entities because they are objective, unproblematically composed of forces, elements, atoms. Only the social, the human realm, is difficult to interpret, we often think, because it is complexly historical and, as they say, "symbolic." But whenever we talk of matter we are really considering, as I am trying to show here, a *package* of former crossovers between social and natural elements, so that what we take to be primitive and pure terms are belated and mixed ones. Already we have seen that matter varies greatly from layer to layer – matter in the layer I have called "political ecology" differs from that in the layers called "technology" and "networks of power." Far from being primitive, immutable, and ahistorical, matter too has a complex genealogy and is handed down to us through a long and convoluted pragmatogony.

The extraordinary feat of what I will call *industry* is to extend to matter a further property that we think of as exclusively social, the capacity to relate to others of one's kind, to conspecifics, so to speak. Nonhumans have this capacity when they are made part of the assembly of actants that we call a machine: an automaton endowed with autonomy of some sort and submitted to regular laws that can be measured with instruments and accounting procedures. From tools held in the hands of human workers, the shift historically was to assemblies of machines, where tools relate to one another, creating a massive array of labor and material relations in factories that Marx described as so many circles of hell. The paradox of this stage of relations between humans and nonhumans is that it has been termed "alienation," dehumanization, as if this were the first time that poor and exploited human weakness was confronted by an all-powerful objective force. However, to relate nonhumans together in an assembly of machines, ruled by laws and accounted for by instruments, is to grant them a sort of social life.

Indeed, the modernist project consists in creating this peculiar hybrid: a fabricated nonhuman that has nothing of the character of society and politics yet builds the body politic all the more effectively because it seems completely estranged from humanity. This famous shapeless matter, celebrated so fervently throughout the eighteenth and nineteenth centuries, which is there for Man's – but rarely Woman's – ingenuity to mold and fashion, is only one of many ways to socialize nonhumans. They have been socialized to such an extent that they now have the capacity to create an assembly of their own, an automaton, checking and surveying, pushing and triggering other automata, as if with full autonomy. In effect, however, the properties of the "megamachine" (see Level 7) have been extended to nonhumans.

It is only because we have not undertaken an anthropology of our modern world that we can overlook the strange and hybrid quality of matter as it is seized and implemented by industry. We take matter as mechanistic, forgetting that mechanism is one half of the modern definition of society. A society of machines? Yes, the eighth meaning of the word sociotechnical, though it seems to designate an unproblematic industry, dominating matter through machinery, is the strangest sociotechnical imbroglio yet. Matter is not a given but a recent historical creation.

Level 7: The Megamachine

But where does industry come from? It is neither a given nor the sudden discovery by capitalism of the objective laws of matter. We have to imagine its genealogy through earlier and more primitive meanings of the term sociotechnical. Lewis Mumford has made the intriguing suggestion that the megamachine – the organization of large numbers of humans via chains of command, deliberate planning, and accounting procedures – represents a change of scale that had to be made before wheels and gears could be developed. At some point in history human interactions come to be mediated through a large, stratified, externalized body politic that keeps track, through a range of "intellectual techniques" (writing and counting, basically), of the many nested subprograms for action. When some, though not all, of these subprograms are replaced by nonhumans, machinery and factories are born. The nonhumans, in this view, enter an organization that is already in place and take on a role rehearsed for centuries

by obedient human servants enrolled in the imperial megamachine.

In this seventh level, the mass of nonhumans assembled in cities by an internalized ecology (I will define this expression shortly) has been brought to bear on empire building. Mumford's hypothesis is debatable, to say the least, when our context of discussion is the history of technology; but the hypothesis makes excellent sense in the context of my pragmatogony. Before it is possible to delegate action to nonhumans, and possible to relate nonhumans to one another in an automaton, it must first be possible to nest a range of subprograms for action into one another without losing track of them. Management, Mumford would say, precedes the expansion of material techniques. More in keeping with the logic of my story, one might say that *whenever we learn something about the management of humans, we shift that knowledge to nonhumans and endow them with more and more organizational properties.* The even-numbered episodes I have recounted so far follow this pattern: industry shifts to nonhumans the management of people learned in the imperial machine, much as technoscience shifts to nonhumans the large-scale management learned through networks of power. In the odd-numbered levels, the opposite process is at work: *what has been learned from nonhumans is reimported so as to reconfigure people.*

Level 6: Internalized Ecology

In the context of layer seven, the megamachine seems a pure and even final form, composed entirely of social relations; but, as we reach layer six and examine what underlies the megamachine, we find the most extraordinary extension of social relations to nonhumans: agriculture and the domestication of animals. The intense socialization, reeducation, and reconfiguration of plants and animals – so intense that they change shape, function, and often genetic makeup – is what I mean by the term "internalized ecology." As with our other even-numbered levels, domestication cannot be described as a sudden access to an objective material realm that exists *beyond* the narrow limits of the social. In order to enroll animals, plants, proteins in the emerging collective, one must first endow them with the social characteristics necessary for their integration. This shift of characteristics results in a manmade landscape for society (villages and cities) that completely alters what was until then meant by social and material life. In describing the sixth level we may speak of urban life, empires, and organizations, but not of society and techniques – or of symbolic representation and infrastructure. So profound are the changes entailed at this level that we pass beyond the gates of history and enter more profoundly those of prehistory, of mythology.

Level 5: Society

What is a society, the starting point of all social explanations, the *a priori* of all social science? If my pragmatogony is even vaguely suggestive, society cannot be part of our final vocabulary, since the term had itself to be made – "socially constructed" as the misleading expression goes. But according to the Durkheimian interpretation, a society is primitive indeed: it precedes individual action, lasts very much longer than any interaction does, dominates our lives; it is that in which we are born, live, and die. It is externalized, reified, more real than ourselves, and hence the origin of all religion and sacred ritual, which for Durkheim are nothing but the return, through figuration and myth, of the transcendent to individual interactions.

And yet society itself is constructed only through such quotidian interactions. However advanced, differentiated, and disciplined society becomes, we still repair the social fabric out of our own, immanent knowledge and methods. Durkheim may be right, but so is Harold Garfinkel. Perhaps the solution, in keeping with the generative principle of my genealogy, is to look for nonhumans. (This explicit principle is: look for nonhumans when the emergence of a social feature is inexplicable; look to the state of social relations when a new and inexplicable type of object enters the collective.) What Durkheim mistook for the effect of a *sui generis* social order is simply the effect of having brought so many techniques to bear on our social relations. It was from techniques, that is, the ability to nest several subprograms, that we learned what it means to subsist and expand, to accept a role

and discharge a function. By reimporting this competence into the definition of society, we taught ourselves to reify it, to make society stand independent of fast-moving interactions. We even learned how to delegate to society the task of relegating us to roles and functions. Society exists, in other words, *but is not socially constructed.* Nonhumans proliferate below the bottom line of social theory.

Level 4: Techniques

By this stage in our speculative genealogy we can no longer speak of humans, of anatomically modern humans, but only of social prehumans. At last we are in a position to define technique, in the sense of a *modus operandi*, with some precision. Techniques, we learn from archaeologists, are articulated subprograms for actions that subsist (in time) and extend (in space). Techniques imply not society (that late-developing hybrid) but a semisocial organization that brings together nonhumans from very different seasons, places, and materials. A bow and arrow, a javelin, a hammer, a net, an article of clothing are composed of parts and pieces that require recombination in sequences of time and space that bear no relation to their original settings. Techniques are what happen to tools and nonhuman actants when they are processed through an organization that extracts, recombines, and socializes them. Even the simplest techniques are sociotechnical; even at this primitive level of meaning, forms of organization are inseparable from technical gestures.

Level 3: Social Complication

But what form of organization can explain these recombinations? Recall that at this stage there is no society, no overarching framework, no dispatcher of roles and functions; there are merely interactions among prehumans. Shirley Strum and I call this third layer of meaning *social complication*. Here complex interactions are marked and followed by nonhumans enrolled for a specific purpose. What purpose? Nonhumans stabilize social negotiations. Nonhumans are at once pliable and durable; they can be shaped very quickly but, once shaped, last far longer than the interactions that fabricated them. Social interactions are extremely labile and transitory. More precisely, either they are negotiable but transient or, if they are encoded (for instance) in the genetic makeup, they are extremely durable but difficult to renegotiate. The involvement of nonhumans resolves the contradiction between durability and negotiability. It becomes possible to follow (or "blackbox") interactions, to recombine highly complicated tasks, to nest subprograms into one another. What was impossible for complex social animals to accomplish becomes possible for prehumans – who use tools not to acquire food but to fix, underline, materialize, and keep track of the social realm. Though composed only of interactions, the social realm becomes visible and attains through the enlistment of nonhumans – tools – some measure of durability.

Level 2: The Basic Tool Kit

The tools themselves, wherever they came from, offer the only testimony on behalf of hundreds of thousands of years. Many archaeologists proceed on the assumption that the basic tool kit (as I call it) and techniques are directly related by an evolution of tools into composite tools. But there is no *direct* route from flints to nuclear power plants. Further, there is no direct route, as many social theorists presume there to be, from social complication to society, megamachines, networks. Finally, there is not a set of parallel histories, the history of infrastructure and the history of superstructure, but only one sociotechnical history.

What, then, is a tool? The extension of social skills to nonhumans. Machiavellian monkeys and apes possess little in the way of techniques, but can devise social tools (as Hans Kummer has called them) through complex strategies of manipulating and modifying one another. If you grant the prehumans of my own mythology the same kind of social complexity, you grant as well that they may generate tools by *shifting* that competence to nonhumans, by treating a stone, say, as a social partner, modifying it, then using it to act on a second stone. Prehuman tools, in contrast to the ad hoc implements of other primates, also represent the extension of a skill rehearsed in the realm of social interactions.

Level 1: Social Complexity

We have finally reached the level of the Machiavellian primates, the last circumvolution in Daedalus's maze. Here they engage in social interactions to repair a constantly decaying social order. They manipulate one another to survive in groups, with each group of conspecifics in a state of constant mutual interference. We call this state, this level, social complexity. I will leave it to the ample literature of primatology to show that this stage is no more free of contact with tools and techniques than any of the later stages.

An Impossible But Necessary Recapitulation

I know I should not do it. I more than anyone ought to see that it is madness, not only to peel away the different meanings of sociotechnical, but also to recapitulate all of them in a single diagram, as if we could read off the history of the world at a glance. And yet it is always surprising to see how few alternatives we have to the grandiose scenography of progress. We may tell a lugubrious countertale of decay and decadence as if, at each step in the extension of science and technology, we were stepping down, away from our humanity. This is what Heidegger did, and his account has the somber and powerful appeal of all tales of decadence. We may also abstain from telling any master narrative, under the pretext that things are always local, historical, contingent, complex, multiperspectival, and that it is a crime to hold them all in one pathetically poor scheme. But this ban on master narratives is never very effective, because, in the back of our minds, no matter how firmly we are convinced of the radical multiplicity of existence, something surreptitiously gathers everything into one little bundle which may be even cruder than my diagrams – including the postmodern scenography of multiplicity and perspective. This is why, against the ban on master narratives, I cling to the right to tell a "servant" narrative. My aim is not to be reasonable, respectable, or sensible. It is to fight modernism by finding the hideout in which science has been held since being kidnapped for political purposes I do not share.

If we gather in one table the different layers I have briefly outlined – one of my other excuses is how brief the survey, covering so many millions of years, has been! – we may give some sense to a story in which the further we go the more articulated are the collectives we live in (see Figure 4.4). To be sure, we are not ascending toward a future made of more subjectivity and more objectivity. But neither are we descending, chased ever further from the Eden of humanity and *poesis*.

Even if the speculative theory I have outlined is entirely false, it shows, at the very least, the possibility of imagining a genealogical alternative to the dualist paradigm. We are not forever trapped in a boring alternation between objects or matter and subjects or symbols. We are not limited to "not only . . . but also" explanations. My little origin myth makes apparent the impossibility of having an artifact that does not incorporate social relations, as well as the impossibility of defining social structures without accounting for the large role played in them by nonhumans.

Second, and more important, the genealogy demonstrates that it is false to claim, as so many do, that once we abandon the dichotomy between society and techniques we are faced with a seamless web of factors in which all is included in all. The properties of humans and nonhumans cannot be swapped haphazardly. Not only is there an order in the exchange of properties, but in each of the eleven layers the meaning of the word "sociotechnical" is clarified if we consider the exchange: that which has been learned from nonhumans and reimported into the

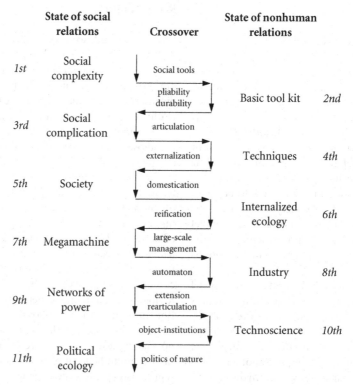

Figure 4.4 If the successive crossovers are summed up, a pattern emerges: relations among humans are made out of a previous set of relations that related nonhumans to one another; these new skills and properties are then reused to pattern new types of relations among nonhumans, and so on; at each (mythical) stage the scale and the entanglement increase. The key feature of this myth, is that, at the final stage, the definitions we can make of humans and nonhumans should recapitulate all the earlier layers of history. The further we go, the less pure are the definitions of humans and nonhumans.

social realm, that which has been rehearsed in the social realm and exported back to the nonhumans. Nonhumans too have a history. They are not material objects or constraints. Sociotechnical[1] is different from sociotechnical[6] or[7] or[8] or[11]. By adding superscripts we are able to qualify the meanings of a term that until now has been hopelessly confused. In place of the great vertical dichotomy between society and techniques, there is conceivable (in fact, now, available) a range of horizontal distinctions between very different meanings of the sociotechnical hybrids. It is possible to have our cake and eat it too – to be monists and make distinctions.

All this is not to claim that the old dualism, the previous paradigm, had nothing to say for itself. We do indeed alternate between states of social and states of nonhuman relations, but this is not the same as alternating between humanity and objectivity. The mistake of the dualist paradigm was its definition of humanity. Even the shape of humans, our very body, is composed to a great extent of sociotechnical negotiations and artifacts. To conceive of humanity and techno-logy as polar opposites is, in effect, to wish away humanity: we are sociotechnical animals, and each human interaction is sociotechnical. We are never limited to social ties. We are never faced only with objects. This final diagram relo-cates humanity right where we belong – in the crossover, the central column, the articulation, the possibility of mediating between mediators.

But my main point is that, in each of the eleven episodes I have traced, an increasingly large number of humans are mixed with an increasingly large number of nonhumans, to the point that, today, the whole planet is engaged in the making of politics, law, and soon, I suspect, morality. The illusion of modernity was to believe that the more we grew, the more separate objectivity and subjectivity would become, thus creating a future radically different from our past. After the paradigm shift in our conception of science and technology, we now know that this will never be the case, indeed that this has never *been* the case. Objectivity and subjectivity are not opposed, they grow together, and they do so irreversibly. At the very least, I hope I have convinced the reader that, if we are to meet our challenge, we will not meet it by considering artifacts as things. They deserve better. They deserve to be housed in our intellectual culture as full-fledged social actors. Do they mediate our actions? No, they are us. The goal of our philosophy, social theory, and morality is to invent political institutions that can absorb this much history, this vast spiraling movement, this labyrinth, this fate.

The nasty problem we now have to deal with is that, unfortunately, we do *not* have a defin-ition of politics that can answer the specifications of this nonmodern history. On the contrary, every single definition we have of politics comes from the modernist settlement and from the polemical definition of science that we have found so wanting. Every one of the weapons used in the science wars, including *the very distinction* between science and politics, has been handed down to the combatants by the side we want to oppose. No wonder we always lose and are accused of politicizing science! It is not only the practice of science and technology that epis-temology has rendered opaque, but also that of politics. As we shall soon see, the fear of mob rule, the proverbial scenography of might versus right, is what holds the old settlement together, is what has rendered us modern, is what has kidnapped the practice of science, all for the most implausible political project: that of doing away with politics.

5

Technology and Ethics

Kristin Shrader-Frechette

Kristin Shrader-Frechette has been since 1998 the O'Neill Family Professor of Philosophy and Biological Sciences at the University of Notre Dame. This selection is one of the best-regarded accounts of the ethical evaluation of technology, and this essay was originally published in the *Encyclopedia of Ethics*. Shrader-Frechette notes that there is considerable overlap of concerns between philosophers, scientists, and policy-makers. For instance, engineers, policy analysts, and philosophers share an interest in risk assessment that echoes Jonas's call for a new ethics of technology in the face of uncertainty about outcomes. In *Risk and Rationality: Philosophical Foundations for Populist Reforms*, Shrader-Frechette discusses what she calls "five dilemmas of risk assessment." Public policy is formulated within the context of risk perceptions, which, in turn, reflect the public's opinions of the quality of risk assessment. Risk assessors must estimate as closely as possible the potential hazards of new technology. These assessments are evaluated by a public that desires that risk assessments make use of social and ethical as well as technical criteria. Risk assessors and the public then face dilemmas in their attempts to balance technical evaluations with the public's desires for nontechnical input into risk assessment.

1 *The Fact-Value Dilemma*: Risk evaluations cannot be both wholly factual/scientific and wholly sanctioned via democratic processes because lay persons want ethical/moral considerations to be taken into account.
2 *The Standardization Dilemma*: Risk assessors seek standardization of evaluation criteria to avoid the appearance of arbitrariness. But local groups want them to take into account their special conditions.

Originally: "Technology," in Lawrence C. Becker and Charlotte B. Becker, eds., *Encyclopedia of Ethics*, vol. 2 (New York: Garland Publishing, 1992), pp. 1231–4. © 1992. Reproduced by permission of Routledge, Inc., a division of Informa plc.

3 *The Contributors Dilemma*: Risk assessment can be costly to undertake. Thus, many potential hazards are not examined. In addition, aggregate effects of subthreshold levels of risk often go untested. Pesticides A and B might be considered safe, for example, but are these pesticides safe when they both are applied to the same crop?

4 *The De Minimis Dilemma*: How safe is safe enough? Declaring a threshold level at which to define negligible risk is a difficult task when citizens hold different expectations of safety.

5 *The Consent Dilemma*: Persons most affected by risk are sometimes those persons least able to give consent. Persons who lack the economic means and political access to challenge public policy might also be the ones who bear most of the burden of potential hazards.

As illustrated in Paul Thompson's essay on transgenic foods (chapter 28), no perfect solution exists to resolve these dilemmas. Shrader-Frechette's interest is in bringing these contradictions to light so the public can openly discuss competing agendas as part of the policy-making process. She shares with the pragmatists Dewey and Hickman, and also with Feenberg, a belief that technology assessment need not be the province of experts alone. In fact, a democratic technology requires that philosophers and other scholars place their expertise at the use of public understanding to facilitate informed discussions and decisions.

Aristotle (384–322 BC) pointed out in Book III of the *Nicomachean Ethics* that one can deliberate only about what is within one's power to do. Technologies such as gene splicing and nuclear fission were not within the power of the Greeks, so there was no ethical deliberation about them until centuries later. Throughout history, technology (knowledge associated with the industrial arts, applied sciences, and various forms of engineering) has opened new possibilities for actions. As a result, it has also raised new ethical questions.

Most of these questions have not generated new ethical concepts; instead they have expanded the scope of existing ones. For example, because hazardous technologies threaten those who live nearby, ethicists have expanded the notion of "equal treatment" to include "geographical equity," equal treatment of persons located different distances from dangerous facilities.

Because new developments force the expansion of ethical concepts, those who investigate technology and ethics need both technical and philosophical skills. To assess the ethical desirability of using biological (versus chemical) pest control, for example, one must know the relevant biology and chemistry, as well as the economic constraints on the choice. Although such factual knowledge does not determine the ethical decision, it constrains it in important and unavoidable ways.

Since policymakers evaluate virtually all technologies, at least in part, by methods such as benefit-cost or benefit-risk analysis, knowledge of economics is essential for informed discussions of technology and ethics. Philosophers investigate both the ethical constraints on developing or implementing particular technologies and the ethical acceptability of various economic and policy methods used to evaluate technology.

Philosophical questions about technology and ethics generally fall into one of at least five categories. These are (1) conceptual or *metaethical* questions; (2) *general normative* questions; (3) *particular normative* questions about specific technologies; (4) questions about the ethical *consequences* of technological developments; and (5) questions about the ethical justifiability of various *methods* of technology assessment.

Examples of (1) are: "how ought one to define 'free, informed consent' to risks imposed by sophisticated technologies?" or "how ought one to define 'equal protection' from such risks?" Examples of (2) are: "does one have a right, as Alan Gewirth argues, not to be caused to contract cancer?" or "are there duties to future generations potentially harmed by various technologies?" Examples of (3) are: "should commercial nuclear power licensees, contrary to the Price–Anderson Act, be subject to strict and full liability?" or "should the US continue to export banned pesticides to developing nations?" Examples of (4) are: "would development of a plutonium-based energy technology threaten civil liberties?" or "would deregulation of the airline industry result in less safe air travel?" Examples of (5) are: "does benefit-cost analysis ignore noneconomic components of human welfare?" or "do Bayesian methods of technology assessment ignore the well-being of minorities likely to be harmed by a technological development?"

The leading philosophical issues concerning technology and ethics are the following:

How to Define Technological Risk

Engineers and technical experts tend to define "technological risk" as a probability of physical harm, usually as "average annual probability of fatality." Philosophers and other humanistic critics claim both that technological risk cannot be defined purely quantitatively, and that it includes more than physical harm. Instead they argue that technology often threatens other goods, such as civil liberties, personal autonomy, or rights such as due process.

The technologists argue for a quantitative definition of risk, claiming that we need a common denominator for evaluating diverse technological hazards. They also claim that it is impossible to evaluate nonquantitative notions, such as the technological threat to democracy. Those who oppose the quantitative definition

argue not only that it excludes qualitative factors (like equity of risk distribution) affecting welfare, but also that the nonquantitative factors are sometimes more important than the quantitative ones. Hence they argue, for example, that an equitably distributed technological risk could be more desirable than a quantitatively smaller one (in terms of probability of fatality) that is inequitably distributed.

How to Evaluate Technologies in the Face of Uncertainty

Whether one technological risk is quantitatively greater than another, in terms of average annual probability of fatality, however, is often difficult to determine. Most evaluations of technology are conducted in the face of probabilistic uncertainty about the magnitude of potential hazards. Typically this uncertainty ranges from two to four orders of magnitude. It arises because the developments most needing evaluation, e.g., biotechnology, are new. We have limited experience with them and hence limited data about their accident frequency.

How should one evaluate technologies whose level of risk is uncertain? According to John Harsanyi, the majority position is that, in such situations, one should either use subjective probabilities of experts, or assume that all uncertain events are equally probable. The desirable technological choice is then the one having the highest "average expected utility," as measured by the probability and utility of the outcomes associated with each choice. Critics of the majority position, like John Rawls, maintain that it has all the flaws of utilitarianism. It fails, they say, to take adequate account of minorities likely to be harmed by high-consequence, low-probability risks. Rawls argues instead that we should use a maximin rule in situations of probabilistic uncertainty. Such a rule, like the difference principle, would direct us to avoid the outcome having the worst possible consequences, regardless of its alleged probability.

Critics of the Rawlsian position claim that it is irrational to choose so as to avoid worst-case technological accidents. They claim that taking small chances with technology often brings great economic benefits for everyone. Opponents of the majority Bayesian position respond, however, that such benefits are neither assured nor worth the risk, and that the subjective probabilities of experts often exhibit an "overconfidence bias" that there will be no serious accidents or negative health effects from a given technology.

Technological Threats to Due Process

Ethicists also charge that technology threatens due-process rights. To the extent that hazardous technologies cause (what Judith Thomson calls) "incompensable risks," like death, due process is impossible because the victim cannot be compensated.

One of the most controversial due-process debates concerns commercial nuclear fission for generating electricity. Current United States law limits liability of the nuclear licensees to less than three percent of total possible losses in a catastrophic accident. Critics maintain that this law (the Price–Anderson Act) violates citizens' due-process rights. Defenders argue that it is needed to protect the industry from possible bankruptcy, and that a catastrophic nuclear accident is unlikely. Critics respond that if a catastrophic nuclear accident is unlikely, then industry needs no protection from bankruptcy caused by such an event.

How Safe Is Safe Enough?

Because a zero-risk society is impossible, philosophers and policy makers debate both how much risk is acceptable and how it ought to be distributed. The distribution controversies raise all the classical problems associated with utilitarian *versus* egalitarian ethical schemes. Conflicts over how much technological risk is acceptable typically raise issues of whether the public has certain welfare rights, like the right to breathe clean air. The controversies also focus on how much economic progress can be traded for the negative health consequences of technology-induced risks.

Philosophers are particularly divided about how to evaluate numerous negligible risks, from a variety of technologies, that together pose a serious hazard. Small cancer risks that are singly harmless, but cumulatively and synergistically harmful, provide a good example of such cases. They raise the classical ethical problem of the contributor's dilemma. This dilemma occurs because the benefit of avoiding imposing a single small technological risk is imperceptible, although the cumulative benefit of everyone's doing so is great. Some philosophers view such small risks as ethically insignificant, while others claim they are important. Those in the latter group argue that agents are responsible for the effects of *sets* of acts (that together cause harm) of which their individual act is only one member.

Consent to Risk

The sophistication of many technologies, from genetically engineered organisms to the latest nuclear weapons, makes it questionable whether many individuals understand them. If they do not, then it is likewise questionable whether persons are able to give free, informed consent to the risks that they impose. Critics of some contemporary technologies point out that those persons most likely to take technological risks (e.g., blue-collar workers in chemical or radiation-related industries) are precisely those who are least able to give free, informed consent to them. This is because they are often persons with limited education and no alternative job skills.

Those who claim that both workers and the public have given consent to technological risks use notions like "compensating wage differential" to defend their position. They say that, since workers in hazardous technologies receive correspondingly higher pay because of the greater risks that they face, they are compensated. Likewise they maintain that accepting a risky job constitutes a form of consent. They also claim that society's acceptance of the economic benefits created by hazardous technologies constitutes implicit acceptance of the technologies.

In response, more conservative ethicists argue both that economic analysis does not show the existence of a compensating wage differential in all cases, and that mere acceptance of a job in a risky technology does not constitute consent to the hazard, especially if the worker has no other realistic employment alternatives. They also argue that acceptance of the benefits of hazardous technologies does not constitute acceptance of the technologies themselves since many people are inadequately informed about such risks.

Bibliography

Baier, Kurt, and Nicholas Rescher, eds. *Values and the Future: The Impact of Technological Change on American Values*. New York: Free Press, 1969. Philosophical analysis of problems associated with technological development.

Durbin, Paul T., ed. *Philosophy and Technology*. Boston: Kluwer, 1985–. Annual publication of the Society for Philosophy and Technology. Essays, many of which deal with ethics and technology, written from a variety of philosophical perspectives.

Ellul, Jacques. *The Technological Society*. Translated by John Wilkinson. New York: Knopf, 1964. General overview of problems associated with technology; widely viewed as anti-technology.

Ellul, Jacques. *The Technological System*. Translated by Joachim Neugroschel. New York: Continuum, 1980. General overview of problems associated with technology; widely viewed as anti-technology.

Goodpaster, Kenneth, and Kenneth Sayre, eds. *Ethics and the Problems of the 21st Century*. South Bend, Ind.: University of Notre Dame Press, 1979. Original philosophical essays on various problems associated with technological development.

Harsanyi, John C. "Can the Maximin Principle Serve as a Basis for Morality? A Critique of John Rawls' Theory." *American Political Science Review* 59, no. 2 (1975): 594–605.

Jonas, Hans. *The Imperative of Responsibility*. Chicago: University of Chicago Press, 1985. Criticism of contemporary technological society and proposals for reform.

Kasperson, Roger, and Mimi Berberian, eds. *Equity Issues in Radioactive Waste Management*. Boston: Oelgeschlager, 1983. Descriptive analysis of equity issues related to radwaste management.

MacLean, Douglas. *Values at Risk*. Totowa, N.J.: Rowman and Allanheld, 1984. Consideration of the ethical problems posed by issues of nuclear security and deterrence.

MacLean, Douglas, and Peter Brown, eds. *Energy and the Future*. Totowa, N.J.: Rowman and Allanheld, 1983. Essays on ethical issues associated with energy technologies.

Rescher, Nicholas. *Unpopular Essays on Technological Progress*. Pittsburgh, Pa.: University of Pittsburgh Press, 1980. Widely ranging essays on a variety of topics related to technology.

Shrader-Frechette, Kristin. *Nuclear Power and Public Policy*. Boston: Reidel, 1983. Criticism of nuclear technology on grounds that it violates various ethical principles.

Shrader-Frechette, Kristin. *Science Policy, Ethics, and Economic Methodology*. Boston: Reidel, 1984. Analysis of problems with benefit-cost methods and proposals for amending them.

Shrader-Frechette, Kristin. *Risk Analysis and Scientific Method*. Boston: Reidel, 1984. Analysis of problems associated with assessment of technological risk assessment and proposals for amending them.

Thomson, Judith Jarvis. *Rights, Restitution, and Risk*. Cambridge: Harvard University Press, 1986. Treats a variety of ethical issues related to technological risk.

Part II
Considering the Autonomy of Technology

6

The Autonomy of Technology

Jacques Ellul

Jacques Ellul (1912–94) was a French sociologist, theologian, and philosopher. Ellul's work is one of the most important examples of a humanist philosophy of technology, and offers us one of the most forceful and compelling accounts of the possible autonomy of technology. This selection includes excerpts from his influential 1954 book *The Technological Society*. In the prefatory notes, Ellul argues that to understand the nature of technology and technological society, one must not be sidetracked by empirical investigations of the surface of the phenomena, but must develop an account of technology itself, as a whole. This position is an example of the transcendental approach to philosophy of technology. He then proceeds to argue that technology, what he calls "technique" has become autonomous of other social forces and values, and further that technology has become the determining force, or prime mover, of contemporary society. Ellul defines "technique" as the "totality of methods rationally arrived at and having absolute efficiency." Thus, for Ellul, technology consists not only, or even most importantly, of artifacts, but is a totalizing system of methods that envelops all of human existence. As such, technique increasingly erases mystery as a dimension of human experience, and shapes everything – politics, economics, personal relationships, education, love, sex, religion – as processes aimed at the efficient realization of clearly identifiable, and ideally quantifiable, goals.

This erasure of mystery and humanness is, for Ellul, the problem posed by modern technology. He argues that all too often attention is paid to what he calls "False Problems" of technology, such as pollution or changing values or challenges to existing forms of artistic production. For each of these, technology both presents a problem and a solution, in an ongoing dialectic. Thus, photography presents a challenge to painting, and not only opens new forms of creativity in its own right, but plays a role in transforming painting. Or, technologies of smoke-stack-scrubbing and emissions trading stand as a response to acid rain caused by power plant emissions.

The primary aspect of autonomy is perfectly expressed by Frederick Winslow Taylor, a leading technician. He takes, as his point of departure, the view that the industrial plant is a whole in itself, a "closed organism," an end in itself. Giedion adds: "What is fabricated in this plant and what is the goal of its labor – these are questions outside its design." The complete separation of the goal from the mechanism, the limitation of the problem to the means, and the refusal to interfere in any way with efficiency; all this is clearly expressed by Taylor and lies at the basis of technical autonomy.

Autonomy is the essential condition for the development of technique, as Ernst Kohn-Bramstedt's study of the police clearly indicates. The police must be independent if they are to become efficient. They must form a closed, autonomous organization in order to operate by the most direct and efficient means and not be shackled by subsidiary considerations. And in this autonomy, they must be self-confident in respect to the law. It matters little whether police action is legal, if it is efficient. The rules obeyed by a technical organization are no longer rules of justice or injustice. They are "laws" in a purely technical sense. As far as the police are concerned, the highest stage is reached when the legislature legalizes their independence of the legislature itself and recognizes the primacy of technical laws. This is the opinion of Best, a leading German specialist in police matters.

The autonomy of technique must be examined in different perspectives on the basis of the different spheres in relation to which it has this characteristic. First, technique is autonomous with respect to economics and politics. We have already seen that, at the present, neither economic nor political evolution conditions technical progress. Its progress is likewise independent of the social situation. The converse is actually the

case, a point I shall develop at length. Technique elicits and conditions social, political, and economic change. It is the prime mover of all the rest, in spite of any appearance to the contrary and in spite of human pride, which pretends that man's philosophical theories are still determining influences and man's political regimes decisive factors in technical evolution. External necessities no longer determine technique. Technique's own internal necessities are determinative. Technique has become a reality in itself, self-sufficient, with its special laws and its own determinations.

Let us not deceive ourselves on this point. Suppose that the state, for example, intervenes in a technical domain. Either it intervenes for sentimental, theoretical, or intellectual reasons, and the effect of its intervention will be negative or nil; or it intervenes for reasons of political technique, and we have the combined effect of two techniques. There is no other possibility. The historical experience of the last years shows this fully.

To go one step further, technical autonomy is apparent in respect to morality and spiritual values. Technique tolerates no judgment from without and accepts no limitation. It is by virtue of technique rather than science that the great principle has become established: *chacun chez soi*. Morality judges moral problems; as far as technical problems are concerned, it has nothing to say. Only technical criteria are relevant. Technique, in sitting in judgment on itself, is clearly freed from this principal obstacle to human action. (Whether the obstacle is valid is not the question here. For the moment we merely record that it is an obstacle.) Thus, technique theoretically and systematically assures to itself that liberty which it has been able to win practically. Since it has put itself beyond good and evil, it need fear no limitation whatever. It was long claimed that technique was neutral. Today this is no longer a useful distinction. The power and autonomy of technique are so well secured that it, in its turn, has become the judge of what is moral, the creator of a new morality. Thus, it plays the role of creator of a new civilization as well. This morality – internal to technique – is assured of not having to suffer from technique. In any case, in respect to traditional morality, technique affirms itself as an independent power. Man alone is subject, it would seem, to moral judgment. We no longer live in that primitive epoch in which things were good or bad in themselves. Technique in itself is neither, and can therefore do what it will. It is truly autonomous.

However, technique cannot assert its autonomy in respect to physical or biological laws. Instead, it puts them to work; it seeks to dominate them.

Giedion, in his probing study of mechanization and the manufacture of bread, shows that "wherever mechanization encounters a living substance, bacterial or animal, the organic substance determines the laws." For this reason, the mechanization of bakeries was a failure. More subdivisions, intervals, and precautions of various kinds were required in the mechanized bakery than in the nonmechanized bakery. The size of the machines did not save time; it merely gave work to larger numbers of people. Giedion shows how the attempt was made to change the nature of the bread in order to adapt it to mechanical manipulations. In the last resort, the ultimate success of mechanization turned on the transformation of human taste. Whenever technique collides with a natural obstacle, it tends to get around it either by replacing the living organism by a machine, or by modifying the organism so that it no longer presents any specifically organic reaction.

The same phenomenon is evident in yet another area in which technical autonomy asserts itself: the relations between techniques and man. We have already seen, in connection with technical self-augmentation, that technique pursues its own course more and more independently of man. This means that man participates less and less actively in technical creation, which, by the automatic combination of prior elements, becomes a kind of fate. Man is reduced to the level of a catalyst. Better still, he resembles a slug inserted into a slot machine: he starts the operation without participating in it.

But this autonomy with respect to man goes much further. To the degree that technique must attain its result with mathematical precision, it has for its object the elimination of all human variability and elasticity. It is a commonplace to say that the machine replaces the human being. But it replaces him to a greater degree than has been believed.

Industrial technique will soon succeed in completely replacing the effort of the worker, and it would do so even sooner if capitalism were not an obstacle. The worker, no longer needed to guide or move the machine to action, will be required merely to watch it and to repair it when it breaks down. He will not participate in the work any more than a boxer's manager participates in a prize fight. This is no dream. The automated factory has already been realized for a great number of operations, and it is realizable for a far greater number. Examples multiply from day to day in all areas. Man indicates how this automation and its attendant exclusion of men operates in business offices; for example, in the case of the so-called tabulating machine. The machine itself interprets the data, the elementary bits of information fed into it. It arranges them in texts and distinct numbers. It adds them together and classifies the results in groups and subgroups, and so on. We have here an administrative circuit accomplished by a single, self-controlled machine. It is scarcely necessary to dwell on the astounding growth of automation in the last ten years. The multiple applications of the automatic assembly line, of automatic control of production operations (so-called cybernetics) are well known. Another case in point is the automatic pilot. Until recently the automatic pilot was used only in rectilinear flight; the finer operations were carried out by the living pilot. As early as 1952 the automatic pilot effected the operations of takeoff and landing for certain supersonic aircraft. The same kind of feat is performed by automatic direction finders in antiaircraft defense. Man's role is limited to inspection. This automation results from the development of servomechanisms which act as substitutes for human beings in more and more subtle operations by virtue of their "feedback" capacity.

This progressive elimination of man from the circuit must inexorably continue. Is the elimination of man so unavoidably necessary? Certainly!

Freeing man from toil is in itself an ideal. Beyond this, every intervention of man, however educated or used to machinery he may be, is a source of error and unpredictability. The combination of man and technique is a happy one only if man has no responsibility. Otherwise, he is ceaselessly tempted to make unpredictable choices and is susceptible to emotional motivations which invalidate the mathematical precision of the machinery. He is also susceptible to fatigue and discouragement. All this disturbs the forward thrust of technique.

Man must have nothing decisive to perform in the course of technical operations; after all, he is the source of error. Political technique is still troubled by certain unpredictable phenomena, in spite of all the precision of the apparatus and the skill of those involved. (But this technique is still in its childhood.) In human reactions, howsoever well calculated they may be, a "coefficient of elasticity" causes imprecision, and imprecision is intolerable to technique. As far as possible, this source of error must be eliminated. Eliminate the individual, and excellent results ensue. Any technical man who is aware of this fact is forced to support the opinions voiced by Robert Jungk, which can be summed up thus: "The individual is a brake on progress." Or: "Considered from the modern technical point of view, man is a useless appendage." For instance, ten per cent of all telephone calls are wrong numbers, due to human error. An excellent use by man of so perfect an apparatus!

Now that statistical operations are carried out by perforated-card machines instead of human beings, they have become exact. Machines no longer perform merely gross operations. They perform a whole complex of subtle ones as well. And before long – what with the electronic brain – they will attain an intellectual power of which man is incapable.

Thus, the "great changing of the guard" is occurring much more extensively than Jacques Duboin envisaged some decades ago. Gaston Bouthoul, a leading sociologist of the phenomena of war, concludes that war breaks out in a social group when there is a "plethora of young men surpassing the indispensable tasks of the economy." When for one reason or another these men are not employed, they become ready for war. It is the multiplication of men who are excluded from working which provokes war. We ought at least to bear this in mind when we boast of the continual decrease in human participation in technical operations.

However, there are spheres in which it is impossible to eliminate human influence. The autonomy of technique then develops in another direction. Technique is not, for example, autonomous in respect of clock time. Machines, like abstract technical laws, are subject to the law of speed, and coordination presupposes time adjustment. In his description of the assembly line, Giedion writes: "Extremely precise time tables guide the automatic cooperation of the instruments, which, like the atoms in a planetary system, consist of separate units but gravitate with respect to each other in obedience to their inherent laws." This image shows in a remarkable way how technique became simultaneously independent of man and obedient to the chronometer. Technique obeys its own specific laws, as every machine obeys laws. Each element of the technical complex follows certain laws determined by its relations with the other elements, and these laws are internal to the system and in no way influenced by external factors. It is not a question of causing the human being to disappear, but of making him capitulate, of inducing him to accommodate himself to techniques and not to experience personal feelings and reactions.

No technique is possible when men are free. When technique enters into the realm of social life, it collides ceaselessly with the human being to the degree that the combination of man and technique is unavoidable, and that technical action necessarily results in a determined result. Technique requires predictability and, no less, exactness of prediction. It is necessary, then, that technique prevail over the human being. For technique, this is a matter of life or death. Technique must reduce man to a technical animal, the king of the slaves of technique. Human caprice crumbles before this necessity; there can be no human autonomy in the face of technical autonomy. The individual must be fashioned by techniques, either negatively (by the techniques of understanding man) or positively (by the adaptation of man to the technical framework), in order to wipe out the blots his personal determination introduces into the perfect design of the organization.

But it is requisite that man have certain precise inner characteristics. An extreme example is the atomic worker or the jet pilot. He must be of calm temperament, and even temper, he must be phlegmatic, he must not have too much initiative, and he must be devoid of egotism. The ideal jet pilot is already along in years (perhaps thirty-five) and has a settled direction in life. He flies his jet in the way a good civil servant goes to his office. Human joys and sorrows are fetters on technical aptitude. Jungk cites the case of a test pilot who had to abandon his profession because "his wife behaved in such a way as to lessen his capacity to fly. Every day, when he returned home, he found her shedding tears of joy. Having become in this way accident conscious, he dreaded catastrophe when he had to face a delicate situation." The individual who is a servant of technique must be completely unconscious of himself. Without this quality, his reflexes and his inclinations are not properly adapted to technique.

Moreover, the physiological condition of the individual must answer to technical demands. Jungk gives an impressive picture of the experiments in training and control that jet pilots have to undergo. The pilot is whirled on centrifuges until he "blacks out" (in order to measure his toleration of acceleration). There are catapults, ultrasonic chambers, etc., in which the candidate is forced to undergo unheard-of tortures in order to determine whether he has adequate resistance and whether he is capable of piloting the new machines. That the human organism is, technically speaking, an imperfect one is demonstrated by the experiments. The sufferings the individual endures in these "laboratories" are considered to be due to "biological weaknesses," which must be eliminated. New experiments have pushed even further to determine the reactions of "space pilots" and to prepare these heroes for their roles of tomorrow. This has given birth to new sciences, biometry for example; their one aim is to create the new man, the man adapted to technical functions.

It will be objected that these examples are extreme. This is certainly the case, but to a greater or lesser degree the same problem exists everywhere. And the more technique evolves, the more extreme its character becomes. The object of all the modern "human sciences" (which I will examine later on) is to find answers to these problems.

The enormous effort required to put this technical civilization into motion supposes that all individual effort is directed toward this goal alone and that all social forces are mobilized to attain the mathematically perfect structure of the edifice. ("Mathematically" does not mean "rigidly." The perfect technique is the most adaptable and, consequently, the most plastic one. True technique will know how to maintain the illusion of liberty, choice, and individuality; but these will have been carefully calculated so that they will be integrated into the mathematical reality merely as appearances!) Henceforth it will be wrong for a man to escape this universal effort. It will be inadmissible for any part of the individual not to be integrated in the drive toward technicization; it will be inadmissible that any man even aspire to escape this necessity of the whole society. The individual will no longer be able, materially or spiritually, to disengage himself from society. Materially, he will not be able to release himself because the technical means are so numerous that they invade his whole life and make it impossible for him to escape the collective phenomena. There is no longer an uninhabited place, or any other geographical locale, for the would-be solitary. It is no longer possible to refuse entrance into a community to a highway, a high-tension line, or a dam. It is vain to aspire to live alone when one is obliged to participate in all collective phenomena and to use all the collective's tools, without which it is impossible to earn a bare subsistence. Nothing is gratis any longer in our society; and to live on charity is less and less possible. "Social advantages" are for the workers alone, not for "useless mouths." The solitary is a useless mouth and will have no ration card – up to the day he is transported to a penal colony. (An attempt was made to institute this procedure during the French Revolution, with deportations to Cayenne.)

Spiritually, it will be impossible for the individual to disassociate himself from society. This is due not to the existence of spiritual techniques which have increasing force in our society, but rather to our situation. We are constrained to be "engaged," as the existentialists say, with technique. Positively or negatively, our spiritual attitude is constantly urged, if not determined, by this situation. Only bestiality, because it is unconscious, would seem to escape this situation, and it is itself only a product of the machine.

Every conscious being today is walking the narrow ridge of a decision with regard to technique. He who maintains that he can escape it is either a hypocrite or unconscious. The autonomy of technique forbids the man of today to choose his destiny. Doubtless, someone will ask if it has not always been the case that social conditions, environment, manorial oppression, and the family conditioned man's fate. The answer is, of course, yes. But there is no common denominator between the suppression of ration cards in an authoritarian state and the family pressure of two centuries ago. In the past, when an individual entered into conflict with society, he led a harsh and miserable life that required a vigor which either hardened or broke him. Today the concentration camp and death await him; technique cannot tolerate aberrant activities.

Because of the autonomy of technique, modern man cannot choose his means any more than his ends. In spite of variability and flexibility according to place and circumstance (which are characteristic of technique) there is still only a single employable technique in the given place and time in which an individual is situated. We have already examined the reasons for this.

At this point, we must consider the major consequences of the autonomy of technique. This will bring us to the climax of this analysis.

Technical autonomy explains the "specific weight" with which technique is endowed. It is not a kind of neutral matter, with no direction, quality, or structure. It is a power endowed with its own peculiar force. It refracts in its own specific sense the wills which make use of it and the ends proposed for it. Indeed, independently of the objectives that man pretends to assign to any given technical means, that means always conceals in itself a finality which cannot be evaded. And if there is a competition between this intrinsic finality and an extrinsic end proposed by man, it is always the intrinsic finality which carries the day. If the technique in question is not exactly adapted to a proposed human end, and if an individual pretends that he is adapting the technique to this end, it is generally quickly evident that it is the end which is being modified, not the technique. Of course, this statement must be qualified by what has already been said concern-

ing the endless refinement of techniques and their adaptation. But this adaptation is effected with reference to the techniques concerned and to the conditions of their applicability. It does not depend on external ends. Perrot has demonstrated this in the case of judicial techniques, and Giedion in the case of mechanical techniques. Concerning the overall problem of the relation between the ends and the means, I take the liberty of referring to my own work, *Présence au monde moderne.*

Once again we are faced with a choice of "all or nothing." If we make use of technique, we must accept the specificity and autonomy of its ends, and the totality of its rules. Our own desires and aspirations can change nothing.

The second consequence of technical autonomy is that it renders technique at once sacrilegious and sacred. (*Sacrilegious* is not used here in the theological but in the sociological sense.) Sociologists have recognized that the world in which man lives is for him not only a material but also a spiritual world; that forces act in it which are unknown and perhaps unknowable; that there are phenomena in it which man interprets as magical; that there are relations and correspondences between things and beings in which material connections are of little consequence. This whole area is mysterious. Mystery (but not in the Catholic sense) is an element of man's life. Jung has shown that it is catastrophic to make superficially clear what is hidden in man's innermost depths. Man must make allowance for a background, a great deep above which lie his reason and his clear consciousness. The mystery of man perhaps creates the mystery of the world he inhabits. Or perhaps this mystery is a reality in itself. There is no way to decide between these two alternatives. But, one way or the other, mystery is a necessity of human life.

Man cannot live without a sense of the secret. The psychoanalysts agree on this point. But the invasion of technique desacralizes the world in which man is called upon to live. For technique nothing is sacred, there is no mystery, no taboo. Autonomy makes this so. Technique does not accept the existence of rules outside itself, or of any norm. Still less will it accept any judgment upon it. As a consequence, no matter where it penetrates, what it does is permitted, lawful, justified.

To a great extent, mystery is desired by man. It is not that he cannot understand, or enter into, or grasp mystery, but that he does not desire to do so. The sacred is what man decides unconsciously to respect. The taboo becomes compelling from a social standpoint, but there is always a factor of adoration and respect which does not derive from compulsion and fear.

Technique worships nothing, respects nothing. It has a single role: to strip off externals, to bring everything to light, and by rational use to transform everything into means. More than science, which limits itself to explaining the "how," technique desacralizes because it demonstrates (by evidence and not by reason, through use and not through books) that mystery does not exist. Science brings to the light of day everything man had believed sacred. Technique takes possession of it and enslaves it. The sacred cannot resist. Science penetrates to the great depths of the sea to photograph the unknown fish of the deep. Technique captures them, hauls them up to see if they are edible – but before they arrive on deck they burst. And why should technique not act thus? It is autonomous and recognizes as barriers only the temporary limits of its action. In its eyes, this terrain, which is for the moment unknown but not mysterious, must be attacked. Far from being restrained by any scruples before the sacred, technique constantly assails it. Everything which is not yet technique becomes so. It is driven onward by itself, by its character of self-augmentation. Technique denies mystery a priori. The mysterious is merely that which has not yet been technicized.

Technique advocates the entire remaking of life and its framework because they have been badly made. Since heredity is full of chance, technique proposes to suppress it so as to engender the kind of men necessary for its ideal of service. The creation of the ideal man will soon be a simple technical operation. It is no longer necessary to rely on the chances of the family or on the personal vigor which is called virtue. Applied biogenetics is an obvious point at which technique desacralizes; but we must not forget psychoanalysis, which holds that dreams, visions, and the psychic life in general are nothing more than objects. Nor must we forget the penetration and exploitation of the earth's secrets. Crash programs, particularly in the United States, are attempting to reconstruct the soil which massive exploitation and the use of chemical fertilizers have impaired. We shall soon discover the functions of chlorophyll and thus entirely transform the conditions of life. Recent investigations in electronic techniques applied to biology have emphasized the importance of DNA and will possibly result in the discovery of the link between the living and the nonliving.

Nothing belongs any longer to the realm of the gods or the supernatural. The individual who lives in the technical milieu knows very well that there is nothing spiritual anywhere. But man cannot live without the sacred. He therefore transfers his sense of the sacred to the very thing which has destroyed its former object: to technique itself. In the world in which we live, technique has become the essential mystery, taking widely diverse forms according to place and race. Those who have preserved some of the notions of magic both admire and fear technique. Radio presents an inexplicable mystery, an obvious and recurrent miracle. It is no less astonishing than the highest manifestations of magic once were, and it is worshiped as an idol would have been worshiped, with the same simplicity and fear.

But custom and the recurrence of the miracle eventually wear out this primitive adoration. It is scarcely found today in European countries; the proletariat, workers and peasants alike, with their motorcycles, radios, and electrical appliances, have an attitude of condescending pride toward the jinn who is their slave. Their ideal is incarnated in certain things which serve them. Yet they retain some feeling of the sacred, in the sense that life is not worth the trouble of living unless a man has these jinns in his home. This attitude goes much further in the case of the conscious segment of the proletariat, among whom technique is seen as a whole and not merely in its occasional aspects. For them, technique is the instrument of liberation for the proletariat. All that is needed is for technique to make a little more headway, and they will be freed proportionately from their chains. Stalin pointed to industrialization as the sole condition for the realization of Communism. Every gain made by technique is a gain for the proletariat. This represents indeed a belief in the sacred. Technique is the god which brings salvation. It is good in its essence. Capitalism is an abomination because on occasion it opposes technique. Technique is the hope of the proletarians; they can have faith in it because its miracles are visible and progressive. A great part of their sense of the mysterious remains attached to it. Karl Marx may have been able to explain rationally how technique would free the proletariat, but the proletariat itself is scarcely equal to a full understanding of this "how." It remains mysterious for them. They retain merely the formula of faith. But their faith addresses itself with enthusiasm to the mysterious agent of their liberation.

The nonintellectual classes of the bourgeoisie are perhaps less caught up in this worship of technique. But the technicians of the bourgeoisie are without doubt the ones most powerfully taken with it. For them, technique *is* sacred, since they have no reason to feel a passion for it. Technical men are always disconcerted when one asks them the motives for their faith. No, they do not expect to be liberated; they expect nothing, yet they sacrifice themselves and devote their lives with frenzy to the development of industrial plants and the organization of banks. The happiness of the human race and suchlike nonsense are the commonplaces they allege. But these are no longer of any service even as justifications, and they certainly have nothing at all to do with man's passion for technique.

The technician uses technique perhaps because it is his profession, but he does so with adoration because for him technique is the locus of the sacred. There is neither reason nor explanation in his attitude. The power of technique, mysterious though scientific, which covers the whole earth with its networks of waves, wires, and paper, is to the technician an abstract idol which gives him a reason for living and even for joy. One sign, among many, of the feeling of the sacred that man experiences in the face of technique is the care he takes to treat it with familiarity. Laughter and humor are common human reactions in the presence of the sacred. This is true for primitive peoples; and for the same reason the first atomic bomb was called "Gilda," the giant cyclotron of Los Alamos "Clementine," the atomic piles "water pots," and radioactive contamination "scalding." The technicians of Los Alamos have banned the word *atom* from their vocabulary. These things are significant.

In view of the very different forms of technique, there is no question of a technical religion. But there is associated with it the feeling of the sacred, which expresses itself in different ways. The way differs from man to man, but for all men the feeling of the sacred is expressed in this marvelous instrument of the power instinct which is always joined to mystery and magic. The worker brags about his job because it offers him joyous confirmation of his superiority. The young snob speeds along at 100 m.p.h. in his Porsche. The technician contemplates with satisfaction the gradients of his charts, no matter what their reference is. For these men, technique is in every way sacred: it is the common expression of human power without which they would find themselves poor, alone, naked, and stripped of all pretentions. They would no longer be the heroes, geniuses, or archangels which a motor permits them to be at little expense.

What shall we say of the outburst of frenzy when the Sputnik went into orbit? What of the poems of the Soviets, the metaphysical affirmations of the French, the speculations on the conquest of the universe? What of the identification of this artificial satellite with the sun, or of its invention with the creation of the earth? And, on the other side of the Atlantic, what was the real meaning of the excessive consternation of the Americans? All these bore witness to a marked social attitude with regard to a simple technical fact.

Even people put out of work or ruined by technique, even those who criticize or attack it (without daring to go so far as to turn worshipers against them) have the bad conscience of all iconoclasts. They find neither within nor without themselves a compensating force for the one they call into question. They do not even live in despair, which would be a sign of their freedom. This bad conscience appears to me to be perhaps the most revealing fact about the new sacralization of modern technique.

The characteristics we have examined permit me to assert with confidence that there is no common denominator between the technique of today and that of yesterday. Today we are dealing with an utterly different phenomenon. Those who claim to deduce from man's technical situation in past centuries his situation in this one show that they have grasped nothing of the technical phenomenon. These deductions prove that all their reasonings are without foundation and all their analogies are astigmatic.

The celebrated formula of Alain has been invalidated: "Tools, instruments of necessity, instruments that neither lie nor cheat, tools with which necessity can be subjugated by obeying her, without the help of false laws; tools that make it possible to conquer by obeying." This formula is true of the tool which puts man squarely in contact with a reality that will bear no excuses, in contact with matter to be mastered, and the only way to use it is to obey it. Obedience to the plow and the plane was indeed the only means of dominating earth and wood. But the formula is not true for our techniques. He who serves these techniques enters another realm of necessity. This new necessity is not natural necessity; natural necessity, in fact, no longer exists. It is technique's necessity, which becomes the more constraining the more nature's necessity fades and disappears. It cannot be escaped or mastered. The tool was not false. But technique causes us to penetrate into the innermost realm of falsehood, showing us all the while the noble face of objectivity of result. In this innermost recess, man is no longer able to recognize himself because of the instruments he employs.

The tool enables man to conquer. But, man, dost thou not know there is no more victory which is thy victory? The victory of our days belongs to the tool. The tool alone has the power and carries off the victory. Man bestows on himself the laurel crown, after the example of Napoleon III, who stayed in Paris to plan the strategy of the Crimean War and claimed the bay leaves of the victor.

But this delusion cannot last much longer. The individual obeys and no longer has victory which is his own. He cannot have access even to his apparent triumphs except by becoming himself the object of technique and the offspring of the mating of man and machine. All his accounts are falsified. Alain's definition no longer corresponds to anything in the modern world. In writing this, I have, of course, omitted innumerable facets of our world. There are still artisans, petty tradesmen, butchers, domestics, and small agricultural landowners. But theirs are the faces of yesterday, the more or less hardy survivals of our past. Our world is not made of these static residues of history, and I have attempted to consider only moving forces. In the complexity of the present world, residues do exist, but they have no future and are consequently disappearing.

Only the things which have a future interest us. But how are we to discern them? By making a comparison of three planes of civilization which coexist today: India, Western Europe, and the United States. And by considering the line of historical progression from one to the other – all of this powerfully reinforced by the evolution of the Soviet Union, which is causing history to boil.

Artifice and Order

Langdon Winner

Langdon Winner is the Thomas Phelan Chair of Humanities and Social Sciences at Rensselaer Polytechnic Institute. Winner rejects the thesis that technology is value-neutral, but his relation to the thesis of technological autonomy is complex. He rejects the notion that technology is necessarily autonomous. But technological development since the Industrial Revolution has largely taken place without conscious political interventions or evaluations of the value implications. Winner characterizes this as "somnambulism," or sleepwalking, a state similar to Ortega's notion that we have come to relate to technology as a "second nature." As such, technology increasingly disciplines our lives, determining the order and pace of our activities. Those who have attempted to question new technologies are typically dismissed as fringe romantics (consider the environmental movement until quite recently) or wacky religious fundamentalists (hesitations about new genetic biotechnologies or human embryonic stem cell research). Going further, Winner notes that in discussions about the place of technology in our lives we tend to group into extremes of what Drengson calls technophilia and technophobia.

It is important to notice, first of all, the conception of society which takes shape in the technological perspective. Absolutely fundamental is the view that modern technology is a way of organizing the world and that, potentially, there is no limit to the extent of this organization. In the end, literally everything within human reach can or will be rebuilt, resynthesized, reconstructed, and incorporated into the system of technical instrumentality. In this all-encompassing arrangement, human society – the total range of relationships among persons – is one segment.

From Langdon Winner, *Autonomous Technology: Technics-out-of-control as a Theme in Political Thought* (Cambridge, MA: MIT Press, 1977), pp. 191–209. © 1977 Massachusetts Institute of Technology, by permission of MIT Press.

"Technological society" is actually a subsystem of something much larger, the technological order. Social relationships are merely one sort of connection. Individuals and social groups are merely one variety of component. The connections and groupings of inanimate parts are equally crucial to the functioning of the whole.

This is not to say that any existing society has been integrated in all its parts into a purely technological order. There are some kinds of social relationships, those involving love and friendship for example, that have not yet been fully adapted to the demands of technical routine. The position of the theory is that a strong tendency toward order of this kind is highly pronounced in all spheres of Western society and that its development will in all likelihood proceed rapidly on a worldwide scale.

An apt comparison can be found in the notions of order and society of the medieval Christian world view. In the great chain of being, with its hierarchy of God created things, men and women in their various social positions occupied merely one of several levels and by no means the most important stratum at that. Each being had its own "degree" or grade of perfection, and it was mandatory that each level in the hierarchy keep its established place. For humans to aspire to anything more than their appointed position was an act of sinful pride and defiance, an invitation to chaos.[1] In the present view, then, the liberatory quality of technology must be weighed against what its system of order imposes upon and requires of man. Attempts to deal with this side of the technical situation have engendered two distinctive themes in modern social philosophy: first, the mechanization of human activity and social relationships; second, the more thorough conditioning of individuals through their contact with technical systems and apparatus.

The idea of a mechanized humanity was a prominent part of nineteenth-century literature in Europe and America. Thinkers looked at the advance of an advanced, industrialized world and wondered openly about the capacity of man to retain his integrity in the face of such marvelous instrumentation. Thomas Carlyle, to whom we owe the idea that "man is a tool using animal. . . . Without tools he is nothing, with tools he is all," wrote of the possibility that men would internalize the external reality of mechanization and become themselves thoroughly mechanical in thought and behavior.[2] Predictions of this sort were not uncommon. The scientific world view of the nineteenth century was still centered in the Newtonian vision of a universe running like a colossal clock.[3] Since this model had served well in the development of physics and since the industrial machine had indeed proven to be an astounding force, the image of man as a machine was widely thought a natural. For some thinkers this meant that homo sapiens was quite literally *l'homme machine* of the sort described by the eighteenth-century philosopher Julien Offray de La Mettrie: "The human body is a machine which winds its own springs. It is the living image of perpetual movement."[4] In a totally mechanical universe, a mechanical man is an appropriate microcosm. To this day, views of this kind live on in the writings of Skinner, Woolridge, and others, accompanied by the peculiar conviction that if human beings are *in any way* like machines, then it follows that they either do or very well ought to behave totally like mechanical devices.[5] In the nineteenth century many of those who took violent objection to this conception of humanity were convinced that the masses of men had already become greatly mechanized in their personal and social existence. Man's relationship to the industrial process, particularly to the machinery, organization, and techniques of the factory, would eventually bring La Mettrie's prophecy to fruition in a remade human environment.[6]

The mid-twentieth century has brought the eclipse of the machine as a model for everything under the sun. Too many recent developments in science and technology – quantum physics, relativity, modern chemistry and biology, the alloys, plastics, the transistor – simply do not match the two primary images of the older mechanical tradition: Newton's clockwork universe and the cog and wheel machine of nineteenth-century industry. Artifice has become more subtle. Many devices properly called machines are no longer truly mechanical. Even Lewis Mumford, who emphasizes the idea of society as machine, has changed his emphasis to something called the "Power Complex." What needs expression is the idea of a set of large-scale, complex, interdependent, functioning networks which form the basis of modern life; for this, "the machine" will no longer suffice.[7]

But the decline of a metaphor does not mean that concerns it represented vanish. The possibility that man faces an unwitting bondage in his relationships with technical systems is still a living hypothesis. Father of this question in modern social philosophy was Jean Jacques Rousseau. In the *Discourses* Rousseau placed himself in open disagreement with the prevailing opinion of his time by arguing that the advance of the "arts and sciences" was a degenerative rather than a progressive movement in history. Human existence, he believed, had long ago reached something of a golden age in which men were free and happy. But then there occurred a "fatal accident"[8] (a chance discovery or invention?) that brought a great revolution in metallurgy and agriculture and thereby "civilized men and ruined the human race."[9] "As soon as some men were needed to smelt and forge iron, other men were needed to feed them. The more the number of workers was multiplied, the fewer hands were engaged in furnishing the common substance . . . and since some needed foodstuffs in exchange for their iron, others finally found the secret of using iron in order to multiply foodstuffs. From this arose husbandry and agriculture. . . . From the cultivation of land, its division necessarily followed; and from property once recognized, the first rules of justice."[10] Rousseau held that by adapting their lives to this early technological revolution and to subsequent ones, men had given up their original freedom and entered the enslaving web of dependencies involved in a complex economic society.

A more familiar conception of how freedom is lost emphasizes the presence of external restraints. The activity of an individual or group is limited by an outside factor, such as the presence of a stronger force or the restrictions of law. Much of liberal political philosophy attempts to delineate the conditions under which such restraints are or are not justified. The theory of technological politics, however, follows Rousseau in seeing the loss of freedom in the modern world as preeminently a situation in which individuals become caught up in webs of relationships which have a pathological completeness. Conditions of life for all persons come to be inextricably tied to systems of transportation, communication, material production, energy, and food supply for which there are no readily available alternatives. An automobile or mass transit network carries the person to work at a particular time every day. The food he or she eats is grown by agribusiness concerns and shipped in from a great distance and distributed by large chain supermarkets. Information about the world is available, prepackaged like the food one eats, in television news programmed at a central source. For all manner of day-to-day activities, apparatus like the telephone become absolutely indispensable. Because in most instances the working of such systems does not include the application of restraints, the problem of human freedom as it involves extreme dependency and helplessness seldom comes up.

Closely tied to the phenomenon of dependency is the situation of servitude within technological relationships. The human encounter with artificial means cannot be summarized solely (or even primarily) as a matter of "use." One must notice that certain kinds of regularized service must be rendered to an instrument before it has any utility at all. One must be aware of the patterns of behavior demanded of the individual or of society in order to accommodate the instrument within the life process. There are, to put it differently, subtle but important costs as well as obvious benefits. These costs, usually forgotten or thought "inevitable" by those who must bear them, are in the aggregate truly staggering.

An early analysis of such circumstances in microcosm appears in Ralph Waldo Emerson's *Works and Days*. "Many facts concur to show," he says, "that we must look deeper for our salvation than to steam, photographs, balloons, or astronomy. These tools have some questionable properties. They are reagents. Machinery is aggressive. The weaver becomes a web, the machinist a machine. If you do not use the tools, they use you."[11] Although Emerson employs the image of the machine, his main point is not actually a standard "mechanization of man" thesis. It is instead the idea that all tools are "reagents"; they are not a passive presence in a human situation but instead evoke a necessary reaction from the person using them.[12] Attachment to apparatus not only requires that men behave in certain ways, it also gives them a positive responsibility and criterion of performance they must meet. "A man builds a fine house; and now he has a master, and a task for life: he is to furnish, watch, show it and keep it in repair, the rest of his days."[13]

Bruno Bettelheim offers a similar report from recent twentieth-century experience. "In my daily work with psychotic children, and in my efforts to create an institutional setting that will induce them to return to sanity, I have come face to face with this problem of how to take best advantage of all the conveniences of a technological age, . . . and to do it without entering a bondage to science and technology."[14] Bettelheim goes on to describe how he learned to study the shaping effect that any new device had upon the work of his institution. "Whenever we introduced a new technological convenience, we had to examine its place in the life of our institution most carefully. The advantages we could enjoy from any new machine were always quite obvious; the bondage we entered by using it was much harder to assess, and much more elusive. Often we were unaware of its negative effects until after long use. By then we had come to rely on it so much, that small disadvantages that came with the use of any one contrivance seemed too trivial to warrant giving it up, or to change the pattern we had fallen into by using it. Nevertheless, when combined with the many other small disadvantages of all the other devices, it added up to a significant and undesirable change in the pattern of our life and work."[15]

Bettelheim's statement reflects a level of awareness altogether rare in contemporary writing, not to mention the practices of everyday life. His eminently sensible conclusion, therefore, smacks of a certain radicalism: "The most careful thinking and planning is needed to enjoy the good use of any technical contrivance without paying a price for it in human freedom."[16] Such sentiments are generally thought to be antiprogressive and are ignored in most polite company. Rousseau's views, similarly, are widely believed to have been those of a romantic fool of history who tried to build a wall against the ineluctable forces of modernity.

The theme of technical servitude becomes a major point in the macrocosmic social theories of, among others, Thorstein Veblen and Jacques Ellul. In such theories, the presence of modern technics is seen to have both subtle and very obvious shaping effects on the whole range of human behavior, consciousness, and social structure. Seen in their totality, these effects do in fact constitute most of what is important in the life of the individual and in all social relationships whatsoever. Still using the machine model of this state of affairs, Veblen spoke of "the cultural incidence of the machine process,"[17] an incidence he took to be completely overwhelming. "The machine pervades the modern life and dominates it in a mechanical sense. Its dominance is seen in the enforcement of precise mechanical measurements and adjustments and the reduction of all manner of things, purposes and acts, necessities, conveniences, and amenities of life, to standard units."[18] In Veblen's eyes the most important single fact about this state of affairs was that it brought a new and stringent "discipline" to all human activities. A society based on the machine process took on a rigid set of rules, responsibilities, and performance criteria much more demanding of human substance and of social relationships than anything known in previous history. Speaking of the effect on the workman, Veblen notes: "It remains true, of course, . . . that he is the intelligent agent concerned in the process, while the machine, furnace, roadway, or retort are inanimate structures devised by man and subject to the workman's supervision. But the process comprises him and his intelligent motions, and it is by virtue of his necessarily taking an intelligent part in what is going forward that the mechanical process has its chief effect upon him. The process standardizes his supervision and guidance of the machine. Mechanically speaking, the machine is not his to do with it as his fancy may suggest. His place is to take thought of the machine and its work in terms given him by the process that is going forward. His thinking in the premises is reduced to standard units of gauge and grade. If he fails of the precise measure, by more or less, the exigencies of the process check the aberration and drive home the need of conformity."[19]

Veblen argued that the advance of the new technological civilization would displace all previous forms of culture. "The machine discipline," he observed, "acts to disintegrate the institutional heritage, of all degrees of antiquity and authenticity – whether it be the institutions that embody the principles of natural liberty or those that comprise the residue of more archaic principles of conduct still current in civilized life."[20] While there is a lament implied in his observations, Veblen certainly did not wish to stop the movement of an increasingly technologized society. He saw it as an inevitable development, sanctioned by the fact that it was, after all, the true center of the modern condition and an improvement in man's material circumstances. As we have already seen, his criticisms came to rest on the contradictions present in this evolving system, namely, that the persons best suited to operate the machine culture – the engineers – were still subordinate to businessmen.

If one substitutes for the concept of "machine" that of "technique," the position Veblen announced is entirely similar to Ellul's. Both men hold that the technological element has outgrown and absorbed the shell of civilization that once enclosed it.[21] Both men assert that individuals and societies do not rule technical means so much as accept with strict obedience the rule that technical means themselves impose. To describe the technological system, therefore, is to describe the true system of governance under which men live.

Now, it is clear that the condition described by Emerson (the effects of apparatus on the behavior and consciousness of the individual) and Bettelheim (the effects of apparatus on a small group) are not entirely analogous. The macrocosmic social theory is not merely a microcosmic insight writ large. What is similar in these cases, however, is an emphasis upon the context in which tools and instruments operate and have their utility. Such statements ask us to consider what is *required* as well as what is received in the activity of technical "use." Technologies, we noted earlier, are commonly thought to be neutral. The important consideration is how they are used, and this is what permits us to judge them. But, as Ellul points out, the matter of "use" may be entirely settled before one can raise the question at all. "Technique *is* a use," he observes. "There is no difference at all between technique and its use. The individual is faced with an exclusive choice, either to use the technique as it should be used according to the technical rules, or not to use it at all."[22] In this assertion, the broader significance of Emerson's notion that tools are reagents becomes clearer. Complex instruments come equipped with certain rules for their employment, which must be obeyed. People are not at liberty to "use" the instruments in an arbitrary manner but must see to it that the appropriate operating procedures and techniques are followed and that all of the material conditions for operation are met. In modern civilization and its various parts, great amounts of time, energy, and resources are expended in making certain that the procedures are followed and that the conditions are met. Of the meanings of autonomous technology that we have encountered so far, this is the most significant. The technological version of Kantian heteronomy – the governance of human activity by external rules or conditions – is present here as a thoroughgoing yet entirely mundane phenomenon.

There is, of course, a vast multiplicity of such rules and conditions. One might ask, Where are they stated and analyzed in detail? An appropriate place to start would be to examine all of the textbooks in engineering, economics, management, and the various technical skill groups. One obvious but perhaps unavoidable source of incompleteness in the theory of technological politics is that much of the real substance of what it tries to account for is buried in diverse teachings of this sort. It is difficult to footnote or discuss that which is known, practiced, and obeyed in thousands upon thousands of technical specialties.

Indeed, anyone seriously critical of conditions in the technological society soon meets up with the demand from technically trained persons that in order to speak at all, one must first "learn technology." A version of the mode of legitimation through expert knowledge, this advice is, in my experience, usually less a plea for understanding than an urging to compliance. Suggestions for learning of this sort are often made to me at the institute of technology where I teach. They range from the study of calculus, physics, or one of the branches of engineering proper – electrical, civil, mechanical – to a mastery of the techniques of cost-benefit analysis, systems theory, and econometrics. I concede the usefulness of knowing the real activities of such domains of practice and have tried whenever possible to achieve the grasp appropriate to an informed outsider. But given the fact that the specialized fields show important differences in approach and content, the mastery of a particular representative specialty seems only superficially helpful to the effort to comprehend broader situations in which technology is problematic. Those who suggest a technological education of this kind implicitly ask that one undergo a process of socialization. "If you knew what we're doing, then you could not make such criticisms." One comes to appreciate and trust one's professional brotherhood. One comes to accept the virtue of such procedures as quantitive cost-benefit analysis. And, above all, one learns to accept the grand wisdom of the view that the world is a set of "problems" awaiting technically refined "solutions."

In summary, life in a sophisticated technological order supposes that each collection of technical rules, procedures, and trained persons outside one's own sphere of competence must be accepted as given. People are content in the knowledge that things are as they are and that they are in good working condition, and that in society everyone and everything has a certain job to do and does it. And everyone, like everything, does not find occasion to inquire into this condition or to dispute the manner in which it structures life.

Seen in this light – the ways in which technical rules and preconditions influence human behavior – the traditional notion that technologies are merely neutral tools becomes problematic. Individuals may still retain the noble idea that they can upon sudden inspiration direct the technical means to whatever ends they choose. They tend to see complex technologies as if they were handsaws or egg beaters. Give me a board and I will saw it in half; give me the eggs and flour and I will whip up a chocolate cake. But in highly developed technologies the conditions that make the tool-use notion tenable seldom hold. The technical equivalent of the Archimedean point – a place to put the lever so that one can move the mechanism – is often missing.

Reasons for this state of affairs are apparent in the very nature of modern technologies. Twentieth-century technical devices, as we have described them here, are characterized by enormous size, complex interconnection, and systemic interdependence. In terms of their own internal structure, most of them require precise coordination of the three major elements in our earlier definition. Apparatus almost always requires refined technique: an elaborate, knowledgeable

kind of human practice to guarantee its successful working. In the great majority of cases, however, both apparatus and technique require the presence of well-developed, rational, social organization. The world of craftsmanship – the world of technique plus apparatus alone – has vanished. Apparatus, technique, and organization are interdependent, that is, reciprocally necessary for each other's successful operation. This condition has become the sine qua non of all higher technologies of manufacturing, communications, transportation, agriculture, and others. And while there are still small pockets where this kind of interconnected technology is not the rule, such cases are now out of the ordinary.

Such circumstances are of special interest in the theory of technological politics, for two parts of the systemic arrangement – technique and apparatus – require that persons in large numbers be induced to behave according to precise technical principles. Through their "employment" such persons serve a specific function in an organization of many coordinated functions. Since this employment is usually their sole livelihood, there is strong pressure toward strict discipline and obedience. One appears at a preestablished time, for precisely determined work, for an exactly designated reward. This situation is so thoroughly normal in the twentieth century that any sense of how it might be otherwise is largely forgotten. In particular, persons so employed have no sense that the design of the work situation or the character of its operating procedures might be changed through their own conscious intervention. Even more than the employer, technology itself is seen as completely authoritative. It is an *authority* that asks for compliance only and never anything more.

What meaning can the traditional tool-use conception have in this context? Indeed, the structure of men and apparatus is "used" to produce something: goods or services for society. It can also be said that the employees "use" vast technical networks to earn a living. But beyond that, the idea that such networks are merely neutral tools under the control of men, to be used for chosen ends, begins to wear thin. Seen as a way of ordering human activity, the total order of networks is anything but neutral or tool-like. In its centrality to the daily activity and consciousness of the "employee," the function-serving human component, the technical order is more properly thought of as a *way of life*. Whatever else it may be, a way of life is certainly not neutral. Opportunities for "use" or "control" that the human components have within this system are minimal, for what kind of "control" is it that at every step requires strict obedience to technique or the necessities of technical organization? One can say that the "control" is exercised from the center or apex of the system; this is true, although we shall soon see that even this has a paradoxical character. But in terms of the functioning of individual components and the complex social interconnections, "control" in the sense of autonomous individuals directing technical means to predetermined ends has virtually no significance. "Control" and "use" simply do not describe anything about relationships of this kind. The direction of governance flows from the technical conditions to people and their social arrangements, not the other way around. What we find, then, is not a tool waiting passively to be used but a technical ensemble that demands routinized behavior.

In this way of seeing, therefore, the tool-use model is a source of illusions and misleading cues. We do not *use* technologies so much as *live* them. One begins to think differently about tools when one notices that the tools include persons as functioning parts. Highly developed, complex technologies are tools without handles or, at least, with handles of extremely remote access. Yet we continue to talk as if telephone and electric systems were analogous in their employment to a simple hand drill, as if an army were similar to an egg beater.

I am not saying that men and women never use technology or that all ideas of use are nonsense, only that many of our most prevalent conceptions here are primarily nostalgia. There was a golden age when the hand was on the handle and alchemy was the queen of the sciences. But except for the world of small-scale appliances, that time has passed.

The question addressed here can also be posed in terms of the idea of extension. In Emerson's words a full century before Marshall McLuhan, "The human body is the magazine of inventions, the patent-office, where are the models from which every hint was taken. All the tools and engines are only extensions of its limbs and senses."[23] But remembering Emerson's thoughts on tools as reagents we encounter a puzzle. What is an extension of what? Looking at contemporary technologies one sees massive aggregations of human and nonhuman parts, rationally ordered, working in precisely coordinated actions and transactions. Men do indeed claim to be in control and to be the instigators of all the motion. But if one considers the structure and behavior necessary for such systems to exist at all, such claims are either incorrect or ambiguous. When one discovers that people are subtly conditioned by their apparatus, when one learns that their conduct is largely determined by preestablished function and learned technique, when one finds that important social relationships are established according to organizational rationality alone – then the idea of technology as controlled extension becomes entirely misleading. Marx concluded that men had become appendages of the machine in the factory system. In a technical environment that is now more massive, more complete, and more intricate, the conclusion takes on new poignancy.

One can appreciate how it is from this vantage point that automation – the displacement of men and women by machines – diminishes as a problem. The crucial difficulty with the existing technological order is not so much that individuals are "unemployed" by automatic processes (though, certainly, this is a source of grief for a significant minority) but that they are overemployed in ways destructive to their humanity. Marcuse argues that advanced industrial societies do in fact suffer from incomplete and imperfect automation. The productive system still employs human beings as its prime components. "This is," he says, "the pure form of servitude: to exist as an instrument, as a thing."[24] If truly complete automized apparatus were introduced into society, this servitude could be ended for all time. "Automation, once it became *the* process of material production, would revolutionize the whole society. The reification of human labor power, driven to perfection would shatter the reified form by cutting the chain that ties the individual to the machinery – the mechanisms through which his own labor enslaves him."[25] Under the existing system of things, however, automation stands as a threat that the managers of the industrial order hold over their employees as a way of extracting even more toil from them. "At the present stage of advanced capitalism," Marcuse writes, "organized labor rightly opposes automation without compensating employment." Labor is thus forced to struggle against its ultimate means of liberation. In this fashion, "The enslavement of man by the instruments of his labor continues in a highly rationalized and vastly efficient and promising form."[26]

Observations and arguments of this kind form the basis of one of the more surprising themes in the literature of technological politics: the myth of labor-saving technology. No one denies that techniques and instruments save time and effort in the performance of specific tasks. And no one denies that the collectivity of such devices enables a society to do things it otherwise could not accomplish or to accomplish them more economically. But one can ask whether the technical innovations added to civilization in the last two hundred years or so have in every case "saved labor" in the sense of lightening man's toil.

Since Marx we have known why this question must be asked in terms of "whose labor?" and "under what conditions in society?" Eli Whitney's cotton gin was a marvelous labor-saving device. Yet its introduction into the system of production in the South actually prolonged slavery and increased the degree of toil extracted from the slaves. The factory system was also a labor-saving innovation, yet there is good evidence that it increased the hours, exertion, and suffering of the workers in the early years of the industrial revolution.[27] Marx's analysis of how such things occur is well known to the reader. What is of interest here is that there exists a distinctive explanation of this phenomenon from the technological perspective. Many have argued that the very nature of advanced technologies – putting aside the matter of ownership and class structure – demands much more of the human being than any previous productive arrangement. The technological order, no matter who *owns* it, is not very efficient at allowing men and women to bank the labor that techniques and instruments have saved. A classic, albeit excessive, statement of this view is found in Friedrich Georg Juenger's *The Failure of Technology*. "Never and nowhere," Juenger argues, "does machine labor reduce the amount of manual labor, however large may be the number of workers tending machines. The machine replaces the worker only where the work can be done in a mechanical fashion. But the burden of which the worker is thus relieved does not vanish at the command of the technical magician. It is merely shifted to areas where work cannot be done mechanically. And, of course, this burden grows apace with the increase of mechanical work."[28]

Juenger's account goes astray in its exclusive emphasis upon the situation of manual labor. Developments in machinery have not had the uniform impact he supposes. Labor-saving devices and automation have eliminated certain tasks, refined others, and created whole new vocational categories. But the broader point that underlies his contention is both valid and significant. Under the relentless pressure of technological processes, the activities of human life in modern society take place at an extremely demanding cadence. Highly productive, fast-moving, intensive, precision systems require highly productive, fast-moving, intense, and precise human participants. The computer has been an especially powerful goad in this direction. Its capacity to do prodigious amounts of work in a very short time puts the humans in the "interface" in a frantic struggle to keep up. The virtues of slow information processing and labor done at a leisurely pace have long since been sacrificed to the norms of work appropriate to the electronic exemplar. The idea that a task is something to be pondered or even savored is entirely foreign to this mode of activity. A telephone call and instantaneous computer check can reserve a room in any of thousands of hotels and motels in a particular network. Since dozens of similar transactions are completed each hour, the employee who does the job cannot spend more than a few moments on any particular request, although certain superficially courteous catchwords may still be part of the rationalized process. What one no longer expects is the innkeeper's handwritten note, received after a characteristic three weeks' delay, which remembers some small detail of your last year's visit.

Pressures of pace are not, however, limited to work environments. They now include the full spectrum of activity involving travel, communication, leisure time, and consumption. Hannah Arendt notes a shift in the focus of human energy that has taken place in the twentieth century. "The two stages through which the ever-recurrent cycle of biological life must pass, the stages of labor and consumption, may change their proportion even to the point where nearly all human 'labor power' is spent in consuming."[29] It is the intensity of the combined activities of labor and leisure that Jules Henry has labeled the "technological drivenness" of modern culture.[30] With Kenneth Keniston, Henry links this phenomenon to the rapidity of technological change. But the connection between pace of work and rate of innovation is by no means a necessary one. Technologies need not be changing rapidly to demand high performance and a rapid tempo of existence.

It is true that technological society is not the first kind of organization to have placed heavy demands upon its members. There is always a price to be paid for culture, for social order, and material well-being. A question raised by the theory of technological politics, however, is whether the price now extracted goes far beyond reasonable limits. How much servitude to technical means is too much? At what point does dependency upon complex systems become a condition of virtual enslavement? The search for criteria upon which one might begin making judgments on these questions offers a rich but as yet relatively poorly explored field of inquiry. Marcuse addresses the issue head on from one direction. He argues that the burdens of civilization as measured by "surplus repression" – "the restrictions necessitated by social domination" above and beyond "the modifications of the instincts necessary for the perpetuation of the human race in civilization" – are now greater than in any previous historical period.[31] The lid upon man's erotic instincts has been screwed down tighter than ever, far beyond any reasonable obedience to the reality principle. Ellul renounces Marcuse's Freudian outlook on the matter.[32] But he agrees that the technological order subjects man to pressures and limitations that are clearly pathological. Not without nostalgia, Ellul describes a condition "common to all civilizations up to the eighteenth century" in which techniques were local and limited, a part of culture rather than its whole.[33] "Man worked as little as possible and was content with a restricted consumption of goods . . . a prevalent attitude, which limits both techniques of production and

techniques of consumption."[34] This state was abandoned when men in Western society perceived the inestimable boon that *la technique* promised.

Whether explicitly stated or strongly implied by those who adopt its vantage point, the theory of technological politics always proceeds with an understanding of limits. Its criticisms point to a boundary beyond which technical artifice no longer enables or liberates mankind. In its evolution, technology arrives at a turning point after which it tends to thwart rather than facilitate the building of an emancipated society. The problem of specifying more clearly what the conditions of human liberation and social emancipation might be is an ambitious project, one that I shall only be able to touch upon briefly in the last chapter. Most of the analysis at this point must attend, perhaps even to a fault, to circumstances of pathology and excess – a corrective to the dewy-eyed traditional assumptions about tools, mastery, and endless benefit.

One is entitled to ask how much of the condition described enters the awareness of persons who live in this world. Do they notice the costs? The answer must be a qualified "yes." In one sense, there is nothing that men and women who live and work in the technological society understand better than the basic conditions which enable this system to function. But their awareness has an intuitive, largely passive quality. The influence of large-scale technical networks is so pervasive and indelible that few of us find occasion to wonder at their effects. We know that "this is how things work." We know that "this is how I do my job." The technological order includes a notion of *citizenship*, which consists in serving one's own function well and not meddling with the mechanism.

In general, then, we live with the costs and do not make the connections as to their origin. Thus, a yawning crevasse opens between the dream of progress and its fulfillment. Men convince themselves, as Ellul points out, that they are about to enter a paradise "in which everything would be at the disposal of everyone, in which men, replaced by the machine, would have only pleasures and play."[35] "In practice, things have not turned out to be so simple. Man is not yet relieved of the brutal fate which pursues him."[36] As a consequence, the citizen of technological society feels a growing frustration and begins the dangerous business of seeking a scapegoat, "the foe who stands in his way and who alone has barred Paradise to him."[37] "He is seized," Ellul continues, "by a sacred delirium when he sees the shining track of a supersonic jet or visualizes the vast granaries stocked for him. He projects this delirium into the myths through which he can control, explain, direct, and justify his actions … and his new slavery. The myth of destruction and the myth of action have their roots in this encounter of man with the promise of technique, and in his wonder and admiration."[38] Rather than question the myths of technics, modern man prefers to conjure a malevolent "other" to account for his difficulties.

Notes

1 See Hiram Haydn's *The Counter-Renaissance* (New York: Harcourt, Brace & World, 1950), pp. 295–6, and Michael Walzer, *The Revolution of the Saints* (London: Weidenfeld and Nicolson, 1966), chap. 1. Both works give an interesting comparison of medieval and modern notions of order and membership.

2 Thomas Carlyle, *Sartor Resartus*, in Carlyle's Complete Works, The Vellum Edition, Vol. I (Boston: Dana Estes and Charles E. Lauriat, 1884), pp. 31–2.

3 E. J. Dijksterhuis, *The Mechanization of the World Picture*, trans. Dikshoorn (Oxford: Oxford University Press, 1961).

4 Julien Offray de La Mettrie, *Man a Machine* (La Salle, Ill.: Open Court Publishing Company, 1961), p. 93.

5 For the major arguments now marshalled in this continuing debate, see Dean E. Woolridge, *Mechanical Man: The Physical Basis of Intelligent Life* (New York: McGraw-Hill, 1968); Floyd W. Matson, *The Broken Image: Man, Science and Society* (New York: Doubleday & Company, Anchor Books, 1966); B. F. Skinner, *Beyond Freedom and Dignity* (New York: Alfred A. Knopf, 1971), p. 204. Woolridge, p. 204, explains in defense of the mechanistic view that "men who know they are machines should be able to bring a higher degree of objectivity to bear on their problems than machines that think they are Men."

6 See Leo Marx, *The Machine in the Garden: Technology and the Pastoral Ideal in America* (New York: Oxford University Press, 1964).

7 Lewis Mumford, *The Myth of the Machine: The Pentagon of Power* (New York: Harcourt Brace Jovanovich, 1970), pp. 163–9.

8 Jean Jacques Rousseau, *First and Second Discourses*, ed. Roger D. Masters, trans. Roger D. and Judith R. Masters (New York: St. Martin's Press, 1964), p. 151.

9 Ibid., p. 152.

10 Ibid., pp. 153–4.

11 Ralph Waldo Emerson, *Works and Days*, in *Of Men and Machines*, ed. Arthur O. Lewis (New York: E. P. Dutton, 1963), p. 68.

12 "Civilization is a reagent and eats away the old traits." Emerson in *English; Traits*, cited by *Oxford English Dictionary*, "reagent."

13 Emerson, *Works and Days*, p. 68.

14 Bruno Bettelheim, *The Informed Heart: Autonomy in a Mass Age* (New York: Free Press, 1960), p. 48.

15 Ibid., p. 49.

16 Ibid.

17 Thorstein Veblen, *The Theory of Business Enterprise* (New York: New American Library, Mentor Books, 1970), p. 144.

18 Ibid., p. 146.

19 Ibid., pp. 146–7.

20 Ibid., p. 177.

21 "Without exception in the course of history, *technique belonged to a civilization* and was merely a single element among a host of nontechnical activities. Today *technique has taken over the whole of civilization*. Certainly, technique is no longer the simple machine substitute for human labor. It has come to be the 'intervention into the very substance not only of the inorganic but also of the organic.'" J. Ellul, *The Technological Society*, (New York: Random House, 1964), p. 128.

22 Ibid., p. 98.

23 Emerson, *Works and Days*, p. 64.

24 H. Marcuse, *One-Dimensional Man: Studies in the Ideology of Advanced Industrial Society* (Boston: Beacon Press, 1968), p. 33.

25 Ibid., pp. 36–7.

26 Ibid., p. 37.

27 Ibid., p. 42.

28 Friedrich Georg Juenger, *The Failure of Technology* (Chicago: Henry Regnery, 1956), p. 8.

29 Hannah Arendt, *The Human Condition* (Chicago: University of Chicago Press, 1958), p. 131.

30 Jules Henry, *Culture Against Man* (New York: Random House, Vintage Books, 1965) pp. 15–24.

31 H. Marcuse, *Eros and Civilization* (Boston: Beacon Press, 1969), pp. 32–4, and *One-Dimensional Man*, chaps. 1–3.

32 Ellul's assessment of Marcuse's position in the debate is offered in his *Autopsy of Revolution*,

trans. Patricia Wolf (New York: Alfred A. Knopf, 1971), pp. 287–90. One of the kinder things he says is the following: "Although Freud's work suggests a revolutionary approach to sexual repression, which begins in the family and is perpetuated by a network of social relationships, he was cautious and never indulged in the acrobatic fantasies of Marcuse, not because he was hopelessly bourgeois, but out of recognition of the unreliability of the unconscious, which made repression necessary, and out of a somewhat skeptical view of revolution. He thought that revolution could not 'change life,' as the reinforced patterns of servility, guilt, and repression would be likely to reappear in seemingly different social surroundings. I accept the logic of that view, whereas 'Freudian-Marxist syntheses' strike me as so much haphazard verbiage – but dangerous, still, as all meaningless verbiage is, for they shunt the revolutionary impulse into dead storage, identifying the sexual explosion with revolution, and giving sterile and brutish expression to the whole legacy of revolution" (ibid., p. 287). To the best of my knowledge, Herbert Marcuse has not yet published a response to the work of Jacques Ellul.

33 Ellul, *Technological Society*, pp. 64–79.
34 Ibid., p. 65.
35 Ibid., p. 191.
36 Ibid.
37 Ibid., p. 192.
38 Ibid.

8

The Autonomy of Technology

Joseph Pitt

Joseph Pitt, one of the founders of the Science, Technology, and Society program at Virginia Tech, has long argued that philosophers should pay more attention to technology. Much of his work has focused on epistemological concerns, and he argues that such concerns are not separable from the positions we take in creating our visions of the good life, and of the future. Further, these visions have been a direct function of what we have thought is possible; that is, such visions are a function of our technological development. Thus, Pitt argues, if the set of technologies we command are central to our way of life and to our future, if they reflect our value system, or even if they merely affect the economic structure of our society, we need to know what this means and how it happens.

In his book, *Thinking About Technology: Foundations of the Philosophy of Technology*, Joseph Pitt attempts to do three things: (1) disentangle science and technology, (2) promote an epistemological rather than ideological basis for the philosophy of technology, and (3) provide a foundation for that epistemological basis. Unlike Ellul, Pitt embraces the empirical approach to understanding technology, and rejects the notion that technology is any one thing. As he writes, "Try as I may, I cannot find the one thing." In this selection, he directly responds to Ellul, and through the careful examination of specific cases challenges the notion that technology has become autonomous.

It might seem that it is but one step from the view that technology is ideologically neutral to the view that technology is autonomous. If, as we noted in Chapter 5, a tool or system can contribute to the decision-making process by forcing changes in values, then surely, it might be suggested, the

From Joseph Pitt, *Thinking about Technology: Foundations of Philosophy of Technology* (New York: Seven Bridges Press, 2000), pp. 87–99.

system itself becomes an independent actor in the process. Maybe so, but probably not. But the view that technology is autonomous is a popular one. Consider what Jacques Ellul has to say on the subject:

> – Technique is autonomous with respect to economics and politics –Technique elicits and conditions social, political and economic change. It is the prime mover of all the rest, in spite of any appearance to the contrary and in spite of human pride, which pretends that man's philosophical theories are still determining influences and man's political regimes are decisive factors in technical evolution. (Ellul 1964, 133)

Ellul may be right about the role philosophical theories and political regimes play in technical evolution, but his claims also sound somewhat exaggerated. More important, the kind of claim he makes for the autonomy of technology makes it sound as if it were unfalsifiable, especially given assertions such as "in spite of any appearance to the contrary."

Unfortunately, claims like Ellul's have become commonplace. They amount to treating technology as a kind of "thing," and in so doing they reify it, attributing causal powers to it and endowing it with a mind and intentions of its own. In addition to the fact that it is empirically false that *Technology* has these characteristics, reifying Technology moves the discussion, and hence any hope of philosophical progress, down blind alleys. The profit in treating Technology in this way, to the extent there is any, is only negative. It lies in removing the responsibility from human shoulders for the way in which we make our way around in the world. Now we can blame all the terrible things that happen to us on Technology! It is only after the first moves have been made toward reifying Technology that we hear about such things as the "threat" of technology taking over our lives. Likewise, reification leads to misleading talk about technology being the handmaiden to science, or some variant on that theme. In other words, reification makes talk about autonomy possible. But, I will argue, it is a major mistake to think there is any *useful* sense in which we could conceive of technology as autonomous.[1]

It is important to stress the "useful" here. It is no doubt possible to contrive outrageous examples to show there is something called autonomous technology. But before we allow misdirected philosophical analysis to take us into the world of science fiction, we can at least take the time to understand what is really going on. Technology, even understood in its more popular-culture sense as new gadgets and electronics, among other things, is such an integral part of our society and culture that unless we ferret out the ways in which these devices are actually embedded in our lives, we may fall victim to a kind of intellectual hysteria that makes successful dealings with the real world impossible. The first step to take if we are to avoid this danger is to clarify the kinds of issues that can reasonably be addressed. To a large degree this means separating the significant from the trivial.

Section 1. Trivial Autonomy

There are at least two cases of talk about the autonomy of technology that are non-starters. That is, if these popular topics of discussion are considered carefully, they easily can be shown to be irrelevant to serious consideration of the issue, since the kind of autonomy they address is trivial.

In the first case, some version of the following account is given of what it means for technology to be autonomous: technology is autonomous when the inventor of a technology, once the technology is made available, loses control over his or her invention. The development of the digital computer can be used as such an example. Once computers entered the public domain, it was impossible for anyone to call them back. The rapid increase in their sophistication and the all-pervasiveness of their employment in society made it impossible to avoid them once they entered the marketplace. Surely, the story goes, this is a case of autonomous technology.

Well, yes and no. Yes, it is autonomous, if by that is meant only that the inventor alone can no longer control the development of the technology. But this is a trivial sense of "autonomy," since it is true of all aspects of our society. Once in the public domain, each item is beyond the control of its inventor in some sense or other. But that does not make the item autonomous. Its further development is a direct function of how people employ it and extend it. To the extent that people are necessarily involved in that process, the invention cannot be autonomous. Rather than being conceived of as an independent agent that acts on its own, the invention is seized opportunistically as a means to an end. It is used, changed, augmented, or discarded, depending on the goals of the agents. That these various uses were not envisioned or intended by its inventor does not make the invention autonomous in any interesting sense.

The second trivial case of autonomous technology concerns the consequences of innovation. Here it might be claimed, for instance, that because the inventor of a device or system failed to see the consequences of employing it in a certain way, the item has a life of its own and is autonomous. Thus it would appear on this scenario that the use of nuclear plants to generate electricity is evidence for the autonomy of nuclear energy, since this use was not foreseen by Einstein in his famous letter to President Roosevelt informing him of the wartime potential for nuclear energy. This, too, is an incorrect conclusion. The fact of the matter is that no one can foresee all the consequences of any act. That fact, however, does not entail that once some action is taken, the consequences of that action are autonomous. That the full consequences of introducing large-scale manufacturing techniques for the production of automobiles were not anticipated by Henry Ford does not mean that those consequences were due to the automobile or to the processes, economic, social, and engineering, that produced it.

The key to understanding this second point lies in realizing that once an invention or innovation leaves the hands of its inventor, it also leaves behind the circumstances in which the actions of only one person can affect its development and employment. Once it enters the public domain, its diffusion generally will be the result of community decisions; and as we noted, these are the kinds of decisions that are the results of compromises. That there is no logical order to the patterns these decisions take should come as no surprise. Compromise is a function of a variety of factors, and it is impossible to tell in advance which of them will be persuasive in any given situation. Furthermore, it may be that *it is this lack of absolute predictability with respect to the outcome of community decisions that itself produces the illusion of the autonomy of technology*. But the fact that the role an innovation acquires in a society is a function of complicated community decisions, which decisions are at best compromises (at worst they are the results of collusion and corruption, which themselves involve compromise), does not entail that the innovation is autonomous. *Quite the contrary*. Given the kind of buffeting and manipulation this process involves, it would appear that it would be anything but autonomous!

Thus arguments from the eventual lack of control of the inventor and the failure to foresee all the consequences fail to secure the case for the "autonomy of technology." But there are also other arguments we need to consider.

Section 2. The Process of Technology

Well-intentioned writers and critics have commented on various aspects of technology which they see as raising the possibility of a serious sense of autonomous technology and, along with it, the specter of apocalypse. One of the best examples of the kind of worry expressed by these authors can be found in John McDermott's essay review of Emmanuel Mesthene's *Technological Change*, "Technology: The Opiate of the Intellectuals" (McDermott 1969). In that review McDermott speaks of a kind of momentum certain devices or systems acquire, thereby providing the appearance of autonomy.

Consider the following McDermottian scenario. A growing retail company located in Fairbanks has just hired a fancy up-to-date accountant with an MBA to manage the financial records of the company, which records are currently in a condition closely resembling chaos. Our accountant is a bright young urban professional. Given the size of the company and its projected growth, she argues persuasively that in the long run it will be cheaper and more efficient to buy a couple of computers than to hire additional staff and to continue handling the books in the traditional way, with ledgers entered by hand, etc. She produces a report showing the projected costs of people versus machines, calculating only for the long run the cost of benefits and retirement for the people and maintenance for the machines. She wins her case and the computers are purchased. But once the computers are introduced, air conditioning is not far behind, because the computers need a cool environment to function optimally. But, our fictional tale continues, air conditioning simply can't be added on to the current structure housing the company offices. Either we redesign the old building to handle air flow and pressure, or we look for a new one. Finally, our storyteller says with a knowing look, the president of the company is totally confused and dismayed and yells: "How did we get into this fix? The old building is perfectly good, we really don't need air conditioning in Alaska; since we introduced those machines, things have gotten out of control!"

This is a typical story – one often told and perhaps even representing a situation often experienced. But just because such stories are told, and some people may interpret their experiences in this fashion, it doesn't follow that they have lost control to some autonomous technology that has taken over their company. What the tale allows us to see is that despite the fact that machines play a prominent role in the unfolding sequence of events, the major overlooked fact is that people often tend to forget the reasons for which they introduced a certain kind of tool or procedure. Instead of taking time to assess critically the impact of making further accommodations to the tools, possibly even concluding that it may be time to reexamine the whole situation, people often simply "go with the flow" and take what appears to be the course of least resistance. Still, from the fact that people sometimes tend to react to the circumstances of a situation in certain ways, perhaps accommodating a new procedure at first, rather than either replacing it with another or eliminating it altogether, it does not follow that the procedure is autonomous.

A basic point we sometimes tend to forget is that *there is no getting rid of tools, written large.* Humanity making its way around in the world is humanity using tools of wide variety and complexity, e.g., hammers, automobiles, governments, electricity. The tools we invent to help us survive and go beyond are essential – perhaps even to the concept of humanity. It isn't as if we can remove tools altogether and continue without them. When we introduce an implement or

a complex system, it is to help us achieve a goal. If we find that the device produces results or side effects in conflict with other goals and/or values, we may replace it or modify it. Whichever we choose, devices, tools, and systems remain with us; they are part of how we go about making our way in the world. What McDermott overlooked (when he spoke of how technologies become so ingrained in our procedures that in accommodating the requirements of the technology we lose our independence of action) was that it is the *perception, or lack of it, that people have of the usefulness of a new product that determines the extent to which they are willing to make concessions in its direction.* They may also lose sight of the goal that first guided their actions and, therefore, may react blindly to the circumstance with which they are now faced. But that is not to say that the product has "taken over." For nothing *in principle* rules our later modifications and, if necessary, replacements. What is required is that the individuals involved keep their objectives in mind and be strong enough to act in their own best interests.

Section 3. Common Sense

Phrased as I have put it, technology conceived of as humanity at work represents the results of the systematic application of common sense; common sense is how people first gain experience and then knowledge by acting on that experience. Nor should this result come as a surprise. Since, if we acknowledge that the concept of a tool lies at the commonsense heart of technology, and if we accept the rather obvious point that not all tools are physical tools, i.e., that there are conceptual tools, social tools, economic tools, etc., then it is not difficult to agree that knowledge is a tool, and if knowledge is constantly being updated, the tool is constantly being honed. In other words, if science produces knowledge, then the knowledge science produces is constantly being upgraded and changed by virtue of the impact of various other tools on the efforts of science to discover more and more about the world. Or to put it differently, quite aside from the resolution of the question of the independence or interdependence of technology and science, if science produces knowledge, and if that knowledge is sometimes used to develop tools that are used in the world, then what those tools produce should generate a form of knowledge that ought to have a bearing in turn on the original knowledge that produced the tools. In addition, it follows that what we do and how we do it is also constantly changing in the face of these developments, and that is as it should be. The bottom line is that, on this account, once a relation between a science and some tool or procedure is established, neither can lay further claims to autonomy – the interdependence is an essential aspect of the process of science itself.[2] But this point of view cannot be established only by *a priori* argument. We need to look at what actually goes on; and I have selected an historical case study to illustrate my points. This is not to say that the analysis of one historical example will settle the issue, but it should help clarify some matters.

Indeed, the case I want to look at, Galileo and the telescope, ought to help exhibit just the issues relevant to sorting out some of the confusions surrounding the interrelations between the development of science and the use of tools and systems of tools. Furthermore, there is a punch line. The general thesis, as already expressed, is that science and technology – where "technology" should now be read as tools, techniques, and systems of tools and techniques – where they interact at all, are mutually nurturing. There is also a caveat, to wit, in point of fact some technologies are science-independent, e.g., the roads of Rome. This is not to say that they are autonomous, since those technologies were responses to needs and goals also; just not the needs and goals of some scientific theory. And some science generates no technology, e.g., Aristotelian biology. The punch line is this: once that is said, something of a paradox emerges. For the history of science is the history of failed theories. But the failure of theory most often does not force a discarding of whatever technology that theory generated or was involved with, nor does the failure of the theory force the abandoning of the technology if a technology was responsible for that theory. To oversimplify: sciences come and go, but their technologies remain. But oversimplification is what got us into trouble at the start, so a more accurate claim would be: scientific theories come and go, but some technologies with which they are in one way or another associated remain. It

is also the case that some technologies associated with specific scientific descriptions disappear when they are replaced or superseded by new techniques.

But there is one sense in which the transient character of scientific theories becomes somewhat problematic. That is, if, as I put it earlier, technology is an integral part of science and partially responsible for changing the science, then the failure of the particular theories could be construed as a failure of the technology involved as well. This may in fact be true. But we should also emphasize our model, MT, in which technology is seen as a process of policy formation, implement-system implementation, assessment, and updating, which process functions at a variety of levels and with varying degrees of significance for technologies further up and down the line; e.g., the initial failure of the Hubble to produce clear pictures of the heavens did not spell disaster for the entire project. Goal-achieving activities are nested within one another and, as we shall see, as a matter of historical and physical accident the nesting will have different degrees of importance depending on the case. Thus placing the blame for a failed scientific theory on its associated technology once again oversimplifies the situation.

Section 4. Galileo and the Telescope

To illustrate some of the notions introduced here, let us turn to an examination of the development of the telescope by Galileo and its effect on some of the theoretical problems he faced in his efforts to show that Copernicus's theory was worthy of serious scientific consideration. As we shall see, the story is not a simple one, and the issue takes on an increasing degree of complexity as the tale proceeds.

To begin with, we need to be perfectly clear that Galileo did not begin his work on the telescope in order to prove anything about Copernicus. The full story of how Galileo came to construct his first telescope is clearly and succinctly put forth by Drake in his *Galileo at Work*. There, quoting from a number of Galileo's letters and published works, Drake makes it clear that Galileo was first drawn to the idea of constructing a telescope out of financial need. To summarize the account: in

July 1609 Galileo was in poor health and, as always, if not nearly broke at least bothered by his lack of money. Having heard of the telescope, Galileo claims to have thought out the principles on which it worked by himself, "my basis being the theory of refraction" (as quoted in Drake 1978, 139). Drake acknowledges that there was no theory of refraction at the time, but excuses Galileo's claim on the grounds that this was not the first time that Galileo arrived at a correct result by reasoning from false premises. (Historians of the logic of discovery, take note.) Once having reconstructed the telescope, Galileo writes: "Now having known how useful this would be for maritime as well as land affairs, and seeing it desired by the Venetian government, I resolved on the 25th of this month [August] to appear in the College and make a free gift of it to his Lordship" (as quoted in Drake 1978, 141; Galileo's letter to his brother-in-law Benedetto Landucci). The result of this gift was the offer of a lifetime appointment with a nice salary increase from 520 to 1000 florins per year. What was unclear at the time, and later became the source of major annoyance on Galileo's part, was that along with the stipend came the provision that there was also to be no further increase for life. So he reinitiated his efforts, eventually successful, to return to Florence.

Now there are some problems here that need not delay us, but they ought to be mentioned in passing. How Galileo managed to reconstruct the telescope from just having heard reports of its existence in Holland remains something of a mystery. Galileo provides us with his own account of the reasoning he followed; but, as Drake notes, his description has been ridiculed by historians because, despite the fact that the telescope he constructed worked, he did not quite think it through correctly. Nevertheless, Drake's observation, that "the historical question of discovery (or in this case, rediscovery) relates to results, not to rigorous logic," seems to the point (Drake 1978, 140). Despite the fact that a telescope using two convex lenses can be made to exceed the power of one using a convex and a concave lens, the truth of the matter is that Galileo's telescope worked. On the other hand, this point about faulty reasoning leading to good results seems to tie into the paradoxical way in which technologies (thought of as artifacts of varying

degrees of complexity and abstractness) emerge and remain with us. But more of this later.

We can now turn to the question of the impact of the telescope on Galileo's work. As he reports it, Galileo first turned his original eight-power telescope toward the moon in the presence of Cosimo, the Grand Duke of Florence. He and Cosimo apparently discussed the mountainous nature of the moon, and shortly after his return to Padua in late 1609, Galileo built a twenty-power telescope, apparently to confirm his original observations of the moon. He did so and then wrote to the Grand Duke's secretary to announce his results. So far then, Galileo has constructed the telescope for profit and is continuing to use it to advance his own position by courting Cosimo.

Galileo, never retiring about his work, continued to use the telescope and to make his new discoveries known through letters to close friends. Consequently, he also began to attract attention. But others such as Clavius now also had access to telescopes. That meant Galileo had to put his results before the public in order to establish his priority of discovery. Therefore, in March 1610 Galileo published *The Starry Messenger*, reporting his lunar observations as well as accounts of the Medicean stars and the hitherto unobserved density of the heavens. At this point controversy enters the picture. These reports of Galileo essentially challenge one of the fundamental assumptions of the Aristotelian theory of the nature of the heavenly sphere: its perfection and immutability. While the rotation of the Medicean stars around Jupiter can be shown to be compatible with both the Copernican and the Tychonian mathematical astronomies, it conflicts with the philosophical and metaphysical view that demands that the planets be carried about a stationary earth embedded in crystalline spheres. And to be clear about the way the battle lines were drawn, remember that Galileo's major opposition came primarily from the philosophers, not from the proto-scientists and other astronomers of his time.

The consequences of Galileo's telescopic observations were more far-reaching than even Copernicus's mathematical model. For the problems Copernicus set were problems in astronomical physics and, as such, had to do with meeting the observational restraints represented by detailed records of celestial activity. Galileo's results, however, and his further arguments concerning the lack of an absolute break between terrestrial and celestial phenomena, maintaining as they did the similarities between the moon and the earth, etc., forced the philosophers to the wall. It was the philosophers' theories that were being challenged when the immutability of the heavens was confronted with the Medicean stars, the phases of Venus, sunspots, and new comets. One might conclude, then, that this represented something akin to a radical Kuhnian paradigm switch.

Much has been written about the extent to which Kuhn's paradigm shifts and their purported likeness to Gestalt switches actually commit someone who experiences one to seeing a new and completely different world. But to see mountains on the moon in a universe in which celestial bodies are supposed to be perfectly smooth comes pretty close to making sense of what this extreme interpretation of Kuhn might mean. Prior to the introduction of the telescope, observations of the heavens, aside from providing inspiration for poets and lovers, were limited to supporting efforts to plot the movements of the planets against the rotation of the heavenly sphere. Furthermore, metaphysical considerations derived from Aristotle interfered with the conceptual possibility of learning much more, given the absence of alternatives. The one universally accepted tool that was employed in astronomical calculation was geometry, and its use was not predicated on any claims of realism for the mathematical models that were developed, another point derived from Aristotelian methodology. The acceptable problem for mathematical astronomy was to plot the relative positions of various celestial phenomena, not to try to explain them. Nor were astronomers expected to astound the world with new revelations about the population of the heavens, since that was assumed to be fixed and perfect. So whatever else astronomers were to do, it was not to discover new facts; there were not supposed to be any.

But the telescope revealed new facts. And for Galileo this meant that some way had to be found to accommodate them. Furthermore, to make the new telescopic findings acceptable, Galileo had to do more than merely let people look

and see for themselves. The strategy he adopted was to link the telescopic data to something already secure in the minds of the community: geometry. This, however, was not as simple as it sounds. He had to build a case for extending geometry as a tool for physics, thereby releasing it from the restrictions under which it labored when used only as a modeling device for descriptive astronomy. In other words, Galileo had to advance the case of Archimedean mechanics. To this end he was forced to do two different things: (1) emphasize rigor in proof – extolling the virtues of geometry and decrying the lack of demonstrations by his opposition; and (2) de-emphasize the appeal to causes in providing explanations of physical phenomena (since abandoning the Aristotelian universe entailed abandoning the metaphysics of causes and teleology – without which the physics was empty).

Section 5. Geometry as a Technology

This is not the place to detail the actual way in which Galileo employed geometry to radicalize the notions of proof, explanation, and evidence.[3] Suffice it to say that he did and that it met with mixed success. The general maneuver was to begin by considering a problem of terrestrial physics, proceed to "draw a little picture," analyze the picture using the principles of Euclidean geometry, and (1) interpret the geometric proof in terrestrial terms, just as a logical positivist would interpret an axiomatic system *via* a "neutral" observation language, and then (2) extend the terrestrial interpretation to celestial phenomena. This is how he proceeded with his account of mountains on the moon, namely by establishing an analogy with terrestrial mountains. This process took place in stages. He first subjected the terrestrial phenomena to geometric analysis and then he extended that analysis to the features of the moon. Not all of Galileo's efforts at explanation using this method succeeded, e.g., his account of the tides. Nevertheless, the central role of geometry cannot be denied.

While Galileo used geometry for most of his career, it was not until he was forced to support publicly his more novel observations and hypotheses that we find in his writings the beginnings of what was eventually to become a very sophisticated methodological process. This procedure is most clearly evident in his last two works, the *Dialogue on the Two Chief World Systems* and his *Discourses on Two New Sciences*. But in the end the *geometric method* as employed by Galileo, or to put it more specifically, Galilean science, dies with Galileo. No one significant carried on his research program using his methods. Whatever impetus he gives to mathematics in science, his mathematics, geometry, very quickly gives way to Newton's calculus and the mathematics of the modern era.

Galileo's use of geometry was as much the employing of a technology conceived of as a tool/technique as was his use of the telescope. Furthermore, it represents the first major step toward the mathematization of what today we would call science. This much is commonplace. The challenging part comes in two sections. (1) The telescope was a new technology, whose introduction for primarily nonscientific reasons, i.e., money, was in fact science-independent, i.e., its invention by the Dutch was theory-independent. (The inventor, Hans Lipperhey, was a lens grinder; the invention was apparently the result of simply fooling around with a couple of lenses, the basic properties of which were known through Lipperhey's daily experience.) In many ways, the use of this new technology by Galileo can be held responsible for the extension of the *geometric method* as a radical method of supporting knowledge claims. (2) Geometry was also theory-independent. But, unlike the telescope, geometry was a very old technology. It was called upon to rescue, as it were, the new technology. It was a very different kind of technology from the telescope, being a method for providing justifications, i.e., proofs, of abstract conclusions regarding spatial relations, not a physical thing. Furthermore, despite the fact that this old technology was required to establish the viability of the new, the old was soon to become obsolete with respect to the justificatory role it was to play in science. That it was to be replaced also had nothing to do with any significant relation between the telescope and the development of the theory Newton outlined in his *Principia*. In other words, the telescope itself had little direct bearing on the development of the calculus, and yet it was the calculus that superseded geometry (but did not

completely eliminate it) as the mathematical basis for scientific proof.

Section 6. Technology and the Dynamics of Change: Autonomy Socialized

If we try to sort it all out, the results are uncomfortable for standard views of technology and the growth of knowledge. The two technologies remain, the two sciences have been replaced. Furthermore, in one of the truly nice bits of irony that history reveals, one of the superseded technologies, geometry, after being replaced by a different kind of mathematical system for justificatory functions, experiences a resurrection in the nineteenth century and ends up playing a crucial role (but not a justificatory role) in the development of yet another physics, having been modified and expanded in the process.

Where is the autonomy here? Both Galileo's physics and the telescope, while capable of being viewed as independent products of one man's creative energy, can also be seen performing an intricate *pas de deux* of motivation and justification when the process of inquiry is examined. It is getting difficult to determine which view ought to take priority. A resolution of the problem might be found if we stop looking at the history and examine the concept of "autonomy" itself.

If we define "autonomous" as "free from influence in both its development and its use," then technology cannot be autonomous since it is inherently something used to accomplish specific goals. But what happens if we try to define "technology" so as to allow technology to have an impact on us as well as on our environment? Are we then committed to the view that, given a technology in use, there emerges from its use a self-propagating process outside the control of humankind? If (1) technology is a product, and (2) we do not add some additional properties to technology beyond its being a thing we manipulate, then (3) there is no reason why we should even begin to think of technology as not within our control.

In other words, we can talk of Galileo being forced to employ geometry and to develop novel methods of justification in order to defend his telescopic discoveries, for what sense does "forced" carry here? The telescope did not with logical necessity precipitate him headlong into battle. Much of what Galileo did to defend his claims and insure his priority of discovery was the product of his flamboyant personality. This was a man who loved fights and being in the public eye. How these features of Galileo's personality can be factored into the tool so as to make it appear that the tool itself is responsible for the action of the man is beyond serious consideration. Given the tool, we can plot its history. What that history amounts to is how it is used. How it is used is a complicated process, for it can entail more than intentional application of a device. "Use" may also mean "rely on," and it may be the case that what we rely on we take for granted, never giving thought to the cost. But this does not thereby entail that, in the absence of human deliberation, the tool by default acquires intentionality and, along with it, control of human affairs.

An alternative would be to endorse the idea that both the telescope and geometry used Galileo. This suggests a science fiction scenario in which as soon as any technology is used by a person, it "takes over" that individual. In the case of populations adopting constitutions that establish governments, all freedom of human action is lost since the government "takes over." Surely this amounts to a *reductio*. For the tool used to adopt government is reason. Is reason, too, going to be something sufficiently alien that we should fear it? The image really does become Mephistophelian enough that we ought to worry about the extent to which we have lost touch with reality.

Further, the existence of a technology does not entail that it will be used. We all know people who refuse to use computers today, not because they cannot, but simply because they feel more comfortable with the old technology of pen and paper. Surely we do not want to say that these individuals are controlled by pencil and paper. The decision to employ a certain means to an end requires thought, information, a determination of the nature and desirability of the end, assessment of the long- and short-term costs and benefits, as well as constant updating of the database. What if, in his declining years, our pen-and-pencil advocate changes his mind and opts for the computer, having decided that time is running out and he has too many things to finish by hand? Do we really want to say that the machine won out over man? Surely not; the man initiated the process that led to the machine, so why not include him in that process?

We are at the point where, in closing, we might ask: Why are we so quick to point to the machines and wag our finger? Well, the long and the short of it is that those who fear reified technology really fear men. It is not the machine that is frightening, but what some men will do with the machine; or, given the machine, what we fail to do by way of assessment and planning. It may be only a slogan, but there is a ring of truth to: "Guns don't kill, people do." There is no problem about the autonomy of technology. Pogo was right: "We have met the enemy and he is us" (Quoted in Kelly 1985, 114). The tools by themselves do nothing. That, I propose, is the only significant sense of autonomy you can find for technology.

Notes

1 For examples of this "style" of philosophizing about technology, see all of Ellul 1964; Winner 1977.
2 See Dewey 1929, for the development of a similar argument.
3 I have worked on the topic; see Pitt 1978, 1982, 1986, 1991; as have McMullin 1968; Shea 1972; Wallace 1992; among others.

Bibliography

Drake, S. *Galileo at Work* (Chicago: University of Chicago Press, 1978).

Kelly, W. *Outrageously Pogo*, ed. W. Kelly and B. Crouch, Jr. (New York: Simon and Schuster, Inc., 1985).

McDermott, J. "Technology: The Opiate of the Intellectuals." *New York Review of Books* (July 31, 1969): 25–35.

McMullin, E. *Galileo, Man of Science* (New York: Basic Books, 1968).

Pitt, J. C. "Galileo: Causation and the Use of Geometry." In *New Perspectives on Galileo*, ed. R. E. Butts and J. C. Pitt (Dordrecht: D. Reidel, 1978), 181–92.

Pitt, J. C. "The Role of Inductive Generalizations in Sellars' Theory of Explanation," *Theory and Decision* 13 (1982): 345–56.

Pitt, J. C. "The Character of Galilean Evidence." *PSA* (1986): 125–34.

Pitt, J. C. *Galileo, Human Knowledge and The Book of Nature: Method Replaces Metaphysics* (Dordrecht: Kluwer, 1991).

Shea, W. *Galileo's Intellectual Revolution* (London: Macmillan, 1972).

Wallace, W. A. *Galileo's Logic of Discovery and Proof: The Background Content and Use of His Appropriated Treatises on Aristotle's Posterior Analytics* (Boston: Kluwer Academic Publishers, 1992).

Part III

Existential and Phenomenological Considerations

9

The Question Concerning Technology

Martin Heidegger

Martin Heidegger (1889–1976) was one of the most influential, innovative, and controversial thinkers of the twentieth century. Throughout his life, he remained concerned with understanding our ontological condition. His work was influential in many areas of thought, including: phenomenology, hermeneutics, political theory, existentialism, theology, psychology, and postmodernism. Heidegger's essay, "The Question Concerning Technology," is perhaps the single most important work in philosophy of technology, and certainly one of the most controversial. This work is a most important touchstone in what Carl Mitcham calls "humanities philosophy of technology," and has been a wellspring for a great deal of pessimism about technology. This is the start of a tradition that includes Ihde and Borgmann.

Heidegger seeks the essence of technology. In this, his work is similar to that of Ellul. In the ordinary sense of essence it is clear that technology is a human activity and a means to an end. In order to identify the essence of technology, Heidegger works to answer three questions: "What kind of activity is technology?" "What are the ends of technology?" and, "What are the means used to achieve technological ends?" Consider that we can say of this activity and its products that they profoundly influence every area of human life so that it is not surprising that people call this age the age of technology. We may think that technology is entirely beneficial to mankind, or that it brings both benefits and harms. We might come to believe that, although it is a human activity, it is one which is beyond the control of individuals, perhaps even of the human race. This may lead us to demonize technology. But this is not what interests Heidegger.

From Martin Heidegger, *The Question Concerning Technology and Other Essays*, trans. William Lovitt (New York: Harper & Row, 1977), pp. 287–317. English language translation © 1977 by Harper & Row, Publishers, Inc. Reprinted by permission of HarperCollins Publishers.

There is a sense in which, for Heidegger, technology is the supreme danger to man. The *essence* of technology, in the Heideggerian sense, presents a supreme danger because it prevents us from having a proper understanding of our own being, of our own essence. The essence of technology, as it shapes the ways in which we can understand ourselves, our essence, is such as to exclude other non-technological ways of understanding being. For example, purely technological ways of thinking exclude those involved in creating and engaging with works of art. Heidegger argues that it is not merely understanding being, not only self-understanding, but understanding being in many ways that makes us human.

His thesis is that technology is a kind of thinking as a revelation of being, or a way of thinking that reveals to us one, and only one, way of existing. The essence of this way of thinking is to seek more and more efficiency simply for its own sake. On his account, technology is a way of conceiving and acting that will consider only instrumental considerations. As a result, we have come to see nature as merely a resource and people as human resources. Technology is thus not neutral. Rather, modern technology presents, or reveals, a world in which everyone and everything is part of a "standing reserve" awaiting use. This process he calls "enframing." Heidegger invites us to question technology in order to free ourselves from such a limiting way of experiencing the world. Once we question technology, we come to understand that we have a technological understanding of ourselves and the world, and in this realization we have begun to move beyond the technological framework. We can then call on values other than efficiency in shaping our lives.

In what follows we shall be *questioning* concerning technology. Questioning builds a way. We would be advised, therefore, above all to pay heed to the way, and not to fix our attention on isolated sentences and topics. The way is one of thinking. All ways of thinking, more or less perceptibly, lead through language in a manner that is extraordinary. We shall be questioning concerning *technology*, and in so doing we should like to prepare a free relationship to it. The relationship will be free if it opens our human existence to the essence of technology. When we can respond to this essence, we shall be able to experience the technological within its own bounds.

Technology is not equivalent to the essence of technology. When we are seeking the essence of "tree," we have to become aware that what pervades every tree, as tree, is not itself a tree that can be encountered among all the other trees.

Likewise, the essence of technology is by no means anything technological. Thus we shall never experience our relationship to the essence of technology so long as we merely conceive and push forward the technological, put up with it, or evade it. Everywhere we remain unfree and chained to technology, whether we passionately affirm or deny it. But we are delivered over to it in the worst possible way when we regard it as something neutral; for this conception of it, to which today we particularly like to do homage, makes us utterly blind to the essence of technology.

According to ancient doctrine, the essence of a thing is considered to be *what* the thing is. We ask the question concerning technology when we ask what it is. Everyone knows the two statements that answer our question. One says: Technology is a means to an end. The other says: Technology is a human activity. The two definitions of technology belong together. For to posit ends and procure and utilize the means to them is a human activity. The manufacture and utilization of equipment, tools, and machines, the manu-factured and used things themselves, and the needs and ends that they serve, all belong to what technology is. The whole complex of these contrivances is technology. Technology itself is a contrivance – in Latin, an *instrumentum*.

The current conception of technology, according to which it is a means and a human activity, can therefore be called the instrumental and anthropological definition of technology.

Who would ever deny that it is correct? It is in obvious conformity with what we are envi-sioning when we talk about technology. The instrumental definition of technology is indeed so uncannily correct that it even holds for modern technology, of which, in other respects, we maintain with some justification that it is, in contrast to the older handwork technology, something completely different and therefore new. Even the power plant with its turbines and generators is a man-made means to an end established by man. Even the jet aircraft and the high-frequency apparatus are means to ends. A radar station is of course less simple than a weather vane. To be sure, the construction of a high-frequency apparatus requires the interlock-ing of various processes of technical-industrial production. And certainly a sawmill in a secluded valley of the Black Forest is a primitive means com-pared with the hydroelectric plant on the Rhine River.

But this much remains correct: modern technology too is a means to an end. This is why the instrumental conception of technology con-ditions every attempt to bring man into the right relation to technology. Everything depends on our manipulating technology in the proper manner as a means. We will, as we say, "get" technology "spiritually in hand." We will master it. The will to mastery becomes all the more urgent the more technology threatens to slip from human control.

But suppose now that technology were no mere means, how would it stand with the will to master it? Yet we said, did we not, that the instrumental definition of technology is correct? To be sure. The correct always fixes upon some-thing pertinent in whatever is under consideration. However, in order to be correct, this fixing by no

means needs to uncover the thing in question in its essence. Only at the point where such an uncovering happens does the true come to pass. For that reason the merely correct is not yet the true. Only the true brings us into a free rela-tionship with that which concerns us from its essence. Accordingly, the correct instrumental definition of technology still does not show us technology's essence. In order that we may arrive at this, or at least come close to it, we must seek the true by way of the correct. We must ask: What is the instrumental itself? Within what do such things as means and end belong? A means is that whereby something is effected and thus attained. Whatever has an effect as its consequence is called a cause. But not only that by means of which something else is effected is a cause. The end in keeping with which the kind of means to be used is determined is also considered a cause. Wherever ends are pursued and means are employed, wherever instrumentality reigns, there reigns causality.

For centuries philosophy has taught that there are four causes: (1) the *causa materialis*, the material, the matter out of which, for example, a silver chalice is made; (2) the *causa formalis*, the form, the shape into which the material enters; (3) the *causa finalis*, the end, for example, the sacrificial rite in relation to which the chalice required is determined as to its form and matter; (4) the *causa efficiens*, which brings about the effect that is the finished, actual chalice, in this instance, the silversmith. What technology is, when represented as a means, discloses itself when we trace instrumentality back to fourfold causality.

But suppose that causality, for its part, is veiled in darkness with respect to what it is? Certainly for centuries we have acted as though the doctrine of the four causes had fallen from heaven as a truth as clear as daylight. But it might be that the time has come to ask, why are there just four causes? In relation to the afore-mentioned four, what does "cause" really mean? From whence does it come that the causal char-acter of the four causes is so unifiedly determined that they belong together?

So long as we do not allow ourselves to go into these questions, causality, and with it instru-mentality, and with this the accepted definition of technology, remain obscure and groundless.

For a long time we have been accustomed to representing cause as that which brings something about. In this connection, to bring about means to obtain results, effects. The *causa efficiens*, but one among the four causes, sets the standard for all causality. This goes so far that we no longer even count the *causa finalis*, telic finality, as causality. *Causa, casus*, belongs to the verb *cadere*, to fall, and means that which brings it about that something turns out as a result in such and such a way. The doctrine of the four causes goes back to Aristotle. But everything that later ages seek in Greek thought under the conception and rubric "causality," in the realm of Greek thought and for Greek thought per se has simply nothing at all to do with bringing about and effecting. What we call cause [*Ursache*] and the Romans call *causa* is called *aition* by the Greeks, that to which something else is indebted [*das, was ein anderes verschuldet*]. The four causes are the ways, all belonging at once to each other, of being responsible for something else. An example can clarify this.

Silver is that out of which the silver chalice is made. As this matter (*hyle*), it is co-responsible for the chalice. The chalice is indebted to, i.e., owes thanks to, the silver for that of which it consists. But the sacrificial vessel is indebted not only to the silver. As a chalice, that which is indebted to the silver appears in the aspect of a chalice, and not in that of a brooch or a ring. Thus the sacred vessel is at the same time indebted to the aspect (*eidos*) of chaliceness. Both the silver into which the aspect is admitted as chalice and the aspect in which the silver appears are in their respective ways co-responsible for the sacrificial vessel.

But there remains yet a third that is above all responsible for the sacrificial vessel. It is that which in advance confines the chalice within the realm of consecration and bestowal. Through this the chalice is circumscribed as sacrificial vessel. Circumscribing gives bounds to the thing. With the bounds the thing does not stop; rather, from within them it begins to be what after production it will be. That which gives bounds, that which completes, in this sense is called in Greek *telos*, which is all too often translated as "aim" and "purpose," and so misinterpreted. The *telos* is responsible for what as matter and what as aspect are together co-responsible for the sacrificial vessel.

Finally there is a fourth participant in the responsibility for the finished sacrificial vessel's lying before us ready for use, i.e., the silversmith – but not at all because he, in working, brings about the finished sacrificial chalice as if it were the effect of a making; the silversmith is not a *causa efficiens*.

The Aristotelian doctrine neither knows the cause that is named by this term, nor uses a Greek word that would correspond to it.

The silversmith considers carefully and gathers together the three aforementioned ways of being responsible and indebted. To consider carefully [*überlegen*] is in Greek *legein, logos*. *Legein* is rooted in *apophainesthai*, to bring forward into appearance. The silversmith is co-responsible as that from whence the sacred vessel's bringing-forth and subsistence take and retain their first departure. The three previously mentioned ways of being responsible owe thanks to the pondering of the silversmith for the "that" and the "how" of their coming into appearance and into play for the production of the sacrificial vessel.

Thus four ways of owing hold sway in the sacrificial vessel that lies ready before us. They differ from one another, yet they belong together. What unites them from the beginning? In what does this playing in unison of the four ways of being responsible play? What is the source of the unity of the four causes? What, after all, does this owing and being responsible mean, thought as the Greeks thought it?

Today we are too easily inclined either to understand being responsible and being indebted moralistically as a lapse, or else to construe them in terms of effecting. In either case we bar to ourselves the way to the primal meaning of that which is later called causality. So long as this way is not opened up to us we shall also fail to see what instrumentality, which is based on causality, actually is.

In order to guard against such misinterpretations of being responsible and being indebted, let us clarify the four ways of being responsible in terms of that for which they are responsible. According to our example, they are responsible for the silver chalice's lying ready before us as a sacrificial vessel. Lying before and lying ready (*hypokeisthai*) characterize the presencing of something that is present. The four ways of being responsible bring something into appearance.

They let it come forth into presencing [*Anwesen*]. They set it free to that place and so start it on its way, namely, into its complete arrival. The principal characteristic of being responsible is this starting something on its way into arrival. It is in the sense of such a starting something on its way into arrival that being responsible is an occasioning or an inducing to go forward [*Ver-an-lassen*]. On the basis of a look at what the Greeks experienced in being responsible, in *aitia*, we now give this verb "to occasion" a more inclusive meaning, so that it now is the name for the essence of causality thought as the Greeks thought it. The common and narrower meaning of "occasion." in contrast, is nothing more than striking against and releasing, and means a kind of secondary cause within the whole of causality.

But in what, then, does the playing in unison of the four ways of occasioning play? These let what is not yet present arrive into presencing. Accordingly, they are unifiedly governed by a bringing that brings what presences into appearance. Plato tells us what this bringing is in a sentence from the *Symposium* (205b): *hē gar toi ek tou mē ontos eis to on ionti hotōioun aitia pasa esti poiēsis*. "Every occasion for whatever passes beyond the nonpresent and goes forward into presencing is *poiēsis*, bringing-forth [*Hervor-bringen*]."

It is of utmost importance that we think bringing-forth in its full scope and at the same time in the sense in which the Greeks thought it. Not only handicraft manufacture, not only artistic and poetical bringing into appearance and concrete imagery, is a bringing-forth, *poiēsis*. *Physis* also, the arising of something from out of itself, is a bringing-forth, *poiēsis*. *Physis* is indeed *poiēsis* in the highest sense. For what presences by means of *physis* has the bursting open belonging to bringing-forth, e.g., the bursting of a blossom into bloom, in itself (*en heautōi*). In contrast, what is brought forth by the artisan or the artist, e.g., the silver chalice, has the bursting open belonging to bringing-forth, not in itself, but in another (*en allōi*), in the craftsman or artist.

The modes of occasioning, the four causes, are at play, then, within bringing-forth. Through bringing-forth the growing things of nature as well as whatever is completed through the crafts and the arts come at any given time to their appearance.

But how does bringing-forth happen, be it in nature or in handwork and art? What is the bringing-forth in which the fourfold way of occasioning plays? Occasioning has to do with the presencing [*Anwesen*] of that which at any given time comes to appearance in bringing-forth. Bringing-forth brings out of concealment into unconcealment. Bringing-forth comes to pass only insofar as something concealed comes into unconcealment. This coming rests and moves freely within what we call revealing [*das Entbergen*]. The Greeks have the word *alētheia* for revealing. The Romans translate this with *veritas*. We say "truth" and usually understand it as correctness of representation.

But where have we strayed to? We are questioning concerning technology, and we have arrived now at *alētheia*, at revealing. What has the essence of technology to do with revealing? The answer: everything. For every bringing-forth is grounded in revealing. Bringing-forth, indeed, gathers within itself the four modes of occasioning – causality – and rules them throughout. Within its domain belong end and means as well as instrumentality. Instrumentality is considered to be the fundamental characteristic of technology. If we inquire step by step into what technology, represented as means, actually is, then we shall arrive at revealing. The possibility of all productive manufacturing lies in revealing.

Technology is therefore no mere means. Technology is a way of revealing. If we give heed to this, then another whole realm for the essence of technology will open itself up to us. It is the realm of revealing, i.e., of truth.

This prospect strikes us as strange. Indeed, it should do so, as persistently as possible and with so much urgency that we will finally take seriously the simple question of what the name "technology" means. The word stems from the Greek. *Technikon* means that which belongs to *technē*. We must observe two things with respect to the meaning of this word. One is that *technē* is the name not only for the activities and skills of the craftsman, but also for the arts of the mind and the fine arts. *Technē* belongs to bringing-forth, to *poiēsis*; it is something poetic.

The other thing that we should observe with regard to *technē* is even more important. From earliest times until Plato the word *technē* is linked with the word *epistēmē*. Both words are terms for knowing in the widest sense. They mean to be entirely at home in something, to understand and be expert in it. Such knowing provides an opening up. As an opening up it is a revealing. Aristotle, in a discussion of special importance (*Nicomachean Ethics*, Bk. VI, chaps. 3 and 4), distinguishes between *epistēmē* and *technē* and indeed with respect to what and how they reveal. *Technē* is a mode of *alētheuein*. It reveals whatever does not bring itself forth and does not yet lie here before us, whatever can look and turn out now one way and now another. Whoever builds a house or a ship or forges a sacrificial chalice reveals what is to be brought forth, according to the terms of the four modes of occasioning. This revealing gathers together in advance the aspect and the matter of ship or house, with a view to the finished thing envisioned as completed, and from this gathering determines the manner of its construction. Thus what is decisive in *technē* does not lie at all in making and manipulating nor in the using of means, but rather in the revealing mentioned before. It is as revealing, and not as manufacturing, that *technē* is a bringing-forth.

Thus the clue to what the word *technē* means and to how the Greeks defined it leads us into the same context that opened itself to us when we pursued the question of what instrumentality as such in truth might be.

Technology is a mode of revealing. Technology comes to presence in the realm where revealing and unconcealment take place, where *alētheia*, truth, happens.

In opposition to this definition of the essential domain of technology, one can object that it indeed holds for Greek thought and that at best it might apply to the techniques of the hand-craftsman, but that it simply does not fit modern machine-powered technology. And it is precisely the latter and it alone that is the disturbing thing, that moves us to ask the question concerning technology per se. It is said that modern technology is something incomparably different from all earlier technologies because it is based on modern physics as an exact science. Meanwhile we have come to understand more clearly that the reverse holds true as well: modern physics,

as experimental, is dependent upon technical apparatus and upon progress in the building of apparatus. The establishing of this mutual relationship between technology and physics is correct. But it remains a merely historiographical establishing of facts and says nothing about that in which this mutual relationship is grounded. The decisive question still remains: Of what essence is modern technology that it thinks of putting exact science to use?

What is modern technology? It too is a revealing. Only when we allow our attention to rest on this fundamental characteristic does that which is new in modern technology show itself to us.

And yet, the revealing that holds sway throughout modern technology does not unfold into a bringing-forth in the sense of *poiēsis*. The revealing that rules in modern technology is a challenging [*Herausfordern*], which puts to nature the unreasonable demand that it supply energy which can be extracted and stored as such. But does this not hold true for the old windmill as well? No. Its sails do indeed turn in the wind; they are left entirely to the wind's blowing. But the windmill does not unlock energy from the air currents in order to store it.

In contrast, a tract of land is challenged in the hauling out of coal and ore. The earth now reveals itself as a coal mining district, the soil as a mineral deposit. The field that the peasant formerly cultivated and set in order appears different from how it did when to set in order still meant to take care of and maintain. The work of the peasant does not challenge the soil of the field. In sowing grain it places seed in the keeping of the forces of growth and watches over its increase. But meanwhile even the cultivation of the field has come under the grip of another kind of setting-in-order, which *sets upon* nature. It sets upon it in the sense of challenging it. Agriculture is now the mechanized food industry. Air is now set upon to yield nitrogen, the earth to yield ore, ore to yield uranium, for example; uranium is set upon to yield atomic energy, which can be released either for destruction or for peaceful use.

This setting-upon that challenges the energies of nature is an expediting, and in two ways. It expedites in that it unlocks and exposes. Yet that expediting is always itself directed from the beginning toward furthering something else, i.e., toward driving on to the maximum yield at the minimum expense. The coal that has been hauled out in some mining district has not been produced in order that it may simply be at hand somewhere or other. It is being stored; that is, it is on call, ready to deliver the sun's warmth that is stored in it. The sun's warmth is challenged forth for heat, which in turn is ordered to deliver steam whose pressure turns the wheels that keep a factory running.

The hydroelectric plant is set into the current of the Rhine. It sets the Rhine to supplying its hydraulic pressure, which then sets the turbines turning. This turning sets those machines in motion whose thrust sets going the electric current for which the long-distance power station and its network of cables are set up to dispatch electricity. In the context of the interlocking processes pertaining to the orderly disposition of electrical energy, even the Rhine itself appears to be something at our command. The hydroelectric plant is not built into the Rhine River as was the old wooden bridge that joined bank with bank for hundreds of years. Rather, the river is dammed up into the power plant. What the river is now, namely, a water-power supplier, derives from the essence of the power station. In order that we may even remotely consider the monstrousness that reigns here, let us ponder for a moment the contrast that is spoken by the two titles: "The Rhine," as dammed up into the *power* works, and "The Rhine," as uttered by the *art* work, in Hölderlin's hymn by that name. But, it will be replied, the Rhine is still a river in the landscape, is it not? Perhaps. But how? In no other way than as an object on call for inspection by a tour group ordered there by the vacation industry.

The revealing that rules throughout modern technology has the character of a setting-upon, in the sense of a challenging-forth. Such challenging happens in that the energy concealed in nature is unlocked, what is unlocked is transformed, what is transformed is stored up, what is stored up is, in turn, distributed, and what is distributed is switched about ever anew. Unlocking, transforming, storing, distributing, and switching about are ways of revealing. But the revealing never simply comes to an end. Neither does it run off into the indeterminate. The revealing reveals to itself its own manifoldly interlocking paths, through regulating their course. This regulating itself is, for its part,

everywhere secured. Regulating and securing even become the chief characteristics of the revealing that challenges.

What kind of unconcealment is it, then, that is peculiar to that which results from this setting-upon that challenges? Everywhere everything is ordered to stand by, to be immediately on hand, indeed to stand there just so that it may be on call for a further ordering. Whatever is ordered about in this way has its own standing. We call it the standing-reserve [*Bestand*]. The word expresses here something more, and something more essential, than mere "stock." The word "standing-reserve" assumes the rank of an inclusive rubric. It designates nothing less than the way in which everything presences that is wrought upon by the revealing that challenges. Whatever stands by in the sense of standing-reserve no longer stands over against us as object.

Yet an airliner that stands on the runway is surely an object. Certainly. We can represent the machine so. But then it conceals itself as to what and how it is. Revealed, it stands on the taxi strip only as standing-reserve, inasmuch as it is ordered to insure the possibility of transportation. For this it must be in its whole structure and in every one of its constituent parts itself on call for duty, i.e., ready for takeoff. (Here it would be appropriate to discuss Hegel's definition of the machine as an autonomous tool. When applied to the tools of the craftsman, his characterization is correct. Characterized in this way, however, the machine is not thought at all from the essence of technology within which it belongs. Seen in terms of the standing-reserve, the machine is completely unautonomous, for it has its standing only from the ordering of the orderable.)

The fact that now, wherever we try to point to modern technology as the revealing that challenges, the words "setting-upon," "ordering," "standing-reserve," obtrude and accumulate in a dry, monotonous, and therefore oppressive way, has its basis in what is now coming to utterance.

Who accomplishes the challenging setting-upon through which what we call the real is revealed as standing-reserve? Obviously, man. To what extent is man capable of such a revealing? Man can, indeed, conceive, fashion, and carry through this or that in one way or another. But man does not have control over unconcealment itself, in which at any given time the real shows itself or withdraws. The fact that the real has been showing itself in the light of Ideas ever since the time of Plato, Plato did not bring about. The thinker only responded to what addressed itself to him.

Only to the extent that man for his part is already challenged to exploit the energies of nature can this revealing which orders happen. If man is challenged, ordered, to do this, then does not man himself belong even more originally than nature within the standing-reserve? The current talk about human resources, about the supply of patients for a clinic, gives evidence of this. The forester who measures the felled timber in the woods and who to all appearances walks the forest path in the same way his grandfather did is today ordered by the industry that produces commercial woods, whether he knows it or not. He is made subordinate to the orderability of cellulose, which for its part is challenged forth by the need for paper, which is then delivered to newspapers and illustrated magazines. The latter, in their turn, set public opinion to swallowing what is printed, so that a set configuration of opinion becomes available on demand. Yet precisely because man is challenged more originally than are the energies of nature, i.e., into the process of ordering, he never is transformed into mere standing-reserve. Since man drives technology forward, he takes part in ordering as a way of revealing. But the unconcealment itself, within which ordering unfolds, is never a human handiwork, any more than is the realm man traverses every time he as a subject relates to an object.

Where and how does this revealing happen if it is no mere handiwork of man? We need not look far. We need only apprehend in an unbiased way that which has already claimed man so decisively that he can only be man at any given time as the one so claimed. Wherever man opens his eyes and ears, unlocks his heart, and gives himself over to meditating and striving, shaping and working, entreating and thanking, he finds himself everywhere already brought into the unconcealed. The unconcealment of the unconcealed has already come to pass whenever it calls man forth into the modes of revealing allotted to him. When man, in his way, from within unconcealment reveals that which presences, he merely responds to the call of unconcealment even when he contradicts it. Thus when man, investigating,

observing, pursues nature as an area of his own conceiving, he has already been claimed by a way of revealing that challenges him to approach nature as an object of research, until even the object disappears into the objectlessness of standing-reserve.

Modern technology, as a revealing which orders, is thus no mere human doing. Therefore we must take that challenging, which sets upon man to order the real as standing-reserve, in accordance with the way it shows itself. That challenging gathers man into ordering. This gathering concentrates man upon ordering the real as standing-reserve.

That which primordially unfolds the mountains into mountain ranges and courses through them in their folded togetherness is the gathering that we call "Gebirg" [mountain chain].

That original gathering from which unfold the ways in which we have feelings of one kind or another we name "Gemüt" [disposition].

We now name that challenging claim which gathers man thither to order the self-revealing as standing-reserve: "Ge-stell" [enframing].

We dare to use this word in a sense that has been thoroughly unfamiliar up to now.

According to ordinary usage, the word Gestell [frame] means some kind of apparatus, e.g., a bookrack. Gestell is also the name for a skeleton. And the employment of the word Gestell [enframing] that is now required of us seems equally eerie, not to speak of the arbitrariness with which words of a mature language are so misused. Can anything be more strange? Surely not. Yet this strangeness is an old custom of thought. And indeed thinkers follow this custom precisely at the point where it is a matter of thinking that which is highest. We, late born, are no longer in a position to appreciate the significance of Plato's daring to use the word eidos for that which in everything and in each particular thing endures as present. For eidos, in the common speech, meant the outward aspect [Ansicht] that a visible thing offers to the physical eye. Plato exacts of this word, however, something utterly extraordinary: that it name what precisely is not and never will be perceivable with physical eyes. But even this is by no means the full extent of what is extraordinary here. For idea names not only the nonsensuous aspect of what is physically visible. Aspect (idea) names and also is that which constitutes the essence in the audible, the tasteable, the tactile, in everything that is in any way accessible. Compared with the demands that Plato makes on language and thought in this and in other instances, the use of the word *Gestell* as the name for the essence of modern technology, which we are venturing, is almost harmless. Even so, the usage now required remains something exacting and is open to misinterpretation.

Enframing means the gathering together of that setting-upon that sets upon man, i.e., challenges him forth, to reveal the real, in the mode of ordering, as standing-reserve. Enframing means that way of revealing that holds sway in the essence of modern technology and that is itself nothing technological. On the other hand, all those things that are so familiar to us and are standard parts of assembly, such as rods, pistons, and chassis, belong to the technological. The assembly itself, however, together with the aforementioned stockparts, falls within the sphere of technological activity. Such activity always merely responds to the challenge of enframing, but it never comprises enframing itself or brings it about.

The word *stellen* [to set upon] in the name *Ge-stell* [enframing] not only means challenging. At the same time it should preserve the suggestion of another *Stellen* from which it stems, namely that producing and presenting [*Her- und Dar-stellen*], which, in the sense of *poiēsis*, lets what presences come forth into unconcealment. This producing that brings forth, e.g., erecting a statue in the temple precinct, and the ordering that challenges now under consideration are indeed fundamentally different, and yet they remain related in their essence. Both are ways of revealing, of *alētheia*. In enframing that unconcealment comes to pass in conformity with which the work of modern technology reveals the real as standing-reserve. This work is therefore neither only a human activity nor a mere means within such activity. The merely instrumental, merely anthropological definition of technology is therefore in principle untenable. And it may not be rounded out by being referred back to some metaphysical or religious explanation that undergirds it.

It remains true, nonetheless, that man in the technological age is, in a particularly striking way, challenged forth into revealing. That revealing concerns nature, above all, as the chief storehouse of the standing energy reserve. Accordingly, man's ordering attitude and behavior display themselves first in the rise of modern physics as an exact science. Modern science's way of representing pursues and entraps nature as a calculable coherence of forces. Modern physics is not experimental physics because it applies apparatus to the questioning of nature. The reverse is true. Because physics, indeed already as pure theory, sets nature up to exhibit itself as a coherence of forces calculable in advance, it orders its experiments precisely for the purpose of asking whether and how nature reports itself when set up in this way.

But after all, mathematical science arose almost two centuries before technology. How, then, could it have already been set upon by modern technology and placed in its service? The facts testify to the contrary. Surely technology got underway only when it could be supported by exact physical science. Reckoned chronologically, this is correct. Thought historically, it does not hit upon the truth.

The modern physical theory of nature prepares the way not simply for technology but for the essence of modern technology. For such gathering-together, which challenges man to reveal by way of ordering, already holds sway in physics. But in it that gathering does not yet come expressly to the fore. Modern physics is the herald of enframing, a herald whose origin is still unknown. The essence of modern technology has for a long time been concealed, even where power machinery has been invented, where electrical technology is in full swing, and where atomic technology is well underway.

All coming to presence, not only modern technology, keeps itself everywhere concealed to the last. Nevertheless, it remains, with respect to its holding sway, that which precedes all: the earliest. The Greek thinkers already knew of this when they said: That which is earlier with regard to its rise into dominance becomes manifest to us men only later. That which is primally early shows itself only ultimately to men. Therefore, in the realm of thinking, a painstaking effort to think through still more primally what was primally thought is not the absurd wish to revive what is past, but rather the sober readiness to be astounded before the coming of the dawn.

Chronologically speaking, modern physical science begins in the seventeenth century. In contrast, machine-power technology develops only in the second half of the eighteenth century. But modern technology, which for chronological reckoning is the later, is, from the point of view of the essence holding sway within it, historically earlier.

If modern physics must resign itself ever increasingly to the fact that its realm of representation remains inscrutable and incapable of being visualized, this resignation is not dictated by any committee of researchers. It is challenged forth by the rule of enframing, which demands that nature be orderable as standing-reserve. Hence physics, in its retreat from the kind of representation that turns only to objects, which has been the sole standard until recently, will never be able to renounce this one thing: that nature reports itself in some way or other that is identifiable through calculation and that it remains orderable as a system of information. This system is then determined by a causality that has changed once again. Causality now displays neither the character of the occasioning that brings forth nor the nature of the *causa efficiens*, let alone that of the *causa formalis*. It seems as though causality is shrinking into a reporting – a reporting challenged forth – of standing-reserves that must be guaranteed either simultaneously or in sequence. To this shrinking would correspond the process of growing resignation that Heisenberg's lecture depicts in so impressive a manner.[1]

Because the essence of modern technology lies in enframing, modern technology must employ exact physical science. Through its so doing the deceptive illusion arises that modern technology is applied physical science. This illusion can maintain itself only so long as neither the essential origin of modern science nor indeed the essence of modern technology is adequately found out through questioning.

We are questioning concerning technology in order to bring to light our relationship to its essence. The essence of modern technology shows itself in what we call enframing. But simply to point to this is still in no way to answer the question concerning technology, if to answer means to respond, in the sense of correspond, to the essence of what is being asked about.

Where do we find ourselves if now we think one step further regarding what enframing itself actually is? It is nothing technological, nothing on the order of a machine. It is the way in which the real reveals itself as standing-reserve. Again we ask: Does such revealing happen somewhere beyond all human doing? No. But neither does it happen exclusively *in* man, or definitively *through* man.

Enframing is the gathering together which belongs to that setting-upon which challenges man and puts him in position to reveal the real, in the mode of ordering, as standing-reserve. As the one who is challenged forth in this way, man stands within the essential realm of enframing. He can never take up a relationship to it only subsequently. Thus the question as to how we are to arrive at a relationship to the essence of technology, asked in this way, always comes too late. But never too late comes the question as to whether we actually experience ourselves as the ones whose activities everywhere, public and private, are challenged forth by enframing. Above all, never too late comes the question as to whether and how we actually admit ourselves into that wherein enframing itself comes to presence.

The essence of modern technology starts man upon the way of that revealing through which the real everywhere, more or less distinctly, becomes standing-reserve. "To start upon a way" means "to send" in our ordinary language. We shall call the sending that gathers [*versammelnde Schicken*], that first starts man upon a way of revealing, *destining* [*Geschick*]. It is from this destining that the essence of all history [*Geschichte*] is determined. History is neither simply the object of written chronicle nor merely the process of human activity. That activity first becomes history as something destined.[2] And it is only the destining into objectifying representation that makes the historical accessible as an object for historiography, i.e., for a science, and on this basis makes possible the current equating of the historical with that which is chronicled.

Enframing, as a challenging-forth into ordering, sends into a way of revealing. Enframing is an ordaining of destining, as is every way of revealing. Bringing-forth, *poiēsis*, is also a destining in this sense.

Always the unconcealment of that which is goes upon a way of revealing. Always the destining of revealing holds complete sway over men. But that destining is never a fate that compels. For man becomes truly free only insofar as he belongs to the realm of destining and so becomes one who listens, though not one who simply obeys.

The essence of freedom is *originally* not connected with the will or even with the causality of human willing.

Freedom governs the open in the sense of the cleared and lighted up, i.e., the revealed. To the occurrence of revealing, i.e., of truth, freedom stands in the closest and most intimate kinship. All revealing belongs within a harboring and a concealing. But that which frees – the mystery – is concealed and always concealing itself. All revealing comes out of the open, goes into the open, and brings into the open. The freedom of the open consists neither in unfettered arbitrariness nor in the constraint of mere laws. Freedom is that which conceals in a way that opens to light, in whose lighting shimmers that veil that hides the essential occurrence of all truth and lets the veil appear as what veils. Freedom is the realm of the destining that at any given time starts a revealing on its way.

The essence of modern technology lies in enframing. Enframing belongs within the destining of revealing. These sentences express something different from the talk that we hear more frequently, to the effect that technology is the fate of our age, where "fate" means the inevitableness of an unalterable course.

But when we consider the essence of technology we experience enframing as a destining of revealing. In this way we are already sojourning within the open space of destining, a destining that in no way confines us to a stultified compulsion to push on blindly with technology or, what comes to the same, to rebel helplessly against it and curse it as the work of the devil. Quite to the contrary, when we once open ourselves expressly to the *essence* of technology we find ourselves unexpectedly taken into a freeing claim.

The essence of technology lies in enframing. Its holding sway belongs within destining. Since destining at any given time starts man on a way of revealing, man, thus underway, is continually approaching the brink of the possibility of pursuing and pushing forward nothing but what is revealed in ordering, and of deriving all his standards on this basis. Through this the other possibility is blocked, that man might be admitted more and sooner and ever more primally to the essence of what is unconcealed and to its unconcealment, in order that he might experience as his essence the requisite belonging to revealing.

Placed between these possibilities, man is endangered by destining. The destining of revealing is as such, in every one of its modes, and therefore necessarily, *danger*.

In whatever way the destining of revealing may hold sway, the unconcealment in which everything that is shows itself at any given time harbors the danger that man may misconstrue the unconcealed and misinterpret it. Thus where everything that presences exhibits itself in the light of a cause-effect coherence, even God, for representational thinking, can lose all that is exalted and holy, the mysteriousness of his distance. In the light of causality, God can sink to the level of a cause, of *causa efficiens*. He then becomes even in theology the God of the philosophers, namely of those who define the unconcealed and the concealed in terms of the causality of making, without ever considering the essential origin of this causality.

In a similar way the unconcealment in accordance with which nature presents itself as a calculable complex of the effects of forces can indeed permit correct determinations; but precisely through these successes the danger may remain that in the midst of all that is correct the true will withdraw.

The destining of revealing is in itself not just any danger, but *the* danger.

Yet when destining reigns in the mode of enframing, it is the supreme danger. This danger attests itself to us in two ways. As soon as what is unconcealed no longer concerns man even as object, but exclusively as standing-reserve, and man in the midst of objectlessness is nothing but the orderer of the standing-reserve, then he comes to the very brink of a precipitous fall, that is, he comes to the point where he himself will have to be taken as standing-reserve. Meanwhile, man, precisely as the one so threatened, exalts himself to the posture of lord of the earth. In this way the illusion comes to prevail that everything man encounters exists only insofar as it is his construct. This illusion gives rise in turn to one final delusion: it seems as though man everywhere and always encounters only himself. Heisenberg has with complete correctness pointed out that the real must present itself to contemporary man in this way.[3] *In truth, however, precisely nowhere does man today any longer encounter himself, i.e., his essence.* Man stands so decisively in attendance on the challenging-forth of enframing that he does not grasp enframing as a claim, that he fails to see himself as the one spoken to, and hence also fails in every way to hear in what respect he ek-sists, from out of his essence, in the realm of an exhortation or address, so that he *can never* encounter only himself.

But enframing does not simply endanger man in his relationship to himself and to everything that is. As a destining, it banishes man into that kind of revealing that is an ordering. Where this ordering holds sway, it drives out every other possibility of revealing. Above all, enframing conceals that revealing which, in the sense of *poiēsis*, lets what presences come forth into appearance. As compared with that other revealing, the setting-upon that challenges forth thrusts man into a relation to whatever is that is at once antithetical and rigorously ordered. Where enframing holds sway, regulating and securing of the standing-reserve mark all revealing. They no longer even let their own fundamental characteristic appear, namely, this revealing as such.

Thus the challenging-enframing not only conceals a former way of revealing, bringing-forth, but it conceals revealing itself and with it that wherein unconcealment, i.e., truth, comes to pass.

Enframing blocks the shining-forth and holding sway of truth. The destining that sends into ordering is consequently the extreme danger. What is dangerous is not technology. Technology is not demonic; but its essence is mysterious. The essence of technology, as a destining of revealing, is the danger. The transformed meaning of the word "enframing" will perhaps become somewhat more familiar to us now if we think enframing in the sense of destining and danger.

The threat to man does not come in the first instance from the potentially lethal machines and apparatus of technology. The actual threat has already afflicted man in his essence. The rule of enframing threatens man with the possibility that it could be denied to him to enter into a more original revealing and hence to experience the call of a more primal truth.

Thus where enframing reigns, there is *danger* in the highest sense.

> But where danger is, grows
> The saving power also.

Let us think carefully about these words of Hölderlin. What does it mean to "save"? Usually we think that it means only to seize hold of a thing threatened by ruin in order to secure it in its former continuance. But the verb "to save" says more. "To save" is to fetch something home into its essence, in order to bring the essence for the first time into its genuine appearing. If the essence of technology, enframing, is the extreme danger, if there is truth in Hölderlin's words, then the rule of enframing cannot exhaust itself solely in blocking all lighting-up of every revealing, all appearing of truth. Rather, precisely the essence of technology must harbor in itself the growth of the saving power. But in that case, might not an adequate look into what enframing is, as a destining of revealing, bring the upsurgence of the saving power into appearance?

In what respect does the saving power grow also there where the danger is? Where something grows, there it takes root, from thence it thrives. Both happen concealedly and quietly and in their own time. But according to the words of the poet we have no right whatsoever to expect that there where the danger is we should be able to lay hold of the saving power immediately and without preparation. Therefore we must consider now, in advance, in what respect the saving power does most profoundly take root and thence thrive even where the extreme danger lies – in the holding sway of enframing. In order to consider this, it is necessary, as a last step upon our way, to look with yet clearer eyes into the danger. Accordingly, we must once more question concerning technology. For we have said that in technology's essence roots and thrives the saving power.

But how shall we behold the saving power in the essence of technology so long as we do not consider in what sense of "essence" it is that enframing is actually the essence of technology?

Thus far we have understood "essence" in its current meaning. In the academic language of philosophy "essence" means *what* something is;

in Latin, *quid*. *Quidditas*, whatness, provides the answer to the question concerning essence. For example, what pertains to all kinds of trees – oaks, beeches, birches, firs – is the same "treeness." Under this inclusive genus – the "universal" – fall all real and possible trees. Is then the essence of technology, enframing, the common genus for everything technological? If this were the case then the steam turbine, the radio transmitter, and the cyclotron would each be an enframing. But the word "enframing" does not mean here a tool or any kind of apparatus. Still less does it mean the general concept of such resources. The machines and apparatus are no more cases and kinds of enframing than are the man at the switchboard and the engineer in the drafting room. Each of these in its own way indeed belongs as stockpart, available resource, or executor, within enframing; but enframing is never the essence of technology in the sense of a genus. Enframing is a way of revealing which is a destining, namely the way that challenges forth. The revealing that brings forth (*poiēsis*) is also a way that has the character of destining. But these ways are not kinds that, arrayed beside one another, fall under the concept of revealing. Revealing is that destining which, ever suddenly and inexplicably to all thinking, apportions itself into the revealing that brings forth and the revealing that challenges, and which allots itself to man. The revealing that challenges has its origin as a destining in bringing-forth. But at the same time enframing, in a way characteristic of a destining, blocks *poiēsis*.

Thus enframing, as a destining of revealing, is indeed the essence of technology, but never in the sense of genus and *essentia*. If we pay heed to this, something astounding strikes us: it is technology itself that makes the demand on us to think in another way what is usually understood by "essence." But in what way?

If we speak of the "essence of a house" and the "essence of a state" we do not mean a generic type; rather we mean the ways in which house and state hold sway, administer themselves, develop, and decay – the way in which they "develop" [*wesen*]. Johann Peter Hebel in a poem, "Ghost on Kanderer Street," for which Goethe had a special fondness, uses the old word *die Weserei*. It means the city hall, inasmuch as there the life of the community gathers and village existence is

constantly in play, i.e., comes to presence. It is from the verb *wesen* that the noun is derived. *Wesen* understood as a verb is the same as *währen* [to last or endure], not only in terms of meaning, but also in terms of the phonetic formation of the word. Socrates and Plato already think the essence of something as what essences, what comes to presence, in the sense of what endures. But they think what endures as what remains permanently (*aei on*). And they find what endures permanently in what persists throughout all that happens in what remains. That which remains they discover, in turn, in the aspect (*eidos, idea*), for example the Idea "house."

The Idea "house" displays what anything is that is fashioned as a house. Particular, real, and possible houses, in contrast, are changing and transitory derivatives of the Idea and thus belong to what does not endure.

But it can never in any way be established that enduring is based solely on what Plato thinks as *idea* and Aristotle thinks as *to ti ēn einai* (that which any particular thing has always been), or what metaphysics in its most varied interpretations thinks as *essentia*.

All essencing endures. But is enduring only permanent enduring? Does the essence of technology endure in the sense of the permanent enduring of an Idea that hovers over everything technological, thus making it seem that by technology we mean some mythological abstraction? The way in which technology essences lets itself be seen only on the basis of that permanent enduring in which enframing comes to pass as a destining of revealing. Goethe once uses the mysterious word *fortgewähren* [to grant permanently] in place of *fortwähren* [to endure permanently].[4] He hears *währen* [to endure] and *gewähren* [to grant] here in one unarticulated accord. And if we now ponder more carefully than we did before what it is that actually endures and perhaps alone endures, we may venture to say: *Only what is granted endures. What endures primally out of the earliest beginning is what grants.*

As the essencing of technology, enframing is what endures. Does enframing hold sway at all in the sense of granting? No doubt the question seems a horrendous blunder. For according to everything that has been said, enframing is, rather, a destining that gathers together into the revealing that challenges forth. Challenging is anything but a granting. So it seems, so long as we do not notice that the challenging-forth into the ordering of the real as standing-reserve still remains a destining that starts man upon a way of revealing. As this destining, the coming to presence of technology gives man entry into something which, of himself, he can neither invent nor in any way make. For there is no such thing as a man who exists singly and solely on his own.

But if this destining, enframing, is the extreme danger, not only for man's coming to presence, but for all revealing as such, should this destining still be called a granting? Yes, most emphatically, if in this destining the saving power is said to grow. Every destining of revealing comes to pass from a granting and as such a granting. For it is granting that first conveys to man that share in revealing that the coming-to-pass of revealing needs. So needed and used, man is given to belong to the coming-to-pass of truth. The granting that sends one way or another into revealing is as such the saving power. For the saving power lets man see and enter into the highest dignity of his essence. This dignity lies in keeping watch over the unconcealment – and with it, from the first, the concealment – of all coming to presence on this earth. It is precisely in enframing, which threatens to sweep man away into ordering as the supposed single way of revealing, and so thrusts man into the danger of the surrender of his free essence – it is precisely in this extreme danger that the innermost indestructible belongingness of man within granting may come to light, provided that we, for our part, begin to pay heed to the essence of technology.

Thus the coming to presence of technology harbors in itself what we least suspect, the possible upsurgence of the saving power.

Everything, then, depends upon this: that we ponder this arising and that we, recollecting, watch over it. How can this happen? Above all through our catching sight of what comes to presence in technology, instead of merely gaping at the technological. So long as we represent technology as an instrument, we remain transfixed in the will to master it. We press on past the essence of technology.

When, however, we ask how the instrumental comes to presence as a kind of causality, then we experience this coming to presence as the destining of a revealing.

When we consider, finally, that the coming to presence of the essence of technology comes to pass in the granting that needs and uses man so that he may share in revealing, then the following becomes clear:

The essence of technology is in a lofty sense ambiguous. Such ambiguity points to the mystery of all revealing, i.e., of truth.

On the one hand, enframing challenges forth into the frenziedness of ordering that blocks every view into the coming-to-pass of revealing and so radically endangers the relation to the essence of truth.

On the other hand, enframing comes to pass for its part in the granting that lets man endure – as yet inexperienced, but perhaps more experienced in the future – that he may be the one who is needed and used for the safekeeping of the essence of truth. Thus does the arising of the saving power appear.

The irresistibility of ordering and the restraint of the saving power draw past each other like the paths of two stars in the course of the heavens. But precisely this, their passing by, is the hidden side of their nearness.

When we look into the ambiguous essence of technology, we behold the constellation, the stellar course of the mystery.

The question concerning technology is the question concerning the constellation in which revealing and concealing, in which the coming to presence of truth comes to pass.

But what help is it to us to look into the constellation of truth? We look into the danger and see the growth of the saving power.

Through this we are not yet saved. But we are thereupon summoned to hope in the growing light of the saving power. How can this happen? Here and now and in little things, that we may foster the saving power in its increase. This includes holding always before our eyes the extreme danger.

The coming to presence of technology threatens revealing, threatens it with the possibility that all revealing will be consumed in ordering and that everything will present itself only in the unconcealedness of standing-reserve. Human activity can never directly counter this danger. Human achievement alone can never banish it. But human reflection can ponder the fact that all saving power must be of a higher essence than what is endangered, though at the same time kindred to it.

But might there not perhaps be a more primally granted revealing that could bring the saving power into its first shining-forth in the midst of the danger that in the technological age rather conceals than shows itself?

There was a time when it was not technology alone that bore the name *technē*. Once that revealing which brings forth truth into the splendor of radiant appearance was also called *technē*.

Once there was a time when the bringing-forth of the true into the beautiful was called *technē*. The *poiēsis* of the fine arts was also called *technē*.

At the outset of the destining of the West, in Greece, the arts soared to the supreme height of the revealing granted them. They illuminated the presence [*Gegenwart*] of the gods and the dialogue of divine and human destinings. And art was simply called *technē*. It was a single, manifold revealing. It was pious, *promos*, i.e., yielding to the holding sway and the safekeeping of truth.

The arts were not derived from the artistic. Art works were not enjoyed aesthetically. Art was not a sector of cultural activity.

What was art – perhaps only for that brief but magnificent age? Why did art bear the modest name *technē*? Because it was a revealing that brought forth and made present, and therefore belonged within *poiēsis*. It was finally that revealing which holds complete sway in all the fine arts, in poetry, and in everything poetical that obtained *poiēsis* as its proper name.

The same poet from whom we heard the words

> But where danger is, grows
> The saving power also . . .

says to us:

> . . . poetically dwells man upon this earth.

The poetical brings the true into the splendor of what Plato in the *Phaedrus* calls *to ekphanestaton*, that which shines forth most purely. The poetical thoroughly pervades every art, every revealing of coming to presence into the beautiful.

Could it be that the fine arts are called to poetic revealing? Could it be that revealing lays claim to the arts most primally, so that they for their part may expressly foster the growth of the saving power, may awaken and found anew our vision of that which grants and our trust in it?

Whether art may be granted this highest possibility of its essence in the midst of the extreme danger, no one can tell. Yet we can be astounded. Before what? Before this other possibility: that the frenziedness of technology may entrench itself everywhere to such an extent that someday, throughout everything technological, the essence of technology may come to presence in the coming-to-pass of truth.

Because the essence of technology is nothing technological, essential reflection upon technology and decisive confrontation with it must happen in a realm that is, on the one hand, akin to the essence of technology and, on the other, fundamentally different from it.

Such a realm is art. But certainly only if reflection upon art for its part, does not shut its eyes to the constellation of truth concerning which we are *questioning*.

Thus questioning, we bear witness to the crisis that in our sheer preoccupation with technology we do not yet experience the coming to presence of technology, that in our sheer aesthetic-mindedness we no longer guard and preserve the coming to presence of art. Yet the more questioningly we ponder the essence of technology, the more mysterious the essence of art becomes.

The closer we come to the danger, the more brightly do the ways into the saving power begin to shine and the more questioning we become. For questioning is the piety of thought.

Notes

1   W. Heisenberg, "Das Naturbild in der heutigen Physik," in *Die Künste im technischen Zeitalter* (Munich, 1954), pp. 43 ff.

2   See "On the Essence of Truth" (1930), first edition 1943, pp. 16 ff.

3   "Das Naturbild," pp. 60 ff.

4   "Die Wahlverwandtschaften," pt. 2, chap. 10, in the novel *Die Wunderlichen Nachbarskinder*.

10

Man the Technician

José Ortega Y Gasset

José Ortega y Gasset (1883–1955) was a Spanish philosopher, journalist, and political activist who was active in resisting the dictatorship of Primo de Rivera during the 1920s and in leading the
short-lived Spanish republic of the 1930s. These selections from
Ortega y Gasset's "Man the Technician" present an existentialist
approach to the Philosophy of Technology. Like Dewey (chapter 16),
Ortega argues that we are in large part responsible for the character
of our experiences because we make the things and situations within
which we find ourselves. He examines technological history as a story
of human self-consciousness, and argues that at our present stage,
what he calls the stage of the "technician," we have potentially the
greatest amount of self-determination, but also potentially the most
empty of lives.

For Ortega, as for Sartre, or de Beauvoir, the unique challenge of
human life is "autofabrication" or self-making. Doing so responsibly
requires a critical relation to self and world, which includes understanding evidence, weighing and evaluating options, articulating
goals and reasons. In a position echoed by Winner, Ortega argues that
contemporary humans have become so enmeshed in technology that
it has become a "second nature." As such, most of us take it for granted,
we make few attempts to understand or evaluate it, and consequently we fail to take responsibility for the shape and directions of
technological development. If he is correct that we make ourselves
through our relations to self and world, and if those relations are largely
mediated by technology, then we also fail to take responsibility for
ourselves. This is both a striking existential critique, and unveils an

From José Ortega y Gasset, *History as a System and Other Essays Toward a Philosophy of History* (New York: W. W.
Norton & Company), pp. 109–55. © 1941, 1961, by W. W. Norton & Company, Inc. Used by permission of W. W.
Norton & Company Inc.

attitude that is all too present in developed world business, politics, and consumption.

Excursion to the Substructure of Technology

The answers which have been given to the question, what is technology, are appallingly superficial; and what is worse, this cannot be blamed on chance. For the same happens to all questions dealing with what is truly human in human beings. There is no way of throwing light upon them until they are tackled in those profound strata from which everything properly human evolves. As long as we continue to speak of the problems that concern man as though we knew what man really is, we shall only succeed in invariably leaving the true issue behind. That is what happens with technology. We must realize into what fundamental depths our argument will lead us. How does it come to pass that there exists in the universe this strange thing called technology, the absolute cosmic fact of man the technician? If we seriously intend to find an answer, we must be ready to plunge into certain unavoidable profundities.

We shall then come upon the fact that an entity in the universe, man, has no other way of existing than by being in another entity, nature or the world. This relation of being one in the other, man in nature, might take on one of three possible aspects. Nature might offer man nothing but facilities for his existence in it. That would mean that the being of man coincides fully with that of nature or, what is the same, that man is a natural being. That is the case of the stone, the plant, and, probably, the animal. If it were that of man, too, he would be without necessities, he would lack nothing, he would not be needy. His desires and their satisfaction would be one and the same. He would wish for nothing that did not exist in the world and, conversely, whatever he wished for would be there of itself, as in the fairy tale of the magic wand. Such an entity could not experience the world as something alien to himself; for the world would offer him no resistance. He would be in the world as though he were in himself.

Or the opposite might happen. The world might offer to man nothing but difficulties, i.e., the being of the world and the being of man might be completely antagonistic. In this case the world would be no abode for man; he could not exist in it, not even for the fraction of a second. There would be no human life and, consequently, no technology.

The third possibility is the one that prevails in reality. Living in the world, man finds that the world surrounds him as an intricate net woven of both facilities and difficulties. Indeed, there are not many things in it which, potentially, are not both. The earth supports him, enabling him to lie down when he is tired and to run when he has to flee. A shipwreck will bring home to him the advantage of the firm earth – a thing grown humble from habitude. But the earth also means distance. Much earth may separate him from the spring when he is thirsty. Or the earth may tower above him as a steep slope that is hard to climb. This fundamental phenomenon – perhaps the most fundamental of all – that we are surrounded by both facilities and difficulties gives to the reality called human life its peculiar ontological character.

For if man encountered no facilities it would be impossible for him to be in the world, he would not exist, and there would be no problem. Since he finds facilities to rely on, his existence is possible. But this possibility, since he also finds difficulties, is continually challenged, disturbed, imperiled. Hence, man's existence is no passive being in the world; it is an unending struggle to accommodate himself in it. The stone is given its existence; it need not fight for being what it is – a stone in the field. Man has to be himself in spite of unfavorable circumstances; that means he has to make his own existence at every single moment. He is given the abstract possibility of existing, but not the reality. This he has to conquer hour after hour. Man must earn his life, not only economically but metaphysically.

And all this for what reason? Obviously – but this is repeating the same thing in other words –

because man's being and nature's being do not fully coincide. Because man's being is made of such strange stuff as to be partly akin to nature and partly not, at once natural and extranatural, a kind of ontological centaur, half immersed in nature, half transcending it. Dante would have likened him to a boat drawn up on the beach with one end of its keel in the water and the other in the sand. What is natural in him is realized by itself; it presents no problem. That is precisely why man does not consider it his true being. His extranatural part, on the other hand, is not there from the outset and of itself; it is but an aspiration, a project of life. And this we feel to be our true being; we call it our personality, our self. Our extra- and antinatural portion, however, must not be interpreted in terms of any of the older spiritual philosophies. I am not interested now in the so-called spirit (*Geist*), a pretty confused idea laden with speculative wizardry.

If the reader reflects a little upon the meaning of the entity he calls his life, he will find that it is the attempt to carry out a definite program or project of existence. And his self – each man's self – is nothing but this devised program. All we do we do in the service of this program. Thus man begins by being something that has no reality, neither corporeal nor spiritual; he is a project as such, something which is not yet but aspires to be. One may object that there can be no program without somebody having it, without an idea, a mind, a soul, or whatever it is called. I cannot discuss this thoroughly because it would mean embarking on a course of philosophy. But I will say this: although the project of being a great financier has to be conceived of in an idea, "being" the project is different from holding the idea. In fact, I find no difficulty in thinking this idea but I am very far from being this project.

Here we come upon the formidable and unparalleled character which makes man unique in the universe. We are dealing – and let the disquieting strangeness of the case be well noted – with an entity whose being consists not in what it is already, but in what it is not yet, a being that consists in not-yet-being. Everything else in the world is what it is. An entity whose mode of being consists in what it is already, whose potentiality coincides at once with his reality, we call a "thing." Things are given their being ready-made.

In this sense man is not a thing but an aspiration, the aspiration to be this or that. Each epoch, each nation, each individual varies in its own way the general human aspiration.

Now, I hope, all terms of the absolute phenomenon called "my life" will be clearly understood. Existence means, for each of us, the process of realizing, under given conditions, the aspiration we are. We cannot choose the world in which to live. We find ourselves, without our previous consent, embedded in an environment, a here and now. And my environment is made up not only by heaven and earth around me, but by my own body and my own soul. I am not my body; I find myself with it, and with it I must live, be it handsome or ugly, weak or sturdy. Neither am I my soul; I find myself with it and must use it for the purpose of living although it may lack will power or memory and not be of much good. Body and soul are things; but I am a drama, if anything, an unending struggle to be what I have to be. The aspiration or program I am, impresses its peculiar profile on the world about me, and that world reacts to this impress, accepting or resisting it. My aspiration meets with hindrance or with furtherance in my environment.

At this point one remark must be made which would have been misunderstood before. What we call nature, circumstance, or the world is essentially nothing but a conjunction of favorable and adverse conditions encountered by man in the pursuit of this program. The three names are interpretations of ours; what we first come upon is the experience of being hampered or favored in living. We are wont to conceive of nature and world as existing by themselves, independent of man. The concept "thing" likewise refers to something that has a hard and fast being and has it by itself and apart from man. But I repeat, this is the result of an interpretative reaction of our intellect upon what first confronts us. What first confronts us has no being apart from and independent of us; it consists exclusively in presenting facilities and difficulties, that is to say, in what it is in respect to our aspiration. Only in relation to our vital program is something an obstacle or an aid. And according to the aspiration animating us the facilities and difficulties, making up our pure and fundamental environment, will be such or such, greater or smaller.

This explains why to each epoch and even to each individual the world looks different. To the particular profile of our personal project, circumstance answers with another definite profile of facilities and difficulties. The world of the businessman obviously is different from the world of the poet. Where one comes to grief, the other thrives; where one rejoices, the other frets. The two worlds, no doubt, have many elements in common, viz., those which correspond to the generic aspiration of man as a species. But the human species is incomparably less stable and more mutable than any animal species. Men have an intractable way of being enormously unequal in spite of all assurances to the contrary.

Life as Autofabrication – Technology and Desires

From this point of view human life, the existence of man, appears essentially problematic. To all other entities of the universe existence presents no problem. For existence means actual realization of an essence. It means, for instance, that "being a bull" actually occurs. A bull, if he exists, exists as a bull. For a man, on the contrary, to exist does not mean to exist at once as the man he is, but merely that there exists a possibility of, and an effort towards, accomplishing this. Who of us is all he should be and all he longs to be? In contrast to the rest of creation, man, in existing, has to make his existence. He has to solve the practical problem of transferring into reality the program that is himself. For this reason "my life" is pure task, a thing inexorably to be made. It is not given to me as a present; I have to make it. Life gives me much to do; nay, it is nothing save the "to do" it has in store for me. And this "to do" is not a thing, but action in the most active sense of the word.

In the case of other beings the assumption is that somebody or something, already existing, acts; here we are dealing with an entity that has to act in order to be; its being presupposes action. Man, willy-nilly, is self-made, autofabricated. The word is not unfitting. It emphasizes the fact that in the very root of his essence man finds himself called upon to be an engineer. Life means to him at once and primarily the effort to

bring into existence what does not exist offhand, to wit: himself. In short, human life "is" production. By this I mean to say that fundamentally life is not, as has been believed for so many centuries, contemplation, thinking, theory, but action. It is fabrication; and it is thinking, theory, science only because these are needed for its autofabrication, hence secondarily, not primarily. To live . . . that is to find means and ways for realizing the program we are.

[. . .]

It is, therefore, a fundamental error to believe that man is an animal endowed with a talent for technology, in other words, that an animal might be transmuted into a man by magically grafting on it the technical gift. The opposite holds: because man has to accomplish a task fundamentally different from that of the animal, an extranatural task, he cannot spend his energies in satisfying his elemental needs, but must stint them in this realm so as to be able to employ them freely in the odd pursuit of realizing his being in the world.

We now see why man begins where technology begins.

[. . .]

Insight into the specific character of modern technology itself will best be gained by deliberately setting off its peculiar silhouette against the background of the whole of man's technical past. This means that we must give a sketch, if only the briefest, of the great changes undergone by the technical function itself; in other words, that we must define various stages in the evolution of technology. In this way, drawing some border lines and underlining others, we shall see the hazy past take on relief and perspective, revealing the forms from which technology has set out and those to which it has been coming.

The Stages of Technology – Technology of Chance

The subject is difficult. It took me some time to decide upon the principle best suited to distinguish periods of technology. I do not hesitate to reject the one readiest to hand, viz., that we should divide the evolution according to the appearance of certain momentous and characteristic inventions. All I have said in this essay aims to correct the current error of regarding such or such a definite invention as the thing which matters in technology. What really matters and what can bring about a fundamental advance is a change in the general character of technology. No single invention is of such caliber as to bear comparison with the tremendous mass of the integral evolution. We have seen that magnificent advances have been achieved only to be lost again, whether they disappeared completely or whether they had to be rediscovered. [. . .]

The best principle of delimiting periods in technical evolution is, to my judgment, furnished by the relation between man and technology, in other words by the conception which man in the course of history held, not of this or that particular technology but of the technical function as such. In applying this principle we shall see that it not only clarifies the past, but also throws light on the question we have asked before: how could modern technology give birth to such radical changes, and why is the part it plays in human life unparalleled in any previous age?

Taking this principle as our point of departure we come to discern three main periods in the evolution of technology: technology of chance; technology of the craftsman; technology of the technician.

What I call technology of chance, because in it chance is the engineer responsible for the invention, is the primitive technology of pre- and protohistoric man. [. . .]

How does primitive man conceive technology? The answer is easy. He is not aware of his technology as such; he is unconscious of the fact that there is among his faculties one which enables him to refashion nature after his desires.

The repertory of technical acts at the command of primitive man is very small and does not form a body of sufficient volume to stand out against, and be distinguished from, that of his natural acts, which is incomparably more important. That is to say, primitive man is very little man and almost all animal. His technical acts are scattered over and merged into the totality of his natural acts and appear to him as part of his natural life. He finds himself with the ability to light a fire as he finds himself with the ability to walk, swim, use his arms . . . His natural acts are a

given stock fixed once and for all; and so are his technical. It does not occur to him that technology is a means of virtually unlimited changes and advances.

The simplicity and scantiness of these pristine technical acts account for their being executed indiscriminately by all members of the community, who all light fires, carve bows and arrows, and so forth. The one differentiation noticeable very early is that women perform certain technical functions and men certain others. But that does not help primitive man to recognize technology as an isolated phenomenon. For the repertory of natural acts is also somewhat different in men and women. That the woman should plow the field – it was she who invented agriculture – appears as natural as that she should bear the children.

Nor does technology at this stage reveal its most characteristic aspect, that of invention. Primitive man is unaware that he has the power of invention; his inventions are not the result of a premeditated and deliberate search. He does not look for them; they seem rather to look for him. In the course of his constant and fortuitous manipulation of objects he may suddenly and by mere chance come upon a new useful device. While for fun or out of sheer restlessness he rubs two sticks together a spark springs up, and a vision of new connections between things will dawn upon him. The stick, which hitherto has served as weapon or support, acquires the new aspect of a thing producing fire. Our savage will be awed, feeling that nature has inadvertently loosed one of its secrets before him. Since fire had always seemed a godlike power, arousing religious emotions, the new fact is prone to take on a magic tinge. All primitive technology smacks of magic. In fact, magic, as we shall shortly see, is nothing but a kind of technology, albeit a frustrated and illusory one.

Primitive man does not look upon himself as the inventor of his inventions. Invention appears to him as another dimension of nature, as part of nature's power to furnish him – nature furnishing man, not man nature – with certain novel devices. He feels no more responsible for the production of his implements than for that of his hands and feet. He does not conceive of himself as *homo faber*. He is therefore very much in the same situation as Mr. Koehler's monkey when it suddenly notices that the stick in his hands may serve an unforeseen purpose. Mr. Koehler calls this the "aha-impression" after the exclamation of surprise a man utters when coming upon a startling new relation between things. It is obviously a case of the biological law of trial and error applied to the mental sphere. The infusoria "try" various movements and eventually find one with favorable effects on them which they consequently adopt as a function.

The inventions of primitive man, being, as we have seen, products of pure chance, will obey the laws of probability. Given the number of possible independent combinations of things, a certain possibility exists of their presenting themselves some day in such an arrangement as to enable man to see preformed in them a future implement.

Technology as Craftsmanship – Technology of the Technician

We come to the second stage, the technology of the artisan. This is the technology of Greece, of preimperial Rome, and of the Middle Ages. Here are in swift enumeration some of its essential features.

The repertory of technical acts has grown considerably. But – and this is important – a crisis and setback, or even the sudden disappearence of the principal industrial arts, would not yet be a fatal blow to material life in these societies. The life people lead with all these technical comforts and the life they would have to lead without them are not so radically different as to bar, in case of failures or checks, retreat to a primitive or almost primitive existence. The proportion between the technical and the non-technical is not yet such as to make the former indispensable for the supporting of life. Man is still relying mainly on nature. At least, and that is what matters, so he himself feels. When technical crises arise he does therefore not realize that they will hamper his life, and consequently fails to meet them in time and with sufficient energy.

Having made this reservation we may now state that technical acts have by this time enormously increased both in number and in complexity. It has become necessary for a definite group of people to take them up systematically and make a full-time job of them. These people are the artisans. Their existence is bound to help man become conscious of technology as an independent entity. He sees the craftsman at work – the cobbler, the blacksmith, the mason, the saddler – and therefore comes to think of technology in terms and in the guise of the technician, the artisan. That is to say, he does not yet know that there is technology, but he knows that there are technicians who perform a peculiar set of activities which are not natural and common to all men.

[...]

At the second stage of technology everybody knows shoemaking to be a skill peculiar to certain men. It can be greater or smaller and suffer slight variations as do natural skills, running for instance, or swimming or, better still, the flying of a bird, the charging of a bull. That means shoemaking is now recognized as exclusively human and not natural, i.e., animal; but it is still looked upon as a gift granted and fixed once and for all. Since it is something exclusively human it is extranatural, but since it is something fixed and limited, a definite fund not admitting of substantial amplification, it partakes of nature; and thus technology belongs to the nature of man. As man finds himself equipped with the unexchangeable system of his bodily movements, so he finds himself equipped with the fixed system of the "arts." For this is the name technology bears in nations and epochs living on the technical level in question; and this also is the original meaning of the Greek word *techne*.

The way technology progresses might disclose that it is an independent and, in principle, unlimited function. But, oddly enough, this fact becomes even less apparent in this than in the primitive period. After all, the few primitive inventions, being so fundamental, must have stood out melodramatically against the workaday routine of animal habits. But in craftsmanship there is no room whatever for a sense of invention. The artisan must learn thoroughly in long apprenticeship – it is the time of masters and apprentices – elaborate usages handed down by long tradition. He is governed by the norm that man must bow to tradition as such. His mind is turned towards the past and closed to novel possibilities. He follows the established routine. Even such modifications and improvements as may be brought about in his craft through continuous and therefore imperceptible shifts present themselves not as fundamental novelties, but rather as differences of personal style and skill. And these styles of certain masters again will spread in the forms of schools and thus retain the outward character of tradition.

We must mention another decisive reason why the idea of technology is not at this time separated from the idea of the person who practices it. Invention has as yet produced only tools and not machines. The first machine in the strict sense of the word – and with it I anticipate the third period – was the weaving machine set up by Robert in 1825. It is the first machine because it is the first tool that works by itself, and by itself produces the object. Herewith technology ceases to be what it was before, handiwork, and becomes mechanical production. In the crafts the tool works as a complement of man; man with his natural actions continues to be the principal agent. In the machine the tool comes to the fore, and now it is no longer the machine that serves man but man who waits on the machine. Working by itself, emancipated from man, the machine, at this stage, finally reveals that technology is a function apart and highly independent of natural man, a function which reaches far beyond the bounds set for him. What a man can do with his fixed animal activities we know beforehand; his scope is limited. But what the machine man is capable of inventing may do, is in principle unlimited.

One more feature of craftsmanship remains to be mentioned which helps to conceal the true character of technology. I mean this: technology implies two things. First, the invention of a plan of activity, of a method or procedure – *mechane*, said the Greeks – and, secondly, the execution of this plan. The former is technology strictly speaking, the latter consists merely in handling the raw material. In short, we have the technician and the worker who between them, performing very different functions, discharge the technical job. The craftsman is both technician and worker; and what appears first is a man at work with his hands, and what appears last, if at all, is the technology behind him. The dissociation of the artisan into his two ingredients, the worker and

the technician, is one of the principal symptoms of the technology of the third period.

We have anticipated some of the traits of this technology. We have called it the technology of the technician. Man becomes clearly aware that there is a capacity in him which is totally different from the immutable activities of his natural or animal part. He realizes that technology is not a haphazard discovery, as in the primitive period; that it is not a given and limited skill of some people, the artisans, as in the second period; that it is not this or that definite and therefore fixed "art"; but that it is a source of practically unlimited human activity.

This new insight into technology as such puts man in a situation radically new in his whole history and in a way contrary to all he has experienced before. Hitherto he has been conscious mainly of all the things he is unable to do, i.e., of his deficiencies and limitations. But the conception our time holds of technology – let the reader reflect a moment on his own – places us in a really tragicomic situation. Whenever we imagine some utterly extravagant feat, we catch ourselves in a feeling almost of apprehension lest our reckless dream – say a voyage to the stars – should come true. Who knows but that tomorrow morning's paper will spring upon us the news that it has been possible to send a projectile to the moon by imparting to it a speed great enough to overcome the gravitational attraction. That is to say, present-day man is secretly frightened by his own omnipotence. And this may be another reason why he does not know what he is. For finding himself in principle capable of being almost anything makes it all the harder for him to know what he actually is.

In this connection I want to draw attention to a point which does not properly belong here, that technology for all its being a practically unlimited capacity will irretrievably empty the lives of those who are resolved to stake everything on their faith in it and it alone. To be an engineer and nothing but an engineer means to be potentially everything and actually nothing. Just because of its promise of unlimited possibilities technology is an empty form like the most formalistic logic and is unable to determine the content of life. That is why our time, being the most intensely technical, is also the emptiest in all human history.

Relation Between Man and Technology in our Time – the Engineer in Antiquity

This third stage of technical evolution, which is our own, is characterized by the following features:

Technical acts and achievements have increased enormously. Whereas in the Middle Ages – the era of the artisan – technology and the nature of man counterbalanced each other and the conditions of life made it possible to benefit from the human gift of adapting nature to man without denaturalizing man, in our time the technical devices outweigh the natural ones so gravely that material life would be flatly impossible without them. This is no manner of speaking, it is the literal truth. In *The Revolt of the Masses* I drew attention to the most noteworthy fact that the population of Europe between 500 and 1800 AD, i.e., for thirteen centuries, never exceeded 180 millions; whereas by now, in little over a century, it has reached 500 millions, not counting those who have emigrated to America. In one century it has grown nearly three and a half times its size. If today 500 million people can live well in a space where 180 lived badly before, it is evident that, whatever the minor causes, the immediate cause and most necessary condition is the perfection of technology. Were technology to suffer a setback, millions of people would perish.

Such fecundity of the human animal could occur only after man had succeeded in interposing between himself and nature a zone of exclusively technical provenance, solid and thick enough to form something like a supernature. Present-day man – I refer not to the individual but to the totality of men – has no choice of whether to live in nature or to take advantage of this supernature. He is as irremediably dependent on, and lodged in, the latter as primitive man is in his natural environment. And that entails certain dangers. Since present-day man, as soon as he opens his eyes to life, finds himself surrounded by a superabundance of technical objects and procedures forming an artificial environment of such compactness that primordial nature is hidden behind it, he will tend to believe that all these things are there in the same way as nature itself is there without further effort on his part: that aspirin and automobiles grow on trees like apples. That is to say, he may easily lose sight of technology and of the conditions – the moral conditions, for example – under which it is produced and return to the primitive attitude of taking it for the gift of nature which is simply there. We thus have the curious fact that, at first, the prodigious expansion of technology made it stand out against the sober background of man's natural activities and allowed him to gain full sight of it, whereas by now its fantastic progress threatens to obscure it again.

Another feature helping man to discover the true character of his own technology we found to be the transition from mere tools to machines, i.e., mechanically working apparatus. A modern factory is a self-sufficient establishment waited on occasionally by a few persons of very modest standing. In consequence, the technician and the worker, who were united in the artisan, have been separated and the technician has grown to be the live expression of technology as such – in a word, the engineer.

Today technology stands before our mind's eye for what it is, apart, unmistakable, isolated, and unobscured by elements other than itself. And this enables certain persons, called engineers, to devote their lives to it. In the paleolithic age or in the Middle Ages technology, that is invention, could not have been a profession because man was ignorant of his own inventive power. Today the engineer embraces as one of the most normal and firmly established forms of activity the occupation of inventor. In contrast to the savage, he knows before he begins to invent that he is capable of doing so, which means that he has "technology" before he has "a technology." To this degree and in this concrete sense our previous assertion holds that technologies are nothing but concrete realizations of the general technical function of man. The engineer need not wait for chances and favorable odds; he is sure to make discoveries. How can he be?

The question obliges us to say a word about the technique of technology. To some people technique and nothing else is technology. They are right in so far as without technique – the intellectual method operative in technical creation – there is no technology. But with technique alone there is none either. As we have seen before, the existence of a capacity is not enough to put that capacity into action.

11

Focal Things and Practices

Albert Borgmann

Albert Borgmann is Regents Professor of Philosophy at the University of Montana. This excerpt is a chapter from Borgmann's *Technology and the Character of Contemporary Life*. Borgmann draws on Heidegger and Aristotle to highlight the difference between the "device paradigm" of modern technology and "focal things and practices" of pre-modern technology. The "promise of modern technology" is to free us from the misery and work imposed on us by nature and social existence. Technology will not only help us tame the forces of nature, but make our lives better by liberating us and enriching our lives. Like all modern thought, however, it leaves out the traditional (ancient Greek and Judeo-Christian) concern for the Good Life. Technology has failed to live up to its promise of liberation, because it is silent on ends, purposes, and goods. The device paradigm describes the situation of those for whom technological devices have become omnipresent. That includes most in the developed world. A device is distinguished from a "thing" in that a device requires little human intervention (and little knowledge) to operate. The entire trend toward making technological artifacts and systems "user-friendly" involves de-skilling, making them easier to use. Borgmann argues that these devices distance us from ourselves and from the world around us. A central example in his work is preparing and sharing a meal. Pre-prepared meals that can be heated in a microwave and eaten alone at anytime and anywhere disengage us from experience and knowledge of food, nutrition, agriculture, cooking, and human community. In contrast he proposes that we nurture "focal things and practices." For example, we should prepare a meal for friends or family from fresh foodstuffs, sharing conversation, and perhaps the labor, with those with whom we will share the meal. Reviving focal

From Albert Borgmann, *Technology and the Character of Contemporary Life* (Chicago: University of Chicago Press, 1984), pp. 196–210. Reprinted by permission of the publisher, The University of Chicago Press, and by Albert Borgmann.

practices, which contain within them a vision of the Good Life, can reform modern technology. Borgmann echoes the Communitarian-neo-Aristotelian concerns over the failure of liberalism to provide an adequate framework for moral-political institutions.

To see that the force of nature can be encountered analogously in many other places, we must develop the general notions of focal things and practices. This is the first point of this chapter. The Latin word *focus*, its meaning and etymology, are our best guides to this task. But once we have learned tentatively to recognize the instances of focal things and practices in our midst, we must acknowledge their scattered and inconspicuous character too. Their hidden splendor comes to light when we consider Heidegger's reflections on simple and eminent things. But an inappropriate nostalgia clings to Heidegger's account. It can be dispelled, so I will argue, when we remember and realize more fully that the technological environment heightens rather than denies the radiance of genuine focal things and when we learn to understand that focal things require a practice to prosper within. These points I will try to give substance in the subsequent parts of this chapter by calling attention to the focal concerns of running and of the culture of the table.

The Latin word *focus* means hearth. [. . .] [I]n a pretechnological house the fireplace constituted a center of warmth, of light, and of daily practices. For the Romans the *focus* was holy, the place where the housegods resided. In ancient Greece, a baby was truly joined to the family and household when it was carried about the hearth and placed before it. The union of a Roman marriage was sanctified at the hearth. And at least in the early periods the dead were buried by the hearth. The family ate by the hearth and made sacrifices to the housegods before and after the meal. The hearth sustained, ordered, and centered house and family.[1] Reflections of the hearth's significance can yet be seen in the fireplace of many American homes. The fireplace often has a central location in the house. Its fire is now symbolical since it rarely furnishes sufficient warmth. But the radiance, the sounds, and the fragrance of living fire consuming logs that are split, stacked, and felt in their grain have retained their force. There are no longer images of the ancestral gods placed by the fire; but there often are pictures of loved ones on or above the mantel, precious things of the family's history, or a clock, measuring time.[2]

The symbolical center of the house, the living room with the fireplace, often seems forbidding in comparison with the real center, the kitchen with its inviting smells and sounds. Accordingly, the architect Jeremiah Eck has rearranged homes to give them back a hearth, "a place of warmth and activity" that encompasses cooking, eating, and living and so is central to the house whether it literally has a fireplace or not.[3] Thus we can satisfy, he says, "the need for a place of focus in our family lives."[4]

"Focus," in English, is now a technical term of geometry and optics. Johannes Kepler was the first so to use it, and he probably drew on the then already current sense of focus as the "burning point of lens or mirror."[5] Correspondingly, an optic or geometric focus is a point where lines or rays converge or from which they diverge in a regular or lawful way. Hence "focus" is used as a verb in optics to denote moving an object in relation to a lens or modifying a combination of lenses in relation to an object so that a clear and well-defined image is produced.

These technical senses of "focus" have happily converged with the original one in ordinary language. Figuratively they suggest that a focus gathers the relations of its context and radiates into its surroundings and informs them. To focus on something or to bring it into focus is to make it central, clear, and articulate. It is in the context of these historical and living senses of "focus" that I want to speak of focal things and practices. Wilderness on this continent, it now appears, is a focal thing. It provides a center of orientation; when we bring the surrounding technology into it, our relations to technology become clarified and well-defined. But just how strong its gathering and radiating force is requires further reflection. And surely there will be other focal things and practices: music, gardening, the culture of the table, or running.

We might in a tentative way be able to see these things as focal; what we see more clearly and readily is how inconspicuous, homely, and dispersed they are. This is in stark contrast to the focal things of pretechnological times, the Greek temple or the medieval cathedral that we have mentioned before. Martin Heidegger was deeply impressed by the orienting force of the Greek temple. For him, the temple not only gave a center of meaning to its world but had orienting power in the strong sense of first originating or establishing the world, of disclosing the world's essential dimensions and criteria.[6] Whether the thesis so extremely put is defensible or not, the Greek temple was certainly more than a self-sufficient architectural sculpture, more than a jewel of well-articulated and harmoniously balanced elements, more, even, than a shrine for the image of the goddess or the god. As Vincent Scully has shown, a temple or a temple precinct gathered and disclosed the land in which they were situated. The divinity of land and sea was focused in the temple.[7]

To see the work of art as the focus and origin of the world's meaning was a pivotal discovery for Heidegger. He had begun in the modern tradition of Western philosophy where, as suggested in the first chapter of this book, the sense of reality is to be grasped by determining the antecedent and controlling conditions of all there is (the *Bedingungen der Möglichkeit* as Immanuel Kant has it). Heidegger wanted to outdo this tradition in the radicality of his search for the fundamental conditions of being. Perhaps it was the relentlessness of his pursuit that disclosed the ultimate futility of it. At any rate, when the universal conditions are explicated in a suitably general and encompassing way, what truly matters still hangs in the balance because everything depends on how the conditions come to be actualized and instantiated.[8] The preoccupation with antecedent conditions not only leaves this question unanswered; it may even make it inaccessible by leaving the impression that, once the general and fundamental matters are determined, nothing of consequence remains to be considered. Heidegger's early work, however, already

contained the seeds of its overcoming. In his determination to grasp reality in its concreteness, Heidegger had found and stressed the inexorable and unsurpassable givenness of human existence, and he had provided analyses of its pretechnological wholeness and its technological distraction though the significance of these descriptions for technology had remained concealed to him.[9] And then he discovered that the unique event of significance in the singular work of art, in the prophet's proclamation, and in the political deed was crucial. This insight was worked out in detail with regard to the artwork. But in an epilogue to the essay that develops this point, Heidegger recognized that the insight comes too late. To be sure, our time has brought forth admirable works of art. "But," Heidegger insists, "the question remains: is art still an essential and necessary way in which that truth happens which is decisive for historical existence, or is art no longer of this character?"[10]

Heidegger began to see technology (in his more or less substantive sense) as the force that has eclipsed the focusing powers of pretechnological times. Technology becomes for him the final phase of a long metaphysical development. The philosophical concern with the conditions of the possibility of whatever is now itself seen as a move into the oblivion of what finally matters. But how are we to recover orientation in the oblivious and distracted era of technology when the great embodiments of meaning, the works of art, have lost their focusing power? Amidst the complication of conditions, of the *Bedingungen*, we must uncover the simplicity of things, of the *Dinge*.[11] A jug, an earthen vessel from which we pour wine, is such a thing. It teaches us what it is to hold, to offer, to pour, and to give. In its clay, it gathers for us the earth as it does in containing the wine that has grown from the soil. It gathers the sky whose rain and sun are present in the wine. It refreshes and animates us in our mortality. And in the libation it acknowledges and calls on the divinities. In these ways the thing (in agreement with its etymologically original meaning) gathers and discloses what Heidegger calls the fourfold, the interplay of the crucial dimensions of earth and sky, mortals and divinities.[12] A thing, in Heidegger's eminent sense, is a focus; to speak of focal things is to emphasize the central point twice.

Still, Heidegger's account is but a suggestion fraught with difficulties. When Heidegger described the focusing power of the jug, he might have been thinking of a rural setting where wine jugs embody in their material, form, and craft a long and local tradition; where at noon one goes down to the cellar to draw a jug of table wine whose vintage one knows well; where at the noon meal the wine is thoughtfully poured and gratefully received.[13] Under such circumstances, there might be a gathering and disclosure of the fourfold, one that is for the most part understood and in the background and may come to the fore on festive occasions. But all of this seems as remote to most of us and as muted in its focusing power as the Parthenon or the Cathedral of Chartres. How can so simple a thing as a jug provide that turning point in our relation to technology to which Heidegger is looking forward? Heidegger's proposal for a reform of technology is even more programmatic and terse than his analysis of technology.[14] Both, however, are capable of fruitful development.[15] Two points in Heidegger's consideration of the turn of technology must particularly be noted. The first serves to remind us of arguments already developed which must be kept in mind if we are to make room for focal things and practices. Heidegger says, broadly paraphrased, that the orienting force of simple things will come to the fore only as the rule of technology is raised from its anonymity, is disclosed as the orthodoxy that heretofore has been taken for granted and allowed to remain invisible.[16] As long as we overlook the tightly patterned character of technology and believe that we live in a world of endlessly open and rich opportunities, as long as we ignore the definite ways in which we, acting technologically, have worked out the promise of technology and remain vaguely enthralled by that promise, so long simple things and practices will seem burdensome, confining, and drab. But if we recognize the central vacuity of advanced technology, that emptiness can become the opening for focal things. It works both ways, of course. When we see a focal concern of ours threatened by technology, our sight for the liabilities of mature technology is sharpened.

A second point of Heidegger's is one that we must develop now. The things that gather the fourfold, Heidegger says, are inconspicuous and humble. And when we look at his litany of things, we also see that they are scattered and of yesterday: jug and bench, footbridge and plow, tree and pond, brook and hill, heron and deer, horse and bull, mirror and clasp, book and picture, crown and cross.[17] That focal things and practices are inconspicuous is certainly true; they flourish at the margins of public attention. And they have suffered a diaspora; this too must be accepted, at least for now. That is not to say that a hidden center of these dispersed focuses may not emerge some day to unite them and bring them home. But it would clearly be a forced growth to proclaim such a unity now. A reform of technology that issues from focal concerns will be radical not in imposing a new and unified master plan on the technological universe but in discovering those sources of strength that will nourish principled and confident beginnings, measures, i.e., which will neither rival nor deny technology.

But there are two ways in which we must go beyond Heidegger. One step in the first direction has already been taken. It led us to see in the preceding chapter that the simple things of yesterday attain a new splendor in today's technological context. The suggestion in Heidegger's reflections that we have to seek out pretechnological enclaves to encounter focal things is misleading and dispiriting. Rather we must see any such enclave itself as a focal thing heightened by its technological context. The turn to things cannot be a setting aside and even less an escape from technology but a kind of affirmation of it. The second move beyond Heidegger is in the direction of practice, into the social and, later, the political situation of focal things.[18] Though Heidegger assigns humans their place in the fourfold when he depicts the jug in which the fourfold is focused, we scarcely see the hand that holds the jug, and far less do we see of the social setting in which the pouring of the wine comes to pass. In his consideration of another thing, a bridge, Heidegger notes the human ways and works that are gathered and directed by the bridge.[19] But these remarks too present practices from the viewpoint of the focal thing. What must be shown is that focal things can prosper in human practices only. Before we can build a bridge, Heidegger suggests, we must be able to dwell.[20] But what does that mean concretely?

The consideration of the wilderness has disclosed a center that stands in a fruitful counterposition to technology. The wilderness is beyond the procurement of technology, and our response to it takes us past consumption. But it also teaches us to accept and to appropriate technology. We must now try to discover if such centers of orientation can be found in greater proximity and intimacy to the technological everyday life. And I believe they can be found if we follow up the hints that we have gathered from and against Heidegger, the suggestions that focal things seem humble and scattered but attain splendor in technology if we grasp technology properly, and that focal things require a practice for their welfare. Running and the culture of the table are such focal things and practices. We have all been touched by them in one way or another. If we have not participated in a vigorous or competitive run, we have certainly taken walks; we have felt with surprise, perhaps, the pleasure of touching the earth, of feeling the wind, smelling the rain, of having the blood course through our bodies more steadily. In the preparation of a meal we have enjoyed the simple tasks of washing leaves and cutting bread; we have felt the force and generosity of being served a good wine and homemade bread. Such experiences have been particularly vivid when we came upon them after much sitting and watching indoors, after a surfeit of readily available snacks and drinks. To encounter a few simple things was liberating and invigorating. The normal clutter and distraction fall away when, as the poet says,

> there, in limpid brightness shine,
> on the table, bread and wine.[21]

If such experiences are deeply touching, they are fleeting as well. There seems to be no thought or discourse that would shelter and nurture such events; not in politics certainly, nor in philosophy where the prevailing idiom sanctions and applies equally to lounging and walking, to Twinkies, and to bread, the staff of life. But the reflective care of the good life has not withered away. It has left the profession of philosophy and sprung up among practical people. In fact, there is a tradition in this country of persons who are engaged by life in its concreteness and simplicity and who are so filled with this engagement that they have reached for the pen to become witnesses and teachers, speakers of deictic discourse. Melville and Thoreau are among the great prophets of this tradition. Its present health and extent are evident from the fact that it now has no overpowering heroes but many and various more or less eminent practitioners. Their work embraces a spectrum between down-to-earth instruction and soaring speculation. The span and center of their concerns vary greatly. But they all have their mooring in the attention to tangible and bodily things and practices, and they speak with an enthusiasm that is nourished by these focal concerns. Pirsig's book is an impressive and troubling monument in this tradition, impressive in the freshness of its observations and its pedagogical skill, troubling in its ambitious and failing efforts to deal with the large philosophical issues. Norman Maclean's *A River Runs through It* can be taken as a fly-fishing manual, a virtue that pleases its author.[22] But it is a literary work of art most of all and a reflection on technology inasmuch as it presents the engaging life, both dark and bright, from which we have so recently emerged. Colin Fletcher's treatise of *The Complete Walker* is most narrowly a book of instruction about hiking and backpacking.[23] The focal significance of these things is found in the interstices of equipment and technique; and when the author explicitly engages in deictic discourse he has "an unholy awful time" with it.[24] Roger B. Swain's contemplation of gardening in *Earthly Pleasures* enlightens us in cool and graceful prose about the scientific basis and background of what we witness and undertake in our gardens.[25] Philosophical significance enters unbidden and easily in the reflections on time, purposiveness, and the familiar. Looking at these books, I see a stretch of water that extends beyond my vision, disappearing in the distance. But I can see that it is a strong and steady stream, and it may well have parts that are more magnificent than the ones I know.[26]

To discover more clearly the currents and features of this, the other and more concealed, American mainstream, I take as witnesses two books where enthusiasm suffuses instruction vigorously, Robert Farrar Capon's *The Supper of the Lamb* and George Sheehan's *Running and Being*.[27] Both are centered on focal events, the great run and the great meal. The great run, where one exults in the strength of one's body, in the ease and the length of the stride, where nature speaks powerfully in the hills, the wind, the heat, where one takes endurance to the breaking point, and where one is finally engulfed by the good will of the spectators and the fellow runners.[28] The great meal, the long session as Capon calls it, where the guests are thoughtfully invited, the table has been carefully set, where the food is the culmination of tradition, patience, and skill and the presence of the earth's most delectable textures and tastes, where there is an invocation of divinity at the beginning and memorable conversation throughout.[29]

Such focal events are compact, and if seen only in their immediate temporal and spatial extent they are easily mistaken. They are more mistakable still when they are thought of as experiences in the subjective sense, events that have their real meaning in transporting a person into a certain mental or emotional state. Focal events, so conceived, fall under the rule of technology. For when a subjective state becomes decisive, the search for a machinery that is functionally equivalent to the traditional enactment of that state begins, and it is spurred by endeavors to find machineries that will procure the state more instantaneously, ubiquitously, more assuredly and easily. If, on the other hand, we guard focal things in their depth and integrity, then, to see them fully and truly, we must see them in context. Things that are deprived of their context become ambiguous.[30] The letter "a" by itself means nothing in particular. In the context of "table" it conveys or helps to convey a more definite meaning. But "table" in turn can mean many things. It means something more powerful in the text of Capon's book where he speaks of "The Vesting of the Table."[31] But that text must finally be seen in the context and texture of the world. To say that something becomes ambiguous is to say that it is made to say less, little, or nothing. Thus to elaborate the context of focal events is to grant them their proper eloquence.

"The distance runner," Sheehan says, "is the least of all athletes. His sport the least of all sports."[32] Running is simply to move through time and space, step-by-step. But there is splendor in that simplicity. In a car we move of course much faster, farther, and more comfortably. But we are not moving on our own power and in our own right. We cash in prior labor for present motion. Being beneficiaries of science and engineering and having worked to be able to pay for a car, gasoline, and roads, we now release what has been earned and stored and use it for transportation. But when these past efforts are consumed and consummated in my driving, I can at best take credit for what I have done. What I am doing now, driving, requires no effort, and little or no skill or discipline. I am a divided person; my achievement lies in the past, my enjoyment in the present. But in the runner, effort and joy are one; the split between means and ends, labor and leisure is healed.[33] To be sure, if I have trained conscientiously, my past efforts will bear fruit in a race. But they are not just cashed in. My strength must be risked and enacted in the race which is itself a supreme effort and an occasion to expand my skill.

This unity of achievement and enjoyment, of competence and consummation, is just one aspect of a central wholeness to which running restores us. Good running engages mind and body. Here the mind is more than an intelligence that happens to be housed in a body. Rather the mind is the sensitivity and the endurance of the body.[34] Hence running in its fullness, as Sheehan stresses over and over again, is in principle different from exercise designed to procure physical health. The difference between running and physical exercise is strikingly exhibited in one and the same issue of the New York Times Magazine. It contains an account by Peter Wood of how, running the New York City Marathon, he took in the city with body and mind, and it has an account by Alexandra Penney of corporate fitness programs where executives, concerned about their Coronary Risk Factor Profile, run nowhere on treadmills or ride stationary bicycles.[35] In another issue, the Magazine shows executives exercising their bodies while busying their dissociated minds with reading.[36] To be sure, unless a runner concentrates on bodily performance, often in an effort to run the best possible race, the mind wanders as the body runs. But as in free association we range about the future and the past, the actual and the possible, our mind, like our breathing, rhythmically gathers itself to the here and now, having spread itself to distant times and faraway places.

It is clear from these reflections that the runner is mindful of the body because the body is intimate with the world. The mind becomes relatively disembodied when the body is severed from the depth of the world, i.e., when the world is split into commodious surfaces and inaccessible machineries. Thus the unity of ends and means, of mind and body, and of body and world is one and the same. It makes itself felt in the vividness with which the runner experiences reality. "Somehow you feel more in touch," Wood says, "with the realities of a massive inner-city housing problem when you are running through it slowly enough to take in the grim details, and, surprisingly, cheered on by the remaining occupants."[37] As this last remark suggests, the wholeness that running establishes embraces the human family too. The experience of that simple event releases an equally simple and profound sympathy. It is a natural goodwill, not in need of drugs nor dependent on a common enemy. It wells up from depths that have been forgotten, and it overwhelms the runners ever and again.[38] As Wood recounts his running through streets normally besieged by crime and violence, he remarks: "But we can only be amazed today at the warmth that emanates from streets usually better known for violent crime." And his response to the spectators' enthusiasm is this: "I feel a great proximity to the crowd, rushing past at all of nine miles per hour; a great affection for them individually; a commitment to run as well as I possibly can, to acknowledge their support."[39] For George Sheehan, finally, running discloses the divine. When he runs, he wrestles with God.[40] Serious running takes us to the limits of our being. We run into threatening and seemingly unbearable pain. Sometimes, of course, the plunge into that experience gets arrested in ambition and vanity. But it can take us further to the point where in suffering our limits we experience our greatness too. This, surely, is a hopeful place to escape technology, metaphysics, and the God of the philosophers and reach out to the God of Abraham, Isaac, and Jacob.[41]

If running allows us to center our lives by taking in the world through vigor and simplicity, the culture of the table does so by joining simplicity with cosmic wealth. Humans are such complex and capable beings that they can fairly comprehend the world and, containing it, constitute a cosmos in their own right. Because we are standing so eminently over against the world, to come in touch with the world becomes for us a challenge and a momentous event. In one sense, of course, we are always already in the world, breathing the air, touching the ground, feeling the sun. But as we can in another sense withdraw from the actual and present world, contemplating what is past and to come, what is possible and remote, we celebrate correspondingly our intimacy with the world. This we do most fundamentally when in eating we take in the world in its palpable, colorful, nourishing immediacy. Truly human eating is the union of the primal and the cosmic. In the simplicity of bread and wine, of meat and vegetable, the world is gathered.

The great meal of the day, be it at noon or in the evening, is a focal event par excellence. It gathers the scattered family around the table. And on the table it gathers the most delectable things nature has brought forth. But it also recollects and presents a tradition, the immemorial experiences of the race in identifying and cultivating edible plants, in domesticating and butchering animals; it brings into focus closer relations of national or regional customs, and more intimate traditions still of family recipes and dishes. It is evident from the preceding chapters how this living texture is being rent through the procurement of food as a commodity and the replacement of the culture of the table by the food industry. Once food has become freely available, it is only consistent that the gathering of the meal is shattered and disintegrates into snacks, T.V. dinners, bites that are grabbed to be eaten; and eating itself is scattered around television shows, late and early meetings, activities, overtime work, and other business. This is increasingly the normal condition of technological eating. But it is within our power to clear a central space amid the clutter and distraction. We can begin with the simplicity of a meal that has a beginning, a middle, and an end and that breaks through the superficiality of convenience food in the simple steps of beginning with raw ingredients, preparing and transforming them, and bringing them to the table. In this way we can again become freeholders of our culture. We are disfranchised from world citizenship when the foods we eat are mere commodities. Being essentially opaque surfaces, they repel all efforts at extending our sensibility and competence into the deeper reaches of the world. A Big Mac and a Coke can overwhelm our tastebuds and accommodate our hunger. Technology is not, after all, a children's crusade but a principled and skillful enterprise of defining and satisfying human needs. Through the diversion and busyness of consumption we may have unlearned to feel constrained by the shallowness of commodities. But having gotten along for a time and quite well, it seemed, on institutional or convenience food, scales fall from our eyes when we step up to a festively set family table. The foods stand out more clearly, the fragrances are stronger, eating has once more become an occasion that engages and accepts us fully.

To understand the radiance and wealth of a festive meal we must be alive to the interplay of things and humans, of ends and means. At first a meal, once it is on the table, appears to have commodity character since it is now available before us, ready to be consumed without effort or merit. But though there is of course in any eating a moment of mere consuming, in a festive meal eating is one with an order and discipline that challenges and ennobles the participants. The great meal has its structure. It begins with a moment of reflection in which we place ourselves in the presence of the first and last things. It has a sequence of courses; it requires and sponsors memorable conversation; and all this is enacted in the discipline called table manners. They are warranted when they constitute the respectful and skilled response to the great things that are coming to pass in the meal. We can see how order and discipline have collapsed when we eat a Big Mac. In consumption there is the pointlike and inconsequential conflation of a sharply delimited human need with an equally contextless and closely fitting commodity. In a Big Mac the sequence of courses has been compacted into one object and the discipline of table manners has been reduced to grabbing and eating. The social context reaches no further than the pleasant faces and quick hands of the people who run the fast-food outlet. In a festive meal, however, the food is served, one of the most generous gestures human beings are capable of. The serving is of a piece with garnishing; garnishing is the final phase of cooking, and cooking is one with preparing the food. And if we are blessed with rural circumstances, the preparation of food draws near the harvesting and the raising of the vegetables in the garden close by. This context of activities is embodied in persons. The dish and the cook, the vegetable and the gardener tell of one another. Especially when we are guests, much of the meal's deeper context is socially and conversationally mediated. But that mediation has translucence and intelligibility because it extends into the farther and deeper recesses without break and with a bodily immediacy that we too have enacted or at least witnessed firsthand. And what seems to be a mere receiving and consuming of food is in fact the enactment of generosity and gratitude, the affirmation of mutual and perhaps religious obligations. Thus eating in a focal setting differs sharply from the social and cultural anonymity of a fast-food outlet.

The pretechnological world was engaging through and through, and not always positively. There also was ignorance, to be sure, of the final workings of God and king; but even the unknown engaged one through mystery and awe. In this web of engagement, meals already had focal character, certainly as soon as there was anything like a culture of the table.[42] Today, however, the great meal does not gather and order a web of thoroughgoing relations of engagement; within the technological setting it stands out as a place of profound calm, one in which we can leave behind the narrow concentration and one-sided strain of labor and the tiring and elusive diversity of consumption. In the technological setting, the culture of the table not only focuses our life; it is also distinguished as a place of healing, one that restores us to the depth of the world and to the wholeness of our being.

As said before, we all have had occasion to experience the profound pleasure of an invigorating walk or a festive meal. And on such occasions we may have regretted the scarcity of such events; we might have been ready to allow such events a more regular and central place in our lives. But for the most part these events remain occasional, and indeed the ones that still grace us may be slipping from our grasp. But why are we acting against our better insights and aspirations? This at first seems all the more puzzling as the engagement in a focal activity is for most citizens of the technological society an instantaneous and ubiquitous possibility. On any day I can decide to run or to prepare a meal after work. Everyone has some sort of suitable equipment. At worst one has to stop on the way home to pick up this or that. It is of course technology that has opened up these very possibilities. But why are they lying fallow for the most part? There is a convergence of several factors. Labor is exhausting, especially when it is divided. When we come home, we often feel drained and crippled. Diversion and pleasurable consumption appear to be consonant with this sort of disability. They promise to untie the knots and to soothe the aches. And so they do at a shallow level of our existence. At any rate, the call for exertion and engagement seems like a cruel and unjust demand. We have sat in the easy chair, beer at hand and television before us; when we felt stirrings of ambition, we found it easy to ignore our superego.[43] But we also may have had our alibi refuted on occasion when someone to whom we could not say no prevailed on us to put on our coat and to step out into cold and windy weather to take a walk. At first our indignation grew. The discomfort was worse than we had thought. But gradually a transformation set in. Our gait became steady, our blood began to flow vigorously and wash away our tension, we smelled the rain, began thoughtfully to speak with our companion, and finally returned home settled, alert, and with a fatigue that was capable of restful sleep.

But why did such occurrences remain episodes also? The reason lies in the mistaken assumption that the shaping of our lives can be left to a series of individual decisions. Whatever goal in life we entrust to this kind of implementation we in fact surrender to erosion. Such a policy ignores both the frailty and strength of human nature. On the spur of the moment, we normally act out what has been nurtured in our daily practices as they have been shaped by the norms of our time. When we sit in our easy chair and contemplate what to do, we are firmly enmeshed in the framework of technology with our labor behind us and the blessings of our labor about us, the diversions and enrichments of consumption. This arrangement has had our lifelong allegiance, and we know it to have the approval and support of our fellows. It would take superhuman strength to stand up to this order ever and again. If we are to challenge *the rule of technology*, we can do so only through *the practice of engagement*.

The human ability to establish and commit oneself to a practice reflects our capacity to comprehend the world, to harbor it in its expanse as a context that is oriented by its focal points. To found a practice is to guard a focal concern, to shelter it against the vicissitudes of fate and our frailty. John Rawls has pointed out that there is decisive difference between the justification of a practice and of a particular action falling under it.[44] Analogously, it is one thing to decide for a focal practice and quite another to decide for a particular action that appears to have focal character.[45] Putting the matter more clearly, we must say that without a practice an engaging action or event can momentarily light up our life, but it cannot order and orient it focally. Competence, excellence, or virtue, as Aristotle first saw, come into being as an *éthos*, a settled disposition and a way of life.[46] Through a practice, Alasdaire MacIntyre says accordingly, "human powers to achieve excellence, and human conceptions of the ends and goods involved, are systematically extended."[47] Through a practice we are able to accomplish what remains unattainable when aimed at in a series of individual decisions and acts.

How can a practice be established today? Here, as in the case of focal things, it is helpful to consider the foundation of pretechnological practices. In mythic times the latter were often established through the founding and consecrating act of a divine power or mythic ancestor. Such an act set up a sacred precinct and center that gave order to a violent and hostile world. A sacred practice, then, consisted in the regular reenactment of the founding act, and so it renewed and sustained the order of the world. Christianity came into being this way; the eucharistic meal, the Supper of the Lamb, is its central event, established with the instruction that it be reenacted. Clearly a focal practice today should have centering and orienting force as well. But it differs in important regards from its grand precursors. A mythic focal practice derived much force from the power of its opposition. The alternative to the preservation of the cosmos was chaos, social and physical disorder and collapse. It is a reduction to see mythic practices merely as coping behavior of high survival value. A myth does not just aid survival; it defines what truly human life is. Still, as in the case of pretechnological morality, economic and social factors were interwoven with mythic practices. Thus the force of brute necessity supported, though it did not define, mythic focal practices. Since a mythic focal practice united in itself the social, the economic, and the cosmic, it was naturally a prominent and public affair. It rested securely in collective memory and in the mutual expectations of the people.

This sketch, of course, fails to consider many other kinds of pretechnological practices. But it does present one important aspect of them and more particularly one that serves well as a backdrop for focal practices in a technological setting. It is evident that technology is itself a sort of practice, and it procures its own kind of order and security. Its history contains great moments of innovation, but it did not arise out of a founding event that would have focal character; nor has it produced focal things. Thus it is not a *focal* practice, and it has indeed, so I have urged, a debilitating tendency to scatter our attention and to clutter our surroundings. A focal practice today, then, meets no tangible or overtly hostile opposition from its context and is so deprived of the wholesome vigor that derives from such opposition. But there is of course an opposition at a more profound and more subtle level. To feel the support of that opposing force one must have experienced the subtly debilitating character of technology, and above all one must understand, explicitly or implicitly, that the peril of technology lies not in this or that of its manifestations but in *the pervasiveness and consistency of its pattern*. There are always occasions where a Big Mac, an exercycle, or a television program are unobjectionable and truly helpful answers to human needs. This makes a case-by-case appraisal of technology so inconclusive. It is when we attempt to take the measure of technological life in its normal totality that we are distressed by its shallowness. And I believe that the more strongly we sense and the more clearly we understand the coherence and the character of technology, the more evident it becomes to us that technology must be countered by an equally patterned and social commitment, i.e., by a practice.

At this level the opposition of technology does become fruitful to focal practices. They can now be seen as restoring a depth and integrity to our lives that are in principle excluded within the paradigm of technology. MacIntyre, though his foil is the Enlightenment more than technology, captures this point by including in his definition of practice the notion of "goods internal to a practice."[48] These are one with the practice and can only be obtained through that practice. The split between means and ends is healed. In contrast "there are those goods externally and contingently attached" to a practice; and in that case there "are always alternative ways for achieving such goods, and their achievement is never to be had *only* by engaging in some particular kind of practice."[49] Thus practices (in a looser sense) that serve external goods are subvertible by technology. But MacIntyre's point needs to be clarified and extended to include or emphasize not only the essential unity of human being and a particular sort of doing but also the tangible things in which the world comes to be focused. The importance of this point has been suggested by the consideration of running and the culture of the table. There are objections to this suggestion that will be examined in the next chapter. Here I want to advance the thesis by considering Rawls's contention that a practice is defined by rules. We can take a rule as an instruction for a particular domain of life to act in a certain way under specified circumstances. How important is the particular character of the tangible setting of the rules? Though Rawls does not address this question directly he suggests in using baseball for illustration that "a peculiarly shaped piece of wood" and a kind of bag become a bat and base only within the confines defined by the rules of baseball.[50] Rules and the practice they define, we might argue in analogy to what Rawls says about their relation to particular cases, are logically prior to their tangible setting. But the opposite contention seems stronger to me. Clearly the possibilities and challenges of baseball are crucially determined by the layout and the surface of the field, the weight and resilience of the ball, the shape and size of the bat, etc. One might of course reply that there are rules that define the physical circumstances of the game. But this is to take "rule" in a broader sense. Moreover it would be more accurate to say that the rules of this latter sort reflect and protect the identity of the original tangible circumstances in which the game grew up. The rules, too, that circumscribe the actions of the players can be taken as ways of securing and ordering the playful challenges that arise in the human interplay with reality. To be sure there are developments and innovations in sporting equipment. But either they quite change the nature of the sport as in pole vaulting, or they are restrained to preserve the identity of the game as in baseball.

It is certainly the purpose of a focal practice to guard in its undiminished depth and identity the thing that is central to the practice, to shield it against the technological diremption into means and end. Like values, rules and practices are recollections, anticipations, and, we can now say, guardians of the concrete things and events that finally matter. Practices protect focal things not only from technological subversion but also against human frailty. The ultimately significant things to which we respond in deictic discourse cannot be possessed or controlled. Hence when we reach out for them, we miss them occasionally and sometimes for quite some time. Running becomes unrelieved pain and cooking a thankless chore. If in the technological mode we insisted on assured results or if more generally we estimated the value of future efforts on the basis of recent experience, focal things would vanish from our lives. A practice keeps faith with focal things and saves for them an opening in our lives. To be sure, eventually the practice needs to be empowered again by the reemergence of the great thing in its splendor. A practice that is not so revived degenerates into an empty and perhaps deadening ritual.

We can now summarize the significance of a focal practice and say that such a practice is required to counter technology in its patterned pervasiveness and to guard focal things in their depth and integrity. Countering technology through a practice is to take account of our susceptibility to technological distraction, and it is also to engage the peculiarly human strength of comprehension, i.e., the power to take in the world in its extent and significance and to respond through an enduring commitment. Practically a focal practice comes into being through resoluteness, either an explicit resolution where one vows regularly to engage in a focal activity from this day on or in a more implicit resolve that is nurtured by a focal thing in favorable circumstances and matures into a settled custom.

In considering these practical circumstances we must acknowledge a final difference between focal practices today and their eminent pretech-nological predecessors. The latter, being public and prominent, commanded elaborate social and physical settings: hierarchies, offices, ceremonies, and choirs; edifices, altars, implements, and vestments. In comparison our focal practices are humble and scattered. Sometimes they can hardly be called practices, being private and limited. Often they begin as a personal regimen and mature into a routine without ever attaining the social richness that distinguishes a practice. Given the often precarious and inchoate nature of focal practices, evidently focal things and practices, for all the splendor of their simplicity and their fruitful opposition to technology, must be further clarified in their relation to our everyday world if they are to be seen as a foundation for the reform of technology.

Notes

1 See *Paulys Realencyclopedie der classischen Altertumswissenschaft* (Stuttgart, 1893–1963), 15: 615–17; See also Fustel de Coulanges, "The Sacred Fire," in *The Ancient City*, trans. Willard Small (Garden City, NY, n.d. [first published in 1864]), pp. 25–33.

2 See Kent C. Bloomer and Charles W. Moore, *Body, Memory, and Architecture* (New Haven, 1977), pp. 2–3 and 50–1.

3 See Jeremiah Eck, "Home Is Where the Hearth Is," *Quest* 3 (April 1979): 12.

4 Ibid., p. 11.

5 See *The Oxford English Dictionary*.

6 See Martin Heidegger, "The Origin of the Work of Art," in *Poetry, Language, Thought*, trans. Albert Hofstadter (New York, 1971), pp. 15–87.

7 See Vincent Scully, *The Earth, the Temple, and the Gods* (New Haven, 1962).

8 See my *The Philosophy of Language* (The Hague, 1974), pp. 126–31.

9 See Heidegger, *Being and Time*, trans. John Macquarrie and Edward Binson (New York, 1962), pp. 95–107, 163–8, 210–24.

10 See Heidegger, "The Origin of the Work of Art," p. 80.

11 See Heidegger, "The Thing," in *Poetry, Language, Thought*, pp. 163–82. Heidegger alludes to the turn from the *Bedingungen* to the *Dinge* on p. 179 of the original, "Das Ding," in *Vorträge und Aufsätze* (Pfullingen, 1959). He alludes to the turn from technology to (focal) things in "The Question Concerning Technology," in *The Question Concerning Technology and Other Essays*, trans. William Lovitt (New York, 1977), p. 43.

12 See Heidegger, "The Thing."

13 See M. F. K. Fisher, *The Cooking of Provincial France* (New York, 1968) p. 50.

14 Though there are seeds for a reform of technology to be found in Heidegger as I want to show, Heidegger insists that "philosophy will not be able to effect an immediate transformation of the present condition of the world. Only a god can save us." See "Only a God Can Save Us: Der Spiegel's Interview with Martin Heidegger," trans. Maria P. Alter and John D. Caputo, *Philosophy Today*, 20 (1976): 277.

15 I am not concerned to establish or defend the claim that my account of Heidegger or my development of his views are authoritative. It is merely a matter here of acknowledging a debt.

16 See Heidegger, "The Question Concerning Technology," p. 43; Langdon Winner makes a similar point in "The Political Philosophy of Alternative Technology," in *Technology and Man's Future*. ed. Albert H. Teich, 3d edn. (New York, 1981), pp. 369–73.

17 See Heidegger, "The Thing," pp. 180–2.

18 The need of complementing Heidegger's notion of the thing with the notion of practice was brought home to me by Hubert L. Dreyfus's essay, "Holism and Hermeneutics," *Review of Metaphysics*, 34 (1980): 22–3.

19 See Heidegger, "Building Dwelling Thinking," in *Poetry, Language, Thought*, pp. 152–3.

20 Ibid., pp. 148–9.

21 Georg Trakl, quoted by Heidegger in "Language," in *Poetry, Language, Thought*, pp. 194–5 (I have taken some liberty with Hofstadter's translation).

22 See Norman Maclean, *A River Runs through It and Other Stories* (Chicago, 1976). Only the first of the three stories instructs the reader about fly fishing.

23 See Colin Fletcher, *The Complete Walker* (New York, 1971).

24 Ibid., p. 9.

25 See Roger B. Swain, *Earthly Pleasures: Tales from a Biologist's Garden* (New York, 1981).

26 Here are a few more: Wendell Berry, *Farming: A Handbook* (New York, 1970); Stephen Kiesling, *The Shell Game: Reflections on Rowing and the Pursuit of Excellence* (New York, 1982); John Richard Young, *Schooling for Young Riders* (Norman, Okla., 1970); W. Timothy Gallwey, *The Inner Game of Tennis* (New York, 1974); Ruedi Bear, *Pianta Su: Ski Like the Best* (Boston, 1976). Such books must be sharply distinguished from those that promise to teach accomplishments without effort and in no time. The latter kind of book is technological in intent and fraudulent in fact.

27 See Robert Farrar Capon, *The Supper of the Lamb: A Culinary Reflection* (Garden City, NY, 1969); and George Sheehan, *Running and Being: The Total Experience* (New York, 1978).

28 See Sheehan, pp. 211–20 and elsewhere.

29 See Capon, pp. 167–81.

30 See my "Mind, Body, and World," *Philosophical Forum*, 8 (1976): 76.

31 See Capon, pp. 176–7.

32 See Sheehan, p. 127.

33 On the unity of achievement and enjoyment, see Alasdair Macintyre, *After Virtue* (Notre Dame, Ind., 1981), p. 184.

34 See my "Mind, Body, and World," pp. 68–86.

35 See Peter Wood, "Seeing New York on the Run," *New York Times Magazine*, 7 October 1979; Alexandra Penney, "Health and Grooming: Shaping Up the Corporate Image," ibid.

36 See *New York Times Magazine*, 3 August 1980, pp. 20–1.

37 See Wood, p. 112.

38 See Sheehan, pp. 211–17.

39 See Wood, p. 116.

40 See Sheehan, pp. 221–31 and passim.

41 There is substantial anthropological evidence to show that running has been a profound focal practice in certain pretechnological cultures. I am unable to discuss it here. Nor have I discussed the problem, here and elsewhere touched upon, of technology and religion. The present study, I believe, has important implications for that issue, but to draw them out would require more space and circumspection than are available now. I have made attempts to provide an explication in "Christianity and the Cultural Center of Gravity," *Listening*, 18 (1983): 93–102; and in "Prospects for the Theology of Technology," *Theology and Technology*, ed. Carl Mitcham and Jim Grote (Lanham, Md., 1984), pp. 305–22.

42 See M. F. K. Fisher, pp. 9–31.

43 Some therapists advise lying down till these stirrings go away.

44 See John Rawls, "Two Concepts of Rules," *Philosophical Review*, 64 (1955): 3–32.

45 Conversely, it is one thing to break a practice and quite another to omit a particular action. For we define ourselves and our lives in our practices; hence to break a practice is to jeopardize one's identity while omitting a particular action is relatively inconsequential.

46 See Aristotle's *Nicomachean Ethics*, the beginning of Book Two in particular.

47 See MacIntyre, p. 175.

48 Ibid., pp. 175–7.

49 Ibid., p. 176.

50 See Rawls, p. 25.

12

A Phenomenology of Technics

Don Ihde

Don Ihde, a distinguished philosopher who has taught for decades at the State University of New York, Stony Brook, argues, in these excerpts from *Technology and the Lifeworld*, that human life has always been technologically embedded, and a return to a pre-technological state of nature is at best a thought experiment. He then examines two aspects of human–technology relations. The first are phenomenological investigations to uncover the invariant structures of human–technology relations; the second are hermeneutical investigations to reveal variant cultural patterns of the embeddedness of technology.

First Ihde describes four distinct patterns of human–technology relations, each being a form of mediation between a person and the world. These are: an embodiment relation, a hermeneutic relation, an alterity relation, and a background relation. In an embodiment relation, technology stands between self and world as an extension of self, as my reading glasses extend the power of my eyes. A hermeneutic relation is one in which we experience some aspect of the world through experiencing some aspect of a technology, as when your cell phone tells you that it can obtain a strong signal. Alterity relations are characterized by a relation between self and technology, where the world may be present only as a distant referent. For example, when I watch a movie, I am aware of the movie and the world it creates. There may well be a human narrative behind the movie narrative, and there is certainly a human and technological story behind the existence of the movie as an artifact. But, my relation here is with a technological artifact as "other." Finally, as Ortega noted, technologies increasingly form the background of daily life, especially within the developed world. We take for granted, except when the systems fail, electricity, heating, cooling, and so on.

From Don Ihde, *Technology and the Lifeworld* (Bloomington: Indiana University Press, 1990), pp. 72–100, 105–12. Reprinted by permission of the publisher, Indiana University Press.

In the second part of this piece, Ihde notes that human–technology relations are both structurally cross-cultural yet realized differently in each culture. Ihde uncovers a variety of technological experiences, overcoming the tendency of earlier theorists to conceive of it as single, totalizing technique of control. Instead, he emphasizes its multiple transformative powers that, within the perceptual–cultural dimensions of human life, create the possibility of what he calls "pluriculturality."

Ihde's work is influenced by Heidegger's work on technology. But a stronger influence on him was Edmund Husserl, and Ihde's most common focus has been on the way culture – including technological culture and its many instruments and gadgets – shapes perception. Ihde is generally considered one of the leading figures among phenomenological-hermeneutic philosophers of technology; he now considers his work to be "post-phenomenological."

The task of a phenomenology of human-technology relations is to discover the various structural features of those ambiguous relations. In taking up this task, I shall begin with a focus upon experientially recognizable features that are centered upon the ways we are bodily engaged with technologies. The beginning will be within the various ways in which I-as-body interact with my environment by means of technologies.

A. Technics Embodied

If much of early modern science gained its new vision of the world through optical technologies, the process of embodiment itself is both much older and more pervasive. To embody one's praxis *through* technologies is ultimately an *existential* relation with the world. It is something humans have always – since they left the naked perceptions of the Garden – done.

I have previously and in a more suggestive fashion already noted some features of the visual embodiment of optical technologies. Vision is technologically transformed through such optics. But while the fact *that* optics transform vision may be clear, the variants and invariants of such a transformation are not yet precise. That becomes the task for a more rigorous and structural phenomenology of embodiment. I shall begin by drawing from some of the previous features mentioned in the preliminary phenomenology of visual technics.

Within the framework of phenomenological relativity, visual technics first may be located within the intentionality of seeing.

I see – through the optical artifact – the world

This seeing is, in however small a degree, at least minimally distinct from a direct or naked seeing.

I see – the world

I call this first set of existential technological relations with the world *embodiment relations*, because in this use context I take the technologies *into* my experiencing in a particular way by way of perceiving *through* such technologies and through the reflexive transformation of my perceptual and body sense.

In Galileo's use of the telescope, he embodies his seeing through the telescope thusly:

Galileo – telescope – Moon

Equivalently, the wearer of eyeglasses embodies eyeglass technology:

I – glasses – world

The technology is actually *between* the seer and the seen, in a *position of mediation*. But the referent of the seeing, that towards which sight is directed, is "on the other side" of the optics. One sees *through* the optics. This, however, is not enough to specify this relation as an embodiment one. This is because one first has to determine *where* and *how*, along what will be described as a continuum of relations, the technology is experienced.

There is an initial sense in which this positioning is doubly ambiguous. First, the technology must be *technically* capable of being seen through; it must be transparent. I shall use the term *technical* to refer to the physical characteristics of the technology. Such characteristics may be designed or they may be discovered. Here the disciplines that deal with such characteristics are informative, although indirectly so for the philosophical analysis per se. If the glass is not transparent enough, seeing-through is not possible. If it is transparent enough, approximating whatever "pure" transparency could be empirically attainable, then it becomes possible to embody the technology. This is a material condition for embodiment.

Embodying as an activity, too, has an initial ambiguity. It must be learned or, in phenomenological terms, constituted. If the technology is good, this is usually easy. The very first time I put on my glasses, I see the now-corrected world. The adjustments I have to make are not usually focal irritations but fringe ones (such as the adjustment to backglare and the slight changes in spatial motility). But once learned, the embodiment relation can be more precisely described as one in which the technology becomes maximally "transparent." It is, as it were, taken into my own perceptual-bodily self experience thus:

(I-glasses)-world

My glasses become part of the way I ordinarily experience my surroundings; they "withdraw" and are barely noticed, if at all. I have then actively embodied the technics of vision. Technics is the symbiosis of artifact and user within a human action.

Embodiment relations, however, are not at all restricted to visual relations. They may occur for any sensory or microperceptual dimension. A hearing aid does this for hearing, and the blind man's cane for tactile motility. Note that in these corrective technologies *the same structural features of embodiment* obtain as with the visual example. Once learned, cane and hearing aid "withdraw" (if the technology is good – and here we have an experiential clue for the perfecting of technologies). I hear the world through the hearing aid and feel (and hear) it through the cane. The juncture (I-artifact)-world is through the technology and brought close by it.

Such relations *through* technologies are not limited to either simple or complex technologies. Glasses, insofar as they are engineered systems, are much simpler than hearing aids. More complex than either of these monosensory devices are those that entail whole-body motility. One such common technology is automobile driving. Although driving an automobile encompasses more than embodiment relations, its pleasurability is frequently that associated with embodiment relations.

One experiences the road and surroundings *through* driving the car, and motion is the focal activity. In a finely engineered sports car, for example, one has a more precise feeling of the road and of the traction upon it than in the older, softer-riding, large cars of the fifties. One embodies the car, too, in such activities as parallel parking: when well embodied, one feels rather than sees the distance between car and curb – one's bodily sense is "extended" to the parameters of the driver-car "body." And although these embodiment relations entail larger, more complex artifacts and entail a somewhat longer, more complex learning process, the bodily tacit knowledge that is acquired is perceptual-bodily.

Here is a first clue to the polymorphous sense of bodily extension. The experience of one's "body image" is not fixed but malleably extendable and/or reducible in terms of the material or technological mediations that may be embodied. I shall restrict the term embodiment, however, to those types of mediation that can be so experienced. The same dynamic polymorphousness can also be located in non-mediational or direct experience. Persons trained in the martial arts, such as karate, learn to feel the vectors and trajectories of the opponent's moves within the space of the combat. The near space around one's material body is charged.

Embodiment relations are a particular kind of use-context. They are technologically relative in a double sense. First, the technology must "fit" the use. Indeed, within the realm of embodiment relations one can develop a quite specific set of qualities for design relating to attaining the requisite technological "withdrawal." For example, in handling highly radioactive materials at a distance, the mechanical arms and hands which are designed to pick up and pour glass tubes inside the shielded enclosure have to "feed back" a delicate sense of touch to the operator. The closer to invisibility, transparency, and the extension of one's own bodily sense this technology allows, the better. Note that the design perfection is not one related to the machine alone but to the combination of machine and human. The machine is perfected along a bodily vector, molded to the perceptions and actions of humans.

And when such developments are most successful, there may arise a certain romanticizing of technology. In much anti-technological literature there are nostalgic calls for returns to simple tool technologies. In part, this may be because long-developed tools are excellent examples of bodily expressivity. They are both direct in actional terms and immediately experienced; but what is missed is that such embodiment relations may take any number of directions. Both the sports car driver within the constraints of the racing route and the bulldozer driver destroying a rainforest may have the satisfactions of powerful embodiment relations.

There is also a deeper desire which can arise from the experience of embodiment relations. It is the doubled desire that, on one side, is a wish for *total transparency*, total embodiment, for the technology to truly "become me." Were this possible, it would be equivalent to there being no technology, for total transparency would *be* my body and senses; I desire the face-to-face that I would experience without the technology. But that is only one side of the desire. The other side is the desire to have the power, the transformation that the technology makes available. Only by using the technology is my bodily power enhanced and magnified by speed, through distance, or by any of the other ways in which technologies change my capacities. These capacities are always *different* from my naked capacities.

The desire is, at best, contradictory. I want the transformation that the technology allows, but I want it in such a way that I am basically unaware of its presence. I want it in such a way that it becomes me. Such a desire both secretly *rejects* what technologies are and overlooks the transformational effects which are necessarily tied to human-technology relations. This illusory desire belongs equally to pro- and anti-technology interpretations of technology.

The desire is the source of both utopian and dystopian dreams. The actual, or material, technology always carries with it only a partial or quasi-transparency, which is the price for the extension of magnification that technologies give. In extending bodily capacities, the technology also transforms them. In that sense, all technologies in use are non-neutral. They change the basic situation, however subtly, however minimally; but this is the other side of the desire. The desire is simultaneously a desire for a change in situation – to inhabit the earth, or even to go beyond the earth – while sometimes inconsistently and secretly wishing that this movement could be without the mediation of the technology.

The direction of desire opened by embodied technologies also has its positive and negative thrusts. Instrumentation in the knowledge activities, notably science, is the gradual extension of perception into new realms. The desire is to see, but seeing is seeing through instrumentation. Negatively, the desire for pure transparency is the wish to escape the limitations of the material technology. It is a platonism returned in a new form, the desire to escape the newly extended body of technological engagement. In the wish there remains the contradiction: the user both wants and does not want the technology. The user wants what the technology gives but does not want the limits, the transformations that a technologically extended body implies. There is a fundamental ambivalence toward the very human creation of our own earthly tools.

The ambivalence that can arise concerning technics is a reflection of one kind upon the *essential ambiguity* that belongs to technologies in use. But this ambiguity, I shall argue, has its own distinctive shape. Embodiment relations display an essential magnification/reduction structure which has been suggested in the instrumentation examples. Embodiment relations simultaneously magnify or amplify and reduce or place aside what is experienced through them.

The sight of the mountains of the moon, through all the transformational power of the telescope, removes the moon from its setting in the expanse of the heavens. But if our technologies were only to replicate our immediate and bodily experience, they would be of little use and ultimately of little interest. A few absurd examples might show this:

In a humorous story, a professor bursts into his club with the announcement that he has just invented a reading machine. The machine scans the pages, reads them, and perfectly reproduces them. (The story apparently was written before the invention of photocopying. Such machines might be said to be "perfect reading machines" in actuality.) The problem, as the innocent could see, was that this machine leaves us with precisely the problem we had prior to its invention. To have reproduced through mechanical "reading" all the books in the world leaves us merely in the library.

A variant upon the emperor's invisible clothing might work as well. Imagine the invention of perfectly transparent clothing through which we might technologically experience the world. We could see through it, breathe through it, smell and hear through it, touch through it. Indeed, it effects no changes of any kind, since it is *perfectly* invisible. Who would bother to pick up such clothing (even if the presumptive wearer could find it)? Only by losing some invisibility – say, with translucent coloring – would the garment begin to be usable and interesting. For here, at least, fashion would have been invented – but at the price of losing total transparency – by becoming that through which we relate to an environment.

Such stories belong to the extrapolated imagination of fiction, which stands in contrast to even the most minimal actual embodiment relations, which in their material dimensions simultaneously extend and reduce, reveal and conceal.

In actual human-technology relations of the embodiment sort, the transformational structures may also be exemplified by variations: In optical technologies, I have already pointed out how spatial significations change in observations through lenses. The entire gestalt changes. When the apparent size of the moon changes, along with it the apparent position of the observer changes.

Relativistically, the moon is brought "close"; and equivalently, this optical near-distance applies to both the moon's appearance and my bodily sense of position. More subtly, every dimension of spatial signification also changes. For example, with higher and higher magnification, the well-known phenomenon of depth, instrumentally mediated as a "focal plane," also changes. Depth diminishes in optical near-distance.

A related phenomenon in the use of an optical instrument is that it transforms the spatial significations of vision in an instrumentally focal way. But my seeing without instrumentation is a full bodily seeing – I see not just with my eyes but with my whole body in a unified sensory experience of things. In part, this is why there is a noticeable irreality to the apparent position of the observer, which only diminishes with the habits acquired through practice with the instrument. But the optical instrument cannot so easily transform the entire sensory gestalt. The focal sense that is magnified through the instrument is monodimensioned.

Here may be the occasion (although I am not claiming a cause) for a certain interpretation of the senses. Historians of perception have noted that, in medieval times, not only was vision not the supreme sense but sound and smell may have had greatly enhanced roles so far as the interpretation of the senses went. Yet in the Renaissance and even more exaggeratedly in the Enlightenment, there occurred the reduction to sight as the favored sense, and within sight, a certain reduction *of* sight. This favoritism, however, also carried implications for the other senses.

One of these implications was that each of the senses was interpreted to be clear and distinct from the others, with only certain features recognizable through a given sense. Such an interpretation impeded early studies in echo location.

In 1799 Lazzaro Spallanzani was experimenting with bats. He noticed not only that they could locate food targets in the dark but also that they could do so blindfolded. Spallanzani wondered if bats could guide themselves by their ears rather than by their eyes. Further experimentation, in which the bats' ears were filled with wax, showed that indeed they could not guide themselves without their ears. Spallanzani surmised that either bats locate objects through hearing or they had some sense of which humans knew nothing.

Given the doctrine of separate senses and the identification of shapes and objects through vision alone, George Montagu and Georges Guvier virtually laughed Spallanzani out of the profession.

This is not to suggest that such an interpretation of sensory distinction was due simply to familiarity with optical technologies, but the common experience of enhanced vision through such technologies was at least the standard practice of the time. Auditory technologies were to come later. When auditory technologies did become common, it was possible to detect the same amplification/reduction structure of the human-technology experience.

The telephone in use falls into an auditory embodiment relation. If the technology is good, I hear *you* through the telephone and the apparatus "withdraws" into the enabling background:

(I-telephone)-you

But as a monosensory instrument, your phenomenal presence is that of a voice. The ordinary multidimensioned presence of a face-to-face encounter does not occur, and I must at best imagine those dimensions through your vocal gestures. Also, as with the telescope, the spatial significations are changed. There is here an auditory version of visual near-distance. It makes little difference whether you are geographically near or far, none at all whether you are north or south, and none with respect to anything but your bodily relation to the instrument. Your voice retains its partly irreal near-distance, reduced from the full dimensionality of direct perceptual situations. This telephonic distance is different both from immediate face-to-face encounters and from visual or geographical distance as normally taken. Its distance is a mediated distance with its own identifiable significations.

While my primary set of variations is to locate and demonstrate the invariance of a magnification/reduction structure to any embodiment relation, there are also secondary and important effects noted in the histories of technology. In the very first use of the telephone, the users were fascinated and intrigued by its auditory transparency. Watson heard and recognized Bell's *voice*, even though the instrument had a high ratio of noise to message. In short, the fascination attaches to magnification, amplification,

enhancement. But, contrarily, there can be a kind of forgetfulness that equally attaches to the reduction. What is *revealed* is what excites; what is concealed may be forgotten. Here lies one secret for technological trajectories with respect to development. There are *latent telics* that occur through inventions.

Such telics are clear enough in the history of optics. Magnification provided the fascination. Although there were stretches of time with little technical progress, this fascination emerged from time to time to have led to compound lenses by Galileo's day. If some magnification shows the new, opens to what was poorly or not at all previously detected, what can greater magnification do? In our own time, the explosion of such variants upon magnification is dramatic. Electron enhancement, computer image enhancement, CAT and NMR internal scanning, "big-eye" telescopes – the list of contemporary magnificational and visual instruments is very long.

I am here restricting myself to what may be called a *horizontal* trajectory, that is, optical technologies that bring various micro- or macrophenomena to vision through embodiment relations. By restricting examples to such phenomena, one structural aspect of embodiment relations may be pointed to concerning the relation to microperception and its Adamic context. While *what* can be seen has changed dramatically – Galileo's New World has now been enhanced by astronomical phenomena never suspected and by micro-phenomena still being discovered – there remains a strong phenomenological constant in *how* things are seen. All lenses and optical technologies of the sort being described bring what is to be seen into a normal bodily space and distance. Both the macroscopic and the microscopic appear within the same near-distance. The "image size" of galaxy or amoeba is the *same*. Such is the existential condition for visibility, the counterpart to the technical condition, that the instrument makes things visually present.

The mediated presence, however, must fit, be made close to my actual bodily position and sight. Thus there is a reference within the instrumental context to my face-to-face capacities. These remain primitive and central within the new mediational context. Phenomenological theory claims that for every change in what is seen (the object correlate), there is a noticeable change in how (the experiential correlate) the thing is seen.

In embodiment relations, such changes retain both an equivalence and a difference from non-mediated situations. What remains constant is the bodily focus, the reflexive reference back to my bodily capacities. What is seen must be seen from or within my visual field, from the apparent distance in which discrimination can occur regarding depth, etc., just as in face-to-face relations. But the range of what can be brought into this proximity is transformed by means of the instrument.

Let us imagine for a moment what was never in fact a problem for the history of instrumentation: If the "image size" of both a galaxy and an amoeba is the "same" for the observer using the instrument, how can we tell that one is macrocosmic and the other microcosmic? The "distance" between us and these two magnitudes, Pascal noted, was the same in that humans were interpreted to be between the infinitely large and the infinitely small.

What occurs through the mediation is not a problem *because our construction of the observation presupposes ordinary praxical spatiality.* We handle the paramecium, placing it on the slide and then under the microscope. We aim the telescope at the indicated place in the sky and, before looking through it, note that the distance is at least that of the heavenly dome. But in our imagination experiment, what if our human were *totally immersed* in a technologically mediated world? What if, from birth, all vision occurred only through lens systems? Here the problem would become more difficult. But in our distance from Adam, it is precisely the presumed difference that makes it possible for us to see both nakedly *and* mediately – and thus to be able to locate the difference – that places us even more distantly from any Garden. It is because we retain this ordinary spatiality that we have a reflexive point of reference from which to make our judgments.

The noetic or bodily reflexivity implied in all vision also may be noticed in a magnified way in the learning period of embodiment. Galileo's telescope had a small field, which, combined with early hand-held positioning, made it very difficult to locate any particular phenomenon. What must have been noted, however, even if not commented upon, was the exaggerated sense of bodily motion experienced through trying to fix upon a heavenly body – and more, one quickly learns something about the earth's very motion in the attempt to use such primitive telescopes. Despite the apparent fixity of the stars, the hand-held telescope shows the earth-sky motion dramatically. This magnification effect is within the experience of one's own bodily viewing.

This bodily and actional point of reference retains a certain privilege. All experience refers to it in a taken-for-granted and recoverable way. The bodily condition of the possibility for seeing is now twice indicated by the very situation in which mediated experience occurs. Embodiment relations continue to locate that privilege of my being here. The partial symbiosis that occurs in well-designed embodied technologies retains that motility which can be called expressive. Embodiment relations constitute one existential form of the full range of the human-technology field.

B. Hermeneutic Technics

Heidegger's hammer in use displays an embodiment relation. Bodily action through it occurs within the environment. But broken, missing, or malfunctioning, it ceases to be the means of praxis and becomes an obtruding *object* defeating the work project. Unfortunately, that negative derivation of objectness by Heidegger carries with it a block against understanding a second existential human-technology relation, the type of relation I shall term *hermeneutic*.

The term hermeneutic has a long history. In its broadest and simplest sense it means "interpretation," but in a more specialized sense it refers to *textual* interpretation and thus entails *reading*. I shall retain both these senses and take hermeneutic to mean a special interpretive action within the technological context. That kind of activity calls for special modes of action and perception, modes analogous to the reading process.

Reading is, of course, a reading of ____; and in its ordinary context, what fills the intentional blank is a text, something *written*. But all writing entails technologies. Writing has a product. Historically, and more ancient than the revolution brought about by such crucial technologies as the clock or the compass, the invention and development of writing was surely even more revolutionary than clock or compass with respect to human experience. Writing transformed the very perception and understanding we have of language. Writing is a technologically embedded form of language.

There is a currently fashionable debate about the relationship between speech and writing, particularly within current Continental philosophy. The one side argues that speech is primary, both historically and ontologically, and the other – the French School – inverts this relation and argues for the primacy of writing. I need not enter this debate here in order to note the *technological difference* that obtains between oral speech and the materially connected process of writing, at least in its ancient forms.

Writing is inscription and calls for both a process of writing itself, employing a wide range of technologies (from stylus for cuneiform to word processors for the contemporary academic), and other material entities upon which the writing is recorded (from clay tablet to computer printout). Writing is technologically mediated language. From it, several features of hermeneutic technics may be highlighted. I shall take what may at first appear as a detour into a distinctive set of human-technology relations by way of a phenomenology of reading and writing.

Reading is a specialized perceptual activity and praxis. It implicates my body, but in certain distinctive ways. In an ordinary act of reading, particularly of the extended sort, what is read is placed before or somewhat under one's eyes. We read in the immediate context from some miniaturized bird's-eye perspective. What is read occupies an expanse within the focal center of vision, and I am ordinarily in a somewhat rested position. If the object-correlate, the "text" in the broadest sense, is a chart, as in the navigational examples, what is represented retains a representational isomorphism with the natural features of the landscape. The chart represents the land- (or sea)scape and insofar as the features are isomorphic, there is a kind of representational "transparency." The chart in a peculiar way "refers" beyond itself to what it represents.

Now, with respect to the embodiment relations previously traced, such an isomorphic representation is both similar and dissimilar to what would be seen on a larger scale from some observation position (at bird's-eye level). It is similar in that the shapes on the chart are reduced representations of distinctive features that can be directly or technologically mediated in face-to-face or embodied perceptions. The reader can compare these similarities. But chart reading is also different in that, during the act of reading, the perceptual focus is the chart itself, a substitute for the landscape.

I have deliberately used the chart-reading example for several purposes. First, the "textual" isomorphism of a representation allows this first example of hermeneutic technics to remain close to yet differentiated from the perceptual isomorphism that occurs in the optical examples. The difference is at least perceptual in that one sees *through* the optical technology, but now one *sees* the chart as the visual terminus, the "textual" artifact itself.

Something much more dramatic occurs, however, when the representational isomorphism disappears in a printed text. There is no isomorphism between the printed word and what it "represents," although there is some kind of *referential* "transparency" that belongs to this new technologically embodied form of language. It is apparent from the chart example that the chart itself becomes the *object of perception* while simultaneously referring beyond itself to what is not immediately seen. In the case of the printed text, however, the referential transparency is distinctively different from technologically embodied perceptions. *Textual transparency is hermeneutic transparency, not perceptual transparency.*

[. . .]

The movement from embodiment relations to hermeneutic ones can be very gradual, as in the history of writing, with little-noticed differentiations along the human-technology continuum. A series of wide-ranging variants upon readable technologies will establish the point. First, a fairly explicit example of a readable technology: Imagine sitting inside on a cold day. You look out the window and notice that the snow is blowing, but you are toasty warm in front of the fire. You can clearly "see" the cold in Merleau-Ponty's pregnant sense of perception – but you do not actually *feel* it. Of course, you could, were you to go outside. You would then have a full face-to-face verification of what you had seen.

But you might also see the thermometer nailed to the grape arbor post and *read* that it is 28°F. You would now "know" how cold it was, but you still would not feel it. To retain the full sense of an embodiment relation, there must also be retained some isomorphism with the felt sense of the cold – in this case, tactile – that one would get through face-to-face experience. One could invent such a technology; for example, some conductive material could be placed through the wall so that the negative "heat," which is cold, could be felt by hand. But this is not what the thermometer does.

Instead, you read the thermometer, and in the immediacy of your reading you *hermeneutically* know that it is cold. There is an instantaneity to such reading, as it is an already constituted intuition (in phenomenological terms). But you should not fail to note that *perceptually* what you have seen is the dial and the numbers, the thermometer "text." And that text has hermeneutically delivered its "world" reference, the cold.[1]

Such constituted immediacy is not always available. For instance, although I have often enough lived in countries where Centigrade replaces Fahrenheit, I still must translate from my intuitive familiar language to the less familiar one in a deliberate and self-conscious hermeneutic act. Immediacy, however, is not the test for whether the relation is hermeneutic. A hermeneutic relation mimics sensory perception insofar as it is also a kind of seeing as ____; but it is a referential seeing, which has as its immediate perceptual focus seeing the thermometer.

Now let us make the case more complex. In the example cited, the experiencer had both embodiment (seeing the cold) and hermeneutic access to the phenomenon (reading the thermometer). Suppose the house were hermetically sealed, with no windows, and the only access to the weather were through the thermometer (and any other instruments we might include). The hermeneutic character of the relation becomes more obvious. I now clearly have to know how to read the instrumentation and from this reading knowledge get hold of the "world" being referred to.

This example has taken actual shape in nuclear power plants. In the Three Mile Island incident, the nuclear power system was observed only through instrumentation. Part of the delay that caused a near meltdown was *misreadings* of the instruments. There was no face-to-face, independent access to the pile or to much of the machinery involved, nor could there be.

An intentionality analysis of this situation retains the mediational position of the technology:

I-technology-world
(engineer-instruments-pile)

The operator has instruments between him or her and the nuclear pile. But – and here, an essential difference emerges between embodiment and hermeneutic relations – what is immediately perceived is the instrument panel itself. It becomes the object of my microperception, although in the special sense of a hermeneutic transparency, I *read* the pile through it. This situation calls for a different formalization:

I-(technology-world)

The parenthesis now indicates that the immediate *perceptual* focus of my experience *is* the control panel. I read through it, but this reading is now dependent upon the semi-opaque connection between the instruments and the referent object (the pile). This *connection* may now become enigmatic.

In embodiment relations, what allows the partial symbiosis of myself and the technology is the capacity of the technology to become perceptually transparent. In the optical examples, the glass-maker's and lens-grinder's arts must have accomplished this end if the embodied use is to become possible. Enigmas which may occur regarding embodiment-use transparency thus may occur within the parenthesis of the embodiment relation:

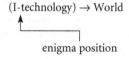

(I-technology) → World

enigma position

(This is not to deny that once the transparency is established, thus making microperception clear, the observer may still fail, particularly at the macroperceptual level. For the moment, however, I shall postpone this type of interpretive problem.) It would be an oversimplification of the history of lens-making were not problems of this sort recognized. Galileo's instrument not only was hard to look through but was good only for certain "middle range" sightings in astronomical terms (it did deliver the planets and even some of their satellites). As telescopes became more powerful, levels, problems with chromatic effects, diffraction effects, etc., occurred. As Ian Hacking has noted,

> Magnification is worthless if it magnifies two distinct dots into one big blur. One needs to resolve the dots into two distinct images. . . . It is a matter of diffraction. The most familiar example of diffraction is the fact that shadows of objects with sharp boundaries are fuzzy. This is a consequence of the wave character of light.[2]

Many such examples may be found in the history of optics, technical problems that had to be solved before there could be any extended reach within embodiment relations. Indeed, many of the barriers in the development of experimental science can be located in just such limitations in instrumental capacity.

Here, however, the task is to locate a parallel difficulty in the emerging new human-technology relation, hermeneutic relations. The location of the technical problem in hermeneutic relations lies in the *connector* between the instrument and the referent. Perceptually, the user's visual (or other) terminus is *upon* the instrumentation itself. To read an instrument is an analogue to reading a text. But if the text does not correctly refer, its reference object or its world cannot be present. Here is a new location for an enigma:

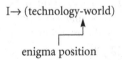

I→ (technology-world)

enigma position

While breakdown may occur at any part of the relation, in order to bring out the graded distinction emerging between embodiment and hermeneutic relations, a short pathology of connectors might be noted.

If there is nothing that impedes my direct perceptual situation with respect to the instrumentation (in the Three Mile Island example, the lights remain on, etc.), interpretive problems in reading a strangely behaving "text" at least occur in the open; but the technical enigma may also occur within the text-referent relation. How

could the operator tell if the instrument was malfunctioning or that to which the instrument refers? Some form of *opacity* can occur within the technology-referent pole of the relation. If there is some independent way of verifying which aspect is malfunctioning (a return to unmediated face-to-face relations), such a breakdown can be easily detected. Both such occurrences are reasons for instrumental redundancy. But in examples where such independent verification is not possible or untimely, the opacity would remain.

Let us take a simple mechanical connection as a borderline case. In shifting gears on my boat, there is a lever in the cockpit that, when pushed forward, engages the forward gear; upward, neutral; and backwards, reverse. Through it, I can ordinarily feel the gear change in the transmission (embodiment) and recognize the simple hermeneutic signification (forward for forward) as immediately intuitive. Once, however, on coming in to the dock at the end of the season, I disengaged the forward gear – and the propeller continued to drive the boat forward. I quickly reversed – and again the boat continued. The hermeneutic significance had failed; and while I also felt a difference in the way the gear lever felt, I did not discover until later that the clasp that retained the lever itself had corroded, thus preventing any actual shifting at all. But even at this level there can be opacity within the technology-object relation.

The purpose of this somewhat premature pathology of human-technology relations is not to cast a negative light upon hermeneutic relations in contrast to embodiment ones but rather to indicate that there are different locations where perceptual and human-technology relations interact. Normally, when the technologies work, the technology-world relation would retain its unique hermeneutic transparency. But if the I-(technology-world) relation is far enough along the continuum to identify the relation as a hermeneutic one, the intersection of perceptual-bodily relations with the technology changes.

Readable technologies call for the extension of my hermeneutic and "linguistic" capacities *through* the instruments, while the reading itself retains its bodily perceptual location as a relation *with* or *towards* the technology. What is emerging here is the first suggestion of an emergence of the technology as "object" but without its negative Heideggerian connotation. Indeed, the type of special capacity as a "text" is a condition for hermeneutic transparency.

The transformation made possible by the hermeneutic relation is a transformation that occurs precisely through *differences* between the text and what is referred to. What is needed is a particular set of textually clear perceptions that "reduce" to that which is immediately readable. To return to the Three Mile Island example, one problem uncovered was that the instrument panel design was itself faulty. It did not incorporate its dials and gauges in an easily readable way. For example, in airplane instrument panel design, much thought has been given to pattern recognition, which occurs as a perceptual gestalt. Thus, in a four-engined aircraft, the four dials indicating r.p.m. will be coordinated so that a single glance will indicate which, if any, engine is out of synchronization. Such technical design accounts for perceptual structures.

There is a second caution concerning the focus upon connectors and pathology. In all the examples I have used to this point, the hermeneutic technics have involved material connections. (The thermometer employs a physical property of a bimetallic spring or mercury in a column; the instrument panel at TMI employs mechanical, electrical, or other material connections; the shift lever, a simple mechanical connection.) If reading does not employ any such material connections, it might seem that its referentiality is essentially different, yet not even all technological connections are strictly material. Photography retains representational isomorphism with the object, yet does not "materially" connect with its object; it is a minimal beginning of action at a distance.

I have been using contemporary or post-scientific examples, but non-material hermeneutic relations do not obtain only for contemporary humans. As existential relations, they are as "old" as post-Garden humanity. Anthropology and the history of religions have long been familiar with a wide variety of shamanistic praxes which fall into the pattern of hermeneutic technics. In what may at first seem a somewhat outrageous set of examples, note the various "reading" techniques employed in shamanism. The reading of animal entrails, of thrown bones, of bodily marks – all are hermeneutic techniques. The patterns of the

entrails, bones, or whatever are taken to *refer* to some state of affairs, instrumentally or textually.

Not only are we here close to a familiar association between magic and the origins of technology suggested by many writers, but we are, in fact, closer to a wider hermeneutic praxis in an intercultural setting. For that reason, the very strangeness of the practice must be critically examined. If the throwing of bones is taken as a "primitive" form of medical diagnosis – which does play a role in shamanism – we might conclude that it is indeed a poor form of hermeneutic relations. What we might miss, however, is that the entire gestalt of what is being diagnosed may differ radically from the other culture and ours.

It may well be that as a focused form of diagnosis upon some particular bodily aliment (appendicitis, for example), the diagnosis will fail. But since one important element in shamanism is a wider diagnosis, used particularly as the occasion of locating certain communal or social problems, it may work better. The sometimes socially contextless emphasis of Western medicine upon a presumably "mechanical" body may overlook precisely the context which the shaman so clearly recognizes. The entire gestalt is different and differently focused, but in both cases there are examples of hermeneutic relations.

In our case, the very success of Western medicine in certain diseases is due to the introduction of technologies into the hermeneutic relation (fever/thermometer; blood pressure/manometer, etc.) The point is that hermeneutic relations are as commonplace in traditional and ancient social groups as in ours, even if they are differently arranged and practiced.

By continuing the intentionality analysis I have been following, one can now see that hermeneutic relations vary the continuum of human-technology-world relations. Hermeneutic relations maintain the general mediation position of technologies within the context of human praxis towards a world, but they also change the variables within the human-technology-world relation. A comparative formalism may be suggestive:

General intentionality relations
Human-technology-world

Variant A: embodiment relations
(I-technology) → world

Variant B: hermeneutic relations
I → (technology-world)

While each component of the relation changes within the correlation, the overall shapes of the variants are distinguishable. Nor are these matters of simply how technologies are experienced.

Another set of examples from the set of optical instruments may illustrate yet another way in which instrumental intentionalities can follow new trajectories. Strictly embodiment relations can be said to work best when there is both a transparency and an isomorphism between perceptual and bodily action within the relation. I have suggested that a trajectory for development in such cases may often be a horizontal one. Such a trajectory not only follows greater and greater degrees of magnification but also entails all the difficulties of a technical nature that go into allowing what is to be seen as though by direct vision. But not all optical technologies follow this strategy. The introduction of hermeneutic possibilities opens the trajectory into what I shall call *vertical* directions, possibilities that rely upon quite deliberate hermeneutic transformations.

It might be said that the telescope and microscope, by extending vision while transforming it, remained *analogue* technologies. The enhancement and magnification made possible by such technologies remain visual and transparent to ordinary vision. The moon remains recognizably the moon, and the microbe – even if its existence was not previously suspected – remains under the microscope a beastie recognized as belonging to the animate continuum. Here, just as the capacity to magnify becomes the foreground phenomenon to the background phenomenon of the reduction necessarily accompanying the magnification, so the similitude of what is seen with ordinary vision remains central to embodiment relations.

Not all optical technologies mediate such perceptions. In gradually moving towards the visual "alphabet" of a hermeneutic relation, deliberate variations may occur which enhance previously undiscernible *differences*:

1) Imagine using spectacles to correct vision, as previously noted. What is wanted is to *return* vision as closely as possible to ordinary perception, not to distort or modify it in any extreme micro- or macroperceptual direction. But now, for snowscapes or sun on the water or desert, we modify the lenses by coloring or polarizing them to cut glare. Such a variation transforms *what* is seen in some degree. Whether we say the polarized lens removes glare or "darkens" the landscape, what is seen is now clearly different from what may be seen through untinted glasses. This difference is a clue which may open a new *telic direction* for development.

2) Now say that somewhere, sometime, someone notes that certain kinds of tinting reveal unexpected results. Such is a much more complex technique now used in infrared satellite photos. (For the moment, I shall ignore the fact that part of this process is a combined embodiment and hermeneutic relation.) If the photo is of the peninsula of Baja California, it will remain recognizable in shape. Geography, whatever depth and height representations, etc., remain but vary in a direction different from any ordinary vision. The infrared photo enhances the difference between vegetation and non-vegetation beyond the limits of any isomorphic color photography. This difference corresponds, in the analogue example, to something like a pictograph. It simultaneously leaves certain analogical structures there and begins to modify the representation into a different, non-perceived "representation."

3) Very sophisticated versions of still representative but non-ordinary forms of visual recognition occur in the new heat-sensitive and light-enhanced technologies employed by the military and police. Night scopes which enhance a person's heat radiation still look like a person but with entirely different regions of what stands out and what recedes. In high-altitude observations, "heat shadows" on the ground can indicate an airplane that has recently had its engines running compared to others which have not. Here visual technologies bring into visibility what was not visible, but in a distinctly now perceivable way.

4) If now one takes a much larger step to spectrographic astronomy, one can see the acceleration of this development. The spectrographic picture of a star no longer "resembles" the star at all. There is no point of light, no disk size, no spatial isomorphism at all – merely a band of differently colored rainbow stripes. The naive reader would not know that this was a picture of a star at all – the reader would have to know the language, the alphabet, that has coded the star. The

astronomer-hermeneut does know the language and "reads" the visual "ABCs" in such a way that he knows the chemical composition of the star, its internal makeup, rather than its shape or external configuration. We are here in the presence of a more fully hermeneutic relation, the star mediated not only instrumentally but in a transformation such that we must now thematically *read* the result. And only the informed reader can do the reading.

There remains, of course, the *reference* to the star. The spectograph is *of* Rigel or *of* Polaris, but the individuality of the star is now made present hermeneutically. Here we have a beginning of a special transformation of perception, a transformation which deliberately enhances differences rather than similarities in order to get at what was previously unperceived.

5) Yet even the spectrograph is but a more radical transformation of perception. It, too, can be transformed by a yet more radical *hermeneutic* analogue to the *digital* transformation which lies embedded in the preferred quantitative praxis of science. The "alphabet" of science is, of course, mathematics, a mathematics that separates itself by yet another hermeneutic step from perception embodied.

There are many ways in which this transformation can and does occur, most of them interestingly involving a particular act of *translation* that often goes unnoticed. To keep the example as simple as possible, let us assume *mechanical* or *electronic* "translation." Suppose our spectrograph is read by a machine that yields not a rainbow spectrum but a set of numbers. Here we would arrive at the final hermeneutic accomplishment, the transformation of even the analogue to a digit. But in the process of hermeneuticization, the "transparency" to the object referred to becomes itself enigmatic. Here more explicit and thematic interpretation must occur.

Hermeneutic relations, particularly those utilizing technologies that permit vertical transformations, move away from perceptual isomorphism. It is the *difference* between what is shown and how something is shown which is informative. In a hermeneutic relation, the world is first transformed into a text, which in turn is read. There is potentially as much flexibility within hermeneutic relations as there is in the various uses of language. Emmanuel Mournier early

recognized just this analogical relationship with language:

> The machine as implement is not a simple material extension of our members. It is of another order, an annex to our language, an auxiliary language to mathematics, a means of penetrating, dissecting and revealing the secret of things, their implicit intentions, their unemployed capacities.[3]

Through hermeneutic relations we can, as it were, *read* ourselves into any possible situation without being there. In science, in contrast to literature, what is important is that the reading retain *some* kind of reference or hermeneutic transparency to what is there. Perhaps that is one reason for the constant desire to reverse what is read back towards what may be perceived. In this reversal, contemporary technologically embodied science has frequently derived what might be called *translation technologies*. I mention two in passing:

(*a*) Digital processes have become *de rigueur* within the perceptual domain. The development of pictures from space probes is such a *double translation* process. The photograph of the surface of Venus is a technological analogue to human vision. At least is a field display of the surface, incorporating the various possible figures and contrasts that would be seen instantaneously in a visual gestalt – but this holistic result cannot be transmitted in this way by the current technologies. Thus it is "translated" into a digital code, which can be transmitted. The "seeing" of the instrument is broken down into a series of digits that are radiographically transmitted to a receiver; then they are reassembled into a spatter pattern and enhanced to reproduce the photograph taken millions of miles away. It would be virtually impossible for anyone to read the digits and tell what was to be seen; only when the linear text of the digits has been retranslated back into the span of an instantaneous visual gestalt can it be seen that the rocks on Venus are or are not like those on the moon. Here the analogues of perception and language are both utilized to extend vision beyond the earth.

(*b*) The same process is used audially in digital recordings. Once again, the double translation process takes place and sound is reduced to digital form, reproduced through the record, and translated back into an auditory gestalt.

Digital and analogue processes blur together in certain configurations. Photos transmitted as points of black on a white ground and reassembled within certain size limits are perceptually gestalted; we see Humphrey Bogart, not simply a mosaic of dots. (Pointillism did the same in painting, although in color. So-called concrete poetry employs the same crossover by placing the words of the poem in a visual pattern so the poem may be both read and seen as a visual pattern.)

Such translation and retranslation processes are clearly transformations from perceptually gestalted phenomena into analogues of writing (serial translation and retranslation processes are clearly transformations from perceptual gestalt phenomena into analogues of writing serial transmissions along a "line," as it were), which are then retranslatable into perceptual gestalts.

I have suggested that the movement from embodiment relations to hermeneutic ones occurs along a human-technology continuum. Just as there are complicated, borderline cases along the continuum from fully haired to bald men, there are the same less-than-dramatic differences here. I have highlighted some of this difference by accenting the bodily-perceptual distinctions that occur between embodiment and hermeneutic relations. This has allowed the difference in perceptual and hermeneutic transparencies to stand out.

There remain two possible confusions that must be clarified before moving to the next step in this phenomenology of technics. First, there is a related sense in which perception and interpretation are intertwined. Perception is primitively already interpretational, in both micro- and macrodimensions. To perceive is already "like" reading. Yet reading is also a specialized act that receives both further definition and elaboration within literate contexts. I have been claiming that one of the distinctive differences between embodiment and hermeneutic relations involves perceptual position, but in the broader sense, interpretation pervades both embodiment and hermeneutic action.

A second and closely related possible confusion entails the double sense in which a technology may be used. It may be used simultaneously both as

something *through* which one experiences and as something *to* which one relates. While this is so, the doubled relation takes shapes in embodiment different from those of hermeneutic relations. Return to the simple embodiment relation illustrated in wearing eyeglasses. *Focally*, my perceptual experience finds its directional aim *through* the lenses, terminating my gaze upon the object of vision; but as a *fringe* phenomenon, I am simultaneously aware of (or can become so) the way my glasses rest upon the bridge of my nose and the tops of my ears. In this fringe sense, I am aware *of* the glasses, but the focal phenomenon is the perceptual transparency that the glasses allow.

In cases of hermeneutic transparency, this doubled role is subtly changed. Now I may carefully read the dials within the core of my visual field and attend to them. But my reading is simultaneously a reading through them, although now the terminus of reference is not necessarily a perceptual object, nor is it, strictly speaking, perceptually present. While the type of transparency is distinct, it remains that the purpose of the reading is to gain hermeneutic transparency.

Both relations, however, at optimum, occur within the familiar acquisitional praxes of the lifeworld. Acute perceptual seeing must be learned and, once acquired, occurs as familiarly as the act of seeing itself. For the accomplished and critical reader, the hermeneutic transparency of some set of instruments is as clear and as immediate as a visual examination of some specimen. The peculiarity of hermeneutic transparency does not lie in either any deliberate or effortful accomplishment of interpretation (although in learning any new text or language, that effort does become apparent). That is why the praxis that grows up within the hermeneutic context retains the same sense of spontaneity that occurs in simple acts of bodily motility. Nevertheless, a more distinctive presence *of* the technology appears in the example. My awareness of the instrument panel is both stronger and centered more focally than the fringe awareness of my eyeglasses frames, and this more distinct awareness is essential to the optimal use of the instrumentation.

In both embodiment and hermeneutic relations, however, the technology remains short of full objectiveness or *otherwise*. It remains the means through which something else is made present. The negative characterization that may occur in breakdown pathologies may return. When the technology in embodiment position breaks down or when the instrumentation in hermeneutic position fails, what remains is an obtruding, and thus negatively derived, object.

Both embodiment and hermeneutic relations, while now distinguished, remain basic existential relations between the human user and the world. There is the danger that my now-constant and selective use of scientific instrumentation could distort the full impact of the existential dimension.

[. . .]

C. Alterity Relations

Beyond hermeneutic relations there lie *alterity relations*. The first suggestions of such relations, which I shall characterize as relations *to* or *with* a technology, have already been suggested in different ways from within the embodiment and hermeneutic contexts. Within embodiment relations, were the technology to intrude upon rather than facilitate one's perceptual and bodily extension into the world, the technology's objectness would necessarily have appeared negatively. Within hermeneutic relations, however, there emerged a certain positivity to the objectness of instrumental technologies. The bodily-perceptual focus *upon* the instrumental text is a condition of its own peculiar hermeneutic transparency. But what of a positive or presentential sense of relations with technologies? In what phenomenological senses can a technology be *other*?

The analysis here may seem strange to anyone limited to the habits of objectivist accounts, for in such accounts technologies as objects usually come first rather than last. The problem for a phenomenological account is that objectivist ones are non-relativistic and thus miss or submerge what is distinctive about human-technology relations.

A naive objectivist account would likely begin with some attempt to circumscribe or define technologies by object characteristics. Then, what I have called the technical properties of technologies would become focal. Some combination of physical and material properties would be

taken to be definitional. (This is an inherent tendency of the standard nomological positions such as those of Bunge and Hacking). The definition will often serve a secondary purpose by being stipulative: only those technologies that are obviously dependent upon or strongly related to contemporary scientific and industrial productive practices will count.

This is not to deny that objectivist accounts have their own distinctive strengths. For example, many such accounts recognize that technological or "artificial" products are different from the simply found object or the natural object. But the submergence of the human-technology relation remains hidden, since either object may enter into praxis and both will have their material, and thus limited, range of technical usability within the relation. Nor is this to deny that the objectivist accounts of types of technologies, types of organization, or types of designed purposes should be considered. But the focus in this first program remains the phenomenological derivation of the set of human-technology relations.

There is a tactic behind my placing alterity relations last in the order of focal human-technology relations. The tactic is designed, on the one side, to circumvent the tendency succumbed to by Heidegger and his more orthodox followers to see the otherness of technology only in negative terms or through negative derivations. The hammer example, which remains paradigmatic for this approach, is one that derives objectness from breakdown. The broken or missing or malfunctioning technology could be *discarded*. From being an obtrusion it could become *junk*. Its objectness would be clear – but only partly so. Junk is not a focal object of use relations (except in certain limited situations). It is more ordinarily a background phenomenon, that which has been put out of use.

Nor, on the other side, do I wish to fall into a naively objectivist account that would simply concentrate upon the material properties of the technology as an object of knowledge. Such an account would submerge the relativity of the intentionality analysis, which I wish to preserve here. What is needed is an analysis of the positive or presentential senses in which humans relate to technologies as relations *to* or *with* technologies, to technology-as-other. It is this sense which is included in the term "alterity."

Philosophically, the term "alterity" is borrowed from Emmanuel Levinas. Although Levinas stands within the traditions of phenomenology and hermeneutics, his distinctive work, *Totality and Infinity*, was "anti-Heideggerian." In that work, the term "alterity" came to mean the radical difference posed to any human by another human, an *other* (and by the ultimately other, God). Extrapolating radically from within the tradition's emphasis upon the non-reducibility of the human to either objectness (in epistemology) or as a means (in ethics), Levinas poses the otherness of humans as a kind of *infinite* difference that is concretely expressed in an ethical, face-to-face encounter.

I shall retain but modify this radical Levinasian sense of human otherness in returning to an analysis of human-technology relations. How and to what extent do technologies become other or, at least, *quasi-other*? At the heart of this question lie a whole series of well-recognized but problematic interpretations of technologies. On the one side lies the familiar problem of anthropomorphism, the personalization of artifacts. This range of anthropomorphism can reach from serious artifact-human analogues to trivial and harmless affections for artifacts.

An instance of the former lies embedded in much AI research. To characterize computer "intelligence" as human-like is to fall into a peculiarly contemporary species of anthropomorphism, however sophisticated. An instance of the latter is to find oneself "fond" of some particular technofact as, for instance, a long-cared-for automobile which one wishes to keep going and which may be characterized by quite deliberate anthropomorphic terms. Similarly, in ancient or non-Western cultures, the role of sacredness attributed to artifacts exemplifies another form of this phenomenon.

The religious object (idol) does not simply "represent" some absent power but is endowed with the sacred. Its aura of sacredness is spatially and temporally present within the range of its efficacy. The tribal devotee will defend, sacrifice to, and care for the sacred artifact. Each of these illustrations contains the seeds of an alterity relation.

A less direct approach to what is distinctive in human-technology alterity relations may perhaps better open the way to a phenomenologically relativistic analysis. My first example comes from a comparison to a technology and to an animal "used" in some practical (although possibly sporting) context: the spirited horse and the spirited sports car.

To ride a spirited horse is to encounter a lively animal *other*. In its pre- or nonhuman context, the horse has a life of its own within the environment that allowed this form of life. Once domesticated, the horse can be "used" as an "instrument" of human praxis – but only to a degree and in a way different from counterpart technologies; in this case, the "spirited" sports car.

There are, of course, analogues which may at first stand out. Both horse and car give the rider/driver a magnified sense of power. The speed and the experience of speed attained in riding/driving are dramatic extensions of my own capacities. Some prominent features of embodiment relations can be found analogously in riding/driving. I experience the trail/road through horse/car and guide/steer the mediating entity under way. But there are equally prominent differences. No matter how well trained, no horse displays the same "obedience" as the car. Take malfunction: in the car, a malfunction "resists" my command – I push the accelerator, and because of a clogged gas line, there is not the response I expected. But the animate resistance of a spirited horse is more than such a mechanical lack of response – the response is more than malfunction, it is *disobedience*. (Most experienced riders, in fact, prefer spirited horses over the more passive ones, which might more nearly approximate a mechanical obedience.) This life of the other in a horse may be carried much further – it may live without me in the proper environment; it does not need the *deistic* intervention of turning the starter to be "animated." The car will not shy at the rabbit springing up in the path any more than most horses will obey the "command" of the driver to hit the stone wall when he is too drunk to notice. The horse, while approximating some features of a mediated embodiment situation, never fully enters such a relation in the way a technology does. Nor does the car ever attain the sense of animation to be found in horseback riding. Yet the analogy is so deeply embedded in our contemporary consciousness (and perhaps the lack of sufficient experience with horses helps) that we might be tempted to emphasize the similarities rather than the differences.

Anthropomorphism regarding the technology on the one side and the contrast with horseback riding on the other point to a first approximation to the unique type of otherness that relations to technologies hold. Technological otherness is a *quasi-otherness*, stronger than mere objectness but weaker than the otherness found within the animal kingdom or the human one; but the phenomenological derivation must center upon the positive experiential aspects outlining this relation.

In yet another familiar phenomenon, we experience technologies as *toys* from childhood. A widely cross-cultural example is the spinning top. Prior to being put into use, the top may appear as a top-heavy object with a certain symmetry of design (even early tops approximate the more purely functional designs of streamlining, etc.), but once "deistically" animated through either stick motion or a string spring, the now spinning top appears to take on a life of its own. On its tip (or "foot") the top appears to defy its top-heaviness and gravity itself. It traces unpredictable patterns along its pathway. It is an object of *fascination*.

Note that once the top has been set to spinning, what was imparted through an embodiment relation now exceeds it. What makes it fascinating is this property of quasi-animation, the life of its own. Also, of course, once "automatic" in its motion, the top's movements may be entered into a whole series of possible contexts. I might enter a game of warring tops in which mine (suitably marked) represents me. If I-as-top am successful in knocking down the other tops, then this game of hermeneutics has the top winning for me. Similarly, if I take its quasi-autonomous motion to be a hermeneutic predictor, I may enter a divination context in which the path traced or the eventual point of stoppage indicates some fortune. Or, entering the region of scientific instrumentation, I may transform the top into a gyroscope, using its constancy of direction within its now-controlled confines as a better-than-magnetic compass. But in each of these cases, the top may become the focal center of attention as a quasi-other to which I may relate. Nor need the object of fascination carry either an embodiment or hermeneutic referential transparency.

To the ancient and contemporary top, compare briefly the fascination that occurs around video games. In the actual use of video games, of course, the embodiment and hermeneutic relational dimensions are present. The joystick that embodies hand and eye coordination skills extends the player into the displayed field. The field itself displays some hermeneutic context (usually either some "invader" mini-world or some sports analogue), but this context does not refer beyond itself into a worldly reference.

[...]

When I first introduced the notion of hermeneutic relations, I employed what could be called a "static" technology: writing. The long and now ancient technologies of writing result in fixed texts (books, manuscripts, etc., all of which, barring decay or destruction, remain stable in themselves). With film, the "text" remains fixed only in the sense that one can repeat, as with a written text, the seeing and hearing of the cinema text. But the mode of presentation is dramatically different. The "characters" are now animate and theatrical, unlike the fixed alphabetical characters of the written text. The dynamic "world" of the cinema-text, while retaining many of the functional features of writing, also now captures the semblance of real-time, action, etc. It remains to be "read" (viewed and heard), but the object-correlate necessarily appears more "life-like" than its analogue – written text. This factor, naively experienced by the current generations of television addicts, is doubtless one aspect in the problems that emerge between television watching habits and the state of reading skills. James Burke has pointed out that "the majority of the people in the advanced industrialized nations spend more time watching television than doing anything else beside work."[4] The same balance of time use also has shown up in surveys regarding students. The hours spent watching television among college and university students, nationally, are equal to or exceed those spent in doing homework or out-of-class preparation.

Film, cinema, or television can, in its hermeneutic dimension, refer in its unique way to a "world." The strong negative response to the Vietnam War was clearly due in part to the virtually unavoidable "presence" of the war in virtually everyone's living room. But films, like readable technologies, are also *presentations*, the focal terminus of a perceptual situation. In that emergent sense, they are more dramatic forms of perceptual immediacy in which the presented display has its own characteristics conveying quasi-alterity. Yet the engagement with the film normally remains short of an engagement with an *other*. Even in the anger that comes through in outrage about civilian atrocities or the pathos experienced in seeing starvation epidemics in Africa, the emotions are not directed to the screen but, indirectly, through it, in more appropriate forms of political or charitable action. To this extent there is retained a hermeneutic reference elsewhere than at the technological instrument. Its quasi-alterity, which is also present, is not fully focal in the case of such media technologies.

A high-technology example of breakdown, however, provides yet another hint at the emergence of alterity phenomena. Word processors have become familiar technologies, often strongly liked by their users (including many philosophers who fondly defend their choices, profess knowledge about the relative abilities of their machines and programs, etc.). Yet in breakdown, this quasi-love relationship reveals its quasi-hate underside as well. Whatever form of "crash" may occur, particularly if some fairly large section of text is involved, it occasions frustration and even rage. Then, too, the programs have their idiosyncrasies, which allow or do not allow certain movements; and another form of human-technology competition may emerge. (Mastery in the highest sense most likely comes from learning to program and thus overwhelm the machine's previous brainpower. "Hacking" becomes the game-like competition in which an entire system is the alterity correlate.) Alterity relations may be noted to emerge in a wide range of computer technologies that, while failing quite strongly to mimic bodily incarnations, nevertheless display a quasi-otherness within the limits of linguistics and, more particularly, of logical behaviors. Ultimately, of course, whatever contest emerges, its sources lie opaquely with other humans as well but also with the transformed technofact, which itself now plays a more obvious role within the overall relational net.

I have suggested that the computer is one of the stronger examples of a technology which may be positioned within alterity relations. But its otherness remains a quasi-otherness, and its genuine usefulness still belongs to the borders of its hermeneutic capacities. Yet in spite of this, the tendency to fantasize its quasi-otherness into an authentic otherness is pervasive. Romanticizations such as the portrayal of the emotive, speaking "Hal" of the movie *2001: A Space Odyssey*, early fears that the "brain power" of computers would soon replace human thinking, fears that political or military decisions will not only be informed by but also made by computers – all are symptoms revolving around the positing of otherness to the technology.

These romanticizations are the alterity counterparts to the previously noted dreams that wish for total embodiment. Were the technofact to be genuinely an other, it would both be and not be a *technology*. But even as quasi-other, the technology falls short of such totalization. It retains its unique role in the human-technology continuum of relations as the medium of transformation, but as a recognizable medium.

The wish-fulfillment desire occasioned by embodiment relations – the desire for a fully transparent technology that would *be* me while at the same time giving me the powers that the use of the technology makes available – here has its counterpart fantasy, and this new fantasy has the same internal contradiction: It both reduces or, here, extrapolates the technology into that which is not a technology (in the first case, the magical transformation is *into* me; in this case, *into the other*), and at the same time, it desires what is not identical with me or the other. The fantasy is for the transformational effects. Both fantasies, in effect, deny technologies playing the roles they do in the human-technology continuum of relations; yet it is only on the condition that there be some detectable differentiation within the relativity that the unique ways in which technologies transform human experience can emerge.

In spite of the temptation to accept the fantasy, what the quasi-otherness of alterity relations does show is that humans may relate positively or presententially *to* technologies. In that respect and to that degree, technologies emerge as focal entities that may receive the multiple attentions humans give the different forms of the other. For this reason, a third formalization may be employed to distinguish this set of relations:

I → technology-(-world)

I have placed the parentheses thusly to indicate that in alterity relations there may be, but need not be, a relation through the technology to the world (although it might well be expected that the *usefulness* of any technology will necessarily entail just such a referentiality). The world, in this case, may remain context and background, and the technology may emerge as the foreground and focal quasi-other with which I momentarily engage.

This disengagement of the technology from its ordinary-use context is also what allows the technology to fall into the various disengaged engagements which constitute such activities as play, art, or sport.

A first phenomenological itinerary through direct and focal human-technology relations may now be considered complete. I have argued that the three sets of distinguishable relations occupy a continuum. At the one extreme lie those relations that approximate technologies to a quasi-me (embodiment relations). Those technologies that I can so take into my experience that through their semi-transparency they allow the world to be made immediate thus enter into the existential relation which constitutes my self. At the other extreme of the continuum lie alterity relations in which the technology becomes quasi-other, or technology "as" other *to* which I relate. Between lies the relation with technologies that both mediate and yet also fulfill my perceptual and bodily relation with technologies, hermeneutic relations. The variants may be formalized thus:

> Human-technology-World Relations
> Variant 1, Embodiment Relations
> (Human-technology) → World
> Variant 2, Hermeneutic Relations
> Human → (technology-World)
> Variant 3, Alterity Relations
> Human → technology-(-World)

Although I have characterized the three types of human-technology relations as belonging to a continuum, there is also a sense in which the elements within each type of relation are differently distributed. There is a *ratio* between the objectness of the technology and its transparency

in use. At the extreme height of embodiment, a background presence of the technology may still be detected. Similarly but with a different ratio, once the technology has emerged as a quasi-other, its alterity remains within the domain of human invention through which the world is reached. Within all the types of relations, technology remains artifactual, but it is also its very artifactual formation which allows the transformations affecting the earth and ourselves.

All the relations examined heretofore have also been focal ones. That is, each of the forms of action that occur through these relations have been marked by an implicated self-awareness. The engagements through, with, and to technologies stand within the very core of praxis. Such an emphasis, while necessary, does not exhaust the role of technologies nor the experiences of them. If focal activities are central and foreground, there are also fringe and background phenomena that are no more neutral than those of the foreground. It is for that reason that one final foray in this phenomenology of technics must be undertaken. That foray must be an examination of technologies in the background and at the horizons of human-technology relations.

D. Background Relations

With background relations, this phenomenological survey turns from attending to technologies in a foreground to those which remain in the background or become a kind of near-technological environment itself. Of course, there are discarded or no-longer-used technologies, which in an extreme sense occupy a background position in human experience – junk. Of these, some may be recuperated into non-use but focal contexts such as in technology museums or in the transformation into junk art. But the analysis here points to specifically functioning technologies which ordinarily occupy background or field positions.

First, let us attend to certain individual technologies designed to function in the background – automatic and semiautomatic machines, which are so pervasive today – as good candidates for this analysis. In the mundane context of the home, lighting, heating, and cooling systems, and the plethora of semiautomatic appliances are good examples. In each case, there is some necessity for an instant of deistic intrusion to program or set the machinery into motion or to its task. I set the thermostat; then, if the machinery is high-tech, the heating/cooling system will operate independently of ongoing action. It may employ time-temperature changes, external sensors to adjust to changing weather, and other cybernetic operations. (While this may function well in the home situation, I remain amused at the still-primitive state of the art in the academic complex I occupy. It takes about two days for the system to adjust to the sudden fall and spring weather changes, thus making offices which actually have opening windows – a rarity – highly desirable.) Once operating, the technology functions as a barely detectable background presence; for example, in the form of background noise, as when the heating kicks in. But in operation, the technology does not call for focal attention.

Note two things about this human-technology relation: First, the machine activity in the role of background presence is not displaying either what I have termed a transparency or an opacity. The "withdrawal" of this technological function is phenomenologically distinct as a kind of "absence." The technology is, as it were, "to the side." Yet as a present absence, it nevertheless becomes part of the experienced field of the inhabitant, a piece of the immediate environment.

Somewhat higher on the scale of semiautomatic technologies are task-oriented appliances that call for explicit and repeated deistic interventions. The washing machine, dryer, microwave, toaster, etc., all call for repeated programming and then for dealing with the processed product (wash, food, etc.). Yet like the more automated systems, the semiautomatic machine remains in the background while functioning.

In both systems and appliances, however, one also may detect clues to the ways in which background relations texture the immediate environment. In the electric home, there is virtually a constant hum of one sort or the other, which is part of the technological texture. Ordinarily, this "white noise" may go unnoticed, although I am always reassured that it remains part of fringe awareness, as when guests visit my mountain home in Vermont. The inevitable comment is about the silence of the woods. At once, the absence of background hum becomes noticeable.

Technological texturing is, of course, much deeper than the layer of background noise which signals its absent presence. Before turning to further implications, one temptation which could occur through the too-narrow selection of contemporary examples must be avoided. It might be thought that only, or predominantly, the high-technology contemporary world uses and experiences technologies as backgrounds. That is not the case, even with respect to automated or semi-automatic technologies.

The scarecrow is an ancient "automated" device. Its mimicry of a human, with clothes flapping in the breeze, is a specifically designed automatic crow scarer, made to operate in the absence of humans. Similarly, in ancient Japan there were automated deer scarers, made of bamboo tubes, pivoted on a pin and placed so that a waterfall or running stream would slowly fill the tube. When it is full enough, the device would trip and its other end strike a sounding board or drum, the noise of which would frighten away any marauding deer. We have already noted the role automation plays in religious rituals (prayer wheels and worship representations thought to function continuously).

Interpreted technologically, there are even some humorous examples of "automation" to be found in ancient religious praxes. The Hindu prayer windmill "automatically" sends its prayers when the wind blows; and in the ancient Sumerian temples there were idols with large eyes at the altars (the gods), and in front of them were smaller, large-eyed human statues representing worshipers. Here was an ancient version of an "automated" worship. (Its contemporary counterpart would be the joke in which the professor leaves his or her lecture on a tape recorder for the class – which students could also "automatically" hear, by leaving their own cassettes to tape the master recording.)

While we do not often conceptualize such ancient devices in this way, part of the purpose of an existential analysis is precisely to take account of the identity of function and of the "ancientness" of all such existential relations. This is in no way to deny the differences of context or the degree of complexity pertaining to the contemporary, as compared to the ancient, versions of automation.

Another form of background relation is associated with various modalities of the technologies that serve to insulate humans from an external environment. Clothing is a borderline case. Clothing clearly insulates our bodies from temperature, wind, and other external weather phenomena that could become dangerous to life; but clothing experienced is borderline with embodiment relations, for we do feel the external environment through clothing, albeit in a particularly damped-down mode. Clothing is not designed, in most cases, to be "transparent" in the way the previous instrument examples were but rather to have a certain opacity without restricting movement. Yet clothing is part of a fringe awareness in most of our daily activities (I am obviously not addressing fashion aspects of clothing here).

A better example of a background relation is a shelter technology. Although shelters may be found (caves) and thus enter untransformed into human praxis, most are constructed, as are most technological artifacts; but once constructed and however designed to insulate or account for external weather, they become a more field-like background phenomenon. Here again, human cultures display an amazing continuum from minimalist to maximalist strategies with respect to this version of a near-background.

Many traditional cultures, particularly in Southern Hemisphere areas, practice an essentially open shelter technology, perhaps with primarily a roof to keep off rain and sun. Such peoples frequently find distasteful such items as windows and, particularly, glassed windows. They do not wish to be too isolated or insulated from the elements. At the other extreme is the maximalist strategy, which most extremely wishes to totalize shelter technology into a virtual life-support system, autonomous and enclosed. I shall call this a technological cocoon.

A contemporary example of a near-cocoon is the nuclear submarine. Its crew lives inside, and the vessel is designed to remain at sea for prolonged periods, even underwater for long stretches of time. There are sophisticated recycling systems for waste, water, and air. Contact with the outside, obviously important in this case, is primarily through monitoring equally sophisticated hermeneutic devices (sonar, low-frequency radio, etc.). All ordinary duties take place in the cocoon-like interior. A multibillion-dollar projection to a greater degree of cocoonhood is the long-term space station now under debate.

Part of the very purpose of the space station is to experiment with creating a mini-environment, or artificial "earth," which would be totally technologically mediated. Yet contemporary high-tech suburban homes show similar features. Fully automated for temperature and humidity, tight air structures, some with glass that adjusts to glare, all such homes lie on the same trajectory of self-containment. But while these illustrations are uniquely high-technology textured, there remain, as before, differently contexted but similar examples from the past.

Totally enclosed spaces have frequently been associated with ritual and religious praxis. The Kiva of past southwestern native American cultures was dug deep into the ground, windowless and virtually sealed. It was the site for important initiatory and secret societies, which gathered into such ancient cocoons for their own purposes. The enclosure bespeaks different kinds of totalization.

What is common to the entire range of examples pointed to here is the position occupied by such technology, background position, the position of an absent presence as a part of or a total field of immediate technology.

In each of the examples, the background role is a field one, not usually occupying focal attention but nevertheless conditioning the context in which the inhabitant lives. There are, of course, great differences to be detailed in terms of the types of contexts which such background technologies play. Breakdown, again, can play a significant indexical role in pointing out such differences.

The involvement implications of contemporary, high-technology society are very complex and often so interlocked as to fall into major disruption when background technology fails. In 1985 Long Island was swept by Hurricane Gloria with massive destruction of power lines. Most areas went without electricity for at least a week, and in some cases, two. Lighting had to be replaced by older technologies (lanterns, candles, kerosene lamps), supplies for which became short immediately. My own suspicion is that a look at birth statistics at the proper time after this radical change in evening habits will reveal the same glitch which actually did occur during the blackouts of earlier years in New York.

Similarly, with the failure of refrigeration, eating habits had to change temporarily. The example could be expanded quite indefinitely; a mass purchase of large generators by university buyers kept a Minnesota company in full production for several months after, to be prepared the "next time." In contrast, while the same effects on a shorter-term basis were experienced in the grid-wide blackouts of 1965, I was in Vermont at my summer home, which is lighted by kerosene lamps and even refrigerated with a kerosene refrigerator. I was simply unaware of the massive disruption until the Sunday *Times* arrived. Here is a difference between an older, loose-knit and a contemporary, tight-knit system.

Despite their position as field or background relations, technologies here display many of the same transformational characteristics found in the previous explicit focal relations. Different technologies texture environments differently. They exhibit unique forms of non-neutrality through the different ways in which they are interlinked with the human lifeworld. Background technologies, no less than focal ones, transform the gestalts of human experience and, precisely because they are absent presences, may exert more subtle indirect effects upon the way

a world is experienced. There are also involvements both with wider circles of connection and amplification/reduction selectivities that may be discovered in the roles of background relations; and finally, the variety of minimalist to maximalist strategies remains as open to this dimension of human-technology relations as each of the others.

Notes

1 This illustration is my version of a similar one developed by Patrick Heelan in his more totally hermeneuticized notion of perception in *Space Perception and the Philosophy of Science* (Berkeley: University of California Press, 1989).

2 Ian Hacking, *Representing and Intervening* (Cambridge: Cambridge University Press, 1983), p. 195. Hacking develops a very excellent and suggestive history of the use of microscopes. His focus, however, is upon the technical properties that were resolved before microscopes could be useful in the sciences. He and Heelan, however, along with Robert Ackermann, have been among the pioneers dealing with perception and instrumentation in instruments. Cf. also my *Technics and Praxis* (Dordrecht: Reidel Publishers, 1979).

3 Emmanuel Mournier, *Be Not Afraid*, trans. Cynthia Rowland (London: Rockcliffe, 1951), p. 195.

4 James Burke, *Connections. Alternative History of Technology*, New York: MacMillan, 1978, p. 5.

Part IV
Critical Theory

Part IV

Critical Theory

13

The New Forms of Control

Herbert Marcuse

Marcuse (1898–1979) was a leading advocate of Critical Theory, and along with Theodor Adorno and Max Horkheimer, an early member of the Frankfurt School. He wrote his *Habilitation* with Martin Heidegger, and the influence of Heidegger's concern with essences is present throughout Marcuse's work. However, Marcuse's work takes a materialist turn, and he developed one of the most insightful critiques of mid-twentieth century industrial and consumer society.

The key method of Critical Theory is immanent critique. Critical theory as immanent critique focuses on the internal tensions of the theory or social form under analysis. Using immanent critique, critical theorists identify the internal contradictions of society and of thought, with the aim of analyzing and identifying (i) prospects for progressive social change and (ii) those structures of society and consciousness that contribute to human domination. When applying immanent critique to science and technology, Critical Theorists identify both the oppressive and the libratory potentials.

Regarding science and technology, all critical theorists hold that science and technology are intertwined into a single complex, or realm of human activity, that we can call technoscience. Further, they believe that technoscience is not neutral with respect to human values, but rather it creates and bears value. They agree that the tools we use shape our way of life in modern societies where technoscience has become all-pervasive. Hence, how we do things determines who and what we are, and technological development transforms what it is to be human. Critical theorists agree that the apparently neutral formulations of science and technology often hide oppressive or repressive interests. They differ, however, in their ideas about whether technoscience is of necessity a force of dehumanization, and if it is not, why and how it might be a force for greater freedom.

From Herbert Marcuse, *One Dimensional Man* (Boston: Beacon Press, 1964), pp. 1–18.

For Theodor Adorno and Max Horkheimer, technoscientific development brings with it increasing dehumanization. Modern institutions and ideas, including transnational organizations and democracy, are shaped and guided by instrumental rationality, and exist primarily to preserve their own continuation. It is no longer possible to ask about, or critically evaluate, ends; they are taken for granted. Because only questions about means can be considered by instrumental rationality, questions about ends are now considered irrational. So, the progress of Enlightenment reason, restricted to instrumental rationality, contradicts the very goal sought by the Enlightenment – the increasing liberation of human beings. And, modern technoscience that should contribute to greater human freedom increasingly becomes a cage of our own making.

Marcuse rejects this critique of technoscience insofar as it suggests that science and technology are necessarily oppressive. Instead, he argues that technoscience is oppressive under capitalism, but might be otherwise under a different social order, and hence might embody different values. His *One Dimensional Man* is one of the central works in Critical Theory of technology. Here, Marcuse argues that advanced industrial societies create false needs that integrate individuals into the existing system of production and consumption via mass media, advertising, industrial management, and scientific-technological modes of thought. Technological rationality homogenizes nature and people into neutral objects of manipulation. The result is a "one-dimensional" universe of thought and behavior in which our capacity for critical thinking and oppositional behavior withers away. Marcuse's view shares many similarities with that of Ellul in understanding technology as absorbing religion, politics, the economy, and culture, and in thinking that technological rationality obeys its own internal logic. Like Heidegger, Marcuse views technology not as some particular thing or system, but as an all-encompassing logic that shapes also how we think of ourselves. Unlike these thinkers, Marcuse is more sanguine about the possibilities of change. The rapid pace of change of a capitalist-technological order will always displace many, and those who are displaced or excluded, those persons at the margins (including those who for intellectual or aesthetic reasons choose to be there) are repositories of alternatives and oppositions to a one-dimensional society.

A comfortable, smooth, reasonable, democratic unfreedom prevails in advanced industrial civilization, a token of technical progress. Indeed, what could be more rational than the suppression of individuality in the mechanization of socially necessary but painful performances; the concentration of individual enterprises in more effective, more productive corporations; the regulation of free competition among unequally equipped economic subjects; the curtailment of prerogatives and national sovereignties which impede the international organization of resources. That this technological order also involves a political and intellectual coordination may be a regrettable and yet promising development.

The rights and liberties which were such vital factors in the origins and earlier stages of industrial society yield to a higher stage of this society: they are losing their traditional rationale and content. Freedom of thought, speech, and conscience were – just as free enterprise, which they served to promote and protect – essentially *critical* ideas, designed to replace an obsolescent material and intellectual culture by a more productive and rational one. Once institutionalized, these rights and liberties shared the fate of the society of which they had become an integral part. The achievement cancels the premises.

To the degree to which freedom from want, the concrete substance of all freedom, is becoming a real possibility, the liberties which pertain to a state of lower productivity are losing their former content. Independence of thought, autonomy, and the right to political opposition are being deprived of their basic critical function in a society which seems increasingly capable of satisfying the needs of the individuals through the way in which it is organized. Such a society may justly demand acceptance of its principles and institutions, and reduce the opposition to the discussion and promotion of alternative policies *within* the status quo. In this respect, it seems to make little difference whether the increasing satisfaction of needs is accomplished by an authoritarian or a non-authoritarian system. Under the conditions of a rising standard of living, non-conformity with the system itself appears to be socially useless, and the more so when it entails tangible economic and political disadvantages and threatens the smooth operation of the whole. Indeed, at least in so far as the necessities of life are involved, there seems to be no reason why the production and distribution of goods and services should proceed through the competitive concurrence of individual liberties.

Freedom of enterprise was from the beginning not altogether a blessing. As the liberty to work or to starve, it spelled toil, insecurity, and fear for the vast majority of the population. If the individual were no longer compelled to prove himself on the market, as a free economic subject, the disappearance of this kind of freedom would be one of the greatest achievements of civilization. The technological processes of mechanization and standardization might release individual energy into a yet uncharted realm of freedom beyond necessity. The very structure of human existence would be altered; the individual would

be liberated from the work world's imposing upon him alien needs and alien possibilities. The individual would be free to exert autonomy over a life that would be his own. If the productive apparatus could be organized and directed toward the satisfaction of the vital needs, its control might well be centralized; such control would not prevent individual autonomy, but render it possible.

This is a goal within the capabilities of advanced industrial civilization, the "end" of technological rationality. In actual fact, however, the contrary trend operates: the apparatus imposes its economic and political requirements for defense and expansion on labor time and free time, on the material and intellectual culture. By virtue of the way it has organized its technological base, contemporary industrial society tends to be totalitarian. For "totalitarian" is not only a terroristic political coordination of society, but also a non-terroristic economic-technical coordination which operates through the manipulation of needs by vested interests. It thus precludes the emergence of an effective opposition against the whole. Not only a specific form of government or party rule makes for totalitarianism, but also a specific system of production and distribution which may well be compatible with a "pluralism" of parties, newspapers, "countervailing powers," etc.

Today political power asserts itself through its power over the machine process and over the technical organization of the apparatus. The government of advanced and advancing industrial societies can maintain and secure itself only when it succeeds in mobilizing, organizing, and exploiting the technical, scientific, and mechanical productivity available to industrial civilization. And this productivity mobilizes society as a whole, above and beyond any particular individual or group interests. The brute fact that the machine's physical (only physical?) power surpasses that of the individual, and of any particular group of individuals, makes the machine the most effective political instrument in any society whose basic organization is that of the machine process. But the political trend may be reversed; essentially the power of the machine is only the stored-up and projected power of man. To the extent to which the work world is conceived of as a machine and mechanized accordingly, it becomes the *potential* basis of a new freedom for man.

Contemporary industrial civilization demonstrates that it has reached the stage at which "the free society" can no longer be adequately defined in the traditional terms of economic, political, and intellectual liberties, not because these liberties have become insignificant, but because they are too significant to be confined within the traditional forms. New modes of realization are needed, corresponding to the new capabilities of society.

Such new modes can be indicated only in negative terms because they would amount to the negation of the prevailing modes. Thus economic freedom would mean freedom *from* the economy – from being controlled by economic forces and relationships; freedom from the daily struggle for existence, from earning a living. Political freedom would mean liberation of the individuals *from* politics over which they have no effective control. Similarly, intellectual freedom would mean the restoration of individual thought now absorbed by mass communication and indoctrination, abolition of "public opinion" together with its makers. The unrealistic sound of these propositions is indicative, not of their utopian character, but of the strength of the forces which prevent their realization. The most effective and enduring form of warfare against liberation is the implanting of material and intellectual needs that perpetuate obsolete forms of the struggle for existence.

The intensity, the satisfaction and even the character of human needs, beyond the biological level, have always been preconditioned. Whether or not the possibility of doing or leaving, enjoying or destroying, possessing or rejecting something is seized as a *need* depends on whether or not it can be seen as desirable and necessary for the prevailing societal institutions and interests. In this sense, human needs are historical needs and, to the extent to which the society demands the repressive development of the individual, his needs themselves and their claim for satisfaction are subject to overriding critical standards.

We may distinguish both true and false needs. "False" are those which are superimposed upon the individual by particular social interests in his repression: the needs which perpetuate toil, aggressiveness, misery, and injustice. Their satisfaction might be most gratifying to the individual, but this happiness is not a condition which has to be maintained and protected if it serves to arrest the development of the ability (his own and others) to recognize the disease of the whole and grasp the chances of curing the disease. The result then is euphoria in unhappiness. Most of the prevailing needs to relax, to have fun, to behave and consume in accordance with the advertisements, to love and hate what others love and hate, belong to this category of false needs.

Such needs have a societal content and function which are determined by external powers over which the individual has no control; the development and satisfaction of these needs is heteronomous. No matter how much such needs may have become the individual's own, reproduced and fortified by the conditions of his existence; no matter how much he identifies himself with them and finds himself in their satisfaction, they continue to be what they were from the beginning – products of a society whose dominant interest demands repression.

The prevalence of repressive needs is an accomplished fact, accepted in ignorance and defeat, but a fact that must be undone in the interest of the happy individual as well as all those whose misery is the price of his satisfaction. The only needs that have an unqualified claim for satisfaction are the vital ones – nourishment, clothing, lodging at the attainable level of culture. The satisfaction of these needs is the prerequisite for the realization of *all* needs, of the unsublimated as well as the sublimated ones.

For any consciousness and conscience, for any experience which does not accept the prevailing societal interest as the supreme law of thought and behavior, the established universe of needs and satisfactions is a fact to be questioned – questioned in terms of truth and falsehood. These terms are historical throughout, and their objectivity is historical. The judgment of needs and their satisfaction, under the given conditions, involves standards of *priority* – standards which refer to the optimal development of the individual, of all individuals, under the optimal utilization of the material and intellectual resources available to man. The resources are calculable. "Truth" and "falsehood" of needs designate objective conditions to the extent to which the universal satisfaction of vital needs and, beyond it, the progressive alleviation of toil and poverty, are universally valid standards. But as historical standards, they do not only vary according to area and stage of development, they also can be defined only in (greater or lesser) *contradiction* to the prevailing ones. What tribunal can possibly claim the authority of decision?

In the last analysis, the question of what are true and false needs must be answered by the individuals themselves, but only in the last analysis; that is, if and when they are free to give their own answer. As long as they are kept incapable of being autonomous, as long as they are indoctrinated and manipulated (down to their very instincts), their answer to this question cannot be taken as their own. By the same token, however, no tribunal can justly arrogate to itself the right to decide which needs should be developed and satisfied. Any such tribunal is reprehensible, although our revulsion does not do away with the question: how can the people who have been the object of effective and productive domination by themselves create the conditions of freedom?

The more rational, productive, technical, and total the repressive administration of society becomes, the more unimaginable the means and ways by which the administered individuals might break their servitude and seize their own liberation. To be sure, to impose Reason upon an entire society is a paradoxical and scandalous idea – although one might dispute the righteousness of a society which ridicules this idea while making its own population into objects of total administration. All liberation depends on the consciousness of servitude, and the emergence of this consciousness is always hampered by the predominance of needs and satisfactions which, to a great extent, have become the individual's own. The process always replaces one system of preconditioning by another; the optimal goal is the replacement of false needs by true ones, the abandonment of repressive satisfaction.

The distinguishing feature of advanced industrial society is its effective suffocation of those needs which demand liberation – liberation also from that which is tolerable and rewarding and comfortable – while it sustains and absolves the destructive power and repressive function of the affluent society. Here, the social controls exact the overwhelming need for the production and consumption of waste; the need for stupefying work where it is no longer a real necessity; the need for modes of relaxation which soothe and prolong this stupefication; the need for maintaining such deceptive liberties as free competition at administered prices, a free press which censors itself, free choice between brands and gadgets.

Under the rule of a repressive whole, liberty can be made into a powerful instrument of domination. The range of choice open to the individual is not the decisive factor in determining the degree of human freedom, but *what* can be chosen and what *is* chosen by the individual. The criterion for free choice can never be an absolute one, but neither is it entirely relative. Free election of masters does not abolish the masters or the slaves. Free choice among a wide variety of goods and services does not signify freedom if these goods and services sustain social controls over a life of toil and fear – that is, if they sustain alienation. And the spontaneous reproduction of superimposed needs by the individual does not establish autonomy; it only testifies to the efficacy of the controls.

Our insistence on the depth and efficacy of these controls is open to the objection that we overrate greatly the indoctrinating power of the "media," and that by themselves the people would feel and satisfy the needs which are now imposed upon them. The objection misses the point. The preconditioning does not start with the mass production of radio and television and with the centralization of their control. The people enter this stage as preconditioned receptacles of long standing; the decisive difference is in the flattening out of the contrast (or conflict) between the given and the possible, between the satisfied and the unsatisfied needs. Here, the so-called equalization of class distinctions reveals its ideological function. If the worker and his boss enjoy the same television program and visit the same resort places, if the typist is as attractively made up as the daughter of her employer, if the Negro owns a Cadillac, if they all read the same newspaper, then this assimilation indicates not the disappearance of classes, but the extent to which the needs and satisfactions that serve the preservation of the Establishment are shared by the underlying population.

Indeed, in the most highly developed areas of contemporary society, the transplantation of social into individual needs is so effective that the difference between them seems to be purely theoretical. Can one really distinguish between the mass media as instruments of information and entertainment, and as agents of manipulation and indoctrination? Between the automobile as nuisance and as convenience? Between the horrors and the comforts of functional architecture? Between the work for national defense and the work for corporate gain? Between the private pleasure and the commercial and political utility involved in increasing the birth rate?

We are again confronted with one of the most vexing aspects of advanced industrial civilization: the rational character of its irrationality. Its productivity and efficiency, its capacity to increase and spread comforts, to turn waste into need, and destruction into construction, the extent to which this civilization transforms the object world into an extension of man's mind and body makes the very notion of alienation questionable. The people recognize themselves in their commodities; they find their soul in their automobile, hi-fi set, split-level home, kitchen equipment. The very mechanism which ties the individual to his society has changed, and social control is anchored in the new needs which it has produced.

The prevailing forms of social control are technological in a new sense. To be sure, the technical structure and efficacy of the productive and destructive apparatus have been a major instrumentality for subjecting the population to the established social division of labor throughout the modern period. Moreover, such integration has always been accompanied by more obvious forms of compulsion: loss of livelihood, the administration of justice, the police, the armed forces. It still is. But in the contemporary period, the technological controls appear to be the very embodiment of Reason for the benefit of all social groups and interests – to such an extent that all contradiction seems irrational and all counteraction impossible.

No wonder then that, in the most advanced areas of this civilization, the social controls have been introjected to the point where even individual protest is affected at its roots. The intellectual and emotional refusal "to go along" appears neurotic and impotent. This is the socio-psychological aspect of the political event that marks the contemporary period: the passing of the historical forces which, at the preceding stage of industrial society, seemed to represent the possibility of new forms of existence.

But the term "introjection" perhaps no longer describes the way in which the individual by himself reproduces and perpetuates the external controls exercised by his society. Introjection suggests a variety of relatively spontaneous processes by which a Self (Ego) transposes the "outer" into the "inner." Thus introjection implies the existence of an inner dimension distinguished from and even antagonistic to the external exigencies – an individual consciousness and an individual unconscious *apart from* public opinion and behavior.[1] The idea of "inner freedom" here has its reality: it designates the private space in which man may become and remain "himself."

Today this private space has been invaded and whittled down by technological reality. Mass production and mass distribution claim the *entire* individual, and industrial psychology has long since ceased to be confined to the factory. The manifold processes of introjection seem to be ossified in almost mechanical reactions. The result is, not adjustment but *mimesis*: an immediate identification of the individual with *his* society and, through it, with the society as a whole.

This immediate, automatic identification (which may have been characteristic of primitive forms of association) reappears in high industrial civilization; its new "immediacy," however, is the product of a sophisticated, scientific management and organization. In this process, the "inner" dimension of the mind in which opposition to the status quo can take root is whittled down. The loss of this dimension, in which the power of negative thinking – the critical power of Reason – is at home, is the ideological counterpart to the very material process in which advanced industrial society silences and reconciles the opposition. The impact of progress turns Reason into submission to the facts of life, and to the dynamic capability of producing more and bigger facts of the same sort of life. The efficiency of the system blunts the individuals' recognition that it contains no facts which do not communicate the repressive power of the whole. If the individuals find themselves in the things which shape their life, they do so, not by giving, but by accepting the law of things – not the law of physics but the law of their society.

I have just suggested that the concept of alienation seems to become questionable when the individuals identify themselves with the existence which is imposed upon them and have in it their own development and satisfaction. This identification is not illusion but reality. However, the reality constitutes a more progressive stage of alienation. The latter has become entirely objective; the subject which is alienated is swallowed up by its alienated existence. There is only one dimension, and it is everywhere and in all forms. The achievements of progress defy ideological indictment as well as justification; before their tribunal, the "false consciousness" of their rationality becomes the true consciousness.

This absorption of ideology into reality does not, however, signify the "end of ideology." On the contrary, in a specific sense advanced industrial culture is *more* ideological than its predecessor, inasmuch as today the ideology is in the process of production itself.[2] In a provocative form, this proposition reveals the political aspects of the prevailing technological rationality. The productive apparatus and the goods and services which it produces "sell" or impose the social system as a whole. The means of mass transportation and communication, the commodities of lodging, food, and clothing, the irresistible output of the entertainment and information industry carry with them prescribed attitudes and habits, certain intellectual and emotional reactions which bind the consumers more or less pleasantly to the producers and, through the latter, to the whole. The products indoctrinate and manipulate; they promote a false consciousness which is immune against its falsehood. And as these beneficial products become available to more individuals in more social classes, the indoctrination they carry ceases to be publicity; it becomes a way of life. It is a good way of life – much better than before – and as a good way of life, it militates against qualitative change. Thus emerges a pattern of *one-dimensional thought and behavior* in which

ideas, aspirations, and objectives that, by their content, transcend the established universe of discourse and action are either repelled or reduced to terms of this universe. They are redefined by the rationality of the given system and of its quantitative extension.

The trend may be related to a development in scientific method: operationalism in the physical, behaviorism in the social sciences. The common feature is a total empiricism in the treatment of concepts; their meaning is restricted to the representation of particular operations and behavior. The operational point of view is well illustrated by P. W. Bridgman's analysis of the concept of length:[3]

> We evidently know what we mean by length if we can tell what the length of any and every object is, and for the physicist nothing more is required. To find the length of an object, we have to perform certain physical operations. The concept of length is therefore fixed when the operations by which length is measured are fixed: that is, the concept of length involves as much and nothing more than the set of operations by which length is determined. In general, we mean by any concept nothing more than a set of operations; *the concept is synonymous with the corresponding set of operations.*

Bridgman has seen the wide implications of this mode of thought for the society at large:[4]

> To adopt the operational point of view involves much more than a mere restriction of the sense in which we understand 'concept,' but means a far-reaching change in all our habits of thought, in that we shall no longer permit ourselves to use as tools in our thinking concepts of which we cannot give an adequate account in terms of operations.

Bridgman's prediction has come true. The new mode of thought is today the predominant tendency in philosophy, psychology, sociology, and other fields. Many of the most seriously troublesome concepts are being "eliminated" by showing that no adequate account of them in terms of operations or behavior can be given. The radical empiricist onslaught thus provides the methodological justification for the debunking of the mind by the intellectuals – a positivism which, in its denial of the transcending elements of Reason, forms the academic counterpart of the socially required behavior.

Outside the academic establishment, the "far-reaching change in all our habits of thought" is more serious. It serves to coordinate ideas and goals with those exacted by the prevailing system, to enclose them in the system, and to repel those which are irreconcilable with the system. The reign of such a one-dimensional reality does not mean that materialism rules, and that the spiritual, metaphysical, and bohemian occupations are petering out. On the contrary, there is a great deal of "Worship together this week," "Why not try God," Zen, existentialism, and beat ways of life, etc. But such modes of protest and transcendence are no longer contradictory to the status quo and no longer negative. They are rather the ceremonial part of practical behaviorism, its harmless negation, and are quickly digested by the status quo as part of its healthy diet.

One-dimensional thought is systematically promoted by the makers of politics and their purveyors of mass information. Their universe of discourse is populated by self-validating hypotheses which, incessantly and monopolistically repeated, become hypnotic definitions or dictations. For example, "free" are the institutions which operate (and are operated on) in the countries of the Free World; other transcending modes of freedom are by definition either anarchism, communism, or propaganda. "Socialistic" are all encroachments on private enterprises not undertaken by private enterprise itself (or by government contracts), such as universal and comprehensive health insurance, or the protection of nature from all too sweeping commercialization, or the establishment of public services which may hurt private profit. This totalitarian logic of accomplished facts has its Eastern counterpart. There, freedom is the way of life instituted by a communist regime, and all other transcending modes of freedom are either capitalistic, or revisionist, or leftist sectarianism. In both camps, non-operational ideas are non-behavioral and sub-

versive. The movement of thought is stopped at barriers which appear as the limits of Reason itself.

Such limitation of thought is certainly not new. Ascending modern rationalism, in its speculative as well as empirical form, shows a striking contrast between extreme critical radicalism in scientific and philosophic method on the one hand, and an uncritical quietism in the attitude toward established and functioning social institutions. Thus Descartes' *ego cogitans* was to leave the "great public bodies" untouched, and Hobbes held that "the present ought always to be preferred, maintained, and accounted best." Kant agreed with Locke in justifying revolution *if and when* it has succeeded in organizing the whole and in preventing subversion.

However, these accommodating concepts of Reason were always contradicted by the evident misery and injustice of the "great public bodies" and the effective, more or less conscious rebellion against them. Societal conditions existed which provoked and permitted real dissociation from the established state of affairs; a private as well as political dimension was present in which dissociation could develop into effective opposition, testing its strength and the validity of its objectives.

With the gradual closing of this dimension by the society, the self-limitation of thought assumes a larger significance. The interrelation between scientific-philosophical and societal processes, between theoretical and practical Reason, asserts itself "behind the back" of the scientists and philosophers. The society bars a whole type of oppositional operations and behavior; consequently, the concepts pertaining to them are rendered illusory or meaningless. Historical transcendence appears as metaphysical transcendence, not acceptable to science and scientific thought. The operational and behavioral point of view, practiced as a "habit of thought" at large, becomes the view of the established universe of discourse and action, needs and aspirations. The "cunning of Reason" works, as it so often did, in the interest of the powers that be. The insistence on operational and behavioral concepts turns against the efforts to free thought and behavior *from* the given reality and *for* the suppressed alternatives. Theoretical and practical Reason, academic and social behaviorism meet on common ground: that of an advanced society which makes scientific and technical progress into an instrument of domination.

"Progress" is not a neutral term; it moves toward specific ends, and these ends are defined by the possibilities of ameliorating the human condition. Advanced industrial society is approaching the stage where continued progress would demand the radical subversion of the prevailing direction and organization of progress. This stage would be reached when material production (including the necessary services) becomes automated to the extent that all vital needs can be satisfied while necessary labor time is reduced to marginal time. From this point on, technical progress would transcend the realm of necessity, where it served as the instrument of domination and exploitation which thereby limited its rationality; technology would become subject to the free play of faculties in the struggle for the pacification of nature and of society.

Such a state is envisioned in Marx's notion of the "abolition of labor." The term "pacification of existence" seems better suited to designate the historical alternative of a world which – through an international conflict which transforms and suspends the contradictions within the established societies – advances on the brink of a global war. "Pacification of existence" means the development of man's struggle with man and with nature, under conditions where the competing needs, desires, and aspirations are no longer organized by vested interests in domination and scarcity – an organization which perpetuates the destructive forms of this struggle.

Today's fight against this historical alternative finds a firm mass basis in the underlying population, and finds its ideology in the rigid orientation of thought and behavior to the given universe of facts. Validated by the accomplishments of science and technology, justified by its growing productivity, the status quo defies all transcendence. Faced with the possibility of pacification on the grounds of its technical and intellectual achievements, the mature industrial society closes itself against this alternative. Operationalism, in theory and practice, becomes the theory and practice of *containment*. Underneath its obvious dynamics, this society is a thoroughly static system of life: self-propelling in its oppressive productivity and in its beneficial coordination. Containment of technical progress goes hand in hand with its growth in the established direction. In spite of the political fetters imposed by the status quo, the more technology appears capable of creating the conditions for pacification, the more are the minds and bodies of man organized against this alternative.

The most advanced areas of industrial society exhibit throughout these two features: a trend toward consummation of technological rationality, and intensive efforts to contain this trend within the established institutions. Here is the internal contradiction of this civilization: the irrational element in its rationality. It is the token of its achievements. The industrial society which makes technology and science its own is organized for the ever-more-effective domination of man and nature, for the ever-more-effective utilization of its resources. It becomes irrational when the success of these efforts opens new dimensions of human realization. Organization for peace is different from organization for war; the institutions which served the struggle for existence cannot serve the pacification of existence. Life as an end is qualitatively different from life as a means.

Such a qualitatively new mode of existence can never be envisaged as the mere by-product of economic and political changes, as the more or less spontaneous effect of the new institutions which constitute the necessary prerequisite. Qualitative change also involves a change in the *technical* basis on which this society rests – one which sustains the economic and political institutions through which the "second nature" of man as an aggressive object of administration is stabilized. The techniques of industrialization are political techniques; as such, they prejudge the possibilities of Reason and Freedom.

To be sure, labor must precede the reduction of labor, and industrialization must precede the development of human needs and satisfactions. But as all freedom depends on the conquest of alien necessity, the realization of freedom depends on the *techniques* of this conquest. The highest productivity of labor can be used for the perpetuation of labor, and the most efficient industrialization can serve the restriction and manipulation of needs.

When this point is reached, domination – in the guise of affluence and liberty – extends to all spheres of private and public existence, integrates all authentic opposition, absorbs all alternatives. Technological rationality reveals its political character as it becomes the great vehicle of better domination, creating a truly totalitarian universe in which society and nature, mind and body are kept in a state of permanent mobilization for the defense of this universe.

Notes

1 The change in the function of the family here plays a decisive role: its "socializing" functions are increasingly taken over by outside groups and media. See my *Eros and Civilization* (Boston: Beacon Press, 1955), p. 96 ff.

2 Theodor W. Adorno, *Prismen. Kulturkritik und Gesellschaft.* (Frankfurt: Suhrkamp, 1955), p. 24 f.

3 P. W. Bridgman, *The Logic of Modern Physics* (New York: Macmillan, 1928), p. 5. The operational doctrine has since been refined and qualified. Bridgman himself has extended the concept of "operation" to include the "paper-and-pencil" operations of the theorist (in Philipp J. Frank, *The Validation of Scientific Theories* [Boston: Beacon Press, 1954], Chap. II). The main impetus remains the same: it is "desirable" that the paper-and-pencil operations "be capable of eventual contact, although perhaps indirectly, with instrumental operations."

4 P. W. Bridgman, *The Logic of Modern Physics*, loc. cit., p. 31.

14

Technical Progress and the Social Life-World

Jürgen Habermas

Habermas is the most important contemporary representative of Critical Theory, and one of the more important thinkers of the latter part of the twentieth century. A student of Adorno's, and later director of the Institute for Social Research in Frankfurt, Habermas has devoted his work to understanding and advocating for the continued relevance of Enlightenment ideals along with the importance and possibilities of democracy. Technoscience presents both possibilities and threats to both of these. This selection offers a response to Marcuse. Habermas criticizes Marcuse's position as hopeless romanticism, and one that dangerously will restrict the careful use of instrumental reasoning in the areas where it is appropriate to use it.

Habermas also rejects the cynicism of Adorno and Horkheimer, which is often present in philosophy of technology. Habermas agrees that technology is a form of instrumental reason, but disagrees with the claim that it will subsume all forms of thought and action. Drawing on T. H. Huxley's distinction between a domain of literature and a domain of science, he distinguishes between a legitimate realm of technical control and the "social life-world" which is free, open, and communicative. Huxley argued that progress would occur once science and literature are newly assimilated to each other. C. P. Snow, making a similar distinction between a culture of science and a culture of the humanities, suggests that progress demands the primacy of a scientific worldview. Habermas rejects both of these positions. Science and technology appropriately operate according to their own internal principles and goals; the same is true for literature and the humanities. Neither, he claims, should have priority over the other; each represents a human interest. Technology has merged with science, capitalism, and bureaucracy, forming a new "technocratic consciousness" that has become the leading productive

From Jürgen Habermas, *Toward a Rational Society*, trans. Jeremy Shapiro (Boston: Beacon Press, 1970), pp. 50–61.

force in contemporary life. The trend noted by Marcuse, Ellul, Ortega, and others, toward treating technology as a second nature and allowing technological values of order, quantification, and efficiency to trump all other concerns, is a very real threat to other human values (such as love, care, respect, or moral autonomy). The task is to prevent the further colonization of the lifeworld by technological rationality.

When C. P. Snow published *The Two Cultures* in 1959, he initiated a discussion of the relation of science and literature which has been going on in other countries as well as in England. Science in this connection has meant the strictly empirical sciences, while literature has been taken more broadly to include methods of interpretation in the cultural sciences. The treatise with which Aldous Huxley entered the controversy, however, *Literature and Science*, does limit itself to confronting the natural sciences with the belles lettres.

Huxley distinguishes the two cultures primarily according to the specific experiences with which they deal: literature makes statements mainly about private experiences, the sciences about intersubjectively accessible experiences. The latter can be expressed in a formalized language, which can be made universally valid by means of general definitions. In contrast, the language of literature must verbalize what is in principle unrepeatable and must generate an intersubjectivity of mutual understanding in each concrete case. But this distinction between private and public experience allows only a first approximation to the problem. The element of ineffability that literary expression must overcome derives less from a private experience encased in subjectivity than from the constitution of these experiences within the horizon of a life-historical environment. The events whose connection is the object of the lawlike hypotheses of the sciences can be described in a spatio-temporal coordinate system, but they do not make up a world:

The world with which literature deals is the world in which human beings are born and live and finally die; the world in which they love and hate, in which they experience triumph and humiliation, hope and despair; the world of

sufferings and enjoyments, of madness and common sense, of silliness, cunning and wisdom; the world of social pressures and individual impulses, of reason against passion, of instincts and conventions, of shared language and unsharable feelings and sensations....[1]

In contrast, science does not concern itself with the contents of a life-world of this sort, which is culture-bound, ego-centered, and pre-interpreted in the ordinary language of social groups and socialized individuals:

... As a professional chemist, say, a professional physicist or physiologist, [the scientist] is the inhabitant of a radically different universe – not the universe of given appearances, but the world of inferred fine structures, not the experienced world of unique events and diverse qualities, but the world of quantified regularities.[2]

Huxley juxtaposes the *social life-world* and the *worldless universe of facts*. He also sees precisely the way in which the sciences transpose their information about this worldless universe into the life-world of social groups:

Knowledge is power and, by a seeming paradox, it is through their knowledge of what happens in this unexperienced world of abstractions and inferences that scientists have acquired their enormous and growing power to control, direct, and modify the world of manifold appearances in which human beings are privileged and condemned to live.[3]

But Huxley does not take up the question of the relation of the two cultures at this juncture, where the sciences enter the social life-world through the technical exploitation of their information. Instead he postulates an immediate relation. Literature should assimilate scientific statements as such, so that science can take on "flesh and blood."

> ... Until some great artist comes along and tells us what to do, we shall not know how the muddled words of the tribe and the too precise words of the textbooks should be poetically purified, so as to make them capable of harmonizing our private and unsharable experiences with the scientific hypotheses in terms of which they are explained.[4]

This postulate is based, I think, on a misunderstanding. Information provided by the strictly empirical sciences can be incorporated in the social life-world only through its technical utilization, as technological knowledge, serving the expansion of our power of technical control. Thus, such information is not on the same level as the action-orienting self-understanding of social groups. Hence, without mediation, the information content of the sciences cannot be relevant to that part of practical knowledge which gains expression in literature. It can only attain significance through the detour marked by the practical results of technical progress. Taken for itself, knowledge of atomic physics remains without consequence for the interpretation of our life-world, and to this extent the cleavage between the two cultures is inevitable. Only when with the aid of physical theories we can carry out nuclear fission, only when information is exploited for the development of productive or destructive forces, can its revolutionary practical results penetrate the literary consciousness of the life-world: poems arise from consideration of Hiroshima and not from the elaboration of hypotheses about the transformation of mass into energy.

The idea of an atomic poetry that would elaborate on hypotheses follows from false premises. In fact, the problematic relation of literature and science is only one segment of a much broader problem: *How is it possible to translate technically exploitable knowledge into the practical consciousness of a social life-world?* This question obviously sets a new task, not only or even primarily for literature. The skewed relation of the two cultures is so disquieting only because, in the seeming conflict between the two competing cultural traditions, a true life-problem of scientific civilization becomes apparent: namely, how can the relation between technical progress and the social life-world, which today is still clothed in a primitive, traditional, and unchosen form, be reflected upon and brought under the control of rational discussion?

To a certain extent practical questions of government, strategy, and administration had to be dealt with through the application of technical knowledge even at an earlier period. Yet today's problem of transposing technical knowledge into practical consciousness has changed not merely its order of magnitude. The mass of technical knowledge is no longer restricted to pragmatically acquired techniques of the classical crafts. It has taken the form of scientific information that can be exploited for technology. On the other hand, behavior-controlling traditions no longer naively define the self-understanding of modern societies. Historicism has broken the natural-traditional validity of action-orienting value systems. Today, the self-understanding of social groups and their worldview as articulated in ordinary language is mediated by the hermeneutic appropriation of traditions as traditions. In this situation questions of life conduct demand a rational discussion that is not focused exclusively either on technical means or on the application of traditional behavioral norms. The reflection that is required extends beyond the production of technical knowledge and the hermeneutical clarification of traditions to the employment of technical means in historical situations whose objective conditions (potentials, institutions, interests) have to be interpreted anew each time in the framework of a self-understanding determined by tradition.

This problem-complex has only entered consciousness within the last two or three generations. In the nineteenth century one could still maintain that the sciences entered the conduct of life through two separate channels: through the technical exploitation of scientific information and through the processes of individual education and culture during academic study. Indeed, in the German university system, which goes back to Humboldt's reform, we still maintain the fiction that the sciences develop their action-orienting power through educational processes within the life history of the individual student. I should like to show that the intention designated by Fichte as a "transformation of knowledge into works" can no longer be carried out in the private sphere of education, but rather can be realized only on the politically relevant level at which technically exploitable knowledge is translatable into the context of our life-world. Though literature participates in this, it is primarily a problem of the sciences themselves.

At the beginning of the nineteenth century, in Humboldt's time, it was still impossible, looking at Germany, to conceive of the scientific transformation of social life. Thus, the university reformers did not have to break seriously with the tradition of practical philosophy. Despite the profound ramifications of revolutions in the political order, the structures of the preindustrial work world persisted, permitting for the last time, as it were, the classical view of the relation of theory to practice. In this tradition, the technical capabilities employed in the sphere of social labor are not capable of immediate direction by theory. They must be pragmatically practiced according to traditional patterns of skill. Theory, which is concerned with the immutable essence of things beyond the mutable region of human affairs, can obtain practical validity only by molding the manner of life of men engaged in theory. Understanding the cosmos as a whole yields norms of individual human behavior, and it is through the actions of the philosophically educated that theory assumes a positive form. This was the only relation of theory to practice incorporated in the traditional idea of university education. Even where Schelling attempts to provide the physician's practice with a scientific basis in natural philosophy, the medical *craft* is unexpectedly transformed into a medical *praxiology*. The physician must orient himself to Ideas derived from natural philosophy in the same way that the subject of moral action orients itself through the Ideas of practical reason.

Since then it has become common knowledge that the scientific transformation of medicine succeeds only to the extent that the pragmatic doctrine of the medical art can be transformed into the control of isolated natural processes, checked by scientific method. The same holds for other areas of social labor. Whether it is a matter of rationalizing the production of goods, management and administration, construction of machine tools, roads, or airplanes, or the manipulation of electoral, consumer, or leisure-time behavior, the professional practice in question will always have to assume the form of technical control of objectified processes.

In the early nineteenth century, the maxim that scientific knowledge is a source of culture required a strict separation between the university and the technical school because the preindustrial forms of professional practice were impervious to theoretical guidance. Today, research processes are coupled with technical conversion and economic exploitation, and production and administration in the industrial system of labor generate feedback for science. The application of science in technology and the feedback of technical progress to research have become the substance of the world of work. In these circumstances, unyielding opposition to the decomposition of the university into specialized schools can no longer invoke the old argument. Today, the reason given for delimiting study on the university model from the professional sphere is not that the latter is still foreign to science, but conversely, that science – to the very extent that it has penetrated professional practice – has estranged itself from humanistic culture. The philosophical conviction of German idealism that scientific knowledge is a source of culture no longer holds for the strictly empirical scientist. It was once possible for theory, via humanistic culture, to become a practical force. Today, theories can become technical power while remaining unpractical, that is, without being expressly oriented to the interaction of a community of human beings. Of course, the sciences now transmit a specific capacity: but the capacity for control, which they teach, is not the same capacity for life and action that was to be expected of the scientifically educated and cultivated.

The cultured possessed orientation in action. Their culture was universal only in the sense of the universality of a culture-bound horizon of a world in which scientific experiences could be interpreted and turned into practical abilities, namely, into a reflected consciousness of the practically necessary. The only type of experience which is admitted as scientific today according to positivistic criteria is not capable of this transposition into practice. The capacity for *control* made possible by the empirical sciences is not to be confused with the capacity for *enlightened action*. But is science, therefore, completely discharged of this task of action-orientation, or does the question of academic education in the framework of a civilization transformed by scientific means arise again today as a problem of the sciences themselves?

First, production processes were revolutionized by scientific methods. Then expectations of technically correct functioning were also transferred to those areas of society that had become independent in the course of the industrialization of labor and thus supported planned organization. The power of technical control over nature made possible by science is extended today directly to society: for every isolatable social system, for every cultural area that has become a separate, closed system whose relations can be analyzed immanently in terms of presupposed system goals, a new discipline emerges in the social sciences. In the same measure, however, the problems of technical control solved by science are transformed into life problems. For the scientific control of natural and social processes – in a word, technology – does not release men from action. Just as before, conflicts must be decided, interests realized, interpretations found – through both action and transaction structured by ordinary language. Today, however, these practical problems are themselves in large measure determined by the system of our technical achievements.

But if technology proceeds from science, and I mean the technique of influencing human behavior no less than that of dominating nature, then the assimilation of this technology into the practical life-world, bringing the technical control of particular areas within the reaches of the communication of acting men, really requires scientific reflection. The prescientific horizon of experience becomes infantile when it naively incorporates contact with the products of the most intensive rationality.

Culture and education can then no longer indeed be restricted to the ethical dimension of personal attitude. Instead, in the political dimension at issue, the theoretical guidance of action must proceed from a scientifically explicated understanding of the world.

The relation of technical progress and social life-world and the translation of scientific information into practical consciousness is not an affair of private cultivation.

I should like to reformulate this problem with reference to political decision-making. In what follows we shall understand "technology" to mean scientifically rationalized control of objectified processes. It refers to the system in which research and technology are coupled with feedback from the economy and administration. We shall understand "democracy" to mean the institutionally secured forms of general and public communication that deal with the practical question of how men can and want to live under the objective conditions of their ever-expanding power of control. Our problem can then be stated as one of the relation of technology and democracy: how can the power of technical control be brought within the range of the consensus of acting and transacting citizens?

I should like first to discuss two antithetical answers. The first, stated in rough outline, is that of Marxian theory. Marx criticizes the system of capitalist production as a power that has taken on its own life in opposition to the interests of productive freedom, of the producers. Through the private form of appropriating socially produced goods, the technical process of producing use values falls under the alien law of an economic process that produces exchange values. Once we trace this self-regulating character of the accumulation of capital back to its origins in private property in the means of production, it becomes possible for mankind to comprehend economic compulsion as an alienated result of its own free productive activity and then abolish it. Finally, the reproduction of social life can be rationally planned as a process of producing use values; society places this process under its technical control. The latter is exercised democratically in accordance with the will and insight of the associated individuals. Here Marx equates the practical insight of a political public with successful technical control. Meanwhile we have learned that even a well-functioning planning bureaucracy with scientific control of the production of goods and services is not a sufficient condition for realizing the associated material and intellectual productive forces in the interest of the enjoyment and freedom of an emancipated society. For Marx did not reckon with the possible emergence at every level of a discrepancy between scientific control of the material conditions of life and a democratic decision-making process. This is the philosophical reason why socialists never anticipated the authoritarian welfare state, where social wealth is relatively guaranteed while political freedom is excluded.

Even if technical control of physical and social conditions for preserving life and making it less burdensome had attained the level that Marx expected would characterize a communist stage of development, it does not follow that they would be linked automatically with social emancipation of the sort intended by the thinkers of the Enlightenment in the eighteenth century and the Young Hegelians in the nineteenth. For the techniques with which the development of a highly industrialized society could be brought under control can no longer be interpreted according to an instrumental model, as though appropriate means were being organized for the realization of goals that are either presupposed without discussion or clarified through communication.

Hans Freyer and Helmut Schelsky have outlined a counter-model which recognizes technology as an independent force. In contrast to the primitive state of technical development, the relation of the organization of means to given or preestablished goals today seems to have been reversed. The process of research and technology – which obeys immanent laws – precipitates in an unplanned fashion new methods for which we then have to find purposeful application. Through progress that has become automatic, Freyer argues, abstract potential continually accrues to us in renewed thrusts. Subsequently, both life interests and fantasy that generates meaning have to take this potential in hand and expend it on concrete goals. Schelsky refines and simplifies this thesis to the point of asserting that technical progress produces not only unforeseen methods but the unplanned goals and applications themselves: technical potentialities command their own practical realization. In particular, he puts

forth this thesis with regard to the highly complicated objective exigencies that in political situations allegedly prescribed solutions without alternatives.

> Political norms and laws are replaced by objective exigencies of scientific-technical civilization, which are not posited as political decisions and cannot be understood as norms of conviction or weltanschauung. Hence, the idea of democracy loses its classical substance, so to speak. In place of the political will of the people emerges an objective exigency, which man himself produces as science and labor.

In the face of research, technology, the economy, and administration – integrated as a system that has become autonomous – the question prompted by the neohumanistic ideal of culture, namely, how can society possibly exercise sovereignty over the technical conditions of life and integrate them into the practice of the life-world, seems hopelessly obsolete. In the technical state such ideas are suited at best for "the manipulation of motives to help bring about what must happen anyway from the point of view of objective necessity."

It is clear that this thesis of the autonomous character of technical development is not correct. The pace and *direction* of technical development today depend to a great extent on public investments: in the United States the defense and space administrations are the largest sources of research contracts. I suspect that the situation is similar in the Soviet Union. The assertion that politically consequential decisions are reduced to carrying out the immanent exigencies of disposable techniques and that therefore they can no longer be made the theme of practical considerations, serves in the end merely to conceal preexisting, unreflected social interests and prescientific decisions. As little as we can accept the optimistic convergence of technology and democracy, the pessimistic assertion that technology excludes democracy is just as untenable.

These two answers to the question of how the force of technical control can be made subject to the consensus of acting and transacting citizens are inadequate. Neither of them can deal appropriately with the problem with which we are objectively confronted in the West and East, namely, how we can actually bring under control the preexisting, unplanned relations of technical progress and the social life-world. The tensions between productive forces and social intentions that Marx diagnosed and whose explosive character has intensified in an unforeseen manner in the age of thermonuclear weapons are the consequence of an ironic relation of theory to practice. The direction of technical progress is still largely determined today by social interests that arise autochthonously out of the compulsion of the reproduction of social life without being reflected upon and confronted with the declared political self-understanding of social groups. In consequence, new technical capacities erupt without preparation into existing forms of life-activity and conduct. New potentials for expanded power of technical control make obvious the disproportion between the results of the most organized rationality and unreflected goals, rigidified value systems, and obsolete ideologies.

Today, in the industrially most advanced systems, an energetic attempt must be made consciously to take in hand the mediation between technical progress and the conduct of life in the major industrial societies, a mediation that has previously taken place without direction, as a mere continuation of natural history. This is not the place to discuss the social, economic, and political conditions on which a long-term central research policy would have to depend. It is not enough for a social system to fulfill the conditions of technical rationality. Even if the cybernetic dream of a virtually instinctive self-stabilization could be realized, the value system would have contracted in the meantime to a set of rules for the maximization of power and comfort; it would be equivalent to the biological base value of survival at any cost, that is, ultrastability. Through the unplanned sociocultural consequences of technological progress, the human species has challenged itself to learn not merely to affect its social destiny, but to control it. This challenge of technology cannot be met with technology alone. It is rather a question of setting into motion a politically effective discussion that rationally brings the social potential constituted by technical knowledge and ability into a defined and controlled relation to our practical knowledge and will. On the one hand, such discussion could enlighten those who act politically about the tradition-bound self-understanding of their interests in relation to what is technically possible and feasible. On the other hand, they would be able to judge practically, in the light of their now articulated and newly interpreted needs, the direction and the extent to which they want to develop technical knowledge for the future.

This *dialectic of potential and will* takes place today without reflection in accordance with interests for which public justification is neither demanded nor permitted. Only if we could elaborate this dialectic with political consciousness could we succeed in directing the mediation of technical progress and the conduct of social life, which until now has occurred as an extension of natural history; its conditions being left outside the framework of discussion and planning. The fact that this is a matter for reflection means that it does not belong to the professional competence of specialists. The substance of domination is not dissolved by the power of technical control. To the contrary, the former can simply hide behind the latter. The irrationality of domination, which today has become a collective peril to life, could be mastered only by the development of a political decision-making process tied to the principle of general discussion free from domination. Our only hope for the rationalization of the power structure lies in conditions that favor political power for thought developing through dialogue. The redeeming power of reflection cannot be supplanted by the extension of technically exploitable knowledge.

Notes

1 Aldous Huxley, *Literature and Science*, New York, 1963, p. 8.
2 *Ibid.*
3 *Ibid.*, p. 9.
4 *Ibid.*, p. 107.

15

The Critical Theory of Technology

Andrew Feenberg

Andrew Feenberg studied with Marcuse and is currently Canada Research Chair in Philosophy of Technology in the School of Communication, Simon Fraser University. His work stands in the Western-Marxist tradition, yet unlike his predecessors he finds ambiguity and possibilities of freedom and democracy within technological development itself. Feenberg thus goes beyond Marcuse's arguments that change could only arise from those marginalized, from oppositional intellectuals, who might create a new science and a new technology, or from philosophers and artists, who could also further change through their work. And he also disagrees with Habermas, who argues that a techno-scientific sphere properly constrained should have relative autonomy. Feenberg's position is that we can only understand technology and the possibilities for increasing human control and freedom by viewing technology as fully enmeshed within its social use and development context. This means looking at everyday uses and transformations of technology. In this chapter excerpted from his 2002 book *Transforming Technology*, he rejects the notion that technology is either neutral or autonomous; rather, it must be understood as a combination of technical and social factors, both of which shape its meaning. Technological development is contingent, not deterministic. It often responds to public pressures, market forces, user-determined uses and meanings, aesthetic considerations, and other non-technical concerns. A further democratizing of technology requires increased attention to, and perhaps intervention in, planning and design processes. Feenberg argues against both essentialism and its cousin – determinism – to put forward a political theory of technology that embraces the social dimensions of technological systems, including their impact on the environment and

From Andrew Feenberg, *Transforming Technology: A Critical Theory Revisited* (Oxford: Oxford University Press, 2002), pp. 162–90. By permission of Oxford University Press, Inc.

workers' skills and their role on the distribution of power. Feenberg constructs a multi-level analysis of the essence of technology in order to do justice to the complex, historic, indeterminate character of technological design, use, and transformation. Only then, he argues, can we begin to criticize our society's culture of technology and imagine alternatives that would bring out a more democratic, meaningful and livable environment.

The Critique of Scientific-Technical Rationality

Modernity and critique

Modernity is the affirmation of autonomy against every traditional or social authority (Pippin, 1991). Modern societies organize apparently neutral mediations such as markets, elections, administrations, and technical systems for the expression of an unlimited variety of contingent interests and visions of life that cannot and need not be justified, reconciled, or ranked. This system does not favor this or that substantive value but maximizes autonomy in general, promising liberation of the human essence from fixed definitions. Rationality enters this scheme only at the level of means, both the means individuals employ to achieve particular ends and the means instituted by society to mediate their relations. These means fall under formal norms of efficiency and equity.

Capitalist democracy is the most successful modern political institution. As a specific instance of modernity, capitalism is subject to critique either as all too modern or as not modern enough. The first type of critique is usually conservative. Heidegger, for example, condemns modern society as nihilistic and attempts to conceive a philosophical alternative to autonomy. Traditionalist reactions to modernity are of course commonplace today under the guise of ethnic or national identity. More interesting for our argument are those progressive critiques of capitalism which address it as a failed instance of modernity. Such arguments generally contrast the ideal of autonomy with capitalist realities, identify interests capitalism is structurally blocked from serving, or denounce the substantive goals it imposes in the course of structuring social life around neutral mediations.

The progressive critical strategy can be pursued in two rather different ways. One approach argues that capitalism interferes with the neutral media – markets, elections, administrations, and technical systems – through which modern individuals pursue their interests. This is the logic of suspicion, the demystifying attack on vested interests that manipulate the public from behind the scenes. The product and process critique of technology is of this type. The other style of critique argues that "the medium is the message," that the media distort the contents they express. For example, not every good can find a place on the market. Markets are not therefore neutral arbiters of the community's values but prejudice choice wherever they are instituted. The question is not just who profits but what way of life is determined by the market. Since it appears to be essentially transparent and universal, however imperfect particular practical realizations may be, the critique must undermine the standard of rationality that defines it. The design critique of technology and the related theory of the technical code follow this general approach in the technical domain.

The critique of rationality also characterizes critical theory from its origins in Hegel and Marx, down through the early Marxist Lukács, Ernst Bloch, and the Frankfurt School. Today, some feminists and ecologists find resources in this tradition that they seem to be practically alone in continuing. Yet far more work of this type is needed in a society in which scientific-technical rationality has become the principal legitimating discourse. This chapter attempts to contribute to such a revival of radical social critique.

This critique usually contains at least an implicit reference to what Bloch called "Left Aristotelianism."[1] In one late essay, Bloch defines the agenda of a critical theory of scientific-technical rationality in terms of the still viable heritage of premodern, qualitative images of nature (Bloch, 1988: 59). Conceding that nature has the reified dimension attributed to it by modern science, Bloch argues that a modern holistic ontology must relativize that dimension with respect to other dimensions science ignores. These other dimensions are manifested in ecological crisis which, like economic crisis, demonstrates the limits of scientific-technical rationality (Bloch, 1988: 67). Bloch offers here a typical Hegelian-Marxist critique of the formalistic character of modern reason that fails adequately to grasp its "content" (nature).

But today, Bloch's formulation appears excessively optimistic. It is not the heritage of the premodern conception of nature that needs saving but the heritage of classical Critical Theory itself. The waves appear to be rapidly closing over that tradition under the combined attack of Habermas and postmodernism. What both have in common, despite their many obvious differences, is a rejection of that tradition's dialectical concept of reason, which is now identified with a nostalgic organicism that seeks a utopia in the past, in nature, in the immediate (Jay, 1984: chap. 15, epilogue). On this account, Critical Theory would be a regression behind the level of rationality achieved by modernity rather than a transcendence of its capitalist forms. A vigorous modernity or postmodernity, as the case may be, looks forward without illusion and affirms a culture based on fragmentation in which wholeness is at best a regulative ideal for the conversation of fractured identities.

One might object to the polemical exaggeration in these characterizations of Critical Theory. There is a certain arrogance in assuming that such profound students of Hegel as Lukács or Marcuse were mere romantics haunted by Rousseauian reveries. But the argument can be advanced more rapidly by accepting at the outset the necessary choice forced on us by the polemic against dialectics. It is true that these critical theorists retain a romantic reference to an original immediacy as a symbol of the dialectical reunification of what analysis has fragmented. They thus attempt to place romanticism within a more or less Hegelian framework rather than rejecting it outright. It is difficult to accurately characterize a position that hovers "dialectically" between alternatives it hopes to redeem rather than

select. Is it possible to reformulate the critique without playing on these ambiguities, without opening a flank to attack by today's sober censors of intellectual nostalgia?

The task is complicated by a second problem. Because natural science and technology share a fundamentally similar form of rationality, Critical Theory tends to identify them. The critique of what has come to be called "technoscience" unveils the secret complicity between the apparently innocent activity of the researcher and the horrifying military applications. Science is undoubtedly influenced by society in all sorts of ways and can no more claim to be socially neutral than can technology. But despite their growing interconnections, science and technology are very different institutions (Goldman, 1990). The difference shows up in the reform programs that sound plausible in the two cases: political reform for technology and reform from within for science. Yet if technoscience is a single phenomenon, on what basis can one make this strategic distinction? In fact, critical theorists tend to waver uncomfortably between a utopian politics of technoscience (Marcuse) and acceptance of the neutrality of technoscience in its proper sphere (Habermas). Both positions are mistaken, but until we discriminate conceptually between science and technology, we will be unable to put forward a credible case for a critique and transformation of modern forms of rationality. Indeed, we will be easy targets for the charge of irrationalism.

The rest of this chapter attempts to resolve these problems. I first reconstruct several of the core arguments of the Critical Theory tradition and discuss similar arguments in contemporary feminism. In the second half of the chapter, I develop the critique of technology in a new way that avoids romantic subtexts and opens positive perspectives on the future. Along the way, I attempt to clarify the issues raised above and to show that Critical Theory reconceptualizes reason rather than rejecting it.

Reason and domination

Critical Theory attacks capitalism by attacking its forms of rationality. The approach appears strangely circuitous. Why not solve the problem of poverty through redistribution? Why drag in a critique of the rationality of the market? Similarly, if one is opposed to deskilling, why not use regulation to protect the skill content of jobs, much as one now protects endangered species? Why complicate the issue with a critique of scientific-technical rationality? No such critique was required to introduce affirmative action, food stamps, and welfare. In Weberian terms, the argument would be that reforms motivated by substantively rational ends can soften the hard edges of a formally rationalized society. These proposals place us on the familiar terrain of dual compliance.

Such *moral reformism* has the advantage of assuming the self-evidences of the age. The formal mediations introduced by capitalism are not challenged, but their effects are compensated. Technical reason is not criticized but subordinated to humanistic objectives. The gradual moralization of social life can create a better world, trading off certain economic for human values. What is wrong with this approach?

In fact, modern critical theory grows out of the work of two thinkers who rejected it, Marx and Weber. They formulated some of the earliest social theories of formal rational systems such as markets and technology. These theories emphasize the self-expanding character of formal mediation. The dynamic of rationalization inherent in the system conflicts with substantive correctives. Since these correctives are by nature formally irrational, they create social tensions likely to be resolved at a later stage through the sacrifice of "ideals" for practical efficiency. Hence the political oscillations of the welfare state, caught in unresolvable goal conflicts. Both Marx and Weber are therefore skeptical of moralistic reformism, although they draw very different political conclusions.

Marx attempts to establish a coherent strategy of civilizational change based on a critique of the class bias of capitalist rationality. He analyzes the mechanisms by which market rationality reproduces the class structure and reinforces capitalist hegemony. In identifying these limits of capitalist rationality, Marx situates himself beyond them in a higher dialectical rationality. Socialism is then described as a new form of rational order rather than as a regression to premodern conditions or an irrational and inefficient excrescence on the market.

Marx's general approach was anticipated by Hegel. In Hegel, dialectical *reason* overcomes the tendency of analytic *understanding* to split objects up into abstractly separated parts. Hegel does not regress to the immediate givenness of the objects of the understanding but believes that reason can recapture totality at a higher level through mediating the fragmented parts. Hegel thus proposes moving forward from fragmentation to totality rather than backward to an original unity. But Marx's version of dialectics falls short because he fails to explain the dialectical rationality of the planned society he wants to substitute for capitalism.

Weber rejects dialectics and does not propose an alternative to capitalism. Although aware of its social bias, Weber has no philosophical critique of formal rationality; for him, as for most modern social theory, the rise to power of specific social strata in the course of rationalization is ultimately no more than an unavoidable side effect of progress. Thus, he overlooks the connection between the limitations of formal rationality and the problems of capitalism and bureaucracy.[2]

Lukács's early theory of reification first makes that connection explicit and sketches a theory of dialectical rationality. Lukács introduces the term "reification" to describe Marx's "fetishism" and Weber's "rationalization." He argues that the structure of both market and bureaucracy is essentially related to the structure of formal rationality, and brings to light the congruence of modes of thought and action that rest on the fragmentation of society, analytic thinking, technology, and the autonomization of production units under the control of private owners. Lukács thus explains the preestablished harmony between a particular organization of society and a historically concrete form of rationality, unifying in the same concept social facts that remain separate for Marx and Weber. Where Marx had foreseen a recovery of wholeness at the economic level, Lukács offers a similar argument at the level of culture, attacking capitalist fragmentation not from the standpoint of premodern organicism but in terms of a dialectical concept of the mediated totality (Feenberg, 1986: chap. 3).

[...] His theory is intended to show how, starting out from the specific degradation of its life and work under the reign of the law of value, the proletariat can break with capitalist forms of thought and action and realize the potentialities for a very different type of society contained and repressed in capitalism. Lukács argues that the standpoint of the proletariat is not merely immanent to capitalism but opens up a broader view of the most fundamental limitations of that system. Lukács calls that broader view, in which capitalism is relativized with respect to its own potentialities, the "totality." Totality is thus not a synoptic view or a conceptual myth as critics contend but the basis of an immanent critique.

In Lukács, formal rationality is the basis of capitalist culture, and dialectical reason, by contrast, supports a socialist society. Thus, the same relation holds between formal rationality and capitalism as between dialectics and socialism. And, just as socialism does not reject the capitalist heritage but employs it as an ambivalent basis of development, so dialectics encompasses formal rationality in a larger framework that determines its limits and significance. This approach goes beyond dual compliance to suggest the possibility of founding socialism as an alternative civilization, as coherent and rational in its own way as capitalism.

[. . .]

Marcuse goes beyond Lukács and attempts to explain the growing political role of science and technology in advanced capitalism. Continuing along the path opened in Adorno and Horkheimer's *Dialectic of Enlightenment*, he aims at nothing less than a general theory of the link between formalism and class domination throughout history, and on that basis he anticipates the main outlines of a new society, including its scientific and technical practice (Adorno and Horkheimer, 1972; Marcuse, 1964).

Like Lukács, Marcuse considers the universality of bias in the rationalization process to be a *problem* and not simply an accident of world-historical scope. He writes:

Scientific-technical rationality and manipulation are welded together into new forms of social control. Can one rest content with the assumption that this unscientific outcome is the result of a specific societal *application* of science? I think that the general direction in which it came to be applied was inherent in pure science even where no practical purposes were intended,

and that the point can be identified where theoretical Reason turns into social practice. (Marcuse, 1964: 146)

We can rephrase Marcuse's point by asking what it means that formal systems are generally available for applications biased to favor domination. Is there something about their very structure that opens them to such applications? What happened "originally" in the initial construction of the formal mode of abstraction that rendered it pliable in this particular way?

It is difficult to follow Marcuse's argument to this point because we do not normally think of such formal systems as mathematics or technology as essentially implicated in their own applications. Rather, they appear neutral in themselves. Of course, one can make a bad use of them just as one might pick up a rock and throw it at a passerby. It would be comical to suggest that the rock is "biased" a priori toward such uses, that its hardness is the essential precondition by which it lends itself to violence. Marcuse's very question reverses our normal assumptions and connects formal neutrality and domination as moments in a dialectical totality. This is perhaps admissible to the extent that formal systems, unlike rocks, are human inventions created for a purpose in specific social contexts.

Marcuse's treatment of this problem depends on his dialectical ontology, which, in turn, presupposes the distinction between "substantive" and "logico-mathematical" or "formal" universals (Marcuse, 1964: chap. 5). This distinction separates a holistic approach to human and natural systems from the mechanistic breakdown of these systems into their reified parts.[3] Substantive universals are essences constructed through an abstractive process that brings to the fore the internal coherence and potentialities of their objects. These objects are not isolated and self-subsistent things but contextually dependent "wholes" developing in essential interaction with an environment. Formal thinking, on the contrary, abstracts from the whole not toward its potentialities but rather toward its "form." By "form," Marcuse intends abstract properties that are isolated from each other and from the inner order of the objects from which they are abstracted. These properties include colors, shapes, number, and so on.

Formal universals decontextualize their objects in both time and space, evacuating their "content" and abstracting from their developmental dynamics. Instead of transcending the given toward its essential potentialities, this type of universality classifies or quantifies objects in terms of the function they can be made to serve in an instrumental system imposed on them from without. Although apparently neutral and value free, in suppressing the dimensions of contextual relatedness and potentiality, the *decontextualizing practice* of formal abstraction transforms its objects into mere means, an operation that prejudices their status as much as any valuative choice.

Here is the core of Marcuse's argument. Formal universals are indeed "value-free" in the sense that they do not prescribe the ends of the objects they conceive as means; however, they are value-laden in systematically overlooking the difference between the extrinsic values of an instrumental subject and the intrinsic telos of an independent, self-developing object. Insofar as formal thinking considers its objects only in terms of their utility, it treats their potentialities as no different from the outcome of a technical manipulation. The essential difference between self-development and control is obscured, and a founding bias is thereby introduced. The very conception of value from which formal universals are "free" is itself a product of the abstractive process in which formalism obscures the nature of potentiality. Despite, or rather because of, its neutrality as between potentialities and utilitarian values, formal reason is biased toward the actual, what is already realized and available for technical control.

Methodologically, this bias appears in the inability to grasp history and social contexts as the scene of development. Formal abstraction works with the immediate appearance of its artificially isolated object. It accepts this appearance as truth and in so doing comes under the horizon of the existing reified society and its modes of practice. The range of manipulation opened by formal abstraction is the uncritically accepted horizon of domination under which its objects lie. These objects can be used, but not transformed, adapted to the dominant social purposes, but not transcended toward the realization of their potentialities in the context of a better society.

This is why formal systems are intrinsically available as a power base. In cutting the essential connections between objects and their history and contexts, formal abstraction ignores the inner tensions in reality that open possibilities of progressive development. Instead, objects are conceptualized as fixed and frozen, unchanging in themselves but available for manipulation from above.

This construction of objectivity comes back to haunt formal thinking in the biased application of its products. Repressive applications arise as soon as its abstractions are reintegrated to a real world of historical contingencies. Then it becomes clear that "formalization and functionalization are, *prior* to all application, the 'pure form' of a concrete societal practice" (Marcuse, 1964: 157).

The hypothetical system of forms and functions becomes dependent on another system – a preestablished universe of ends, in which and *for* which it develops. What appeared extraneous, foreign to the theoretical project, shows forth as part of its very structure (method and concepts); pure objectivity reveals itself as object *for* a *subjectivity* which provides the Telos, the ends. In the construction of the technological reality, there is no such thing as a purely rational scientific order; the process of technological rationality is a political process. (Marcuse, 1964: 168)

According to Marcuse, such formal abstraction is the technical "a priori" of modern capitalist society and its communist imitators.

Toward a successor technoscience?

Marcuse's theory of potentiality implies a participatory epistemology and a holistic ontology. The potentialities of objects come into focus in active involvement with them as wholes, rather than through calculative contemplation of their manipulable components: "creative receptivity versus repressive productivity" (Marcuse, 1974: II, 286). Marcuse conceives this receptivity under the categories of the erotic and the aesthetic, which he generalizes beyond the spheres of sexuality and art to include a dereified relationship to nature. Nature is not merely an object of technical conquest but can be an active partner of human beings. We should stand in "a 'human relation' to matter . . . [which] is part of the *life* environment and thus assumes traits of a living object" (Marcuse, 1972: 65).

These ideas have an affinity with certain strands of feminist theory, and in the early 1970s Marcuse formulated his concept of socialism in feminist terms. In his view, capitalist patriarchy shelters women to some degree from the full force of reification by confining them to subordinate roles in the home. In the struggle between "eros and aggression," women are inclined to the former as a consequence of the very oppression they suffer. Marcusean socialism would generalize the "female" traits of "tenderness, receptivity, sensuousness" in creating a society freed of male domination (Marcuse, 1972: 74–78). A new science would emerge from these changes, incorporating human values into its very structure.

This convergence of Critical Theory and feminism is less surprising than it may seem at first. From Aristotle to Hegel to the Frankfurt School, holistic ontologies have offered a powerful alternative to the mechanistic worldview. Feminists who privilege modes of knowing based on involvement and receptivity find resources in this tradition (Bordo, 1987: 103–105). Their gendered epistemologies have inspired a whole contemporary literature that has striking similarities to certain positions of the Frankfurt School. As Sandra Harding writes,

> The feminist standpoint epistemologies ground a distinctive feminist science in a theory of gendered activity and social experience. They simultaneously privilege women or feminists (the accounts vary) epistemically and yet also claim to overcome the dichotomizing that is characteristic of the Enlightenment/bourgeois world view and its science. It is useful to think of the standpoint epistemologies, like the appeals to feminist empiricism, as "successor science" projects: in significant ways, they aim to reconstruct the original goals of modern science. (Harding, 1986: 142)

Marcuse's critique of the repressive implications of modern scientific-technical thinking also culminates in a successor science project. He rejects scientific pretensions to value neutrality and

argues for science "becoming political" in order to recognize the suppressed dimensions of inner and outer nature (Marcuse, 1964: 233–234). Similarly, Harding summarizes one feminist account as demanding "an epistemology which holds that appeals to the subjective are legitimate, that intellectual and emotional domains must be united, that the domination of reductionism and linearity must be replaced by the harmony of holism and complexity" (Harding, 1986: 144).

The idea of an alternative science parallels at a more fundamental level the similar notion of an alternative technology. If, like machines, facts and theories are social constructions, how can they be innocent and neutral? Once social criticism shows how deeply these supposedly autonomous fields have been marked by politics, they can be treated as ambivalent institutions subject to reconstruction in the context of a new hegemony (Marcuse, 1964: 233–234).

This parallel raises a delicate question. What is the role of politics in the transformation of technoscience? Despite ritual disclaimers, the critique of scientific-technical rationality appears to lead straight to political control of research not just through familiar external manipulations such as grants, but far more profoundly at the level of fundamental epistemological choices. After all, if science is completely colonized by a false rationality, then it is difficult to see how it could reform itself (even with a boost from a reformed NSF). Indeed, why should its fate differ fundamentally from that of other oppressive superstructures such as law? Earlier chapters in this book have in fact discussed the transformation of technology as a political affair, and unless one distinguishes science from technology, it, too, would seem to fall under an external practical critique.

But there are warning signs posted along this path. Shortly after the Russian Revolution, an organization called Proletcult argued for the substitution of a new proletarian culture for the reactionary inheritance of bourgeois technology, science, and even language (Claudin-Urondo, 1975: 47–60). Up to this point Marxists had generally considered these phenomena as nonideological. The exemption of science from political critique was a foundational assumption of orthodox Marxism. Following Engels, most

Marxists connected the genesis of modern science with early bourgeois society, while insisting that this historical background in no way diminished the universality of modern scientific achievements. Proletcult resolved this split between genesis and validity, treating science as Marxism had always treated the superstructures. The embarrassing residue of transhistorical truth was eliminated from the system.

Although both Lenin and Stalin opposed this view in theory, Lysenko was able to introduce political criteria into the actual institutions of Russian science. His genetic theories won state support while most of his scientific adversaries were executed. The catastrophic failure of this experiment in "proletarian" thought continues to inspire widespread fear of any ideology critique of natural science (Graham, 1998).

Even those unaware of this history are likely to be affected by it, so deeply did it discredit the project of politicizing science. For the most part, current social criticism of science responds to this dangerous precedent by arguing against political interference and instead calls for the "reclamation from within of science" (Keller, 1985: 178). Civilizational change would eventually promote scientific change without the risk of further Lysenko affairs. Not political power but scientists' own evolving categories and perceptions in a radically new social environment would inspire new types of questions and new theories generated spontaneously in the course of research by scientists themselves. As Marcuse writes, scientific "hypotheses, without losing their rational character, would develop in an essentially different experimental context (that of a pacified world); consequently, science would arrive at essentially different concepts of nature and establish essentially different facts" (Marcuse, 1964: 166–167).

This view of scientific progress and its likely course makes sense, however, noninterventionism is incompatible with the statement of clear guidelines for a successor science. One must choose between affirming the self-reconstructive powers of science, which will surely yield an unexpected outcome, or devising an extrinsic program anticipating a future state of science that would have to be implemented politically. The first alternative allows us only to contest premature totalizations, such as reductionist

paradigms in sociobiology or "neurophilosophy"; it does not dictate theoretical developments. Social critique of science cannot contain the future, but only hold it open.

These qualifications raise questions about the extrinsic ontological and epistemological criteria used to evaluate current science. What, one might ask, guarantees that in a "pacified" world, a holistic science would discover ways to overcome the split between value and fact, emotion and reason, part and whole? How can we foresee today the general outline of the results of future research? Perhaps scientific method will change far less than we imagine and instead new theories will address the problems that concern us today. What is more, holism itself is politically controversial. There is no lack of evidence that it can be accommodated to repressive ends (Haraway, 1989: 256). Thus, Donna Haraway writes, "Evaluations and critiques cannot leap over the crafted standards for producing credible accounts in the natural sciences because neither the critiques nor the objects of their discourse have any place to stand 'outside' to legitimate such an arrogant overview. To insist on value and story-ladenness at the heart of the production of scientific knowledge is not equivalent to standing nowhere talking about nothing but one's biases – quite the opposite" (Haraway, 1989: 13).

Haraway's doubts about the successor science project are reasonable, and one does not appear to pay a high political price for the caution she recommends. But similar doubts might be raised about politically motivated reform in every sphere. For example, one might argue that technological change cannot be anticipated from outside the engineering profession, legal change from outside the legal profession, and so on. That would result in the dismissal of political criteria for sociotechnical transformation that have emerged laboriously from generations of struggle and analysis.

There is another way to look at the difficulty. The holistic critique of modern science is perhaps misdirected. Alienated objectivism has an obvious *venue* in our experience other than natural science, with which few people have any direct contact. Rather, the living source of the critique is our participation in technically mediated social institutions. The operational autonomy these institutions support founds an epistemological standpoint that is congruent with the detached analytic standpoint of science but that has neither scientific purpose nor institutional context. It is as though the discursive framework of scientific rationality had escaped the confines of inquiry to become a cultural principle and a basis of social organization. This is in fact the original insight of Lukács's theory of reification: "What is important is to recognize clearly that all human relations (viewed as the objects of social activity) assume increasingly the objective forms of the abstract elements of the conceptual systems of natural science and of the abstract substrata of the laws of nature. And also, the subject of this 'action' likewise assumes increasingly the attitude of the pure observer of these – artificially abstract – processes, the attitude of the experimenter" (Lukács, 1971: 131).[4]

Unlike the successor science project, *techno-logical holism* cannot be accused of extrinsic political interference because ordinary people are intrinsic participants in technical processes. They can transform technology through enlarging the margin of maneuver they already enjoy in the technical networks in which they are enrolled. The extrapolation of the logic of that transformation to the domain of the sciences is a different story. The point is not that science is purer than technology but that social contradictions traverse science differently. Because science does not form the life-world of ordinary people, but only affects them through technology, it remains a specialized activity. The holistic criteria of change relevant to the critique of technology do not therefore apply to science, or at any rate not in the same way.[5]

Distinguishing between the critique of natural science and the critique of technology has both strategic and theoretical consequences. The idea of a successor technoscience combines a plausible approach to technological change with speculative and politically charged proposals for scientific change. The entire enterprise risks foundering because of the connection. Conservative objections to technological critique can shelter behind the self-righteous defense of scientific freedom. The only effective response is to clearly separate a nonteleological critique of science from the teleological critique of technology based on notions of human, social, and natural potential.

These strategic considerations raise a larger problem. Richard Bernstein argues that to define "true human potentiality" we must be prepared to defend the supposedly outdated ontologies of Aristotle and Hegel. He points out, "This is not a rarefied philosophical or intellectual problem when we remind ourselves that however much we condemn totalitarianism and fascism as 'untrue' and 'evil,' they are *also* realizations of human potentialities" (Bernstein, 1988: 24). In Bernstein's view, one can oppose totalitarianism and fascism from the standpoint of a modern formal concept of freedom, but the old teleological approach is no longer intellectually respectable.

This argument challenges Critical Theory – either to find a nonontological formulation of the notion of potentiality or to come forward openly in defense of a holistic ontology of some sort. In the following I attempt to carry out the first program; however, I can conceive of no way of including science as well as technology in this project. A holistic conception of nature as such is by definition a speculative ontological project until such time as science, on its own terms, gives a scientific content to the notion.

Ontological holism is of course an interesting notion, but the critique of technological rationality does not require it. A nonontological formulation of a critical theory of technology is possible on terms that leave natural science out of account. I believe this is the best way to counter the undifferentiated defense of technoscience in the writings of the many philosophers and social theorists who see a threat to rationality as a whole in any critique of technology.

Instrumentalization Theory

Two types of instrumentalization

The holistic technology critique I propose depends on an analytic distinction between what I call the primary and secondary instrumentalizations.[6] The primary instrumentalization is the technical orientation toward reality that Heidegger identified as the technological "mode of revealing." However, as we have seen, the technical involves not just an orientation but also action in the world, and that action is socially conditioned through and through. Hence the need for a theory of secondary instrumentalizations through which the skeletal primary instrumentalization takes on body and weight in actual devices and systems in a social context.

An analogy with literature explains how these two levels together form a single "essence of technology." Literature depends on an imaginative orientation toward reality. Yet it is obvious that a definition of literature that included only that orientation would be incomplete. What about genres such as the novel or tragedy? What about composition and performance? Markets and careers? Surely all this belongs to literature too. The essence of literature must include a reference to imagination, to be sure, but it must include a lot more besides, and this carries us into social territory we must explore if we really want to understand it.

Technology is similar. A complete definition must show how the orientation toward reality characteristic of technology is combined with the realization of technology in the social world. A véry simple example can illustrate this point. Carpentry involves perceiving wood as a resource and grasping the affordances it offers. In phenomenological language, we could say that the world reveals itself to the carpenter as such a resource, as such affordances. Without this primary instrumentalization of wood, no one would have thought to make a saw, but a saw is not just an "application" of a technical orientation toward wood. Rather, it is a concrete object produced in a specific society according to a social logic. Even such basic facts about saw design as whether it will cut on the push or the pull are socially relative. To understand the form of the saw, its manufacture, its symbolic status, and so on, we need more than a theory of technical orientation. Furthermore, a theory of technical orientation will not tell us what becomes of persons whose lives are dedicated to working wood, how that activity will shape their hands, their reflexes, their language and personality so that it will make sense to call someone a carpenter. All these are secondary instrumentalizations, inseparable from the essence of technology.

My intent in analyzing technology at these two levels is to combine essentialist insights into the technical orientation toward the world with critical and constructivist insights into the social nature of technology. I will show that what are usually presented as competing theories are in fact analytically distinguishable levels in a complex object.

For example, Marcuse's critical account focuses on the primary instrumentalization of the object of technical practice. In the next section, I will break this conception down into the various moments through which the object is isolated and exposed to external manipulation. These moments are the basis for formal bias which works with the technical elements released from the instrumentalized objects.

But as they develop, technologies reappropriate aspects of contextual relatedness and self-development from which abstraction was originally made in establishing the technical object relation. It is only because technology has these integrative potentialities that it can be enlisted to repair the damage it does, for example, by redesigning technical processes to take into account their effects on workers, users, and the environment. The description of "informating" technology attempts to conceptualize such potentialities of the computer, and a later section of this chapter discusses the theoretical implications of integrative technical development.

On the basis of this concept of integration, I argue that technique is dialectical. A full definition of it must include a secondary instrumentalization that works with dimensions of the object denied at the primary level. This dialectical account of technology breaks with the overly negative evaluation of technology in the Frankfurt School. On the other hand, it continues Critical Theory's search for a positive moment in Enlightenment that compensates for the disaster of modernity. That moment surfaces in concepts like Adorno's "mindfulness of nature" or Marcuse's notion of "potentiality." Instrumentalization theory identifies resources in the technical sphere through which it can be realized in a redeemed modernity.

The complementarity of primary and secondary instrumentalization is a normal aspect of the technical sphere. Secondary instrumentalizations lie at the intersection of technical action and the other action systems with which technique is inextricably linked insofar as it is a social enterprise. The dialectics of technology is thus not a mysterious "new concept of reason" but an ordinary aspect of the technical sphere, familiar to all who work with machines if not to all who write about them.

But capitalism has a unique relation to these aspects of technique. Because its hegemony rests on formal bias, it strives to reduce technique to the primary level of decontextualization, calculation, and control. The definition of "technique" is narrowed as much as possible to the primary instrumentalization, and other aspects of technique are considered nontechnical. Suppressed are the integrative potentialities of technique that compensate for some of the negative effects of the primary instrumentalization.

The dialectic of technology is short-circuited under capitalism in one especially important domain: the technical control of the labor force. Special obstacles to secondary instrumentalization are encountered wherever integrative technical change would threaten that control. These obstacles are not merely ideological but are incorporated into technical codes that determine formally biased designs. As we have seen, the integration of skill and intelligence into production is often arrested by the fear that the firm will become dependent on its workers. The larger context of work, which includes these suppressed potentialities, is uncovered in a critique of the formal bias of existing designs. The critical theory of technology exposes the obstacles to the release of technology's integrative potential and thus serves as the link between political and technical discourse.

The dialectic of technology

In traditional societies, technique is always embedded in a larger framework of social relations. Not only does technical practice serve extratechnical values – it does that in all societies, including capitalist ones – but more than that, it is contextualized by practices that define its place in an encompassing nontechnical action system. One finds remnants of such a structure today in child rearing and artistic production. The parent who employs modern medicine, the artist who welds a sculpture or uses videotape, integrate these technologies into a larger framework of nurturing or aesthetic practices. Although the actors may rationalize the technologies they employ, the larger system in which these technologies are embedded resists rationalization and

does not fall under the norm of efficiency (Feenberg, 1995: chap. 7).

Capitalist labor organization is no longer embedded in the various social subsystems it serves, controlled by nontechnical forms of action such as religious or paternal moral authority. Capitalism liberates technique from such internal controls and organizes work and an ever enlarging share of the rest of the social system in pursuit of efficiency and power. Thus even though technique in itself has many similar traits in precapitalist and capitalist societies, only in the latter is it a universal human destiny (Habermas, 1970: 94–98).

This destiny can be summarized as four reifying moments of technical practice that have always characterized the object relation in the small technical enclaves of social life but that embrace society as a whole for the first time under capitalism. To each of these reifying moments, there corresponds a compensating integrative moment that, as we will see, is severely restricted as it is accommodated to capitalism.

1 *Decontextualization and Systematization*: the separation of the technical object from its immediate context, and a corresponding systematization through which the decontextualized objects are connected with each other, with human users, and nature to form devices and technical organizations.

2 *Reductionism and Mediation*: the separation of primary from secondary qualities, that is, the reduction of objects to their useful aspects, and a corresponding mediation of technical devices by aesthetic and ethical qualities that are incorporated into their design.

3 *Autonomization and Vocation*: the separation of subject from object, that is, the protection of the autonomized technical actor from the immediate consequences of its actions, and a corresponding vocational investment of the actor who is shaped as a person with an occupation by the technical actions in which he or she engages.

4 *Positioning and Initiative*: the subject situates or positions itself strategically to navigate among its objects and control them, and a corresponding sphere of initiative in which those of the "objects" which are in fact

subordinated human beings, workers and consumers, enjoy a certain tactical free play.

Capitalism applies the four primary moments most broadly, while partially suppressing secondary moments of the technical relation. The remainder of this section will show how these characteristics of technical action apply to both the collective laborer and to nature as the object of production under capitalism.

Decontextualization and systematization Capitalist technology is based on the *reified decontextualization* of the objects it constructs. It is because basic technical elements are abstracted from all particular contexts that they can be combined in devices and reinserted into any context whatsoever to further a hegemonic interest. Capitalism emerges from the generalization of this feature of technology at the expense of labor and the natural environment. Communist societies imitated these aspects of the capitalist inheritance and so offered no alternative in this respect.

The construction of abstract labor power under capitalism is unique in achieving a properly technical decontextualization of human capacities. All earlier societies employed human labor in the context of the social conditions of its reproduction, such as the family and community. The creative powers of labor were developed through vocations such as crafts transmitted from one generation to the next. Thus, however impoverished and exploited, the worker always remained the organizer of technical action, not its object.

Under capitalism, on the contrary, the hand, back, and elbow are required to release their schemas of action on exactly the same terms as tree trunks, fire, or oil. To get at these technical potentials, workers must be split off from institutions such as community and family and reduced to pure instrumentalities. Workers on the assembly line are not essentially members of a community, nor are they merely a source of muscle power as a slave might be: insofar as possible, they are components of the machinery. The computer can be used to extend this logic to education, reducing human inputs to mechanical routines. The reifying extraction of technical elements thus harmonizes with the requirements of the capitalist division of labor, as they are both based on decontextualizing practice.

Decontextualization is of course only the starting point in technical development since the decontextualized elements must be combined with each other to be useful. The resulting device must then be related to other devices and to its natural environment. "Systematization" is the secondary instrumentalization in which these connections are established. The process of systematization has the potential for overcoming the mutilating effects of decontextualization where technical design addresses a sufficiently wide range of contexts. Capitalism does greatly enlarge that range insofar as devices form each others' contexts, integrating enormous numbers of them in tightly coupled networks. This gives rise to what I call "system-centered design," the typical design strategy of modern societies.[7] However, where the wellbeing of workers and nature are concerned, it limits those contexts as much as possible for the sake of control and profits. A socialist technology would not impose such limits on systematization but would reach out to embrace the widest range of contexts in all areas.

Reductionism and mediation Technical means are "abstracted" by reducing complex totalities to those of their elements through which they are exposed to control from above. I will call these controlling elements "primary qualities," not in Locke's epistemological sense but in terms of their essential place in particular technical projects. "Secondary qualities" include everything else about the object, everything that is unimportant to the technical project in which it is enrolled. To the extent that all of reality comes under the sign of technique, the real is progressively reduced to primary qualities.

For example, a valley chosen as a roadbed presents itself to technical reason as a certain concatenation of (primary) geographical and geological qualities subject to manipulation in the interest of transport. Other secondary qualities, such as the valley's plant and animal life or its historical and aesthetic associations, can be overlooked in reconstructing it. A reduction of this sort is unfortunate in the case of a green valley, but it is tragic in the case of a human being. The essential object of capitalist action is the worker.

Since he is located "above" the social subsystems he commands, the manager cannot rely on means that emerge spontaneously within those sub-systems, such as the moral or sentimental social controls of the family. Formal abstraction, which produces technical knowledge by decontextualizing its objects and reducing them to their primary qualities, supplies means to this decontextualized subject as well.

The reduction of the technical object to primary qualities is compensated to some extent in all societies by aesthetic and ethical investments that enrich it once again and adapt it to its environment. All traditional craftsmen apply ethical or religious rules in the course of their work in order to adjust their technical interventions to the requirements of meaning and social stability. They also produce and ornament simultaneously in order to reinsert the object extracted from nature into its new social context. This "mediation-centered design" process disappears in modern societies.[8] They are unique in distinguishing production from aesthetics and ethical regulation. They are heedless of the social insertion of their objects, substitute packaging for an inherent aesthetic elaboration, and are indifferent to the unintended consequences of technology for human beings and nature. Various system crises result from this artificial separation of technique, ethics, and aesthetics.

Autonomization and vocation These reflections on capitalism as a quasitechnical system suggest a metaphoric application to society of Newton's third law: "For every action there is an equal and opposite reaction." In mechanics, actor and object belong to the same system and so every effect is simultaneously a cause, every object simultaneously a subject. In technical action, however, the subject is unaffected by the object on which it acts, thus forming an exception to Newton's law. Technical action autonomizes the subject through dissipating or deferring feedback from the object of action to the actor.[9] This autonomization of the subject has momentous social implications under capitalism, where subject (manager) and object (worker) are both human beings.

Ordinary human relations have a "Newtonian" character. Every action one friend, lover, or family member directs toward another provokes a comparable reaction that promptly affects the initiator of the exchange. The human relations involved in the organization of traditional work are similar. For example, the father, as leader of the familial work group, is exposed by his treatment of his dependent coworkers to consequences fully in proportion to his effects on them. If he drives his "workers" too hard, he suffers in his family, which must aid them to recover. Here action is caught up in a short feedback loop, returning promptly to the acting subject in an "equal and opposite reaction."

The case is different in the technical sphere. The driver of an automobile accelerates to high speeds while experiencing only a slight pressure and small vibrations; the marksman shoots and experiences only a small force transmitted to his shoulder by the stock of the gun. By the same token, management controls workers while minimizing and channeling resistance so far as possible. The absolute disproportion between the "reaction" experienced by the actor and the effect of his action distinguishes these activities as technical. The feedback loop is extended here as far as possible to isolate the subject from the effects of his or her action. Extrapolating this disposition to the limit, one arrives at the ideal of the god, external to the system on which it operates and omnipotent in relation to it.

In fact, of course, human beings are not gods but finite beings. As such, they are part of every system on which they act. The strategic manipulation of people appears to require independence on the part of the actor and passivity on the part of the human object on which he acts. But in fact this polarity is an illusion masking reciprocal interactions. One cannot affect other people without approaching them and becoming in some measure vulnerable to them. *The nearest approximation to being truly "above" the social system to which the actor belongs is for that system itself to reproduce the actor's operational autonomy within it.* This is the nature of capitalist leadership. The capitalist's operational autonomy provides opportunities to place workers in a dependent position where they need precisely the sort of leadership the capitalist supplies. The capitalist enterprise consists in such loops of circular causality through which the enterprise reproduces itself in response to internal tensions and encounters with the outside world.[10]

Once established in this way, the collective laborer can be organized only through external coordination, which gradually comes to seem like just one of the many technical conditions of cooperative production. So normal does it become to exercise control from above that management functions are transferred first from owners to hired executives and eventually, under state socialism, to civil servants, without fundamentally altering the labor process.

In precapitalist societies the autonomization of the technical subject with respect to its objects is overcome in the acquisition of a craft, a vocation. Here what I have called the "Newtonian" character of action, the reciprocity of the relation of subject to object, is recovered in a technical context at a higher level. In vocation, the subject is no longer isolated from objects but is transformed by its own technical relation to them. This relation exceeds passive contemplation or external manipulation and involves the worker as bodily subject and member of a community. It is precisely this quality of traditional technical practice that is eliminated in deskilling and that must be recaptured in a modern context to create a socialist technology. A vocation centered design process would preserve faculty skills by supporting their application in the online environment.

Positioning and initiative In a sense all technique is navigation. Just as the sailor uses the "law" of the winds to reach a destination and the trader anticipates the movements of the market and rides them to success, so too the technical subject falls in with the object's own tendencies to extract a desired outcome. By positioning itself strategically with respect to its objects, the technical subject turns their inherent properties to account. Lukács calls this a "contemplative" form of practice because it changes the "form" of its objects but not their nature (Feenberg, 2000).

The capitalist, like the bureaucrat who inherits his powers in state socialist societies, has established an interiority from which to *act on* social reality, rather than *acting out of* a reality in which he is essentially engaged. Situated in this ideal social locus "above" social processes, he "positions" himself advantageously with respect to the "things" into which his world is fragmented, including the human communities in which he

works and lives. Capitalist practice thus has a strategic aspect: it is based not on a substantive role *within* a given social group but rather on an external relationship to groups in general. The operational autonomy the capitalist enjoys once he enters a social system is the trace of his quasi-externality. Operational autonomy is the occupation of a strategic position with respect to a reified reality.

Capitalist management and product design aims to limit and channel the little initiative that remains to workers and consumers. Their margin of maneuver is reduced to occasional tactical gestures. But the enlargement of margin of maneuver in a socialist trajectory of development would lead to voluntary cooperation in the coordination of effort. It seems appropriate to call this praxis "collegial" since individuals participate in it only insofar as they share responsibility for an institution. In precapitalist societies, such cooperation was often regulated by tradition or paternal authority exercised within moral limits that represented interests of the work group and the craft. In modern societies collegiality is an alternative to traditional bureaucracy with widespread, if imperfect, applications in the organization of professionals such as teachers and doctors. Reformed and generalized, it has the potential for reducing alienation through substituting conscious cooperation for control from above.

Technological Holism

Recontextualizing practice

The hegemony of capital does not rest on a particular technique of social control but more fundamentally on the technical reconstruction of the entire field of social relations within which it operates. The power of the businessman or bureaucrat is already present in the fragmentation of the various social spheres of production, management and labor, family and home life, economics and politics, and so on. The fragmented individuals and institutions can be organized only by agents who dominate them from above.

The secondary instrumentalizations support the reintegration of object with context, primary with secondary qualities, subject with object, and leadership with group. In today's industrial societies, technical practice supports these progressive forms of integration only to the extent that political protest or competitive pressures impose them, but under socialism, technique could incorporate integrative principles and procedures in its basic modus operandi. This new form of technical practice would be characterized by the movement *through reification to reintegration*. It would be adapted to the requirements of a socialist society much as contemporary technique is adapted to the requirements of capitalism.

Since decontextualization predestines technology to serve capitalist power, socialism must recover some of those contextual elements lost in the narrowing of technology to class-specific applications. This requires a *recontextualizing practice* oriented toward a wide range of interests that capitalism represents only partially, interests that reflect human and natural potentialities capitalism ignores or suppresses.

These interests correspond to the lost contexts from which technology is abstracted and the "secondary qualities" of its objects, the sacrificed dimensions of society and nature that bear the burden of technical action. In an earlier period, the socialist movement brought the existence of such interests to light through labor's resistance to total instrumentalization by capital. More recently, feminism and ecology have familiarized us with other suppressed dimensions.

A socialist technical code would be oriented toward the reintegration of the contexts and secondary qualities of both the subjects and objects of capitalist technique. These include ecological, medical, aesthetic, urbanistic, and work-democratic considerations that capitalist and communist societies encounter as "problems," "externalities," and "crises." Health and environmental considerations, the enrichment of work and industrial democracy, must all be internalized as *engineering* objectives. This can be accomplished by multiplying the technical systems that are brought to bear on design to take into account more and more of the essential features of the object of the technology, the needs of operators, consumers, and clients, and the requirements of the environment.

There are probably limits to how far one can go in this direction in the existing industrial civilization. The point is not that capitalism is incapable of dealing with many of its current problems through reactive crisis avoidance.[11] It does usually meet crises with solutions of some sort. But the solutions are often so flawed that they provoke public resistance, as in the case of costly environmental regulation. Deeper problems, such as the pernicious dependence on the automobile, cannot even be posed in the framework of the system.

The need for a general overhaul of technology is ever more apparent, and that overhaul is incompatible with the continued existence of a system of control from above based on social fragmentation. So long as environmental hazards or job dissatisfaction appear as "externalities" they cannot be fundamentally overcome. In this respect, the capitalist or communist bureaucrat cannot claim to be the neutral agent of society's choices because the system that places him or her in a position to represent society has immense substantive consequences.

The underlying problem is the reified separation of labor, consumption, and social decision-making in all modern industrial societies. Given the authoritarian structure of the industrial enterprise, labor has no direct influence on the design of technology but instead manifests its wishes in union strife. Because they do not participate in the original networks of design choice, workers' interests can only be incorporated later through a posteriori regulations that sometimes appear to conflict with the direction of technical progress. But labor is not so much opposed to the advance of technology as to a system in which it is the object rather than the subject of progress.[12] In another social system where it had more influence at an earlier stage in design, it could return to the technical elements and recombine them in conformity with the requirements of a different technical code.

A similar observation applies to environmental problems. These problems appear as such to individuals in social roles remote from industrial decision-making. The very same person who, as a decision maker, accepts the environmentally destructive implications of the dominant technical code flees privately with his family to distant suburbs to find a safe haven from the consequences of decisions such as his. The political protest against pollution returns to haunt the design process in the form of external regulation once flawed technologies have been unleashed on society.

Soviet-style planning offered no improvement over capitalist regulation (O'Connor, 1989). Soviet production depended on transferred technology designed according to capitalist technical codes. No socialist innovation process addressed the inherent flaws of this technology, and a management system based on production quotas left the imported technological base essentially intact. Regulation and planning are thus not so much alternatives to reification as ways of achieving a partial recognition of the totality under the horizon of reification, that is to say, in a social order based on mastery through fragmentation.

The external character of regulation in both capitalist and communist economies introduces inefficiencies into the operation of industrial processes. The problem is not the expense of serving needs such as health, safety, clean air and water, aesthetic goals, and full employment. There is nothing inherently inefficient about such expenditures so long as a proportionate benefit is received. Rather, the essential problem lies in the cascading impacts of the various ex post facto "fixes" imposed on technologies, the workplace, and the environment.

Because technology is designed in abstraction from these so-called soft values, including them at a later stage has highly visible costs. These costs appear to represent essential trade-offs inscribed in the very nature of industrial society when in reality they are side effects of a reified design process. The design of the automobile engine, for example, is complicated by the addition of inelegant pollution control devices, such as catalytic converters. The design of cities is compromised, in turn, by attempts to adapt them to ever more automobiles, and so on. It would be easy to multiply such examples of the social construction of the dilemma of environmental values versus technical efficiency.

The process in which capitalism assembled a collective laborer and supplied it with tools was essentially fragmenting. The mark of that origin can be removed through a new process of sociotechnical integration. The technical heritage must be *overcome* insofar as it reflects the social requirements of capitalism. The many

connections that industrial societies today treat as external must be internalized as technology is reproduced under the aegis of a new dereifying technical code. This is why the integration of the social and technical subsectors requires more than a central plan: it will take technical progress to reform the technology inherited from capitalism.

Concretization

That progress can be theorized in terms of Gilbert Simondon's concept of the "concretization" of technology (Simondon, 1958: chap. 1). Concretization is the discovery of synergisms between technologies and their various environments. Recall that Simondon situates technologies along a continuum that runs from less to more structurally integrated designs. He describes loose designs, in which each part performs a separate function, as "abstract." In the course of technical progress, parts are redesigned to perform multiple functions and structural interactions take on functional roles. These integrative changes yield a more "concrete" technical object that is in fact a system rather than a bunch of externally related elements. For example, a typical concretization occurs in engine design when the surfaces used for the dissipation of heat are merged with those used to reinforce the engine case: two separate structures and their distinct functions are combined in a single structure with two functions.

Simondon argues that technical objects are adapted to their multiple milieus by concretizing advances. Technologies must be compatible with the major constraints of their technical and natural environments: a car's metal skin must protect it from the weather while also reducing air drag to increase effective power; the base of a light bulb must seal it for operation within a certain range of temperatures and pressures while also fitting in standard sockets. All developed technologies exhibit more or less elegant condensations aimed at achieving compatibilities of this sort.

The most sophisticated technologies employ synergies between their various milieus to create a semiartificial environment that supports their own functioning. Simondon calls the combined technical and natural conditions these technologies generate an "associated milieu." It forms a niche with which the technology is in continual recursive causal interaction. The associated milieu

> is that by which the technical object conditions itself in its functioning. This milieu is not manufactured, or at least not totally manufactured; it is a certain order of natural elements surrounding the technical object and linked to a certain order of elements constituting the technical object. The associated milieu mediates the relation between the manufactured technical elements and the natural elements within which the technical object functions. (Simondon, 1958: 57, my trans.)

This higher level of "organic" concreteness is achieved where the technology itself generates the environmental conditions to which it is adapted, as when the heat generated by a motor supplies a favorable operating environment. Energy-efficient housing design offers another example of a technical system that is not simply compatible with environmental constraints but that internalizes them, making them in some sense part of the "machinery." In this case, factors that are only externally and accidentally related in most homes, such as the direction of sunlight and the distribution of glass surfaces, are purposefully combined to achieve a desired effect. The niche in which the house operates is constituted by its angle with respect to the sun.

Human beings are also an operating environment. The craftsman is actually the most important associated milieu of traditional tools, which are adapted primarily to their human users. Although modern machines are organized as technical "individuals" and do not depend on human operators to the same degree, it is still possible to adapt them to an environment of intelligence and skill. But the capitalist technical code militates against solutions to technical problems that place workers once again at the center of the technical system.

The idea of a "concrete technology," which includes nature in its very structure, contradicts the commonplace notion that technical progress "conquers" nature. In Simondon's theory the most advanced forms of progress create complex synergies of technical and natural forces.

Such synergies are achieved by creative acts of invention that transcend apparent constraints or trade-offs and generate a relatively autonomous system out of elements that at first seem opposed or disconnected. *The passage from abstract technical beginnings to concrete outcomes is a general integrative tendency of technological development* that overcomes the reified heritage of capitalist industrialism.

The theory of concretization shows how technical progress might be able to address contemporary social problems through advances that incorporate the wider contexts of human and environmental needs into the structure of machines. While there is no strictly technological imperative dictating such an approach, strategies of concretization could embrace these contexts as they do others in the course of technical development. Where these contexts include environmental considerations, the technology is reintegrated or adapted to nature; where they include the capacities of the human operators, the technology progresses beyond deskilling to become the basis for vocational self-development.

The argument shows that socialist demands for environmentally sound technology and humane, democratic, and safe work are not extrinsic to the logic of technology but respond to the inner tendency of technical development to construct synergistic totalities of natural, human, and technical elements. Nor would the incorporation of socialist requirements into the structure of technology diminish productive efficiency so long as it was achieved through further concretization rather than through multiplying external controls in ever more abstract designs.

All modern industrial societies stand today at the crossroads, facing two different directions of technical development. They can either intensify the exploitation of human beings and nature, or they can take a new path in which the integrative tendencies of technology support emancipatory applications. This choice is essentially political. The first path yields a formally biased system that consistently reinforces elite power. The second path requires a concretizing application of technical principles, taking into account the many larger contexts on which technology has impacts. These contexts reflect potentialities – values – that can be realized only through a new organization of society.

Forward to nature

Some environmentalists argue that the problems caused by modern technology can be solved only by returning to more primitive conditions. This position belongs to a long tradition of antitechnological critique that denounces the alienation of modern society from nature. The "nature" in question is the immediacy from which the objects of technical practice are originally decontextualized, including naturelike elements of culture such as the family. But the price of a return to immediate "naturalness" is the reduction of individuals to mere functions of the whole, absorbed in service to its goals. Such a return to nature would be a reactionary retreat behind the level of emancipation achieved by modernity.

Is there a way of restoring the broken unity of society and nature while avoiding the moral cost of romantic retreat? Or are we destined to oscillate forever between the poles of primitive and modern, solidarity and individuality, domination by nature and domination of nature? This is the ultimate question that a critical theory of technology must address. I have shown that an implausible return to nature is not the only alternative to contemporary industrial society.

Although a new civilization cannot be extracted out of nostalgia for the old, nostalgia is a significant symbolic articulation of interests that are ignored today. These interests point not backward but *forward to nature*, toward a *totality* consciously composed in terms of a wide range of human needs and concerns. This conception of totality as the goal of a process of mediation rather than as an organic presupposition suggests a reply to some common objections to radical arguments for social reconstruction.

We cannot recover what reification has lost by regressing to pretechnological conditions, to some prior unity irrelevant to the contemporary world. The solution is neither a romantic return to the primitive, qualitative, and natural, nor a speculative leap into a "new age" and a whole "new technology." On the contrary, the critical concept of totality aids in identifying the *contingency* of the existing technological system, the points at which it can be invested with new values and bent to new purposes. Those points are to be found where the fragmentation of the established system maintains an alienated power.

The reified systems constructed by capitalist technology must be resituated in the larger contexts from which they are abstracted *today*, not in the past. A partial return to craft labor might be desirable, but it is no solution to the alienation of industrial labor; a further technical advance is needed to reduce alienation through empowering the kind of workers employed in today's society. The horse and plow are not the "context" to which modern agriculture needs to be related but rather the actual environmental and health considerations from which it is abstracted in being constituted as a technological enterprise according to the prevailing technical codes.

We take the reification of technology for granted today, but the present system is completely artificial. Never before have human beings organized their practice in fragments and left the integration of the bits and pieces to chance. The technical environment of capitalism is essentially fragile, constantly at risk from externalities and conflicts, and unable to adjust to the ecological and social problems it causes. As industry becomes ever more powerful, the fragility of the system as a whole increases despite our best efforts to regulate sanity into an insane process of development.

In the past, tradition and custom accomplished a many-sided integration of society and nature. Premodern societies had an organic quality like all other living things on the surface of the earth. Unlike our Promethean assaults on nature, their technologies, however primitive, *conquered time* by constantly reproducing a viable relationship between society and nature.[13] This is the one "conquest" our vaunted technology seems unable to achieve. We must recover the lost art of survival formerly contained in tradition and custom.

That goal cannot be achieved by a regression to traditional forms of personal identity, however comforting these may be in an anomic society. What is required, rather, is a rational recognition of the natural and human constraints on technical development. Such recognition should not be confused with passive submission to external necessity. That confusion arises from the capitalist fixation on the paradigm of primary instrumentalization in terms of which the objects of technique appear simply as raw materials in service to extrinsic goals. Synergisms by which the environment can be enlisted in the structure of appropriate technology are overlooked. These are captured at the level of secondary instrumentalization, which determines a different paradigm of technical practice.

This conception of practice conforms with our current understanding of biological adaptation. From an evolutionary standpoint, living things relate to their environment actively as well as passively, selecting out that dimension of the world around them to which they adapt. This process of selection is of course unconscious, but it is formally quite similar to the way in which a human society might choose to treat the variety of natural limits it confronts.

In adapting, living things engage in concretizing strategies not so different from the technical developments discussed here. They, too, incorporate environmental constraints into their structure, something that human societies must also learn to do through redesigning technology in more concrete forms.[14] No social system can be natural, but a socialist society would have at least some of the essential interdependence with its environment that characterizes organic beings. It would therefore represent an advance to a higher level of integration between humanity and nature (Moscovici, 1968: 562).

Nature as a context of development is not a final purpose but a dialectical limitation that invites transcendence through adaptation. To conceptualize a totality once again, we need not know in advance precisely in what way human beings will confront the limitations they meet. We need only gain insight into the *form of the process* of mediation. As the structure of a new social practice, this mediating activity opens infinite possibilities rather than foreclosing the future in some preconceived utopia. Adaptation maintains the formal character of the modern concept of freedom and therefore does not reduce individuals to mere functions of society. Freedom lies in this lack of determinacy.

Notes

1 I am grateful to John Ely for pointing out this connection. For accounts of the tension in Marxism between naturalistic holism and theory of the social construction of nature, see Ely (1988), Ely (1989), and Vogel (1995).

2 See Schluchter (1979: 57, 117–18).

3 The contribution of the Frankfurt School to the age-old debate on the problem of universals deserves a study. I would guess that such a study would find considerable agreement, if not a doctrine. For example, Marcuse's position and Adorno's have more in common than is usually recognized. Michael Ryan points out that in contrast to Marcuse, who claims that universals like "freedom" contain more content than is ever realized in particular institutions, Adorno claims that it is the particular that contains an excess of content with respect to the universal (Ryan 1982: 73). But the universal at issue is different. The surplus to which Adorno refers is precisely the basis on which Marcuse refuses to identify limited realizations of freedom with the universal (Marcuse, 1964: 105–6).

4 The passivity of the experimenter to which he refers is only apparent: the experimenter actively constructs the observed object but, at least in Lukács's view, is not aware of having done so and interprets the experiment as the voice of nature. While this does not criticize the epistemological consequences of this illusion in natural science, in the social arena it defines reification.

5 Is there still a distinction between science and technology? Not if you believe certain science studies scholars who talk about a single unified "technoscience." The concept of "scaling up" is supposed to get us from the laboratory to society. But that concept can mask the tremendously complex and differentiated processes involved in applying new scientific ideas to production. There is a significant gap here that justifies the distinction.

6 For a complementary exposition of instrumentalization theory, see Feenberg (1999: chap. 9).

7 I called this "system congruent design" in the conclusion to *Alternative Modernity* (Feenberg, 1995: 228).

8 I called this "expressive design" in the conclusion to *Alternative Modernity* (Feenberg, 1995: 225).

9 For the environmental consequences of autonomization, see O'Connor (1989) and Beck (1992).

10 This analysis can be clarified in terms of Jean-Pierre Dupuy's system theoretic interpretation of the concept of alienation. Dupuy defines "autonomy" as the ability of a system to reproduce certain stable characteristics under a variety of conditions. These stable characteristics can be considered "system effects," emergent behaviors proper to the system itself. Dupuy's analysis of panic illustrates this notion by showing that leadership in crowds is a system effect: the power that apparently flows down from the leader is in fact based on the relations governing the interactions

of the mass. The leader is an "endogenous fixed point . . . produced by the crowd although the crowd believes itself to have been produced by it. Such a tangling of different levels is . . . a distinguishing feature of autonomous systems" (Dupuy, n.d.: 23).

11 The theory of reactive crisis avoidance as the general form of movement of the capitalist state can be extended to the domain of technology using Simondon's categories of concrete and abstract design. See Habermas (1975); Offe (1987); O'Connor (1984).

12 See "A Technology Bill of Rights," in Shaiken (1984). On the role of workers in innovation, see Wilkinson (1983: chap. 9).

13 It is true that some premodern societies destroyed their own natural environments, for example, through overgrazing. However, one can hardly compare destructive processes that took many centuries to show their effects with modern environmental problems and the threat of nuclear weapons.

14 Levins and Lewontin (1985: 104). Merleau-Ponty (1963) expressed the idea clearly in an early book: he signifies that the organism itself measures the action of things upon it and itself delimits its milieu by a circular process which is without analogy in the physical world.

References

Adorno, T. and M. Horkheimer. *The Dialectic of Enlightenment* (New York: Herder and Herder, 1972).

Beck, U. *Risk Society* (London: Sage, 1992).

Bernstein, R. "Negativity: Theme and Variations." In R. Pippin, A. Feenberg, and C. Weber, eds., *Marcuse: Critical Theory and the Promise of Utopia* (South Hadley, MA: Bergin and Garvey, 1988).

Bloch, E. "Art and Society". In *The Utopian Function of Art and Literature*. Trans. J. Zipes and F. Mecklenberg (Cambridge, MA: MIT Press, 1988).

Bordo, S. *The Flight to Objectivity: Essays on Cartesianism and Culture* (Albany: SUNY Press, 1987).

Claudin-Urondo, C. *Lénine et la révolution culturelle* (Paris: Mouton, 1998).

Dupuy, J.-P. *The Autonomy of Social Reality: On the Contributions of the Theory of Systems to the Theory of Society* (Unpublished manuscript, n.d.).

Ely, J. "Lukác's Construction of Nature". *Capitalism, Nature, Socialism* 1, 1988.

Ely, J. "Ernst Bloch and the Second Contradiction of Capitalism". *Capitalism, Nature, Socialism* 2, 1989.

Feenberg, A. *Lukács, Marx, and The Source of Critical Theory* (New York: Oxford University Press, 1986).

Feenberg, A. "A Fresh Look at Lukács: On Steven Vogel's *Against Nature*." *Rethinking Marxism*, Vol. 11, No. 4, 2000.

Goldman, S. "Philosophy, Engineering, and Western Culture." In P. Durbin, ed., *Broad and Narrow Interpretations of Philosophy of Technology* (Dordrecht: Kluwer Academic Publishers, 1990).

Graham, L. *What We Have Learned About Science and Technology From The Russian Experience* (Stanford: Stanford University Press, 1998).

Harding, S. *The Science Question in Feminism* (Ithaca: Cornell University Press, 1986).

Jay, M. *Marxism and Totality: The Adventures of a Concept from Lukács to Habermas* (Berkeley: University of California Press, 1984).

Levins, R. and Lewontin, R. *The Dialectical Biologist* (Cambridge, MA: Harvard University Press, 1985).

Lukács, G. *History and Class Consciousness* (Cambridge, MA: MIT Press, 1971).

Marcuse, H. *One Dimensional Man: Studies in the Ideology of Advanced Industrial Society* (Boston: Beacon Press, 1964).

Marcuse, H. "Marxism and Feminism." *Women's Studies*, Vol. 2. 1974.

O'Connor, J. *The Meaning of Crisis* (Oxford: Basil Blackwell, 1989).

Offe, C. *Contradictions of the Welfare State*, ed. J. Keane (Cambridge, MA: MIT Press, 1987).

Pippin, R. *Modernism as a Philosophical Problem: On The Dissatisfactions of European High Culture* (New York: Basil Blackwell, 1991).

Ryan, M. *Marxism and Deconstruction: A Critical Appraisal* (Baltimore, MD: Johns Hopkins University Press, 1982).

Schluchter, W. *The Rise of Western Rationalism: Max Weber's Developmental History*, trans. G. Roth (Berkeley, CA: University of California Press, 1979).

Shaiken, H. *Work Transformed* (Lexington, MA: D. C. Heath, 1984).

Simondon, G. *Du mode d'existence des objets techniques* (Paris: Aubier, 1958).

Vogel, S. *Against Nature: The Concept of Nature in Critical Theory* (Unpublished manuscripts, 1995).

Wilkinson, B. *The shepfloor Politics of New Technology* (London: Heineman Educational Books, 1983).

Part V

Pragmatic Considerations

16

Science and Society

John Dewey

John Dewey (1859–1952) was one of the foremost developers of American Pragmatism. Along with Charles Sanders Peirce and William James, he took the open, experimental, and practical nature of technoscientific inquiry to be the paradigmatic example of all inquiry. For Dewey, all inquiry is similar in form to technoscientific inquiry in that it is fallibilistic, resolves in practice some initial question through an experimental method, but provides no final absolute answer.

Dewey argues that humans are constantly developing through transactions with their environment, and a central aspect of that environment is technology. Thus, we are a prime source of those environments through which we develop, and either thrive or fail. In this essay, he argues that many of the problems laid at the feet of technology are actually the fault of a human failing to be sufficiently attentive and critical as we create technologies, including human social institutions. He further argues that successful technology is democratic, as both terms embody the method of intelligence and both are forms of human inquiry. He challenges us to think carefully about the relations between technology and democracy, and about our responsibilities for consciously directing technological change.

Three features of Dewey's conception of inquiry are relevant: inquiry as problem-solving, as historical and progressive, and as communal. We engage in inquiry, Dewey thought, as part of a struggle with an objectively precarious but improvable environment. Inquiry is demanded by what he calls an incomplete or problematic situation, that is, one in which something must be done. The goal of inquiry is not simply a change in the beliefs of the inquirers but

From *John Dewey: The Later Works, 1925–1953*, vol. 6: *1931–1932*, ed. Jo Ann Boydston (Carbondale, IL: Southern Illinois University Press, 1985), pp. 53–63.

the resolution of the problematic situation. Technology is, on this understanding, our interventions in these problematic situations.

The significant outward forms of the civilization of the western world are the product of the machine and its technology. Indirectly, they are the product of the scientific revolution which took place in the seventeenth century. In its effect upon men's external habits, dominant interests, the conditions under which they work and associate, whether in the family, the factory, the state, or internationally, science is by far the most potent social factor in the modern world. It operates, however, through its undesigned effects rather than as a transforming influence of men's thoughts and purposes. This contrast between outer and inner operation is the great contradiction in our lives. Habits of thought and desire remain in substance what they were before the rise of science, while the conditions under which they take effect have been radically altered by science.

When we look at the external social consequences of science, we find it impossible to apprehend the extent or gauge the rapidity of their occurrence. Alfred North Whitehead has recently called attention to the progressive shortening of the time-span of social change. That due to basic conditions seems to be of the order of half a million years; that due to lesser physical conditions, like alterations in climate, to be of the order of five thousand years. Until almost our own day the time-span of sporadic technological changes was of the order of five hundred years; according to him, no great technological changes took place between, say, 100 A.D. and 1400 A.D. With the introduction of steam-power, the fifty years from 1780 to 1830 were marked by more changes than are found in any previous thousand years. The advance of chemical techniques and in use of electricity and radio-energy in the last forty years makes even this last change seem slow and awkward.

Domestic life, political institutions, international relations and personal contacts are shifting with kaleidoscopic rapidity before our eyes. We cannot appreciate and weigh the changes; they occur too swiftly. We do not have time to take

them in. No sooner do we begin to understand the meaning of one such change than another comes and displaces the former. Our minds are dulled by the sudden and repeated impacts. Externally, science through its applications is manufacturing the conditions of our institutions at such a speed that we are too bewildered to know what sort of civilization is in process of making.

Because of this confusion, we cannot even draw up a ledger account of social gains and losses due to the operation of science. But at least we know that the earlier optimism which thought that the advance of natural science was to dispel superstition, ignorance, and oppression, by placing reason on the throne, was unjustified. Some superstitions have given way, but the mechanical devices due to science have made it possible to spread new kinds of error and delusion among a larger multitude. The fact is that it is foolish to try to draw up a debit and credit account for science. To do so is to mythologize; it is to personify science and impute to it a will and an energy on its own account. In truth science is strictly impersonal; a method and a body of knowledge. It owes its operation and its consequences to the human beings who use it. It adapts itself passively to the purposes and desires which animate these human beings. It lends itself with equal impartiality to the kindly offices of medicine and hygiene and the destructive deeds of war. It elevates some through opening new horizons; it depresses others by making them slaves of machines operated for the pecuniary gain of owners.

The neutrality of science to the uses made of it renders it silly to talk about its bankruptcy, or to worship it as the usherer in of a new age. In the degree in which we realize this fact, we shall devote our attention to the human purposes and motives which control its application. Science is an instrument, a method, a body of technique. While it is an end for those inquirers who are engaged in its pursuit, in the large human sense it is a means, a tool. For what ends shall it be used? Shall it be used deliberately, systematically, for the promotion of social well-being, or shall it be employed primarily for private aggrandizement, leaving its larger social results to chance? Shall the scientific attitude be used to create new mental and moral attitudes, or shall it continue to be subordinated to the service of desires, purposes and institutions which were formed before science came into existence? Can the attitudes which control the use of science be themselves so influenced by scientific technique that they will harmonize with its spirit?

The beginning of wisdom is, I repeat, the realization that science itself is an instrument which is indifferent to the external uses to which it is put. Steam and electricity remain natural forces when they operate through mechanisms; the only problem is the purposes for which men set the mechanisms to work. The essential technique of gunpowder is the same whether it be used to blast rocks from the quarry to build better human habitations, or to hurl death upon men at war with one another. The airplane binds men at a distance in closer bonds of intercourse and understanding, or it rains missiles of death upon hapless populations. We are forced to consider the relation of human ideas and ideals to the social consequences which are produced by science as an instrument.

The problem involved is the greatest which civilization has ever had to face. It is, without exaggeration, the most serious issue of contemporary life. Here is the instrumentality, the most powerful, for good and evil, the world has ever known. What are we going to do with it? Shall we leave our underlying aims unaffected by it, treating it merely as a means by which uncooperative individuals may advance their own fortunes? Shall we try to improve the hearts of men without regard to the new methods which science puts at our disposal? There are those, men in high position in church and state, who urge this course. They trust to a transforming influence of a morals and religion which have not been affected by science to change human desire and purpose, so that they will employ science and machine technology for beneficent social ends. The recent Encyclical of the Pope is a classic document in expression of a point of view which would rely wholly upon inner regeneration to protect society from the injurious uses to which science may be put. Quite apart from any ecclesiastical

connection, there are many "intellectuals" who appeal to inner "spiritual" concepts, totally divorced from scientific intelligence, to effect the needed work. But there is another alternative: to take the method of science home into our own controlling attitudes and dispositions, to employ the new techniques as means of directing our thoughts and efforts to a planned control of social forces.

Science and machine technology are young from the standpoint of human history. Though vast in stature, they are infants in age. Three hundred years are but a moment in comparison with thousands of centuries man has lived on the earth. In view of the inertia of institutions and of the mental habits they breed, it is not surprising that the new technique of apparatus and calculation, which is the essence of science, has made so little impression on underlying human attitudes. The momentum of traditions and purposes that preceded its rise took possession of the new instrument and turned it to their ends. Moreover, science had to struggle for existence. It had powerful enemies in church and state. It needed friends and it welcomed alliance with the rising capitalism which it so effectively promoted. If it tended to foster secularism and to create predominantly material interests, it could still be argued that it was in essential harmony with traditional morals and religion. But there were lacking the conditions which are indispensable to the serious application of scientific method in reconstruction of fundamental beliefs and attitudes. In addition, the development of the new science was attended with so many internal difficulties that energy had to go to perfecting the instrument just as an instrument. Because of all these circumstances the fact that science was used in behalf of old interests is nothing to be wondered at.

The conditions have now changed, radically so. The claims of natural science in the physical field are undisputed. Indeed, its prestige is so great that an almost superstitious aura gathers about its name and work. Its progress is no longer dependent upon the adventurous inquiry of a few untrammeled souls. Not only are universities organized to promote scientific research and learning, but one may almost imagine the university laboratories abolished and still feel confident of the continued advance of science. The development of industry has compelled the inclusion of scientific inquiry within the processes of production and distribution. We find in the public prints as many demonstrations of the benefits of science from a business point of view as there are proofs of its harmony with religion.

It is not possible that, under such conditions, the subordination of scientific techniques to purposes and institutions that flourished before its rise can indefinitely continue. In all affairs there comes a time when a cycle of growth reaches maturity. When this stage is reached, the period of protective nursing comes to an end. The problem of securing proper use succeeds to that of securing conditions of growth. Now that science has established itself and has created a new social environment, it has (if I may for the moment personify it) to face the issue of its social responsibilities. Speaking without personification, we who have a powerful and perfected instrument in our hands, one which is determining the quality of social changes, must ask what changes we want to see achieved and what we want to see averted. We must, in short, plan its social effects with the same care with which in the past we have planned its physical operation and consequences. Till now we have employed science absentmindedly as far as its effects upon human beings are concerned. The present situation with its extraordinary control of natural energies and its totally unplanned and haphazard social economy is a dire demonstration of the folly of continuing this course.

The social effects of the application of science have been accidental, even though they are intrinsic to the private and unorganized motives which we have permitted to control that application. It would be hard to find a better proof that such is the fact than the vogue of the theory that such unregulated use of science is in accord with "natural law," and that all effort at planned control of its social effects is an interference with nature. The use which has been made of a peculiar idea of personal liberty to justify the dominion of accident in social affairs is another convincing proof. The doctrine that the most potent instrument of widespread, enduring, and objective social changes must be left at the mercy of purely private desires for purely personal gain is a doctrine of anarchy. Our present insecurity of life is the fruit of the adoption in practice of this anarchic doctrine.

The technologies of industry have flowed from the intrinsic nature of science. For that is itself essentially a technology of apparatus, materials and numbers. But the pecuniary aims which have decided the social results of the use of these technologies have not flowed from the inherent nature of science. They have been derived from institutions and attendant mental and moral habits which were entrenched before there was any such thing as science and the machine. In consequence, science has operated as a means for extending the influence of the institution of private property and connected legal relations far beyond their former limits. It has operated as a device to carry an enormous load of stocks and bonds and to make the reward of investment in the way of profit and power one out of all proportion to that accruing from actual work and service.

Here lies the heart of our present social problem. Science has hardly been used to modify men's fundamental acts and attitudes in social matters. It has been used to extend enormously the scope and power of interests and values which anteceded its rise. Here is the contradiction in our civilization. The potentiality of science as the most powerful instrument of control which has ever existed puts to mankind its one outstanding present challenge.

There is one field in which science has been somewhat systematically employed as an agent of social control. Condorcet, writing during the French Revolution in the prison from which he went to the guillotine, hailed the invention of the calculus of probabilities as the opening of a new era. He saw in this new mathematical technique the promise of methods of insurance which should distribute evenly and widely the impact of the disasters to which humanity is subject. Insurance against death, fire, hurricanes and so on have in a measure confirmed his prediction. Nevertheless, in large and important social areas, we have only made the merest beginning of the method of insurance against the hazards of life and death. Insurance against the risks of maternity, of sickness, old age, unemployment, is still rudimentary; its idea is fought by all reactionary forces. Witness the obstacles against which social insurance with respect to accidents incurred in industrial employment had to contend. The anarchy called natural law and personal liberty still operates with success against a planned social use of the resources of scientific knowledge.

Yet insurance against perils and hazards is the place where the application of science has gone the furthest, not the least, distance in present society. The fact that motor cars kill and maim more persons yearly than all factories, shops, and farms is a fair symbol of how backward we are in that province where we have done most. Here, however, is one field in which at least the idea of planned use of scientific knowledge for social welfare has received recognition. We no longer regard plagues, famine and disease as visitations of necessary "natural law" or of a power beyond nature. By preventive means of medicine and public hygiene as well as by various remedial measures we have in idea, if not in fact, placed technique in the stead of magic and chance and uncontrollable necessity in this one area of life. And yet, as I have said, here is where the socially planned use of science has made the most, not least, progress. Were it not for the youth of science and the historically demonstrated slowness of all basic mental and moral change, we could hardly find language to express astonishment at the situation in which we have an extensive and precise control of physical energies and conditions, and in which we leave the social consequences of their operation to chance, laissez-faire, privileged pecuniary status, and the inertia of tradition and old institutions.

Condorcet thought and worked in the Baconian strain. But the Baconian ideal of the systematic organization of all knowledge, the planned control of discovery and invention, for the relief and advancement of the human estate, remains almost as purely an ideal as when Francis Bacon put it forward centuries ago. And this is true in spite of the fact that the physical and mathematical technique upon which a planned control of social results depends has made in the meantime incalculable progress. The conclusion is inevitable. The outer arena of life has been transformed by science. The effectively working mind and character of man have hardly been touched.

Consider that phase of social action where science might theoretically be supposed to have taken effect most rapidly, namely, education. In dealing with the young, it would seem as if scientific methods might at once take effect in transformation of mental attitudes, without meeting the obstacles which have to be overcome in dealing with adults. In higher education, in universities and technical schools, a great amount of research is done and much scientific knowledge is imparted. But it is a principle of modern psychology that the basic attitudes of mind are formed in the earlier years. And I venture the assertion that for the most part the formation of intellectual habits in elementary education, in the home and school, is hardly affected by scientific method. Even in our so-called progressive schools, science is usually treated as a side line, an ornamental extra, not as the chief means of developing the right mental attitudes. It is treated generally as one more body of ready-made information to be acquired by traditional methods, or else as an occasional diversion. That it is the method of all effective mental approach and attack in all subjects has not gained even a foothold. Yet if scientific method is not something esoteric but is a realization of the most effective operation of intelligence, it should be axiomatic that the development of scientific attitudes of thought, observation, and inquiry is the chief business of study and learning.

Two phases of the contradiction inhering in our civilization may be especially mentioned. We have long been committed in theory and words to the principle of democracy. But criticism of democracy, assertions that it is failing to work and even to exist are everywhere rife. In the last few months we have become accustomed to similar assertions regarding our economic and industrial system. Mr. Ivy Lee, for example, in a recent commencement address, entitled "This Hour of Bewilderment," quoted from a representative clergyman, a railway president, and a publicist, to the effect that our capitalistic system is on trial. And yet the statements had to do with only one feature of that system: the prevalence of unemployment and attendant insecurity. It is not necessary for me to invade the territory of economics and politics. The essential fact is that if both democracy and capitalism are on trial, it is in reality our collective intelligence which is on trial. We have displayed enough intelligence in the physical field to create the new and powerful instrument of science and technology. We have not as yet had enough intelligence to use this instrument deliberately and systematically to control its social operations and consequences.

The first lesson which the use of scientific method teaches is that control is coordinate with knowledge and understanding. Where there is technique there is the possibility of administering forces and conditions in the region where the technique applies. Our lack of control in the sphere of human relations, national, domestic, international, requires no emphasis of notice. It is proof that we have not begun to operate scientifically in such matters. The public press is full of discussion of the five-year plan and the ten-year plan in Russia. But the fact that the plan is being tried by a country which has a dictatorship foreign to all our beliefs tends to divert attention from the fundamental consideration. The point for us is not this political setting nor its communistic context. It is that by the use of all available resources of knowledge and experts an attempt is being made at organized social planning and control. Were we to forget for the moment the special Russian political setting, we should see here an effort to use coordinated knowledge and technical skill to direct economic resources toward social order and stability.

To hold that such organized planning is possible only in a communistic society is to surrender the case to communism. Upon any other basis, the effort of Russia is a challenge and a warning to those who live under another political and economic regime. It is a call to use our more advanced knowledge and technology in scientific thinking about our own needs, problems, evils, and possibilities so as to achieve some degree of control of the social consequences which the application of science is, willy-nilly, bringing about. What stands in the way is a lot of outworn traditions, moth-eaten slogans and catchwords, that do substitute duty for thought, as well as our entrenched predatory self-interest. We shall only make a real beginning in intelligent thought when we cease mouthing platitudes; stop confining our idea to antitheses of individualism and socialism, capitalism and communism, and realize that the issue is between chaos and order, chance and control: the haphazard use and the planned use of scientific techniques.

Thus the statement with which we began, namely, that we are living in a world of change extraordinary in range and speed, is only half true.

It holds of the outward applications of science. It does not hold of our intellectual and moral attitudes. About physical conditions and energies we think scientifically; at least, some men do, and the results of their thinking enter into the experiences of all of us. But the entrenched and stubborn institutions of the past stand in the way of our thinking scientifically about human relations and social issues. Our mental habits in these respects are dominated by institutions of family, state, church, and business that were formed long before men had an effective technique of inquiry and validation. It is this contradiction from which we suffer to-day.

Disaster follows in its wake. It is impossible to overstate the mental confusion and the practical disorder which are bound to result when external and physical effects are planned and regulated, while the attitudes of mind upon which the direction of external results depends are left to the medley of chance, tradition, and dogma. It is a common saying that our physical science has far outrun our social knowledge; that our physical skill has become exact and comprehensive while our humane arts are vague, opinionated, and narrow. The fundamental trouble, however, is not lack of sufficient information about social facts, but unwillingness to adopt the scientific attitude in what we do know. Men floundered in a morass of opinion about physical matters for thousands of years. It was when they began to use their ideas experimentally and to create a technique or direction of experimentation that physical science advanced with system and surety. No amount of mere fact-finding develops science nor the scientific attitude in either physics or social affairs. Facts merely amassed and piled up are dead; a burden which only adds to confusion. When ideas, hypotheses, begin to play upon facts, when they are methods for experimental use in action, then light dawns; then it becomes possible to discriminate significant from trivial facts, and relations take the place of isolated scraps. Just as soon as we begin to use the knowledge and skills we have to control social consequences in the interest of shared abundant and secured life, we shall cease to complain of the backwardness of our social knowledge. We shall take the road which leads to the assured building up of social science just as men built up physical science when they actively used the techniques of tools and numbers in physical experimentation.

In spite, then, of all the record of the past, the great scientific revolution is still to come. It will ensue when men collectively and cooperatively organize their knowledge for application to achieve and make secure social values; when they systematically use scientific procedures for the control of human relationships and the direction of the social effects of our vast technological machinery. Great as have been the social changes of the last century, they are not to be compared with those which will emerge when our faith in scientific method is made manifest in social works. We are living in a period of depression. The intellectual function of trouble is to lead men to think. The depression is a small price to pay if it induces us to think about the cause of the disorder, confusion, and insecurity which are the outstanding traits of our social life. If we do not go back to their cause, namely our half-way and accidental use of science, mankind will pass through depressions, for they are the graphic record of our unplanned social life. The story of the achievement of science in physical control is evidence of the possibility of control in social affairs. It is our human intelligence and human courage which are on trial; it is incredible that men who have brought the technique of physical discovery, invention, and use to such a pitch of perfection will abdicate in the face of the infinitely more important human problem.

17

Technology and Community Life

Larry Hickman

*When the machine age has thus perfected its machinery it will be a means of life
and not its despotic master. Democracy will come into its own, for democracy is a
name for a life of free and enriching communion. It had its seer in Walt Whitman.
It will have its consummation when free social inquiry is indissolubly wedded to
the art of full and moving communication.*
— John Dewey, *The Public and Its Problems* (LW.2.350)

Larry Hickman has been since 1993 Director of the Center for
Dewey Studies, and Professor of Philosophy at Southern Illinois
University, Carbondale. He is one of the leading developers of pragmatic philosophy of technology. His work takes Dewey's as a starting
point, and then, often by responding to criticisms of Dewey, he
expands and deepens Dewey's insights. Two important criticisms have
been offered against Dewey's view of technology. One is the claim
that everything becomes technology, and the second is the argument
that pragmatism is apologetics for consumerism and the status quo
and calls for a technocracy. In response to each, Hickman offers a
revisioned pragmatism, what he calls "productive pragmatism," that
draws on Dewey but is distinctly his own.

Addressing the first, Hickman provides us with a four-part typology of human activities. The four types of activities are those that
involve tool use, the technological (tool use and cognitive activity)
and the technical (tool use but little or no cognitive activity), and
those that do not, the non-instrumental but cognitive, and the
non-instrumental and non-cognitive. The first two of these involve
tool use, but only the first is technology properly understood. On this
Hickman has commented, "I argued that technology is a term that
should be treated as analogous to biology or geology. Technology is

From Larry Hickman, *Philosophical Tools for Technological Culture* (Bloomington, IN: Indiana University Press, 2001),
pp. 44–64. Reprinted by permission of the publisher, Indiana University Press.

inquiry into our tools and techniques." Important in this typology is the clear fact that a significant portion of human activity is not technological as understood on this typology, and, as Hickman notes, much of it falls into the fourth category, the non-cognitive and non-instrumental. "The greatest part of life," he writes, "is what is immediate and habitual."

The critique of Dewey's work as calling for a technocracy has been heard from the political left and right, and Hickman responds to charges from both quarters. The fundamentalist religious right attacks Dewey and science, often in the same breath. And, the critical theorists, starting with Horkheimer, find Dewey's talk of "instrumentalism" indistinguishable from the instrumental rationality critiqued in works such as *The Dialectic of Enlightenment* (Adorno and Horkheimer, 1976). These are no mere problems of theory, as almost daily we learn of some attempt to roll back the serious presentation of science in public education or policy, or yet another technological advance so complicated that it is presumed beyond the understanding of most people. Hickman argues that precisely because it excludes democratic participation and denies the importance of individual experience, technocracy is a social dead-end. "Wherever individuals are not free to articulate problems and to attack them experimentally, then growth within the society is greatly diminished." Hickman thus wrestles with (and against) all the transcendental thinkers mentioned earlier, arguing that the right kind of technology can avoid the excesses of both the left and the right in dealing with technosocial problems.

I. Dewey's Productive Pragmatism in Context

One of the central lessons of the history of technology is that change generally entails displacement and conflict. At the level of community life, new institutions and methods compete with older ones for public acceptance. Familiar jobs and occupations disappear and are replaced by others that require new skills. Geographic migrations occur. Forms of social organization and practice that seemed satisfactory for parents and grandparents no longer seem appropriate or even possible.

From early in his life, Dewey was impressed with the effects of rapid technological change upon community life. During a visit to his father, who was then in the Union army in Virginia, the seven-year-old Dewey witnessed the devastation that had been brought about by the new technologies and techniques of the American Civil War. As he began his academic career, his world exploded with new inventions. During his decade at the University of Michigan (1884–94), Americans received news of the first steam turbine engine, the first single-cylinder automobile engine, the first pneumatic tire, and the first wireless telegraph. The following decade, which Dewey spent at the University of Chicago, was a time of even more rapid technological innovation. The cinema, X rays, magnetic recording of sound, radio transmission, airplane flight, and many other inventions contributed to a period of breathtaking change. Chicago was irrevocably altered by changing technological factors. During his decade there, Dewey's city absorbed massive waves of immigration from Europe and migration from the southern United States. It was also the site of bloody confrontations between workers and industrialists.

Dewey's association with the Hull House experiments of social reformer Jane Addams provided him with an important vantage point from which to view these phenomena. He became acutely aware that technological change not only produces the difficulties of displacement, but also offers opportunities for new forms of cooperation and communication. Even more important, he developed the view that philosophy could play a positive role in the transformation of American culture. In 1925, two decades after he left Chicago for New York and Columbia University, he would write that the "proper task" of philosophy is the liberation and clarification of meanings, including those that have been generated and propagated by advances made in scientific technology (Later Works.1.307).

In Dewey's time, as in our own, many philosophers have argued that their discipline should be above the fray of ordinary, everyday concerns. These arguments have taken two principal forms. The first, espoused by some of the heirs of the logical positivism of the Vienna Circle tradition, has been that philosophy should be modeled after the theoretical sciences,[1] and that its proper focus should thus be limited to what is narrowly empirical or rigorously abstract. The second, most recently espoused by some "post-modern" philosophers and even by some neo-pragmatists,[2] has been that philosophy is primarily an imaginative or literary art, and that its focus should thus be limited to exploration of the aesthetic dimensions of experience. Although these two approaches appear to have little else in common, they do share the view that philosophy has little or no role to play in the public sphere with respect to the reform of technological culture.

It was Dewey's position that both of these views are defective as they are usually articulated, but that each nevertheless contains an element of truth. He thought that any philosophical activity worthy of the name aims at transforming inchoate or confused human experiences by helping to generate the conditions and implements necessary for their enlargement and clarification. Consequently, it is a part of the task of philosophy to facilitate the construction and use of abstract entities insofar as such entities can function as tools of inquiry. But Dewey also thought that a central aim of philosophy is criticism of received values, or, as he put it, the promotion of "a heightened consciousness of deficiencies and corruptions in the scheme and distribution of values that obtains at any period" (Later Works.1.308). Consequently, it is a part of the task of philosophy to promote active and refined aesthetic appreciation as a tool for the development of critical consciousness.

An important source of Dewey's activism with respect to the reform of technological culture in general, and his own American culture in particular, was his commitment to what has been termed "evolutionary naturalism." [. . .] This is the view that humans are biological organisms who live their lives interacting with and evolving within the rest of nature. As such, we humans are not ontologically separate from nature, nor can we ever fully escape the existential pushes and pulls of its facilities and constraints. We are, however, able to enhance some of those facilities and mitigate some of those constraints through the use of tools. As Dewey wrote in his lecture notes during 1926, "[t]ools are the expression of the man/environment interaction; by their way means and consequences of action are adapted to each other."[3]

As I also indicated in the last chapter, when Dewey used the term "tools" he was referring to more than just tangible objects such as hammers or computers. He thought that other tools, such as personal habits, shared ideas, and even social institutions, are not any less involved in what is technical and technological just because they are abstract or intangible. He argued that the traditional separation of what is concrete from what is abstract, with its tendency to award the abstract a place of special honor above the concrete, has been the source of considerable confusion and a brake to social progress.

It is by means of the use of many sorts of tools – tangible as well as intangible, and concrete as well as abstract – that human beings are able both to alter their environing conditions and to accommodate themselves to those conditions. This adjustment occurs as a result of inquiry that enables human beings to project themselves forward in time beyond the preoccupations of the present moment in ways that are unavailable to less complex organisms. In populations of such organisms, adjustment occurs as a result of multiple factors that include sexual selection, genetic mutations, and the demise of individuals that are not fitted to changing conditions. Adjustment within human populations also occurs in these ways, but humans are also able to enhance their adjustment by means of an ongoing invention, development, and use of tools of all sorts. The most important tool that humans have at their disposal is language, which Dewey called "the tool of tools" (Later Works.1.134).

Unlike many current philosophers of language, however, Dewey thought that the domain of language is much wider than spoken or written expression. He argued that the many varieties of expression in the plastic and visual arts, for example, not to mention music, are also instances of communication or language. They are among the tools that human beings use to assess the meanings of their experience and to effect ongoing adjustment within their changing environments. Dewey thought that knowledge itself – in the arts as well as in the sciences – is both a technologically[4] produced artifact and a tool that human beings use in order to make other artifacts.

In the last chapter I explicated some of the meanings of "technology" along Deweyan lines. In terms that are more specific to the philosophical movement of which he was one of the founding members, however, Dewey's productive version of pragmatism is committed to the use of tools of all sorts within experimental situations in order to effect forward-looking adjustment to environing conditions by means of consideration of practical effects. As he put it in his well-known (1925) essay "The Development of American Pragmatism," productive pragmatism, or what he called "instrumentalism," involves "an attempt to establish a precise logical theory of concepts, of judgments and inferences in their various forms, by considering primarily how thought functions in the experimental determinations of future consequences. . . . It aims to constitute a theory of the general forms of conception and reasoning, and not of this or that particular judgment or concept related to its own content, or to its particular implications" (Later Works.2.14).

Implicit in these remarks is the claim that "forms of conception" and "reasoning" take place not just "in the mind," but as features of a fully fleshed out involvement of the organism within its environment. In Dewey's view, reasoning takes place in the literary and plastic arts, in engineering, in jurisprudence, in the writing of history, in agriculture, in music, in the culinary and vintner's arts, and wherever else systematic, self-conscious, creative, forward-looking adjustment occurs.

Dewey's productive pragmatism has been sharply attacked from two fronts. From one side it has been dismissed as being too weak to provide adequate guidance for difficult decisions.[5] Some of these critics, such as the partisans of fundamentalist Christianity, have argued on the basis of their personal or institutional religious commitments that decisions must be grounded on absolute truths that are revealed by God and applicable to all times and places.[6] Other critics, such as some of the first generation of the Frankfurt School, have argued on more strictly philosophical grounds that truth is much more than a tool of action.[7] In their view it is the absolute, unshakable bedrock of certainty in an otherwise uncertain and dangerous world.

From the other side, the position advanced as a part of Dewey's productive pragmatism has been assailed by those who think its claims too strong. Some of these critics have objected to the claim that there are grounds for assessing one judgment or form of life as better than another.[8] Others have argued that productive pragmatism cannot be a participant in community life as long as it holds that its own method is superior to all others.[9] For these critics, truth remains tightly bound to the context of cultural or individual behavior in ways that render objective assessments of such forms of behavior impossible.

Dewey met both of these attacks head on. First, he recognized that a major feature of the history of philosophy has been what he called its "quest for certainty." Ever since Plato, he noted, philosophers have attempted to discover ideals or rules that could serve humankind as foundations, or absolutes. The quest for certainty has also been a dominant feature of most institutional religious thought. But productive pragmatism breaks with this long tradition. It treats ideals and rules as artifacts, and it holds that neither artifacts nor the tools that are used to produce them are absolute. Productive pragmatism rejects the idea that there are absolutes in two of the leading senses of the word.

First, if the term "absolute" means "unmoved or unconditioned by anything else," in the sense in which some theologians have held that the Christian God is absolute, then even if there were anything of that sort it would not be possible to know it. This is because whatever is known comes by that very fact to be related to whoever knows it. More specifically, it is related to the interests and attitudes of the knower. As Dewey argued in his classic essay "The Reflex Arc Concept in Psychology" (Early Works.5.96–109), it is the interests and attitudes of the knower that lead to the selection of data from an indefinitely large field of possible experience, and it is also interests and attitudes that contribute to the reworking and reconfiguration of that data into objects of knowledge. "The fact is," Dewey wrote, "that stimulus and response are not distinctions of existence, but teleological distinctions, that is, distinctions of function, or part played, with reference to reaching or maintaining an end" (Early Works.5.104). Knowing is thus constructed by relating data to other data on the basis of context and interest. Knowing invokes comparison, contrast, measurement, and assessment of one thing in relation to another. In short, it involves experimentation that results in the alteration of something relative to something else.

Knowing is also relative in the sense that it involves connections to other knowers. Knowing is sharpened and extended by taking the stances or viewpoints of others within a community of inquiry, that is, by considering a problem from as many different perspectives as possible. Thinking, language, and knowledge are all community enterprises, both in terms of their historical development and in terms of their ongoing function of construction and reconstruction.

Second, Dewey rejected the notion of an "absolute" in the related sense of "absolutely certain or immutable knowledge." Because we live forward in time and can never be sure what the future will bring, there is nothing in our experience that is totally impervious to change. Physics textbooks, for example, have tended to treat the speed of light as fixed and certain. That may be true in a perfect vacuum, but in the existential world where perfect vacuums do not exist, the speed of light is anything but fixed and certain. Harvard physicist Lene Vestergaard Hau and her team, for example, were able to slow a beam of laser light to just thirty-eight miles per hour. They transmitted laser light through a cloud of ultra-cold sodium atoms, thus reducing its speed

to what one report termed "a pace slower than her bicycle."[10]

The commitment of Dewey and the other pragmatists to experimentalism – and thus experiments of this sort – led them to develop the view that they termed "fallibilism" – the view that knowing is a project that is open to continual review and revision, and that knowing advances by means of such adjustments.

It might be objected that some statements (even if they are not the ones that report the speed of light) are beyond revision, and therefore that Dewey's version of fallibilism suffers from fatal difficulties. The statement "2 + 2 = 4," it might be argued, could never prove to be false and so must count as an example of absolutely certain knowledge. If anything is certain, it would seem, then this equation must be.

Dewey took up this difficult issue in his 1938 *Logic: The Theory of Inquiry*. In mathematical propositions such as "2 + 2 = 4," he argued, "the interpretation to be put upon the contents is irrelevant to any material considerations whatever" (Later Works.12.395). It is this feature that distinguishes mathematical propositions from the laws of physics. Physical laws require what Dewey called "preferred or privileged interpretation," such as "operating in a perfect vacuum." On the other hand,

> the contents of a mathematical proposition, *qua* mathematical, are free from the conditions that require any limited interpretation. They have no meaning or interpretation save that which is formally imposed by the need of satisfying the condition of transformability within the system, with no extra-systemic reference whatever. In the sense which "meaning" bears in any conception having even indirect existential reference, the terms have no meaning – a fact which accounts, probably, for the view that mathematical subject-matter is simply a string of arbitrary marks. But in the wider logical sense, they have a meaning constituted exclusively and wholly by their relations to one another as determined by satisfaction of the condition of transformability. (Later Works.12.395–96)

The question that Dewey considered in this passage is "what do we know when we know that 2 + 2 = 4"? His answer is that we know the "meaning" of a mathematical proposition, insofar as it has a meaning, relative to, or as a function of, its role in a mathematical system, which is itself a construct.

Further, insofar as the proposition "2 + 2 = 4" may be said to be "true" or "false," it refers to an existential situation. Our knowledge in this sense is not absolute either, but dependent upon the existential facts of the case. If we place two apples beside two other apples at a certain point in time on a table, then we can have empirical knowledge that there are exactly four apples on the table. If, on the other hand, we add two cups of milk to a beaker containing two cups of popped popcorn, then we can have empirical knowledge that at that moment the beaker contains two cups of soggy popcorn. (This is one of several examples provided in *The Mathematical Experience* by Philip J. Davis and Reuben Hersh.)[11] In one case, it is true that 2 + 2 = 4. In the other, it is false that 2 + 2 = 4. Here is another example, one that involves division. If I divide 100 by 3, I get 33.333. ... The string of 3's on the right side of the decimal, of course, goes on indefinitely. But if I go to the grocery store to buy one of the cans of tomato sauce that cost three for a dollar, how much do I pay? I pay 34 cents. It will do me no good to inform the clerk that it is false that 100 divided by 3 is 34. In a world in which I get my food from grocery stores, which is the existential world in which I live, 100 divided by 3 is sometimes 34.

Put another way, Dewey distinguished between a proposition that refers "to *each and every individual* [that] has certain characteristics ... and a proposition that refers in its own content to *no* individual" (Later Works.12.256). If we take "2 + 2 = 4" as an example of the former type of proposition, then the matter is an empirical one, that is, it requires that we examine the objects being added together in order to determine whether they have such characteristics that when two of them are taken together with two others, the sum will be four. As is obvious from the examples just given, sometimes this is the case, and sometimes it is not. If we take "2 + 2 = 4" as an example of the latter type of proposition, however, then it refers to *no* individual, but functions merely as a part of a system that has been constructed in order to effect certain types of transformations among abstract objects of knowledge. Knowledge of the meaning of this type of proposition is not absolute, but only relative to its restricted domain.

Even though Dewey viewed knowledge as relative, however, he rejected the type of relativism advanced by some "deconstructionist" philosophers who claim that there is no way to decide between "alternative readings of a text," whether that "text" be a written one, a "text" of nature, or the information that we have about our artifactual world. This position holds that conflicting views about a particular matter do not really prove to be better or worse, but are just expressions of different cultural biases (cultural relativism) or individual preferences or emotions (subjectivism or emotivism) that exist on an equal footing because there is no independent or context-free basis on which to decide among them.

Dewey thought this family of views faulty for several reasons. First, he thought it possible to articulate a general method of intelligence that takes into account successful inquiry in many different areas of human activity, such as the various sciences, the arts, politics, and jurisprudence. He thought that this method undergoes continual revision as it takes new cases into account, and that it has proven the best method so far devised for making decisions. And although technoscience has made major contributions to the development of this general method of intelligence, it is only one of the many sources that continue to nourish that general method and to help it evolve. In this way Dewey avoided the charge that he had accepted "scientism," or the view that the methods of the sciences should be the paradigm for all other forms of inquiry.

Second, Dewey believed that there have been numerous areas of technological and social life in which it is possible to point to examples of objective progress. In the domain of astronomy, for example, the geocentric model of the solar system that was accepted before the Copernican heliocentric model has been shown to be false. Among biologists, the pre-Darwinian view that all living things were created simultaneously is now regarded as quaint. Even though it is true that some individuals, and even some cultures, still hold these discredited views, it is an objective technological fact that such views have been tested and found to be inferior to the views that replaced them.

Dewey thought that objectivity is a function of experimentation within a community of candid and committed inquiry. Not all hypotheses carry equal weight within such communities of inquiry: some of them have been shown to be of little or no value as starting points for getting further knowledge. Others have even proven to be a barrier to getting further knowledge.

Of course there are also social hypotheses that were at one time widely held but that have since proven false. There was a time, for example, when attempts were made on "Biblical" grounds to justify the idea that one race is inferior to another. The same was true of the view that armed combat is the best method of solving disputes about cultural superiority. But both of these hypotheses have been shown to be faulty. Examples of social hypotheses that have proven to be true, in Dewey's sense of enhancing adjustment through what he called "warranted assertibility," include women's suffrage, universal public education, and social security for the elderly.

But it might be objected that this claim, that certain hypotheses should be privileged above others, violates the principle of fallibilism, that is, the principle that holds that our knowledge is never finished and certain. In fact, Dewey would have agreed that our current models of the solar system and the origin of plant and animal species are open to revision. This is true because these are matters that involve abstraction from concrete experience, transformation at the level of abstraction, and then application and reconstruction with respect to existential situations. In other words, these are matters that are subject to ongoing experience. It is conceivable that current models of the solar system or evolutionary biology might someday be replaced by other views, but it is also highly unlikely that they will be subject to substantial correction. The same might be said of women's suffrage, universal education, and social security. In Dewey's view, these hypotheses have been tested and have proven to be "true," that is, to constitute relatively stable platforms for further action.

Dewey's position with respect to these matters might be termed "objective relativism."[12] His position is objective because of his belief that when individual interests and goals are subjected to public, objective, falsifiable experimentation, then they can and do yield concrete results that can become broadly accepted and utilized within communities that take seriously the methods and results of experimentation.

Another way of putting this is to say that Dewey's productive pragmatism involves a moderate form of relativism that is based on the observation that intelligent action takes into account the different legitimate interests of the various groups that make up larger communities of discourse. It is also based on the observation that action tends to be unintelligent when it is based upon private or subjective viewpoints that have not been adequately tested. He thought that subjective viewpoints are appropriate places for inquiry to begin, but that they are not very good places for it to end. He thought that the history of human progress is a history of men and women coming together to form communities of discussion, inquiry, and activity and then constructing new tools: new ideas and new habits of action that are based on careful experimentation and held in common. When this fails to occur, parties to disagreement tend to remain intransigent with respect to received ideas. The development of new tools and new options is cut short.

It was evident to Dewey that cultural, political, and religious differences may sometimes be the sources of sharp disagreement. But a part of his faith in the methods of science and democracy was his belief that even the most serious disagreements could be transcended if subjected to the application of the proper tools.

Dewey therefore rejected the version of what has come to be known as the "incommensurability thesis,"[13] which states that there are unbridgeable chasms of understanding between different cultures or between different cultural groups within complex societies. He thought such a view unsatisfactory on two counts. First, the incommensurability thesis prejudices the very issue it purports to investigate. The assumption that cross-cultural communication is ineffective tends to be an instrument of its own validation.

Dewey's second objection rests on his view of human communication and the role of philosophy within human communities. Even though it is an observable fact that cultural and religious misunderstandings occur, and even that the practitioners of the various scientific-technical disciplines utilize different conceptual models and sometimes fail to understand one another, it is also an observable fact that properly controlled inquiry has been capable of bridging even the most profoundly recalcitrant differences between competing interest groups by identifying common interests and generating common goals.

Dewey regarded communication as one of the most wonderful of human activities, and he thought that wherever enhanced communication is held honestly as a goal and an ideal, then new areas of agreement can be constructed and community life rendered more satisfactory for all concerned.

But what does productive pragmatism have to say to those who hold the general method of intelligence and democratic ideals in contempt? What support can productive pragmatism offer for its claim that its method is superior to the methods of religious or political dogmatism? The answer to dogmatists of these sorts is that the methods of productive pragmatism have proven themselves to be the best methods so far devised for settling disputes and enhancing cooperation between conflicting factions. To privilege these methods is just to recognize that they are the only methods that have proven capable of avoiding the intransigence of absolutism on one side, and the drift of radical relativism, subjectivism, and emotivism on the other. And to opt for either intransigence or drift is to return to methods that have been tried and have failed. Productive pragmatism thus neither appeals to absolute values nor admits that there are no grounds for decision. At the same time, however, it treats its own ideals of inquiry and democracy as tools that are in need of continuing refinement. In other words, it is not just that its methods are the best so far devised, but, more importantly, that they are the best hope for settling future difficulties.

Dewey's rejection of these extremes – absolutism on one side and unqualified relativism on the other – led him to reject some of the most persistent dogmas of the history of philosophy.

He rejected Plato's ideal, perfect, immutable Forms and sought to supplant them with ideals and goals that are artifacts and therefore temporary, provisional, and in need of periodic tune-ups. He also rejected Aristotle's view that there is a fixed order of nature and that we can know natural kinds by making copies of them in our own minds. The model of nature advanced by productive pragmatism is based instead on multiple taxonomies that interact with one another and that are based on an awareness that the kind of information we get from natural events depends on the types of questions we ask and the types of tools we employ.

In fact, productive pragmatism doesn't treat nature as a "thing" at all, but as a human social artifact constructed from many different received ideas, interests, test results, points of view, and working hypotheses. Dewey thought that one of the most serious difficulties associated with Aristotle's view of nature was that he had treated natural science as an empirical activity rather than as an experimental one. The principal difference between the "empirical" and the "experimental" in this context is that the latter involves the intervention of technical and technological artifacts as tools to enhance and extend knowing, whereas the former is primarily a matter of observation and single-model classification. Dewey thought that the great advances in human knowledge since the seventeenth century have been the result of the experimental methods of technoscience. Or put another way, science as we now know it would have proven impossible without the application of instrumentation to the selection and testing of what is observed, as data of experimentation.

Dewey's productive pragmatism also rejected Descartes's idea that thinking substance (mind) and extended substance (matter) are experienced as the most basic ontological categories of things. It holds instead that we experience neither "mind" nor "matter" directly or in the absence of the other, and that both "mind" and "matter" are concepts or tools that we use to divide up our gross, immediate experiences, to operate on them, and to render them more manageable.

Consequently, productive pragmatism holds that the uniqueness of human beings is not attributable to a non-empirical "spirit" or "soul" that is qualitatively different from the rest of the animal world.[14] Human uniqueness is instead the result of an extremely high order of complexity that enables us to take control of the ways by which we form our own habits and therefore to take charge of our own development as individuals and our own evolution as a species. Productive pragmatism regards "mind" and "body" as different focal points within experience, or as different phases in the act of getting knowledge and adjusting to changing environmental conditions. In Dewey's book, "mind" and "body" are not natural ontological entities, but instead important functional tools of knowing.

All of this means that Dewey took a very broad view of what counts as technology. He held that technology is the *invention, development, and cognitive deployment of tools and other artifacts, brought to bear on raw materials and intermediate stock parts, to resolve perceived problems.* This means that tractors and televisions count as technological artifacts, but so do individual habits, the social habits that we call institutions, and even working hypotheses. Sports skills, universities, political parties, and pi are as much technological artifacts as is a hammer. This is so because neither individual habits, nor social institutions, nor shared concepts, are just "given" to us by a god or by nature. They are instead artifacts that are constructed in much the same way that hardware is constructed – not out of nothing, but out of various raw materials and previously constructed artifacts. And since it holds that goals and plans are also technological constructs, productive pragmatism takes very seriously individual and collective responsibility for the future. Unlike less complex animals, we human beings construct our own futures; they are among the artifacts that we continually build and rebuild.

In "The Development of American Pragmatism," Dewey summed up his view of productive pragmatism, or "instrumentalism." "It is therefore not the origin of a concept, it is its application which becomes the criterion of its value. ... The function of intelligence is therefore not that of copying the objects of the environment, but rather of taking account of the way in which more effective and more profitable relations with these objects may be established in the future" (Later Works.2.16–17).

II. The Individual, Publics, and Community Life

Because of its view of the ways in which human beings interact with their environments, continually adjusting and readjusting with respect to them, the term "community life" takes on great significance for productive pragmatism.

Dewey thought that both of the principal ways in which social philosophers since the seventeenth century have conceived of community life have proven faulty. The first view is what has generally been known as "classical liberalism" and is now termed "conservatism." This was the view of thinkers such as John Locke, Jeremy Bentham, and John Stuart Mill. This view holds that each of us is born and develops as an individual in the strict sense of the word, that is, as a more or less complete social atom. At some point in prehistory, according to some versions of this thesis, individuals joined together with other more or less equally independent and complete individuals by agreeing among themselves to the terms of some version of a "social contract."

Now Dewey did not object to the claim of the conservatives that social groups are purposely and consciously formed, since he thought that such groups are technological artifacts just as surely as are hammers and saws. But he thought it absurd that anyone would think that individuality and self-conscious personhood could arise in the absence of social interaction. He thought that it is only by means of communication and shared experiences that human beings can become self-conscious individuals in the first place. He consequently regarded education as one of the most important human activities: it is the means by which children are enabled to develop their own talents and interests in ways that take into account environing social conditions.

Dewey also thought that most versions of the "social contract" worked out by political and social philosophers were faulty. Some proponents of this view argued that there had been an actual historical moment in which individuals must have come together to form such a contract. But Dewey pointed out that no such historical moment had ever been demonstrated. Still others attempted to support their view by listing the specific rules that must have been agreed upon and then followed by all the parties to the new social contract. But Dewey thought that such lists of rules were fiction at best. He thought that lists of rules are just as often abstractions from actual living human communities, and not historically prior to them. Human communities develop rules on the basis of practice, and their practice is developed on the basis of rules as those rules indicate and serve as solutions to common difficulties.

It was Dewey's view that a second group of social and political philosophers have also misunderstood community life. These philosophers hold that the individual is *only* or *primarily* a function of his or her society, and that the needs of the community should always take precedence over the needs of its constituents. This view was popular in the Soviet Union during most of its history. It was also a feature of European fascism during the 1930s, and in our own time it is the guiding principle of some "theocratic" states, such as certain Islamic republics, and some religious groups as well. The adjectives "totalitarian" and "authoritarian" are often used to refer to such social arrangements because of their failure to take into account relevant differences between human beings.

Dewey thought this view of community life faulty for several reasons. First, it tended to stifle human growth and development. Each of us has unique interests, talents, and outlooks that can be focused and developed through well-thought-out and carefully articulated educational practice. But when indoctrination is substituted for education, as is usually the case in closed societies, then development is cut short. Second, closed societies also cut short the open discussion and social experimentation that are necessary for determining the nature of common problems and seeking solutions to them. As a consequence, such societies tend to stagnate and eventually to decay.

The major source of invention and insight, in Dewey's view, lies with individuals as they strive to overcome some experienced difficulty. "Every *new* idea," he wrote, "every conception of things differing from that authorized by current belief, must have its origin in an individual" (Middle Works.9.305). Wherever individuals are not free to articulate problems and to attack them experimentally, then growth within the society is greatly diminished. At the same time, however, problems exist within social and cultural contexts, and they are best articulated and refined by means of discourse within communities.

Underlying his critique of these two extreme views of community life was Dewey's contention that both of them had failed to realize that the concepts "individual" and "society" are just abstractions, or tools to be used, and not absolutes. Dewey thought that the problems of community life would not be solved by deciding whether individual or society should have priority over the other, but by recasting the discussion in terms of the relation between what was properly private and what was properly public. What is private, he argued, does not have consequences beyond the boundaries of narrow actions and associations. What is public has wider applicability, and should therefore be a matter of common involvement and oversight.

The real problems in community life are thus for Dewey not the consequences of conflicts between individuals and society, or even conflicts between individuals. Instead, he thought that the problems of community life are due to conflicts within two broad and vague areas: one of them takes place where the public encounters the private, and the other is where various publics encounter one another. In the first of these areas, space must be scrupulously maintained for private activities that have little or no public consequence. Sexual interaction between consenting adults, for example, including the choice of partner and means of contraception, would normally constitute just such a domain of private action.

In the second area of potential conflict, ways must be found to arbitrate and adjudicate between legitimate but conflicting interests of various publics. Dewey thought that both of these areas of discourse can be made more manageable and productive through active planning undertaken at the level of the comprehensive public we call "the state." Dewey thus saw the state as more than just referee or traffic cop that ensures that the rules of the road are observed by all. The state must be more than simply neutral. He thought that because the state is the most comprehensive of publics in which most of us participate, it is also the most comprehensive means available for developing common interests and directing energies toward change that results in the growth of individuals and communities.

At the same time, however, Dewey did not think that action by the state should be a substitute for concrete action within the various publics that it comprehends. The proper role of the state is to aid the liberation of individual talents and resources and to enable and empower various interacting groups so that they can develop new goals and ideals and thus make their best case within a meritocracy of ideas.

In all this Dewey urged us not to forget that publics are technological products. There is no such thing as a "natural" public. Publics are artifacts created and maintained by human effort. One of the primary tools utilized to create publics is the dissemination of information. Books, newspapers, television, and electronic networks are among the tools by means of which publics are formed and held together for common purposes. But disinformation and propaganda, as well as information that is warranted and useful, are also among the tools by which publics are formed. And although there is no place within a free society for the censorship of even what a majority perceives as disinformation, it is also important that public education should foster the development of the tools of critical intelligence by which information of all types can be evaluated.

III. Technological Design for Community Life

Productive pragmatism rejects technological determinism. Unlike Marx, Dewey did not think that forms of technological artifacts uniquely determine social arrangements. In one of his more famous statements, for example, Marx had pronounced that the hand mill produces a society with the feudal lord, whereas the steam mill produces a society with the industrial capitalist. For Dewey, however, tools do not have the last say. Instead, technological innovations tend to rearrange existing alliances, tip balances of power, render some forms of community life obsolete, and encourage the development of others.

Dewey did argue, however, that we can learn from the past. The history of technological innovation teaches us that after initial periods of invention, experimentation, and decentralization, new forms of technology tend to become centralized and monopolized as economic interests are reorganized and consolidated. And the danger of centralization and monopoly is that those publics that are the most vulnerable tend to be ignored and thus to become politically marginalized. In our own time, children, the mentally ill, and the homeless have been excellent candidates

for marginalization, and this because they do not have deep-pocket lobbyists and they are either unable to vote or not qualified to do so.

A part of Dewey's rejection of technological determinism was his view that new forms of technoscience do not carry a single value on their faces. He thought that new techniques and technologies are multi-valent, that is, that they offer all sorts of new possibilities and that it is the obligation of those who use them to choose the best of those possibilities and then to rework them in order to render them more valuable. Electronic technology, for example, produces artifacts that are inherently neither pro- nor anti-democratic. They are both – and neither. They are what their users will, and can, make of them.

If Dewey were alive today, he would surely realize that the phase of the electronic revolution that we are now entering will radically alter the ways in which publics are formed and the means by which they interact with one another. Because productive pragmatism is democratic at its core, and because it is experimental in the sense that the technological disciplines are experimental, it always asks whether new forms of technology will tend to support or undercut democratic procedures and institutions.

The much-discussed "information superhighway" offers an excellent case by which to test whether the tools of productive pragmatism can play an effective role in social reconstruction. The first thing to notice about this new form of technology is that it is still in an incipient stage of development. Although its bases such as the Internet have been in existence for some time, they have only recently begun to be available beyond relatively narrow educational, research, military, and corporate circles. As the general public has expressed increasing interest in computer-to-computer communication, however, its use has begun to experience enormous growth. As a consequence, various commercial and public interest groups have begun to recognize its potential and to compete with one another to define its parameters for their own benefit.

What would be the results of applying Dewey's productive pragmatism to the problems associated with the configuration and use of this new public medium?

First, because productive pragmatism is a form of evolutionary naturalism, it takes into account the genetic development of tools as well as organic structures. In pragmatic terms, then, a successful design strategy would take into account the fact that new forms of technological methods and artifacts tend to incorporate elements of older techniques and artifacts as their content. New technologies do not arise out of nothing, but are built on the basis of more or less viable institutions, customs, and habits. Nevertheless, painful discontinuities and displacements sometimes occur when accepted institutional practice is altered. One of the goals of intelligent design should be to mitigate the pain of such displacements. What kinds of displacement might the information superhighway involve, and how could they be taken into account at the design stage?

One form of displacement involves what has been called the "electronic sweatshop." Once work becomes decentralized, for example, there is a tendency to return to the outmoded practice of piecework. Because laptops can be used at home and on trips away from the office, even salaried workers are beginning to find themselves working longer hours than they did when most of their work was done at centralized locations.

A second form of displacement involves the de facto exclusion of various segments of the population from participation in the electronic network because they lack the skills or the resources with which to access it. Children who attend schools with inadequate computer facilities or instruction, for example, suffer a profound disadvantage with respect to peers whose education affords computerized instruction. A study released by the U.S. Department of Commerce in July of 1999, for example, indicated that 47 percent of whites own computers, whereas only 19 percent of blacks do. Even worse, it found that a child in a white low-income family was three times more likely than a child in a black low-income family to have Internet access.[15]

A third form of displacement is the result of novel stresses experienced by individuals accustomed to "hands-on" work, but who have been reassigned to work with more abstract tools such as computers. In the previous chapter I recalled Shoshana Zuboff's detailed study of just such a situation in the bleach plant of a pulp mill. Zuboff has argued that the transition from "action-centered" to "intellective" job skills not only requires massive retraining, but can be the source of high levels of job-related stress.[16]

A fourth form of displacement results from the facility with which specialized, interest-specific communities develop within "information space." Will the point-to-point communication features of the information superhighway contribute to a splintering of comprehensive community life into smaller and smaller communities that reinforce their own eccentricities and insulate themselves from the methods and forces that serve to promote the coherence of the wider community?

In its own unique fashion, each of these forms of displacement harbors potential threats to community life.

How would intelligent design based on Dewey's productive pragmatism deal with these potential problems? First, it would seek to avoid the techniques of "technology assessment" that were popular in the United States during the 1950s and early 1960s. This usually involved top-down, expert-based assessments of the features of proposed technologies, and there was a tendency to take into account narrow "technical" considerations or to engage in behind-the-scenes social engineering. At the same time, input from the publics most affected by the proposed changes was usually ignored. The massive "urban renewal" and public housing programs of that era continue to serve as reminders of the limitations of such approaches.

Because productive pragmatism exhibits a core belief in the methods of democracy, its methods dictate that all affected parties be heard from not only during the planning stages of significant public projects, but during the stages of their implementation as well. As Dewey put it, productive pragmatism calls not so much for a "planned" society as for one that is continually "planning." The role of the "expert" within productive pragmatism is thus to draw on the energies and sources of information within affected publics and to formulate scenarios for action, but not to exercise ultimate decisions regarding the determination and execution of public policy.

Dewey was well aware that not every affected party or public has the ability to appreciate or articulate its own best interests. It was for this reason that he thought that the interaction between "expert" and affected publics should be a "transaction" that would serve to educate experts with respect to the needs of publics, as well as to educate publics with respect to the assessment and articulation of their own needs and goals.

On the pragmatic model, then, existing forms of mass and point-to-point communication would be examined for strengths and weaknesses in the performance of their various roles such as entertainment, communication, and education. Media theorists and media historians would be commissioned to address the strengths and weaknesses of broadcast radio and television, cable, the Internet in its current form, and even the telephone, so that past mistakes could be avoided and past successes could be used as models for new developmental strategies. The results of their research would most likely have important consequences for ownership of media, public access, and other regulatory matters.

As new forms of communication alter the nature of work, new agreements would need to be forged among government, industry, and workers in order to avoid new patterns of exploitation. As I write, for example, negotiations of this type are being demanded of the World Trade Organization by members of various non-governmental organizations (NGOs) such as environmental groups and critics of child labor. More equitable income distribution (through more progressive forms of taxation and other means such as raising the minimum wage) would probably be required to ensure that no segment of the larger society – such as the low-income black children just mentioned – is excluded from participation in what promises to become the new social and cultural "central nervous system." Public access to the new information superhighway might be modeled on the successes of rural electrification during the 1930s and the network of interstate highways constructed during the years immediately after World War II.

Government, business, industry, workers' groups, and private foundations would need to enter into new cooperative ventures in order to ensure that every member of the community has access to basic information services. This would in turn require that a realistic accessibility baseline be established and reevaluated on a periodic basis. Historically, new forms of technology have required significant realignments of existing social structures. Whenever this has occurred in a haphazard way and without intelligent planning, the results have usually been disastrous for those publics that are the most fragile.

Second, there would need to be a continuing commitment to the methods of the technosciences and the methods of democracy that have proven in the past to be highly effective tools for the improvement of community life. One of the pillars of the experimental method is fallibilism, and one of the pillars of democracy is its rejection of the intransigence of absolutism, at one extreme, and the drift of radical relativism, subjectivism, and emotivism, at the other. What this means in practical terms is that the information superhighway would need to be designed so as to leave open as many parallel paths of development as are practicable, so that if one path were to prove undesirable or unworkable, another could be taken up and pursued. Fallibilism dictates strategies of design and implementation that remain flexible by maximizing options and creating redundancies wherever possible. Objective relativism, in its turn, takes into account the unique perspectives of the various groups that have a legitimate interest in the development of the new technology. At the same time, however, democratic processes such as negotiation and third-party arbitration would need to be employed, and their techniques continually improved, in order to ascertain the extent to which the actions of a particular public should be judged unreasonably obstructionist to the good of other publics and to the wider community.

Third, because one of the central tenets of productive pragmatism is its belief in the importance of education, one of the central aims of the information superhighway would be to increase educational options. Dewey argued that education is neither indoctrination nor unregulated change. Instead, he suggested, it is a process by which a teacher enters into a transaction with a learner with the aim of developing the learner's talents and interests and enhancing transaction between the learner and the institutional features of his or her society. Good teachers are capable of appreciating the wide range of talents and interests that they find among learners. They also understand that education is a source of personal development for teacher and learner alike, as well as a source of novel ideas about the ways in which social problems can be articulated and addressed.

One of the most interesting features of the emerging information environment is the rapidity with which new tools are becoming available. As a consequence of this, education is already becoming much less age-graded. Younger people often find themselves educating their elders in the use of new electronic tools and techniques. At the same time, however, older and more mature individuals are able to offer the young insights developed during a lifetime of learning to work within the context of complex relationships. Design and implementation of the educational dimension of the information superhighway would thus call for a reconstruction of educational practice that would take these factors into account with a view to establishing new patterns of interaction between teacher and learner. In short, new types of conversation will need to be constructed to take into account these new units of discourse.

Finally, if cultural splintering is to be avoided, educators will need to reevaluate their methods for leading children to an appreciation of the historical and cultural contexts of the wider society in which they will live and work as adults. Dewey argued that educators have a double task in this regard: it is their task to help children develop their interests and talents in ways that enable them to be individuals in the best sense of the word, and it is also their task to help children find a place within the broader society and to envision their own role in its reform.[17]

argued that productive pragmatism is uncritically technophilic and elitist, and that it is therefore incapable of taking into account cultural values that have emerged outside of what they regard as its own liberal-democratic circle. A third group of critics has argued that productive pragmatism is overly sanguine about the prospects for progress; that it does not take into sufficient account the darker, dysfunctional aspects of human life.

It was with the first type of critic in mind that I have suggested some of the ways in which democratic inclusiveness can continue to strengthen community life. And it was in response to the second type of critic that I have argued that technology comprehends more than just the tangible implements that occupy the spaces of human life. It also involves the methods and ideals by which men and women can organize themselves into mutually beneficial overlapping publics and thereby enhance the life of the wider community. To the third group of critics it must be admitted that human methods and institutions are unlikely to reach perfection; that it appears that there will continue to be serious constraints on human progress; and that pathologies sometimes deflect the best-intentioned attempts to build community life.

In the face of all this, however, the productive pragmatist continues to exhibit faith in the methods of democracy and technology because she is convinced that they offer the best alternative so far devised for overcoming what is mean, debilitating, and even pathological in human life.

Productive pragmatism offers many more tools for the design and implementation of new forms of technology and the reconstruction of community life than I have been able to detail in this chapter. In the chapters that follow, I shall also discuss its applicability to some of the aesthetic dimensions of community life and its treatment of religious issues.

The methods of productive pragmatism are by no means uncontroversial. Some of its critics have charged that its commitment to democratic inclusiveness runs the risk of undermining the very democratic processes and institutions that it holds so dear. Critics of another sort have

Notes

1 This position was a part of the program of the logical positivists during the 1930s and 1940s. Some of their followers continued to advance it well into the 1960s. As I write, there are still attempts to rehabilitate portions of the positivist program. See, for example, Michael Friedman, *Reconsidering Logical Positivism* (Cambridge: Cambridge University Press, 1999).

2 This view is usually attributed to neo-pragmatist Richard Rorty. See, for example, his essay "Private Irony and Liberal Hope," in *Contingency, Irony, and Solidarity* (New York: Cambridge University Press, 1989), 73–95.

3   Dewey in lecture notes for 10 April 1926, Special
    Collections at Morris Library, Southern Illinois
    University at Carbondale, John Dewey Papers,
    65/1. I owe thanks to Dario Segato for drawing my
    attention to this reference.
4   I want here to distinguish between technically
    produced artifacts and technologically produced
    artifacts. A technically produced artifact is one
    that is made with very little or no cognitive inter-
    action with materials. A technologically produced
    artifact is accordingly one that is the result of
    deliberate cognitive intervention into materials.
    Although this distinction fuzzes at its borders,
    it is nevertheless helpful in meeting the types of
    objections to my general account of technology.
    In *Art as Experience* (1934), Dewey made the
    point that even the most utilitarian of domestic
    objects often enjoy cognitive attention in the
    form of decorative elements added.
5   Dewey was attacked on just these grounds by
    Mortimer Adler during a speech in New York
    in 1940. Adler charged that Dewey, because of
    his claim that there are no absolute values, had
    become a "serious threat to Democracy" (*Vital
    Speeches*, 1 December 1940, p. 100). See Robert
    Westbrook, *John Dewey and American Democracy*
    (Ithaca, NY: Cornell University Press, 1991), esp.
    pp. 519–20, for more on this subject.
6   Televangelist Pat Robertson has often repeated
    Adler's attack on Dewey, and school board mem-
    bers across the country have joined the attack as
    well. Page 2 of the July–August issue of the *Illinois
    School Board Journal*, for example, contained a
    letter from the vice president of a suburban
    Chicago school district in which Dewey was
    blamed for the shootings at Columbine High
    School in Littleton, Colorado. According to the
    writer, "pragmatism proposes that there are no
    unchanging truths, no fixed standard of morality."
    From this the author concludes that "the seemingly
    mindless slaughter at Littleton was the acting out
    of the pragmatic view. If it works, if it feels good,
    do it. They did!" The writer's primary fallacy lies
    in his essentialist coupling of absolute values,
    which for him are of supernatural origin, and
    morality. This is precisely what Dewey wanted
    to de-couple, since he thought that morality
    could flourish in a world without supernatural
    influences.
7   Max Horkheimer attacked Dewey as having
    abandoned the notion of "objective truth:" See
    Horkheimer, *The Eclipse of Reason* (New York:
    Oxford University Press, 1947; reprint, New
    York: Seabury Press, 1974), 45.
8   Richard Rorty has claimed that there is no fun-
    damental difference between his own view and that
    of Dewey (see Giovanna Borradori, *The American
    Philosopher* (Chicago: University of Chicago Press,
    1994), 106). Some have claimed, however, that
    Rorty advances a "fuzzy" version of relativism
    that is not only at odds with Dewey's more
    objective relativism, but that serves as an implicit
    criticism of it (see Richard Bernstein, *The New
    Constellation* (Cambridge: MIT Press, 1992), 233 ff.).
    The problem, in the view of Bernstein and others,
    is Rorty's deep sympathy with deconstructionist
    philosophers such as Jacques Derrida, who is well
    known for his view that human beings are caught
    in an infinite regress of textual interpretations.
    See Derrida, *Of Grammatology,* trans. G. C.
    Spivak (Baltimore: Johns Hopkins University
    Press, 1976). Christopher Norris is one of the
    many critics of Derrida who see him as a kind
    of half-hearted pragmatist. "So Derrida becomes
    a kind of half-way honorary pragmatist, having
    deconstructed a great deal of surplus ontological
    baggage but then fallen victim to the lure of his
    own negative metaphysics or systematized anti-
    philosophy" (Norris, *Derrida* (Cambridge: Harvard
    University Press, 1987), 151).
9   See William Andrew Paringer, *John Dewey and the
    Paradox of Liberal Reform* (Albany: SUNY Press,
    1990), 130.
10 Malcolm W. Browne, "She Puts the Brakes on
    Light;" *New York Times*, 30 March 1999, p. D1.
11 P. I. Davis and R. Hersh, *The Mathematical
    Experience* (New York: Houghton Mifflin, 1981),
    p. 71.
12 Donald Koch has taken exception to this term. He
    thinks that Dewey avoided relativism altogether,
    since he thinks that the emphasis should be on the
    situation that is problematic. The situation, of
    course, includes the individual who is attempting
    to solve the problem, but much more as well.
    And it is not the satisfaction of the individual
    problem solver that interests but the resolution of
    the objective situation. Of course Koch is correct
    in this assessment. I have used the term "objec-
    tive relativism;" however, as a way of accounting
    for what I take to be Dewey's emphasis on the per-
    spectival nature of perception and his innocuous
    form of *cultural* relativism.
13 For a discussion of incommensurability in sci-
    ence, see T. Kuhn, *The Structure of Scientific
    Revolutions* (Chicago: University of Chicago Press,
    1962).
14 Max Scheler, for example, held the view that the
    big gap that separated humans from non-human
    animals was due to the addition of a human
    "soul".
15 Jeri Clausing, "Push to Narrow Disparities in
    Training and Access to Web;" *New York Times*, 10

December 1999, p. A25. The NPRJ Kaiser/ Kennedy School Survey found that the digital divide is the greatest between Americans over and under sixty years of age. The survey found a twenty-two-point gap, 73% to 51%, between blacks and whites who have a computer in the home. Perhaps even more importantly, however, its authors concluded that the schools appear to be a major factor in equalizing computer access for children of different economic and ethnic groups. The results of this poll were reported on NPR and obtained from their Internet site, at www.npr.org.

16   Shoshana Zuboff, *In the Age of the Smart Machine* (New York: Basic Books, 1989).

17   In the "Around Alone" event, a round-the-world race in which sailboat captains competed solo, schoolchildren followed the progress of the race on the event's Internet site. The sailors communicated with the children via e-mail, responding to their questions about geography, marine life, ocean currents, and so on. In another project, schoolchildren were sent out to find out about the "roadkill," or dead animals on the streets around their school. Using the Internet, they researched the habitats and feeding habits of the animals. They concluded that the number of squirrels run over by automobiles in the neighborhood of their school could be reduced if they stopped throwing apple cores out of the windows of the school bus. These are precisely the kind of activities that Dewey was undertaking at his laboratory school at the University of Chicago in the 1890s. Dewey would, I think, have loved the Internet.

Part VI

Feminist Considerations

18

A Cyborg Manifesto: Science, Technology, and Socialist-Feminism in the Late Twentieth Century

Donna Haraway

Donna Haraway is currently a Professor and Chair of the History of Consciousness Program at the University of California, Santa Cruz. In this groundbreaking postmodern feminist essay, she argues that contemporary culture is creating humans as cyborgs, as creatures who have no single constitutive identity. Drawing extensively on the Critical Theory tradition, Haraway's "Cyborg Manifesto" champions the metaphor of human–machine hybrids as a model for overcoming the dualisms of human and animal, machine and organism, physical and non-physical, and thus as a means to rethink the bounds of science, technology, and socialist-feminism.

Pursuing a line of argument similar to that of Latour, she suggests that to imagine ourselves as cyborgs, and to think and experience fully the ways that humans and technology shape each other, can help us overcome totalizing theories that make false dichotomies and privilege one side of a dualism over the other. She also argues that it helps us see the connections among science, technology, and social relations so that we may take responsibility for them and communicate better with one another. Instead of trying to make our world more humane, we should instead try to become more like cyborgs.

An Ironic Dream of a Common Language for Women in the Integrated Circuit

[...]

A cyborg is a cybernetic organism, a hybrid of machine and organism, a creature of social reality as well as a creature of fiction. Social reality is lived social relations, our most important political construction, a world-changing fiction. The international women's movements have constructed 'women's experience', as well as uncovered or discovered this crucial collective object. This experience is a fiction and fact of the

From Donna Haraway, *Simians, Cyborgs and Women: The Reinvention of Nature* (New York: Routledge, 1991), pp. 149–81.

most crucial, political kind. Liberation rests on the construction of the consciousness, the imaginative apprehension, of oppression, and so of possibility. The cyborg is a matter of fiction and lived experience that changes what counts as women's experience in the late twentieth century. This is a struggle over life and death, but the boundary between science fiction and social reality is an optical illusion.

Contemporary science fiction is full of cyborgs – creatures simultaneously animal and machine, who populate worlds ambiguously natural and crafted. Modern medicine is also full of cyborgs, of couplings between organism and machine, each conceived as coded devices, in an intimacy and with a power that was not generated in the history of sexuality. Cyborg 'sex' restores some of the lovely replicative baroque of ferns and invertebrates (such nice organic prophylactics against heterosexism). Cyborg replication is uncoupled from organic reproduction. Modern production seems like a dream of cyborg colonization work, a dream that makes the nightmare of Taylorism seem idyllic. And modern war is a cyborg orgy, coded by C^3I, command-control-communication-intelligence, an $84 billion item in 1984's US defence budget. I am making an argument for the cyborg as a fiction mapping our social and bodily reality and as an imaginative resource suggesting some very fruitful couplings. Michael Foucault's biopolitics is a flaccid premonition of cyborg politics, a very open field.

By the late twentieth century, our time, a mythic time, we are all chimeras, theorized and fabricated hybrids of machine and organism; in short, we are cyborgs. The cyborg is our ontology; it gives us our politics. The cyborg is a condensed image of both imagination and material reality, the two joined centres structuring any possibility of historical transformation. In the traditions of 'Western' science and politics – the tradition of racist, male-dominant capitalism; the tradition of progress; the tradition of the appropriation of nature as resource for the productions of culture; the tradition of reproduction of the self from the reflections of the other – the relation between organism and machine has been a border war. The stakes in the border war have been the territories of production, reproduction, and imagination. This chapter is an argument for *pleasure* in the confusion of boundaries and for *responsibility* in their construction. It is also an effort to contribute to socialist-feminist culture and theory in a post-modernist, non-naturalist mode and in the utopian tradition of imagining a world without gender, which is perhaps a world without genesis, but maybe also a world without end. The cyborg incarnation is outside salvation history. Nor does it mark time on an oedipal calendar, attempting to heal the terrible cleavages of gender in an oral symbiotic utopia or post-oedipal apocalypse. As Zoe Sofoulis argues in her unpublished manuscript on Jacques Lacan, Melanie Klein, and nuclear culture, *Lacklein*, the most terrible and perhaps the most promising monsters in cyborg worlds are embodied in non-oedipal narratives with a different logic of repression, which we need to understand for our survival.

The cyborg is a creature in a pot-gender world; it has no truck with bisexuality, pre-oedipal symbiosis, unalienated labour, or other seductions to organic wholeness through a final appropriation of all the powers of the parts into a higher unity. In a sense, the cyborg has no origin story in the Western sense – a 'final' irony since the cyborg is also the awful apocalyptic *telos* of the 'West's' escalating dominations of abstract individuation, an ultimate self untied at last from all dependency, a man in space. An origin story in the 'Western', humanist sense depends on the myth of original unity, fullness, bliss and terror, represented by the phallic mother from whom all humans must separate, the task of individual development and of history, the twin potent myths inscribed most powerfully for us in psychoanalysis and Marxism. Hilary Klein has argued that both Marxism and psychoanalysis, in their concepts of labour and of individuation and gender formation, depend on the plot of original unity out of which difference must be produced and enlisted in a drama of escalating domination of woman/nature. The cyborg skips the step of original unity, of identification with nature in the Western sense. This is its illegitimate promise that might lead to subversion of its teleology as star wars.

The cyborg is resolutely committed to partiality, irony, intimacy, and perversity. It is oppositional, utopian, and completely without innocence. No longer structured by the polarity of public and private, the cyborg defines a technological polis based partly on a revolution of social relations in the *oikos*, the household. Nature and culture are reworked; the one can no longer be the resource for appropriation or incorporation by the other. The relationships for forming wholes from parts, including those of polarity and hierarchical domination, are at issue in the cyborg world. Unlike the hopes of Frankenstein's monster, the cyborg does not expect its father to save it through a restoration of the garden; that is, through the fabrication of a heterosexual mate, through its completion in a finished whole, a city and cosmos. The cyborg does not dream of community on the model of the organic family, this time without the oedipal project. The cyborg would not recognize the Garden of Eden; it is not made of mud and cannot dream of returning to dust. Perhaps that is why I want to see if cyborgs can subvert the apocalypse of returning to nuclear dust in the manic compulsion to name the Enemy. Cyborgs are not reverent; they do not re-member the cosmos. They are wary of holism, but needy for connection – they seem to have a natural feel for united front politics, but without the vanguard party. The main trouble with cyborgs, of course, is that they are the illegitimate offspring of militarism and patriarchal capitalism, not to mention state socialism. But illegitimate offspring are often exceedingly unfaithful to their origins. Their fathers, after all, are inessential.

I will return to the science fiction of cyborgs at the end of this chapter, but now I want to signal three crucial boundary breakdowns that make the following political-fictional (political-scientific) analysis possible. By the late twentieth century in United States scientific culture, the boundary between human and animal is thoroughly breached. The last beachheads of uniqueness have been polluted if not turned into amusement parks – language, tool use, social behaviour, mental events, nothing really convincingly settles the separation of human and animal. And many people no longer feel the need for such a separation; indeed, many branches of feminist culture affirm the pleasure of connection of human and other living creatures. Movements for animal rights are not irrational denials of human uniqueness; they are a clear-sighted recognition of connection across the discredited breach of nature and culture. Biology and evolutionary theory over the last two centuries have simultaneously produced modern organisms as objects of knowledge and reduced the line between humans and animals to a faint trace re-etched in ideological struggle or professional disputes between life and social science. Within this framework teaching modern Christian creationism should be fought as a form of child abuse.

Biological-determinist ideology is only one position opened up in scientific culture for arguing the meanings of human animality. There is much room for radical political people to contest the meanings of the breached boundary.[1] The cyborg appears in myth precisely where the boundary between human and animal is transgressed. Far from signalling a walling off of people from other living beings, cyborgs signal disturbingly and pleasurably tight coupling. Bestiality has a new status in this cycle of marriage exchange.

The second leaky distinction is between animal-human (organism) and machine. Pre-cybernetic machines could be haunted; there was always the spectre of the ghost in the machine. This dualism structured the dialogue between materialism and idealism that was settled by a dialectical progeny, called spirit or history, according to taste. But basically machines were not self-moving, self-designing, autonomous. They could not achieve man's dream, only mock it. They were not man, an author to himself, but only a caricature of that masculinist reproductive dream. To think they were otherwise was paranoid. Now we are not so sure. Late twentieth-century machines have made thoroughly ambiguous the difference between natural and artificial, mind and body, self-developing and externally designed, and many other distinctions that used to apply to organisms and machines. Our machines are disturbingly lively, and we ourselves frighteningly inert.

Technological determination is only one ideological space opened up by the reconceptions of machine and organism as coded texts through which we engage in the play of writing and reading the world.[2] 'Textualization' of everything in poststructuralist, postmodernist theory has been damned by Marxists and socialist feminists for its utopian disregard for the lived relations of domination that ground the 'play' of arbitrary reading.[3] It is certainly true that postmodernist strategies, like my cyborg myth, subvert myriad organic wholes (for example, the poem, the primitive culture, the biological organism). In short, the certainty of what counts as nature – a source of insight and promise of innocence – is undermined, probably fatally. The transcendent authorization of interpretation is lost, and with it the ontology grounding 'Western' epistemology. But the alternative is not cynicism or faithlessness, that is, some version of abstract existence, like the accounts of technological determinism destroying 'man' by the 'machine' or 'meaningful political action' by the 'text'. Who cyborgs will be is a radical question; the answers are a matter of survival. Both chimpanzees and artefacts have politics, so why shouldn't we (de Waal, 1982; Winner, 1980)?

The third distinction is a subset of the second: the boundary between physical and non-physical is very imprecise for us. Pop physics books on the consequences of quantum theory and the indeterminacy principle are a kind of popular scientific equivalent to Harlequin romances [the US equivalent of Mills & Boon] as a marker of radical change in American white heterosexuality: they get it wrong, but they are on the right subject. Modern machines are quintessentially microelectronic devices: they are everywhere and they are invisible. Modern machinery is an irreverent upstart god, mocking the Father's ubiquity and spirituality. The silicon chip is a surface for writing; it is etched in molecular scales disturbed only by atomic noise, the ultimate interference for nuclear scores. Writing, power, and technology are old partners in Western stories of the origin of civilization, but miniaturization has changed our experience of mechanism. Miniaturization has turned out to be about power; small is not so much beautiful as pre-eminently dangerous, as in cruise missiles. Contrast the TV sets of the 1950s or the news cameras of the 1970s with the TV wrist bands or hand-sized video cameras now advertised. Our best machines are made of sunshine; they are all light and clean because they are nothing but signals, electromagnetic waves, a section of a spectrum, and these machines are eminently portable, mobile – a matter of immense human pain in Detroit and Singapore. People are nowhere near so fluid, being both material and opaque. Cyborgs are ether, quintessence.

The ubiquity and invisibility of cyborgs is precisely why these sunshine-belt machines are so deadly. They are as hard to see politically as materially. They are about consciousness – or its simulation.[4] They are floating signifiers moving in pickup trucks across Europe, blocked more effectively by the witch-weavings of the displaced and so unnatural Greenham women, who read the cyborg webs of power so very well, than by the militant labour of older masculinist politics, whose natural constituency needs defence jobs. Ultimately the 'hardest' science is about the realm of greatest boundary confusion, the realm of pure number, pure spirit, C^3I, cryptography, and the preservation of potent secrets. The new machines are so clean and light. Their engineers are sun-worshippers mediating a new scientific revolution associated with the night dream of post-industrial society. The diseases evoked by these clean machines are 'no more' than the minuscule coding changes of an antigen in the immune system, 'no more' than the experience of stress. The nimble fingers of 'Oriental' women, the old fascination of little Anglo-Saxon Victorian girls with doll's houses, women's enforced attention to the small take on quite new dimensions in this world. There might be a cyborg Alice taking account of these new dimensions. Ironically, it might be the unnatural cyborg women making chips in Asia and spiral dancing in Santa Rita jail [A practice at once both spiritual and political that linked guards and arrested anti-nuclear demonstrators in the Alameda County jail in California in the early 1980s] whose constructed unities will guide effective oppositional strategies.

So my cyborg myth is about transgressed boundaries, potent fusions, and dangerous possibilities which progressive people might explore as one part of needed political work. One of my premises is that most American socialists and feminists see deepened dualisms of mind and body, animal and machine, idealism and materialism in the social practices, symbolic formulations, and physical artefacts associated with 'high technology' and scientific culture. From One-Dimensional Man (Marcuse, 1964) to The Death of Nature (Merchant, 1980), the analytic resources developed by progressives have insisted on the necessary domination of technics and recalled us to an imagined organic body to integrate our resistance. Another of my premises is that the need for unity of people trying to resist world-wide intensification of domination has never been more acute. But a slightly perverse shift of perspective might better enable us to contest for meanings, as well as for other forms of power and pleasure in technologically mediated societies.

From one perspective, a cyborg world is about the final imposition of a grid of control on the planet, about the final abstraction embodied in a Star Wars apocalypse waged in the name of defence, about the final appropriation of women's bodies in a masculinist orgy of war (Sofia, 1984). From another perspective, a cyborg world might be about lived social and bodily realities in which people are not afraid of their joint kinship with animals and machines, not afraid of permanently partial identities and contradictory standpoints. The political struggle is to see from both perspectives at once because each reveals both dominations and possibilities unimaginable from the other vantage point. Single vision produces worse illusions than double vision or many-headed monsters. Cyborg unities are monstrous and illegitimate; in our present political circumstances, we could hardly hope for more potent myths for resistance and recoupling. I like to imagine LAG, the Livermore Action Group, as a kind of cyborg society, dedicated to realistically converting the laboratories that most fiercely embody and spew out the tools of technological apocalypse, and committed to building a political form that actually manages to hold together witches, engineers, elders, perverts, Christians, mothers, and Leninists long enough to disarm the state. Fission Impossible is the name of the affinity group in my town. (Affinity: related not by blood but by choice, the appeal of one chemical nuclear group for another, avidity.)[5]

[. . .]

The Informatics of Domination

In this attempt at an epistemological and political position, I would like to sketch a picture of possible unity, a picture indebted to socialist and feminist principles of design. The frame for my sketch is set by the extent and importance of rearrangements in world-wide social relations tied to science and technology. I argue for a politics rooted in claims about fundamental changes in the nature of class, race, and gender in an emerging system of world order analogous in its novelty and scope to that created by industrial capitalism; we are living through a movement from an organic, industrial society to a polymorphous, information system – from all work to all play, a deadly game. Simultaneously material and ideological, the dichotomies may be expressed in the following chart of transitions from the comfortable old hierarchical dominations to the scary new networks I have called the informatics of domination (see Table 18.1).

This list suggests several interesting things.[6] First, the objects on the right-hand side cannot be coded as 'natural', a realization that subverts naturalistic coding for the left-hand side as well. We cannot go back ideologically or materially. It's not just that 'god' is dead; so is the 'goddess'. Or both are revivified in the worlds charged with microelectronic and biotechnological politics. In relation to objects like biotic components, one must think not in terms of essential properties, but in terms of design, boundary constraints, rates of flows, systems logics, costs of lowering constraints. Sexual reproduction is one kind of reproductive strategy among many, with costs and benefits as a function of the system environment. Ideologies of sexual reproduction can no longer reasonably call on notions of sex and sex role as organic aspects in natural objects like organisms and families. Such reasoning will be unmasked as irrational, and ironically corporate executives reading *Playboy* and anti-porn radical feminists will make strange bedfellows in jointly unmasking the irrationalism.

Likewise for race, ideologies about human diversity have to be formulated in terms of frequencies of parameters, like blood groups or intelligence scores. It is 'irrational' to invoke concepts like primitive and civilized. For liberals and radicals, the search for integrated social systems gives way to a new practice called 'experimental ethnography' in which an organic object dissipates in attention to the play of writing. At the level of ideology, we see translations of racism and colonialism into languages of development and under-development, rates and constraints of modernization. Any objects or persons can be reasonably thought of in terms of disassembly and reassembly; no 'natural'

Table 18.1

Representation	Simulation
Bourgeois novel, realism	Science fiction, postmodernism
Organism	Biotic component
Depth, integrity	Surface, boundary
Heat	Noise
Biology as clinical practice	Biology as inscription
Physiology	Communications engineering
Small group	Subsystem
Perfection	Optimization
Eugenics	Population Control
Decadence, *Magic Mountain*	Obsolescence, *Future Shock*
Hygiene	Stress Management
Microbiology, tuberculosis	Immunology, AIDS
Organic division of labour	Ergonomics / cybernetics of labour
Functional specialization	Modular construction
Reproduction	Replication
Organic sex role specialization	Optimal genetic strategies
Biological determinism	Evolutionary inertia, constraints
Community ecology	Ecosystem
Racial chain of being	Neo-imperialism, United Nations humanism
Scientific management in home / factory	Global factory / Electronic cottage
Family / Market / Factory	Women in the Integrated Circuit
Family wage	Comparable worth
Public / Private	Cyborg citizenship
Nature / Culture	Fields of difference
Co-operation	Communications enhancement
Freud	Lacan
Sex	Genetic engineering
Labour	Robotics
Mind	Artificial Intelligence
Second World War	Star Wars
White Capitalist Patriarchy	Informatics of Domination

architectures constrain system design. The financial districts in all the world's cities, as well as the export-processing and free-trade zones, proclaim this elementary fact of 'late capitalism'. The entire universe of objects that can be known scientifically must be formulated as problems in communications engineering (for the managers) or theories of the text (for those who would resist). Both are cyborg semiologies.

One should expect control strategies to concentrate on boundary conditions and interfaces, on rates of flow across boundaries – and not on the integrity of natural objects. 'Integrity' or 'sincerity' of the Western self gives way to decision procedures and expert systems. For example, control strategies applied to women's capacities to give birth to new human beings will be developed in the languages of population control and maximization of goal achievement for individual decision-makers. Control strategies will be formulated in terms of rates, costs of constraints, degrees of freedom. Human beings, like any other component or subsystem, must be localized in a system architecture whose basic modes of operation are probabilistic, statistical. No objects, spaces, or bodies are sacred in themselves; any component can be interfaced with any other if the proper standard, the proper code, can be constructed for processing signals in a common language. Exchange in this world transcends the universal translation effected by capitalist markets that Marx analysed so well.

The privileged pathology affecting all kinds of components in this universe is stress – communications breakdown (Hogness, 1983). The cyborg is not subject to Foucault's biopolitics; the cyborg simulates politics, a much more potent field of operations.

This kind of analysis of scientific and cultural objects of knowledge which have appeared historically since the Second World War prepares us to notice some important inadequacies in feminist analysis which has proceeded as if the organic, hierarchical dualisms ordering discourse in 'the West' since Aristotle still ruled. They have been cannibalized, or as Zoe Sofia (Sofoulis) might put it, they have been 'techno-digested'. The dichotomies between mind and body, animal and human, organism and machine, public and private, nature and culture, men and women, primitive and civilized are all in question ideologically. The actual situation of women is their integration/exploitation into a world system of production/reproduction and communication called the informatics of domination. The home, workplace, market, public arena, the body itself – all can be dispersed and interfaced in nearly infinite, polymorphous ways, with large consequences for women and others – consequences that themselves are very different for different people and which make potent oppositional international movements difficult to imagine and essential for survival. One important route for reconstructing socialist-feminist politics is through theory and practice addressed to the social relations of science and technology, including crucially the systems of myth and meanings structuring our imaginations. The cyborg is a kind of disassembled and reassembled, postmodern collective and personal self. This is the self feminists must code.

Communications technologies and biotechnologies are the crucial tools recrafting our bodies. These tools embody and enforce new social relations for women world-wide. Technologies and scientific discourses can be partially understood as formalizations, i.e., as frozen moments, of the fluid social interactions constituting them, but they should also be viewed as instruments for enforcing meanings. The boundary is permeable between tool and myth, instrument and concept, historical systems of social relations and historical anatomies of possible bodies, including objects of knowledge. Indeed, myth and tool mutually constitute each other.

Furthermore, communications sciences and modern biologies are constructed by a common move – *the translation of the world into a problem of coding*, a search for a common language in which all resistance to instrumental control disappears and all heterogeneity can be submitted to disassembly, reassembly, investment, and exchange.

In communications sciences, the translation of the world into a problem in coding can be illustrated by looking at cybernetic (feedback-controlled) systems theories applied to telephone technology, computer design, weapons deployment, or data base construction and maintenance. In each case, solution to the key questions rests on a theory of language and control; the key operation is determining the rates, directions, and probabilities of flow of a quantity called information. The world is subdivided by boundaries differentially permeable to information. Information is just that kind of quantifiable element (unit, basis of unity) which allows universal translation, and so unhindered instrumental power (called effective communication). The biggest threat to such power is interruption of communication. Any system breakdown is a function of stress. The fundamentals of this technology can be condensed into the metaphor C^3I, command-control-communication-intelligence, the military's symbol for its operations theory.

In modern biologies, the translation of the world into a problem in coding can be illustrated by molecular genetics, ecology, sociobiological evolutionary theory, and immunobiology. The organism has been translated into problems of genetic coding and read-out. Biotechnology, a writing technology, informs research broadly.[7] In a sense, organisms have ceased to exist as objects of knowledge, giving way to biotic components, i.e., special kinds of information-processing devices. The analogous moves in ecology could be examined by probing the history and utility of the concept of the ecosystem. Immunobiology and associated medical practices are rich exemplars of the privilege of coding and recognition systems as objects of knowledge, as constructions of bodily reality for us. Biology here is a kind of cryptography. Research is necessarily a kind of intelligence activity. Ironies abound. A stressed system goes awry; its communication processes break down; it fails to recognize the difference between self and other. Human babies with baboon hearts evoke national ethical perplexity – for animal rights activists at least as much as for the guardians of human purity. In the US gay men and intravenous drug users are the 'privileged' victims of an awful immune system disease that marks (inscribes on the body) confusion of boundaries and moral pollution (Treichler, 1987).

But these excursions into communications sciences and biology have been at a rarefied level; there is a mundane, largely economic reality to support my claim that these sciences and technologies indicate fundamental transformations in the structure of the world for us. Communications technologies depend on electronics. Modern states, multinational corporations, military power, welfare state apparatuses, satellite systems, political processes, fabrication of our imaginations, labour-control systems, medical constructions of our bodies, commercial pornography, the international division of labour, and religious evangelism depend intimately upon electronics. Microelectronics is the technical basis of simulacra; that is, of copies without originals.

Microelectronics mediates the translations of labour into robotics and word processing, sex into genetic engineering and reproductive technologies, and mind into artificial intelligence and decision procedures. The new biotechnologies concern more than human reproduction. Biology as a powerful engineering science for redesigning materials and processes has revolutionary implications for industry, perhaps most obvious today in areas of fermentation, agriculture, and energy. Communications sciences and biology are constructions of natural-technical objects of knowledge in which the difference between machine and organism is thoroughly blurred; mind, body, and tool are on very intimate terms. The 'multinational' material organization of the production and reproduction of daily life and the symbolic organization of the production and reproduction of culture and imagination seem equally implicated. The boundary-maintaining images of base and superstructure, public and private, or material and ideal never seemed more feeble.

I have used Rachel Grossman's (1980) image of women in the integrated circuit to name the situation of women in a world so intimately restructured through the social relations of science and technology.[8] I used the odd circumlocution, 'the social relations of science and technology', to indicate that we are not dealing with a technological determinism, but with a historical system depending upon structured relations among people. But the phrase should also indicate that science and technology provide fresh sources of power, that we need fresh sources of analysis and political action (Latour, 1984). Some of the rearrangements of race, sex, and class rooted in high-tech-facilitated social relations can make socialist-feminism more relevant to effective progressive politics.

The 'Homework Economy' Outside 'The Home'

The 'New Industrial Revolution' is producing a new world-wide working class, as well as new sexualities and ethnicities. The extreme mobility of capital and the emerging international division of labour are intertwined with the emergence of new collectivities, and the weakening of familiar groupings. These developments are neither gender- nor race-neutral. White men in advanced industrial societies have become newly vulnerable to permanent job loss, and women are not disappearing from the job rolls at the same rates as men. It is not simply that women in Third World countries are the preferred labour force for the science-based multinationals in the export-processing sectors, particularly in electronics. The picture is more systematic and involves reproduction, sexuality, culture, consumption, and production. In the prototypical Silicon Valley, many women's lives have been structured around employment in electronics-dependent jobs, and their intimate realities include serial heterosexual monogamy, negotiating childcare, distance from extended kin or most other forms of traditional community, a high likelihood of loneliness and extreme economic vulnerability as they age. The ethnic and racial diversity of women in Silicon Valley structures a microcosm of conflicting differences in culture, family, religion, education, and language.

Richard Gordon has called this new situation the 'homework economy'.[9] Although he includes the phenomenon of literal homework emerging in connection with electronics assembly, Gordon intends 'homework economy' to name a restructuring of work that broadly has the characteristics formerly ascribed to female jobs, jobs literally done only by women. Work is being redefined as both literally female and feminized, whether performed by men or women. To be feminized means to be made extremely vulnerable; able to be disassembled, reassembled, exploited as a reserve labour force; seen less as workers than as servers; subjected to time arrangements on and off the paid job that make a mockery of a limited work day; leading an existence that always borders on being obscene, out of place, and reducible to sex. Deskilling is an old strategy newly applicable to formerly privileged workers. However, the homework economy does not refer only to large-scale deskilling, nor does it deny that new areas of high skill are emerging, even for women and men previously excluded from skilled employment. Rather, the concept indicates that factory, home, and market are integrated on a new scale and that the places of women are crucial – and need to be analysed for differences among women and for meanings for relations between men and women in various situations.

The homework economy as a world capitalist organizational structure is made possible by (not caused by) the new technologies. The success of the attack on relatively privileged, mostly white, men's unionized jobs is tied to the power of the new communications technologies to integrate and control labour despite extensive dispersion and decentralization. The consequences of the new technologies are felt by women both in the loss of the family (male) wage (if they ever had access to this white privilege) and in the character of their own jobs, which are becoming capital-intensive; for example, office work and nursing.

The new economic and technological arrangements are also related to the collapsing welfare state and the ensuing intensification of demands on women to sustain daily life for themselves as well as for men, children, and old people. The feminization of poverty – generated by dismantling the welfare state, by the homework economy where stable jobs become the exception, and sustained by the expectation that women's wages will not be matched by a male income for the support of children – has become an urgent focus. The

causes of various women-headed households are a function of race, class, or sexuality; but their increasing generality is a ground for coalitions of women on many issues. That women regularly sustain daily life partly as a function of their enforced status as mothers is hardly new; the kind of integration with the overall capitalist and progressively war-based economy is new. The particular pressure, for example, on US black women, who have achieved an escape from (barely) paid domestic service and who now hold clerical and similar jobs in large numbers, has large implications for continued enforced black poverty *with* employment. Teenage women in industrializing areas of the Third World increasingly find themselves the sole or major source of a cash wage for their families, while access to land is ever more problematic. These developments must have major consequences in the psychodynamics and politics of gender and race.

Within the framework of three major stages of capitalism (commercial/early industrial, monopoly, multinational) – tied to nationalism, imperialism, and multinationalism, and related to Jameson's three dominant aesthetic periods of realism, modernism, and postmodernism – I would argue that specific forms of families dialectically relate to forms of capital and to its political and cultural concomitants. Although lived problematically and unequally, ideal forms of these families might be schematized as (1) the patriarchal nuclear family, structured by the dichotomy between public and private and accompanied by the white bourgeois ideology of separate spheres and nineteenth-century Anglo-American bourgeois feminism; (2) the modern family mediated (or enforced) by the welfare state and institutions like the family wage, with a flowering of a-feminist heterosexual ideologies, including their radical versions represented in Greenwich Village around the First World War; and (3) the 'family' of the homework economy with its oxymoronic structure of women-headed households and its explosion of feminisms and the paradoxical intensification and erosion of gender itself. This is the context in which the projections for worldwide structural unemployment stemming from the new technologies are part of the picture of the homework economy. As robotics and related technologies put men out of work in 'developed' countries and exacerbate failure to generate male jobs in Third World 'development', and as the automated office becomes the rule even in labour-surplus countries, the feminization of work intensifies. Black women in the United States have long known what it looks like to face the structural underemployment ('feminization') of black men, as well as their own highly vulnerable position in the wage economy. It is no longer a secret that sexuality, reproduction, family, and community life are interwoven with this economic structure in myriad ways which have also differentiated the situations of white and black women. Many more women and men will contend with similar situations, which will make cross-gender and race alliances on issues of basic life support (with or without jobs) necessary, not just nice.

The new technologies also have a profound effect on hunger and on food production for subsistence world-wide. Rae Lessor Blumberg (1983) estimates that women produce about 50 per cent of the world's subsistence food.[10] Women are excluded generally from benefiting from the increased high-tech commodification of food and energy crops, their days are made more arduous because their responsibilities to provide food do not diminish, and their reproductive situations are made more complex. Green Revolution technologies interact with other high-tech industrial production to alter gender divisions of labour and differential gender migration patterns.

The new technologies seem deeply involved in the forms of 'privatization' that Ros Petchesky (1981) has analysed, in which militarization, right-wing family ideologies and policies, and intensified definitions of corporate (and state) property as private synergistically interact.[11] The new communications technologies are fundamental to the eradication of 'public life' for everyone. This facilitates the mushrooming of a permanent high-tech military establishment at the cultural and economic expense of most people, but especially of women. Technologies like video games and highly miniaturized televisions seem crucial to production of modern forms of 'private life'. The culture of video games is heavily orientated to individual competition and extraterrestrial warfare. High-tech, gendered imaginations are produced here, imaginations that can contemplate destruction of the planet and

a sci-fi escape from its consequences. More than our imaginations is militarized; and the other realities of electronic and nuclear warfare are inescapable. These are the technologies that promise ultimate mobility and perfect exchange – and incidentally enable tourism, that perfect practice of mobility and exchange, to emerge as one of the world's largest single industries.

The new technologies affect the social relations of both sexuality and of reproduction, and not always in the same ways. The close ties of sexuality and instrumentality, of views of the body as a kind of private satisfaction- and utility-maximizing machine, are described nicely in sociobiological origin stories that stress a genetic calculus and explain the inevitable dialectic of domination of male and female gender roles.[12] These sociobiological stories depend on a high-tech view of the body as a biotic component or cybernetic communications system. Among the many transformations of reproductive situations is the medical one, where women's bodies have boundaries newly permeable to both 'visualization' and 'intervention'. Of course, who controls the interpretation of bodily boundaries in medical hermeneutics is a major feminist issue. The speculum served as an icon of women's claiming their bodies in the 1970s; that handcraft tool is inadequate to express our needed body politics in the negotiation of reality in the practices of cyborg reproduction. Self-help is not enough. The technologies of visualization recall the important cultural practice of hunting with the camera and the deeply predatory nature of a photographic consciousness.[13] Sex, sexuality, and reproduction are central actors in high-tech myth systems structuring our imaginations of personal and social possibility.

Another critical aspect of the social relations of the new technologies is the reformulation of expectations, culture, work, and reproduction for the large scientific and technical work-force. A major social and political danger is the formation of a strongly bimodal social structure, with the masses of women and men of all ethnic groups, but especially people of colour, confined to a homework economy, illiteracy of several varieties, and general redundancy and impotence, controlled by high-tech repressive apparatuses ranging from entertainment to surveillance and disappearance. An adequate socialist-feminist politics should address women in the privileged occupational categories, and particularly in the production of science and technology that constructs scientific-technical discourses, processes, and objects.[14]

This issue is only one aspect of enquiry into the possibility of a feminist science, but it is important. What kind of constitutive role in the production of knowledge, imagination, and practice can new groups doing science have? How can these groups be allied with progressive social and political movements? What kind of political accountability can be constructed to tie women together across the scientific-technical hierarchies separating us? Might there be ways of developing feminist science/technology politics in alliance with anti-military science facility conversion action groups? Many scientific and technical workers in Silicon Valley, the high-tech cowboys included, do not want to work on military science.[15] Can these personal preferences and cultural tendencies be welded into progressive politics among this professional middle class in which women, including women of colour, are coming to be fairly numerous?

Women in the Integrated Circuit

Let me summarize the picture of women's historical locations in advanced industrial societies, as these positions have been restructured partly through the social relations of science and technology. If it was ever possible ideologically to characterize women's lives by the distinction of public and private domains – suggested by images of the division of working-class life into factory and home, of bourgeois life into market and home, and of gender existence into personal and political realms – it is now a totally misleading ideology, even to show how both terms of these dichotomies construct each other in practice and in theory. I prefer a network ideological image, suggesting the profusion of spaces and identities and the permeability of boundaries in the personal body and in the body politic. 'Networking' is both a feminist practice and a multinational corporate strategy – weaving is for oppositional cyborgs.

So let me return to the earlier image of the informatics of domination and trace one vision of women's 'place' in the integrated circuit, touching only a few idealized social locations seen primarily from the point of view of advanced capitalist societies: Home, Market, Paid Work Place, State, School, Clinic-hospital, and Church. Each of these idealized spaces is logically and practically implied in every other locus, perhaps analogous to a holographic photograph. I want to suggest the impact of the social relations mediated and enforced by the new technologies in order to help formulate needed analysis and practical work. However, there is no 'place' for women in these networks, only geometrics of difference and contradiction crucial to women's cyborg identities. If we learn how to read these webs of power and social life, we might learn new couplings, new coalitions. There is no way to read the following list from a standpoint of 'identification', of a unitary self. The issue is dispersion. The task is to survive in the diaspora.

Home: Women-headed households, serial monogamy, flight of men, old women alone, technology of domestic work, paid homework, re-emergence of home sweat-shops, home-based businesses and telecommuting, electronic cottage, urban homelessness, migration, module architecture, reinforced (simulated) nuclear family, intense domestic violence.

Market: Women's continuing consumption work, newly targeted to buy the profusion of new production from the new technologies (especially as the competitive race among industrialized and industrializing nations to avoid dangerous mass unemployment necessitates finding ever bigger new markets for ever less clearly needed commodities); bimodal buying power, coupled with advertising targeting of the numerous affluent groups and neglect of the previous mass markets; growing importance of informal markets in labour and commodities parallel to high-tech, affluent market structures; surveillance systems through electronic funds transfer; intensified market abstraction (commodification) of experience, resulting in ineffective utopian or equivalent cynical theories of community; extreme mobility (abstraction) of marketing/financing systems; interpenetration of sexual and labour markets; intensified sexualization of abstracted and alienated consumption.

Paid Work Place: Continued intense sexual and racial division of labour, but considerable growth of membership in privileged occupational categories for many white women and people of colour; impact of new technologies on women's work in clerical, service, manufacturing (especially textiles), agriculture, electronics; international restructuring of the working classes; development of new time arrangements to facilitate the homework economy (flex time, part time, over time, no time); homework and out work; increased pressures for two-tiered wage structures; significant numbers of people in cash-dependent populations world-wide with no experience or no further hope of stable employment; most labour 'marginal' or 'feminized'.

State: Continued erosion of the welfare state; decentralizations with increased surveillance and control; citizenship by telematics; imperialism and political power broadly in the form of information rich/information poor differentiation; increased high-tech militarization increasingly opposed by many social groups; reduction of civil service jobs as a result of the growing capital intensification of office work, with implications for occupational mobility for women of colour; growing privatization of material and ideological life and culture; close integration of privatization and militarization, the high-tech forms of bourgeois capitalist personal and public life; invisibility of different social groups to each other, linked to psychological mechanisms of belief in abstract enemies.

School: Deepening coupling of high-tech capital needs and public education at all levels, differentiated by race, class, and gender; managerial classes involved in educational reform and refunding at the cost of remaining progressive educational democratic structures for children and teachers; education for mass ignorance and repression in technocratic and militarized culture; growing anti-science mystery cults in dissenting and radical political movements; continued relative scientific illiteracy among white women and people of colour; growing industrial direction of education (especially higher education) by science-based multinationals (particularly in electronics- and biotechnology-dependent companies); highly educated, numerous élites in a progressively bimodal society.

Clinic-hospital: Intensified machine–body relations; renegotiations of public metaphors which channel personal experience of the body, particularly in relation to reproduction, immune system functions, and 'stress' phenomena; intensification of reproductive politics in response to world historical implications of women's unrealized, potential control of their relation to reproduction; emergence of new, historically specific diseases; struggles over meanings and means of health in environments pervaded by high technology products and processes; continuing feminization of health work; intensified struggle over state responsibility for health; continued ideological role of popular health movements as a major form of American politics.

Church: Electronic fundamentalist 'super-saver' preachers solemnizing the union of electronic capital and automated fetish gods; intensified importance of churches in resisting the militarized state; central struggle over women's meanings and authority in religion; continued relevance of spirituality, intertwined with sex and health, in political struggle.

The only way to characterize the informatics of domination is as a massive intensification of insecurity and cultural impoverishment, with common failure of subsistence networks for the most vulnerable. Since much of this picture interweaves with the social relations of science and technology, the urgency of a socialist-feminist politics addressed to science and technology is plain. There is much now being done, and the grounds for political work are rich. For example, the efforts to develop forms of collective struggle for women in paid work should be a high priority for all of us. These efforts are profoundly tied to technical restructuring of labour processes and reformations of working classes. These efforts also are providing understanding of a more comprehensive kind of labour organization, involving community, sexuality, and family issues never privileged in the largely white male industrial unions.

The structural rearrangements related to the social relations of science and technology evoke strong ambivalence. But it is not necessary to be ultimately depressed by the implications of late twentieth-century women's relation to all aspects of work, culture, production of knowledge, sexuality, and reproduction. For excellent reasons, most Marxisms see domination best and have trouble understanding what can only look like false consciousness and people's complicity in their own domination in late capitalism. It is crucial to remember that what is lost, perhaps especially from women's points of view, is often virulent forms of oppression, nostalgically naturalized in the face of current violation. Ambivalence towards the disrupted unities mediated by high-tech culture requires not sorting consciousness into categories of 'clear-sighted critique grounding a solid political epistemology' versus 'manipulated false consciousness', but subtle understanding of emerging pleasures, experiences, and powers with serious potential for changing the rules of the game.

There are grounds for hope in the emerging bases for new kinds of unity across race, gender, and class, as these elementary units of socialist-feminist analysis themselves suffer protean transformations. Intensifications of hardship experienced world-wide in connection with the social relations of science and technology are severe. But what people are experiencing is not transparently clear, and we lack sufficiently subtle connections for collectively building effective theories of experience. Present efforts – Marxist, psychoanalytic, feminist, anthropological – to clarify even 'our' experience are rudimentary.

I am conscious of the odd perspective provided by my historical position – a PhD in biology for an Irish Catholic girl was made possible by Sputnik's impact on US national science-education policy. I have a body and mind as much constructed by the post-Second World War arms race and cold war as by the women's movements. There are more grounds for hope in focusing on the contradictory effects of politics designed to produce loyal American technocrats, which also produced large numbers of dissidents, than in focusing on the present defeats.

The permanent partiality of feminist points of view has consequences for our expectations of forms of political organization and participation.

We do not need a totality in order to work well. The feminist dream of a common language, like all dreams for a perfectly true language, of perfectly faithful naming of experience, is a totalizing and imperialist one. In that sense, dialectics too is a dream language, longing to resolve contradiction. Perhaps, ironically, we can learn from our fusions with animals and machines how not to be Man, the embodiment of Western logos. From the point of view of pleasure in these potent and taboo fusions, made inevitable by the social relations of science and technology, there might indeed be a feminist science.

Cyborgs: a Myth of Political Identity

I want to conclude with a myth about identity and boundaries which might inform late twentieth-century political imaginations. I am indebted in this story to writers like Joanna Russ, Samuel R. Delany, John Varley, James Tiptree, Jr, Octavia Butler, Monique Wittig, and Vonda McIntyre.[16] These are our story-tellers exploring what it means to be embodied in high-tech worlds. They are theorists for cyborgs. Exploring conceptions of bodily boundaries and social order, the anthropologist Mary Douglas (1966, 1970) should be credited with helping us to consciousness about how fundamental body imagery is to world view, and so to political language. French feminists like Luce Irigaray and Monique Wittig, for all their differences, know how to write the body; how to weave eroticism, cosmology, and politics from imagery of embodiment, and especially for Wittig, from imagery of fragmentation and reconstitution of bodies.[17]

American radical feminists like Susan Griffin, Audre Lorde, and Adrienne Rich have profoundly affected our political imaginations – and perhaps restricted too much what we allow as a friendly body and political language.[18] They insist on the organic, opposing it to the technological. But their symbolic systems and the related positions of ecofeminism and feminist paganism, replete with organicisms, can only be understood in Sandoval's terms as oppositional ideologies fitting the late twentieth century. They would simply bewilder anyone not preoccupied with the machines and consciousness of late capitalism. In that sense they are part of the cyborg world. But there are also great riches for feminists in explicitly embracing the possibilities inherent in the breakdown of clean distinctions between organism and machine and similar distinctions structuring the Western self. It is the simultaneity of breakdowns that cracks the matrices of domination and opens geometric possibilities. What might be learned from personal and political 'technological' pollution?

[. . .]

Writing is pre-eminently the technology of cyborgs, etched surfaces of the late twentieth century. Cyborg politics is the struggle for language and the struggle against perfect communication, against the one code that translates all meaning perfectly, the central dogma of phallogocentrism. That is why cyborg politics insist on noise and advocate pollution, rejoicing in the illegitimate fusions of animal and machine. These are the couplings which make Man and Woman so problematic, subverting the structure of desire, the force imagined to generate language and gender, and so subverting the structure and modes of reproduction of 'Western' identity, of nature and culture, of mirror and eye, slave and master, body and mind. 'We' did not originally choose to be cyborgs, but choice grounds a liberal politics and epistemology that imagines the reproduction of individuals before the wider replications of 'texts'.

From the perspective of cyborgs, freed of the need to ground politics in 'our' privileged position of the oppression that incorporates all other dominations, the innocence of the merely violated, the ground of those closer to nature, we can see powerful possibilities. Feminisms and Marxisms have run aground on Western epistemological imperatives to construct a revolutionary subject from the perspective of a hierarchy of oppressions and/or a latent position of moral superiority, innocence, and greater closeness to nature. With no available original dream of a common language or original symbiosis promising protection from hostile 'masculine' separation, but written into the play of a text that has no finally privileged reading or salvation history, to recognize 'oneself' as fully implicated in the world, frees us of the need to root politics in identification, vanguard parties, purity, and mothering. Stripped of identity, the bastard race teaches about the power of the margins and the importance of a mother like Malinche. Women of colour have transformed her from the evil mother of masculinist fear into the originally literate mother who teaches survival.

This is not just literary deconstruction, but liminal transformation. Every story that begins with original innocence and privileges the return to wholeness imagines the drama of life to be individuation, separation, the birth of the self, the tragedy of autonomy, the fall into writing, alienation; that is, war, tempered by imaginary respite in the bosom of the Other. These plots are ruled by a reproductive politics – rebirth without flaw, perfection, abstraction. In this plot women are imagined either better or worse off, but all agree they have less selfhood, weaker individuation, more fusion to the oral, to Mother, less at stake in masculine autonomy. But there is another route to having less at stake in masculine autonomy, a route that does not pass through Woman, Primitive, Zero, the Mirror Stage and its imaginary. It passes through women and other present-tense, illegitimate cyborgs, not of Woman born, who refuse the ideological resources of victimization so as to have a real life. These cyborgs are the people who refuse to disappear on cue, no matter how many times a 'Western' commentator remarks on the sad passing of another primitive, another organic group done in by 'Western' technology, by writing.[19] These real-life cyborgs (for example, the Southeast Asian village women workers in Japanese and US electronics firms described by Aihwa Ong) are actively rewriting the texts of their bodies and societies. Survival is the stakes in this play of readings.

To recapitulate, certain dualisms have been persistent in Western traditions; they have all been systemic to the logics and practices of domination of women, people of colour, nature, workers, animals – in short, domination of all constituted as others, whose task is to mirror the self. Chief among these troubling dualisms are self/other, mind/body, culture/nature, male/female, civilized/primitive, reality/appearance, whole/part, agent/resource, maker/made, active/passive, right/wrong, truth/illusion, total/partial, God/man. The self is the One who is not dominated, who knows that by the service of the other, the other is the one who holds the future, who knows that by the experience of domination, which gives the lie to the autonomy of the self. To be One is to be autonomous, to be powerful, to be God; but to be One is to be an illusion, and so to be involved in a dialectic of apocalypse with the other. Yet to be other is to be multiple, without clear boundary, frayed, insubstantial. One is too few, but two are too many.

High-tech culture challenges these dualisms in intriguing ways. It is not clear who makes and who is made in the relation between human and machine. It is not clear what is mind and what body in machines that resolve into coding practices. In so far as we know ourselves in both formal discourse (for example, biology) and in daily practice (for example, the homework economy in the integrated circuit), we find ourselves to be cyborgs, hybrids, mosaics, chimeras. Biological organisms have become biotic systems, communications devices like others. There is no fundamental, ontological separation in our formal knowledge of machine and organism, of technical and organic. The replicant Rachel in the Ridley Scott film *Blade Runner* stands as the image of a cyborg culture's fear, love, and confusion.

One consequence is that our sense of connection to our tools is heightened. The trance state experienced by many computer users has become a staple of science-fiction film and cultural jokes. Perhaps paraplegics and other severely handicapped people can (and sometimes do) have the most intense experiences of complex hybridization with other communication devices.[20] Anne McCaffrey's pre-feminist *The Ship Who Sang* (1969) explored the consciousness of a cyborg, hybrid of girl's brain and complex machinery, formed after the birth of a severely handicapped child. Gender, sexuality, embodiment, skill: all were reconstituted in the story. Why should our bodies end at the skin, or include at best other beings encapsulated by skin? From the seventeenth century till now, machines could be animated – given ghostly souls to make them speak or move or to account for their orderly development and mental capacities. Or organisms could be mechanized – reduced to body understood as resource of mind. These machine/organism relationships are obsolete, unnecessary. For us, in imagination and in other practice, machines can be prosthetic devices, intimate components, friendly selves. We don't need organic holism to give impermeable wholeness, the total woman and her feminist variants (mutants?). Let me conclude this point by a very partial reading of the logic of the cyborg monsters of my second group of texts, feminist science fiction.

The cyborgs populating feminist science fiction make very problematic the statuses of man or woman, human, artefact, member of a race, individual entity, or body. Katie King clarifies how pleasure in reading these fictions is not largely based on identification. Students facing Joanna Russ for the first time, students who have learned to take modernist writers like James Joyce or Virginia Woolf without flinching, do not know what to make of *The Adventures of Alyx* or *The Female Man*, where characters refuse the reader's search for innocent wholeness while granting the wish for heroic quests, exuberant eroticism, and serious politics. *The Female Man* is the story of four versions of one genotype, all of whom meet, but even taken together do not make a whole, resolve the dilemmas of violent moral action, or remove the growing scandal of gender. The feminist science fiction of Samuel R. Delany, especially *Tales of Nevèrÿon*, mocks stories of origin by redoing the neolithic revolution, replaying the founding moves of Western civilization to subvert their plausibility. James Tiptree, Jr, an author whose fiction was regarded as particularly manly until her 'true' gender was revealed, tells tales of reproduction based on non-mammalian technologies like alternation of generations of male brood pouches and male nurturing. John Varley constructs a supreme cyborg in his arch-feminist exploration of Gaea, a mad goddess-planet-trickster-old woman-technological device on whose surface an extraordinary array of post-cyborg symbioses are spawned. Octavia Butler writes of an African sorceress pitting her powers of transformation against the genetic manipulations of her rival (*Wild Seed*), of time warps that bring a modern US black woman into slavery where her actions in relation to her white master-ancestor determine the possibility of her own birth (*Kindred*), and of the illegitimate insights into identity and community of an adopted cross-species child who came to know the enemy as self (*Survivor*). In *Dawn* (1987), the first instalment of a series called *Xenogenesis*, Butler tells the story of Lilith Iyapo, whose personal name recalls Adam's first and repudiated wife and whose family name marks her status as the widow of the son of Nigerian immigrants to the US. A black woman and a mother whose child is dead, Lilith mediates the transformation of humanity through genetic exchange with extra-terrestrial lovers/rescuers/destroyers/genetic engineers, who reform earth's habitats after the nuclear holocaust and coerce surviving humans into intimate fusion with them. It is a novel that interrogates reproductive, linguistic, and nuclear politics in a mythic field structured by late twentieth-century race and gender.

Because it is particularly rich in boundary transgressions, Vonda McIntyre's *Superluminal* can close this truncated catalogue of promising and dangerous monsters who help redefine the pleasures and politics of embodiment and feminist writing. In a fiction where no character is 'simply' human, human status is highly problematic. Orca, a genetically altered diver, can speak with killer whales and survive deep ocean conditions, but she longs to explore space as a pilot, necessitating bionic implants jeopardizing her kinship with the divers and cetaceans. Transformations are effected by virus vectors carrying a new developmental code, by transplant surgery, by implants of microelectronic devices, by analogue doubles, and other means. Laenea becomes a pilot by accepting a heart implant and a host of other alterations allowing survival in transit at speeds exceeding that of light. Radu Dracul survives a virus-caused plague in his outerworld planet to find himself with a time sense that changes the boundaries of spatial perception for the whole species. All the characters explore the limits of language; the dream of communicating experience; and the necessity of limitation, partiality, and intimacy even in this world of protean transformation and connection. *Superluminal* stands also for the defining contradictions of a cyborg world in another sense; it embodies textually the intersection of feminist theory and colonial discourse in the science fiction I have alluded to in this chapter. This is a conjunction with a long history that many 'First World' feminists have tried to repress, including myself in my readings of *Superluminal* before being called to account by Zoe Sofoulis, whose different location in the world system's informatics of domination made her acutely alert to the imperialist moment of all science fiction cultures, including women's science fiction. From an Australian feminist

sensitivity, Sofoulis remembered more readily McIntyre's role as writer of the adventures of Captain Kirk and Spock in TV's *Star Trek* series than her rewriting the romance in *Superluminal*.

Monsters have always defined the limits of community in Western imaginations. The Centaurs and Amazons of ancient Greece established the limits of the centred polis of the Greek male human by their disruption of marriage and boundary pollutions of the warrior with animality and woman. Unseparated twins and hermaphrodites were the confused human material in early modern France who grounded discourse on the natural and supernatural, medical and legal, portents and diseases – all crucial to establishing modern identity.[21] The evolutionary and behavioural sciences of monkeys and apes have marked the multiple boundaries of late twentieth-century industrial identities. Cyborg monsters in feminist science fiction define quite different political possibilities and limits from those proposed by the mundane fiction of Man and Woman.

There are several consequences to taking seriously the imagery of cyborgs as other than our enemies. Our bodies, ourselves; bodies are maps of power and identity. Cyborgs are no exception. A cyborg body is not innocent; it was not born in a garden; it does not seek unitary identity and so generate antagonistic dualisms without end (or until the world ends); it takes irony for granted. One is too few, and two is only one possibility. Intense pleasure in skill, machine skill, ceases to be a sin, but an aspect of embodiment. The machine is not an *it* to be animated, worshipped, and dominated. The machine is us, our processes, an aspect of our embodiment. We can be responsible for machines; *they* do not dominate or threaten us. We are responsible for boundaries; we are they. Up till now (once upon a time), female embodiment seemed to be given, organic, necessary; and female embodiment seemed to mean skill in mothering and its metaphoric extensions. Only by being out of place could we take intense pleasure in machines, and then with excuses that this was organic activity after all, appropriate to females. Cyborgs might consider more seriously the partial, fluid, sometimes aspect of sex and sexual embodiment. Gender might not be global identity after all, even if it has profound historical breadth and depth.

The ideologically charged question of what counts as daily activity, as experience, can be approached by exploiting the cyborg image. Feminists have recently claimed that women are given to dailiness, that women more than men somehow sustain daily life, and so have a privileged epistemological position potentially. There is a compelling aspect to this claim, one that makes visible unvalued female activity and names it as the ground of life. But *the* ground of life? What about all the ignorance of women, all the exclusions and failures of knowledge and skill? What about men's access to daily competence, to knowing how to build things, to take them apart, to play? What about other embodiments? Cyborg gender is a local possibility taking a global vengeance. Race, gender, and capital require a cyborg theory of wholes and parts. There is no drive in cyborgs to produce total theory, but there is an intimate experience of boundaries, their construction and deconstruction. There is a myth system waiting to become a political language to ground one way of looking at science and technology and challenging the informatics of domination – in order to act potently.

One last image: organisms and organismic, holistic politics depend on metaphors of rebirth and invariably call on the resources of reproductive sex. I would suggest that cyborgs have more to do with regeneration and are suspicious of the reproductive matrix and of most birthing. For salamanders, regeneration after injury, such as the loss of a limb, involves regrowth of structure and restoration of function with the constant possibility of twinning or other odd topographical productions at the site of former injury. The regrown limb can be monstrous, duplicated, potent. We have all been injured, profoundly. We require regeneration, not rebirth, and the possibilities for our reconstitution include the utopian dream of the hope for a monstrous world without gender.

Cyborg imagery can help express two crucial arguments in this essay: first, the production of universal, totalizing theory is a major mistake that misses most of reality, probably always, but certainly now; and second, taking responsibility for the social relations of science and technology means refusing an anti-science metaphysics, a demonology of technology, and so means

embracing the skilful task of reconstructing the boundaries of daily life, in partial connection with others, in communication with all of our parts. It is not just that science and technology are possible means of great human satisfaction, as well as a matrix of complex dominations. Cyborg imagery can suggest a way out of the maze of dualisms in which we have explained our bodies and our tools to ourselves. This is a dream not of a common language, but of a powerful infidel heteroglossia. It is an imagination of a feminist speaking in tongues to strike fear into the circuits of the super-savers of the new right. It means both building and destroying machines, identities, categories, relationships, space stories. Though both are bound in the spiral dance, I would rather be a cyborg than a goddess.

Notes

1 Useful references to left and/or feminist radical science movements and theory and to biological/biotechnical issues include: Bleier (1984, 1986), Harding (1986), Fausto-Sterling (1985), Gould (1981), Hubbard *et al.* (1982), Keller (1985), Lewontin *et al.* (1984), *Radical Science Journal* (became *Science as Culture* in 1987), 26 Freegrove Road, London N7 9RQ; 897 Main St, Cambridge, MA 02139.

2 Starting points for left and/or feminist approaches to technology and politics include: Cowan (1983), Rothschild (1983), Traweek (1988), Young and Levidow (1981, 1985), Weizenbaum (1976), Winner (1977, 1986), Zimmerman (1983), Athanasiou (1987), Cohn (1987a, 1987b), Winograd and Flores (1986), Edwards (1985). *Global Electronics*, 867 West Dana St, #204, Mountain View, CA 94041; *World*, 55 Sutter St, San Francisco, CA 94104; ISIS, Women's International Information Communication Service, PO Box 50 (Cornavin), 12II Geneva 2, Switzerland, and Via Santa Maria Dell'Anima 30, 00186 Rome, Italy. Fundamental approaches to modern social studies of science that do not continue the liberal mystification that it all started with Thomas Kuhn, include: Knorr-Cetina (1981), Knorr-Cetina and Mulkay (1983), Latour and Woolgar (1979), Young (1979). The 1984 Directory of the Network for the Ethnographic Study of Science, Technology, and Organizations lists a wide range of people and projects crucial to better radical analysis; available from NESSTO, PO Box Stanford, CA 94305.

3 A provocative, comprehensive argument about the politics and theories of 'postmodernism' is made by Fredric Jameson (1984), who argues that postmodernism is not an option, a style among others, but a cultural dominant requiring radical reinvention of left politics from within; there is no longer any place without that gives meaning to the comforting fiction of critical distance. Jameson also makes clear why one cannot be for or against postmodernism, an essentially moralist move. My position is that feminists (and others) need continuous cultural reinvention, postmodernist critique, and historical materialism; only a cyborg would have a chance. The old dominations of white capitalist patriarchy seem nostalgically innocent now: they normalized heterogeneity, into man and woman, white and black, for example. 'Advanced capitalism' and postmodernism release heterogeneity without a norm, and we are flattened, without subjectivity, which requires depth, even unfriendly and drowning depths. It is time to write *The Death of the Clinic*. The clinic's methods required bodies and works; we have texts and surfaces. Our dominations don't work by medicalization and normalization anymore; they work by networking, communications redesign, stress management. Normalization gives way to automation, utter redundancy. Michel Foucault's *Birth of the Clinic* (1963), *of* (1976), *and* (1975) name a form of power at its moment of implosion. The discourse of biopolitics gives way to technobabble, the language of the spliced substantive; no noun is left whole by the multinationals. These are their names, listed from one issue of Tech-Knowledge, Genentech, Allergen, Hybritech, Compupro, Genencor, Syntex, Allelix, Agrigenetics Corp., Syntro, Codon, Repligen, MicroAngelo from Scion Corp., Percom Data, Inter Systems, Cyborg Corp., Statcom Corp., Intertec. If we are imprisoned by language, then escape from that prison-house requires language poets, a kind of cultural restriction enzyme to cut the code; cyborg heteroglossia is one form of radical cultural politics. For cyborg poetry, see Perloff (1984); Fraser (1984). For feminist modernist/postmodernist 'cyborg' writing, see HOW(ever), 871 Corbett Ave, San Francisco, CA 94131.

4 Baudrillard (1983). Jameson (1984, p. 66) points out that Plato's definition of the simulacrum is the copy for which there is no original, i.e., the world of advanced capitalism, of pure exchange. See *Discourse 9, Journal for Theoretical Studies in Media and Culture: On Technology – Cybernetics, Ecology, and the Postmodern Imagination* (Spring/ Summer 1987) for a special issue.

5 For ethnographic accounts and political evaluations, see Epstein (forthcoming), Sturgeon (1986). Without explicit irony, adopting the spaceship earth/whole earth logo of the planet photographed from space, set off by the slogan 'Love Your Mother', the May 1987 Mothers and Others Day action at the nuclear weapons testing facility in Nevada none the less took account of the tragic contradictions of views of the earth. Demonstrators applied for official permits to be on the land from officers of the Western Shoshone tribe, whose territory was invaded by the US government when it built the nuclear weapons test ground in the 1950s. Arrested for trespassing, the demonstrators argued that the police and weapons facility personnel, without authorization from the proper officials, were the trespassers. One affinity group at the women's action called themselves the Surrogate Others; and in solidarity with the creatures forced to tunnel in the same ground with the bomb, they enacted a cyborgian emergence from the constructed body of a large, non-heterosexual desert worm.

6 This chart was published in 1985. My previous efforts to understand biology as a cybernetic command-control discourse and organisms as 'natural-technical objects of knowledge' were Haraway (1979, 1983, 1984).

7 For progressive analyses and action on the biotechnology debates: *Gene Watch, A Bulletin of the Committee for Responsible Genetics*, 5 Doane St, 4th Floor, Boston, MA 02109; Genetic Screening Study Group (formerly the Sociobiology Study Group of Science for the People), Cambridge, MA; Wright (1982, 1986); Yoxen (1983).

8 Starting references for 'women in the integrated circuit': D'Onofrio-Flores and Pfaillin (1982), Fernandez-Kelly (1983), Fuentes and Ehrenreich (1983), Grossman (1980), Nash and Fernandez-Kelly (1983), Ong (1987), Science Policy Research Unit (1982).

9 For the 'homework economy outside the home' and related arguments: Gordon (1983); Gordon and Kimball (1985); Stacey (1987); Reskin Hartmann (1986); *and* (1984); S. Rose (1986); Collins (1982); Burr (1982); Gregory and Nussbaum (1982); Piven and Coward (1982); Microelectronics Group (1980); Stallard *et al.* (1983).

10 The conjunction of the Green Revolution's social relations with biotechnologies like plant genetic engineering makes the pressures on land in the Third World increasingly intense. AID's estimates (*New York Times*, 14 October 1984) used at the 1984 World Food Day are that in Africa, women produce about 90 per cent of rural food supplies,

about 60–80 per cent in Asia, and provide 40 per cent of agricultural labour in the Near East and Latin America. Blumberg charges that world organizations' agricultural politics, as well as those of multinationals and national governments in the Third World, generally ignore fundamental issues in the sexual division of labour. The present tragedy of famine in Africa might owe as much to male supremacy as to capitalism, colonialism, and rain patterns. More accurately, capitalism and racism are usually structurally male dominant. See also Blumberg (1981); Hacker (1984); Hacker and Bovit (1981); Busch and Lacy (1983); Wilfred (1982); Sachs (1983); International Fund for Agricultural Development (1985); Bird (1984).

11   See also Enloe (1981, b).

12   For a feminist version of this logic, see Hardy (1981).

13   For the moment of transition of hunting with guns to hunting with cameras in the construction of popular meanings of nature for an American urban immigrant public, see Haraway (1984–5, 1989b), Nash (1979), Sontag (1977), Preston (1984).

14   For guidance for thinking about the political/cultural/racial implications of the history of women doing science in the United States see: Haas and Perucci (1984); Hacker (1981); Keller (1983); National Science Foundation (1988); Rossiter (198z); Schiebinger (1987); Haraway (1989b).

15   Markoff and Siegel (1983). High Technology Professionals for Peace and Computer Professionals for Social Responsibility are promising organizations.

16   King (1984). An abbreviated list of feminist science fiction underlying themes of this essay: Octavia Buder, *Mind of My Mind, Kindred*; Suzy McKee Charnas, *Motherliness*; Samuel R. Delany, the Neveryon series; Anne McCaffrey, *The Ship Who Sang, Dinosaur*; Vonda McIntyre, *Superluminal*; Joanna Russ, *Adventures of Alix, Female Man*; James Tiptree Jr, *Star Songs of an Old Primate, Up the Walls of* John Varley, *Titan, Demon*.

17   French feminisms contribute to cyborg heteroglossia. Burke (1981); (1977, 1979); Marks and de Courtivron (1980); (Autumn 1981); Wittig (1973); Duchen (1986). For English translation of some currents of francophone feminism see *Feminist Issues: A of Feminist 1980*.

18   But all these poets are very complex, not least in their treatment of themes of lying and erotic, decentred collective and personal identities. Griffin (1978), Lorde (1984), Rich (1978).

19   The convention of ideologically taming militarized high technology by publicizing its applications to speech and motion problems of the disabled/differently abled takes on a special irony in monotheistic, patriarchal, and frequently anti-semitic culture when computer-generated speech allows a boy with no voice to chant the Haftorah at his bar mitzvah. See Sussman (1986). Making the always context-relative social definitions of 'ableness' particularly clear, military high-tech has a way of making human beings disabled by definition, a perverse aspect of much automated battlefield and Star Wars R&D. See Welford (1 July 1986).

20   James Clifford (1985, 1988) argues persuasively for recognition of continuous cultural reinvention, the stubborn non-disappearance of those 'marked' by Western imperializing practices.

21   DuBois (1982), Daston and Park (n.d.), Park and Daston (1981). The noun shares its root with the verb *to demonstrate*.

References

Athanasiou, Tom (1987) 'High-tech politics: the case of artificial intelligence', *Socialist Review* 92: 7–35.

Baudrillard, Jean (1983) *Simulations*, trans. P. Foss, P. Patton, P. Beitchman. New York: Semiotext[e].

Bird, Elizabeth (1984) 'Green Revolution imperialism, I & II', papers delivered at the University of California, Santa Cruz.

Blumberg, Rae Lessor (1981) *Stratification: Socio-economic and Sexual Inequality*. Boston: Brown.

Burke, Carolyn (1981) 'Irigaray through the looking glass', *Feminist Studies* 7(2): 288–306.

Burr, Sara G. (1982) 'Women and work', in Barbara K. Haber, ed. *The Women's Annual, 1981*. Boston: G.K. Hall.

Busch, Lawrence and Lacy, William (1983) *Science, Agriculture, and the Politics of Research*. Boulder, CO: Westview.

Clifford, James (1985) 'On ethnographic allegory', in James Clifford and George Marcus, eds. *Writing Culture: The Poetics and Politics of Ethnography*. Berkeley: University of California Press.

Clifford, James (1988) *The Predicament of Culture: Twentieth-Century Ethnography, Literature, and Art*. Cambridge, MA: Harvard University Press.

Cohn, Carol (1987a) 'Nuclear language and how we learned to pat the bomb', *Bulletin of Atomic Scientists*, pp. 17–24.

Cohn, Carol (1987b) 'Sex and death in the rational world of defense intellectuals', *Signs* 12(4): 687–718.

Collins, Patricia Hill (1982) 'Third World women in America', in Barbara K. Haber, ed. *The Women's Annual, 1981*. Boston: G.K. Hall.

Cowan, Ruth Schwartz (1983) *More Work for Mother: The Ironies of Household Technology from the Open Hearth to the Microwave*. New York: Basic.

Daston, Lorraine and Park, Katherine (n.d.) 'Hermaphrodites in Renaissance France', unpublished paper.

de Wall, F. (1982) *Chimpanzee Politics: Power and Sex Among The Ape*. New York: Harper and Row.

D'Onofrio-Flores, Pamela and Pfafflin, Sheila M., eds (1982) *Scientific-Technological Change and the Role of Women in Development*. Boulder: Westview.

Douglas, M. (1966) *Purity and Danger*. London: Routledge & Kegan Paul.

Douglas, M. (1970) *Natural Symbols*. London: Cresset Press.

DuBois, Page (1982) *Centaurs and Amazons*. Ann Arbor: University of Michigan Press.

Duchen, Claire (1986) *Feminism in France from May '68 to Mitterrand*. London: Routledge & Kegan Paul.

Edwards, Paul (1985) 'Border wars: the science and politics of artificial intelligence', *Radical America* 19(6): 39–52.

Enloe, Cynthia (1983a) 'Women textile workers in the militarization of Southeast Asia', in Nash and Fernandez-Kelly (1983), pp. 407–25.

Enloe, Cynthia (1983b) *Does Khaki Become You? The Militarization of Women's Lives*. Boston: South End.

Epstein, Barbara (forthcoming) *Political Protest and Cultural Revolution: Nonviolent Direct Action in the Seventies and Eighties*. Berkeley: University of California Press.

Fausto-Sterling, Anne (1985) *Myths of Gender: Biological Theories about Women and Men*. New York: Basic.

Fernandez-Kelly, Maria Patricia (1983) *For We Are Sold, I and My People*. Albany: State University of New York Press.

Foucault, Michel (1963) *The Birth of the Clinic: An Archaeology of Medical Perception*, trans. A. M. Smith. New York: Vintage, 1975.

Foucault, Michel (1970) *The Order of Things*. New York: Random House.

Foucault, Michel (1972) *The Archaeology of Knowledge*, trans. Alan Sheridan. New York: Pantheon.

Foucault, Michel (1975) *Discipline and Punish: The Birth of the Prison*, trans. Alan Sheridan. New York: Vintage, 1979.

Foucault, Michel (1976) *The History of Sexuality, Vol. 1: An Introduction*, trans. Robert Hurley. New York: Pantheon, 1978.

Fuentes, Annette and Ehrenreich, Barbara (1983) *Women in the Global Factory*. Boston: South End.

Gordon, Richard (1983) 'The computerization of daily life, the sexual division of labor, and the homework economy', Silicon Valley Workshop conference, University of California at Santa Cruz.

Gordon, Richard, and Kimball, Linda (1985) 'High-technology employment and the challenges of education', Silicon Valley Research Project, Working Paper, no 1.

Gould, Stephen J. (1981) *Mismeasure of Man*. New York: Norton.

Gregory, Judith and Nussbaum, Karen (1982) 'Race against time: automation of the office', *Office: Technology and People I*: 197–236.

Griffin, Susan (1978) *Woman and Nature: The Roaring Inside Her*. New York: Harper & Row.

Grossman, R. (1980) "Women's place in the integrated circuit,' *Radical America* 14(1): 29–50.

Haas, Violet and Perucci, Carolyn, eds (1984) *Women in Scientific and Engineering Professions*. Ann Arbor: University of Michigan Press.

Hacker, Sally (1981) 'The culture of engineering: women, workplace, and machine', *Women's Studies International Quarterly* 4(3): 341–53.

Hacker, Sally (1984) 'Doing it the hard way: ethnographic studies in the agribusiness and engineering classroom', paper delivered at the California American Studies Association, Pomona.

Hacker, Sally, and Bovit, Liza (1981) 'Agriculture to agribusiness: Technical imperatives and changing roles', paper delivered at the Society for the History of Technology, Milwaukee.

Haraway, Donna J. (1979) 'The biological enterprise: sex, mind, and profit from human engineering to sociobiology', *Radical History Review* 20: 206–37.

Haraway, Donna J. (1983) 'Signs of dominance: from a physiology to a cybernetics of primate society', *Studies in History of Biology* 6: 129–219.

Haraway, Donna J. (1984) 'Class, race, sex, scientific objects of knowledge: a socialist-feminist perspective on the social construction of productive knowledge and some political consequences', in Violet Haas and Carolyn Perucci (1984), pp. 212–29.

Haraway, Donna J. (1984–5) 'Teddy bear patriarch: taxidermy in the Garden of Eden, New York City, 1908–36', *Social Text II*: 20.

Haraway, Donna J. (1989b) *Primate Visions: Gender, Race, and Nature in the World of Modern Science*. New York: Routledge.

Harding, Sandra (1986) *The Science Question in Feminism*. Ithaca: Cornell University Press.

Hardy, Sarah Blaffer (1981) *The Woman That Never Evolved*. Cambridge, MA: Harvard University Press.

Hogness, E. 'Why stress? A look at the making of stress, 1936–56.' Unpublished manuscript available from the author, 4437 Mill Creek Rd. Healdsburg, CA 95448.

Hubbard, Ruth, Henifin, Mary Sue, and Fried, Barbara, eds (1979) *Women Look at Biology Looking at Women: A Collection of Feminist Critiques*. Cambridge, MA: Schenkman.

Hubbard, Ruth, Henifin, Mary Sue, and Fried, Barbara, eds (1982) *Biological Women, the Convenient Myth*. Cambridge. MA: Schenkman.

International Fund for Agricultural Development (1985) *IFAD Experience Relating to Rural Women, 1977–84*. Rome: IFAD, 37.

Irigaray, Luce (1977) *Ce sexe n'en est pas un*. Paris: Minuit.

Irigaray, Luce (1979) *Et l'une ne bouge pas sans l'autre*. Paris: Minuit.

Jameson, Fredric (1984) 'Post-modernism, or the cultural logic of late capitalism', *New Left Review* 146: 53–92.

Keller, Evelyn Fox (1983) *A Feeling for the Organism*. San Francisco: Freeman.

Keller, Evelyn Fox (1985) *Reflection on Gender and Science*. New Haven: Yale University Press.

King, Katie (1984) 'The pleasure of repetition and the limits of identification in feminist science fiction: reimaginations of the body after the cyborg', paper delivered at the California American Studies Association, Pomona.

Knorr-Cetina, Karin (1981) *The Manufacture of Knowledge*. Oxford: Pergamon.

Knorr-Cetina, Karin and Mulkay, Michael, eds (1983) *Science Observed: Perspectives on the Social Study of Science*. Beverly Hills: Sage.

Latour, B. (1984) *Les Microbes, guerre et paix, suivi des irréductions*. Paris: Métailié.

Latour, Bruno and Woolgar, Steve (1979) *Laboratory Life: The Social Construction of Scientific Facts*. Beverly Hills: Sage.

Lewontin, R.C., Rose, Steven, and Kamin, Leon J. (1984) *Not in Our Genes: Biology, Ideology, and Human Nature*. New York: Pantheon.

Lorde, Audre (1982) *Zami, a New Spelling of My Name*. Trumansberg, NY: Crossing.

Lorde, Audre (1984) *Sister Outsider*. Trumansberg, NY: Crossing.

Marcuse, H. (1964) *One Dimensional Man: Studies in the Ideology of Advanced Industrial Society*. Boston: Beacon Press.

Markoff, John and Siegel, Lenny (1983) 'Military micros', paper presented at Silicon Valley Research Project conference, University of California at Santa Cruz.

Marks, Elaine, and de Courtivron, Isabelle, eds (1980) *New French Feminisms*. Amherst: University of Massachusetts Press.

Merchant, C. (1980) *The Death of Nature: Women, Ecology, and the Scientific Revolution*. New York: Harper and Row.

Microelectronics Group (1981) *Microelectronics: Capitalist Technology and the Working Class*. London: CSE.

Nash, June and Fernandez-Kelly, Maria Patricia, eds (1983) *Women and Men and the International Division of Labor*. Albany: State University of New York Press.

Nash, Roderick (1979) 'The exporting and importing of nature: nature-appreciation as a commodity, 1985–1980', *Perspectives in American History* 3: 517–60.

National Science Foundation (1988) *Women and Minorities in Science and Engineering*. Washington: NSF.

Ong, Aihwa (1987) *Spirits of Resistance and Capitalist Discipline: Factory Workers in Malaysia*. Albany: State University of New York Press.

Park, Katherine and Daston, Lorraine J. (1981) 'Unnatural conceptions: the study of monsters in sixteenth- and seventeenth-century France and England', *Past and Present* 92: 20–54.

Petchesky, R. P. (1981) 'Abortion, anti-feminism, and the rise of the New-Right,' *Feminist Studies* 7(2): 206–46.

Piven, Frances Fox and Coward, Richard (1982) *The New Class War: Reagan's Attack on the Welfare State and Its Consequences*. New York: Pantheon.

Preston, Douglas (1984) 'Shooting in paradise', *Natural History* 93(12): 14–19.

Reskin, Barbara F. and Hartman, Heidi, eds (1986) *Women's Work, Men's Work*. Washington: National Academy of Sciences.

Rich, Adrienne (1978) *The Dream of a Common Language*. New York: Norton.

Rose, Stephen (1986) *The American Profile Poster: Who Owns What, Who Makes How Much, Who Works Where, and Who Lives With Whom?* New York: Pantheon.

Rothschild, Joan, ed. (1983) *Machina ex Dea: Feminist Perspectives on Technology*. New York: Pergamon.

Sachs, Carolyn (1983) *The Invisible Farmers: Women in Agricultural Production*. Totowa: Rowman.

Schiebinger, Londa (1987) 'The history and philosophy of women in science: a review essay', *Signs* 12(2): 305–32.

Science Policy Research Unit (1982) *Microelectronics and Women's Employment in Britain*. University of Sussex.

Sofia, Z. (1984) 'Exterminating Fetuses: abortion, disarmament, and the sexo-semantics of extra-terrestrialism,' *Diacritics* 14(2): 47–59.

Sontag, Susan (1977) *On Photography*. New York: Dell.

Stacey, Judith (1987) 'Sexism by a subtler name? Postindustrial conditions and post feminist consciousness', *Socialist Review* 96: 7–28.

Stallard, Karin, Ehrenreich, Barbara and Sklar, Holly (1983) *Poverty in the American Dream*. Boston: South End.

Sturgeon, Noel (1986) 'Feminism, anarchism and non-violent direct action politics', University of California at Santa Cruz, PhD qualifying essay.

Traweek, Sharon (1988) *Beamtimes and Lifetimes: The World of High Energy Physics*. Cambridge, MA: Harvard University Press.

Treichler, P. (1987) 'AIDS, Homophobia, and Biomedical Discourse: An Epidemic of Signification,' *October* 43: 31–70

Weizenbaum, Joseph (1976) *Computer Power and Human Reason*. San Francisco: Freeman.

Welford, John Noble (1 July, 1986) 'Pilot's helmet helps interpret high speed world', *New York Times*, pp. 21, 24.

Wilfred, Denis (1982) 'Capital and agriculture, a review of Marzian problematic', *Studies in Political Economy* 7: 127–54.

Winner, Langdon (1977) *Autonomous Technology: Technics out of Control as a Theme in Political Thought*. Cambridge, MA: MIT Press.

Winner, Langdon (1980) 'Do artifacts have politics?', *Daedalus* 109(1): 121–36.

Winner, Langdon (1986) *The Whale and the Reactor*. Chicago: University of Chicago Press.

Wittig, Monique (1973) *The Lesbian Body*, trans. David LeVay. New York: Avon, 1975 (*Le Corps lesbian*, 1973).

'Women and Poverty', Special Issue (1984) *Signs* 10(2).

Wright, Susan (1982, July/August) 'Recombinant DNA: the status of hazards and controls', *Environment* 24(6): 12–20, 51–3.

Wright, Susan (1986) 'Recombinant DNA technology and its social transformation, 1972–82', *Osiris*, 2nd series, 2: 303–60.

Wright, Susan (1979, March) 'Interpreting the production of science', *New Scientist* 29: 1026–8.

Wright, Susan, and Lividow, Les, eds (1981, 1985) *Science, Technology and the Labour Process*, 2 vols. London: CSE and Free Association books.

Yoxen, Edward (1983) *The Gene Business*. New York: Harper & Row.

Zimmerman, Jan, ed. (1983) *The Technological Woman: Interfacing with Tomorrow*. New York: Praeger.

19

Technological Ethics in a Different Voice

Diane P. Michelfelder

Diane Michelfelder is Provost and Dean of the Faculty at MacAlester University. Drawing on the work of Borgmann, Feenberg, and feminist ethics, Michelfelder argues that expanding freedom alone is not a sufficient goal for technological development or evaluation. Rather, we must focus on the broader social and political contexts within which technologies are, or will be, used. She begins by noting that whereas Feenberg argues for greater participation from nonprofessionals in planning and design processes, Borgmann begins his analysis once technological artifacts are already in place. While questioning this emphasis, she notes that it is a salutary turn toward considering the role of technology in everyday life (the theme of Part I of Section 2, "Applied Reflections"). Drawing on a feminist ethics of care that places face-to-face encounters, personal relationships, love, and care at the center of moral philosophy, she asks whether Borgmann's distinction between "thing" and "device" holds up. Michelfelder then turns her attention to communications technologies, technologies that Borgmann classifies as devices because of their deskilling and distancing affects. She notes Feenberg's example of the reception and use of the French Minitel system as evidence of public reconfiguration of technologies, and explores the telephone as a device around which whole structures of focal practices have arisen. In each case, she argues, it is less important whether a technological artifact is subject to public participation in design, or whether it is classified as a device or a thing, than the part it plays in everyday lives.

From Eric Higgs, Andrew Light, and David Strong, eds., *Technology and the Good Life* (Chicago: University of Chicago Press, 2000), pp. 219–33. Reprinted by permission of the publisher, The University of Chicago Press, and by Diane Michelfelder.

The rapid growth of modern forms of technology has brought both a threat and a promise
for liberal democratic society. As we grapple to
understand the implications of new techniques
for extending a woman's reproductive life or the
spreading underground landscape of fiber-optic
communication networks or any of the other
developments of contemporary technology, we
see how these changes conceivably threaten the
existence of a number of primary goods traditionally associated with democratic society, including
social freedom, individual autonomy, and personal
privacy. At the same time, we recognize that
similar hopes and promises have traditionally
been associated with both technology and democracy. Like democratic society itself, technology
holds forth the promise of creating expanded
opportunities and a greater realm of individual
freedom and fulfillment. This situation poses a key
question for the contemporary philosophy of
technology. How can technology be reformed to
pose more promise than threat for democratic life?
How can technological society be compatible
with democratic values?

One approach to this question is to suggest
that the public needs to be more involved with
technology not merely as thoughtful consumers
but as active participants in its design. We can
find an example of this approach in the work of
Andrew Feenberg. As he argues, most notably
in his recent book *Alternative Modernity: The
Technical Turn in Philosophy and Social Theory*,
the advantage of technical politics, of greater public participation in the design of technological
objects and technologically mediated services
such as health care, is to open up this process
to the consideration of a wider sphere of values
than if the design process were to be left up to
bureaucrats and professionals, whose main concern is with preserving efficiency. Democratic
values such as personal autonomy and individual agency are part of this wider sphere. For
Feenberg, the route to technological reform and
the preservation of democracy thus runs directly
through the intervention of nonprofessionals in
the early stages of the development of technology
(Feenberg 1995).

By contrast, the route taken by Albert
Borgmann starts at a much later point. His
insightful explorations into the nature of the

technological device – that "conjunction of
machinery and commodity" (Borgmann 1992b,
296) – do not take us into a discussion of how
public participation in the design process might
result in a device more reflective of democratic
virtues. Borgmann's interest in technology starts
at the point where it has already been designed,
developed, and ready for our consumption. Any
reform of technology, from his viewpoint, must
first pass through a serious examination of the
moral status of material culture. But why must it
start here, rather than earlier, as Feenberg suggests?
In particular, why must it start here for the sake
of preserving democratic values?

In taking up these questions in the first part of
this paper, I will form a basis for turning in the
following section to look at Borgmann's work
within the larger context of contemporary moral
theory. With this context in mind, in the third
part of this paper I will take a critical look from
the perspective of feminist ethics at Borgmann's
distinction between the thing and the device, a
distinction on which his understanding of the
moral status of material culture rests. Even if
from this perspective this distinction turns out to
be questionable, it does not undermine, as I will
suggest in the final part of this paper, the wisdom
of Borgmann's starting point in his evaluation of
technological culture.

Public Participation and
Technological Reform

One of the developments that Andrew Feenberg
singles out in *Alternative Modernity* to back up his
claim that public involvement in technological
change can further democratic culture is the
rise of the French videotext system known as
Teletel (Feenberg 1995, 144–66). As originally
proposed, the Teletel project had all the characteristics of a technocracy-enhancing device. It
was developed within the bureaucratic structure
of the French government-controlled telephone
company to advance that government's desire
to increase France's reputation as a leader in
emerging technology. It imposed on the public
something in which it was not interested: convenient access from home terminals (Minitels)
to government-controlled information services.

However, as Feenberg points out, the government plan for Teletel was foiled when the public (thanks to the initial assistance of computer hackers) discovered the potential of the Minitels as a means of communication. As a result of these interventions, Feenberg reports, general public use of the Minitels for sending messages eventually escalated to the point where it brought government use of the system to a halt by causing it to crash. For Feenberg, this story offers evidence that the truth of social constructivism is best seen in the history of the computer.

Let us imagine it does offer this evidence. What support, though, does this story offer regarding the claim that public participation in technical design can further democratic culture? In Feenberg's mind, there is no doubt that the Teletel story reflects the growth of liberal democratic values. The effect generated by the possibility of sending anonymous messages to others over computers is, according to Feenberg, a positive one, one that "enhances the sense of personal freedom and individualism by reducing the 'existential' engagement of the self in its communications" (Feenberg 1995,159). He also finds that in the ease of contact and connection building fostered by computer-mediated communication, any individual or group of individuals who is a part of building these connections becomes more empowered (Feenberg 1995, 160).

But as society is strengthened in this way, in other words, as more and more opportunities open up for electronic interaction among individuals, do these opportunities lead to a more meaningful social engagement and exercise of individual freedom? As Borgmann writes in *Technology and the Character of Contemporary Life* (or *TCCL*): "The capacity for significance is where human freedom should be located and grounded" (Borgmann 1984, 102). Human interaction without significance leads to disengagement; human freedom without significance leads to banality of agency. If computer-mediated communications take one where Feenberg believes they do (and there is little about the more recent development of Internet-based communication to raise doubts about this), toward a point where personal life increasingly becomes a matter of "staging . . . personal performances" (Feenberg

1995, 160), then one wonders what effect this has on other values important for democratic culture: values such as self-respect, dignity, community, and personal responsibility.

The Teletel system, of course, is just one example of technological development, but it provides an illustration through which Borgmann's concern with the limits of public participation in the design process as a means of furthering the democratic development of technological society can be understood. Despite the philosophical foundations of liberal democracy in the idea that the state should promote equality by refraining from supporting any particular idea of the human good, in practice, he writes, "liberal democracy is enacted as technology. It does not leave the question of the good life open but answers it along technological lines" (Borgmann 1984, 92). The example we have been talking about illustrates this claim. Value neutral on its surface with respect to the good life, Feenberg depicts the Teletel system as encouraging a play of self-representation and identity that develops at an ever-intensifying pace while simultaneously blurring the distinction between private and public life. The value of this displacement, though, in making life more meaningful, is questionable.

To put it in another way, for technology to be designed so that it offers greater opportunities for more and more people, what it offers has to be put in the form of a commodity. But the more these opportunities are put in the form of commodities, the more banal they threaten to become. This is why, in Borgmann's view, technical politics cannot lead to technical reform.

For there truly to be a reform of technological society, Borgmann maintains, it is not enough only to think about preserving democratic values. One also needs to consider how to make these values meaningful contributors to the good life without overly determining what the good life is. "The good life," he writes, "is one of engagement, and engagement is variously realized by various people" (Borgmann 1984, 214). While a technical politics can influence the design of objects so that they reflect democratic values, it cannot guarantee that these values will be more meaningfully experienced. While a technical politics can lead to more individual freedom, it does not necessarily lead to an enriched sense of freedom. For an object to lead to an enriched sense of freedom, it needs, according to Borgmann, to promote unity over dispersement, and tradition over instantaneity. Values such as these naturally belong to objects, or can be acquired by them, but cannot be designed into them.

To take some of Borgmann's favorite examples, a musical instrument such as a violin can reflect the history of its use in the texture of its wood (Borgmann 1992b, 294); with its seasonal variations, a wilderness area speaks of the natural belonging together of time and space (Borgmann 1984, 191). We need to bring more things like these into our lives, and use technology to enhance our direct experience of them (as in wearing the right kinds of boots for a hike in the woods), for technology to deliver on its promise of bringing about a better life. As Borgmann writes toward the end of TCCL, "So counterbalanced, technology can fulfill the promise of a new kind of freedom and richness" (Borgmann 1984, 248).

Thus for Borgmann the most critical moral choices that one faces regarding material culture are "material decisions" (Borgmann 1992a, 112): decisions regarding whether to purchase or adopt a technical device or to become more engaged with things. These decisions, like the decisions to participate in the process of design of an artifact, tend to be inconspicuous. The second type of decision, as Wiebe E. Bijker, Thomas P. Hughes, and Trevor Pinch have shown (1987), fades from public memory over time. The end result of design turns into a "black box" and takes on the appearance of having been created solely by technical experts. The moral decisions Borgmann describes are just as inconspicuous because of the nature of the context in which they are discussed and made. This context is called domestic life. "Technology," he observes, "has step by step stripped the household of substance and dignity" (Borgmann 1984, 125). Just as Borgmann recalls our attention to the things of everyday life, he also makes us remember the importance of the household as a locus for everyday moral decision making. Thus Borgmann's reflections on how technology might be reformed can also be seen as an attempt to restore the philosophical significance of ordinary life.

Borgmann and the Renewal of Philosophical Interest in Ordinary Life

In this attempt, Borgmann does not stand alone. Over the course of the past two decades or so in North America, everyday life has been making a philosophical comeback. Five years after the publication of Borgmann's TCCL appeared Charles Taylor's Sources of the Self, a fascinating and ambitious account of the history of the making of modern identity. Heard throughout this book is the phrase "the affirmation of everyday life," a life characterized in Taylor's understanding by our nonpolitical relations with others in the context of the material world. As he sees it, affirming this life is one of the key features in the formation of our perception of who we are (Taylor 1989, 13). Against the horizons of our lives of work and play, friendship and family, we raise moral concerns that go beyond the questions of duties and obligations familiar to philosophers. What sorts of lives have the character of good lives, lives that are meaningful and worth living? What does one need to do to live a life that would be good in this sense? What can give my life a sense of purpose? In raising these questions, we affirm ordinary life. This affirmation is so deeply woven into the fabric of our culture that its very pervasiveness, Taylor maintains, serves to shield it from philosophical sight (Taylor 1989, 498).

Other signs point as well to a resurgence of philosophical interest in the moral dimensions of ordinary life. Take, for example, two fairly recent approaches to moral philosophy. In one of these approaches, philosophers such as Lawrence Blum, Christina Hoff Summers, John Hartwig, and John Deigh have been giving consideration to the particular ethical problems triggered by interpersonal relationships, those relationships among persons who know each other as friends or as family members or who are otherwise intimately connected. As George Graham and Hugh LaFollette note in their book Person to Person, these relationships are ones that almost all of us spend a tremendous amount of time and energy trying to create and sustain (Graham and LaFollette 1989, 1). Such activity engenders a significant amount of ethical confusion. Creating new relationships often means making difficult decisions about breaking off relationships in which one is already engaged. Maintaining interpersonal relationships often means making difficult decisions about what the demands of love and friendship entail. In accepting the challenge to sort through some of this confusion in a philosophically meaningful way, those involved with the ethics of interpersonal relationships willingly pay attention to ordinary life. In the process, they worry about the appropriateness of importing the standard moral point of view and standard moral psychology used for our dealings with others in larger social contexts – the Kantian viewpoint of impartiality and the distrust of emotions as factors in moral decision making – into the smaller and more intimate settings of families and friendships.

Another, related conversation about ethics includes thinkers such as Virginia Held, Nel Noddings, Joan Tronto, Rita Manning, Marilyn Friedman, and others whose work has been influenced by Carol Gilligan's research into the development of moral reasoning among women. I will call the enterprise in which these theorists are engaged feminist ethics, since I believe that description would be agreeable to those whom I have just mentioned, all of whom take the analysis of women's moral experiences and perspectives to be the starting point from which to rethink ethical theory.[1] Like interpersonal ethics, feminist ethics (particularly the ethics of care) places particular value on our relationships with those with whom we come into face-to-face contact in the context of familial and friendly relations. Its key insight lies in the idea that the experience of looking out for those immediately around one, an experience traditionally associated with women, is morally significant, and needs to be taken into account by anyone interested in developing a moral theory that would be a satisfactory and useful guide to the moral dilemmas facing us in all areas of life. Thus this approach to ethics also willingly accepts the challenge of paying philosophical attention to ordinary life. This challenge is summed up nicely by Virginia Held: "Instead of importing into the household principles derived from the marketplace, perhaps we should export to the wider society the relations suitable for mothering persons and children" (Held 1987, 122).

On the surface, these three paths of ethical inquiry – Borgmann's ethics of modern technology, the ethics of interpersonal relationships,

and feminist ethics – are occupied with different ethical questions. But they are united, it seems to me, in at least two ways. First, they are joined by their mutual contesting of the values upon which Kantian moral theory in particular and the Enlightenment in general are based. Wherever the modernist project of submitting public institutions and affairs to one's personal scrutiny went forward, certain privileges were enforced: that of reason over emotion, the "naked self" over the self in relation to others, impartiality over partiality, the public realm over the private sphere, culture over nature, procedural over substantive reasoning, and mind over body. In addition to the critique of Kantian ethics already mentioned by philosophers writing within a framework of an ethics of interpersonal relationships, feminist ethics has argued that these privileges led to the construction of moral theories insensitive to the ways in which women represent their own moral experience. Joining his voice to these critiques, Borgmann has written (while simultaneously praising the work of Carol Gilligan), "Universalism neglects . . . ways of empathy and care and is harsh toward the human subtleties and frailties that do not convert into the universal currency. . . . The major liability of moral universalism is its dominance; the consequence of dominance is an oppressive impoverishment of moral life" (Borgmann 1992a, 54–55).

A second feature uniting these relatively new forms of moral inquiry is a more positive one. Each attempts to limit further increases in the "impoverishment of moral life" by calling attention to the *moral* aspects of typical features of ordinary life that have traditionally been overlooked or even denied. The act of mothering (for Virginia Held), the maintenance of friendships (for Lawrence Blum) and the loving preparation of a home-cooked meal (for Borgmann) have all been defended, against the dominant belief to the contrary, as morally significant events.[2]

Despite the similarities and common concerns of these three approaches to moral philosophy, however, little engagement exists among them. Between feminist ethics and the ethics of interpersonal relationships, some engagement can be found: for instance, the "other-centered" model of friendship discussed in the latter is of interest to care ethicists as part of an alternative to

Kantian ethics. However, both of these modes of ethical inquiry have shown little interest in the ethical dimensions of material culture. Nel Noddings, for example, believes that while caring can be a moral phenomenon when it is directed toward one's own self and that of others, it loses its moral dimension when it is directed toward things. In her book *Caring*, she defends the absence of discussion of our relations to things in her work: "as we pass into the realm of things and ideas, we move entirely beyond the ethical. ... My main reason for setting things aside is that we behave ethically only through them and not toward them" (Noddings 1984, 161–62).

And yet in ordinary life ethical issues of technology, gender, and interpersonal relationships overlap in numerous ways. One wonders as a responsible parent whether it is an act of caring to buy one's son a Mighty Morphin Power Ranger. If I wish to watch a television program that my spouse cannot tolerate, should I go into another room to watch it or should I see what else is on television so that we could watch a program together? Is a married person committing adultery if he or she has an affair with a stranger in cyberspace? Seeing these interconnections, one wonders what might be the result were the probing, insightful questioning initiated by Borgmann into the moral significance of our material culture widened to include the other voices mentioned here. What would we learn, for instance, if Borgmann's technological ethics were explored from the perspective of feminist ethics?

In the context of this paper I can do no more than start to answer this question. With this in mind, I would like to look at one of the central claims of *TCCL*: the claim that the objects of material culture fall either into the category of things or devices.

Feminism and the Device Paradigm

As Borgmann describes them, things are machines that, in a manner of speaking, announce their own narratives and as a result are generous in the effects they can produce. For example, we can see the heat of the wood burning in the fireplace being produced in front of our eyes – the heat announces its own story, its own history, in which its relation to the world is revealed. In turn, fireplaces give us a place to focus our attention, to regroup and reconnect with one another as we watch the logs burn. In this regard, Borgmann speaks compellingly not only of the fireplace but also of wine: "Technological wine no longer bespeaks the particular weather of the year in which it grew since technology is at pains to provide assured, i.e. uniform, quality. It no longer speaks of a particular place since it is a blend of raw materials from different places" (Borgmann 1984, 49).

Devices, on the other hand, hide their narratives by means of their machinery and as a result produce only the commodity they were intended to produce. When I key the characters of the words I want to write into my portable computer they appear virtually simultaneously on the screen in front of me. I cannot see the connection between the one event and the other, and the computer does not demand that I know how it works in order for it to function. The commodity we call "processed words" is the result. While things lead to "multi-sided experiences," devices produce "one-sided experiences."[3] Fireplaces provide warmth, the possibility of conviviality, and a closer tie to the natural world; a central heating system simply provides warmth.

What thoughts might a philosopher working within the framework of feminist ethics have about this distinction? To begin with, I think she would be somewhat uneasy with the process of thinking used to make decisions about whether a particular object would be classified as a thing or a device. In this process, Borgmann abstracts from the particular context of the object's actual use and focuses his attention directly on the object itself. The view that some wine is "technological," as the example described above shows, is based on the derivation of the wine, the implication being that putting such degraded wine on the table would lead to a "one-sided experience" and further thwart, albeit in a small way, technology's capability to contribute meaningfully to the good life. In a feminist analysis of the moral significance of material culture, a different methodology would prevail. The analysis of material objects would develop under the assumption that understanding people's actual experiences of these objects, and in particular understanding the actual experiences of women who use them, would be an important source of information in deciding what direction a technological reform of society should take.

The attempt to make sense of women's experience of one specific technological innovation is the subject of communication professor Lana Rakow's book *Gender on the Line: Women, the Telephone, and Community Life* (1992). As its title suggests, this is a study of the telephone practices of the women residents of a particular community, a small midwestern town she called, to protect its identity, Prospect.

Two features of Rakow's study are of interest with regard to our topic. One relates to the discrepancy between popular perceptions of women's use of the telephone, and the use revealed in her investigation. She was well aware at the beginning of her study of the popular perception, not just in Prospect but widespread throughout American culture, of women's use of the telephone. In the popular perception, characterized by expressions such as "Women just like to talk on the phone" and "Women are on the phone all the time"; telephone conversations among women appear as "productivity sinks," as ways of wasting time. Understandably from this perception the telephone could appear as a device used for the sake of idle chatter that creates distraction from the demands of work and everyday life. This is how Borgmann sees it:

> The telephone network, of course, is an early version of hyperintelligent communication, and we know in what ways the telephone has led to disconnectedness. It has extinguished the seemingly austere communication via letters. Yet this austerity was wealth in disguise. To write a letter one needed to sit down, collect one's thoughts and world, and commit them laboriously to paper. Such labor was a guide to concentration and responsibility. (Borgmann 1992a, 105)

Rakow's study, however, did not support the popular perception. She found that the "women-talk" engaged in by her subjects was neither chatter nor gossip. Rather, it was a means to the end of producing, affirming, and reinforcing the familial and community connections that played a very large role in defining these women's lives. Such "phone work," very often consisting of exchanges of stories, was the stuff of which

relations were made: "Women's talk holds together the fabric of the community, building and maintaining relationships and accomplishing important community relations" (Rakow 1992, 34).

Let me suggest some further support for this view from my own experience. While I was growing up, I frequently witnessed this type of phone work on Sunday afternoons as my mother would make and receive calls from other women to discuss "what had gone on at church." Although these women had just seen each other at church several hours before, their phone calls played exactly the role that Rakow discovered they played in Prospect. At the time, they were not allowed to hold any positions of authority within the organizational structure of this particular church. The meaning of these phone calls would be missed by calling them idle talk; at least in part, these phone visits served to strengthen and reinforce their identity within the gendered community to which these women belonged.

Another interesting feature of Rakow's study was its discovery of how women used the telephone to convey care:

> Telephoning functions as a form of care-giving. Frequency and duration of calls... demonstrate a need for caring or to express care (or a lack of it). Caring here has the dual implication of caring *about* and caring *for* – that is, involving both affection and service.... While this [care-giving role] has been little recognized or valued, the caring work of women over the telephone has been even less noted. (Rakow 1992, 57)

As one of the places where the moral status of the care-giving role of women has been most clearly recognized and valued, feminist ethics is, of course, an exception to this last point. Rakow's recognition of the telephone as a means to demonstrate one's caring for speaks directly to Nel Noddings's understanding of why giving care can be considered a moral activity (Noddings 1984). In caring one not only puts another's needs ahead of one's own, but, in reflecting on how to take care of those needs, one sees oneself as being related to, rather than detached from, the self of the other. In commenting that not only checking on the welfare of another woman or phoning her on her birthday but "listening to others who need to talk is also a form of care" (Rakow 1992, 57), Rakow singles out a kind of caring that well reflects Nodding's description. More often one needs to listen to others who call one than one needs to call others; and taking care of the needs of those who call often involves simply staying on the phone while the other talks. As Rakow correctly points out, this makes this particular practice of telephone caring a form of work. Those who criticize the ethics of care for taking up too much of one's time with meeting the needs of individual others might also be critical of Rakow's subjects who reported that

> they spend time listening on the phone when they do not have the time or interest for it. . . . One elderly woman . . . put a bird feeder outside the window by her telephone so she can watch the birds when she has to spend time with these phone calls. "I don't visit; I just listen to others," she said. (Rakow 1992, 57)

As these features of telephone conversations came to light in the interviews she conducted with the women of Prospect, Rakow began to see the telephone as "a gendered, not a neutral, techno-logy" (Rakow 1992, 33). As a piece of gendered technology, the telephone arguably appears more like a thing than like a device, allowing for, in Borgmann's phrase, the "focal practice" of caring to take place. Looking at the telephone from this perspective raises doubts about Borgmann's assessment of the telephone. Has the telephone in fact become a substitute for the thing of the letter, contributing to our widespread feelings of disconnectedness and to our distraction? Rakow's fieldwork provides support for the idea that phone work, much like letter writing, can be "a guide to concentration and responsibility." By giving care over the phone, the development of both these virtues is supported. Thus, on Borgmann's own terms – "The focal significance of a mental activity should be judged, I believe, by the force and extent with which it gathers and illuminates the tangible world and our appropri-ation of it" (Borgmann 1984, 217) – it is difficult see how using the telephone as a means of con-veying care could not count as a focal concern.

Along with the question of whether a particu-lar item of our material culture is or is not a device,

looking at the device paradigm from a feminist point of view gives rise to at least two other issues. One is connected to an assumption on which this paradigm rests: that the moral significance of an object is directly related to whether or not that object is a substitute for the real thing. This issue is also connected to the idea that because technological objects are always substitutes for the real thing, the introduction of new technology tends to be a step forward in the impoverishment of ordinary life.

Certainly technological objects are always substitutes for *something or another*. A washing machine is a substitute for a washing board, dryers are substitutes for the line out back, krab [sic] is often found these days on salad bars, and so forth. In some cases, the older object gradually fades from view, as happened with the type-writer, which (but only as of fairly recently) is no longer being produced. In other cases, however, the thing substituted for is not entirely replaced, but continues to coexist alongside the substitute. In these cases, it is harder to see how the tech-nological object is a substitute *for the real thing*, and thus harder to see how the introduction of the new object threatens our sense of engagement with the world. While it is true that telephones substitute for letter writing, as Borgmann observes, the practice of letter writing goes on, even to the point of becoming intertwined with the use of the telephone. Again, from Rakow:

> The calls these women make and the letters they send literally call families into existence and maintain them as a connected group. A woman who talks daily to her two nearby sisters demonstrated the role women play in keep-ing track of the well-being of family members and changes in their lives. She said, "If we get a letter from any of them (the rest of the family) we always call and read each other the letters." (Rakow 1992, 64)

Perhaps, though, the largest question prompt-ed by Rakow's study has to do with whether Borgmann's distinction itself between things and devices can hold up under close consideration of the experiences and practices of different individuals. There are many devices that can be and are used as the women in this study used the telephone. Stereos, for example, can be a means for someone to share with someone else particular cuts on a record or songs from a CD to which she or he attaches a great deal of personal significance. In this way, stereos can serve as equipment that aid the development of mutual understanding and relatedness, rather than only being mechanisms for disengagement. The same goes for the use of the computer as a communicative device. Empirical investigations into the gendered use of computer-mediated communications suggest that while women do not necessarily use this environment like the telephone, as a means of promoting care, they do not "flame" (send electronic messages critical of another individual) nearly as much as do men, and they are critical of men who do engage in such activity.[4]

In particular, from a feminist perspective one might well wonder whether, in Borgmann's language, the use of those "conjunctions of machinery and commodity" inevitably hamper one's efforts at relating more to others and to the world. Borgmann argues that because devices hide their origins and their connections to the world, they cannot foster our own bodily and social engagement with the world. But as I have tried to show here, this is arguably not the case. Whether or not a material object hides or reveals "its own story" does not seem to have a direct bearing on that object's capacity to bind others together in a narrative web. For instance, older women participating in Rakow's study generally agreed that telephones improved in their ability to serve as a means of social support and care-giving once their machinery became more hidden: when private lines took the place of party lines and the use of an operator was not necessary to place a local call. To generalize, the machinery that clouds the story of a device does not appear to prevent that device from playing a role in relationship building.

Devices and the Promise of Technology

While a child growing up in New Jersey, I looked forward on Friday evenings in the summer to eating supper with my aunt and uncle. I would run across the yard separating my parents' house from theirs to take my place at a chair placed at the corner of the kitchen table. The best part of the meal, I knew, would always be the same, and that was why I looked forward to these evenings. While drinking lemonade from the multicolored aluminum glasses so popular during the 1950s, we would eat Mrs. Paul's fish sticks topped with tartar sauce. With their dubious nutritional as well as aesthetic value, fish sticks are to fresh fish as, in a contrast described eloquently by Borgmann, Kool Whip is to fresh cream (Borgmann 1987, 239–42). One doesn't know the seas in which the fish that make up fish sticks swim. Nearly anyone can prepare them in a matter of minutes. Still, despite these considerations, these meals were marked by family sociability and kindness, and were not hurried affairs.

I recall these meals now with the following point in mind. One might be tempted by the course of the discussion here to say that the objects of material culture should not be divided along the lines proposed in TCCL but divided in another manner. From the perspective of feminist ethics, one might suggest that one needs to divide up contemporary material culture between relational things, things that open up the possibility of caring relations to others, and nonrelational things: things that open up the possibility of experience but not the possibility of relation. Telephones, on this way of looking at things, would count as relational things. Virtual reality machines, such as the running simulator Borgmann imagines in CPD, or golf simulators that allow one to move from the green of the seventeenth hole at Saint Andrews to the tee of the eighteenth hole at Pebble Beach, would be nonrelational things. One can enjoy the experiences a virtual golf course makes possible, but one cannot in turn, for example, act in a caring manner toward the natural environment it so vividly represents. But the drawback of this distinction seem similar to the drawback of the distinction between things and devices: the possibility of using a thing in a relational and thus potentially caring manner seems to depend more on the individual using that thing and less on the thing itself. Depending on who is playing it, a match of virtual golf has the potential of strengthening, rather than undoing, narrative connections between oneself, others and the world.

But if our discussion does not lead in this direction, where does it lead? Let me suggest that although it does not lead one to reject the device paradigm outright, it does lead one to recognize that while any device does use machinery to produce a commodity, the meaning of one's experience associated with this device does not necessarily have to be diminished. And if one can use technology (such as the telephone) to carry out focal practices (such as caregiving), then we might have cause to believe that there are other ways to recoup the promise of technology than Borgmann sees. As mentioned earlier, his hope is that we will give technology more of a supporting role in our lives than it has at present (Borgmann 1984, 247), a role he interprets as meaning that it should support the focal practices centered around focal things. But if devices can themselves support focal practices, then the ways in which technology can assume a supporting role in our lives are enhanced.

But if the idea that devices can support focal practices is in one way a challenge to the device paradigm, in another way it gives additional weight to the notion that there are limits to reforming technology through the process of democratic design. When they are used in a context involving narrative and tradition, devices can help build engagement and further reinforce the cohesiveness of civil society. Robert Putnam has pointed out the importance of trust and other forms of "social capital" necessary for citizens to interact with each other in a cooperative manner. As social capital erodes, democracy itself, he argues, is threatened (Putnam 1995, 67). While this paper has suggested that devices can under some conditions further the development of social capital, it is difficult to see how they can be deliberately designed to do so. In thinking about how to reform technology from a democratic perspective, we need to remember the role of features of ordinary life such as narrative and tradition in making our experience of democratic values more meaningful. Borgmann's reminder to us of this role is, it seems to me, one of the reasons why *TCCL* will continue to have a significant impact in shaping the field of the philosophy of technology.

Notes

1 I am not using "feminist ethics" in a technical sense, but as a way of referring to the philosophical approach to ethics that starts from a serious exam-ination of the moral experience of women. For philosophers such as Alison Jaggar, the term feminist ethics primarily means an ethics that recognizes the patriarchal domination of women and the need for women to overcome this system of male domination. Thus she and others might disagree that the ethics of care, as I take it here, is an enterprise of feminist ethics.

2 For example, Virginia Held has written: "[Feminist moral inquiry] pays attention to the neglected experience of women and to such a woefully neglected though enormous area of human moral experience as that of mothering.... That this whole vast region of human experience can have been dismissed as 'natural' and thus as irrelevant to morality is extraordinary" (Held 1995, 160).

3 The term "multi-sided experiences" is used by Mihaly Csikszentmihalyi and Eugene Rochbert-Halton in their work *The Meaning of Things*, discussed in Borgmann 1992b.

4 See, for example, Susan Herring, "Gender Differences in Computer-Mediated Communication: Bringing Familiar Baggage to the New Frontier" (unpublished paper).

References

Bijker, Wiebe E., Thomas P. Hughes, and Trevor Pinch, eds. 1987. *The Social Construction of Technological Systems*. Cambridge: MIT Press.

Borgmann, Albert. 1984. *Technology and the Character of Contemporary Life: A Philosophical Inquiry*. Chicago: University of Chicago Press.

Borgmann, Albert. 1987. "The Invisibility of Contemporary Culture." *Revue internationale de philosophie* 41:234–49.

Borgmann, Albert. 1992a. *Crossing the Postmodern Divide*. Chicago: University of Chicago Press.

Borgmann, Albert. 1992b. "The Moral Significance of the Material Culture." *Inquiry* 35:291–300.

Feenberg, Andrew. 1995. *Alternative Modernity: The Technical Turn in Philosophy and Social Theory*. Berkeley and Los Angeles: University of California Press.

Graham, George, and Hugh LaFollette, eds. 1989. *Person to Person*. Philadelphia: Temple University Press.

Held, Virginia. 1987. "Non-contractual Society: A Feminist View." In *Science, Morality and Feminist Theory*. Ed. Marsha Hanen and Kai Nielsen. Calgary: University of Calgary Press.

Held, Virginia. 1995. "Feminist Moral Inquiry and the Feminist Future." In *Justice and Care: Essential Readings in Feminist Ethics*. Ed. Virginia Held. Boulder, Colo.: Westview Press.

Noddings, Nel. 1984. *Caring*. Berkeley and Los Angeles: University of California Press.

Putnam, Robert D. 1995. "Bowling Alone: America's Declining Social Capital." *Journal of Democracy* 6:65–78.

Rakow, Lana F. 1992. *Gender on the Line: Women, the Telephone, and Community Life*. Urbana: University of Illinois Press.

Taylor, Charles. 1989. *Sources of the Self*. Cambridge: Harvard University Press.

Section Two

Applied Reflections on Technology and Value

Part VII

Technology and Value in Everyday Life

Part VII
Technology and Value in Everyday Life

Introduction

Studies of technology have tended either to focus on technology abstracted from concrete context, as in the case of classical transcendental philosophers of technology, or to analyze technologies in the public sphere (as in Parts II and III below, "Values and Biotechnologies" and "Urban Values"). However, some of the most important changes wrought by technological change have been in our everyday lives. The three readings in this section turn our attention to technology and value in everyday life.

Drawing on the work of the American Pragmatist philosophers William James and John Dewey, John J. McDermott, Distinguished Professor of Philosophy and Humanities at Texas A&M University, articulates a theory of the aesthetic possibilities of everyday artifacts. He thus rejects aspects of a worldview inherited from Descartes: that mind and body are clearly separable, as are self and world, that knowledge requires detached contemplation, that the most important of human concerns are the most universal and abstract, and a vocabulary that reflects all of these ideas in how we talk and think about ourselves. In this, the Cartesian worldview and its subsequent influence captures many of the reasons why philosophy has not had much to say about technology. McDermott argues that, in contrast to the Cartesian view, we embody ourselves in our technologies, and create our world. As living creatures, we are permeable and incomplete. His metaphor for our situation is "uterine": we are fully immersed in the world, we absorb, grow, reject, transform, construct, and destroy on a constant basis. Thus, it is in and through our relations with everyday artifacts that we make self and world. He claims "the world is made sacred by our *hand*ling of things."

Judy Wajcman, Professor of Sociology in the Political Science Program, Research School of Social Sciences, Australian National University, begins by noting that one of the sites of daily life most changed by new technologies over the past 150 years is the home, and yet until fairly recently household technologies have not been the subject of study. Beginning with the work of feminist scholars such as Ruth Schwartz Cowan (*More Work for Mother*, New York: Basic Books, 1983) and Ann Oakley (*The Sociology of Housework*, London: Martin Robinson, 1974), a picture of the partial-mechanization of the home came into focus. In this piece, Wajcman examines the impacts of domestic technologies on women's lives. She argues that household technologies have partially mechanized the home, but have not lessened the workload in any straightforward manner. She closes with a call for more detailed empirical work that examines the intersections between public and private spheres and the design process.

Finally, Douglas Browning, Professor Emeritus of Philosophy at the University of Texas, drawing on both an aesthetic attention to everydayness that owes much to Dewey and the "what-it-is-likeness" of phenomenology, examines some meanings of automobiles. This essay is an example of the phenomenological account of our experiences of technological artifacts. Browning argues, in keeping with Ortega and Dewey, that through the ways we interact with our cars, and with each other through our cars, we create ourselves as new, automobiled, persons. Through a process he calls "personation," consistent with Ortega's notion of autofabrication, we play at making ourselves. In this way, our relations with automobiles would fall under Ihde's category of embodiment relations. He further argues that as such people most of us now have a category of life experience, the category of "the sudden" that in previous eras only characterized a few people, and then only for limited periods of time. In this way the automobile changes not only living patterns and cities and global political and economic relations, but to be automobiled transforms our relationships with self and the nature of our embodiment.

20

The Aesthetic Drama of the Ordinary

John McDermott

I wish I could see what my eyes see.
VANILLA FUDGE

Traditionally, we think of ourselves as "in the world," as a button is in a box, a marble in a hole, a coin in a pocket, a spoon in a drawer; in, always in something or other. And yet, to the contrary, I seem to carry myself, to lead myself, to have myself hang around, furtive of nose, eye, and hand, all the while spending and wasting, eating and fouling, minding and drifting, engaging in activities more descriptive of a permeable membrane than of a box. To feel is to be felt. To be in the world is to "world" and to be "worlded." No doubt, the accepted language of expository prose severely limits us in this effort to describe our situation experientially. Were I to say, for example, my presence in the world or my being in the world, I would still fall prey to the container theory and once again be "in" as over against "out." Is this not why it is necessary to describe an unusual person, situation, or state of being as being "out of this world," or "spaced out" or simply "out of it." Why is it that ordinary language, or our language as used ordinarily, so often militates against the ways in which we actually have, that is, undergo, our experiencing? Why is it that we turn to the more specialized

forms of discourse such as jokes, fiction, poetry, music, painting, sculpture, and dance, in order to say what we "really" mean? Does this situation entail the baleful judgment that the comparative bankruptcy of our ordinary language justly points to the comparable bankruptcy of our ordinary experience?

In gross and obvious empirical terms, it is difficult to say no to the necessity of this entailment. Surely it is true that we are surrounded by the banal, monumentalized in a miniature and trivial fashion by the American shopping center. And it is equally, yea, painfully true that the "things" of our everyday experience are increasingly de-aestheticized, not only by misuse and failure to maintain, but forebodingly in their very conception of design and choice of material, as witnessed by the recent national scandal in our urban bus fleet, when millions of dollars were spent on buses that were not built for city traffic, roads, or frequency of use. How striking, as well, is the contrast between those Americans at the turn of the century, who built the IRT subway in New York City, complete with a mosaic of inlaid tile, balustrades, and

From John McDermott, *Streams of Experience: Reflections on History and Philosophy in the American Grain* (Amherst, MA: University of Massachusetts Press, 1986), pp. 129–40. © 1986 by The University of Massachusetts Press.

canopied entrances, over against their descen-
dants, our peers, who seem not able to find a way
to eradicate the stink and stain of human urine
from those once proud and promising platforms
and stairwells. So as not to contribute any further
to the offensive and misleading assumption that
our main aesthetic disasters are now found in the
great urban centers of the Northeast, let us point
to one closer to my home.

The city of Houston, in paying homage to a long
outdated frontier myth of every "building" for
itself, proceeds to construct an environment
which buries an urban aesthetic in the wake of
free enterprise. Houston gives rise to tall and
imposing buildings whose eyes of window and
light point to the surrounding plains, but whose
feet are turned inward. These buildings do not
open in a merry Maypole of neighborhood frolic
and function. Houston buildings are truly sky-
buildings, for they look up and out, leaving only
the sneer of a curved lip to waft over the enervated
neighborhoods below, most of them increasingly
grimy and seedy. As an apparent favor to most
of us, Houston provides a way for us to avoid these
neighborhoods, allowing us to career around
the city, looking only at the bellies of the titans
of glass and steel, astride the circular ribbon of
concrete known appropriately as the beltway,
marred only by the dead trees, broken car jacks,
and the intrusive omnipresence of Texas-sized bill-
boards. Perhaps it is just as well that we, too, rise
above the madding crowd, for in that way we miss
the awkwardness of wandering into one of those
walled-off, sometimes covenanted and patrolled,
fancy enclaves which make the city tolerable for
the rich. And as we make our "beltway," we miss
as well that strikingly sad experience of downtown
Houston at 6 P.M. of a weekend evening, when
the loneliness and shabbiness of the streets are
cast into stark relief by the perimeter of empty
skyscrapers and the hollow sounds of the feet of
the occasional snow-belt emigre traveler, emerging from the Hyatt Regency in a futile search for
action. What is startling and depressing about all
of this is that the city of Houston is the nation's
newest and allegedly most promising major city.

Actually, whether it is North, South, East, or
West matters little, for in general the archons
of aesthetic illiteracy have seen to it that on
behalf of whatever other ideology they follow, the
presence of aesthetic sensibility has been either
ruled out or, where traditionally present, allowed
to erode. Further, to the extent that we prehend
ourselves as a thing among things or a functioning item in a box, then we get what we deserve.
Supposing, however, we were to consider the
major metaphorical versions of how we carry on
our human experiencing and, in so doing, avoid
using the imagery of the box. Instead, let us
consider ourselves as being in a uterine situation,
which binds us to nutrition in a distinctively
organic way. James Marston Fitch, a premier
architectural historian, writes about us as follows:

> Life is coexistent with the external natural
> environment in which the body is submerged.
> The body's dependence upon this external
> environment is absolute – in the fullest sense of
> the word – uterine.[1]

No box here. Rather we are floating, gestating
organisms, transacting with our environment,
eating all the while. The crucial ingredient in all
uterine situations is the nutritional quality of the
environment. If our immediate surroundings
are foul, soiled, polluted harbors of disease and
grime, ridden with alien organisms, then we fal-
ter and perish. The growth of the spirit is exactly
analogous to the growth of the organism. It too
must be fed and it must have the capacity to convert its experiences into a nutritious transaction.
In short, the human organism has need of two
livers. The one, traditional and omnipresent,
transforms our blood among its 500 major functions and oversees the elimination from our
body of ammonia, bacteria, and an assortment of
debris, all of which would poison us. The second
is more vague, having no physical analogue. But
its function is similar and crucial. This second liver
eats the sky and the earth, sorts out tones and
colors, and provides a filter through which the
experienced environment enters our consciousness. It is this spiritual liver which generates our
feelings of queasiness, loneliness, surprise, and
celebration. And it is this liver which monitors the
tenuous relationship between expectations and
anticipations on the one hand and realizations,
disappointments, and failures on the other. We
are not simply in the world so much as we are of
and about the world. On behalf of this second type
of livering, let us evoke the major metaphors of
the fabric, of the uterus, through which we have

our natal being. Our context for inquiry shall be the affairs of time and space, as well as the import of things, events, and relations. We shall avoid the heightened and intensified versions of these experiential filters and concentrate on the explosive and implosive drama of their ordinariness.

Time

Time passing is a death knell. With the license of a paraphrase, I ask, For whom does the bell toll? It tolls for thee and me and for ours. We complain about the studied repetition, which striates our lives, and yet, in honesty, we indulge this repetition as a way of hiding from the inexorability of time passing, as a sign equivalent to the imminence of our self-eulogy. Time is a shroud, often opaque, infrequently diaphanous. Yet, from time to time, we are able to bring time into our own self-awareness and to bring time to its knees. On those rare occasions when time is ours rather than we being creatures of time, we feel a burst of singularity, of independence, even perhaps of the eternal import of our being present to ourselves. How has it happened that we have become slaves to time? Surely as children of Kant and Einstein, we should know better. For them and for modern physics, time is a mock-up, an earth phenomenon, no more relevant cosmically than the watches which watch time, supposedly passing. Still, Kant not withstanding, time is the name given to the process of our inevitable dissolution. On the morrow, our kidney is less quick, our liver less conscientious, our lung less pulsatile, and our brain less alert. Is it possible, without indulging ourselves in a Walter Mittyesque self-deception, to turn this erosive quality of time passing to our own advantage?

I suggest that we can beat time at its own game. Having created time, let us obviate it. Time, after all, rushes headlong into the future, oblivious to its damages, its obsoleting, and its imperviousness to the pain it often leaves in its wake. A contrary view is that in its passing, time heals. But it is not time which heals us, it is we who heal ourselves by our retroactive reconstruction of history. It is here that time is vulnerable, for it has no history, no past. Time is ever lurching into the future. We, however, can scavenge its remains and make them part of ourselves. For us, the past is existentially present if we have the will and the attentiveness to so arrange. I offer here that we recover the detritus of time passing and clot its flow with our freighted self-consciousness. We can become like the giant balloons in the Macy's Thanksgiving Day parade, thick with history and nostalgia, forcing time passing to snake around us, assuring that it be incapable of enervating our deepest feelings of continuity. What, for example, could time do to us if every time we met a person, or thought a thought, or dreamt a dream, we involved every person ever met, every thought ever thought, and every dream ever dreamt? What would happen if every event, every place, every thing experienced, resonated all the events, places, and things of our lives? What would happen if we generated a personal environment in which the nostalgic fed into the leads of the present, a self-created and sustained environment with implications fore and aft? In so doing, we would reduce time passing to scratching on the externals of our Promethean presence. Time would revolve around us rather than passing through us. Time would provide the playground for our activities rather than the graveyard of our hopes. We would time the world rather than having the world time us. And we would reverse the old adage, to wit, if you have the place, I have the time, for time is mine to keep and to give. And, in addition to telling our children now is your time, we would tell ourselves, no matter how old, now is our time.

Space

It is equally as difficult to extricate ourselves from the box of space as it is to escape from the penalties of time. Here too, we have failed to listen to Kant and Einstein, for space, just as time, has no existential reality other than our conception of it. Yet we allow the prepossessing character of space to dwarf us. Nowhere is this more apparent than in Texas, where the big sky of Montana is outdone by the scorching presence of a sun that seems never to set, frying our brains in the oven of its arrogance. In the spring of the year, the bluebonnets and Indian paintbrush state our position: fey, lovely, quiet, reserved, and delicate of manner. The Texas sun indulges this temporary human-scaled assertion while hovering in the background with vengeance on its mind. As the flowers fade, the horizon widens and the sun takes its place at the center of our lives, burning us with the downdraft of its rays. Listen to Larry King on the sun and sky in West Texas.

> The land is stark and flat and treeless, altogether as bleak and spare as mood scenes in Russian literature, a great dry-docked ocean with small swells of hummocky tan sand dunes or hump-backed rocky knolls that change colors with the hour and the shadows: reddish brown, slate gray, bruise colored. But it is the sky – God-high and pale, like a blue chenille bedspread bleached by seasons in the sun – that dominates. There is simply *too much* sky. Men grow small in its presence and – perhaps feeling diminished – they sometimes are compelled to proclaim themselves in wild or berserk ways. Alone in those remote voids, one may suddenly half be-lieve he is the last man on earth and go in frantic search of the tribe. Desert fever, the natives call it. . . . The summer sun is as merciless as a loan shark: a blinding, angry orange explosion baking the land's sparse grasses and quickly aging the skin.[2]

Texans pride themselves as being larger than life. But this is just a form of railing against the sun. The centuries-long exodus from the Northeast and the coastal cities was in part an escape from urban claustrophobia. In that regard, the escape was short-lived and self-deceptive, for it soon became apparent that the West presented a claustrophobia of another kind – paradoxically, that of open space. The box was larger, the horizon deeper, but the human self became even more trivialized than it was among the skyscrapers and the crowded alleyways and alcoves of the teeming urban centers. No, to the extent that we are overshadowed by an external overhang, be it artifact or natural, we cower in the presence of an *other* which is larger, more diffuse, still threatening and depersonalizing. In response, just as we must seize the time, so too must we seize the space, and turn it into a place, our place.

The placing of space is the creating of interior space, of personal space, of your space and my space, of our space. I am convinced, painful though it be, that we as human beings have no natural place. We are recombinant organisms in a cosmic DNA chain. Wrapped in the mystery of our origins, we moved from natural places to artifactual ones, from caves to ziggurats to the Eiffel tower. We moved from dunes to pyramids and then to the World Trade Center. The history of our architecture, big and small, functional and grandiloquent, lovely and grotesque, is the history of the extension of the human body into the abyss. We dig and we perch. We level and we raise. We make our places round and square and angular. We make them hard and soft and brittle. We take centuries to make them and we throw them up overnight. In modern America, the new Bedouins repeat the nomadic taste of old and carry their places with them as they plod the highway vascular system of the nation, hooking up here and there.

Some of our idiomatic questions and phrases tell us of our concern for being in place. Do you have a place? Set a place for me. This is my place. Why do we always go to your place? Would you care to place a bet? I have been to that place. Wow, this is *some* place. Win, place, show. The trouble with him is that he never went any place and the trouble with her is that she never got any place. How are you doing? How is it going? Fine, I am getting someplace. Not so well, I seem to be no place.

Recall that poignant scene in *Death of a Salesman* when Willy Loman asks Howard for a place in the showroom rather than on the road. In two lines, Howard tells Willy three times that he has no "spot" for him. I knew your father, Howard, and I knew you when you were an infant. Sorry, Willy! No spot, no place, for you. Pack it in. You are out of time and have no place.

Listen lady, clear out. But this is my place. No lady, this place is to be replaced. The harrowing drama of eviction haunts all of us as we envision our future out of place and on the street.[3] Dorothy Day founded halfway houses, places somewhere between no place and my place, that is, at least, someplace. And, finally, they tell us that we are on the way to our resting place, a place from where there is no return.

These are only anecdotal bare bones, each of them selected from a myriad of other instances which point to our effort to overcome the ontological *angoisse* which accompanies our experience of *Unheimlichkeit*, a deep and pervasive sense of ultimate homelessness. We scratch out a place and we raise a wall. The windows look out but the doors open in. We hang a picture and stick a flower in a vase. We go from cradle and crib to a coffin, small boxes at the beginning and end of journeys through slightly larger boxes. Some of us find ourselves in boxes underneath and on top of other boxes in a form of apartmentalization. Some of our boxes are official boxes and we call them offices, slightly less prestigious than the advantage of a box seat. Everywhere in the nation, the majority of our houses are huddled together, sitting on stingy little pieces of ground, while we ogle the vast stretch of land held by absentees. One recalls here "Little Boxes," a folksong of the 1960s that excoriates the ticky-tacky boxes on the hillsides, as a preface to the yuppiedom of our own time. For the most part, our relation to external space is timid, even craven. From time to time, we send forth a camel, a schooner, a Conestoga wagon, or a space shuttle as probes into the outer reaches of our environ, on behalf of our collective body. Yet these geographical efforts to break out are more symbolic than real, for after our explorations we seem destined to repeat our limited variety of habitat.

The *locus classicus* for an explication of the mortal danger in a sheerly geographical response to space is found in a story by Franz Kafka, "The Burrow." In an effort to protect his food from an assumed intruder, the burrower walls off a series of mazes sure to confuse an opponent. This attempt is executed with such cunning and brilliance that his nonreflective anality is missed as a potential threat. The food is indeed walled off from the intruder – from the burrower as well. He dies of starvation, for he cannot find his own food.

The way out of the box is quite different, for it has to do not with the geography and physicality of space, but rather with our symbolic utilization of space for purposes of the human quest. We manage our ontological dwarfing and trivialization at the hands of infinite space, and the rush of time passing and obsoleting, by our construction, management, placing, and relating of *our* things. It is to our things, to creating our salvation in a world without guarantee of salvation, that we now turn.

Things

Thing, orthographically and pronouncedly, is one of the ugly words in contemporary American usage. Yet it is also, inferentially and historically, one of the most subtle and beautiful of our words. It is lamentable that we do not speak the way Chaucer spoke. From the year 1400 and a work of Lydgate, *Troy-Book*, the text reads: "That thei with Paris to Greece schulde wende, To Brynge this thynge to an ende." The Trojan war was a thing? Of course it was a thing, for thing means concern, assembly, and, above all, an affair. Thing is a woman's menses and a dispute in the town. Thing is a male sex organ and a form of prayer. (The continuity is not intended, although desirable.) Thing is what is to be done or its doing. I can't give you any thing but love, baby. That is the only thing, I have plenty of, baby. When you come, bring your things. I forgot to bring my things. My things are packed away. Everything will be all right. And by the way, I hope that things will be better.

What and who are these things to which we cling? An old pari-mutuel ticket, a stub for game seven of the World Series, a class ring, a mug, a dead Havana cigar, loved but unsmoked. My snuff box, my jewelry drawer, an album, a diary, a yearbook, all tumbled into the box of memories, but transcendent and assertive of me and mine. Do not throw out his things, they will be missed. Put her things in the attic, for someday she will want them as a form of reconnoitering her experienced past. Do you remember those things? I know that we had them. Where are they? They are in my consciousness. Can we find them? We didn't throw them out, did we? How could we?

The making, placing, and fondling of our things is equivalent to the making, placing, and fondling of our world. We are our things. They are personal intrusions into the vast, impersonal reach of space. They are functional clots in the flow of time. They are living memories of experiences had but still viable. They are memorials to experiences undergone and symbolically still present. The renewed handling of a doll, a ticket, a toy soldier, a childhood book, a tea cup, a bubble-gum wrapper, evokes the flood of experiences past but not forgotten.[4] How we strive to say hello, to say here I am, in a cosmos impervious, unfeeling, and dead to our plaintive cry of self-assertion. To make is to be made and to have is to be had. My thing is not anything or something. Your thing is not my thing but it could be our thing. The ancients had it right, bury the things with the person. We should do that again. Bury me with a copy of the *New York Times*, a Willie Mays baseball card, a bottle of Jameson, my William James book, a pipe, some matches, and a package of Seven-Seas tobacco.

The twentieth-century artist Alexander Calder once said that no one is truly human who has not made his or her own fork and knife. Homemade or not, do you have your own fork, your own knife, your own cup, your own bed, desk, chair? You must have your own things! They are you. You are they. As the poet Rilke tells us, "Being here amounts to so much."[5]

Our things are our things. They do not belong to the cosmos or to the gods. They can be had by others only in vicarious terms. Commendable though it may be for those of us who are collectors of other people's things, nonetheless, those who burn their papers or destroy their things just before they die are a testament to both the radical self-presence and transiency of human life. Those of us, myself included, who collect other people's things, are Texas turkey vultures, seizing upon the sacred moments hammered out by transients and eating them in an effort to taste the elixir of memory for our own vapid personal life. Ironically, for the most part their experience of their things were similar efforts, sadly redeemed more by us than by them. Now to the crux of the matter before us.

It is not, I contend, humanly significant to have the primary meaning of one's life as posthumous. We and our things, I and my things, constitute our world. The nectar of living, losing, loving, maintaining, and caring for our things is for us, and for us alone. It is of time but not in time. It is of space but not in space. We and our things make, constitute, arrange, and determine space and time. The elixir garnered by the posthumous is for the survivors. It cannot be of any biological significance to us, although many of us have bartered our present for the ever absent lilt of being remembered. St. Francis of Assisi and John Dewey both taught us the same *thing*: time is sacred, live by the sacrament of the moment and listen to the animals. We may have

a future. It is barely conceivable, although I doubt its existence. We do have, however, a present. It is the present, canopied by our hopefully storied past, that spells the only meaning of our lives. Still, the present would be empty without our things.

You, you out there, you have your things. Take note. Say hello, say hello, things. They are your things. Nay, they are you. No things, no you, or in correct grammar, you become no*thing*. So be it. Space and time are simply vehicles for things, our things, your things, my things. These things do not sit, however, in rows upon rows, like ducks in a shooting gallery. These things make love, hate, and tire. Like us, they are involved. We consider now this involvement of persons, things, things and persons, all struggling to time space and space time, namely, the emergence of events as relations.

Things as Events as Aesthetic Relations

We have been in a struggle to achieve non-derivative presence of ourselves and our things over against the dominating worlds of space and time. Fortunately, for us, space and time do not necessarily speak to each other. Our canniness can play them off, one against the other. The triumph is local, never ultimate, although it does give us staying power in our attempt to say I, me, you, we, us, and other asserted pronominal outrages against the abyss.

A happy phenomenon for human life is that things not only are; they also happen. I like to call these happenings events. The literal meaning of event is intended: a coming out, a party, a debutante dance, a *bar mitzvah*, a hooray for the time, given the circumstance. In my metaphysics, at least, things are bundles of relations, snipped at the edges to be sure. Usually, we give our things a name and this name takes the place of our experience of the thing. It does not take long to teach a child a list of nouns, each bent on obviating and blocking the rich way in which the child first comes upon and undergoes things. It is difficult to overcome this prejudice of language, especially since row upon row of nouns, standing for things, makes perfectly good sense, if you believe that space is a container and time is the measure of external motion. If, however, you believe as I do, that space and time are human instincts, subject to the drama of our inner lives, then things lose their inert form. Emerson says this best when he claims that every fact and event in our private history shall astonish us by "soaring from our body into the empyrean."[6]

The clue here is the presence of a person. Quite aside from the geographical and physical relationships characteristic of things and creatures, we further endow a whole other set of relations, the aesthetic. I refer to the rhythm of how we experience *what* we experience. The most distinctive human activity is the potentially affective dimension of our experiencing ourself, experiencing the world. I say potentially, for some of us all of the time and most of us most of the time are dead to the possible rhythms of our experiences. We are ghouls. We look alive but we are dead, dead to our things and dead even to ourselves. As John Cage warned us, we experience the names of sounds and not the sounds themselves. It is not the things as names, nouns, which are rich. It is how the things do and how they are done to. It is how they marry and divorce, sidle and reject. The aesthetic drama of the ordinary plays itself out as a result of allowing all things to become events, namely, by allowing all things the full run of their implications. This run may fulfill our anticipations and our expectations. This run may disappoint us. This run may surprise us, or blow us out. Implicitness is everywhere and everywhen. Were we to experience an apparently single thing in its full implicitness, as an event reaching out to all its potential relations, then, in fact, we would experience everything, for the leads and the hints would carry us into the nook and cranny of the implicitness of every experience.[7]

We are caught between a Scylla and Charybdis with regard to the drama of the ordinary. The scions of the bland and the anaesthetic convince us that nothing is happening, whereas the arbiters and self-announcers of high culture tell us that only a few can make it happen, so we are reduced to watching. My version is different. The world is already astir with happenings, had we the wit to let them enter our lives in their own way, so that we may press them backward and forward, gathering relations, novelties, all the while. Our affective presence converts the ordinary to the extraordinary. The world is made sacred by our *hand*ling of our things. We are the makers of our world. It is we who praise, lament, and celebrate. Out of the doom of obviousness and repetition shall come the light, a light lit by the fire of our eyes.

Notes

1 Cited in Serge Chermayeff and Christopher Alexander, *Community and Privacy* (New York: Anchor Books, 1965), p. 29.

2 Larry L. King, "The Last Frontier," *The Old Man and Lesser Mortals* (New York: Viking Press, 1975), p. 207.

3 Cf. the moving and poignant scene of "eviction" in Ralph Ellison, *Invisible Man* (New York: Vintage Books, 1972), pp. 261–77.

4 The master of "things" and "boxes" is, of course, Joseph Cornell. Indeed, he is the master of things in boxes, known forever as Cornell boxes. Only those who have experienced these "boxes" can appreciate Cornell's extraordinary ability to merge the surrealism of the imagination and the obviousness of things as a "memorial experience." Cf. Diane Waldman, *Cornell* (New York: George Braziller, Inc., 1977), and Kynaston McShine, *Cornell* (New York: The Museum of Modern Art, 1981). As with Cornell, by "things" we mean, as does William James, bundles of relations. Things are not construed here as Aristotelian essences, much less as conceptually rendered boxes.

5 Rainer Maria Rilke, "The Ninth Elegy," *Duino Elegies* (New York: W. W. Norton, 1939), p. 73.

6 Ralph Waldo Emerson, "The American Scholar," in *Works*, vol. 1 (Boston: Houghton Mifflin, 1903–1904), pp. 96–7.

7 Cf. William Blake, "Auguries of Innocence," *The Poetry and Prose of William Blake*, ed. David V. Erdman (New York: Anchor Books, 1965), p. 481.

> To see a World in a Grain of Sand
> And a Heaven in a Wild Flower,
> Hold Infinity in the palm of your hand
> And Eternity in an hour.

21

Domestic Technology: Labour-saving or Enslaving?

Judy Wajcman

Out there in the land of household work there are small industrial plants which sit idle for the better part of every working day; there are expensive pieces of highly mechanized equipment which only get used once or twice a month; there are consumption units which weekly trundle out to their markets to buy 8 ounces of this nonperishable product and 12 ounces of that one. There are also workers who do not have job descriptions, time clocks, or even paychecks.
Cowan, *From Virginia Dare to Virginia Slims*

The introduction of technology into the home has especially affected women's lives and the work that goes on in the household. Indeed, it has been suggested that we should conceive of an industrial revolution as having occurred in the home too, that 'the change from the laundry tub to the washing machine is no less profound than the change from the hand loom to the power loom' (Cowan, 1976, pp. 8–9). Women's unpaid work in the home, servicing men, children and others, has for a long time been seen by feminists as the key to women's oppression. Relieving women of this burden has been a major project of feminism. As in other spheres, considerable optimism has attached to the possibility that technology may provide the solution to gender inequality in the home.

Since the 1970s housework has finally become the object of serious academic study by historians, sociologists and even a few economists. This was part of a general concern with the relationship between the changing structures of industrial capitalism and the shaping of everyday life within the household. *The Sociology of Housework* by Ann Oakley published in 1974 marked an important break in treating housework as work within the framework of industrial sociology. In the same year, Joann Vanek's article on 'Time Spent in Housework' compared the findings of the US time use studies of housework from the 1920s to the late 1960s. She argued that the aggregate time spent on housework by full-time housewives had remained remarkably constant throughout the period, although there had been some redistribution of time between individual tasks. Her surprising conclusion, that the introduction of domestic technology had practically no effect on the aggregate time spent on housework, soon became the orthodoxy amongst feminists working in the area.

[...]

From Judy Wajcman, *Feminism Confronts Technology* (University Park, PA: The Pennsylvania State University Press), pp. 81–109.

Industrialization of the Home and Creation of the Housewife

What was the relationship between the technological developments in the economy and those in the home? To what extent did new technologies 'industrialize' the home and transform domestic labour? Why, despite massive technological changes in the home, such as running water, gas and electric cookers, central heating, washing machines, refrigerators, do studies show that household work in the industrialized countries still accounts for approximately half of the total working time (Sirageldin, 1969)?

The conventional wisdom is that the forces of technological change and the growth of the market economy have progressively absorbed much of the household's role in production. The classic formulation of this position is to be found in Talcott Parsons' (1956) functionalist sociology of the family. He argues that industrialization removed many functions from the family system, until all that remains is consumption. For Parsons, the wife–mother function is the primary socialization of children and the stabilization of the adult personality; it thus becomes mainly expressive or psychological, as compared with the instrumental male world of 'real' work. More generally, modern technology is seen as having either eliminated or made less arduous almost all women's former household work, thus freeing women to enter the labour force. To most commentators, the history of housework is the story of its elimination.

Although it is true that industrialization transformed households, the major changes in the pattern of household work during this period were not those that the traditional model predicts. Ruth Schwartz Cowan (1983), in her celebrated American study of the development of household technology between 1860 and 1960, argued exactly that.[1] For her, the view that the household has passed from being a unit of production to a unit of consumption, with the attendant assumption that women have nothing left to do at home, is grossly misleading. Rather, the processes by which the American home became industrialized were much more complex and heterogeneous than this.

Cowan provides the following explanations for the failure of the 'industrial revolution in the home' to ease or eliminate household tasks. Mechanization gave rise to a whole range of new tasks which, although not as physically demanding, were as time consuming as the jobs they had replaced. The loss of servants meant that even middle-class housewives had to do all the housework themselves. Further, although domestic technology did raise the productivity of housework, it was accompanied by rising expectations of the housewife's role which generated more domestic work for women. Finally, mechanization has only had a limited effect on housework because it has taken place within the context of the privatized, single-family household.

It is important to distinguish between different phases of industrialization that involved different technologies. Cowan characterizes twentieth-century technology as consisting of eight interlocking systems: food, clothing, health care, transportation, water, gas, electricity, and petroleum products. While some technological systems do fit the model of a shift from production to consumption, others do not.

Food, clothing, and health care systems do fit the 'production to consumption' model. By the beginning of the twentieth century, the purchasing of processed foods and ready-made clothes instead of home production was becoming common. Somewhat later, the healthcare system moved out of the household and into centralized institutions. These trends continued with increasing momentum during the first half of this century.

The transportation system and its relation to changing consumption patterns, however, exemplifies the shift in the other direction. During the nineteenth century, household goods were often delivered, mailorder catalogues were widespread and most people did not spend much time buying goods. With the advent of the motor car after the First World War, all this began to change. By 1930 the automobile had become the prime mode of transportation in the United States. Delivery services of all kinds began to disappear and the burden of providing transportation shifted from the seller to the buyer (Strasser, 1982). Meanwhile women gradually replaced men as the drivers of transport, more and more business converted to the 'self-service' concept, and households became increasingly dependent upon housewives to provide the service. The time spent on shopping tasks expanded until today the average time spent is eight hours per week, the equivalent of an entire working day.

In this way, households moved from the net consumption to the net production of transportation services, and housewives became the transporters of purchased goods rather than the receivers of them. The purchasing of goods provides a classic example of a task that is generally either ignored altogether or considered as 'not work', in spite of the time, energy and skill required, and its essential role in the national economy.

In charting the historical development of the last four household systems, water, gas, electricity, and petroleum, Cowan reveals further deficiencies in the 'production to consumption' model. These technological changes totally reorganized housework yet their impact was ambiguous. On the one hand they radically increased the productivity of housewives: 'modern technology enabled the American housewife of 1950 to produce singlehandedly what her counterpart of 1850 needed a staff of three or four to produce; a middle-class standard of health and cleanliness' (1983, p. 100). On the other hand, while eliminating much drudgery, modern labour-saving devices did not reduce the necessity for time-consuming labour. Thus there is no simple cause and effect relation between the mechanization of homes and changes in the volume and nature of household work.

Indeed the disappearance of paid and unpaid servants (unmarried daughters, maiden aunts, grandparents and children fall into the latter category) as household workers, and the imposition of the entire job on the housewife herself, was arguably the most significant change. The proportion of servants to households in America dropped from 1 servant to every 15 households in 1900, down to 1 to 42 in 1950 (Cowan, 1983, p. 99). Most of this shrinkage took place during the 1920s. The disappearance of domestic servants stimulated the mechanization of homes, which in turn may have hastened the disappearance of servants.

This change in the structure of the household labour force was accompanied by a remodelled ideology of housewifery. The development in the early years of this century of the domestic science

movement, the germ theory of disease and the idea of 'scientific motherhood', led to new exacting standards of housework and childcare.[2] As standards of personal and household cleanliness rose during the twentieth century women were expected to produce clean toilets, bathtubs and sinks. With the introduction of washing machines, laundering increased because of higher expectations of cleanliness. There was a major change in the importance attached to child rearing and mother's role. The average housewife had fewer children, but modern 'child-centred' approaches to parenting involved her in spending much more time and effort. These trends were exploited and further promoted by advertisers in their drive to expand the market for domestic appliances.

Housework began to be represented as an expression of the housewife's affection for her family. The split between public and private meant that the home was expected to provide a haven from the alienated, stressful technological order of the workplace and was expected to provide entertainment, emotional support, and sexual gratification. The burden of satisfying these needs fell on the housewife.

With home and housework acquiring heightened emotional significance, it became impossible to rationalize household production along the lines of industrial production (Ravetz, 1965). Cowan graphically captures the completely 'irrational' use of technology and labour within the home, because of the dominance of single-family residences and the private ownership of correspondingly small-scale amenities. 'Several million American women cook supper each night in several million separate homes over several million stoves' (Cowan, 1979, p. 59). Domestic technology has thus been designed for use in single-family households by a lone and loving housewife. Far from liberating women from the home it has further ensnared them. This is not an inevitable, immutable situation, but one whose transformation depends on the transformation of gender relations.

The relationship between domestic technology and household labour thus provides a good illustration of the general problem of technological determinism, where technology is said to have resulted in social changes. The greatest influences on time spent on housework have in fact come from non-technological changes: the demise of domestic servants, changing standards of hygiene and childcare, as well as the ideology of the housewife and the symbolic importance of the home.[3]

Gender Specialization of Household Technology

If domestic technology has not directly reduced the time spent on housework, has it had any effect on the degree of gender specialization of household labour? Is the general relationship women and men have to technology itself a significant factor in determining the division of labour in the home?

Available evidence suggests that domestic technology has reinforced the traditional sexual division of labour between husbands and wives and locked women more firmly into their traditional roles.[4] Because technologies have been used to privatize work, they have cumulatively hindered a reallocation of household labour. Some household appliances may have been substituted for a more equal allocation of household labour, in particular reducing the amount of time men engage in housework.

The allocation of housework between men and women is in fact much the same in households where the wife is employed and those in which she is not. Husbands in all social classes do little housework. Where men do undertake housework, they usually perform non-routine tasks at intervals rather than continually, and frequently the work is outdoors. This is in marked contrast to women's housework, the dominant characteristic of which is that it is never complete.[5]

Task-specific technologies may develop in such a way that women can take over tasks previously done by other family members. For example, Charles Thrall (1982) found that in families which had a garbage disposal unit, husbands and young children were significantly less involved in taking care of the garbage and wives were more likely to do it exclusively. Similarly with dishwashers, which are cited as one of the few appliances that do the job better and save time, husbands were less likely to help occasionally with the dishes.[6] 'In other words, new technologies may reduce the amount of time men engage in housework and increase the time spent by women, a finding which contradicts conventional wisdom' (Bose, et al., 1984, p. 78). Women have not been the prime beneficiaries of domestic technology.

Women's and men's relationship to domestic technology is a compound of their relationship to housework and their relationship to machines. Men's relationship to technology is defined differently to women's. Cultural notions of masculinity stress competence in the use and repair of machines. Machines are extensions of male power and signal men's control of the environment. Women can be users of machines, particularly those to do with housework, but this is not seen as a competence with technology. Women's use of machines, unlike men's, is not seen as a mark of their skill. Women's identity is not enhanced by their use of machines.

The household division of labour is reflected in the differential use of technologies, as Cockburn's (1985) study confirms. Few of the women in her sample used a hammer or screwdriver for more than hanging the occasional picture or mending the proverbial plug. Fewer still would use an electric drill or even a lawnmower, as 'men were proprietorial about these tools and the role that goes with them' (p. 219). Generally, women used utensils and implements – the dishwasher, vacuum cleaner, car – rather than tools. The skills necessary to handle these utensils and implements are no less than the male skills of their husbands. But, as Cynthia Cockburn points out, women cannot fix these utensils and implements when they go wrong and are therefore dependent on husbands or tradesmen, so that finally 'it is men on the whole who are in control of women's domestic machinery and domestic environment' (p. 220).

Technologies related to housework are not the only technologies to be found in the home. Indeed the extent to which the meanings and uses of domestic technologies have a gendered character is perhaps even more clearly demonstrated with regard to the technology of leisure. While for women the home is primarily defined as a sphere of work, for men it is a site of leisure, an escape from the world of paid work. This sexual division of domestic activities is read onto the artefacts themselves.

For example, television viewing reflects existing structures of power and authority relations between household members. In a study of white, working-class nuclear families in London, David Morley (1986) found that women and men gave constrasting accounts of their experience of television. Men prefer to watch television attentively, in silence, and without interruption; women it seems are only able to watch television distractedly and guiltily, because of their continuing sense of their domestic responsibilities. Male power was the ultimate determinant of programme choice on occasions of conflict. Moreover, in families who had a remote control panel, it was not regularly used by women. Typically, the control device was used almost exclusively by the father (or by the son, in the father's absence) and to some extent symbolized his domestic power.

Video recorders, like remote control panels, are the possessions of fathers and sons. In order to highlight the 'gender' of various household objects, Ann Gray (1987) asked women to imagine pieces of domestic equipment as coloured either pink or blue. Although the uniformly pink irons and blue electric drills were predictable, the mixtures in between were revealing. Home entertainment technologies were not wholly a neutral lilac. Both the timer switch and the remote control switch of video recorders were deep blue, that is, used and controlled by men.

Women's estrangement from the video recorder is no simple matter of the technical difficulty of operating it. 'Although women routinely operate extremely sophisticated pieces of domestic technology, often requiring, in the first instance, the study and application of a manual of instructions, they often feel alienated from operating the VCR.' (Gray, 1987, p. 43) Rather, women's experience with the video has to be understood in terms of the 'gendering' of technology. When a new piece of technology arrives in the home it is already inscribed with gendered meanings and expectations. Assuming himself able to install and operate home equipment, the male of the household will quickly acquire the requisite knowledge. Along with television, the video is incorporated into the principally masculine domain of domestic leisure. Gray also points out, however, that some women may have developed what she calls a 'calculated ignorance'

in relation to video, lest operating the machine should become yet another of the domestic tasks expected of them.

Technological Innovation and Housework Time

Attempts by 'post-industrial utopians' to conceive of the likely shape of the household in the future suffer from many of the intellectual defects that have misled analysts of domestic technology in the past. Much of the work of these theorists is speculative. The British economist and sociologist, Jonathan Gershuny (1978, 1983, 1985, 1988) has made the most sustained attempt to give empirical weight to post-industrial predictions about the household.

Gershuny's starting point has little in common with that of the feminist commentators. His work is directed at theories of post-industrial society which see the economy as being based increasingly on services rather than on manufacturing production. By contrast, Gershuny's main thesis is that the economy is moving toward the provision of services within the household, that is, to being a self-service economy.

Although not drawing on the feminist literature, Gershuny shares with it a recognition that unpaid domestic production is in fact work and takes it seriously as such. He goes on to argue for a reorientation of the way we study technical change. Instead of starting from the workplace, as is typical for example in economics, sociology and economic history, in his view we should start from the household.

Households have a certain range of needs, a set of 'service functions' that they wish to satisfy, such as 'food, shelter, domestic services, entertainment, transport, medicine, education, and, more distantly, government services, 'law and order' and defence' (1983, p. 1). Historically the means by which households satisfy these needs changes. Gershuny describes a shift from the purchase of final services (going to the cinema, travelling by train, sending washing to a commercial laundry) to the purchase of domestic technologies (buying a television, buying a car, buying a washing machine). A degree of unpaid domestic work is necessary in order to use such commodities to provide services. This model is used to explain the economic expansion of the developed economies in the 1950s and 1960s, which was based on the creation of new mass markets in consumer durables, electronics and motor vehicles. In this way domestic technology is of enormous economic significance, affecting the pattern of household expenditure, the industrial distribution of employment and the division of labour between paid and unpaid work.

Like Cowan, Gershuny argues that people make rational decisions in this area. However, whereas her emphasis is on moral values and the social nature of human desires and preferences, his emphasis is on prices. The household will choose between alternative technical means of provision on the basis of the household wage rate, the relative prices of final services and goods, and the amount of unpaid time necessary to use the goods to provide the service functions. However, Gershuny assumes that people have unchanging desires and respond to market signals, making narrowly economic decisions primarily in terms of prices but also in terms of domestic labour time per item. But human beings do change and the introduction of machines alters people's preferences and values. The main weakness in Gershuny's analysis is that he ignores the social and cultural dimensions of human desires.

Implicit in this analysis is the assumption that the household can be treated as a unity of interests, in which household members subordinate their individual goals to the pursuit of common household goals. Gershuny shies away from any attempt to explain decisions as to whether men or women should do domestic labour, instead simply referring to 'the traditional segregation of domestic tasks' and 'people's perception of their roles'. What this approach overlooks is that there are conflicts of interest between family members over the differential distribution of tasks and money, and this may well influence how decisions actually come about.

Let us see how this theory explains the widespread purchase and use of washing machines, as opposed to commercial laundries. Gershuny's account differs quite sharply from Cowan (1983, p. 110) who explicitly considers and rejects an economic-rationality argument on laundry. He argues that as the time needed to use a washing machine has fallen, and the price of washing machines relative to the price of laundry

has fallen so their popularity has increased. These developments are not linear however. A central feature of Gershuny's model is that it predicts first a rise, then a plateau, and then a decline in the time spent on domestic labour.

The first phase constitutes the shift from the service to the goods – for example from commercial laundries to domestic washing machines. According to the model this is a rational decision because it is cheaper, even counting the housewife's labour. But clearly, the domestic time spent on laundry goes up at this point. And precisely because it is a cheaper form of washing clothes, it becomes rational to wash more clothes more often, to satisfy (high) marginal desires for clean laundry.

In the second phase, where washing machines are fairly widely diffused, competition between manufacturers at least partly takes the form of offering more efficient machines, replacing the twin tub with the automatic. At the same time, the desire for clean laundry will begin to stabilize – slowing the rate of growth in clothes to be washed. Hence, eventually, time spent in laundry will start to fall. Thus, Gershuny argues, an effect of this move to a self-service economy, is that the amount of time spent on housework has declined since 1960 (1983, p. 151).

Gershuny is so convinced that new technologies increase the productivity of domestic labour that, in a recent paper with Robinson (1988), he takes issue with the feminist 'constancy of housework' thesis. Whilst conceding that prior to the 1960s the time spent by women on domestic work did remain remarkably constant, he insists that a shift occurred at that point. Drawing on evidence from time–budget surveys in the USA and UK, as well as Canada, Holland, Denmark and Norway, he concludes that domestic work time for women has been declining since the 1960s, and even that men do a little more than previously. It is central to his argument that this is so, even after taking into account the effects of such sociodemographic changes as more women having paid jobs, more men being unemployed, and the decreasing size of families. Therefore the diffusion of domestic equipment into households must have had some effect in reducing domestic work time. As Gershuny comments elsewhere: 'it would seem perverse to refuse to ascribe a substantial part of this reduction to the diffusion of domestic techno-logy' (1985, p. 151).

In fact on closer inspection, these findings are more in line with feminist theories about constancy of domestic work than the authors would lead us to believe. Although the central argument is that domestic work time has been declining for women between the 1960s and the 1980s, this is only the case with respect to 'routine' domestic work. Unpaid work is subdivided into three categories: routine domestic chores (cooking, cleaning, other regular housework), shopping and related travel, and childcare (caring for and playing with children).[7] While routine domestic work has declined, the time spent in childcare and shopping have substantially increased.

This finding, however, is entirely consistent with the feminist emphasis on the added time now devoted to shopping and childcare. Certainly the feminist concern with the constancy of housework has employed a broader notion that includes childcare and shopping. To argue that domestic labour time has reduced is only meaningful if it means that leisure or discretionary free time has increased. If however mechanization results in less physical work but more 'personal services' work in the sense of increased time and quality of childcare, then surely this does not mean a real decrease in work. I presume that Gershuny uses such a narrow definition of domestic work because of his interest in the impact of domestic technology. However, it is difficult to maintain that women's domestic work time has declined because of the diffusion of domestic equipment whilst arguing that men's domestic work time has marginally increased at the same time. Men's increase is explained in terms of changing norms and thus inadvertently Gershuny calls into question any direct connection between the domestic technology and the time spent on housework.

Indeed it seems that the preoccupation with increases in productivity due to technological innovation blinds many analysts to more funda-mental social factors. For example, the presence or absence of children, their age and their number all have significantly greater effects on time spent in housework than any combination of technological developments. Similarly, the presence of men in a household increases women's domestic work time by at least a third. In contrast, for men, living with women means

that they do less domestic work (Wyatt et al., 1985, p. 39). Furthermore, it has repeatedly been found that the amount of time women spend on house-work is reduced in proportion to the amount of time they spend in paid employment.[8]

A major problem with most time–budget research is that it does not recognize that the essence of housework is to combine many things, usually concurrently. This has a pro-found bearing on the interpretation of time spent in childcare and the apparent growth of leisure time. For example, watching television or listening to the radio can be combined with childcare, cooking, ironing and washing laundry. And, as I have pointed out with the case of tele-vision, this data would be particularly revealing with regard to women. Time budgets do not analyse whether activities are undertaken exclu-sively or in combination with another activity. Perhaps, as Michael Bittman (1988) suggests, the private and gendered character of the household promotes the kinds of technological innovations that maximize the number of tasks that can be performed simultaneously. To resolve such issues we would need more detailed information about the extent of use of consumer durables, the material output of services performed in the home and the social significance that these activities have for people. Gershuny's focus on technological innovations and tasks *per se* seems indicative, once again, of a technicist orientation which sees the organization of the household as largely determined by machines.

A technicist orientation is also evident in much of the futuristic literature on 'home infor-matics'. Ian Miles (1988), who collaborated with Gershuny on the research into the self-service economy, has attempted to chart the next wave of technological innovations, the new infor-mation and communication technologies, and their effects on the household. He argues that the new consumer electronic products of the coming decade are of major economic and social significance. There is much speculation about the fully automated home of the future known as a 'smart house' or 'interactive home system', where appliances will be able to communicate with each other and to the house within an integrated system. Miles predicts that home informatics will bring substantial changes to people's ways of life, one of which will be to improve the quality of domestic work both in terms of the convenience and effort required. However, Miles (p. 134) gives no reasons whatsoever for his hope that this will result in 'the sexual redivision of labour between men and women in families'.

The sociological literature on the electronic, self-servicing home of the future remains remarkably insensitive to gender issues. In particular, it ignores the way in which the home means very different things for men and women. Many of the new information and communication technologies are being developed for the increasing trend towards home-centred leisure and entertainment. But leisure is deeply divided along the gender lines. Many of these technologies, such as the home computer, demand that the user spend considerable time and concentration mastering them. But women have a lot less time for play in the home than men and boys. Programming the electronic system for the 'smart house' may enhance men's domestic power. Furthermore, the possibilities of home-based commercial operations, from 'telebanking' and shopping to 'teleworking', are likely to involve more housework for women in catering for other home-based family members. Although Miles' subtitle is 'Information Technology and the Transformation of Everyday Life', what is striking about these new technologies is just how little power they have to transform everyday life within the domestic world.

Alternatives to Individualized Housework

Even the most forward looking of the futurists have us living in households which, in social rather than technological terms, resemble the households of today. A more radical approach would be to transform the social context in which domestic technology applies. In view of what has been said about the shortcomings of domestic technology, one is prompted to ask why so much energy and expertise has been devoted to the mechanization of housework in individual households rather than to its collectivization.

During the first few decades of this century there were a range of alternative approaches to housework being considered and experimented with. These included the development of commercial services, the establishment of alternative communities and co-operatives and the invention of different types of machinery. Perhaps the best known exponent of the socialization of domestic work was the nineteenth-century American feminist Charlotte Perkins Gilman. Rather than men and women sharing the housework, as some early feminists and utopian socialists advocated, she envisaged a completely professionalized system of housekeeping which would free women from the ties of cooking, cleaning and childcare.

The call for the socialization of domestic work was not unique to the early feminist movements. Revolutionary socialists such as Engels, Bebel and Kollontai also saw the socialization and collectivization of housework as a precondition for the emancipation of women. And they embraced the new forces of technology as making this possible. Writing in the 1880s, Bebel saw electricity as the great liberator: 'The small private kitchen is just like the workshop of the small master mechanic, a transition stage, an arrangement by which time, power and material are senselessly squandered and wasted' (1971, pp. 338–9). The socialization of the kitchen would expand to all other domestic work in a large-scale socialist economy.

The modern socialist states of Eastern Europe took up some of these ideas, establishing collective laundry systems in apartment blocks and communal eating facilities. Whilst these initiatives certainly represented a different use of technology, they did not challenge the sexual division of labour insofar as women remained responsible for the housework, albeit collectivized. These policies on domestic labour resulted from the economic necessity of drawing women into the workforce combined with the ideology of equality. It is still the case that communal eating places are used a great deal more in the German Democratic Republic than in the West, and in 1974 it was estimated that families who used these facilities saved nearly two and a half hours per day compared to families who did not (Kuhrig, 1978, p. 311). Saving time, however, is not the sole motive as the housing crisis and overcrowded living conditions also encourage this pattern.

History thus provides us with many examples of alternatives to the single-family residence and the private ownership of household tools. Why then, in the USA in particular, has the individualized household triumphed? In particular, why should women apparently be so complicit in a process that was so damaging to them?

> Shall we believe that millions upon millions of women, for five or six generations, have passively accepted a social system that was totally out of their control and totally contrary to their interest? Surely there must have been at least one or two good reasons that all those women actively chose, when choices were available to them, to reside in single-family dwellings, own their own household tools, and do their own housework. (Cowan, 1983, p. 148)

To argue that women just welcomed the new domestic technologies because they became available is to come perilously close to technological determination. On the other hand, how can women have consciously and freely chosen to embrace the new methods when they have been so discredited as a liberating force? It is tempting in these circumstances to see women as duped, as passive respondents to industrialization, and as victims of advertisers.[9]

Cowan argues that women embraced these new technologies because they made possible an increased material standard of living for substantially unchanged expenditure of the housewife's time. To this extent women were acting rationally in their own and their families' interests.

However, as the following passage illustrates, Cowan seems to find the most convincing explanation of the paths chosen in a set of values to which women subscribed – the 'privacy' and 'autonomy' of the family.

> . . . . when decisions have to be made about spending limited funds, most people will still opt for privacy and autonomy over technical efficiency and community interest . . . Americans have decided to live in apartment houses rather than apartment hotels because they believe that something critical to family life is lost when all meals are eaten in restaurants or all food is prepared by strangers; they have decided to buy washing machines rather than patronize commercial laundries because they prefer to wash their dirty linen at home . . . When given choices, in short, most Americans act so as to preserve family life and family autonomy. The single-family home and the private ownership of

tools are social institutions that act to preserve and to enhance the privacy and autonomy of families.' (ibid., p. 150)

Cowan does here depict women as active agents of their own destiny rather than passive recipients of the process. However, an approach that gives such primacy to values and to the symbolic importance of the home inevitably plays down the material context of women's experience. It may be that the effectiveness of the professional experts in imposing new notions of domestic life on behalf of the ruling class has been overestimated, and the resistance they engendered ignored. Most of the available historical research is based on the rhetoric of the experts and ideologues rather than the reality of working-class women's lives. The domestic science movement was never as fully accepted as its advocates hoped (Reiger, 1985; 1986). The evidence rather suggests that women negotiated the ideology of housewifery and motherhood according to their actual circumstances and that major contradictions underlay this attempt to rationalize domestic life. Similarly, when the advertisers were playing on these ideological elements in marketing the new domestic products, women actively participated in accepting or rejecting this process.

In Britain in the 1930s, there already existed an 'infrastructure' for the communal provision of domestic appliances. There were municipal wash-houses and laundries, communal wash-houses in the old tenement blocks, and at this time several local authorities experimented with building blocks of flats, modelled on those built in Russia, and incorporating wash-houses, crèches and communal leisure areas. However, the communal provision of amenities was not always seen as progressive. It was associated in many people's minds with back-to-back houses with their shared water supply and sanitation, and a characteristic squalid view of rows of dustbins and WCs, and the tap at the end of the street. Interestingly, class differences emerged over this issue on the Women's Housing Sub-Committee, with some of the middle-class feminists on the committee more interested in the possibilities for communal childcare, laundries and other facilities. That working-class women favoured privacy and did not favour communal arrangements may have been based on their own experience of communal living in conditions of poverty.

It is important to recognize the extent to which individual choice is constrained by powerful structured forces. The available alternatives to single-family houses were extremely limited, especially for the working class. In fact, state policy in the area of housing and town planning played a key role in promoting privatism. Without the extensive provision of different options, it is not clear to what extent people freely chose private domestic arrangements.

It is even less clear to what extent women, as opposed to men, exercised the degree of choice available. Oddly Cowan separates this American preference for domestic autonomy from the sexual division of domestic labour. No role is granted to men in choosing this single-family home even though Cowan's own historical findings point to men being well served by the private domestic sphere.

The common feminist stress on the negative effects of domestic technology has contributed to the view that women have been duped. There is a tendency among some feminist scholars to assume an unqualified anti-technological stance and to imply that modern housewives are worse off than their grandmothers (Reiger, 1986, p. 110). This tendency is evident in those authors who stress the increasing isolation of the domestic worker and see domestic labour as having lost much of its creativity and individuality. Once we recognize that the mechanization of the home did bring substantial improvements to women's domestic working conditions, even while it also introduced new pressures, women seem less irrational. 'When manufacturers then, in their own interests, marketed washing machines in terms of "make your automatic your clothes basket and wash every day", they were tapping into women's experience of the problems of organizing laundry and the physical drudgery it entailed. They were also opening up greater flexibility in managing some domestic tasks' (Reiger, 1986, pp. 115–16).

Against this there is no doubt that people can be taken in by false promises, especially where advanced technology is involved. Wanting to save time and improve the quality of their housework and in turn the quality of their home life, housewives are susceptible to well-targeted advertising about the capacity of new appliances to meet their needs. The irony is that women have commonly blamed themselves for the failure of technology to deliver them from domestic toil, rather than realizing that the defects lie in the design of technologies and the social relations within which they operate.

Men's Designs on Technology

Thus far my discussion of the literature on domestic technology reveals a preoccupation with its effects on the organization of the household and women's work in the home. However technologies are both socially constructed and society shaping. At a general level, I have argued that the predominance of the single-family household has profoundly structured the form of technology that has become available. There has been much less attention given to the innovation, development and diffusion processes of specific technologies themselves.

The forms of household equipment are almost always taken as given, rather than being understood in their social and cultural context. Yet there are always technological alternatives and any specific machine is the result of non-technological as well as technological considerations. A society's choices among various possible directions of technological development are highly reflective of the patterns of political, social, and economic power in that society. Is it possible to detect these patterns in the design of domestic technology?

Gender relations are most obviously implicated in the development of domestic technology because of the extent to which the sexual division of labour is institutionalized. Most domestic technology is designed by men in their capacity as scientists and engineers, people remote from the domestic tasks involved, for use by women in their capacity as houseworkers. And, as we have seen, modern household equipment is designed and marketed to reinforce rather than challenge the existing household-family pattern.

It is not only gender relations that influence the structure of domestic technology. Like other technologies, domestic technology is big business. Particular technologies are produced not in relation to specific and objectively defined needs of individuals, but largely because they serve the interests of those who produce them. The design and manufacture of household appliances is carried out with a view to profit on the market. And the economic interests involved are not simply those of the manufacturers, but also those of the suppliers of the energy needed by these appliances.

Household appliances are part of technological systems, such as electricity supply networks. The interests of the owners of these systems have played an important part, along with those of the manufacturers, in shaping domestic technology. There is nothing the owner of an electricity supply system, for example, likes better than the widespread diffusion of an electricity-using household appliance that will be on at times of the day when the big industrial consumers are not using electricity. Residential appliances (including heating and cooling equipment) use about a third of the electricity generated in the US today; the refrigerator alone uses about seven per cent. Unlike most other household appliances, the refrigerator operates twenty-four hours a day throughout its life. In fact, many American kitchens now contain between 12 and 20 electric motors. Indeed the drive to motorize all household tasks – including brushing teeth, squeezing lemons and carving meat – is less a response to need than a reflection of the economic and technical capacity for making motors.[10]

The failure or survival, on the basis of vested interests, of some machines at the expense of others has profoundly affected the way our houses and kitchens are both constructed and experienced. This issue is raised in many feminist histories of housework and Cowan (1983, p. 128) presents a detailed example, that of 'the rivalry between the gas refrigerator (the machine that failed) and the electric refrigerator (the one that succeeded)'. There were initially designs for both and, indeed, until 1925 gas refrigerators were more widespread than the electric models. Cowan argues that electric refrigerators came to dominate the market as a result of deliberate corporate decisions about which machine would yield greater profit. The potential market for refrigerators, as well as the potential revenue for gas and electric utility companies, was enormous. Large corporations, like General Electric, with vast technical and financial resources, were in a position to choose which type of machine to

develop. Not surprisingly, with interests in the entire electricity industry, General Electric decided to perfect the design of the electric refrigerator. The manufacturers of gas refrigerators, although they had a product with real advantages from the consumer's point of view, lacked the resources for developing and marketing their machine.

So the demise of the gas refrigerator was not the result of deficiencies in the machine itself; rather, it failed for social and economic reasons. And in this, it is structurally similar to the cases of many other abandoned devices intended for the household. This story illustrates that we have the household machines which we have, not because of their inherent technical superiority, nor simply because of consumer preference, but also because of their profitability to large companies. In this way economic relations shape domestic technology. 'By itself, the gas refrigerator would not have profoundly altered the dominant patterns of household work in the United States: but a reliable refrigerator, combined with a central vacuum-cleaning system, a household incinerator, a fireless cooker, a waterless toilet, and individually owned fertilizer-manufacturing plants would certainly have gone a long way to altering patterns of household expenditure and of municipal services' (Cowan, 1983, p. 144).

What is so original about Cowan's work is that she goes beyond a general account of technological change to present a concrete historical analysis of contingency in the evolution, design and development of a specific technology. She demonstrates the possibility of alternative machines and examines carefully the reasons for the path taken. However, it is disappointing that many of the wider concerns of her book disappear here. Her account is wholly in terms of the interests of, and the power play between, the companies producing the refrigerators, and the gender dimension is lost. Housewives here are relegated to the role of consumer – 'they bought electric refrigerators because they were cheaper'. Our understanding remains incomplete without research on design alternatives which shows how the form of the household, and the sexual division of labour within it, actively shape artefacts. We need much more work of this kind on what shaped these machines in the first place.[11]

An important dimension glossed over in the literature on the development of domestic equipment is the culture of engineering. After all, engineers do not simply follow the manufacturers' directives; they make decisions about design and the use of new technologies, playing an active role in defining what is technically possible. The masculinity of the engineering world has a profound effect on the artefacts generated. This must be particularly true for the design of domestic technologies, most of which are so clearly designed with female users in mind.

When women have designed technological alternatives to time-consuming housework, little is heard of them. One such example is Gabe's innovative self-cleaning house (Zimmerman, 1983). Frances Gabe, an artist and inventor from Oregon spent 27 years building and perfecting the self-cleaning house. In effect, a warm water mist does the basic cleaning and the floors (with rugs removed) serve as the drains. Every detail has been considered. 'Clothes-freshener cupboards' and 'dish-washer cupboards' which wash and dry, relieve the tedium of stacking, hanging, folding, ironing and putting away. But the costs of the building (electricity and plumbing included) are no more than average since her system is not designed as a luxury item. Gabe was ridiculed for even attempting the impossible, but architects and builders now admit that her house is functional and attractive. One cannot help speculating that the development of an effective self-cleaning house has not been high on the agenda of male engineers.

Domestic Technology: A Commercial Afterthought

The fact is that much domestic technology has anyway not been specifically designed for household use but has its origins in very different spheres. Consumer products can very often be viewed as 'technology transfers' from the production processes in the formal economy to those in the domestic informal economy.

Typically, new products are at first too expensive for application to household activities; they are employed on a large scale by industry only, until continued innovation and economies of scale allow substantial reduction in costs or adaptation of technologies to household circumstances. Many domestic technologies were initially developed for commercial, industrial and even defence purposes and only later, as manufacturers sought to expand their markets, were they adapted for home use. Gas and electricity were available for industrial purposes and municipal lighting long before they were adapted for domestic use. The automatic washing machine, the vacuum cleaner and the refrigerator had wide commercial application before being scaled down for use in the home. Electric ranges were used in naval and commercial ships before being introduced to the domestic market. Microwave ovens are a direct descendant of military radar technology and were developed for food preparation in submarines by the US Navy.[12] They were first introduced to airlines, institutions and commercial premises before manufacturers turned their eyes to the domestic market.

Despite the lucrative market that it represents, the household is not usually the first area of application that is considered when new technologies are being developed. For this reason new domestic appliances are not always appropriate to the household work that they are supposed to perform nor are they necessarily the implements that would have been developed if the housewife had been considered first or indeed if she had had control of the processes of innovation.

It is no accident that most domestic technology originates from the commercial sector, nor that much of the equipment which ends up in the home is somewhat ineffectual. As an industrial designer I interviewed put it, why invest heavily in the design of domestic technology when there is no measure of productivity for housework as there is for industrial work? Commercial kitchens, for example, are simple and functional in design, much less cluttered with complicated gadgets and elaborate fittings than most home kitchens. Reliability is at a premium for commercial purchasers who are concerned to minimize their running costs both in terms of breakdowns and labour-time. By contrast, given that women's labour in the home is unpaid, the same economic considerations do not operate. Therefore, when producing for the homes market, manufacturers concentrated on cutting the costs of manufacturing techniques to enable them to sell reasonably cheap products. Much of the design effort is put into making appliances look attractive or impressively high-tech in the showroom – for example giving them an unnecessary array of buttons and flashing lights. In the case of dishwashers and washing machines, a multitude of cycles is provided although only one or two are generally used; vacuum cleaners have been given loud motors to impress people with their power. Far from being designed to accomplish a specific task, some appliances are designed expressly for sale as moderately priced gifts from husband to wife and in fact are rarely used. In these ways the inequalities between women and men, and the subordination of the private to the public sphere are reflected in the very design processes of domestic technology.

In tracing the history of various domestic appliances, Forty (1986) shows how manufacturers have designed their products to represent prevailing ideologies of hygiene and housework. Thus, in the 1930s and 1940s manufacturers styled appliances in forms reminiscent of factory or industrial equipment to emphasize the labour-saving efficiency which they claimed for their products. At that time, domestic equipment was still intended principally for use by servants. However such designs made housework look disturbingly like real work and in the 1950s, when many of the people who bought these appliances were actually working in factories, the physical appearance of appliances changed. A new kind of aesthetic for domestic appliances emerged which was discreet, smooth, and with the untidy, mechanical workings of the machine covered from view in grey or white boxes.[13] The now standard domestic style of domestic appliances '. . . suited the deceits and contradictions of housework well, for their appearance raised no comparisons with machine tools or office equipment and preserved the illusion that housework was an elevated and noble activity', of housework not being work (Forty, 1986, p. 219).

Throughout this chapter I have been examining the way in which the gender division of our society has affected technological change in the home. A crucial point is that the relationship between technological and social change is fundamentally indeterminate. The designers and promoters of a technology cannot completely predict or control its final uses. Technology may well lead a 'double life' '. . . one which conforms to the intentions of designers and interests of power and another which contradicts them – proceeding behind the backs of their architects to yield unintended consequences and unanticipated possibilities' (Noble, 1984, p. 325).

A good illustration of how this double life might operate, and how women can actively subvert the original purposes of a technology, is provided by the diffusion of the telephone. In a study of the American history of the telephone, Claude Fischer (1988) shows that there was a generation-long mismatch between how the consumers used the telephone and how the industry men thought it should be used. Although sociability (phoning relatives and friends) was and still is the main use of the residential telephone, the telephone industry resisted such uses until the 1920s, condemning this use of the technology for 'trivial gossip'. Until that time the telephone was sold as a practical business and household tool. When the promoters of the telephone finally began to advertise its use for sociability, this was at least partly in response to subscribers' insistent and innovative uses of the technology for personal conversation.

Fischer explains this time lag in the industry's attitude toward sociability in terms of the cultural 'mind-set' of the telephone men. The people who developed, built, and marketed telephone systems were predominantly telegraph men. They therefore assumed that the telephone's main use would be to directly replicate that of the parent technology, the telegraph. In this context, people in the industry reasonably considered telephone 'visiting' to be an abuse or trivialization of the service. It did not fit with their understandings of what the technology was supposed to be used for.

The issue of sociability was also tied up with gender. It was women in particular who were attracted to the telephone to reduce their loneliness and isolation and to free their time from unnecessary travel. When industry men criticized 'frivolous' conversation on the telephone, they almost always referred to the speaker as 'she'. A 1930s survey found that whereas men mainly wanted a telephone for business reasons, women ranked talking to kin and friends first (Fischer, 1988, p. 51).

Women's relationship to the telephone is still different to men's in that women use the telephone more because of their confinement at home with small children, because they have the responsibility for maintaining family and social relations and possibly because of their fear of crime in the streets (Rakow, 1988). A recent Australian survey concluded that 'ongoing telephone communication between female family members constitutes an important part of their support structure and contributes significantly to their sense of well-being, security, stability, and self-esteem' (Moyal, 1989, p. 12). The telephone has increased women's access to each other and the outside world. In this way the telephone may well have improved the quality of women's home lives more than many other domestic technologies.[14]

Conclusion: More Work for Social Scientists?

I started this chapter by noting how belated has been the interest in domestic technology and household relations. There is now a substantial body of literature on the history of housework and the division of labour in the home. In recent years too there has been growing interest in domestic technology both among feminist theorists and, from a different perspective, among post-industrial society theorists. This work is still relatively underdeveloped and much of the literature shares a technicist orientation whether optimistic or pessimistic in outlook. Technology is commonly portrayed as the prime mover in social change, carrying people in its wake, for better or worse. But history is littered with examples of alternative ways of organizing housework and with alternative designs for machines we now take for granted. In retrieving these lost options from obscurity the centrality of people's actions and choices is highlighted and with them the social shaping of technology that furnishes our lives.

An adequate analysis of the social shaping of domestic technology cannot be conducted only at the level of the design of individual technologies. The significance of domestic technology lies in its location at the interface of public and private worlds. The fact that men in the public sphere of industry, invention and commerce design and produce technology for use by women in the private domestic sphere, reflects and embodies a complex web of patriarchal and capitalist relations. Although mechanization has transformed the home, it has not liberated women from domestic drudgery in any straightforward way. Time–budget research leaves us wondering whether technology has led to more flexibility in housework or to its intensification. To further our understanding of these issues we need more qualitative research on how people organize housework and use technology in a variety of household forms. Such research should distinguish between different types of domestic technology and examine the significance of gender in people's affinity with technology. Finally, the designers of domestic technology themselves have so far been subjected to very little investigation; an examination of their backgrounds, interests, and motivation may shed light on the development of particular products. By refusing to take technologies for granted we help to make visible the relations of structural inequality that give rise to them.

This portrait of domestic technology is certainly incomplete. In this chapter I have concentrated on domestic technology as a set of physical objects or artefacts and argued that gendered meanings are encoded in the design process. This process involves not only specifying the user but also the appropriate location of technologies within the house. For example, domestic appliances 'belong' in the kitchen, along with women, and communications technology such as the television are found in the 'family room'. This signals the way in which the physical form and spatial arrangement of housing itself expresses assumptions about the nature of domestic life – an issue to be taken up in the next chapter.

Notes

1 See also Ruth Schwartz Cowan (1976 and 1979).
2 There is now quite an extensive feminist literature on the domestic science movement and its attempt to elevate the status of housekeeping. See, for example, Ehrenreich and English (1975,1979) and Margolis (1985) on America;

Davidoff (1976) and Arnold and Burr (1985) on Britain; and Reiger (1985) for Australia. Reiger's book, *The Disenchantment of the Home*, is the most interesting sociologically as she attempts to combine a feminist analysis of the role of the professional and technical experts of the period with a critique of instrumental reason. The infant welfare and domestic science movements are seen as being part of a general extension of 'technical rationality' in the modern world.

3   I am only referring to domestic technology here, as clearly medical technology is central to demographic changes in life expectancy and to birth control.

4   See Bose et al. (1984), Rothschild (1983), and Thrall (1982).

5   In my own qualitative study (1983) in a small market town in Norfolk, England, I found that men always did the 'outdoor' jobs – mowing the lawn, gardening, fixing the car, household repairs and, to a lesser extent, painting and decorating. While the husbands did have a responsibility for performing certain household tasks, these had very different characteristics from those the women performed. Of course, this contrast is exaggerated and depends partly on conventional conceptions; lawn-mowing, for instance, is just as continuous as window cleaning. Nevertheless, there is a general distinction which is reinforced by popular evaluations. Indeed, these evaluations are intrinsic to the domestic division of labour.

6   The microwave cooker is another interesting case where further research is needed to show whether it results in men being more prepared to take up some cooking activities or whether it increases expectations so that mothers cook separate meals for different members of the family at different times.

7   A fourth residual category, odd jobs, is not considered in the article.

8   This might lead one to expect that women in the paid labour force might use their income to substitute consumer durables for domestic labour. Surprisingly however women in employment have slightly less domestic equipment than full-time housewives. From an analysis of the Northampton household survey data, collected in 1987 as part of the British ESRC Social Change in Economic Life Initiative, Sara Horrell found that there were no significant differences in the ownership of consumer durables between working women and non-working women.

9   In her 1976 essay, Cowan has a tendency to adopt this latter position, seeing the corporate advertisers 'the ideologues of the 1920s' as the agents which encouraged American housewives literally to buy the mechanization of the home. The interest of appliance manufacturers in mass markets coincided exactly with the ideological preoccupations of the domestic science advisers, some of whom even entered into employment with appliance companies. According to W. and D. Andrews (1974), nineteenth-century American women, anxious to elevate their status, believed that technology was a powerful ally.

10   The Australian Consumer Association magazine, *Choice*, recently found that many appliances were useless and that a lot of jobs were better done manually. For example, they found that a simple manual citrus squeezer was overall better than many of the electric gadgets.

11   A notable exception is Hardyment's (1988) book on domestic inventions in Britain which documents a multitude of discarded designs, such as sewing machines, washing machines, ovens, irons, wringers, mangles and vacuum cleaners, invented and developed between 1850 and 1950. Unfortunately, the book contains little analysis of the forces which shaped their development. At one point, the author makes the intriguing argument that it was the small electric motor (introduced in the 1920s) more than any other invention which led to the development of domestic machinery along private rather than communal channels. But Hardyment concludes that 'the potential of any machine should lie in the mind of its user rather than its maker' (p. 199), echoing her earlier statement that women should seize the technological means to liberate themselves. It is disappointing that in a book devoted to the history of domestic machines so little attention is paid to the gender interests involved in their production.

12   This point is made by Megan Hicks, 'Microwave Ovens' (MSc dissertation, University of New South Wales, 1987).

13   One can only speculate as to whether covering up the mechanical workings of appliances assisted in alienating women from understanding these machines and how to mend them.

14   However, the unintended consequences of a technology are not always positive. The diffusion of the telephone has facilitated the electronic intrusion of pornography into the home. Not only are abusive and harassing telephone calls made largely by men to women, but new sexual services are being made available. The French post office's Minitel service, which is a small television screen linked to the telephone, has seen a massive 'pink message service' arise. When it was introduced over ten years ago, the Minitel system was intended to replace the

telephone directory. Since then it has developed thousands of services, the most popular being pornographic conversations and sexual dating via the electronic mail. When complaints have been made the French post office claim that they can do nothing to censor hardcore pornography as it is part of private conversations. One wonders how this might affect gender relations in the home.

References

Andrews, W. and D. "Technology and the Housewife in Nineteenth-century America." *Women's Studies* 2, 1974. pp. 309–28.

Arnold, E. and Burr, L. "Housework and the Application of Science" in Faulkner, W. and Arnold, E. (eds); *Smothered by Invention: Technology in Women's Lives* (London: Pluto Press, 1985).

Bebel, A. *Women Under Socialism* (New York: Shoken Books, 1971).

Bittman, M. "Service Provision, Women and the Future of the Household." Unpublished paper, 1988.

Bose, C., Bereano, P. and Malloy, M. "Household Technology and the Social Construction of Housework." *Technology and Culture* 25. 1984. pp. 53–82.

Cockburn, C. *Machinery of Dominance: Women, Men and Yechnocal Know-How.* (London: Pluto Press, 1985).

Cowan, R. S. "The 'Industrial Revolution' in the Home: Household Technology and Social Change in the Twentieth Century." *Technology and Culture* 17, 1976. pp. 1–23.

Cowan, R. S. "From Virginia Dare to Virginia Slims: Women and Technology in American Life." *Technology and Culture* 20(1) 1979. pp. 181–201.

Davidoff, L. "The Rationalization of Housework," in Barker, D. and Allen, S. (eds); *Dependence and Exploitation in Work and Marriage* (London: Longman, 1976).

Ehrenreich, B. and English, D. "The Manufacture of Housework", *Socialist Revolution*, 26 (1975) pp. 5–40.

Ehrenreich, B. and English, D. *For Her Own Good: 150 Years of Experts' Advice to Women* (London: Pluto Press, 1979).

Fischer, C. "Touch Someone: The Telephone Industry Discovers Sociability," *Technology and Culture*, 29(1) 1988. pp. 32–61.

Forty, A. *Objects of Desire: Design and Society 1750–1980* (London: Thames and Hudson, 1986).

Gershuny, J. *After Industrial Society: The Emerging Self-Service Economy* (London: MacMillan, 1978).

Gershuny, J. *Social Innovation and The Division of Labour.* (Oxford: Oxford University Press, 1983).

Gershuny, J. "Economic Development and Change in the Mode of Provision of Services" in Redclift, N. and Minigione, E. (eds); *Beyond Employment: Household, Gender, and Subsistence.* (Oxford: Basil Blackwell, 1985).

Gershuny, J. and Robinson, J. "Historical Changes in the Household Division of Labour." Unpublished manuscript. 1988.

Gray, A. "Behind Closed Doors: Video Recorders in the Home." In Baehr, H. and Dyer, G. (eds) *Boxed In: Women and Television.* (London: Routledge and Kegan Paul, 1987).

Hardyment, C. *From Mangle to Microwave: The Mechanization of the Household.* (Cambridge: Polity Press, 1988).

Kuhrig, H. *Zur gesellschaftlichen Stellung der Frau in der DDR,* (Leipzig: Verlag Fuer die Frau, 1978).

Margolis, M. *Mothers and Such: Views of American Women and Why They Changed* (Berkeley, CA: University of California Press, 1985).

Miles, I. *Home Informatics: Information Technology and the Transformation of Everyday Life.* (London: Punter Publishers, 1988).

Morley, D. *Family Television: Cultural Power and Domestic Leisure.* (London: Comedia, 1986).

Moyal, A. "The Feminine Culture of the Telephone," *Prometheus,* 7(1) 1989. pp. 5–31.

Noble, D. *Forces of Production: A Social History of Industrial Automation.* (New York: Knopf, 1984).

Parson, T. "The American Family: Its Relations to Personality and the Social Structure." In Parsons, T. and Bales, R. (eds) *Family, Socialisation, and Interaction Process.* (London: Routledge and Kegan Paul, 1956).

Rakow, L. "Women and the Telephone: the Gendering of a Communications Technology." In Kramarae, C. (ed.) *Technology and Women's Voices.* (New York: Routledge & Kegan Paul, 1988).

Ravetz, A. "Modern Technology as An Ancient Occupation: Housework in Present-Day Society". *Technology and Culture,* 6 1965 pp. 256–60.

Reiger, K. *The Disenchantment of the Home: Modernizing the Australian Family 1880–1940.* (Melbourne: Oxford University Press, 1985).

Reiger, K. "At Home With Technology," *Arena* 75, 1986, pp. 109–23.

Rothschild, J. (ed.) *Machina Ex Dea: Feminist Perspectives on Technology.* (New York: Pergamon Press, 1983).

Sirageldin, I. *Non-Market Components of National Income.* (Ann Arbor, MI: University of Michigan Survey Research Center, 1969).

Strasser, S. *Never Done: A History of American Housework.* (New York: Pantheon, 1982).

288 JUDY WAJCMAN

Thrall, C., "The Conservative Use of Modern Household Technology," *Technology and Culture* 23, 1982, pp. 175–94.

Wajcman, J. *Women in Control: Dilemmas of a Workers Co-operative.* (Milton Keynes: Open University Press, 1983).

Wyatt, S., Thomas, G., and Miles, I., "Preliminary Analysis of the ESRC 1983/4 Time Budget Data." Science Policy Research Unit, University of Sussex, 1985.

Zimmerman, J. (ed.) *The Technological Woman: Interfacing with Tomorrow.* (New York: Praeger, 1983).

22

Some Meanings of Automobiles

Douglas Browning

Of course there are many meanings of automobiles. I am interested in just three interdependent meanings which I take to be philosophically instructive yet seldom if ever explicitly formulated. These three are the enlargement of the body, the introduction of the category of the sudden, and the deepening of the concept of anonymous privacy. To sum up these points, the distinctive meaning of automobiles with which I am concerned herein is self-enrichment through the anonymity of speed. With it is involved a new understanding of the human self.

Let me state emphatically that I am not concerned with the meanings of automobiles *to* the human being. If meanings come about only through human beings, still there is the living fact of the automobiled human being, just as there are the facts of the economic human being or the enfamilied human being. Meaning is a fact about the concrete fact of organism in environmental transaction and not a fact about a mind separate and distinct from environment.

The personal automobile, especially one with clutch and gearshift, serves to enlarge the body beyond the skin. This first point is the most obvious. Such enlargement begins with the simple fact of the instrumentality of the automobile as a means of conveyance. The automobile is a vehicle for satisfying demands, and is as such a tool like a hammer. But whereas hammers extend only our hands and arms, automobiles extend our legs, hands, fingers, arms, eyes, ears, and so on. The automobile is a thoroughgoing tool. Moreover, though all tools in a sense enlarge the body, some tools become more than simple means when they are chosen for the enjoyment of their functioning. Among those so chosen, some become especially personated. By the verb "to personate" I mean here the proper correlative to the verb "to impersonate." To impersonate is to play at being another; to personate is to play at being oneself. The automobile in the lives of many is a thoroughgoing tool within which the skinned body is absorbed and enjoyed for its functioning and in terms of which one plays at being a self.

Too, an automobile is like a suit of clothes. It is our traffic habit, our highway wear. Clothes serve the ends of modesty and decoration; the automobile gives us anonymity and status. Both protect us from the elements, natural and social. But as everyone is well aware, some garments come to be identified with one's character. They merge into the personality. They are stuck by some secret glue to one's privacies. They are personated.

In Larry A. Hickman, ed., *Technology as a Human Affair* (New York: McGraw Hill Publishing, 1990), pp. 172–7. Reprinted by permission of Douglas Browning.

The richness of this meaning of the automobile is not exhausted by the consideration of it as a personated, thoroughgoing tool and garment. In order for the automobile to enlarge the body, it must be demachined, for the automobile is a machine indeed while man is not. Many grave men once believed that all of nature was a vast machine and that the human physiological plant was one very marvelous machine within the larger one. Men seldom believe this today. But suppose it were true. It is obvious, is it not, that the physiological machine is amazingly personated, organic, pervaded with feeling, intimate. This could only be by some sleight-of-hand, some arch-trickery, whereby what is a machine is discovered to be functioning as though it were not a machine. This is what I mean by demachining. There is the science-fiction story of the humanoid robot that comes to serve as a personal friend of the hero. There is the companion story of the robot who asserts his rights to life, liberty, and property. There is more here than mere personification, for in each case the machine comes to function as what is no longer a machine. The automobile is not only personated, it is necessarily demachined in the process.

Contrast the power of automobiles as the grand machine of the turnpikes, still incompletely demachined, with the personation of the fine frame, which is so one with the protoplasm as to be uncommanded and unmastered. The swell machines are fun to drive, for we set ourselves up as bronco-busters to master their idiosyncratic powers. In fact any unfamiliar automobile is such a machine. It is the autonomous other for our mastery. However, the demachined automobile is not an automobile under control, as though it were another we have mastered; it functions as uncontrolled as our hands and feet. Who thinks of his hands as a thing to be mastered and controlled? And if one did, would it not thereby be meaningfully other than the body?

From what has been said so far it must be clear that not everyone who drives is an automobiled human being. Some of you may not know what I am talking about. The automobile remains a thing *for* you, merely a vehicle, an object of fear or astonishment, or perhaps a friend like a favorite dog, a child, or a mistress. The meanings of which I speak are, as yet, unrealized potentialities in your experience. You must not be disconcerted,

for it is a simple enough truth that the meaning of things often eludes us. If you have never been a party to the kind of fact of which I speak, yet someday you may, and then perhaps you shall find this meaning within the fact.

I have said that the three meanings of automobiles with which I am concerned are interrelated. The initial point, the enlargement of the body, cannot be made adequately without anticipating the next two. Let me make two transitional comments.

First, the identification of automobile with self, its personations and demachining, seems to be at a peak at high speeds. This is, I suspect, analogous to the fact that one's identity with the flesh-and-bone body is experienced most profoundly in those periods of its peak yet effortless functioning. The demachining of the automobile requires, however, an added inducement, which is satisfied in the transformation of landscape at highway speeds into a uniquely alien environment characterized by anonymity and suddenness. Thus the dualism of self and the world is radical, and clearly the automobile itself exists at the still center which "I am." This experience of peak identification I call "the phenomenon of the Texas highway."

The second comment is closely related to the first. The automobile has an inside and an outside. This is a psychological as well as a spatial structure. What passes as ordinary visual and auditory sensations of externals in walking life becomes, in driving life, split between sensations of landscape and sensations of the automobile interior. The latter sensations take on an introspective aura, especially at high speeds, and the automobile interior becomes a nonpublic place of privacies. The character of one's internal life becomes enriched, and a larger self is realized. The dualism of self and world is compounded. This aspect of personation, whereby interiors of sensation become introspections, is the other side of the processes of certain psychoses in which the internals of phantasy are read wholeheartedly into an external world. I call this aspect introspectation. As an illustration consider the role of the automobile radio. Like our eyes and ears, it links us to the world, for through it we sense the happenings of the day; yet the sounds made are not themselves the link from dashboard to the outside, but the interiorized links from receptor to consciousness and as such fall within the body.

I now pass to the second meaning of automobiles, the introduction of the category of the sudden. I am referring to a new way of seeing things, a fresh principle that has arisen into public consciousness since the coming to be of the automobile and according to which the world takes on character and meaning of which human consciousness was not previously aware. I will not say that the basic structure of the universe has itself changed, for I know nothing of the universe, but I suggest that the world as the object of human consciousness has altered since the coming of the automobile. And the alteration that has come to be is simply this: the world presents itself as a place of the sudden, as an arena wherein the rapidly developing lurks to spring across the background of the slow, methodical passage of days. The basic category of processional growth, the natural way of the coming to be of man, trees, and love, is in no way replaced. But now the promise and fact of the sudden cuts across nature as over the surface of the waves of a patiently toiling sea. Birth, life, love, and death, spring, summer, autumn, and winter are there. The ripening of friendship, the rewards of long labor, digestion, pregnancy, creation are there. But also the scream of tires, the yellow light, the narrowly avoided accident.

I would like for you to make some efforts to understand me. I am not maintaining that the sudden is a fact of automobile experiencing alone. It is, no doubt, more apparent when one drives and clearest when one drives in traffic at high speeds. My suggestion is much more radical than this. I am suggesting that the sudden has become a category of all of our experience. I am also suggesting that the automobiled human being is the source and sanction of the publicity of this category and that this is one important meaning of automobiles. I suggest that contemporary man is the first man to live in a world of suddenness, that daily life, art, religion, philosophy, politics, conversation, education, and indeed all human concerns are infected with experience of and concern for crisis, and that the automobiled human being of traffic and highway is the breeder, infector, living reinforcement, and purest exemplification of the flash happening.

Now is the time to rectify one misunderstanding. Throughout this discussion you have suspected perhaps that the perspective of the sudden was neither new nor concerned with automobiles. What you have done is to confuse the sudden with the abrupt, the suddenly-happening with the suddenly-happened, rapidity with discontinuity. These are by no means the same. The category of the suddenly-happened has been native to thought and experience as long as men have lived in an environment in which important things suddenly-happened. And man always has. One suddenly dies, one suddenly wakes up screaming, tigers suddenly spring from trees. But the suddenly-happening is not the over and done with, the finished, the beginning abruptly or the instantly closed out. It is something going on, transpiring, but at the very limit of human capacities for adjustment or control. It directs our attention to a new facet of man, his reaction time. Man becomes, as does his age, crisis oriented. Perhaps this is not totally new. Maybe horse soldiers and gladiators participated in some such world as ours, but the fact is more pervasive now. It is public domain. It is common property. It is categorical.

A new phenomenon requires a new attitude. The attitude one takes up in the automobile is controlled by its adequacy to the landscape as a speedily shifting anonymous field of lurking suddenness, violence, and crisis. The attitude is an orientation to the appearance of the sudden, though the sudden does not constantly appear. When it does appear, the sudden moves a special object of concern across a background of speed and anonymity, as though it were a vividly red flash on a dull greyish ground. The attitude is directed to the occurrence of such flashes, not in order to react to them, but in order to develop with and control them. This means that the attitude must be a constant tension which is both patient with long stretches of routines and capable of instant transformation into rapid and precise business. The attitude is not one of contemplation of the road, for the absorption in the object of attention precludes action. Nor is this attitude an ordinary tension or anxiety, for these allow neither long inactivity nor coolly perfect execution. In fact there is nothing quite like it. It is best understood by exemplification. Consider driving long distances at high speeds. What happens: There is an immense coolness, almost trancelike, in the performance of routine affairs. The times comes when the automobile is no longer handled; it handles itself. You, the driver, sit and await the exigencies. Now suppose you stop for a sandwich and a cup of coffee. Notice how difficult it is to make the transition. Your head rings: because your eyes are too wide to take in the slow subtle movements of human expressions, you have to give your order twice: you have to take a moment to get your bearings. You have to reorient yourself to a world of greater viscosity.

The category of the sudden is only understood by a glance backward at the meaning of the automobile as an extension of the body. You can see now that the new body serves to increase the scope of the manageable far beyond that in any previous age. At night on the highway the headlights pick out an anonymous landscape that extends approximately the length of control. The automobile moves into a world of secrets, anyone of which may emerge at the extreme of vision as a suddenly unfolding drama of collision. Such things may be managed. The mere punch of a foot serves to brake. The slight twitch of a hand swerves an enormous mass into another lane. The scope of the manageable is increased in distance, speed, mass, and momentum. The sudden as the advent of such management is precisely the initial impetus for the demachining of the automobile.

And now let me make a brief comment in transition to a discussion of the last meaning of automobiles to be considered. The new attitude is controlled by the scope of the manageable and posits a world of exigencies, of potential explosiveness, which is radically other than the automobiled human being himself. The human self, enlarged by the automobile, becomes a fact of mobile privacy. We are now in a position to understand this contemporary man.

Privacy, we sometimes feel, is a function of anonymity. The anonymity of the automobile, its faceless commonness, is mobile and hence it is a constant of human traffic, a mask for all occasions. It does not reveal us as do our face and speech, nor must the automobile assume the frightening fixity of assumed joviality, wittiness, or sternness. The automobiled person is from the outside the only typical specimen of the mass man. On the highway one always travels incognito.

In proportion to the perfection of anonymity in the automobile privacy is deepened. The automobile is spacious, like a spacious soul, and contains room for the greatest and the smallest of intimacies. We cannot sing loudly and with proper flourishes even in our homes, but the roar of engine and the passage through anonymous landscapes allow us such expressions in the automobile. The relaxation of the strains of holding steadily to our pedestrian masks effects release in the private car. These expressions no longer have the character of overt behavior but become as internal, demachined, personated, and introspectated as the automobile itself.

There is thus an inside and an outside to the automobiled human being. From the outside there is anonymity. From the inside there is privacy. The windshield is the frontier between the person and the world. The world is the place of exigencies, the coiled potentialities of violence, the possibilities of the sudden. The person is the autonomous master manipulator with the enriched intimacies of a mobile hideout.

But there is an ambiguity involved. Such utter facelessness on the outside and such complete privacy on the inside are only the typical facts of the flow of traffic; these facts are strained when an intrusion occurs. There are sometimes riders, sometimes love, sometimes death. These are intruders in the privacy and the anonymity, brutally contradicting both. If intrusion could destroy anonymity without destroying privacy, if intrusion could cancel privacy without canceling anonymity, then there would no ambiguity. But intrusion actually serves only for the destruction of the simple inside/outside structure of anonymity/privacy. The bringing of outsides into insides, the world into the person, gives the intruders a peculiar quality of intimacy not otherwise encountered. The peculiarity is due to the fact that the intruders are ambiguous – ultimate strangers in a private land yet ultimately personated interiors from a public landscape. Love, death, and riders are internal-externals, which if not contradictions are at least riddles. But the driver of the automobile is a party to the intrusion also. Without him intrusions would be only entries. The automobile man in his moment of intrusion is an ambiguity.

I will illustrate with three vignettes of intrusion.

Vignette One: The Hitchiker. I step into the automobile bidden and yet, as an outsider in his domain, an intruder. I am welcome; I should make myself at home. I am not welcome; I should take up as little of the privacy as possible. I am a question from the start. He does not know me. He does not trust me. I am bringing into his life the unpredictable, the unassimilated, the faceless, the strange. But now he speaks to me. He is running around on his wife; he hates his job; he fears death by cancer. How privileged I am to be the object of such confidence. I am in his deepest soul, a sharer in his hideout, yet surely no one has ever been so external to his person. Interloper and confidant, I am irremediably ambivalent to him. I become, even to myself, a pun.

Vignette Two: Love-Play. This is a mobile rendezvous. You enter myself as something stolen from the world. The secret is deep indeed. We may not linger, for the anonymity of our place is subject to immediate discovery. Wariness is mixed with the recklessness of intimacy. The outside peers into the inside. The ambiguity of sex is unique and exciting. By invitation the alien world admits to my trespass. A piece of the landscape transgresses my soul. I am myself the one who is most naked and vulnerable. I am at the same time the anonymous manipulator. Desperate passion is one with infinite reserve. The suddenness of the world wraps into our privacies, as though the automobile were turned inside out.

Vignette Three: Death. The traffic fatality is statistical. There is in every highway death the symbolic anonymity of a figure in a column of figures, an announcement on the radio. But the suddenly perceived inevitability of final intrusion unfolds the eternally personal moment. You will participate in your death. There in the stark aloneness of the quiet center of suddenness, you will be totally aware. The very enlargement of the body underlines the vulnerability, the contingency of your automobile. The death is public, for it is an object of traffic and the openly observable recovery of a swollen body by the anonymity of landscape. The automobile is still, no longer personated, but a simple inert piece of the environment. It is a personal death; it is a common end.

These vignettes of intrusion take us beyond the meanings of automobiles and introduce us to the meanings of contemporary men. Therefore they may be labeled metaphysical escapades.

Part VIII
Values and BioTechnologies

Introduction

Until quite recently, philosophical and social questions about technology were about machines of one sort or another. Today, however, some of the most pressing popular concerns about technology involve *biotechnology*, or the manipulation and/or creation of life forms rather than machines. The readings in this section examine biotechnology and values, beginning with Daniel Callahan's general reflections on technologies, three readings articulating differing positions on human biotechnologies, and closing with two examinations of food biotechnologies.

In the first essay, Callahan, co-founder and former president of the Hastings Center, asks two questions about new technologies: "How should we cope with those features of our lives, and thus of our human nature, where what is good gradually turns into something bad, usually inadvertently?" and "Is technology simply a neutral reality, neither good nor bad in itself, but to be evaluated solely on the basis of the uses to which it is put – or does it have a life of its own, subtly shaping our values and ways of life whether we choose that to happen or not?" Ultimately, he rejects the value-neutral thesis regarding technology. Using two examples that bridge our concern with everyday artifacts (the automobile) and biomedical technologies, he claims there is something of a "technological imperative" at work that leads us to embrace some technologies rather uncritically, and ignore other strategies that might achieve the same

ends. We are, in the words of Winner, in a state of somnambulism about technologies. Callahan stops short of endorsing technological determinism, but suggests that substantial change is often rooted in crisis.

Opening our more focused considerations on human biotechnology, Laura Purdy, Professor of Philosophy and Ruth and Albert Koch Professor of Humanities at Wells College, New York, starts by asking when, if ever, it might be immoral to have children. Traditionally, some have argued that to bring children into conditions of violent conflict or extreme deprivation might be immoral. We now have new considerations because of new biomedical technologies. It is now possible to know, in some cases, whether parents are carriers for genetic diseases, or, using technologies from ultrasound to pre-implantation genetic diagnosis, whether a child is likely to be born with a disease or disability that might lessen her life chances. Purdy argues that in cases where we know there is a high risk of a serious genetic disease, it is immoral to reproduce. Her straightforward argument is: (1) there exists a duty to provide all children with a normal opportunity for a good life; (2) we do not harm possible children by not letting them exist; (3) the duty to provide for a normal opportunity for a good life for one's children takes precedence over parents' right to reproduce; (4) therefore, use prenatal screening and abortion or do not conceive.

Leon Kass, Addie Clark Harding Professor in the Committee on Social Thought and the College at the University of Chicago and Chair of the President's Council on Bioethics from 2001 to 2005, makes use of many references to Aldous Huxley's *Brave New World* to develop an indictment of all efforts to clone human beings, and of investigations into human genetic technologies in general. In this essay, Kass advocates for both national and international regulations to prevent such cloning.

He offers four specific objections to cloning. (1) Human cloning involves unethical experiments, because most cloning experiments result in fetal deaths or the birth of deformed infants. (2) It threatens identity and individuality. A clone would be like a twin to its parent/sibling creator, and likely would be burdened with expectations of a life already lived rather than a new life yet to come. (3) Cloning would transform the making of a human being into the building of a manufactured product. This creates an unequal relationship between parent and clone. (4) Parents of clones likely would have an unreasonable set of expectations for their children. If created from someone gifted, such as a famous athlete, the clone would live his or her entire life with comparisons to the person from whom genetic material was harvested. The upshot of all of these considerations is that Kass views human reproductive cloning, along with other forms of genetic research, and existing technologies such as the Pre-implantation Genetic Diagnosis essential to realize Purdy's conclusion, as an instrumentalization of human life. In this, he recalls earlier humanities philosophies of technology, fearing that we are making a dystopia in which what is good and valuable about human beings will be engineered away.

Inmaculada de Melo-Martín, currently Associate Professor of Medical Ethics at Weill Cornell Medical College, and Marin Gillis, Director of Medical Humanities and Ethics, University of Nevada School of Medicine, are bioethicists who argue that good science requires good ethics, and also that good ethics requires good science. On both counts, they find much of the existing literature lacking. In this essay, they examine arguments for and against stem-cell research, and find all of them lacking. Their essay is, thus, a rejoinder to both Purdy and Kass.

Turning to food biotechnologies, Nina V. Federoff, Evan Pugh Professor of Biology and Willaman Professor of Life Sciences, Biology Department at the Pennsylvania State University, and independent historian and writer Nancy Marie Brown advocate for the development of genetically modified foods. In this concluding chapter from their book, *Mendel in the Kitchen*, they argue that genetically modified foods can address many existing and looming problems, such as: malnutrition, feeding growing populations, dependence on chemical (and environmentally damaging) fertilizers, and water consumption. They present a value-neutral position in arguing that whether this technology turns out to be a force for good or ill depends on the choices people make.

In this essay, Paul Thompson, W. K. Kellogg Chair in Agricultural, Food and Community Ethics at Michigan State University, identifies some of the philosophical issues and value judgments associated with the claim that risks from transgenic and conventional crops are comparable from a scientific perspective in the US. He notes that from the introduction of transgenic crops, advocates have argued that the risks are comparable to risks from conventional crops, and hence evaluations of new food technologies should be pursued in exactly the same fashion as the evaluations of previous crop technologies. As such, the discussion about risks associated with transgenic crops seems a clear example of the sorts of ethical evaluations cited by Shrader-Frechette above. Thompson discusses techniques in crop production and provides an assessment of the environmental risks of genetically modified and conventional crops. He argues that even when different groups agree to terms used in the evaluations – "Hazard identification, exposure modeling, and comparison populations," for instance – the meanings of these terms is not fixed, and is not determinable by purely scientific means.

23

How Splendid Technologies Can Go Wrong

Daniel Callahan

I begin with a brief account of my ordinary workday. Professionally I am engrossed in matters of national health policy, and particularly the growing cost of health care, once again into double digit inflation. Those costs are increasing the number of uninsured, putting great pressure on the federal programs of Medicare and Medicaid, and threatening employer-provided health insurance. What is to be done about that?

To get to my place of work, up the Hudson River, I cross the Tappan Zee Bridge, which is choked with cars, well beyond its original projected capacity, and fed by a highway that features daily traffic jams. What is to be done about that?

The main source of rising health care costs is the emergence of new technologies and the intensified use of old technologies, accounting for some 40 percent of the annual increase. The main source of the traffic problem, here and elsewhere, is simply too many cars, ever increasing in number. Not many people see analogies between health care technologies and automobiles, but there is nothing like sitting in a traffic jam to expand one's imagination. The problems are more alike than anyone might guess.

Both problems raise two familiar questions, now in a new guise. One is ancient: how should we cope with those features of our lives, and thus of our human nature, where what is good gradually turns into something bad, usually inadvertently? Medical technology is an uncommon human benefit, but when its pursuit and deployment begin to create economic fits, public and private, its good begins to turn bad. The automobile increases freedom and mobility, and like medical technology is a source of enormous national economic benefit. But when our air is besmogged, our highways jammed, and our commutes a misery, its good turns bad.

The second question my two illustrative technologies raise is this: is technology simply a neutral reality, neither good nor bad in itself, but to be evaluated solely on the basis of the uses to which it is put – or does it have a life of its own, subtly shaping our values and ways of life whether we choose that to happen or not? As opponents of gun control are fond of saying "Guns don't kill; only people kill." That may be narrowly true, but is it the full truth? Is it irrelevant for our safety how many guns we have in our house? And are we fully free – the autonomous creatures we are alleged to be – to use or not use medical technology as we like, or to drive or not drive an automobile, or does each technology draw us to its use in ways often beyond our control?

From *The Hastings Center Report*, vol. 33, no. 2 (March–April 2003): 19–22. Reprinted by permission of The Hastings Center. Reprinted by permission of Dan Callahan.

While there are some people who cleverly manage to avoid the health care system altogether, putting their faith in prayer or austere living or exotic jungle potions, most of us can't get away with that. When we hurt we go to the doctor, and doctors these days turn to technology to diagnose us and to treat us. We expect no less. And while there are some strange creatures in Manhattan who have never learned to drive, few of us have successfully evaded the American car culture.

In short, medical technology and the automobile are social realities most of us cannot imagine living without. In each case we can preach moderation; in each, moderation has proved hard to come by. In each case, the cumulative social impact of the technologies seems beyond our control, in great part precisely because it is a good that we cannot easily do without and which (of its very nature?) leads us to want more of it than we now have.

The Ends of Technology

Technology responds to some deep and enduring human needs: for survival (agriculture, defense); for increased choice, pleasure, and convenience (telephones, movies, escalators); for economic benefit (computers and software, machine tools, airplanes for export); for liberating visions of human possibility (space travel, instant world-wide communication). Medical technology and the automobile have many if not most of these attractions, and serve most of those needs. And because they serve multiple needs in multiply complex, overlapping ways, their grip on our culture is extraordinarily powerful. Medical technology serves:

- our survival (forestalling death, relieving our morbidities, and softening our disabilities);
- our choice and convenience (with contraception and prenatal diagnosis);
- our economy (jobs, investments, exportable medical equipment and drugs); and
- our dreams of a better, more liberated life (through genetic engineering and psychopharmacology, for example).

No wonder our culture swallows medicine whole. Yet I believe that equity and technological progress are on a profound collision course. The recent rise of health care costs, beginning in the late 1990s, tells the story – depending upon how we look at it. It is tempting to see the problem as strictly technical, solvable by better management, more effective cost control methods, evidence-based medicine, or market solutions that force people to make decisions about what they are willing to pay for. But at base there is a fundamental clash of long-standing values: the golden ideal of unlimited medical and technological progress, at the core of American health care values, versus the social ideal of universal coverage of health care costs, guaranteeing every citizen decent health care at an affordable price. The trajectory of health care costs is rapidly undermining the ideal of affordable, equitable health care, putting universal care in the United States further and further from our grasp.

In thinking about this tension consider a few economic and demographic realities: 40 million uninsured Americans, with the number going up; $1.3 trillion in health care expenditures, the highest in the world and the highest per capita; an increasing array of expensive technologies and drugs; and the likelihood that the new genetic-based drugs will be even more expensive than present drugs; inflationary pressures of 10–20 percent per year in health care costs over the past couple of years, with no prospect of change in that pattern; an aging population and the imminent retirement of the baby boom generation, likely to push costs even higher; and rising public demands for the latest and the best in medical technology – as the technologies improve, the standards of good health care rise in lock step.

If we are tempted to think that our American problem is unique, it is worth glancing at European and Canadian health care. Those countries long ago embraced universal health care, and for a long time were able to afford it. But they are now hanging on by their nails, subject to the same cost pressures we are. Bit by bit, universal health care in other countries is being eroded: the queues get longer, the out-of-pocket payments increase, denials of care less than urgent become more common. Just as we have, they have tried the efficiency, cut-the-fat route, embraced evidence-based medicine, and even toyed with the idea of health maintenance organizations. Those are all worthy and necessary efforts, but as in the United States, they have not worked well to control costs.

Consider now the automobile. While it does not meet our need for physical survival in quite the way medical technology does, it has proved itself necessary for our:

- social existence (getting where we need to go);
- choice and convenience (going where we want to go, when we want to go there);
- economy (long our leading industry); and
- vision of a better life (living and going where we like).

Yet as with health care, the social and economic costs mount. Some eight million autos were manufactured in 1980, and twelve million in 1999. Americans drove one trillion miles annually in the 1970s, and two trillion by 1999. More is spent on transportation and related costs in the United States than on medical care, education, and clothing combined. The poor can hardly exist without a car to find decent work, and enough has been written about the urban sprawl problem, a product of the car, that I will say nothing more about it.

There is no shortage of solutions to the automobile, but they all have problems. Building more roads is likely to attract more automobiles, eventually using up the initial gains. Improving auto mileage runs into industry opposition. Public transportation is expensive and might not lure people from their autos. Tax incentives and disincentives to control driving patterns might help, but probably won't make a dramatic difference. As with health reform efforts, auto reforms are necessary and sensible – but their chance of putting much of a dent in the overall problem is not encouraging.

What ones sees when comparing medical technology and the automobile is that both serve functions other than those most closely related to their formal ends – health for health care and mobility for the automobile. Those other functions include their economic benefits and their capacity to evoke liberating visions. That combination makes them a far more potent force than if their formal function only was at stake. But those formal functions have proved

more than sufficient to give both of them a seemingly unstoppable social force.

It appears, in sum, that both technologies have built within them the capacity to endlessly escalate our desires and raise the baseline of acceptability. Where it was once possible to speak of "the family car," an increasing number of families have individual cars for each person; upscale houses now routinely feature three-car garages. Even if decent public transportation is available, people still want automobiles, and more automobiles.

Here one encounters a puzzle. Is the proliferation of automobiles the result of industry pressures and advertising, or would the desire for cars be there even if those pressures were diminished? I don't know, but my guess is that the natural attraction of the automobile – "natural" in the sense that most adults in most countries will want one – is simply enhanced by advertising and by the symbolic role of the automobile as a token of social status.

One can imagine a limit to the number of cars that the United States might have. If every individual over age sixteen already owned an automobile, then, except for population growth, the number of automobiles would surely level off. And traffic problems could grow so horrendous that people would be fearful of using cars. But one can also imagine the automobile surmounting these limits. Families might acquire more cars than household members, and people might simply become habituated to daily traffic jams – treating them much as they treat the weather, where everyone complains but nobody does anything about it. Everything will then get worse, and we will say it is the price we pay for the benefits.

The situation is far worse with health care. There, the need for good health is in principle unlimited. Our bodies are finite – subject to injury, disease, decline, and death. The best that medicine can do is to ward off those evils, prevent some diseases, rescue us from others, rehabilitate us from still others, and put death off for a time. Medicine can take just pride in the thirty-year increase of average life expectancy in the twentieth century and the rapidly growing number of people living into their 90s and 100s. But eventually something or other gets us.

However great the improvement in health, the doctor's office will still be full and the hospital still a going concern. Put another way, however great medical progress has been, it eventually runs out. We may not encounter the frontier of progress, of medical possibilities, until we are 100, but there is an endless frontier.

With medical technology, it is seemingly impossible to envision some ultimate saturation point. In addition to the endless possibilities of lengthening and improving the human life span, medical research and technological innovation themselves stimulate ideas for even greater future improvements; and with those improvements come greater expectations about what counts as a decent level of health. Scientific progress spurs expectation progress, with the bar always set higher once the earlier goals are achieved. The annual cry during appropriation hearings for the annual National Institute of Health is that the research promise and possibilities have never been greater; and that is always, in some sense, true. The more we know, the easier it is to know still more.

Going Too Far

It is easy to understand why we can go too far with the automobile and medical research. In medicine, there is considerable evidence that there is often a "technological imperative" at work, making it hard for physicians, patients, and families to stop treatment with a terminal patient no longer able to benefit from the treatment. So, too, it is possible for individuals to be induced to buy automobiles more expensive than they can afford, or to buy two automobiles when one would do quite nicely.

Medical technology seems to know no boundaries because it is hard to say just what bodily failings and lethal threats we should be willing to accept. The medical research agenda now goes after all lethal diseases, but it also goes after human enhancements and wish fulfillment. Death itself is made to seem an accidental, contingent event. Why do we now die? Not because of the inherent finitude of the human body, as most people thought for most of human history. We die, it is said, because we engage in unhealthy lifestyles, or because research has not yet found a cure for our diseases. And many doctors, when they lose a patient, feel that it is somehow their fault,

even if their brain and their colleagues tell them otherwise. For many of our citizens, health has become not simply a necessary means for living a good life, but itself one of the ends of a good life. Medical technology, like the automobile, is imbedded in our culture. It is part of our picture of modern America.

I own a summer house on a small Maine island. When I first went there thirty years ago, there were no more than twenty cars or so. Now there are close to sixty, although the population has not much changed. The island is small and fully walkable; and not long ago everyone did walk, even when they could have driven. The islanders of thirty years ago believed that one of the charms and assets of the island was the absence of cars; and there seems to have been a built-in, culturally transmissible, taboo – a self-imposed limit that kept the cars out. But once the trend toward cars got started, the earlier limits transgressed, little was to stand in their way.

Technology as a Neutral Force

Given the power of medical technology and the automobile to have such a hold on our lives, I find it hard to say that they are merely neutral tools, to be used or not used as we see fit, just lying about. Except for their pejorative connotations, I find technologies to be much like viruses or germs. They are external agents that can invade our bodies and make a great difference in our fate and health.

But that is to speak in a general way. We also know that even in the worst plagues, not everyone perishes. The social context and the character of individual bodies affect the outcome. This is why we often think of viruses as neutral: they do not behave in any wholly independent, context-free way. Of course, common sense dictates that we treat them with care, avoiding them when possible, but no one is guaranteed to get sick from contact with them.

Medical technology and the automobile are similar. Not everyone will, so to speak, invariably be infected by them, though both are hard to avoid. And just as some diseases that are medical threats to individuals also confer some genetic protection, so too medical technology and the automobile have many valuable traits. Our health depends on the e-coli bacteria in our gut; but outside of that setting, let loose in the world, e-coli can be deadly. I once suggested at a meeting at the Centers for Disease Control and Prevention (CDC) that, in light of the national obesity problem, perhaps a Surgeon General's warning should be affixed to every automobile and TV set, warning of their health dangers if used to excess. Everyone laughed except the director of the CDC. He said that he did not think that was a politically promising way to go, and I came away with the impression that even as a joke it was impermissible. Similarly, when I even suggest to most audiences, much less to research advocates, that we might want to rethink and even slow down and redirect innovation in medical technology, I am guaranteed a chilly, shocked reaction.

In sum, if in some literal sense technology is neutral, in the important sense of the way it affects human lives it is anything but neutral. When the technology is ubiquitous, when it serves important human values and ways of life, and when it is all but impossible to avoid using, then it has captured our lives. Most new technologies are introduced not as enslaving tyrants, but as choice-increasing, society-enhancing developments that we are free to take or leave. But as the rise and dominance of the automobile and medical technology show, freedom and choice can be fleeting when the technology takes up a central place in our cultural gut and we require them as much as our private gut requires e-coli.

I will conclude by noting that with both technologies there are alternatives. A few years ago in Prague I was talking with a doctor who often had to spend her weekends at the hospital where she worked. She always took the tram rather than driving the family car. When I pointed out that it was a ten-minute ride by car, with plentiful parking available, but forty-five minutes by tram, she looked at me in an uncomprehending way and said, in effect, that she could not understand why anyone would drive anywhere when public transportation was available.

In the case of medical technology, it is well known that most of the improvements in health status have come from public health measures and improved socioeconomic status. Much more research is needed on the background conditions and determinants of health, but the big research money goes to genetic and other biomedical research, looking into the depths of biology for cures rather than into the breadth of human societies to determine why some people get sick and others do not. This latter approach – often called a population rather than individual perspective – has powerful theoretical support. What it does not have is glamour, an economic lobby, and an intuitive appeal equivalent to the development of new technologies.

The difficulty with these alternative strategies with the automobile and medical technologies is that they have not sunk into our public and private psyche with the power of the technologies. They do not have behind them the overlapping strands of attraction and profit that the technologies do. They do not – to recur to my language above – seem to have the power to change our way of life, to bury themselves so deeply within our way of life that, however much we worry and complain about their harms, we cannot let go of the good and goods they bring.

Could all of that change? Maybe. It may be that the health care cost problem will become so bad that the now well-insured middle class will begin to hurt, and will consider some serious alternatives. It may be that the traffic jams and commutes will grow so long that the public will revolt against the hegemony of the automobile. In other words, for those of us looking for a change, probably the best we can hope for is a nasty crisis that will *force* a change. But it's not likely to happen. As those of us who have longed for universal health care for many decades long ago learned, the capacity of this country to muddle through what in other places would seem a great crisis is formidable.

24

Genetics and Reproductive Risk: Can Having Children be Immoral?

Laura M. Purdy

Is it morally permissible for me to have children? A decision to procreate is surely one of the most significant decisions a person can make. So it would seem that it ought not be made without some moral soul-searching.

There are many reasons why one might hesitate to bring children into this world if one is concerned about their welfare. Some are rather general, such as the deteriorating environment or the prospect of poverty. Others have a narrower focus, such as continuing civil war in one's country or the lack of essential social support for child-rearing in the United States. Still others may be relevant only to individuals at risk of passing harmful diseases to their offspring.

There are many causes of misery in this world, and most of them are unrelated to genetic disease. In the general scheme of things, human misery is most efficiently reduced by concentrating on noxious social and political arrangements. Nonetheless, we should not ignore preventable harm just because it is confined to a relatively small corner of life. So the question arises, Can it be wrong to have a child because of genetic risk factors?[1]

Unsurprisingly, most of the debate about this issue has focused on prenatal screening and abortion: much useful information about a given fetus can be made available by recourse to prenatal testing. This fact has meant that moral questions about reproduction have become entwined with abortion politics, to the detriment of both. The abortion connection has made it especially difficult to think about whether it is wrong to prevent a child from coming into being, because doing so might involve what many people see as wrongful killing; yet there is no necessary link between the two. Clearly, the existence of genetically compromised children can be prevented not only by aborting already existing fetuses but also by preventing conception in the first place.

Worse yet, many discussions simply assume a particular view of abortion without recognizing other possible positions and the difference they make in how people understand the issues. For example, those who object to aborting fetuses with genetic problems often argue that doing so would undermine our conviction that all humans are in some important sense equal.[2] However, this position rests on the assumption that conception marks the point at which humans are endowed with a right to life. So aborting fetuses with genetic problems looks morally the same as killing "imperfect" people without their consent.

From Laura M. Purdy, *Reproducing Persons: Issues in Feminist Bioethics* (Ithaca, NY: Cornell University Press, 1996), pp. 39–49. © 1996 by Cornell University. Used by permission of the publisher, Cornell University Press.

This position raises two separate issues. One pertains to the legitimacy of different views on abortion. Despite the conviction of many abortion activists to the contrary, I believe that ethically respectable views can be found on different sides of the debate, including one that sees fetuses as developing humans without any serious moral claim on continued life. There is no space here to address the details, and doing so would be once again to fall into the trap of letting the abortion question swallow up all others. However, opponents of abortion need to face the fact that many thoughtful individuals do *not* see fetuses as moral persons. It follows that their reasoning process, and hence the implications of their decisions, are radically different from those envisioned by opponents of prenatal screening and abortion. So where the latter see genetic abortion as murdering people who just don't measure up, the former see it as a way to prevent the development of persons who are more likely to live miserable lives, a position consistent with a world-view that values persons equally and holds that each deserves a high-quality life. Some of those who object to genetic abortion appear to be oblivious to these psychological and logical facts. It follows that the nightmare scenarios they paint for us are beside the point: many people simply do not share the assumptions that make them plausible.

How are these points relevant to my discussion? My primary concern here is to argue that conception can sometimes be morally wrong on grounds of genetic risk, although this judgment will not apply to those who accept the moral legitimacy of abortion and are willing to employ prenatal screening and selective abortion. If my case is solid, then those who oppose abortion must be especially careful not to conceive in certain cases, as they are, of course, free to follow their conscience about abortion. Those like myself who do not see abortion as murder have more ways to prevent birth.

Huntington's Disease

There is always some possibility that reproduction will result in a child with a serious disease or handicap. Genetic counselors can help individuals determine whether they are at unusual risk and,

as the Human Genome Project rolls on, their knowledge will increase by quantum leaps. As this knowledge becomes available, I believe we ought to use it to determine whether possible children are at risk *before* they are conceived.

In this chapter I want to defend the thesis that it is morally wrong to reproduce when we know there is a high risk of transmitting a serious disease or defect. This thesis holds that some reproductive acts are wrong, and my argument puts the burden of proof on those who disagree with it to show why its conclusions can be overridden. Hence it denies that people should be free to reproduce mindless of the consequences.[3] However, as moral argument, it should be taken as a proposal for further debate and discussion. It is not, by itself, an argument in favor of legal prohibitions of reproduction.[4]

There is a huge range of genetic diseases. Some are quickly lethal; others kill more slowly, if at all. Some are mainly physical, some mainly mental; others impair both kinds of function. Some interfere tremendously with normal functioning, others less. Some are painful, some are not. There seems to be considerable agreement that rapidly lethal diseases, especially those, such as Tay-Sachs, accompanied by painful deterioration, should be prevented even at the cost of abortion. Conversely, there seems to be substantial agreement that relatively trivial problems, especially cosmetic ones, would not be legitimate grounds for abortion.[5] In short, there are cases ranging from low risk of mild disease or disability to high risk of serious disease or disability. Although it is difficult to decide where the duty to refrain from procreation becomes compelling, I believe that there are some clear cases. I have chosen to focus on Huntington's Disease to illustrate the kinds of concrete issues such decisions entail. However, the arguments are also relevant to many other genetic diseases.[6]

The symptoms of Huntington's Disease usually begin between the ages of 30 and 50:

> Onset is insidious. Personality changes (obstinacy, moodiness, lack of initiative) frequently antedate or accompany the involuntary choreic movements. These usually appear first in the face, neck, and arms, and are jerky, irregular, and stretching in character. Contractions of the facial muscles result in grimaces; those of the

respiratory muscles, lips, and tongue lead to hesitating, explosive speech. Irregular movements of the trunk are present; the gait is shuffling and dancing. Tendon reflexes are increased. . . . Some patients display a fatuous euphoria; others are spiteful, irascible, destructive, and violent. Paranoid reactions are common. Poverty of thought and impairment of attention, memory, and judgment occur. As the disease progresses, walking becomes impossible, swallowing difficult, and dementia profound. Suicide is not uncommon.[7]

The illness lasts about fifteen years, terminating in death.

Huntington's Disease is an autosomal dominant disease, meaning it is caused by a single defective gene located on a non-sex chromosome. It is passed from one generation to the next via affected individuals. Each child of such an affected person has a 50 percent risk of inheriting the gene and thus of eventually developing the disease, even if he or she was born before the parent's disease was evident.[8]

Until recently, Huntington's Disease was especially problematic because most affected individuals did not know whether they had the gene for the disease until well into their childbearing years. So they had to decide about childbearing before knowing whether they could transmit the disease or not. If, in time, they did not develop symptoms of the disease, then their children could know they were not at risk for the disease. If unfortunately they did develop symptoms, then each of their children could know there was a 50 percent chance that they too had inherited the gene. In both cases, the children faced a period of prolonged anxiety as to whether they would develop the disease. Then, in the 1980s, thanks in part to an energetic campaign by Nancy Wexler, a genetic marker was found that, in certain circumstances, could tell people with a relatively high degree of probability whether or not they had the gene for the disease.[9] Finally, in March 1993, the defective gene itself was discovered.[10] Now individuals can find out whether they carry the gene for the disease, and prenatal screening can tell us whether a given fetus has inherited it. These technological developments change the moral scene substantially.

How serious are the risks involved in Huntington's Disease? Geneticists often think a 10 percent risk is high.[11] But risk assessment also depends on what is at stake: the worse the possible outcome, the more undesirable an otherwise small risk seems. In medicine, as elsewhere, people may regard the same result quite differently. But for devastating diseases such as Huntington's this part of the judgment should be unproblematic: no one wants a loved one to suffer in this way.[12]

There may still be considerable disagreement about the acceptability of a given risk. So it would be difficult in many circumstances to say how we should respond to a particular risk. Nevertheless, there are good grounds for a conservative approach, for it is reasonable to take special precautions to avoid very bad consequences, even if the risk is small. But the possible consequences here *are* very bad: a child who may inherit Huntington's Disease has a much greater than average chance of being subjected to severe and prolonged suffering. And it is one thing to risk one's own welfare, but quite another to do so for others and without their consent.

Is this judgment about Huntington's Disease really defensible? People appear to have quite different opinions. Optimists argue that a child born into a family afflicted with Huntington's Disease has a reasonable chance of living a satisfactory life. After all, even children born of an afflicted parent still have a 50 percent chance of escaping the disease. And even if afflicted themselves, such people will probably enjoy some thirty years of healthy life before symptoms appear. It is also possible, although not at all likely, that some might not mind the symptoms caused by the disease. Optimists can point to diseased persons who have lived fruitful lives, as well as those who seem genuinely glad to be alive. One is Rick Donohue, a sufferer from the Joseph family disease: "You know, if my mom hadn't had me, I wouldn't be here for the life I have had. So there is a good possibility I will have children."[13] Optimists therefore conclude that it would be a shame if these persons had not lived.

Pessimists concede some of these facts but take a less sanguine view of them. They think a 50 percent risk of serious disease such as Huntington's is appallingly high. They suspect that many children born into afflicted families are liable to spend their youth in dreadful anticipation and fear of the disease. They expect that the disease, if it appears, will be perceived as a tragic and painful end to a blighted life. They point out that Rick Donohue is still young and has not experienced the full horror of his sickness. It is also well-known that some young persons have such a dilated sense of time that they can hardly envision themselves at 30 or 40, so the prospect of pain at that age is unreal to them.[14]

More empirical research on the psychology and life history of suffers and potential sufferers is clearly needed to decide whether optimists or pessimists have a more accurate picture of the experiences of individuals at risk. But given that some will surely realize pessimists' worst fears, it seems unfair to conclude that the pleasures of those who deal best with the situation simply cancel out the suffering of those others when that suffering could be avoided altogether.

I think that these points indicate that the morality of procreation in such situations demands further investigation. I propose to do this by looking first at the position of the possible child, then at that of the potential parent.

Possible Children and Potential Parents

The first task in treating the problem from the child's point of view is to find a way of referring to possible future offspring without seeming to confer some sort of morally significant existence on them. I follow the convention of calling children who might be born in the future but who are not now conceived "possible" children, offspring, individuals, or persons.

Now, what claims about children or possible children are relevant to the morality of childbearing in the circumstances being considered? Of primary importance is the judgment that we ought to try to provide every child with something like a minimally satisfying life. I am not altogether sure how best to formulate this standard, but I want clearly to reject the view that it is morally permissible to conceive individuals so long as we do not expect them to be so miserable that they wish they were dead.[15] I believe that this kind of moral minimalism is thoroughly unsatisfactory and that not many people would really want to live

in a world where it was the prevailing standard. Its lure is that it puts few demands on us, but its price is the scant attention it pays to human well-being.

How might the judgment that we have a duty to try to provide a minimally satisfying life for our children be justified? It could, I think, be derived fairly straightforwardly from either utilitarian or contractarian theories of justice, although there is no space here for discussion of the details. The net result of such analysis would be to conclude that neglecting this duty would create unnecessary unhappiness or unfair disadvantage for some persons.

Of course, this line of reasoning confronts us with the need to spell out what is meant by "minimally satisfying" and what a standard based on this concept would require of us. Conceptions of a minimally satisfying life vary tremendously among societies and also within them. De rigueur in some circles are private music lessons and trips to Europe, whereas in others providing eight years of schooling is a major accomplishment. But there is no need to consider this complication at length here because we are concerned only with health as a prerequisite for a minimally satisfying life. Thus, as we draw out what such a standard might require of us, it seems reasonable to retreat to the more limited claim that parents should try to ensure something like normal health for their children. It might be thought that even this moderate claim is unsatisfactory as in some places debilitating conditions are the norm, but one could circumvent this objection by saying that parents ought to try to provide for their children health normal for that culture, even though it may be inadequate if measured by some outside standard.[16] This conservative position would still justify efforts to avoid the birth of children at risk for Huntington's Disease and other serious genetic diseases in virtually all societies.[17]

This view is reinforced by the following considerations. Given that possible children do not presently exist as actual individuals, they do not have a right to be brought into existence, and hence no one is maltreated by measures to avoid the conception of a possible person. Therefore, the conservative course that avoids the conception of those who would not be expected to enjoy a minimally satisfying life is at present the only fair course of action. The alternative is a *laissez-faire* approach that brings into existence the lucky, but only at the expense of the unlucky. Notice that attempting to avoid the creation of the unlucky does not necessarily lead to *fewer* people being brought into being; the question boils down to taking steps to bring those with better prospects into existence, instead of those with worse ones.

I have so far argued that if people with Huntington's Disease are unlikely to live minimally satisfying lives, then those who might pass it on should not have genetically related children. This is consonant with the principle that the greater the danger of serious problems, the stronger the duty to avoid them. But this principle is in conflict with what people think of as the right to reproduce. How might one decide which should take precedence?

Expecting people to forgo having genetically related children might seem to demand too great a sacrifice of them. But before reaching that conclusion we need to ask what is really at stake. One reason for wanting children is to experience family life, including love, companionship, watching kids grow, sharing their pains and triumphs, and helping to form members of the next generation. Other reasons emphasize the validation of parents as individuals within a continuous family line, children as a source of immortality, or perhaps even the gratification of producing partial replicas of oneself. Children may also be desired in an effort to prove that one is an adult, to try to cement a marriage, or to benefit parents economically.

Are there alternative ways of satisfying these desires? Adoption or new reproductive technologies can fulfill many of them without passing on known genetic defects. Sperm replacement has been available for many years via artificial insemination by donor. More recently, egg donation, sometimes in combination with contract pregnancy,[18] has been used to provide eggs for women who prefer not to use their own. Eventually it may be possible to clone individual humans, although that now seems a long way off. All of these approaches to avoiding the use of particular genetic material are controversial and have generated much debate. I believe that tenable moral versions of each do exist.[19]

None of these methods permits people to extend both genetic lines or realize the desire for immortality or for children who resemble both parents; nor is it clear that such alternatives will necessarily succeed in proving that one is an adult, cementing a marriage, or providing economic benefits. Yet, many people feel these desires strongly. Now, I am sympathetic to William James's dictum regarding desires: "Take any demand, however slight, which any creature, however weak, may make. Ought it not, for its own sole sake be satisfied? If not, prove why not."[20] Thus a world where more desires are satisfied is generally better than one where fewer are. However, not all desires can be legitimately satisfied, because as James suggests, there may be good reasons, such as the conflict of duty and desire, why some should be overruled.

Fortunately, further scrutiny of the situation reveals that there are good reasons why people should attempt with appropriate social support to talk themselves out of the desires in question or to consider novel ways of fulfilling them. Wanting to see the genetic line continued is not particularly rational when it brings a sinister legacy of illness and death. The desire for immortality cannot really be satisfied anyway, and people need to face the fact that what really matters is how they behave in their own lifetimes. And finally, the desire for children who physically resemble one is understandable, but basically narcissistic, and its fulfillment cannot be guaranteed even by normal reproduction. There are other ways of proving one is an adult, and other ways of cementing marriages – and children don't necessarily do either. Children, especially prematurely ill children, may not provide the expected economic benefits anyway. Nongenetically related children may also provide benefits similar to those that would have been provided by genetically related ones, and expected economic benefit is, in many cases, a morally questionable reason for having children.

Before the advent of reliable genetic testing, the options of people in Huntington's families were cruelly limited. On the one hand, they could have children, but at the risk of eventual crippling illness and death for them. On the other, they could refrain from child-bearing, sparing their possible children from significant risk of inheriting this disease, perhaps frustrating intense desires to procreate – only to discover, in some cases, that their sacrifice was unnecessary because they did not develop the disease. Or they could attempt to adopt or try new reproductive approaches.

Reliable genetic testing has opened up new possibilities. Those at risk who wish to have children can get tested. If they test positive, they know their possible children are at risk. Those who are opposed to abortion must be especially careful to avoid conception if they are to behave responsibly. Those not opposed to abortion can responsibly conceive children, but only if they are willing to test each fetus and abort those who carry the gene. If individuals at risk test negative, they are home free.

What about those who cannot face the test for themselves? They can do prenatal testing and abort fetuses who carry the defective gene. A clearly positive test also implies that the parent is affected, although negative tests do not rule out that possibility. Prenatal testing can thus bring knowledge that enables one to avoid passing the disease to others, but only, in some cases, at the cost of coming to know with certainty that one will indeed develop the disease. This situation raises with peculiar force the question of whether parental responsibility requires people to get tested.

Some people think that we should recognize a right "not to know." It seems to me that such a right could be defended only where ignorance does not put others at serious risk. So if people are prepared to forgo genetically related children, they need not get tested. But if they want genetically related children, then they must do whatever is necessary to ensure that affected babies are not the result. There is, after all, something inconsistent about the claim that one has a right to be shielded from the truth, even if the price is to risk inflicting on one's children the same dread disease one cannot even face in oneself.

In sum, until we can be assured that Huntington's Disease does not prevent people from living a minimally satisfying life, individuals at risk for the disease have a moral duty to try not to bring affected babies into this world. There are now enough options available so that this duty needn't frustrate their reasonable desires. Society has a corresponding duty to facilitate moral behavior on the part of individuals. Such support ranges from the narrow and concrete (such as making sure that medical testing and counseling is available to all) to the more general social environment that guarantees that all pregnancies are voluntary, that pronatalism is eradicated, and that women are treated with respect regardless of the reproductive options they choose.

Notes

1   I focus on genetic considerations, although with the advent of AIDS the scope of the general question here could be expanded. There are two reasons for sticking to this relatively narrow formulation. One is that dealing with a smaller chunk of the problem may help us to think more clearly, while realizing that some conclusions may nonetheless be relevant to the larger problem. The other is the peculiar capacity of some genetic problems to affect ever more individuals in the future.

2   For example, see Leon Kass, "Implications of Prenatal Diagnosis for the Human Right to Life," in *Ethical Issues in Human Genetics*, ed. Bruce Hilton et al. (New York: Plenum, 1973).

3   This is, of course, a very broad thesis. I defend an even broader version in ch. 2 of *Reproducing Persons*, "Loving Future People."

4   Why would we want to resist legal enforcement of every moral conclusion? First, legal action has many costs, costs not necessarily worth paying in particular cases. Second, legal enforcement tends to take the matter out of the realm of debate and treat it as settled. But in many cases, especially where mores or technology are rapidly evolving, we don't want that to happen. Third, legal enforcement would undermine individual freedom and decision-making capacity. In some cases, the ends envisioned are important enough to warrant putting up with these disadvantages.

5   Those who do not see fetuses as moral persons with a right to life may nonetheless hold that abortion is justifiable in these cases. I argue at some length elsewhere that lesser defects can cause great suffering. Once we are clear that there is nothing discriminatory about failing to conceive particular possible individuals, it makes sense, other things being equal, to avoid the prospect of such pain if we can. Naturally, other things rarely are equal. In the first place, many problems go undiscovered until a baby is born. Second, there are often substantial costs associated with screening programs. Third, although women should be encouraged to consider the moral dimensions of routine pregnancy, we do not want it to be so fraught with tension that it becomes a miserable

experience. (See ch. 2 of *Reproducing Persons*, "Loving Future People.")

6 It should be noted that failing to conceive a single individual can affect many lives: in 1916, 962 cases could be traced from six seventeenth-century arrivals in America. See Gordon Rattray Taylor, *The Biological Time Bomb* (New York: Penguin, 1968), p. 176.

7 *The Merck Manual* (Rahway, NJ: Merck, 1972), pp. 1363, 1346. We now know that the age of onset and severity of the disease are related to the number of abnormal replications of the glutamine code on the abnormal gene. See Andrew Revkin, "Hunting Down Huntington's," *Discover* (December 1993): 108.

8 Hymie Gordon, "Genetic Counseling," *JAMA*, 217, no. 9 (August 30, 1971): 1346.

9 See Revkin, "Hunting Down Huntington's," 99–108.

10 "Gene for Huntington's Disease Discovered," *Human Genome News*, no. 1 (May 1993): 5.

11 Charles Smith, Susan Holloway, and Alan E. H. Emery, "Individuals at Risk in Families – Genetic Disease," *Journal of Medical Genetics*, 8 (1971): 453.

12 To try to separate the issue of the gravity of the disease from the existence of a given individual, compare this situation with how we would assess a parent who neglected to vaccinate an existing child against a hypothetical viral version of Huntington's.

13 *The New York Times* (September 30, 1975), p. 1. The Joseph family disease is similar to Huntington's Disease except that symptoms start appearing in the twenties. Rick Donohue was in his early twenties at the time he made this statement.

14 I have talked to college students who believe that they will have lived fully and be ready to die at those ages. It is astonishing how one's perspective changes over time and how ages that one once associated with senility and physical collapse come to seem the prime of human life.

15 The view I am rejecting has been forcefully articulated by Derek Parfit, *Reasons and Persons* (Oxford: Clarendon, 1984). For more discussion, see ch. 2 of *Reproducing Persons*, "Loving Future People."

16 I have some qualms about this response, because I fear that some human groups are so badly off that it might still be wrong for them to procreate, even if that would mean great changes in their cultures. But this is a complicated issue that needs to be investigated on its own.

17 Again, a troubling exception might be the isolated Venezuelan group Nancy Wexler found, where, because of inbreeding, a large proportion of the population is affected by Huntington's. See Revkin, "Hunting Down Huntington's."

18 Or surrogacy, as it has been popularly known. I think that "contract pregnancy" is more accurate and more respectful of women. Eggs can be provided either by a woman who also gestates the fetus or by a third party.

19 The most powerful objections to new reproductive technologies and arrangements concern possible bad consequences for women. However, I do not think that the arguments against them on these grounds have yet shown the dangers to be as great as some believe. So although it is perhaps true that new reproductive technologies and arrangements should not be used lightly, avoiding the conceptions discussed here is well worth the risk. For a series of viewpoints on this issue, including my own "Another Look at Contract Pregnancy" (ch. 12 of *Reproducing Persons*), see Helen B. Holmes, *Issues in Reproductive Technology I: An Anthology* (New York: Garland, 1992).

20 William James, *Essays in Pragmatism*, ed. A. Castell (New York: Hafner, 1948), p. 73.

25

Preventing a Brave New World

Leon Kass

I.

The urgency of the great political struggles of the twentieth century, successfully waged against totalitarianisms first right and then left, seems to have blinded many people to a deeper and ultimately darker truth about the present age: all contemporary societies are travelling briskly in the same utopian direction. All are wedded to the modern technological project; all march eagerly to the drums of progress and fly proudly the banner of modern science; all sing loudly the Baconian anthem, "Conquer nature, relieve man's estate." Leading the triumphal procession is modern medicine, which is daily becoming ever more powerful in its battle against disease, decay, and death, thanks especially to astonishing achievements in biomedical science and technology – achievements for which we must surely be grateful.

Yet contemplating present and projected advances in genetic and reproductive technologies, in neuroscience and psychopharmacology, and in the development of artificial organs and computer-chip implants for human brains, we now clearly recognize new uses for biotechnical power that soar beyond the traditional medical goals of healing disease and relieving suffering.

Human nature itself lies on the operating table, ready for alteration, for eugenic and psychic "enhancement," for wholesale re-design. In leading laboratories, academic and industrial, new creators are confidently amassing their powers and quietly honing their skills, while on the street their evangelists are zealously prophesying a post-human future. For anyone who cares about preserving our humanity, the time has come to pay attention.

Some transforming powers are already here. The Pill. In vitro fertilization. Bottled embryos. Surrogate wombs. Cloning. Genetic screening. Genetic manipulation. Organ harvesting. Mechanical spare parts. Chimeras. Brain implants. Ritalin for the young, Viagra for the old, Prozac for everyone. And, to leave this vale of tears, a little extra morphine accompanied by Muzak.

Years ago Aldous Huxley saw it coming. In his charming but disturbing novel, *Brave New World* (it appeared in 1932 and is more powerful on each rereading), he made its meaning strikingly visible for all to see. Unlike other frightening futuristic novels of the past century, such as Orwell's already dated *Nineteen Eighty-Four*, Huxley shows us a dystopia that goes with, rather than against, the human grain. Indeed, it is animated by our own most humane and

From *The New Republic* (May 21, 2001), pp. 265–76. ©2001 by Leon R. Kass. Reprinted by permission of the author.

progressive aspirations. Following those aspirations to their ultimate realization, Huxley enables us to recognize those less obvious but often more pernicious evils that are inextricably linked to the successful attainment of partial goods.

Huxley depicts human life seven centuries hence, living under the gentle hand of humanitarianism rendered fully competent by genetic manipulation, psychoactive drugs, hypnopaedia, and high-tech amusements. At long last, mankind has succeeded in eliminating disease, aggression, war, anxiety, suffering, guilt, envy, and grief. But this victory comes at the heavy price of homogenization, mediocrity, trivial pursuits, shallow attachments, debased tastes, spurious contentment, and souls without loves or longings. The Brave New World has achieved prosperity, community, stability, and nigh-universal contentment, only to be peopled by creatures of human shape but stunted humanity. They consume, fornicate, take "soma," enjoy "centrifugal bumble-puppy," and operate the machinery that makes it all possible. They do not read, write, think, love, or govern themselves. Art and science, virtue and religion, family and friendship are all passé. What matters most is bodily health and immediate gratification: "Never put off till tomorrow the fun you can have today." Brave New Man is so dehumanized that he does not even recognize what has been lost.

Huxley's novel, of course, is science fiction. Prozac is not yet Huxley's "soma"; cloning by nuclear transfer or splitting embryos is not exactly "Bokanovskification"; MTV and virtual-reality parlors are not quite the "feelies"; and our current safe and consequenceless sexual practices are not universally as loveless or as empty as those in the novel. But the kinships are disquieting, all the more so since our technologies of bio-psycho-engineering are still in their infancy, and in ways that make all too clear what they might look like in their full maturity. Moreover, the cultural changes that technology has already wrought among us should make us even more worried than Huxley would have us be.

In Huxley's novel, everything proceeds under the direction of an omnipotent – albeit benevolent – world state. Yet the dehumanization that he portrays does not really require despotism or external control. To the contrary, precisely because the society of the future will deliver exactly what we most want – health, safety, comfort, plenty, pleasure, peace of mind and length of days – we can reach the same humanly debased condition solely on the basis of free human choice. No need for World Controllers. Just give us the technological imperative, liberal democratic society, compassionate humanitarianism, moral pluralism, and free markets, and we can take ourselves to a Brave New World all by ourselves – and without even deliberately deciding to go. In case you had not noticed, the train has already left the station and is gathering speed, but no one seems to be in charge.

Some among us are delighted, of course, by this state of affairs: some scientists and biotechnologists, their entrepreneurial backers, and a cheering claque of sci-fi enthusiasts, futurologists, and libertarians. There are dreams to be realized, powers to be exercised, honors to be won, and money – big money – to be made. But many of us are worried, and not, as the proponents of the revolution self-servingly claim, because we are either ignorant of science or afraid of the unknown. To the contrary, we can see all too clearly where the train is headed, and we do not like the destination. We can distinguish cleverness about means from wisdom about ends, and we are loath to entrust the future of the race to those who cannot tell the difference. No friend of humanity cheers for a post-human future.

Yet for all our disquiet, we have until now done nothing to prevent it. We hide our heads in the sand because we enjoy the blessings that medicine keeps supplying, or we rationalize our inaction by declaring that human engineering is inevitable and we can do nothing about it. In either case, we are complicit in preparing for our own degradation, in some respects more to blame than the bio-zealots who, however misguided, are putting their money where their mouth is. Denial and despair, unattractive outlooks in any situation, become morally reprehensible when circumstances summon us to keep the world safe for human flourishing. Our immediate ancestors, taking up the challenge of their time, rose to the occasion and rescued the human future from the cruel dehumanizations of Nazi and Soviet tyranny. It is our more difficult task to find ways to preserve it from the soft dehumanizations of well-meaning but hubristic biotechnical "re-creationism" – and to do it without undermining biomedical science or rejecting its genuine contributions to human welfare.

Truth be told, it will not be easy for us to do so, and we know it. But rising to the challenge requires recognizing the difficulties. For there are indeed many features of modern life that will conspire to frustrate efforts aimed at the human control of the biomedical project. First, we Americans believe in technological automatism: where we do not foolishly believe that all innovation is progress, we fatalistically believe that it is inevitable ("If it can be done, it will be done, like it or not"). Second, we believe in freedom: the freedom of scientists to inquire, the freedom of technologists to develop, the freedom of entrepreneurs to invest and to profit, the freedom of private citizens to make use of existing technologies to satisfy any and all personal desires, including the desire to reproduce by whatever means. Third, the biomedical enterprise occupies the moral high ground of compassionate humanitarianism, upholding the supreme values of modern life – cure disease, prolong life, relieve suffering – in competition with which other moral goods rarely stand a chance. ("What the public wants is not to be sick," says James Watson, "and if we help them not to be sick, they'll be on our side.")

There are still other obstacles. Our cultural pluralism and easygoing relativism make it difficult to reach consensus on what we should embrace and what we should oppose; and moral objections to this or that biomedical practice are often facilely dismissed as religious or sectarian. Many people are unwilling to pronounce judgments about what is good or bad, right and wrong, even in matters of great importance, even for themselves – never mind for others or for society as a whole. It does not help that the biomedical project is now deeply entangled with commerce: there are increasingly powerful economic interests in favor of going full steam ahead, and no economic interests in favor of going slow. Since we live in a democracy, moreover, we face political difficulties in gaining a consensus to direct our future, and we have almost no political experience in trying to curtail

the development of any new biomedical technology. Finally, and perhaps most troubling, our views of the meaning of our humanity have been so transformed by the scientific-technological approach to the world that we are in danger of forgetting what we have to lose, humanly speaking.

But though the difficulties are real, our situation is far from hopeless. Regarding each of the aforementioned impediments, there is another side to the story. Though we love our gadgets and believe in progress, we have lost our innocence regarding technology. The environmental movement especially has alerted us to the unintended damage caused by unregulated technological advance, and has taught us how certain dangerous practices can be curbed. Though we favor freedom of inquiry, we recognize that experiments are deeds and not speeches, and we prohibit experimentation on human subjects without their consent, even when cures from disease might be had by unfettered research; and we limit so-called reproductive freedom by proscribing incest, polygamy, and the buying and selling of babies.

Although we esteem medical progress, biomedical institutions have ethics committees that judge research proposals on moral grounds, and, when necessary, uphold the primacy of human freedom and human dignity even over scientific discovery. Our moral pluralism notwithstanding, national commissions and review bodies have sometimes reached moral consensus to recommend limits on permissible scientific research and technological application. On the economic front, the patenting of genes and life forms and the rapid rise of genomic commerce have elicited strong concerns and criticisms, leading even former enthusiasts of the new biology to recoil from the impending commodification of human life. Though we lack political institutions experienced in setting limits on biomedical innovation, federal agencies years ago rejected the development of the plutonium-powered artificial heart, and we have nationally prohibited commercial traffic in organs for transplantation, even though a market would increase the needed supply. In recent years, several American states and many foreign countries have successfully taken political action, making certain practices illegal and placing others under moratoriums (the creation of human embryos solely for research; human germline genetic alteration). Most importantly, the majority of Americans are not yet so degraded or so cynical as to fail to be revolted by the society depicted in Huxley's novel. Though the obstacles to effective action are significant, they offer no excuse for resignation. Besides, it would be disgraceful to concede defeat even before we enter the fray.

Not the least of our difficulties in trying to exercise control over where biology is taking us is the fact that we do not get to decide, once and for all, for or against the destination of a post-human world. The scientific discoveries and the technical powers that will take us there come to us piecemeal, one at a time and seemingly independent from one another, each often attractively introduced as a measure that will "help [us] not to be sick." But sometimes we come to a clear fork in the road where decision is possible, and where we know that our decision will make a world of difference – indeed, it will make a permanently different world. Fortunately, we stand now at the point of such a momentous decision. Events have conspired to provide us with a perfect opportunity to seize the initiative and to gain some control of the biotechnical project. I refer to the prospect of human cloning, a practice absolutely central to Huxley's fictional world. Indeed, creating and manipulating life in the laboratory is the gateway to a Brave New World, not only in fiction but also in fact.

"To clone or not to clone a human being" is no longer a fanciful question. Success in cloning sheep, and also cows, mice, pigs, and goats, makes it perfectly clear that a fateful decision is now at hand: whether we should welcome or even tolerate the cloning of human beings. If recent newspaper reports are to be believed, reputable scientists and physicians have announced their intention to produce the first human clone in the coming year. Their efforts may already be under way.

The media, gawking and titillating as is their wont, have been softening us up for this possibility by turning the bizarre into the familiar. In the four years since the birth of Dolly the cloned sheep, the tone of discussing the prospect of human cloning has gone from "Yuck" to "Oh?"

to "Gee whiz" to "Why not?" The sentimentalizers, aided by leading bioethicists, have downplayed talk about eugenically cloning the beautiful and the brawny or the best and the brightest. They have taken instead to defending clonal reproduction for humanitarian or compassionate reasons: to treat infertility in people who are said to "have no other choice" to avoid the risk of severe genetic disease, to "replace" a child who has died. For the sake of these rare benefits, they would have us countenance the entire practice of human cloning, the consequences be damned.

But we dare not be complacent about what is at issue, for the stakes are very high. Human cloning, though partly continuous with previous reproductive technologies, is also something radically new in itself and in its easily foreseeable consequences – especially when coupled with powers for genetic "enhancement" and germline genetic modification that may soon become available, owing to the recently completed Human Genome Project. I exaggerate somewhat, but in the direction of the truth: we are compelled to decide nothing less than whether human procreation is going to remain human, whether children are going to be made to order rather than begotten, and whether we wish to say yes in principle to the road that leads to the dehumanized hell of Brave New World.

. . .

For we have here a golden opportunity to exercise some control over where biology is taking us. The technology of cloning is discrete and well defined, and it requires considerable technical know-how and dexterity; we can therefore know by name many of the likely practitioners. The public demand for cloning is extremely low, and most people are decidedly against it. Nothing scientifically or medically important would be lost by banning clonal reproduction; alternative and non-objectionable means are available to obtain some of the most important medical benefits claimed for (nonreproductive) human cloning. The commercial interests in human cloning are, for now, quite limited; and the nations of the world are actively seeking to prevent it. Now may be as good a chance as we will ever have to get our hands on the wheel of the runaway train now headed for a post-human world and to steer it toward a more dignified human future.

II.

What is cloning? Cloning, or asexual reproduction, is the production of individuals who are genetically identical to an already existing individual. The procedure's name is fancy – "somatic cell nuclear transfer" – but its concept is simple. Take a mature but unfertilized egg; remove or deactivate its nucleus; introduce a nucleus obtained from a specialized (somatic) cell of an adult organism. Once the egg begins to divide, transfer the little embryo to a woman's uterus to initiate a pregnancy. Since almost all the hereditary material of a cell is contained within its nucleus, the re-nucleated egg and the individual into which it develops are genetically identical to the organism that was the source of the transferred nucleus.

An unlimited number of genetically identical individuals – the group, as well as each of its members, is called "a clone" – could be produced by nuclear transfer. In principle, any person, male or female, newborn or adult, could be cloned, and in any quantity; and because stored cells can outlive their sources, one may even clone the dead. Since cloning requires no personal involvement on the part of the person whose genetic material is used, it could easily be used to reproduce living or deceased persons without their consent – a threat to reproductive freedom that has received relatively little attention.

Some possible misconceptions need to be avoided. Cloning is not Xeroxing: the clone of Bill Clinton, though his genetic double, would enter the world hairless, toothless, and peeing in his diapers, like any other human infant. But neither is cloning just like natural twinning: the cloned twin will be identical to an older, existing adult; and it will arise not by chance but by deliberate design; and its entire genetic makeup will be preselected by its parents and/or scientists. Moreover, the success rate of cloning, at least at first, will probably not be very high: the Scots transferred two hundred seventy-seven adult nuclei into sheep eggs, implanted twenty-nine clonal embryos, and achieved the birth of only one live lamb clone.

For this reason, among others, it is unlikely that, at least for now, the practice would be very popular; and there is little immediate worry of mass-scale production of multicopies. Still, for the tens of thousands of people who sustain more than three hundred assisted-reproduction clinics in the United States and already avail themselves of in vitro fertilization and other techniques, cloning would be an option with virtually no added fuss. Panos Zavos, the Kentucky reproduction specialist who has announced his plans to clone a child, claims that he has already received thousands of e-mailed requests from people eager to clone, despite the known risks of failure and damaged offspring. Should commercial interests develop in "nucleus-banking," as they have in sperm-banking and egg-harvesting; should famous athletes or other celebrities decide to market their DNA the way they now market their autographs and nearly everything else; should techniques of embryo and germline genetic testing and manipulation arrive as anticipated, increasing the use of laboratory assistance in order to obtain "better" babies – should all this come to pass, cloning, if it is permitted, could become more than a marginal practice simply on the basis of free reproductive choice.

What are we to think about this prospect? Nothing good. Indeed, most people are repelled by nearly all aspects of human cloning: the possibility of mass production of human beings, with large clones of look-alikes, compromised in their individuality; the idea of father-son or mother-daughter "twins"; the bizarre prospect of a woman bearing and rearing a genetic copy of herself, her spouse, or even her deceased father or mother; the grotesqueness of conceiving a child as an exact "replacement" for another who has died; the utilitarian creation of embryonic duplicates of oneself, to be frozen away or created when needed to provide homologous tissues or organs for transplantation; the narcissism of those who would clone themselves, and the arrogance of others who think they know who deserves to be cloned; the Frankensteinian hubris to create a human life and increasingly to control its destiny; men playing at being God. Almost no one finds any of the suggested reasons for human cloning compelling, and almost everyone anticipates its possible misuses and abuses. And the popular belief that human cloning cannot be prevented makes the prospect all the more revolting.

Revulsion is not an argument; and some of yesterday's repugnances are today calmly accepted – not always for the better. In some crucial cases, however, repugnance is the emotional expression of deep wisdom, beyond reason's power completely to articulate it. Can anyone really give an argument fully adequate to the horror that is father–daughter incest (even with consent), or bestiality, or the mutilation of a corpse, or the eating of human flesh, or the rape or murder of another human being? Would anybody's failure to give full rational justification for his revulsion at those practices make that revulsion ethically suspect?

I suggest that our repugnance at human cloning belongs in this category. We are repelled by the prospect of cloning human beings not because of the strangeness or the novelty of the undertaking, but because we intuit and we feel, immediately and without argument, the violation of things that we rightfully hold dear. We sense that cloning represents a profound defilement of our given nature as procreative beings, and of the social relations built on this natural ground. We also sense that cloning is a radical form of child abuse. In this age in which everything is held to be permissible so long as it is freely done, and in which our bodies are regarded as mere instruments of our autonomous rational will, repugnance may be the only voice left that speaks up to defend the central core of our humanity. Shallow are the souls that have forgotten how to shudder.

III.

Yet repugnance need not stand naked before the bar of reason. The wisdom of our horror at human cloning can be at least partially articulated, even if this is finally one of those instances about which the heart has its reasons that reason cannot entirely know. I offer four objections to human cloning: that it constitutes unethical experimentation; that it threatens identity and individuality; that it turns procreation into manufacture (especially when understood as the harbinger of manipulations to come); and that it means despotism over children and perversion of parenthood. Please note: I speak only about so-called reproductive cloning, not about the creation of cloned embryos for research. The objections that may be raised against creating (or using)

embryos for research are entirely independent of whether the research embryos are produced by cloning. What is radically distinct and radically new is reproductive cloning.

Any attempt to clone a human being would constitute an unethical experiment upon the resulting child-to-be. In all the animal experiments, fewer than two to three percent of all cloning attempts succeeded. Not only are there fetal deaths and stillborn infants, but many of the so-called "successes" are in fact failures. As has only recently become clear, there is a very high incidence of major disabilities and deformities in cloned animals that attain live birth. Cloned cows often have heart and lung problems; cloned mice later develop pathological obesity; other live-born cloned animals fail to reach normal developmental milestones.

The problem, scientists suggest, may lie in the fact that an egg with a new somatic nucleus must re-program itself in a matter of minutes or hours (whereas the nucleus of an unaltered egg has been prepared over months and years). There is thus a greatly increased likelihood of error in translating the genetic instructions, leading to developmental defects some of which will show themselves only much later. (Note also that these induced abnormalities may also affect the stem cells that scientists hope to harvest from cloned embryos. Lousy embryos, lousy stem cells.) Nearly all scientists now agree that attempts to clone human beings carry massive risks of producing unhealthy, abnormal, and malformed children. What are we to do with them? Shall we just discard the ones that fall short of expectations? Considered opinion is today nearly unanimous, even among scientists: attempts at human cloning are irresponsible and unethical. We cannot ethically even get to know whether or not human cloning is feasible.

If it were successful, cloning would create serious issues of identity and individuality. The clone may experience concerns about his distinctive identity not only because he will be, in genotype and in appearance, identical to another human being, but because he may also be twin to the person who is his "father" or his "mother" – if one can still call them that. Unaccountably, people treat as innocent the homey case of intra-familial cloning – the cloning of husband or wife (or single mother). They forget about the unique dangers of mixing the twin relation with the parent-child relation. (For this situation, the relation of contemporaneous twins is no precedent; yet even this less problematic situation teaches us how difficult it is to wrest independence from the being for whom one has the most powerful affinity.) Virtually no parent is going to be able to treat a clone of himself or herself as one treats a child generated by the lottery of sex. What will happen when the adolescent clone of Mommy becomes the spitting image of the woman with whom Daddy once fell in love? In case of divorce, will Mommy still love the clone of Daddy, even though she can no longer stand the sight of Daddy himself?

Most people think about cloning from the point of view of adults choosing to clone. Almost nobody thinks about what it would be like to be the cloned child. Surely his or her new life would constantly be scrutinized in relation to that of the older version. Even in the absence of unusual parental expectations for the clone – say, to live the same life, only without its errors – the child is likely to be ever a curiosity, ever a potential source of déjà vu. Unlike "normal" identical twins, a cloned individual – copied from whomever – will be saddled with a genotype that has already lived. He will not be fully a surprise to the world: people are likely always to compare his doings in life with those of his alter ego, especially if he is a clone of someone gifted or famous. True, his nurture and his circumstance will be different; genotype is not exactly destiny. But one must also expect parental efforts to shape this new life after the original – or at least to view the child with the original version always firmly in mind. For why else did they clone from the star basketball player, the mathematician, or the beauty queen – or even dear old Dad – in the first place?

Human cloning would also represent a giant step toward the transformation of begetting into making, of procreation into manufacture (literally, "handmade"), a process that has already begun with in vitro fertilization and genetic testing of embryos. With cloning, not only is the process in hand, but the total genetic blueprint of the cloned individual is selected and determined by the human artisans. To be sure, subsequent development is still according to natural processes; and the resulting children will be recognizably human. But we would be taking a major step into making man himself simply another one of the man-made things.

How does begetting differ from making? In natural procreation, human beings come together to give existence to another being that is formed exactly as we were, by what we are – living, hence perishable, hence aspiringly erotic, hence procreative human beings. But in clonal reproduction, and in the more advanced forms of manufacture to which it will lead, we give existence to a being not by what we are but by what we intend and design.

Let me be clear. The problem is not the mere intervention of technique, and the point is not that "nature knows best." The problem is that any child whose being, character, and capacities exist owing to human design does not stand on the same plane as its makers. As with any product of our making, no matter how excellent, the artificer stands above it, not as an equal but as a superior, transcending it by his will and creative prowess. In human cloning, scientists and prospective "parents" adopt a technocratic attitude toward human children: human children become their artifacts. Such an arrangement is profoundly dehumanizing, no matter how good the product.

Procreation dehumanized into manufacture is further degraded by commodification, a virtually inescapable result of allowing baby-making to proceed under the banner of commerce. Genetic and reproductive biotechnology companies are already growth industries, but they will soon go into commercial orbit now that the Human Genome Project has been completed. "Human eggs for sale" is already a big business, masquerading under the pretense of "donation." Newspaper advertisements on elite college campuses offer up to $50,000 for an egg "donor" tall enough to play women's basketball and with SAT scores high enough for admission to Stanford; and to nobody's surprise, at such prices there are many young coeds eager to help shoppers obtain the finest babies money can buy. (The egg and womb-renting entrepreneurs shamelessly proceed on the ancient, disgusting, misogynist premise that most women will give you access to their bodies, if the price is right.) Even before the capacity for human cloning is perfected, established companies will have invested in the harvesting of eggs from ovaries obtained at autopsy or through ovarian surgery, practiced embryonic genetic alteration, and initiated the stockpiling of prospective donor tissues. Through the rental of surrogate-womb services, and through the buying and selling of tissues and embryos priced according to the merit of the donor, the commodification of nascent human life will be unstoppable.

Finally, the practice of human cloning by nuclear transfer – like other anticipated forms of genetically engineering the next generation – would enshrine and aggravate a profound misunderstanding of the meaning of having children and of the parent-child relationship. When a couple normally chooses to procreate, the partners are saying yes to the emergence of new life in its novelty – are saying yes not only to having a child, but also to having whatever child this child turns out to be. In accepting our finitude, in opening ourselves to our replacement, we tacitly confess the limits of our control.

Embracing the future by procreating means precisely that we are relinquishing our grip in the very activity of taking up our own share in what we hope will be the immortality of human life and the human species. This means that our children are not our children: they are not our property, they are not our possessions. Neither are they supposed to live our lives for us, or to live anyone's life but their own. Their genetic distinctiveness and independence are the natural foreshadowing of the deep truth that they have their own, never-before-enacted life to live. Though sprung from a past, they take an uncharted course into the future.

Much mischief is already done by parents who try to live vicariously through their children. Children are sometimes compelled to fulfill the broken dreams of unhappy parents. But whereas most parents normally have hopes for their children, cloning parents will have expectations. In cloning, such overbearing parents will have taken at the start a decisive step that contradicts the entire meaning of the open and forward-looking nature of parent-child relations. The child is given a genotype that has already lived, with full expectation that this blueprint of a past life ought to be controlling the life that is to

come. A wanted child now means a child who exists precisely to fulfill parental wants. Like all the more precise eugenic manipulations that will follow in its wake, cloning is thus inherently despotic, for it seeks to make one's children after one's own image (or an image of one's choosing) and their future according to one's will.

Is this hyperbolic? Consider concretely the new realities of responsibility and guilt in the households of the cloned. No longer only the sins of the parents, but also the genetic choices of the parents, will be visited on the children – and beyond the third and fourth generation; and everyone will know who is responsible. No parent will be able to blame nature or the lottery of sex for an unhappy adolescent's big nose, dull wit, musical ineptitude, nervous disposition, or anything else that he hates about himself. Fairly or not, children will hold their cloners responsible for everything, for nature as well as for nurture. And parents, especially the better ones, will be limitlessly liable to guilt. Only the truly despotic souls will sleep the sleep of the innocent.

IV.

The defenders of cloning are not wittingly friends of despotism. Quite the contrary. Deaf to most other considerations, they regard themselves mainly as friends of freedom: the freedom of individuals to reproduce, the freedom of scientists and inventors to discover and to devise and to foster "progress" in genetic knowledge and technique, the freedom of entrepreneurs to profit in the market. They want large-scale cloning only for animals, but they wish to preserve cloning as a human option for exercising our "right to reproduce" – our right to have children, and children with "desirable genes." As some point out, under our "right to reproduce" we already practice early forms of unnatural, artificial, and extra-marital reproduction, and we already practice early forms of eugenic choice. For that reason, they argue, cloning is no big deal.

We have here a perfect example of the logic of the slippery slope. The principle of reproductive freedom currently enunciated by the proponents of cloning logically embraces the ethical acceptability of sliding all the way down: to producing children wholly in the laboratory from sperm to term (should it become feasible), and to producing children whose entire genetic makeup will be the product of parental eugenic planning and choice. If reproductive freedom means the right to have a child of one's own choosing by whatever means, then reproductive freedom knows and accepts no limits.

Proponents want us to believe that there are legitimate uses of cloning that can be distinguished from illegitimate uses, but by their own principles no such limits can be found. (Nor could any such limits be enforced in practice: once cloning is permitted, no one ever need discover whom one is cloning and why.) Reproductive freedom, as they understand it, is governed solely by the subjective wishes of the parents-to-be. The sentimentally appealing case of the childless married couple is, on these grounds, indistinguishable from the case of an individual (married or not) who would like to clone someone famous or talented, living or dead. And the principle here endorsed justifies not only cloning but also all future artificial attempts to create (manufacture) "better" or "perfect" babies.

The "perfect baby," of course, is the project not of the infertility doctors, but of the eugenic scientists and their supporters, who, for the time being, are content to hide behind the skirts of the partisans of reproductive freedom and compassion for the infertile. For them, the paramount right is not the so-called right to reproduce, it is what the biologist Bentley Glass called, a quarter of a century ago. "the right of every child to be born with a sound physical and mental constitution, based on a sound genotype . . . the inalienable right to a sound heritage." But to secure this right, and to achieve the requisite quality control over new human life, human conception and gestation will need to be brought fully into the bright light of the laboratory, beneath which the child-to-be can be fertilized, nourished, pruned, weeded, watched, inspected, prodded, pinched, cajoled, injected, tested, rated, graded, approved, stamped, wrapped, sealed, and delivered. There is no other way to produce the perfect baby.

If you think that such scenarios require outside coercion or governmental tyranny, you are mistaken. Once it becomes possible, with the aid of human genomics, to produce or to select for what some regard as "better babies" – smarter, prettier, healthier, more athletic – parents will leap at the opportunity to "improve" their offspring. Indeed, not to do so will be socially regarded as a form of child neglect. Those who would ordinarily be opposed to such tinkering will be under enormous pressure to compete on behalf of their as yet unborn children – just as some now plan almost from their children's birth how to get them into Harvard. Never mind that, lacking a standard of "good" or "better," no one can really know whether any such changes will truly be improvements.

Proponents of cloning urge us to forget about the science-fiction scenarios of laboratory manufacture or multiple-copy clones, and to focus only on the sympathetic cases of infertile couples exercising their reproductive rights. But why, if the single cases are so innocent, should multiplying their performance be so off-putting? (Similarly, why do others object to people's making money from that practice if the practice itself is perfectly acceptable?) The so-called science-fiction cases – say, Brave New World – make vivid the meaning of what looks to us, mistakenly, to be benign. They reveal that what looks like compassionate humanitarianism is, in the end, crushing dehumanization.

V.

Whether or not they share my reasons, most people, I think, share my conclusion: that human cloning is unethical in itself and dangerous in its likely consequences, which include the precedent that it will establish for designing our children. Some reach this conclusion for their own good reasons, different from my own: concerns about distributive justice in access to eugenic cloning; worries about the genetic effects of asexual "inbreeding"; aversion to the implicit premise of genetic determinism; objections to the embryonic and fetal wastage that must necessarily accompany the efforts; religious opposition to "man playing God." But never mind why: the overwhelming majority of our fellow Americans remain firmly opposed to cloning human beings.

For us, then, the real questions are: What should we do about it? How can we best succeed? These questions should concern everyone eager to secure deliberate human control over the powers that could re-design our humanity, even if cloning is not the issue over which they would choose to make their stand. And the answer to the first question seems pretty plain. What we should do is work to prevent human cloning by making it illegal.

We should aim for a global legal ban, if possible, and for a unilateral national ban at a minimum – and soon, before the fact is upon us. To be sure, legal bans can be violated; but we certainly curtail much mischief by outlawing incest, voluntary servitude, and the buying and selling of organs and babies. To be sure, renegade scientists may secretly undertake to violate such a law, but we can deter them by both criminal sanctions and monetary penalties, as well as by removing any incentive they have to proudly claim credit for their technological bravado.

Such a ban on clonal baby-making will not harm the progress of basic genetic science and technology. On the contrary, it will reassure the public that scientists are happy to proceed without violating the deep ethical norms and intuitions of the human community. It will also protect honorable scientists from a public backlash against the brazen misconduct of the rogues. As many scientists have publicly confessed, free and worthy science probably has much more to fear from a strong public reaction to a cloning fiasco than it does from a cloning ban, provided that the ban is judiciously crafted and vigorously enforced against those who would violate it.

. . .

. . . I now believe that what we need is an all-out ban on human cloning, including the creation of embryonic clones. I am convinced that all halfway measures will prove to be morally, legally, and strategically flawed, and – most important – that they will not be effective in obtaining the desired result. Anyone truly serious about preventing human reproductive cloning must seek to stop the process from the beginning. Our changed circumstances, and the now evident defects of the less restrictive alternatives, make an all-out ban by far the most attractive and effective option.

Here's why. Creating cloned human children ("reproductive cloning") necessarily begins by producing cloned human embryos. Preventing the latter would prevent the former, and prudence alone might counsel building such a "fence around the law." Yet some scientists favor embryo cloning as a way of obtaining embryos for research or as sources of cells and tissues for the possible benefit of others. (This practice they misleadingly call "therapeutic cloning" rather than the more accurate "cloning for research" or "experimental cloning," so as to obscure the fact that the clone will be "treated" only to exploitation and destruction, and that any potential future beneficiaries and any future "therapies" are at this point purely hypothetical.

[. . .]

A few moments of reflection show why an anti-cloning law that permitted the cloning of embryos but criminalized their transfer to produce a child would be a moral blunder. This would be a law that was not merely permissively "pro-choice" but emphatically and prescriptively "anti-life." While permitting the creation of an embryonic life, it would make it a federal offense to try to keep it alive and bring it to birth. Whatever one thinks of the moral status or the ontological status of the human embryo, moral sense and practical wisdom recoil from having the government of the United States on record as requiring the destruction of nascent life and, what is worse, demanding the punishment of those who would act to preserve it by (feloniously!) giving it birth.

But the problem with the approach that targets only reproductive cloning (that is, the transfer of the embryo to a woman's uterus) is not only moral but also legal and strategic. A ban only on reproductive cloning would turn out to be unenforceable. Once cloned embryos were produced and available in laboratories and assisted-reproduction centers, it would be virtually impossible to control what was done with them. Biotechnical experiments take place in laboratories, hidden from public view, and, given the rise of high-stakes commerce in biotechnology, these experiments are concealed from the competition. Huge stockpiles of cloned human embryos could thus be produced and bought and sold without anyone knowing it. As we have seen with in vitro embryos created to treat infertility, embryos produced for one reason can be used for another reason: today "spare embryos" once created to begin a pregnancy are now used in research, and tomorrow clones created for research will be used to begin a pregnancy.

Assisted reproduction takes place within the privacy of the doctor-patient relationship, making outside scrutiny extremely difficult. Many infertility experts probably would obey the law, but others could and would defy it with impunity, their doings covered by the veil of secrecy that is the principle of medical confidentiality. Moreover, the transfer of embryos to begin a pregnancy is a simple procedure (especially compared with manufacturing the embryo in the first place), simple enough that its final steps could be self-administered by the woman, who would thus absolve the doctor of blame for having "caused" the illegal transfer. (I have in mind something analogous to Kevorkian's suicide machine, which was designed to enable the patient to push the plunger and the good "doctor" to evade criminal liability.)

Even should the deed become known, governmental attempts to enforce the reproductive ban would run into a swarm of moral and legal challenges, both to efforts aimed at preventing transfer to a woman and – even worse – to efforts seeking to prevent birth after transfer has occurred. A woman who wished to receive the embryo clone would no doubt seek a judicial restraining order suing to have the law overturned in the name of a constitutionally protected interest in her own reproductive choice to clone. (The cloned child would be born before the legal proceedings were complete.) And should an "illicit clonal pregnancy" be discovered, no governmental agency would compel a woman to abort the clone, and there would be an understandable storm of protest should she be fined or jailed after she gives birth. Once the baby is born, there would even be sentimental opposition to punishing the doctor for violating the law – unless, of course, the clone turned out to be severely abnormal.

For all these reasons, the only practically effective and legally sound approach is to block human cloning at the start, at the production of the embryo clone. Such a ban can be rightly characterized not as interference with reproductive freedom, nor even as interference with scientific inquiry, but as an attempt to prevent the unhealthy, unsavory, and unwelcome manufacture of and traffic in human clones.

. . .

I appreciate that a federal legislative ban on human cloning is without American precedent, at least in matters technological. Perhaps such a ban will prove ineffective; perhaps it will eventually be shown to have been a mistake. (If so, it could later be reversed.) If enacted, however, it will have achieved one overwhelmingly important result, in addition to its contribution to thwarting cloning: it will place the burden of practical proof where it belongs. It will require the proponents to show very clearly what great social or medical good can be had only by the cloning of human beings. Surely it is only for such a compelling case, yet to be made or even imagined, that we should wish to risk this major departure – or any other major departure – in human procreation.

Americans have lived by and prospered under a rosy optimism about scientific and technological progress. The technological imperative has probably served us well, though we should admit that there is no accurate method for weighing benefits and harms. And even when we recognize the unwelcome outcomes of technological advance, we remain confident in our ability to fix all the "bad" consequences – by regulation or by means of still newer and better technologies. Yet there is very good reason for shifting the American paradigm, at least regarding those technological interventions into the human body and mind that would surely effect fundamental (and likely irreversible) changes in human nature, basic human relationships, and what it means to be a human being. Here we should not be willing to risk everything in the naive hope that, should things go wrong, we can later set them right again.

Some have argued that cloning is almost certainly going to remain a marginal practice, and that we should therefore permit people to practice it. Such a view is shortsighted. Even if cloning is rarely undertaken, a society in which it is tolerated is no longer the same society – any more than is a society that permits (even small-scale) incest or cannibalism or slavery. A society that allows cloning, whether it knows it or not, has tacitly assented to the conversion of procreation into manufacture and to the treatment of children as purely the projects of our will. Willy-nilly, it has acquiesced in the eugenic redesign of future generations. The humanitarian superhighway to a Brave New World lies open before this society.

But the present danger posed by human cloning is, paradoxically, also a golden opportunity. In a truly unprecedented way, we can strike a blow for the human control of the technological project, for wisdom, for prudence, for human dignity. The prospect of human cloning, so repulsive to contemplate, is the occasion for deciding whether we shall be slaves of unregulated innovation, and ultimately its artifacts, or whether we shall remain free human beings who guide our powers toward the enhancement of human dignity. The humanity of the human future is now in our hands.

26

Ethical Issues in Human Stem Cell Research: Embryos and Beyond

Inmaculada de Melo-Martín and Marin Gillis

1 Introduction

We live in the era of new biotechnological advances. Discussion of the social, legal, ethical, and scientific aspects of genetic therapy, *in vitro* fertilization, genetically engineered food, or cloning, appear everywhere, from prestigious scientific journals, to television programs and the tabloids. In a world where the Human Genome Project hoards millions in public and private monies and thousands of scientists, where infertility seems rampant, and where the search for the perfect human baby occupies people's imagination, one might expect to find this focus on biotechnology quite normal and welcomed.

Not only have these discussions captured the public imagination and the interests of scientists, they seem also to have swept many of the members of the bioethics profession away from more mundane issues, such as questions of access to health care, or just distribution of medical resources. Lately, the issue of human stem cell research seems to be the new kid on the block.

The purpose of this work is to present some of the ethical issues involved with research using embryonic stem cells. First, we will briefly give an account of the scientific and medical background surrounding stem cell research. Next, we will offer an overview of some of the main ethical concerns that have been presented in relation to this kind of research.

2 Scientific Background

2a Some history

Stem cells have two important fundamental characteristics that distinguish them from other types of cells. The first characteristic is that they can give rise to a multiplicity of other specialized cells. In the embryonic state they can literally give rise to all cells and tissues of the body. The second fundamental characteristic is self-renewal. Under certain physiological or experimental conditions, these unspecialized cells can renew themselves for long periods of time through cell division. In the embryonic state, stem cells can propagate indefinitely; they are thus called immortal.

Scientists discovered ways to obtain or derive stem cells from early mouse embryos more than two decades ago. Evan and Kaufman's development of mouse embryonic stem (ES) cells in 1981 has provided the model and much of the technology for the development of human ES cells.[1] However, the concept of a pluripotent embryonic cell evolved out of work on mouse teratocarcinomas, and their embryonal carcinoma (EC) stem cells, which present a malignant surrogate for the normal stem cells of the early embryo.[2] Teratocarcinomas are tumors that arise in the gonads of a few inbred strains, and consist of an array of somatic tissues juxtaposed together in a disorganized fashion. The field of teratocarcinoma research expanded considerably in the 1970s and investigators began to realize the potential value of cultured cell lines from the tumors as models for mammalian development.[3] However, EC cells often contained chromosomal abnormalities, and their ability to differentiate into multiple tissue types was often limited. Because teratocarcinomas can also be induced by grafting blastocysts to ectopic sites, scientists thought that it could be possible to derive pluripotent cell lines directly from blastocysts rather than from tumours. That is what was done in 1981 by Gail Martin and Evans and Kaufman independently.[4]

The direct derivation of embryonic germ stem cells (EG cells) from mouse primordial germ cells was accomplished in 1992.[5] These EG cells have a developmental capacity very similar to that of ES cells, though they differ in their expression of some imprinted genes. These primordial germ cell-derived embryonic germ cells can be induced to differentiate extensively in culture and also form teratocarcinomas when injected into nude mice. Additionally, they contribute to chimeras when injected into host blastocysts.[6]

In 1995, James Thomson's team derived primate ES cells from rhesus monkey blastocysts.[7] The implication of the monkey work and the work on human EC cells was that a pluripotent stem cell could probably be derived from a human blastocyst. And thus, in 1998 Thomson and his colleagues announced the isolation of human embryonic stem cells (hES) from surplus blastocysts donated by couples undergoing infertility treatment.[8] Shamblott and co-workers reported the isolation of human embryonic germ cells (hEG) from embryonic and fetal gonads at 5–9 weeks post-fertilization.[9] In 2000, Reubinoff and colleagues confirmed Thomson's findings and reported that somatic cells could be derived in large numbers in vitro from stem cells.[10]

2b Properties and types of stem cells

All stem cells have three general properties: unlike muscle, or nerve cells, which do not normally replicate themselves, stem cells are capable of dividing and renewing themselves for long periods of time; they are unspecialized, that is, stem cells do not have any tissue-specific structures that allow it to perform specialized functions; finally, stem cells can give rise to specialized cell types.

Human stem cells are classified depending on their source.[11] hES are derived from embryos that develop from eggs that have been donated by couples undergoing in vitro fertilization. The stem cells are obtained from the inner cell mass of embryos that are typically 3–5 days old. hES are isolated by transferring this inner cell mass in a laboratory dish with culture medium. The inner surface of the culture dish is usually coated with a feeder layer made of mouse embryonic skin cells that have been treated so that they cannot divide. Because these systems of support run the risk of cross-transfer of animal pathogens from the animal feeder to the hES cells, thus compromising later clinical application, scientists have recently begun to devise other ways to grow embryonic stem cells.[12] After cells proliferate they are removed and plated into fresh culture dishes. This process of replating can be repeated for many months to yield millions of embryonic stem cells. Human embryonic stem cells can thus grow indefinitely in vitro in the primitive embryonic stage while retaining their pluripotentiality or the ability to differentiate into somatic and extraembryonic cell types.

Adult stem cells, considered to be multipotent, are undifferentiated cells that can be found among differentiated cells in a tissue or organ.[13] They can renew themselves and can differentiate to yield the major specialized cell type of that tissue or organ. Adult stem cells are important cells in maintaining the integrity of tissues like skin, bone and blood. Adult stem cells include haematopoietic stem cells, bone marrow stromal (mesenchymal) stem cells, neural stem cells, dermal (keratinocyte) stem cells, and fetal cord blood stem cells.

Until recently, scientists believed that organ specific stem cells were lineage restricted. Recent work has called this idea into question, and proposed that adult stem cells may have much wider differentiation capabilities. Researchers have showed that bone marrow derived cells could target and differentiate into muscle, liver, kidney, cardiomyocytes, neural cell lineages, and gut.[14]

2c Implications for biomedicine

Although the understanding of human stem cells is at this time limited, it appears that investigations with these cells might have a widespread impact on biomedical research. Some of the areas where stem cell research can yield important results are described in this section.

a. Human developmental biology Studies of human embryonic stem cells can offer information about the genetic processes that help build a human body and can help us to obtain information of the complex events that occur during human development. A better understanding of the genetic and molecular aspects of cell division and differentiation can result in knowledge of medical conditions such as cancer and birth defects. Embryonic stem cells can also be used to identify and study environmental toxins that could cause abnormalities in the differentiation and division of cells.[15] Additionally research on stem cells could help us understand how tissues regenerate.

b. Testing of new drugs Research on stem cells can allow us to develop normal lines of cells that represent specific tissues and organs. This would permit testing of new and already existing drugs and prevent cases of unanticipated liver toxicity that can be fatal.[16] Because testing of new drugs is usually performed on non-human animals, research on stem cells could result in a reduction of such practices.

c. Regenerative medicine Embryonic stem cell research has the potential to provide us with renewable sources of replacement cells and tissues to treat diseases such as Parkinson's, Alzheimer's, spinal cord injury, cancer, atherosclerosis, burns, stroke, diabetes, heart disease, rheumatoid arthritis, and osteoarthritis. Moreover, research on stem cells can provide us with a supply of organs for transplantation that could solve the problems of unavailability that we face today.[17] Some believe that this research may lead to the formation of tissue banks to repair or replace damaged body parts.[18]

In April 2006, for example, 91,846 Americans who are in want of an organ transplant have been registered on UNOS (United Network for Organ Sharing). It is estimated that in the US thousands of people die each year for want of a transplant.[19] Thus, a successful regenerative medicine would meet an important demand in itself, and consequently bring about important medical benefits. In addition, its application would prevent what some believe to be great harms to others, given existing protocols for transplant research and therapy. These include the cultivating of animals, either genetically modified and/or cloned, for the purpose of providing organs for humans (i.e. xenotransplantation). Also, it would greatly diminish if not obliterate the demand for traffic in organs and thus the questionable means of appropriating human organs that have been documented in China and India.[20] Further, the ethical and legal debate about the sale of organs is potentially abated.

d. Somatic cell nuclear transfer (SCNT) One of the most serious medical problems occurring in transplant therapies results from immunological incompatibilities between the donor and the recipient. Currently, host immune responses, which can only be overcome by administering long-term immunosuppressive drug therapy, have frustrated cell replacement therapies using allogeneic hES cells.[21] Thus, the possible generation of stem cells derived from nuclear transfer embryonic stem cell (NT-ESC) has the potential to avoid immunorejection.

Nuclear transfer blastocysts are produced following fusion of a single donor cell from a patient into an enucleated using a pipette to suction its nucleus.[22] Once this is done, the egg cell will have none of its nuclear chromosomes. The chromosomes are then replaced with a nucleus extracted from cultured cells that may have come originally from somatic cells of an embryo, a fetus, or an adult organism. This nucleus, because it is derived from somatic cells, contains a diploid set of chromosomes. Once the new nucleus – sometimes more than just the nucleus – is introduced into the egg cell, it is placed between a couple of electrodes which apply an electric current. The electric shock will allow the egg to fuse with the nucleus or donor cell and behave as if it were now a normally fertilized egg cell. Patient-specific NT-ESCs could then be derived in this way.

The recent retraction of a paper that reported patient-specific embryonic stem cell lines created from a human nuclear transfer blastocyst has constituted a significant setback to this research.[23] Nonetheless, other researchers are actively pursuing the generation of NT-ESCs as an alternate source of cells for cell replacement therapies.[24]

3 Ethical Issues

In many cases, groups within the scientific community, recognizing the promises of stem cell research argue that they should be allowed to pursue these investigations. Nonetheless, the possible medical applications that stem cell research might produce are only part of the equation in the discussion of embryonic stem cells. Another part of the equation comes with the ethical implications of such research. In most cases, the debate on stem cell research centers around the question of whether the ethical consequences of this research are such that they require its prohibition, a strict government regulation, or local institutional guidelines. In this section we describe some of the ethical issues raised by human stem cell research. Some of the concerns, such as the moral status of the embryo and the use of ova, are specific to investigation with embryonic stem cells. Also, these issues arise whether the human embryonic stem cells are derived from embryos created by the union of egg and sperm or whether the embryos result from nuclear transfer. Other ethical problems, such as issues about justice and safety, affect both embryonic and adult stem cell research.

3a Moral status of the embryo

Any research that has to do with the human embryo is bound to be controversial. Consequently, the case of hES cells is not different from other types of research that require the manipulation of human embryos. The main reason for the debate in this case is that, presently, the methods available to derive stem cells from embryos involve their destruction. Thus, the issue of the ontological and moral status of this entity is relevant.[25]

The issue of the ontological status of the embryo concerns the kind of entity embryos are.

Embryos can be considered individual organisms, biological human beings, or persons. Although some authors have suggested that there is no difference between the concept of personhood and that of a biological being, many philosophers and theologians have suggested that the criteria for personhood are more demanding that those simply required for being a biological human. Some of the conditions than have been suggested as required for personhood, include self-consciousness, the capacity to feel pain, capacity to act on reasons, capacity to communicate with others using a language, capacity to act freely, and rationality.[26] Granting that an embryo is a person usually gives the embryo a more significant status than merely saying that it is an individual organism.

Inquiries into the moral status of the embryo attempt to discern whether embryos are the kinds of entities that ought to be given moral and legal rights. Having moral status is to qualify under a range of moral protections. Thus, if human embryos have full moral status, then they possess the same rights as human beings who have been born.

Opponents of hES cell research often argue that early human embryos have the full status of persons and thus, the destruction of embryos during research is equated with human sacrifice in order to obtain scientific knowledge. They argue that because the line between the human and the non-human is appropriately drawn at conception, embryos ought to be considered persons and therefore be attributed full moral status. As humans we have moral value simply because we have a human genotype, no matter what the stage of development might be.[27] Given that present ethical and legal regulations prohibit the sacrifice of human life for the sake of knowledge this would mean that the destruction of embryos in this case would not be morally justifiable. Stem cell research would, on these grounds, need to be prohibited because it violates the embryo's rights.[28]

Some authors who oppose embryonic stem cell research might concede that the embryo is not a person, but that it is a potential person and this is a significant moral property. Because of this, killing an embryo to obtain stem cells must be justified to an extent comparable to the justification required for killing a person.[29]

Given that most people would not justify the killing of a person for the sake of knowledge, then, even when they are not persons, the killing of embryos for that purpose would be equally unjustified.

On the other extreme of the debate, proponents of hES cell research argue that the moral status of an early human embryo is equivalent to any other cell in the human body. At this point of development, embryonic cells are too unspecialized to be a unique entity. Proponents of this view usually argue that the appropriate line between the human and the non-human must be drawn at birth rather than at conception or at some other point of development. They would recognize that the embryo is certainly an individual organism and biologically human, but it is not human in an ontologically significant sense, and therefore it does not have a significant moral status. Hence, there is no ethical need for any protection or regulation in the use of embryonic stem cells for medical research purposes.[30]

In between these views, is the so-called "middle" or "third way", which regards embryos as neither persons nor property. It sees them as being entities worthy of a *profound respect*. Embryos may thus have instrumental value while at the same time they are not simply human tissue. They are special and they cannot be treated or used in any way one might wish.[31] In this context, the protection demanded for an embryo is not absolute and might be weighed against the benefit of research purposes.[32]

This position recognizes the important differences between human embryos and fully developed human beings while at the same time giving weight to the unique relationship they have with each other. Proponents of this view tend to draw the line between the human and the non-human at some point between conception and birth. Thus the embryo has no significant moral status although the fetus might acquire full moral status at some stage of fetal development.

Some in this group argue that we already create spare embryos and sometimes discard them in *in vitro* fertilization practices, and that therefore, nothing is lost and much is gained if, rather than discarding those embryos, we can use them for research that would benefit humanity.[33] This position cannot automatically resolve the problem of regulation of embryo research. Although it does reject outright prohibition of such investigations, or the destruction of embryos as non-problematic, proponents of this middle view need to establish parameters that would balance the interests of embryos with those of scientists and other human beings.

Because the destruction of embryos is regarded by many in the popular media and the research world as virtually the only pressing ethical issue in stem cell technology, many believe that if there were a way around the destruction of embryos, all ethical issues, or at least the most pressing and controversial ones, would dissolve.[34] Indeed, significant time and resources have been devoted to develop new ways of obtaining embryonic stem cells that do not necessitate the destruction of the embryo. These beliefs notwithstanding, there are other important ethical issues in stem cell research beyond the embryo.[35]

3b   Safety concerns

Another ethical issue related to stem cell research (with both embryonic and adult stem cells) is related to the use of these cells for medical purposes. Some scientists argue that the procedures required for implanting cells into the human body put patients at risk.[36] Similarly, researchers have questions about whether these cells will grow normally once inserted into the human body or whether they might become cancerous.[37] Also concerns about immunoreaction from transplant recipients are an issue with this kind of research.

Proponents of stem cell research reply to these arguments maintaining that at this point, given the knowledge, or the lack of it, that we have, it would be unethical to attempt stem cell transplantation or similar techniques on a human being. However, if safety is the issue, it can be proposed that more research on animals be done, and more investigation completed to establish its safety and effectiveness for humans, before we proceed to use this technique on human beings. Thus, opposition to stem cell research on the basis of safety would fail as an argument against this research per se.

Proponents could also argue that once we have reasonable beliefs that regenerative medicine, for example, is safe and effective, it would then be ethically justifiable to use it on human beings who give consent to these techniques. The difficulty with this argument however, is that at least until we have research on human subjects, the issue of safety and efficiency cannot be established. But, if this is so, some critics might argue that given a situation of uncertainty it is unclear how patients can give full informed consent to the use of these techniques.

Shortly after the Nuremberg trials, which presented horrifying accounts of Nazi experimentations on unwilling human subjects, the issue of informed consent began to receive attention.[38] The first sentence of the Nuremberg Codes states that the voluntary consent of human subjects in research is absolutely essential. At Helsinki in 1964, the World Medical Association made consent of patients and subjects a central requirement of ethical research.[39] Since then virtually all prominent medical and research codes as well as institutional rules of ethics dictate that physicians and investigators obtain the free informed consent of patients and subjects prior to any substantial intervention. Procedures for free informed consent have several functions such as the protection of patients and subjects from harm or the promotion of medical responsibility in interactions with patients and subjects. Their more fundamental goal is, however, to enable autonomous choices.

The received approach to the definition of informed consent has been to specify the elements of the concept. Legal, regulatory, medical, psychological, and philosophical literature tend to analyze informed consent in terms of the following elements:[40] (1) disclosure, (2) understanding, (3) voluntariness, and (4) competence. Thus, one gives free informed consent to an intervention if and only if one is competent to act, receives a thorough disclosure about the procedure, understands the disclosure, acts voluntarily, and consents to the intervention. Disclosure refers to the necessity of professionals passing on information to decision-makers and possible risk victims. The professionals' perspectives, opinions, and recommendations are usually essential for a sound decision. They are obligated to disclose a core set of information, including (1) those facts or descriptions that patients usually consider material in deciding whether to consent to or refuse the intervention, (2) information the doc-

tors believe to be material, (3) the professionals' recommendations, (4) the purpose of seeking consent, and (5) the nature and limits of consent as an act of authorization.

Understanding may be the most important component for free informed consent. It requires professionals to help potential risk victims overcome illness, irrationality, immaturity, distorted information, or other factors that can limit their grasp of the situation to which they have the right to give or withhold consent. Thus people understand if they have acquired pertinent information and justified, relevant beliefs about the nature and impacts of their actions. This understanding need not be complete, because a substantial grasp of central facts is generally sufficient. Normally, diagnoses, prognoses, the nature and purpose of the intervention, alternatives, risks, benefits, and recommendations are essential. Patients or subjects also need to share an understanding with professionals about the terms of the authorization before proceeding. Unless agreement exists (about the crucial features of what the patients authorize) there can be no assurance that they have made autonomous decisions. Thus, even if doctor and patients use a word such as 'ovulation induction', their interpretations could be totally different if standard medical definitions have no meaning for the patients.

Another element of informed consent is voluntariness, or being free to act in giving consent. It requires that the subjects act in a way that is free from manipulation and coercion by other persons. Coercion occurs if and only if one person intentionally uses a credible and serious threat of harm or force to control another. Manipulation is convincing people to do what the one wants by means other than direct coercion or rational persuasion. One important form of manipulation in health care is informational manipulation, a deliberate act of handling information that alters patients' understanding of the situation and motivates them to do what the agent of influence plans. Withholding evidence and misleading exaggerations of benefits are instances of manipulation inconsistent with voluntary decisions. The way in which doctors present information by tone of voice, by framing information positively (the therapy is successful most of the time) rather than negatively (the therapy fails in forty percent of the cases) can manipulate patients' perceptions and, therefore, affect understanding.

The criterion of competence refers to the patients' abilities to perform a task. Thus patients or subjects are competent if they have the ability to understand the material information, to make a judgment about the evidence in light of their values, to intend a certain outcome, and freely to communicate their wishes to the professionals.

Given these elements that scholars recognize as necessary for informed consent, to claim that patients can give free informed consent to regenerative medicine, for example, might be questionable. If patients are ignorant of the possible risks involved in these kinds of treatments, then they cannot give genuinely informed consent. Lack of information may seriously hinder people's abilities to make informed choices. A possible way to solve this problem, however, is to inform patients of the uncertainty. It is nevertheless unclear whether people really understand what a situation of uncertainty means for evaluation of risks and benefits.

3c Obtaining ova

The use of research embryos also raises ethical concerns related to the interests of the women whose eggs would be used to make the embryos. There are two potential sources of embryonic stem cells: discarded embryos used in infertility treatments and created embryos for the purpose of research. Given that obtaining eggs from women is far more burdensome and risky than obtaining sperm, the interests of women merit special consideration.

According to empirical evidence, risks to women undergoing IVF treatment vary from simple nausea to death.[41] For example, the hormones that doctors use to stimulate the ovaries are associated with numerous side effects. Some studies assert that ovulation induction may be a risk factor for certain types of hormone-dependent cancers. Researchers have associated excessive estrogen secretion with ovarian and breast carcinoma, and gonadotropin secretion with ovarian cancer. A substantial body of experimental, clinical, and epidemiological evidence indicates that hormones play a major role in the development of several human cancers. The ability of hormones to stimulate cell division in certain organs, such as the breast, endometrium, and the ovary, may lead (following repeated cell divisions) to the accumulation of random genetic errors that ultimately produce cancer. Hormone-related cancers account for more than 30% of all newly diagnosed female cancers in the United States. Hence any technique (like IVF) – that relies on massive doses of hormones to obtain ova – may be quite dangerous for women.

The ovarian hyperstimulation syndrome (OHSS) is another possible iatrogenic (caused by medical treatment) consequence of ovulation induction. Women with the severe form of OHSS may suffer renal impairment, liver dysfunction, thromboembolic phenomena, shock, and even death. The incidence of moderate and severe OHSS in IVF treatment ranges from 3% to 4%.

The procedures that doctors normally use to obtain women's eggs, i.e., laparoscopy and ultrasound-guide oocyte retrieval also pose risks. Although there are no accurate statistical data about hazards associated with these two procedures, risks related to these technologies include postoperative infections, punctures of an internal organ, hemorrhages, ovarian trauma, and intrapelvic adhesions.

If the eggs have been obtained in order to solve a reproductive problem, then, we can say that the potential benefits – the birth of a child – might outweigh the potential risks. Thus, those concerned with issues of pressure and coercion of women might see the use of embryos already created for fertility purposes as less problematic. However, a problem appears when we consider the situation of women who are asked to donate excess embryos for stem cell research. Because of the pressure that women facing reproductive problems suffer in our society it is not unreasonable to believe that women undergoing IVF treatment might feel coerced to produce more embryos than those necessary for implantation. Moreover, in cases where women might be asked to donate eggs for research purposes other problems arise. Here women would be properly considered research subjects. However, these women would not directly experience any direct benefit from this procedure – such as the birth of a child. Similarly, they would experience neither direct nor indirect health benefits from donating their eggs. Furthermore, women presumably could not be compensated for the risks imposed on them by these procedures. Present regulations recommend that it be illegal to sell human embryos because the practice of selling human embryos would be contrary to the respect owed to them. But it can be argued that if such practice ought to be illegal, then consistency would require that the selling of eggs with the sole intention of creating embryos for research be illegal.[42]

The need to obtain embryos for research purposes thus might increase the risk of coercion and exploitation for all women. Exploitation of low-income women would be a clear reality in a system where gametes, embryos, or wombs can be bought and sold. Problems of social and racial discrimination in our society make these concerns even more present and serious. Poor, migrant, refugee, or ethnic minority women might be used as producers of eggs.

Furthermore, in a society where eggs and embryos are treated as properties that can be bought, sold, or rented, problems of commodification arise. "Commodification" refers to the association of a thing or a practice with attitudes and behaviours that accompany typical market transactions.[43] Because eggs and embryos have an intimate connection to personhood, their commodification could contribute to a diminishing sense of human personhood on an individual level, and might damage commitments to human flourishing at the societal level.[44]

The moral objections to the commodification of women's reproductive material echo the most vocal moral objections people often have to human cloning, the patenting of organisms or organic processes in whole or in part, surrogacy, and eugenics. In these activities, it is acceptable to treat some human bodies, some body parts, and some non-human living things in the same manner that manufactured objects are treated. The ethos that accepts such commodification risks fostering the view that the value of some living things, like fertile women, is the same as that of laundry detergent and toasters.[45] This objection combines deontological and consequentialist concerns. On the one hand, there is something inherently valuable about a living thing such that it is wrong to instrumentalize it. On the other hand, thinking of eggs and embryos as commodities shapes our understanding of what it means to be human. It can contribute to a lack of

respect for human dignity or to diminish respect for human life.

Also relevant is the fact that potential egg and embryo donors are encouraged to make such donations to medical research and therapy altruistically. The possibility of saving another person's life (or at least ease his or her suffering in some way) is an act represented as so intrinsically valuable that it would be diminished if the donor were compensated for it. However, as some scholars have pointed out, there is a significant tension between the altruism that individuals exhibit by donating their tissue for research and the current patent system, which encourages companies to stake lucrative property claims in that research.[46] Indeed, others argue even more strongly that donation of eggs and embryos as a gift both masks and legitimizes what is actually the extension of commodification. On the one hand, we have donors who believe that they are demonstrating altruism, but on the other we find biotechnology firms and researchers using the discourse of profit.[47] To call for the supplier to be altruistic when there is no similar call placed on those who would profit greatly from the sale of therapies made from the tissue can give way to exploitative practices.[48]

3d Justice

Another ethical issue resulting from the debate on stem cell research is related to our duties and obligations to promote social justice.[49] Stem cell research and the possible medical application resulting from it raise issues of social justice for several reasons. First, existing inequalities in access to healthcare mean that new technologies affect people quite differently according to ability to pay. Medical resources are not distributed equally and people do not have equal access to new medical procedures. That means that those with insurance will likely be beneficiaries of the new advances in medicine and technology. But not everybody in our society has adequate insurance or insurance at all. Approximately fifty million people at any given time during the year have not insurance at all. There is a fundamental inequality in the options that people have, and medical benefits resulting from stem cell research do nothing to narrow or solve the gap between them.

Second, even if the problem of access is resolved as an economic problem, this might not resolve the problem of inequalities and injustices present in our social context.[50] Some evaluations and reports on medical technologies (mostly European) have considered "economic issues" – but it seems only to the extent that the reports note that there is a problem with some people not having the ability to pay.[51] One response to these concerns has been the recommendation that medical technologies and procedures resulting from stem cell research be available to everyone. However, while following this recommendation would go some way toward addressing issues of social justice, it would not fully and adequately respond to those concerns. For example, we might say that government programs will provide access for those who do not have insurance. This might not solve the problems posed by other factors such as distrust of the healthcare system by certain groups, the role of religious beliefs in directing healthcare decisions, lack of access to information, and lack of education. Each of these factors might prevent people from having access to even free and public healthcare programs. Medical research practices such as involuntary sterilization, hysterectomy, mastectomy, and the Tuskegee study, also show that distrust of government health services might appear reasonable. A conception of medical technologies and practices that ignores these issues when evaluating the implementation of such techniques will then be deficient.

Also, given the fact that many healthcare problems are related to where and how one lives (namely, the relative level of dangers posed by the immediate environment of one's home or work), and that these are differentially distributed throughout the population in ways which track to economic status (and also race and sex), then providing the access to diagnosis and treatment alone will not solve problems of inequality. Thus focusing exclusively on the benefits or risks of stem cell research blinds us to the importance of the environment for preventing or curing diseases.

Furthermore, even if stem cell research were fully privately funded, problems of social justice still arise for several reasons. First, there may be complications due to stem cell procedures when used as medical techniques. Treatment of such complications may result in costs for the public

system. Second, societal interdependencies and professional contracts create and enhance doctors' abilities to use the results of stem cell research. They employ tools and technologies developed in part through societal resources. Also, public money supports physicians through learning, because virtually no student, even in private schools, pays for the full costs of education; taxes or donations usually supplement that cost.

4 Conclusion

Stem cell research is still very new. The potential benefits related to this research have created increasing expectations. However, ethical challenges need also be considered when we are evaluating such benefits. In this work, we have called attention to several ethical issues that deserve our attention in the assessment of stem cell research: the problem of the moral status of the embryo, the issue of safety for patients, concerns about coercion and exploitation of women in order to obtain eggs for research purposes, and issues of social justice related to access to the benefits of new medical technologies and procedures. A complete assessment of stem cell research cannot ignore these ethical issues.

Notes

1 Evans, M. J. and Kaufman, M. (1981) Establishment in culture of pluripotential cell from mouse embryos. *Nature* 292 (5819): 154–6.

2 Andrews, P. W. (2002) From teratocarcinomas to embryonic stem cells. *Philos Trans R Soc Lond B Biol Sci* 357 (1420): 405–17; and Pera, M. F. Reubinoff, B., and Trounson, A. (2000) Human Embryonic Stem Cells. *J Cell Sci* 113 (Pt 1): 5–10.

3 See, for example, Martin, G. R. (1980) Teratocarcinoma and mammalian embryogenesis. *Science* 209 (4458): 768–76.

4 Martin, G. R. (1981) Isolation of a pluripotent cell line from early mouse embryos cultured in medium conditioned by teratocarcinoma stem cell. *Proc. Nat. Acad. Sci.* USA 78 (12): 7634–8; See also Pera, M. F. Reubinoff, B., and Trounson, A. (2000) Human Embryonic Stem Cells. *J Cell Sci* 113 (Pt 1): 5–10.

5 Matsui, Y., Zsebo K., and Hogan B. L. (1992) Derivation of pluripotential embryonic stem cells

from murine primordial germ cells in culture. *Cell* 70 (5): 841–7.

6   Labosky, P. A., Barlow, D. P., and Hogan, B. L. (1994) Embryonic germ cell lines and their derivation from mouse primordial germ cells. *Ciba Found Symp* 182: 157–68; discussion 168–78.

7   Thomson, J. A. et al. (1995) Isolation of a primate embryonic stem cell line. *Proc Natl Acad Sci USA* 92 (17): 7844–8.

8   Thomson, J. A. (1998) Embryonic stem cell lines derived from human blastocysts. *Science* 282 (5391): 1145–7.

9   Shamblott, M. J. (1998) Derivation of pluripotent stem cells from cultured human primordial germ cells. *Proc Natl Acad Sci USA* 95 (23): 13726–31.

10 Reubinoff, B. E. (2000) Embryonic stem cell lines from human blastocysts: somatic differentiation in vitro. *Nat Biotechnol* 18 (4): 399–404.

11 See, for example, National Institutes of Health. (2002) Stem Cells: A Primer. Bethesda, Maryland: National Institutes of Health. Available at www.nih.gov/news/stemcell/primer.htm (accessed April 6, 2003); and Hadjantonakis, A.-K., Papaioannou, V. E. (2001) The stem cells of early embryos. *Differentiation* 68: 159–166.

12 Amit, M. et al. (2003) Human Feeder Layers for Human Embryonic Stem Cells. *Biol Reprod* 68 (6): 2150–6; Lim, J. W. Bodnar, A. (2002) Proteome analysis of conditioned medium from mouse embryonic fibroblast feeder layers which support the growth of human embryonic stem cells. *Proteomics* 2 (9): 1187–203; Richards, M. et al. (2002) Human feeders support prolonged undifferentiated growth of human inner cell masses and embryonic stem cells. *Nat Biotechnol* 20 (9): 933–6.

13 Filip, S., Mokry, J., and Hruska, I. (2003) Adult stem cells and their importance in cell therapy. *Folia Biol* 49 (1): 9–14; Preston, S. L. (2003) The new stem cell biology: something for everyone. *Mol Pathol* 56 (2): 86–96; Vats, A. et al. (2002) Stem cells: sources and applications. *Clin Otolaryngol* 27 (4): 227–32.

14 Preston, S. L. (2003) The new stem cell biology: something for everyone. *J Clin Pathol: Mol Pathol* 56 (2): 86–96; Orlic, D. et al. (2001) Bone marrow cells regenerate infarcted myocardium. *Nature* 410: 701–4; Poulsom, R. et al. (2001) Bone marrow contributes to renal parenchymal turnover and regeneration. *J Pathol* 195: 229–35; Alison, M. R. et al. (2000) Hepatocytes from non-hepatic adult stem cells. *Nature* 406: 257. Ferrari, G., et al. (1998) Muscle regeneration by bone marrow-derived myogenic progenitors. *Science* 279: 1528–30; Eglitis, M. A. and Mezey, E.

(1997) Hematopoietic cells differentiate into both microglia and macroglia in the brains of adult mice. *Proc Natl Acad Sci USA* 94: 4080–5.

15 See, for example, Hamasaki, T. et al. (2003) Neuronal cell migration for the developmental formation of the mammalian striatum. *Brain Res Brain Res Rev* 41 (1): 1–12; and Peault, B., Oberlin, E., and Tavian, M. (2002) Emergence of hematopoietic stem cells in the human embryo. *C R Biol* 325 (10): 1021–6.

16 See, for example, Rich, I. N. (2003) In vitro hematotoxicity testing in drug development: a review of past, present and future applications. *Curr Opin Drug Discov Devel* 6 (1): 100–9; and Margolin, K. (2002) High dose chemotherapy and stem cell support in the treatment of germ cell cancer. *J Urol* 169 (4): 1229–33.

17 See for, example, He, Q. et al. (2003) Embryonic stem cells: new possible therapy for degenerative diseases that affect elderly people. *J Gerontol A Biol Sci Med Sci* 58 (3): 279–87; Ostenfeld. T. and Svendsen, C. N. (2003) Recent advances in stem cell neurobiology. *Adv Tech Stand Neurosurg* 28: 3–89; Lakatos, A., and Franklin, R. J. (2002) Transplant mediated repair of the central nervous system: an imminent solution? *Curr Opin Neurol* 15 (6): 701–5; and Brehm, M., Zeus, T., and Strauer, B. E. (2002) Stem cells–clinical application and perspectives. *Herz* 27 (7): 611–20.

18 Hall, S. (2003) The Recycled Generation. *New York Times Magazine* (January): 30.

19 United Network of Organ Sharing, a US non-profit organization and clearing house. Available at www.unos.org. (accessed April 2006).

20 Rothman, D. (1999) The international organ traffic. In *Moral Issues in a Global Perspective*. ed., Christine Koggel (Peterborough, Ontario: Broadview Press): 611–18; Scheper-Hughes, Nancy (2000) The global traffic in human organs. *Current Anthropology* 41 (April): 2–19; *The Bellagio Task Force Report on Transplantation, Bodily Integrity, and the International Traffic in Organs* Available at www.icrc.org/Web/eng/siteeng0.nsf/iwpList302/87DC95FCA3C3D63EC1256B66005B3F6C (accessed April 2006).

21 Semb, H. (2005) Human embryonic stem cells: origin, properties and applications. *APMIS*. 113 (11–12): 743–50.

22 See, for example, Wilmut, I., Campbell, K., and Tudge, C, 2000 *The Second Creation*. Cambridge, MA: Harvard University Press.

23 Kennedy, D. (2006) Editorial Retraction. *Science* 311 (5759): 335.

24 Hall, V. J., Stojkovic, P., and Stojkovic, M. (2006) Using therapeutic cloning to fight human disease: A conundrum or reality? *Stem Cells*. Mar 30.

25 See, for example, Warren, M. (1997) *Moral Status: Obligations to Persons and Other Living Things.* New York: Oxford University Press.

26 See Warren, M. (1973) On the Moral and Legal Status of Abortion. *The Monist* 57 (1): 43–61.

27 See, for example, Marquis, D. (1989) Why Abortion Is Immoral. *Journal of Philosophy* 86 (4): 470–86.

28 Meyer, J. R. (2002) Human embryonic stem cells and respect for life. *J Med Ethics* 26: 166–70; Meilaender, G. (2001) The point of a ban: or how to think about stem cell research. *Hastings Center Report* 31 (1): 9–16; and Doerflinger, R. M. (1999). The ethics of funding embryonic stem cell research: a Catholic viewpoint. *Kennedy Institute of Ethics Journal* 9: 137–50.

29 See Noonan, J. (1970) An almost absolute value in history. In *The Morality of Abortion: Legal and Historical Perspectives,* ed. J. Noonan. Cambridge, MA: Harvard University Press, pp. 51–9.

30 See, for example, Warren, M. (1997) *Moral Status: Obligations to Persons and Other Living Things.* New York: Oxford University Press.

31 See, for example, Green, R. (2001) *The Human Embryo Research Debates: Bioethics in the Vortex of Controversy.* New York: Oxford University Press; Lauritzen, Paul (2001) Neither person nor property: embryo research and the status of the early embryo. *America* (March 26); Nelson, Lawrence and Meyer, Michael (2001) Respecting what we destroy: reflections on human embryo research. *Hastings Center Report* 31: 16–23; Shanner, L. (2001) Stem cell terminology: practical, theological and ethical implications. *Health Law Review Papers* (September 2): 62–6; Steinbock, B. (2000) What does 'respect for embryos' mean in the context of stem cell research? *Women's Health Issues* 10 (3): 127–30.

32 See, for example, National Research Council and the Institute of Medicine (2002) *Stem Cell and the Future of Regenerative Medicine.* Washington, DC: National Academy Press; McLaren, A. (2001) Ethical and social considerations of stem cell research. *Nature* 414: 129–31; McGee, G. and Caplan, A. (1999) The ethics and politics of small sacrifices in stem cell research. *Kennedy Institute of Ethics Journal* 9 (2): 151–8; and Strong, C. (1997) The moral status of preembryos, embryos, fetuses, and infants. *J Med Philos,* 22:457–78.

33 Outka, G. (2002). The ethics of human stem cell research. *Kennedy Institute of Ethics Journal* 12 (2): 175–214.

34 See, for example, Murray, T. (2005) Will new ways of creating stem cells dodge the objections? *Hastings Center Report* 32 (1): 8–9.

35 See, for example, Chung, Y., Klimanskaya, I., Becker, S., et al. (2005) Embryonic and extraem-bryonic stem cell lines derived from single mouse blastomeres. *Nature* (Oct 16); Meissner, A. and Jaenisch, R. (2005) Generation of nuclear transfer-derived pluripotent ES cells from cloned Cdx2-deficient blastocysts. *Nature* (Oct 16).

36 Marr, K. A. et al. (2002) Invasive aspergillosis in allogeneic stem cell transplant recipients: changes in epidemiology and risk factors. *Blood* 100 (13): 4358–66; Smaglik, P. (1999) Promise and problems loom for stem cell gene therapy. *Scientist* 13:14–15.

37 Solter, D., Gearhart, J. (1999) Putting stem cells to work. *Science* 283 (5407): 1468–70.

38 See, for example, Faden, R. and Beauchamp, T. (1986) *A History and Theory of Informed Consent.* New York: Oxford University Press, ch. 5.

39 See, for example Katz, J. (1972) *Experimenting with Human Beings,* New York: Russell Sage Foundation.

40 See, for example, de Melo-Martín, I. (1997) *Making Babies. Biomedical Technologies, Reproductive Ethics, and Public Policy.* Dordrecht: Kluwer University Press, ch. 4.

41 See, for example, de Melo-Martín, I. (1997) *Making Babies. Biomedical Technologies, Reproductive Ethics, and Public Policy.* Dordrecht: Kluwer University Press, ch. 4.

42 See, for example, Baylis, F. (2000) Our Cells/ Ourselves: Creating Human Embryos for Stem Cell Research. *Women's Health Issues* 10 (3): 140–5.

43 Altman, Scott (1991) (Com)modifying Experience. *Southern California Law Review* 65: 293–340, 293. See, for example, Margaret Jane Raddin (1991) Reflections on objectification. *Southern California Law Review* 65: 341–54; Justice and the market domain. In Roland Pennock and John Chapman, eds. (1989) *Markets and Justice.* New York: New York University Press, pp. 165–97.

44 Holland, S. (2001) Contested commodities at both ends of life: buying and selling gametes, embryos, and body tissues. *Kennedy Inst Ethics J.* 11 (3): 263–84.

45 Statement of the American Humane Association, on behalf of American Society for the Prevention of Cruelty to Animals, Animal Protection Institute, Committee for Humane Legislation, and Massachusetts Society for the Prevention of Cruelty to Animals: "It troubles us that animal patenting reduces the animal kingdom to the same level as laundry detergent and toasters. Animals are not objects." TAPRA '89 Hearings, 288.

46 Knowles, Lori (1999) Property, patents, progeny. *Hastings Center Report* 2: 38–40; Spar, Deborah (2004) The business of stem cells. *New England Journal of Medicine* 351: 211–13; Cahill Lisa Sowle

(2000) Social ethics of embryo and stem cell research. *Women's Health Issues* 10: 131–5.

47   Dickenson, Donna (2001) Property and women's alienation from their own labour. *Bioethics* 15: 204–17; Dodds, Susan (2004) Women, commodification, and embryonic stem cell research. In *Stem Cell Research: Biomedical Ethics Reviews.* James Humber and Robert Almeder, eds. Towota, New Jersey: Humana Press, pp. 152–72; Holland, Suzanne and Davis, Dena eds. (2001) Special Issue: Who's afraid of commodification. *Kennedy Institute of Ethics Journal* 11; Resnik, D. (2001) Regulating the Market for Human Eggs. *Bioethics* 15: 1–25.

48   See, for example, Sample, Ruth (2003) *Exploitation: What It Is and Why It's Wrong.* Rowman and Littlefield; Werthheimer, Allan (1999) *Exploitation.* Princeton: Princeton University Press.

49   See, for example, Rawls, J. (1971) *A Theory of Justice.* Cambridge, MA: Harvard University Press.

50   See, for example, de Melo-Martín, I. Hanks, C. (2001) Genetic technologies and women: the importance of context. *Bulletin of Science, Technology & Society* 21 (5): 354–60.

51   Hoedemaekers, R., ten Have, H. Chadwick, R. (1997) Genetic screening: a comparative analysis of three recent reports. *Journal of Medical Ethics* 23 (3): 135–41.

27

Food for Thought

Nina V. Federoff and Nancy Marie Brown

You people in the developed world are certainly free to debate the merits of genetically modified foods, but can we please eat first?
– Florence Wambugu (2003)

Visitors to Monsanto headquarters in St. Louis in 2003 stopped on their tour to admire a potted YieldgardPlus corn plant, a variety that can fend off both the corn borer and the corn rootworm. According to Eric Sachs, Monsanto's director of Scientific Affairs, "These are the only two corn pests that need pesticide applications." Each is a billion dollar pest for American corn farmers. "Both problems were solved by Monsanto," said Sachs.

Asked why he considered YieldgardPlus such a breakthrough, Sachs explained that traditional methods – even crop rotation – no longer hold the pests in check: "What has emerged over the past five years is that the insects have developed ways to get around crop rotation. Instead of laying their eggs in a cornfield, they lay their eggs in a soybean field, which is the rotation crop. So the larvae hatch in a cornfield. The second way is to lay eggs that overwinter more than one year. They don't hatch in the soybean field, they hatch in the cornfield after the rotation." Eventually, he agreed, the insects will become resistant to

the Bt toxin produced by YieldgardPlus. "We're already working on the second generation plant, producing a second Bt protein that targets the same insect. It makes it much more difficult for the insect to become resistant. It needs to have two rare mutations at the same time."

YieldgardPlus, like earlier versions of Bt corn, will benefit farmers and make money for the company – not trivial accomplishments. But it isn't likely to improve the image of genetically modified foods – or of Monsanto.

A very different variety of corn, casually mentioned by Monsanto vice president Rob Horsch, however, inspired the *St. Louis Post-Dispatch* to print an editorial, "Genetically Modified Crops Feed the World," in which the new corn is called a boon to the world and an example of the technology's greatest promise. Horsch calls it Golden Corn. It has pale kernels with brilliant orange embryos, half moons at the kernel's heart. "It's a white corn with a golden embryo," he explained. "It's really quite beautiful." Like Golden Rice, it has been modified to produce more beta

From Nina V. Federoff and Nancy Marie Brown, eds., *Mendel in the Kitchen* (Washington DC: Joseph Henry Press, 2004), pp. 295–316.

carotene, the precursor to vitamin A. White corn was chosen as the starting material because it is preferred in Africa, where vitamin A deficiency is a problem and corn is a staple crop. "It has an embryo-specific promoter," Horsch said. "But there's enough packed in the embryo that the whole kernel has higher beta carotene than yellow maize."

According to Horsch, Monsanto has said it will donate the rights to Golden Corn to the African Agricultural Technology Foundation (AATF), a nongovernmental organization developed with funding from the Rockefeller Foundation. "Farmers will be able to use the seed without paying Monsanto," the *St. Louis Post-Dispatch* explained.

Initially thought of as a patent bank, the AATF was designed, according to its brochure, "to resolve many of the barriers that have prevented smallholder farmers in Africa from gaining access to existing agricultural technologies that could help relieve food insecurity and alleviate poverty." It describes itself as "the neutral intermediary, a 'responsible' party between owners of proprietary technologies and those that need them."

Florence Wambugu was a member of AATF's Design Advisory Committee, along with representatives from seven African countries. USAID was represented, as were its counterparts in Denmark and the U.K. Gerard Barry represented Monsanto on the committee; he left the company late in 2003 to become head of the Golden Rice project at IRRI in the Philippines.

Yet while Monsanto, Dow AgroSciences, Pioneer Hi-Bred, and Aventis CropScience helped set up the foundation, Horsch points out that "the AATF board of directors doesn't include any of the companies." The companies donate licenses for patented genes or technologies; AATF owns any product that results. "For Golden Corn," said Horsch, "we've donated the basic set of genes to enhance beta carotene." Associates at Ohio State University and at CIMMYT, the international maize and wheat research center in Mexico, will use the genes to transform corn plants, analyze the results, breed the trait into popular varieties, and see them through the regulatory process.

Horsch enthusiastically identified this coalition of public and private entities as "we." "We want to combine beta carotene with the high protein maize already developed," he said of the future. "The problem with the high protein trait is that you can't see it, you can't taste it." But combined with the high beta-carotene trait, the protein trait becomes visible: the plants with orange embryos will also be high in proteins.

Golden Corn is just one example of a new trend in plant breeding, called biofortification. The idea was popularized by economist Howarth Bouis of the International Food Policy Research Institute, one of the 16 Future Harvest Centers under the World Bank's Consultative Group on International Agricultural Research. "The idea is to breed plants for higher nutrition content," said John Beard, a nutritionist at Penn State University. In May 2003 Beard had attended a meeting in Cali, Colombia, of some 70 plant breeders, nutritionists, economists, and community activists. Beard reported: "Ten years ago Bouis had the idea that supplement and fortification programs, from an economics aspect, need constant investment to get a response. He wondered, Is there some way to frontload the system? To make a big investment at first, and then tail off? Why not breed staple crops for higher nutrition?" Each of the international plant breeding centers allied with the Food Policy Research Institute was asked to start screening their varieties for micronutrients, not simply for hardiness or high yield. Said Beard, "You want to shift a whole population to a different plane of nutrition. You won't see benefits right away because the content of the micronutrient is usually quite small; hence, the improvements will occur in small increments over a long time."

Karel Schubert at the Donald Danforth Plant Science Center, for example, is working with nutritionists at Tufts University and the University of Florida to fortify rice with folate. Folate (or vitamin B9) and iron are the most important micronutrients whose deficiency leads to anemia, impaired cognitive development, and neural tube defects such as spina bifida, Schubert explained. "In Asia there are 100,000 cases of neural tube defects a year. In China it's called 'the disease of the winter marriage,' because the mother is deficient in folate prior to conception. It's also the most important birth defect in the U.S." Spinach is a prime source of folate. Legumes and some fruits are high in it, but all cereals, including rice, said Schubert, are "seriously low in folate." So are root and tuber crops, like potatoes. However, unlike the beta-carotene pathway, how plants make folate is not yet well understood. Schubert and his colleagues are studying the folate pathway in the common laboratory plant *Arabidopsis*; they have much to learn before any folate-rich crops will be ready to market.

Plant breeders are increasingly paired with nutritionists in such biofortification projects because simply raising the amount of folate or other nutrients in a plant isn't enough to enrich it: the nutrient must also be in the right form. A bioavailability study is needed to measure how easily the human body can take up the nutrient and make use of it. Penn State's John Beard, along with Jere Haas at Cornell University, became involved in the biofortification of crops after plant breeders at IRRI discovered an iron-rich variety of rice.

The rice, called IR68144, had been developed using conventional breeding techniques to grow well in poor soils and cold temperatures. When Bouis first announced his biofortification challenge, IRRI plant geneticist Glenn Gregario began screening the 80,000 varieties of rice in the IRRI germplasm banks. He grew 2,000 different varieties, harvesting the grain and analyzing it for iron content, before he discovered IR68144 in 1998. Because it wasn't altered by molecular techniques, IR68144 was not subject to any safety testing. It was grown in quantity immediately and used in feeding trials. It advanced through cell culture tests and a pilot study of 27 people to a full feeding trial involving 300 nuns in a convent in Manila. If not for the typhoons that twice destroyed the harvest, proof of the rice's nutritional advantage (if any) would have been available in 2000.

As it was, by 2003 Beard and his collaborators at Cornell University had analyzed 10 different parameters of nutrient status on each of 1,080 blood samples from the 300 nuns. They measured levels of vitamin A, iron, folate, vitamin B12, and zinc. The blood tests were combined with direct dietary analyses. "We weighed everything the sisters consumed three times randomly every two weeks throughout the course of the nine-month feeding trial," Beard explained. "Their meals were analyzed for micronutrient content by using the Philippine Food Tables, which tells us

that so many grams of this food contains so much iron, zinc, etc. Then, based on the amount of vitamin C, coffee, and tea consumed with the meal, we can calculate the bioavailability of the rice. Then we ask, What happens when we switch the forms of rice in the diet? Does the iron status change?"

According to an IRRI publication, "The trial of IR68144 is being widely regarded as an attempt to prove the concept that staple foods enriched with micronutrients directly benefit human nutrition." Said Gregario, "If this new variety is successful, then micronutrient deficiency may be a part of history – like smallpox or polio. But that's still just a dream."

Golden Rice is also still just a dream – although Monsanto's Horsch confidently predicts that it will eventually reach the people it was designed to help. For Golden Rice inventor Ingo Potrykus, bioavailability studies are critical. He agrees that to get governments in developing countries and international humanitarian groups to back Golden Rice as a means of easing vitamin A deficiency, IRRI must prove that the beta carotene in the rice is usable and will make a difference. Until such tests are done, the developers of Golden Rice remain open to attacks from critics who believe they are making "false promises" by claiming that it will help alleviate vitamin A deficiency.

Golden Rice, said Greenpeace, is "fool's gold." Michael Pollan, in a *New York Times Magazine* article, cited a figure that appears to have come from Greenpeace's propaganda: "An 11-year-old would have to eat 15 pounds of cooked golden rice a day – quite a bowlful – to satisfy his minimum daily requirement of vitamin A." Yet this figure is merely a conjecture. Robert Russell, a nutritionist at Tufts University and a specialist in vitamin A nutrition, has calculated a figure of 200 grams per day – 7 ounces – or two out of the three to four bowls full that an adult on a rice-based diet ordinarily eats. Which figure is right? It will take a bioavailability trial to learn. But as long as Golden Rice is confined to the greenhouse, the protocol established for the iron-rich IR68144 cannot be followed: enough rice cannot be grown under glass to feed 300 nuns for nine months.

Lecturing at Yale University in April 2003, Potrykus was asked if Golden Rice, like Borlaug's wheat and the other Green Revolution crops, would not simply contribute more to the problem of malnutrition by encouraging poor farmers to move from nutritious vegetable-based multi-cropping systems to a rice monoculture. "It was not without reason that production moved from high nutritious low-yield crops to low nutritious high-yield crops," answered Potrykus. "You can either die from hunger or from malnutrition. I don't know what is worse. The solution I am offering is to make high production crops like rice or wheat more nutritious."

Asked why he invented Golden Rice, he answered, "I have been asked this before, and I have thought about it. I'm a refugee myself, from a part of Germany generously given to Russia after the war." For Potrykus, a year shy of 70, "the war" is World War II. "I lost my father in the last days of the war," he continued. "He was a medical doctor. My mother had to raise four children without anything. She managed to allow us all higher education. But we have experienced hunger. Between the ages of 12 and 14, much of my brother's and my attention was given to where to find something to eat. You could say we reaped what we did not sow."

After earning a college degree in biology, Potrykus taught high school. He married and had children. He considers himself first "an old-fashioned field biologist, a naturalist. From my mother I have this strong interest in nature," he said. A project he has been working on for many years is to videotape all the birds of North America. "I was teaching in high school when the director of the Max Planck Institute for Plant Breeding, Josef Straub, offered me the possibility of a Ph.D. I was exposed to a lot of breeding practices. I did the Ph.D. while still teaching half time, and then I went back to teaching full time. I came into this idea already when I was teaching high school. I was running courses then on the topic, 'More Food for More People.' This looks strange now, because it was the early sixties and the peak of the Green Revolution."

He was still teaching More Food for More People 20 years later at the university level. "In my student evaluations I heard again and again the complaint that I am talking about transgenic plants and they don't want to hear about it. I felt I had to produce a case that demonstrates to my students that they are wrong, that you can use the technology for a good purpose."

Does the world really need more food? Since 1798, when Malthus published his *Essay on the Principle of Population*, catastrophists have predicted imminent famine. Nearly 200 years later Paul Ehrlich, like Malthus, reduced the problem to its simplest form: too many people, not enough food. In his 1968 book *The Population Bomb*, Ehrlich concludes, "The battle to feed all humanity is over." "At this late date," he says, "nothing can prevent a substantial increase in the world death rate." Ehrlich was writing on the eve of the Green Revolution – the largest expansion in agricultural productivity in the history of human civilization. His prediction, like Malthus's, did not come true.

And yet the earth is finite.

In the course of just one century, the twentieth, the human population doubled twice: from 1.5 billion to 3 billion to 6 billion. It continues to expand by 80 million people a year. It doesn't take a catastrophist to see that humanity is pushing against some planetary limits. The demands of agriculture and industry, human habitation, and transportation are driving to extinction more species per year than at any time since the Cretaceous. In 1998 Dan Simberloff, an ecologist at the University of Tennessee, was quoted in the *Washington Post*: "The speed at which species are being lost is much faster than any we've seen in the past – including those [extinctions] related to meteor collisions."

There's little doubt that the trends of the twentieth century cannot continue. In his 1995 book *How Many People Can the Earth Support?*, Joel Cohen lays out the problem of the earth's "carrying capacity" in all its complexity. There is no single answer, no simple answer to his title's question. The number of people Earth can accommodate depends on how they live and how well they manage the planet's physical, chemical, and biological environments. The choices available to us – and the choices we make – depend on science and technology, on the one hand, and on politics, preferences, and moral judgments, on the other.

There is no single, simple path to a sustainable future, either. To meet people's needs without further harm to the environment requires changes of many kinds. Cohen puts them into three categories: we must "put fewer forks on the table" (decrease population growth and reduce consumption), "make a bigger pie" (grow more food), and "teach better manners" (change how people interact with each other for everyone's benefit).

The rate of population growth *is* declining – more rapidly than the experts predicted even a decade ago, when the population was expected to double yet again before stabilizing. Some of the underlying trends are positive, such as improvements in education and economic development, particularly for women. But some are negative. They are the familiar scourges of too many people: disease, famine, war. As Garrett Hardin, author of the famous essay "The Tragedy of the Commons," points out, no one dies of overpopulation.

Experts now estimate that the number of people will stop growing by the middle of the twenty-first century. Before then, however, some 3 billion more people will be living on Earth than are alive now. This number is nearly 10 times the population of the United States today. Many – probably most – of these people will live in countries that are, even now, unable to provide their people with enough food for good health.

Putting fewer forks on the table doesn't just mean decreasing the population growth rate. It is also about how much we eat, what we eat, and what we waste. An adult needs between 2,000 and 2,200 calories per day – more for men and less for women, more for those who do heavy manual labor and less for those who sit at desks. In *Feeding the World*, Vaclav Smil estimates that the total food available per person in 1990, according to data from the FAO, was about 2,700 calories per day. Averaged worldwide, each person ate about 2,000 calories per day. The rest – about 700 calories per person per day – was lost during harvesting, processing, and distribution, or simply discarded.

These figures say that we already produce enough food to feed the world. What they conceal is the appalling gap between the richest and poorest nations. The amount of food each person in a rich country eats every day is at the high end of the range, about 2,200 calories each day – and sometimes even higher, up to 2,700 calories. In the poorest countries, people eat an average of 1,500 to 1,700 calories per day.

The amount of food available to people in rich and poor countries is also very unequal.

According to FAO statistics, the per person food availability in the United States between 1992 and 1994 was about 3,600 calories per day, some 40 percent more than was eaten. (Available, in this case, assumes a person can afford to buy it, yet even in the United States millions of people use food stamps or go hungry.) In less developed countries, only 10 to 15 percent more food was available than was eaten. Moreover, the people in affluent countries consume between two and four times as much milk and meat as the world's average citizen. Converting plant foodstuffs, like grass and grain, to milk and meat is inefficient. Of the usable energy in animal feed, only 33 percent ends up as food energy in milk, 20 percent in pork, and 6 percent in beef.

But getting everyone to adopt a vegetarian lifestyle is not a likely solution, even if every grain and vegetable is biofortified. As Dennis Avery of the Hudson Institute points out, "No country or culture in history has voluntarily accepted a diet based solely on the relatively low-quality protein found in vegetable sources. Meat and milk consumption is rising by millions of tons per year in China and India right now as their incomes rise." Even among America's 12 million vegetarians, only 4 percent never eat any animal products, according to a survey commissioned by the *Vegetarian Times* in 1992. Fifty percent agreed with the statement, "In order to satisfy my appetite, a main meal must include meat." Nonetheless, Smil argues, cutting down the amount of food rich nations waste, reducing their meat consumption to a healthier 25 percent of total calories, and breeding animals that are more efficient in converting feed to food would go a long way toward feeding the world of the future.

Yet even if food is shared out fairly among all peoples – with no waste – we will soon need more: in 50 years, there might be another 3 billion people who need to eat. To "make a bigger pie," in Cohen's words, we have two choices. We can cultivate more land, knowing that land put under the plow is land taken away from black bears and monarch butterflies, Bengal tigers and tropical birds. Or we can produce more food from the land that is already being farmed.

The huge increases of the Green Revolution came from improvements in the yield of each plant, which made each farmed acre give more food. But in the last decade, yields of the major grains have not increased significantly. Agronomist Ken Cassman has argued that some crops are reaching their yield limits, the maximum amount of grain that they can produce under the very best of weather and fertilization conditions. Where will additional yield gains come from?

When Ingo Potrykus in Switzerland was lecturing his high-school students on the subject of More Food for More People, no one could begin to answer this question. What has changed since then is our knowledge of plant biology. Today we know the sequence of the entire genome of the tiny model plant *Arabidopsis thaliana*, as well as the genome sequences of two rice varieties. Work on sequencing the corn genome is underway, and a start has been made on the genomes of wheat, soybeans, and many other crops. Knowing the genome sequences makes it vastly easier to identify, analyze, change, and reintroduce genes that affect critical processes in plants. Comparing genomes, we've identified the similarities among genes and we've understood that what is learned in one plant can often be applied to another. All of these advances have quickened the pace of discovery and broadened our knowledge of photosynthesis, of how plants use nitrogen, and of how they cope with excess salt, toxic chemicals, and lack of water. Together with the ability to move genes between plants and into plants from other sources, this knowledge lets us begin tackling the barriers that limit agricultural productivity.

The first of these barriers is nitrogen use. Plants, together with the bacteria that live in and with them, convert carbon and nitrogen in the air into sugars and amino acids. These sugars and amino acids are the basic building blocks of the starches and proteins of which plants are made – and which feed both humans and their domestic animals. The amount of nitrogen that can be provided by bacteria or derived from composted animal and plant materials is much less than plants can use. Once people discovered how to fix nitrogen from the air, converting it into fertilizers, it was possible to overcome this limit on the nitrogen supply. Crop yields rose as fertilizer plants were built all over the world.

But beyond a certain point, adding more fertilizer no longer helps. Plants use only about half of the nitrogen applied as fertilizer, even under the best conditions. Much of the rest runs off with the rain into streams, rivers, lakes, and oceans. There it becomes a major pollutant. It acts as a fertilizer for small organisms, particularly algae, whose populations explode. Algae produce oxygen by day and consume it at night, depleting the oxygen available to other animals. When algae die, their decay also uses up oxygen in the water. The end result of an algal bloom is the suffocation of fish and other animals that live deeper down.

One way to solve the problem of nitrogen pollution is to increase the plants' ability to use nitrogen, turning more fertilizer into plant proteins and leaving less to run off the land. By increasing the crop's yield, such an improvement could also benefit farmers who do not use chemical fertilizers, either because they cannot easily afford them or because they are limited by the Organic Rule.

Whether it comes to the plant by means of nitrogen-fixing bacteria, from animal manure or plant compost, or from chemical fertilizer, plants first convert nitrogen to ammonia (NH_3). Then a plant enzyme called glutamine synthetase attaches the nitrogen atom in the ammonia to glutamic acid, an amino acid. Once attached to glutamic acid in the plant, the nitrogen can be moved by other enzymes into a variety of small molecules, including all of the other amino acids. These, in turn, are converted to proteins and other nitrogen-containing compounds. When researchers introduced a bean glutamine synthetase gene into wheat, providing more glutamine synthetase, the wheat plants developed faster, flowered earlier, and produced heavier seeds containing more protein.

Another important nitrogen enzyme is glutamate dehydrogenase. When nitrogen is abundant, glutamate dehydrogenase seems to help the plant redistribute it and make better use of its supply. The enzyme removes the nitrogen from amino acids, leaving keto-acids and NH_3 that are then available to make other needed compounds. Again, preliminary experiments show that plants containing an extra, highly expressed glutamate dehydrogenase gene grow bigger, yielding more biomass than plants that don't carry the extra gene. While these observations are still far from the wheat field, they suggest that changing the expression of a relatively small number of genes might produce substantial gains in crop yields.

Yields don't depend just on nitrogen. They also depend on how efficiently the crop plant makes use of carbon. Plants take carbon from the carbon dioxide (CO_2) in the air and convert it into a sugar molecule, which consists of carbon and the hydrogen and oxygen from water (H_2O). The energy for this reaction comes from photosynthesis. The plant absorbs light and uses it to increase the energy level of electrons. The excited electrons then trickle down through a chain of proteins. There, the energy is extracted and used to drive the reduction of CO_2 by the enzyme RuBP carboxylase, commonly called Rubisco – the most abundant protein on Earth. The net result of this reaction is that the carbon atom is incorporated into a sugar molecule.

Plants differ in their ability to capture CO_2 and in how efficiently they convert it into sugars. The basic photosynthetic process, called "fixing" carbon, captures the carbon in a sugar-like molecule that has three carbon atoms. Plants that can carry out only this basic reaction are called C_3 plants. But some plants, designated C_4 plants, have an additional pathway that makes a four-carbon sugar. In C_3 plants, photosynthesis is always coupled to photorespiration, which is carried out by the same enzyme, Rubisco, and drives the reaction in reverse. That is, photorespiration consumes oxygen and releases CO_2, but it doesn't capture the energy this reaction produces – that energy is wasted. As much as half of the carbon drawn from the air in the first place is released again through photorespiration, using up energy in the process.

Although C_4 plants use a bit more energy to fix carbon in the first place, overall they can be two or even three times as efficient as C_3 plants. C_4 plants have an additional enzyme, called PEP carboxylase, that fixes carbon into a four-carbon sugar. PEP carboxylase, unlike Rubisco, is not bothered by oxygen. In simple terms, the C_4 photosynthetic pathway serves as a CO_2 pump. It concentrates CO_2 near the Rubisco enzyme, suppressing the oxygen-driven photorespiration in favor of carbon fixation. A majority of land plants, including rice, wheat, oats, and rye, are the less-efficient C_3 plants. Among major crops, only corn is a C_4 plant. Although many plant breeders have tried, they have not been able to transfer the C_4 traits to C_3 crop plants using conventional breeding techniques.

In 1999 Maurice Ku of Washington State University, together with a group of Japanese scientists at the National Institute of Agrobiological Resources in Tsukuba and at the Bio-Science Center of Nagoya University, reported that he had successfully transferred the PEP carboxylase gene from maize to rice. The researchers also tested a second maize gene, one that encodes pyruvate orthophosphate dikinase, an enzyme that provides one of the compounds that the PEP carboxylase uses in fixing carbon. The rice plants carrying either maize gene showed higher rates of photosynthesis. Oxygen didn't interfere with photosynthesis in the rice plants carrying the maize PEP carboxylase gene – this oxygen-insensitivity of carbon fixation is the hallmark of the C_4 plant. The gene was expressed at a high level. Indeed, as much as 12 percent of the protein in these plants was PEP carboxylase. The researchers reported that the yields of the rice plants that expressed the added corn PEP carboxylase were 10 to 20 percent higher that those of the parental plants. The rice plants expressing the other maize gene gave yields as much as 35 percent higher.

These results, while they await confirmation and are a long way from being applied to agriculture, make plant breeders optimistic. Molecular techniques might be able to break through limits that have long stymied their best efforts. Indeed, they make it possible to alter the fundamental biochemical reactions that set the upper limit on the yields of crop plants today.

Another limit on yield is water. Lack of water – drought – and too much salt (which has the same effect) reduce crop yields around the world. Both dehydrate plants: water from the inside of the plant's cells moves out by a process called osmosis. This loss of water triggers severe stress reactions. Photosynthesis shuts down. In extreme cases the plant dies. Salty soil can come from irrigating too heavily for too long. Much of the irrigation water simply evaporates, leaving behind whatever salts were dissolved in it. Soil salinization affects more than 700 million acres of otherwise arable land. If plants could better withstand salt stress, yields could increase even on marginal land.

Some plants have evolved to live comfortably under very salty conditions. These plants, called halophytes, have several mechanisms that prevent or limit the damage done by water loss and too much salt. Some plants pump out sodium, the most damaging component of salt. Others accumulate it inside of vacuoles, central compartments in their cells. Still others fill up their cells with compounds called osmoprotectants that keep the water inside. Sometimes these compounds are sugars or amino acids, but plants – and marine algae – use other compounds as well. A number of studies have shown that introducing genes that code for enzymes that produce osmoprotectants increases plants' ability to withstand salt stress.

So far, most of these studies have used genes that come from organisms other than plants, particularly bacteria. Plant genes have yet to be explored for this purpose. But both genes that encode salt pumps and those that code for enzymes that cause osmoprotectants to be made are being identified and studied. Other genes are being investigated as well, such as the regulatory genes that control the plant's overall response to salt.

Still another major factor that limits crop yields is the quality of the soil. Not all soils are hospitable to plants. One major problem in acid soils, for instance, is aluminum. Aluminum is the third most abundant element on Earth. When soil is alkaline or neutral, aluminum is in a form that doesn't harm growing plants. When the soil's acidity increases, the aluminum is converted to a soluble form that is toxic to plants. Soil acidification is exacerbated by some farming practices and by acid rain. It affects an estimated 40 percent of arable land worldwide. In the tropics, aluminum toxicity cuts yields by as much as 80 percent on about half of the arable land. Even at quite low concentrations, aluminum ions inhibit root growth, which in turn affects plant growth and yield.

Plants have developed several mechanisms for aluminum tolerance. Some plants exclude it. Others secrete organic acids, such as citric acid, oxalic acid, and malic acid, that bind tightly to the aluminum and prevent the plant from absorbing it. It is the growing root tips that secrete the acids, forming a protective shield. In 1997 a research group led by Luis Herrera-Estrella at the National Polytechnic Institute in Irapuato, Mexico, reported that the ability of a papaya plant to tolerate aluminum could be enhanced by introducing a bacterial gene coding for citrate synthase, the enzyme that produces citric acid. The plants produced and secreted more citric acid, allowing them to grow in soils that had been toxic to them previously. These experiments establish the principle of using genes to enhance the aluminum tolerance of plants, but the first experiments might not provide the final answers. As more is learned about how plants tolerate aluminum, more genes will be identified.

Limits on nitrogen and carbon use, salt and aluminum toxicity – these are among the major problems that must be overcome if farmers' yields are to double or perhaps even triple to meet the demands of a human population 8 or 9 billion in number and demanding more and better food. It seems unlikely that the future holds another simple breakthrough, like the synergy between dwarfing genes and fertilizer that made the Green Revolution possible. But a breakthrough that enhances either the use of nitrogen or the efficiency of photosynthesis, or both simultaneously, could push yields up dramatically. More likely, the advances will be incremental. Small improvements of many different kinds will be made in many different crops. But it depends. And what it depends on has rather less to do with the science than with people.

In 2002 Zambia's president rejected a shipment of donated corn from the United States, ostensibly because genetically modified food had not been proven safe to eat. According to the *Los Angeles Times*, "Many Zambians in rural areas have resorted to eating leaves, twigs, and even poisonous berries and nuts to cope with the worst food crisis in a decade hitting southern Africa." Zambian president Levy Mwanawasa had declared a food emergency in the nation three months earlier. Yet he refused the American maize, saying, "We would rather starve than get something toxic."

His choice was bewildering. The health consequences of starvation are undeniably terminal – and there is no evidence that genetically modified corn is toxic. As the *Los Angeles Times* reported, "The United States, United Nations, and humanitarian aid groups insist that the U.S.-donated corn is safe and identical to grain eaten daily by people in the United States, Canada, and other countries."

Mwanawasa's logic is indeed incomprehensible – unless one views it from an economic standpoint. The European Union has urged African governments who want to trade in Europe to treat genetically modified crops as a serious biological threat. If the Zambian government were to lose its "GM-free" status, it would lose access to European markets, where its exports include organic baby corn and carrots. And indeed, President Mwanawasa was quoted as saying that he does not want the introduction of genetically modified foods to hurt his export trade with Europe. The government of Zimbabwe, also facing famine in 2002, agreed to accept U.S.-donated corn only if it was first ground into cornmeal "so that the food aid cannot be planted," the BBC reported. "Zimbabwe and some of its neighbors are worried that GM seeds could contaminate locally grown crops, threatening lucrative exports to Europe, which insists that food must be GM-free."

"All across Britain and most of the rest of Europe," the New York Times reported in February 2003, "shoppers would be hard pressed to find any genetically modified, or GM, products on grocery store shelves, and that is precisely how most people want it." At the Happy Apple greengrocer in the small English town of Totnes, "the roasted vegetable pasty is labeled, clearly and proudly, as GM-free," the Times reported. When asked her opinion on GM foods, one shopper replied, "It's a kind of corruption, not the right thing to do, you know?"

The private and personal choices of European shoppers like this one are setting the public policy of African nations. Zambia's decision to refuse American corn was greeted with disbelief around the world. But a thoughtful look reveals it to be a logical, if unintended, consequence of the expression of a preference on the part of Europeans for foods that have not been modified by molecular techniques. African and other less developed nations are caught in a terrible bind. With almost three million people at risk of starvation, they are faced with a choice between immediate suffering and closing the door on future economic prosperity.

The heart of economic development is the ability to grow more crops than growers and their families need to survive. The Rockefeller Foundation's Gordon Conway and Gary Toenniessen note that two-thirds of sub-Saharan Africa's more than 600 million people live on small farms. The food they produce, combined with what they can afford to buy, is insufficient. The result is that 194 million Africans, mostly children, are undernourished. "Africa does not produce enough food to feed itself even with equitable distribution," they wrote in 2003. "Food aid to Africa – currently running at 3.23 million tons annually – helps prevent starvation but can create dependency."

The first step, they say, is to achieve food security, which simply means reliable access to enough food to lead a healthy, active life. "Most African farmers have land assets adequate to provide food security and to rise above subsistence." But to do so, "they need to intensify production with genetic and agro-ecological technologies that require only small amounts of additional labor and capital." In a July 2003 New York Times Op-Ed piece on the same subject, Norman Borlaug argues: "Biotechnology absolutely should be part of African agricultural reform; African leaders would be making a grievous error if they turn their backs on it." He strongly urged African leaders not to follow the lead of Europe, where biotechnology has been "demonized," but to use it for the benefit of their farmers and their people.

How can Africa consider adopting molecular technology if by doing so its farmers are locked out of European markets? How can Africa afford not to adopt approaches that are biology-based, low-cost, and beneficial on both small and large scales? How did we – the industrialized nations that developed these molecular techniques for plant breeding – contribute to this extraordinary and deeply distressing state of affairs?

Calestous Juma, director of the Science, Technology and Globalization Project at Harvard University's Kennedy School of Government, compares it to the persecution of coffee. "In the 1500s," he explains, "Catholic bishops tried to have coffee banned from the Christian world for competing with wine and representing new cultural as well as religious values." Juma continues, "In public smear campaigns similar to those currently directed at biotech products, coffee was rumored to cause impotence and other ills and was either outlawed or its use restricted by leaders in Mecca, Cairo, Istanbul, England, Germany, and Sweden. In a spirited 1674 effort to defend the consumption of wine, French doctors argued that when one drinks coffee: 'The body becomes a mere shadow of its former self; it goes into a decline, and dwindles away. The heart and guts are so weakened that the drinker suffers delusions, and the body receives such a shock that it is as though it were bewitched.'"

In coffeehouses throughout Europe, and increasingly in America, similar campaigns are being waged now against genetically modified foods – using equally exaggerated claims of potential harm. Juma writes: "Debates over biotechnology are part of a long history of social discourse over new products. Claims about the promise of new technology are at times greeted with skepticism, vilification, or outright opposition – often dominated by slander, innuendo, and misinformation. Even some of the most ubiquitous products endured centuries of persecution."

It is sobering, for example, to recollect that vaccinations against smallpox – a disease that kills 30 percent of the people it infects and disfigures the rest – were vilified in editorials and cartoons, publicly protested, and strongly resisted. Fortunately, national governments and the United Nations persisted in vaccinating people – sometimes even with a bit of coercion – and smallpox is gone.

The problem today, suggests Stanford University's Henry Miller, a former FDA official, is compounded by governments that increasingly depend on public opinion in formulating policy involving scientific issues. Noting that in 2003 the British government organized focus groups "to find out what ordinary people really think [about GM foods] once they've heard all the arguments," Miller says: "Getting policy recommendations on an obscure and complex technical question from groups of citizen nonexperts (who are recruited through newspaper ads) is similar to going from your cardiologist's office to a café, explaining to the waitress the therapeutic options for your chest pain, and asking her whether you should have the angioplasty or just take medication."

Even when they've "heard all the arguments," intelligent and inquiring minds not trained in the subject can still be confused. In June 2003, for instance, a Zambian newspaper reported that "Maize is not directly consumed in America." The writer, Simon Mwanza, was one of seven African journalists who had toured the United States to learn about biotechology, for which he used the abbreviation BT. His tour had included stops at Monsanto and Pioneer Hi-Bred, several universities, the Center for Science in the Public Interest, the Pew Initiative on Food and Biotechnology, the National Corn Growers' Association, USAID, and the Washington office of a senator from Iowa. None, apparently, had invited him to try cornflakes, corn chips, or corn-on-the-cob. Or perhaps the problem was one of translation, and the fact that "corn" was the American name for "maize" was not made plain. At the University of Maryland he learned that "most of the maize produced in the USA was for animal feed," he wrote, and that "the US also uses maize to produce ethanol." Later, visiting the National Corn Growers Association, he wrote, "The journalists' eyes popped out when they were shown a wide range of products made from maize – thanks to BT." The Association representatives gave the visiting journalists "t-shirts made from corn to make their BT point abundantly clear." Mwanza's conclusion about biotechnology? "While BT appears a promising solution to agriculture, it is difficult to forget what Dr. Scott Angle of Maryland University said: 'We don't know what we don't know.'"

Most people have not devoted even two weeks, as Mwanza and his colleagues did, to understanding the technology behind genetically modified foods. Still, they have strong opinions. A 2003 survey in America found that 58 percent of the people asked were "unwilling to eat genetically modified (GM) food." That majority response seems to send a clear signal to food producers and seed companies. That is, until you ask the next question: "What food are Americans willing to eat?" A 1993 survey of New Jersey residents found that 41 percent of the respondents would not eat food produced through "traditional hybridization techniques." A full 20 percent said it was "morally wrong to produce plants this way." Yet since 1970, more than 96 percent of the corn grown in America has been hybrid corn. Sweet corn, cornflakes, corn muffins and chips, the corn fed to our beef cows, chickens, and pigs is all hybrid corn, as is the source of the corn sweeteners and corn starch found in mayonnaise, peanut butter, chewing gum, soft drinks, beer, wine, frozen fish, processed meats, all dehydrated foods, all powdered foods, and all granulated foods.

It is perhaps not surprising that the Organic Rule, so heavily influenced by public opinion, forbids the use of irradiation, antibiotics, and molecular genetic modifications in producing "organic" food. But in the end, says Miller, "The goal of policy formulation should be to get the right answers." "Although it may be useful, as well as politic, for governments to consult broadly on high-profile public policy issues," he adds, "after the consultations and deliberations have been completed, government leaders are supposed to *lead*."

Getting the right answers on genetically modified foods matters – profoundly. The science is complex, and advancing daily. As we continue to learn more about how plants grow, and as we become more skillful in transferring useful genes into plant cells, we find ourselves with an opportunity to get it right – not just for the economic benefit of large companies, but for the benefit of ordinary people everywhere in the world.

Getting it "right" will have many local meanings. It will mean virus-resistant tomatoes in Italy, herbicide-resistant wheat in Washington, and insect-resistant Bt corn in Iowa. It will mean aluminum-tolerant crops in the tropics and virus-resistant sweet potatoes in Africa. Some of these crops will be produced by companies because they can return a profit, every company's prerequisite for survival. But others might never make money. These will come only when governments everywhere recognize – and invest much more heavily in – agriculture both as a public good and an environmental necessity. As well, these will come only when regulators and regulations become more responsive to evolving knowledge than to public perceptions and anxieties. Only then will public sector scientists be able to invest their time and knowledge in raising yields in an ecologically sound way.

And yet, in the deepest sense, getting it "right" is the same for all nations: having enough to eat while preserving and protecting the environment.

Every civilization rests on food. Over many thousands of years, humans have devoted a great deal of intellect, energy, and effort to changing wild plants into food plants. These changes – all of them – involved changes in the plants' genes. We have a long history of tinkering with nature. It is no exaggeration to say that our tinkering, our modification of plant genes – and those of domesticated animals – to meet our nutritional needs, has shaped our world.

At the same time, agriculture in its very essence is ecologically destructive, whether it is performed at the subsistence level for a single family or on an industrial scale. The challenge now, as our population pushes against the planet's limits, is to lessen the destructive effects of agriculture on the earth even as we coax it to produce more food.

"To assert that GM techniques are a threat to biodiversity is to state the exact opposite of the truth," writes Peter Raven, director of the Missouri Botanical Garden. "They and other methods and techniques must be used, and used aggressively, to help build sustainable and productive, low-input agricultural systems in many different agricultural zones around the world."

At the International Botanical Congress in 1999, Raven announced: "We are predicting the extinction of about two-thirds of all bird, mammal, butterfly, and plant species by the end of the next century, based on current trends." In a 2003 essay he elaborates, "These organisms are simply beautiful, enriching our lives in many ways and inspiring us every day. By any moral or ethical standard, we simply do not have the right to destroy them, and yet we are doing it savagely, relentlessly, and at a rapidly increasing rate, every day. Many believe, and I agree with them, that we simply do not have the right to destroy what is such a high proportion of the species on Earth. They are, as far as we know, our only living companions in the universe."

And the greatest danger to other species is our own need for food. "Nothing has driven more species to extinction or caused more instability in the world's ecological systems than the development of an agriculture sufficient to feed 6.3 billion people," Raven says. "The less focused and productive this agriculture is, the more destructive its effects will be."

Using our growing knowledge of plants and plant genes, and our increasing skill at modifying them with molecular techniques, we can make agriculture more focused and more productive – if we are careful. The thoughtful choice of genetic modifications can help us become better stewards of the earth. The key words here are careful and thoughtful. Whether the technology will be helpful or harmful, in the long run, depends on how it is used, on the choices people make. The better we understand what this technology is – how it has come to be and what it involves – the wiser will be the decisions we make as a civilization about how it will be used in the future.

28

Value Judgments and Risk Comparisons. The Case of Genetically Engineered Crops

Paul B. Thompson

This paper aims to identify and elucidate some of the philosophical issues and value judgments associated with the claim that risks from transgenic and conventional crops are comparable from a scientific perspective. Crops produced using techniques that insert DNA directly into the genome of the plant, by *Agrobacterium tumefaciens*-mediated transformation or by the gene gun, will be referred to as "transgenic or genetically modified (GM) crops," and the terms "conventionally bred" or simply "conventional crops" will be used to indicate all other agricultural crop varieties. Breeders distinguish farmer-bred land races from conventionally bred crops, but the critical requirement in this context is simply that the crops have been developed without the use of recombinant DNA techniques for gene transfer.

I will not discuss food safety and socioeconomic issues so that I can concentrate on other aspects of GM technology. Although critics of transgenic technology have insisted that these socioeconomic impacts have been unjustly overlooked by scientists and regulators (Krimsky and Wrubel, 1996), they are not material to any substantive differences of opinion about the comparability of risks. For both food safety and environmental safety, U.S. regulatory decision making has been based on the assumption that once transgenes are integrated into the genome of a transformed crop and both gene function and reproductive stability have been verified, risks may be evaluated exclusively with respect to phenotypic traits (Miller, 2000). This assumption is critical to debates over food safety and environmental impact of transgenic crops. However, the scientific basis for the justification of this assumption and subsequent attempts to evaluate and compare risks with respect to food consumption and environmental impact of crop production proceed from different principles. Although biochemistry and toxicology are the primary bases for evaluating food safety risks, environmental risks are evaluated on the basis of ecology, population genetics, and evolutionary biology. Hence, the discussion that follows is not really relevant to food safety.

I will not assess the claim that environmental risks of GM and conventional crops are comparable, either with respect to its accuracy or with respect to regulatory policy for new crops. The aim of this paper is to identify key places in the conceptualization, assessment, and comparison of environmental risk from both types of crops where value judgments would tend to produce contrasting judgments about the equivalence or comparability of these risks. Arguments for or against any of the contrasting value judgments that

From *Plant Physiology* 132 (May 2003): 10–16. © American Society of Plant Biologists.

are identified would prove to be complex, and as such even a philosophical evaluation of these value judgments exceeds the scope of the present discussion.

A "value judgment" in this context is a working assumption for the purposes of risk comparison, often implicit, that involves assumptions about the goodness or badness of an action or outcome or that draws broadly upon philosophical framing assumptions about the nature of environmental risk and the propriety of various human and social responses to it. To say that such assumptions are philosophical is simply to say that differences between judgments based on these assumptions would not be easily settled by data collection or scientific experiment. This does not mean that all parties would regard them as subjective or arbitrary.

The Significance of the Comparative Framework

The claim that risks from transgenic and conventionally bred crops are "comparable" has been central to the debate over agricultural genetic engineering since its inception. Writing in the mid-1980s, Winston Brill suggested that crop scientists' long experience with developing new crops through breeding provides a scientifically valid basis for anticipating environmental hazards associated with the then-novel techniques for introducing new genes using recombinant DNA techniques (Brill, 1985, 1986). Brill argued that the use of rDNA gene transfer techniques would not, in itself, be the source of new or novel types of ecological impact from new crops. His view became the basis for U.S. policy toward environmental risks expressed in the aphorism that evaluations would be based on "product, not process," meaning that the use of recombinant gene transfer (e.g. the process) would not be a rationale for unusual regulatory scrutiny.

Brill's argument was endorsed by a series of reports from the U.S. National Research Council (NRC; 1987, 1989, 2000, 2002). None of these publications offer well-quantified expressions of environmental risk; hence, the statement that the risks of transgenic and conventional crops are comparable both in nature and in magnitude must be understood as making a largely qualita-tive and conceptual claim. The claim that environmental risks from transgenics do not differ from those associated with conventional crops has been important for the debate over agricultural biotechnology in two related respects. First, there has been a general presumption that conventional crops do not pose large environmental risks; hence, to claim or deny that the risks of transgenic crops are comparable has been interpreted respectively as an endorsement or criticism of biotechnology. Second, before the advent of rDNA techniques in plant science, new crop varieties were not being subjected to routine regulatory oversight on environmental grounds. Those who have argued for regulatory evaluation of transgenic crops before commercial release have been in the position of needing to justify special treatment for this class of plants, whereas those who have opposed special regulatory handling have tended to defend the claim that the environmental risks of GM and conventional crops are comparable and that additional regulatory requirements for transgenic crops would be unduly burdensome.

Claims about comparability of risk are important for both normative and rhetorical reasons. Many authors have argued that the acceptability of risks associated with new technology should always be determined through a comparison with feasible alternatives, including and especially the risks associated with nonadoption or rejection of the new technique (see Graham and Wiener, 1995). Even quite risky technologies may prove normatively acceptable in circumstances where the status quo involves a high level of adverse health or environmental impact. Emphasizing claims about the comparability of risk is rhetorically attractive because one appears to be making straightforward factual claims. As such, it is possible to express an argument about the acceptability of a novel technology without making overtly normative or value-laden claims simply by asserting whether risks of the novel technology are greater than, less than, or roughly equal to those of existing technologies. Each of these assertions appears to be making a straightforward factual claim that can be verified or falsified by data collection and scientific analysis. Confining one's rhetoric to such claims has proved to be a particularly effective device for authors and bodies that wish to represent themselves as scientific in origin or method (Thompson, 1988).

Boulter (1997), Stewart et al. (2000), and Saner (2000) have discussed the role of values in comparing transgenic and conventional crops. Boulter reviews some of the procedures proposed for quantifying environmental risk and summarizes the scientific viewpoint as endorsing the "product not process view." The public's views are characterized as being driven by largely qualitative factors such as voluntariness, dread, and trust. Stewart et al. (2000) also characterize the scientific viewpoint as endorsing the comparability of risks and characterizing the public's view in terms of a general lack of knowledge about plant science. Saner suggests that the belief that transgenic and conventional crops have noncomparable risks derives from the application of non-consequentialist ethical values. One common theme of these studies is that those who tend to see transgenic crops as posing novel risks tend to apply broad philosophical conceptualizations of nature and natural processes that are inconsistent with a scientific emphasis on quantifying the probability that harmful outcomes will occur.

However, it has become generally acknowledged that although claims about the relative level of risk for two or more courses of action make claims that are, in principle, refutable with evidence and further analysis, these claims are not at all straightforward. Even within a scientifically oriented and consequence-evaluating approach, conceptualization and comparison of risks involves an array of interpretive judgments. Reasonable and often defensible differences in the interpretation of many parameters in the framing and measurement of risk can result in very different estimates of risk and can produce contradictory comparative estimates (Brunk et al., 1991; Caruso, 2002).

Although it seems reasonable to insist that the comparative evaluation of risks from transgenic and conventional crops should be "based on science," setting up a comparison of the environmental risks from transgenic and conventional crops requires a series of value judgments. Philosophical differences of viewpoint involve both normative judgments about what is and is not harmful from an environmental perspective and pragmatic or working assumptions about problem definition. The latter often involve uneliminable but also unverifiable assumptions about the behavior of relevant phenomena. Thus, for example, to evaluate most agriculturally based environmental risks, one must make assumptions about farmer and farm worker handling of key materials, yet the empirical basis for such human factors in agriculture is virtually nonexistent.

Clearly, other differences of viewpoint refer back to economic interests and power and the more qualitative considerations noted by the three studies mentioned above. Like philosophical differences, they may be shielded from falsifying experiments or data analysis both because of the inherent difficulty in subjecting them to test and because powerful actors may prevent them from being subjected to test. In either case, differences of opinion in any of these three areas cannot be settled by scientific studies. Thus, it is possible to have a view of the environmental risks of transgenic crops that is contrary to science in the sense that it contradicts those elements of the comparative judgment that are well established by theory and data, but it is not possible to have a view that is based wholly on scientific methods and findings.

Hazard Identification

Environmental risks are typically understood as a combination of hazard and exposure. Although the conceptualization and measurement of exposure can make extensive use of scientific theory and methods of quantification, the identification of potential hazards is by all accounts deeply value laden (NRC, 1996). To describe a potential state of affairs as an environmental hazard implies the value judgment that this state of affairs is bad, harmful, unwanted, or in some way less preferred than other possible states of affairs. Some environmental hazards derive their negative value from such non-controversially bad outcomes as human death and disease, but environmental risks from transgenic crops more typically have been characterized in terms of negative effects on the environment itself, effects that eventuate in harm to human health only through extremely indirect, convoluted, and highly contingent further causes. Specifying these environmental hazards is a prime source for differing opinions on the relative risks of transgenic and conventional crops.

Criteria for healthy and well-functioning ecosystems have been acknowledged as inherently value laden. The range of values that can be applied to agriculture's impact on the environment is quite broad. Some philosophers have suggested that agriculture is inherently inimical to ecosystem health or integrity (Westra, 1998), whereas others have argued that traditional notions of agrarian stewardship can supply a model that would be applicable in other areas of environmental ethics (Thompson, 1995). The former view, in particular, would provide a philosophical basis for thinking that any trace of human impact on wild ecosystems from transgenic crops would constitute a hazard to be avoided.

NRC committees have tended to draw upon templates for specifying environmental hazards developed by the Animal and Plant Health Inspection Service (APHIS) of the U.S. Department of Agriculture. Historically, APHIS characterized environmental risk primarily in terms of hazards to U.S. agricultural production in the form of pests. The principal pest hazards have been unintended introduction of plant and animal diseases from products brought into the United States and through unintended effects of invasive species (including insects), some of which were planned introductions. Although APHIS was originally established to guard against such hazards to U.S. agricultural production, APHIS authority and practice have gradually expanded to include protection of uncultivated and natural ecosystems from similar threats of infestation and invasion (McKenzie, 2000).

This expansion permits a broad definition of unwanted environmental effects in terms of rapid and substantial change in the number and composition of species occupying an ecosystem. Thus, an environmental or ecological hazard is the potential to cause a relatively rapid and permanent decline in the number of individual organisms from one or more species currently extant in a natural or agricultural ecosystem. This general characterization of environmental hazards has been utilized by several authors offering scientifically based surveys of the risks associated with transgenic crops (Stewart et al., 2000;

Wolfenbarger and Phifer, 2000; Marvier, 2001; Carpenter et al., 2002).

This definition should not be regarded as stating either necessary or sufficient conditions for an unwanted environmental impact. For example, some large-scale changes in agricultural ecosystems are not regarded as adverse. Nevertheless, it does capture in broad terms the main thrust of scientific thinking on the possible environmental hazards that can be associated with transgenic crops. It encompasses the basis for environmental concerns associated with pesticides and other pollutants and with invasive species and microorganisms. Furthermore, it is capable of recognizing that agriculture itself can be a source of unwanted environmental impact, as when crop or livestock production displaces habitat for native plant and animal species. NRC committees have specified a number of specific hazards in some detail, ranging from the potential for weedy herbicide-resistant plants and the evolution of resistance to the *Bacillus thuringiensis* toxin – two events whose adversity is primarily with respect to the prospects for productive agriculture – to the potential for large-scale displacement of natural ecosystems (NRC, 1989, 2000, 2002).

The recent NRC committees charged with revisiting environmental risks from transgenic crops in general and pest-protected crops (e.g. Bacillus thuringiensis (Bt) maize [*Zea mays*]) in particular both concluded that hazards involving both the number and composition of species in an ecosystem can be posed by phenotypic traits of specific GM crops and by attendant cultural practices and technologies (such as the use of chemical insecticides or herbicides) that may be associated with specific GM crops. The process through which these traits are introduced into the crop, be it through recombinant gene transfer or conventional methods, was not deemed to be material. Hence, the NRC committees effectively concluded that transgenic and conventional crops have comparable environmental risks, though certainly some specific crop traits could have a relatively high likelihood of causing environmental damage; however, by whatever method these traits had been introduced into the crop (NRC, 2000, 2002).

Hazard identification is subject to a systematic ambiguity that plagues many forms of risk analysis. In common parlance, the word "hazard" is often used to describe situations in which the likelihood of a harmful outcome is substantially greater than might normally be the case, but where there may be dispute that any harmful events have actually materialized. Such situations qualify as hazards in the terminology used in risk analysis only if the analyst has judged that they constitute a form of harm or damage in themselves. Although the NRC committee views the presence of a transgene or transgenic plant "in the wrong place" as an event contributing to exposure and to the probability that harm will occur, it is also possible to view such gene flow as a form of pollution and as an event that is adverse in itself without regard to any further effects on other plant and animal species. If after gene flow has occurred a transgene is not maintained in the population because positive selection pressure is absent, is this a harmful event? Discussions after the discovery of Bt maize growing in the fields of small-scale Mexican farmers growing open-pollinated land races often took the tone that this event was itself the materialization of an environmental hazard, irrespective of the potential for further impact on the agro-ecosystem of Mexican maize cultivation (Dalton, 2001; Fig 1). Similarly, some have characterized the possibility for hybridization or gene flow to non-transgenic crops or wild relatives as a hazard (Rissler and Mellon, 1996; Löfstedt et al., 2002), rather than as a mechanism that might promote the kind of ecological change that has been viewed as a hazard in the NRC studies.

It should be evident that on conceptual grounds alone, if the presence of a transgene or transgenic plant anywhere outside the field in which transgenic crops are intentionally planted (including within the genome of other crops or wild relatives) is considered to be an environmental hazard, then the environmental risks of transgenic and conventional crops cannot be comparable in any of the senses alluded to above. Transgenic crops have transgenes and conventional crops don't; therefore, only transgenic crops can be associated with hazards defined as a transgene in the wrong place. Conventional crops may have phenotypic traits capable of disrupting the composition of species in an ecosystem, but they do not have transgenes. Hence, the chance that a transgene or transgenic plant will be found as a result of growing a conventional crop is vanishingly

small. Only a crop with a transgene can pose the risk of transgenes in the environment, and this hazard can materialize without regard to whether transgenes produce traits that do pose hazards to the number and composition of species.

Exposure Pathways

Those who have argued against the comparability of environmental risks from transgenic and conventional crops have stressed the way that recombinant techniques involve gene insertions that can disrupt proper functioning of plant genes and have the potential to produce quite anomalous behavior on the part of modified plants. Conventional crops, in contrast, are portrayed as crosses between genetically related and sexually compatible varieties, resulting in the kind of normal gene functioning we associate with standard forms of sexual reproduction (Palumbi, 2001; VIB, 2001). The typical response to this criticism has been to note that genetic engineering results in a much more precise and well-characterized introduction of genetic novelty than conventional approaches do, hence suggesting that variability and unexpected results should be even less in transgenic than in conventionally bred crops. This unhelpful exchange of views actually conceals a difference of perspective in terms of the way that the sequence of events that can lead to an unwanted event is being understood.

As the 2002 NRC report notes, analytic modeling approaches to risk assessment quantify exposure by modeling the sequence of events and causal processes that contribute to the occurrence of hazards, and the overall probability is derived from probabilities that can be assigned to each sequence in the series of events. Those who find risks of transgenic and conventional crops to be noncomparable are noting an acknowledged point of difference in the reproductive success of plants undergoing standard forms of crossbreeding on the one hand and genetic engineering on the other. Having reached a point in characterizing sequence of events leading to a possible environmental hazard at which transgenic plants are behaving far less predictably than conventional plants, they conclude that the risks cannot be comparable. To say that a transgenic plant improvement process is "more precise" seems to be a form of dissembling, for what matters if a form of plant reproduction that results in dysfunctional and unpredictable performance of the individual organisms does so with greater precision?

However, defenders of genetic engineering see crop development as an extended process that only begins with the introduction of a novel trait, either through genetic engineering, crossbreeding, or any of the other techniques currently used in crop development. From the perspective of a plant breeder, this step almost always results in a plant that has a less desirable overall profile of agronomic traits than do current commercial varieties. The subsequent steps of crop development involve further back crossing of the improved plant with established seed lines, generally through six or more generations, finally resulting in a viable crop variety. In the case of genetic engineering, the first step – introduction of genetic novelty – produces far fewer viable offspring than does ordinary crossbreeding (though not necessarily fewer than other forms of inducing mutations by chemicals or irradiation). This is the difference that the critic of genetic engineering seizes upon in deriving a judgment of risk. However, the few stable individuals produced by genetic engineering can and will be backcrossed with founder stock to produce reproductively stable, genetically functional, and agronomically valuable varieties for commercial release. Because it is these plants that will be released into the environment, and not the dysfunctional individuals that are discarded after the initial attempt at transformation, the model for exposure is based solely on the performance of these apparently healthy and genetically stable plants (NRC, 2002).

In short, those who defend the comparability of risks from transgenic and conventional crops do not see the creation of reproductively and genetically unstable individuals as part of the sequence of events leading to the release of a commercial variety. Anomalous individuals associated with the disruption of a genome from the process of plant transformation are removed from the sequence of events leading to the hazard, and their various dysfunctions are not seen as material to it. In contrast, those who defend the noncomparability of these classes see anomalous individuals as part of a data set highly relevant to the range of uncertainty associated with predicting the behavior of transgenic crops in the environment.

In what sense is this difference of perspective the result of a value judgment? There are several possibilities. One is simply that the creation of a model for exposure involves a number of framing judgments about the systems in question and that for whatever reason, critics and proponents of biotechnology have envisioned the relevant sequence of events very differently. Another is that there is a legitimate difference of scientific opinion about the relevance of the many anomalous results occurring as an immediate consequence of transformation for the probability that varieties developed from apparently stable transformation events will continue to behave in a standard and predictable fashion when subjected to the wide variety of different environmental conditions and stimuli associated with widespread commercial production. This is a value judgment somewhat akin to the kind of theoretical clashes and paradigmatic disputes characteristic of science, and one that may be resolved as experience, data, and theoretical developments clarify the issues. A third possibility, compatible with the first two, is that both ways of characterizing the model for exposure are open possibilities at present and that political or economic interests thrive upon such situations where key questions in framing the mechanisms of risk are underdetermined by existing scientific consensus.

Comparison Populations

When transgenic crops were being discussed in the 1980s, there was a general presumption that recombinant DNA techniques would eventually become the preferred method for introducing genetic novelty into plants. In some instances, the presumption went so far as a belief that the traditional skills of plant breeders would soon be obsolete and that all new crops would be developed using transgenic methods (Busch et al., 1991). A corollary to this belief that was seldom articulated in the literature of the 1980s and early 1990s was that the traits being introduced using transgenic methods would be the same kind of agronomically valuable traits that had long been the aim of conventional breeding programs. Such traits were introduced to enhance yield, either in general or in conjunction with specific conditions such as pests or disease, and drought, climatic variation, and both soil-based and externally applied chemicals. Thus, those who asserted the comparability risks of transgenic and conventional crops were assuming that the crops developed with transgenic and conventional techniques would be broadly similar with respect to their traits and purpose.

The regulatory posture toward environmental risk that developed on the U.S.-coordinated framework clearly reflects such an assumption. The U.S. Department of Agriculture APHIS developed a regulatory approach that permitted small-scale production of transgenic crops with relatively little oversight or prior approval. The intent was to permit field trials needed both for backcrossing and generating data, with the expectation that seed companies would need to go through the approval process before releasing a transgenic variety for commercial production. However, by 2002 it was clear that some biotechnology companies were capable of servicing the entire production run for some high-value crops developed to service non-commodity markets on field-trial sized plots. The lack of appropriate regulatory oversight for these crops was noted by the NRC (2002).

The potential for using plants as systems to produce an entirely new class of products, including pharmaceuticals and industrial biologics in cropping systems, was clearly recognized early on. Yet, it is doubtful that anyone who asserted the comparable risk hypothesis had these crops in mind. However, by the time that the most recent NRC committee convened, a number of things had changed. First, the plant science community had more experience with transgenic techniques and a more realistic understanding of their potential when compared with conventional techniques. More importantly, consumer resistance to GM crops has created an economic environment that has reduced the attractiveness of transgenic techniques for food crops. As a result, a much larger than expected percentage of the transgenic crops that are likely to be used in a production setting during the next decade are those in which a food crop, often maize, has been transformed to produce nonfood pharmaceuticals or biologics.

The key value judgments from a risk assessment perspective have to do with establishing comparison populations or defining what is meant by a transgenic crop. If T is the population of all transgenic crops, T_h is the harmful subset, C is the population of conventional crops, and C_h is the harmful subset, then the risks of transgenic and conventional crops are comparable if and only if:

$$T_h/T = C_h/C$$

In this formula, T and C represent reference populations for expressing risk in statistical terms. Although quantification of this formula would be very difficult given existing data, it does provide an apparently clear statement of what it means to say that the risks of transgenic and conventional crops are equivalent. However, the clarity of the expression depends upon unambiguous and agreed-upon definitions of the relevant classes. Disagreements about hazard identification or exposure characterization can produce different interpretations of T_h, as discussed above, but changing assumptions about the kinds of traits being introduced involve the interpretation of T.

The whole point of analogizing risks from transgenic and conventional crops is to gain predictive insight into the risks of transgenic crops. This demands that C be interpreted historically as the class of all crops developed for commercial release using conventional techniques and that T be understood to include some crops that are yet to be developed. Clearly, T does not include all conceivable plants that could be developed using transgenic technology, for any competent biologist will concede that it is possible to produce some very dangerous transformation events not suitable for agricultural production. However, one should not confine T to those crops that will actually be commercially released because to do so begs precisely the regulatory assessment questions that a risk comparison is intended to address. Although the reference population is not specifically defined in NRC reports, these committees have implicitly understood T to include all and only crops that are prospective candidates for commercial production. This would include crops intended to be grown under proprietary management and not released for commercial sale and conventional registered varieties made available to farmers.

Given the expectation that transgenic techniques would be used to produce crops with agronomic traits such as alternative coloring, disease resistance, climatic variation tolerances, and increase yield, T and C were expected to be roughly similar in terms of their phenotypic profile, and the ratios T_h to T and C_h to C were expected to be similar as well. Although herbicide-tolerant and Bt crops are notable examples of crops with fairly standard agronomic traits, transgenic plants produced for pharmaceutical or industrial chemical production have phenotypic characteristics that are quite unlike those of crops historically produced through conventional means. As such, it no longer seems reasonable to expect that the phenotypic characteristics of plants in T will be similar to those of plants in C and because the parallelism in the makeup of the reference population shifts, so does the comparability of risks. It is, in part, this difference in comparison populations that accounts for the 2002 report's more precautionary stance when compared with previous NRC committees.

Conclusion

Hazard identification, exposure modeling, and comparison populations each involve value judgments that are ethical or pragmatic but in either case cannot be characterized as following from established scientific findings or theories. It may be the case that many plant scientists share common views with respect to these value judgments. Nevertheless, taking different viewpoints on any of the value-oriented questions is, absent more extensive argument at least, fully consistent with taking a scientific view on the comparison of environmental risk from transgenic and conventional crops. Because of the technical sophistication implicit in the foregoing analysis, it is unlikely that the specific value judgments identified above contribute strongly to nonscientific resistance to transgenic crops. However, resolving the conceptual and definitional ambiguities noted herein, and providing a clear and straightforward rationale for such resolution, are critical to the credibility of risk assessment.

Literature Cited

Boulter D (1997) Scientific and public perception of plant genetic manipulation: a critical review. *Crit Rev Plant Sci* 16: 231–51

Brill W (1985) Safety concerns and genetic engineering in agriculture. *Science* 227: 381–4

Brill WJ (1986) The impact of biotechnology and the future of agriculture. *In* K Byrne, ed, *Responsible Science: The Impact of Technology on Society*. Harper & Row, New York, pp 31–48

Brunk C, Haworth L, Lee B (1991) *Values in Risk Assessment*. Wilfrid Laurier University Press, Waterloo, CA

Busch L, Lacy W, Burkhardt J, Lacy L (1991) *Plants, Power and Profit: Social, Economic and Ethical Consequences of the New Biotechnologies*. Basil Blackwell, Cambridge, UK

Carpenter J, Felsot A, Goode T, Hammig M, Onstad D, Sankula S (2002) *Comparative Environmental Impacts of Biotechnology-Derived and Traditional Soybean, Corn and Cotton Crops*. Council for Agricultural Science and Technology, Ames, IA

Caruso D (2002) Risk: *The Art and Science of Choice*. The Hybrid Vigor Institute, San Francisco

Dalton R (2001) Transgenic corn found growing in Mexico. *Nature* 413: 337

Graham JD, Wiener JB (1995) *Risk against Risk: Tradeoffs in Protecting Health and the Environment*. Harvard University Press, Cambridge, MA

Krimsky S, Wrubel R (1996) *Agricultural Biotechnology and the Environment: Science, Policy and Social Issues*. University of Illinois Press, Urbana

Löfstedt RE, Fischhoff B, Fischhoff IR (2002) Precautionary principles: General definitions and specific applications to genetically modified organisms. *J Policy Sci Manag* 21: 381–407

Marvier M (2001) Ecology of transgenic crops. *Am Sci* 89: 160–7

McKenzie D (2000) Agricultural biotechnology, law, APHIS regulation. *In* TH Murray, MJ Mehlman, eds, *Encyclopedia of Ethical, Legal and Policy Issues in Biotechnology*. John Wiley & Sons, New York, pp 56–66

Miller HI (2000) Agricultural biotechnology, law, and food biotechnology regulation. *In* TH Murray, MJ Mehlman, eds, *Encyclopedia of Ethical, Legal and Policy Issues in Biotechnology*. John Wiley & Sons, New York, pp 37–46

NRC (1987) *Agricultural Biotechnology: Strategies for National Competitiveness*. National Academy Press, Washington, DC

NRC (1989) *Field Testing Genetically Modified Organisms: Framework for Decision*. National Academy Press, Washington, DC

NRC (1996) *Understanding Risk: Informing Decisions in a Democratic Society.* National Academy Press, Washington, DC

NRC (2000) *Genetically Modified Pest-Protected Plants: Science and Regulation.* National Academy Press, Washington, DC

NRC (2002) *Environmental Effects of Transgenic Plants: The Scope and Adequacy of Regulation.* National Academy Press, Washington, DC

Palumbi SR (2001) *The Evolution Explosion: How Humans Cause Rapid Evolutionary Change.* W.W. Norton, New York

Rissler J, Mellon M (1996) *The Ecological Risks of Transgenic Crops.* The MIT Press, Cambridge, MA

Saner MA (2000) Ethics as problem and ethics as a solution. *Int J Biotechnol* 2: 219–56

Stewart CN Jr, Richards IVHA, Halfhill MD (2000) Transgenic plants and biosafety: science, miscon-

ceptions and public perceptions, *Bio Techniques* 29: 832–42

Thompson PB (1988) Agriculture, biotechnology, and the political evaluation of risk. *Policy Studies J* 17: 97–108

Thompson PB (1995) *The Spirit of the Soil: Agriculture and Environmental Ethics.* Routledge Publishing Co., London

VIB (Flanders Interuniversity Institute for Biotechnology) (2001) *In* R. Custers, ed, *Safety of Genetically Engineered Crops.* VIB, Zwijnaarde, Belgium, pp 8–12

Westra L (1998) *Living in Integrity: A Global Ethic to Restore a Fragmented Earth.* Rowman & Littlefield, Lanham, MA

Wolfenbarger LL, Phifer PR (2000) The ecological risks and benefits of genetically engineered plants. *Science* 290: 2088–203

Part X

Environmental Values

Introduction

The domain of technology has seen major advances in recent years, both in the numbers and kinds of technology, and in the range of their uses and the materials needed for their creation. A common response to these developments among many environmentalists, who see themselves as "protectors of nature," has been to construe these advances as posing a "threat" to nature. Environmental philosophers, whose task is to articulate and comment upon the concepts and arguments used in debates such as this, have been quick to characterize this confrontation in terms of the contrast between the "natural" and the "artificial." Some, wielding a strong normative conviction that privileges the value of the natural above the value of the artificial, conclude that the advance of technology signals a corresponding diminution of value. These essays present a wide range of responses to the relations of natural environment and value.

Michael Pollan is the Knight Professor of Journalism at the Graduate School of Journalism at UC-Berkeley and director of the Knight Program in Science and Environmental Journalism and a frequent contributor to the *New York Times*, and former editor of *Harper's*. He here reflects on the technological and cultural imperative to mow one's yard, tracing the history of the private lawn in the United States to the nineteenth-century planning of Olmstead, and the growing suburbanization of the country from the late nineteenth through the twentieth centuries. The

private lawn expresses democratic tendencies, and becomes a location of the demand for conformism. He argues that while a lawn might be democratic with respect to neighbors, "with respect to nature it is authoritarian." This essay is an examination of the environmental consequences of something many in the United States take for granted, as well as ruminations on the relations between humans and non-human nature.

Lori Gruen, Associate Professor of Philosophy, and Associate Professor and Chair of Feminist, Gender, and Sexuality Studies at Wesleyan University, takes a consequentialist approach to evaluating the general place of technology in an environmental philosophy. Next, Rajni Kothari, political scientist and founder of the Center for the Study of Developing Societies, argues that we need to rethink the notion of "sustainable development." We are at present, he argues, merely exporting the pending ecological crisis to the future and to the third world. He argues that we should nurture a "sense of sanctity about the Earth."

A central debate in environmental philosophy has been between anthropocentric views that place humans at the center as the primary, or only source of value, and biocentric views that claim there are intrinsic values worthy of our consideration in nonhuman nature. In one of the most important essays in recent environmental ethics, Baird Callicott, Regents Professor of Philosophy,

University of North Texas, draws out the philosophical implications of Aldo Leopold's land ethic, and places Leopold in the same tradition of moral philosophy as David Hume and Smith. Leopold was an ecologist, forester, founder of the land ethic, and father of conservation management. The land ethic claims that we must expand our circle of ethical concern beyond humans to include other members of the "biotic community," including animals, soil, water, and so on. Leopold calls this larger community "the land." Leopold's work has been criticized as lending support to "environmental fascism," the notion that the building environmental crisis can only be solved by a benign authoritarian government. According to the land ethic, a morally responsible technology would be one that respects the ecosystem as a whole, and allows humans to flourish as but one member of a thriving biotic community. This is reasonable, he argues, because individual human lives and human communities only exist and flourish as part of a larger flourishing ecological, or land, community. The Humean moral tradition that places sympathy at the roots of our moral lives explains how we can understand our larger connections with both reason and feeling. In arguing that human lives can only be understood as always already in relations with the non-human, Callicott's work recalls Ortega y Gassett and Jonas, and in holding that in the final instance a human moral community has priority over the biotic community, his argument stands in contrast to the deep ecology we will encounter below.

Drawing on the work of Arne Naess, Bill Devall, Professor Emeritus in Sociology at Humboldt State University in Arcata, California, and George Sessions, former chair of the Philosophy Department at Sierra College in Rocklin, California, develop a comprehensive framework for a new biocentric ethic. Arne Naess, Professor Emeritus at the University of Oslo, is the founder of "deep ecology." The central claim of deep ecology is that mainstream ecological and environmental movements do not go far enough to address the problems that need to be addressed, the very problems the mainstream movements claim to be addressing. Their critique is still apt today. Consider the increase in "green consumption," ranging from a green product line at the hardware store, to gas-electric hybrid SUVs. In each of these cases, consumers can feel, and perhaps even think, that they are addressing some environmental issue (global warming, local water or air pollution, and so on), but they do so in terms of an existing system of expropriation and consumption that is itself unsustainable. Further, this mainstream approach is grounded on an anthropocentric assumption, namely that humans are the only source of value and the only features of the world worthy of moral consideration. Given this approach and this assumption, deep ecology argues that we will continue to dominate nonhuman nature, and that we will continue to lead lives that are destructive to all ecologies. What is needed is a new, deeper ecology, one that recognizes the deep interrelationships between the human and nonhuman aspects of reality (echoing some concerns of Latour), as well as one that abandons the privileged position given to human values in traditional ethics and in mainstream environmentalism.

This section closes with two responses to deep ecology. In response to some strains of the environmental movement in North America, Ramachandra Guha, Indian sociologist, writer, and activist, considers whether the ethic of deep ecology, and its demand that technological development be guided by the criterion of preserving biotic integrity, is useful in the third world. Guha argues that the biocentric emphasis in deep ecology ignores the pressing needs of some of the world's poorest citizens. Peter Wenz, Emeritus Professor of Philosophy at the University of Illinois at Springfield, University Scholar of the University of Illinois, and Adjunct Professor of Medical Humanities at the Southern Illinois University School of Medicine, addresses one of the most pressing, and often invisible issues of our technologized culture – what to do with waste. Drawing on deep ecology, Wenz argues that the burdens and benefits of industrial culture should be equally shared, but often they are not. The burdens fall disproportionately on those at social margins – those in poverty and members of ethnic and racial minorities. Justice about our garbage, he claims, requires rethinking not only about what we do with waste, but industrialization and consumption as well.

29

The Highway and the City

Lewis Mumford

When the American people, through their Congress, voted last year for a twenty-six-billion-dollar highway program, the most charitable thing to assume about this action is that they hadn't the faintest notion of what they were doing. Within the next fifteen years they will doubtless find out; but by that time it will be too late to correct all the damage to our cities and our countryside, to say nothing of the efficient organization of industry and transportation, that this ill-conceived and absurdly unbalanced program will have wrought.

Yet if someone had foretold these consequences before this vast sum of money was pushed through Congress, under the specious guise of a national defense measure, it is doubtful whether our countrymen would have listened long enough to understand; or would even have been able to change their minds if they did understand. For the current American way of life is founded not just on motor transportation but on the religion of the motorcar, and the sacrifices that people are prepared to make for this religion stand outside the realm of rational criticism. Perhaps the only thing that could bring Americans to their senses would be a clear demonstration of the fact that their highway program will, eventually, wipe out the very area of freedom that the private motorcar promised to retain for them.

As long as motorcars were few in number, he who had one was a king: he could go where he pleased and halt where he pleased; and this machine itself appeared as a compensatory device for enlarging an ego which had been shrunken by our very success in mechanization. That sense of freedom and power remains a fact today only in low-density areas, in the open country; the popularity of this method of escape has ruined the promise it once held forth. In using the car to flee from the metropolis the motorist finds that he has merely transferred congestion to the highway; and when he reaches his destination, in a distant suburb, he finds that the countryside he sought has disappeared: beyond him, thanks to the motorway, lies only another suburb, just as dull as his own. To have a minimum amount of communication and sociability in this spread-out life, his wife becomes a taxi driver by daily occupation, and the amount of money it costs to keep this whole system running leaves him with shamefully overcrowded, understaffed schools, inadequate police, poorly serviced hospitals, under-spaced recreation areas, ill-supported libraries.

In short, the American has sacrificed his life as a whole to the motorcar, like someone who,

From Lewis Mumford, *The Urban Prospect* (New York: Harcourt Brace Jovanovich, 1968), pp. 92–107.

demented with passion, wrecks his home in order to lavish his income on a capricious mistress who promises delights he can only occasionally enjoy.

For most Americans, progress means accepting what is new because it is new, and discarding what is old because it is old. This may be good for a rapid turnover in business, but it is bad for continuity and stability in life. Progress, in an organic sense, should be cumulative, and though a certain amount of rubbish-clearing is always necessary, we lose part of the gain offered by a new invention if we automatically discard all the still valuable inventions that preceded it. In transportation, unfortunately, the old-fashioned linear notion of progress prevails. Now that motorcars are becoming universal, many people take for granted that pedestrian movement will disappear and that the railroad system will in time be abandoned; in fact, many of the proponents of highway building talk as if that day were already here, or if not, they have every intention of making it dawn quickly. The result is that we have actually crippled the motorcar, by placing on this single means of transportation the burden for every kind of travel. Neither our cars nor our highways can take such a load. This overconcentration, moreover, is rapidly destroying our cities, without leaving anything half as good in their place.

What's transportation for? This is a question that highway engineers apparently never ask themselves: probably because they take for granted the belief that transportation exists for the purpose of providing suitable outlets for the motorcar industry. To increase the number of cars, to enable motorists to go longer distances, to more places, at higher speeds has become an end in itself. Does this overemployment of the motorcar not consume ever larger quantities of gas, oil, concrete, rubber, and steel, and so provide the very groundwork for an expanding economy? Certainly, but none of these make up the essential purpose of transportation, which is to bring people or goods to places where they are needed, and to concentrate the greatest variety of goods and people within a limited area, in order to widen the possibility of choice without making it necessary to travel. A good transportation system minimizes unnecessary transportation; and in any event, it offers a change of speed and mode to fit a diversity of human purposes.

Diffusion and concentration are the two poles of transportation: the first demands a closely articulated network of roads – ranging from a footpath to a six-lane expressway and a transcontinental railroad system. The second demands a city. Our major highway systems are conceived, in the interests of speed, as linear organizations, that is to say, as arteries. That conception would be a sound one, provided the major arteries were not overdeveloped to the exclusion of all the minor elements of transportation. Highway planners have yet to realize that these arteries must not be thrust into the delicate tissue of our cities; the blood they circulate must, rather, enter through an elaborate network of minor blood vessels and capillaries.

In many ways, our highways are not merely masterpieces of engineering, but consummate works of art: a few of them, like the Taconic State Parkway in New York, stand on a par with our highest creations in other fields. Not every highway, it is true, runs through country that offers such superb opportunities to an imaginative highway builder as this does; but then not every engineer rises to his opportunities as the planners of this highway did, routing the well-separated roads along the ridgeways, following the contours, and thus, by this single stratagem, both avoiding towns and villages and opening up great views across country, enhanced by a lavish planting of flowering bushes along the borders. If this standard of comeliness and beauty were kept generally in view, highway engineers would not so often lapse into the brutal assaults against the landscape and against urban order that they actually give way to when they aim solely at speed and volume of traffic, and bulldoze and blast their way across country to shorten their route by a few miles without making the total journey any less depressing.

Perhaps our age will be known to the future historian as the age of the bulldozer and the exterminator; and in many parts of the country the building of a highway has about the same result upon vegetation and human structures as the passage of a tornado or the blast of an atom bomb. Nowhere is this bulldozing habit of mind so disastrous as in the approach to the city. Since the engineer regards his own work as more important than the other human functions it serves, he does not hesitate to lay waste to woods, streams, parks, and human neighborhoods in order to carry his roads straight to their supposed destination. As a consequence the 'cloverleaf' has become our national flower and 'wall-to-wall concrete' the ridiculous symbol of national affluence and technological status.

The fatal mistake we have been making is to sacrifice every other form of transportation to the private motorcar – and to offer as the only long-distance alternative the airplane. But the fact is that each type of transportation has its special use; and a good transportation policy must seek to improve each type and make the most of it. This cannot be achieved by aiming at high speed or continuous flow alone. If you wish casual opportunities for meeting your neighbors, and for profiting by chance contacts with acquaintances and colleagues, a stroll at two miles an hour in a relatively concentrated area, free from vehicles, will alone meet your need. But if you wish to rush a surgeon to a patient a thousand miles away, the fastest motorway is too slow. And again, if you wish to be sure to keep a lecture engagement in winter, railroad transportation offers surer speed and better insurance against being held up than the airplane. There is no one ideal mode or speed: human purpose should govern the choice of the means of transportation. That is why we need a better transportation *system*, not just more highways. The projectors of our national highway program plainly had little interest in transportation. In their fanatical zeal to expand our highways, the very allocation of funds indicates that they are ready to liquidate all other forms of land and water transportation.

In order to overcome the fatal stagnation of traffic in and around our cities, our highway engineers have come up with a remedy that actually expands the evil it is meant to overcome. They create new expressways to serve cities that are already overcrowded within, thus tempting people who had been using public transportation to reach the urban centers to use these new private facilities. Almost before the first day's tolls on these expressways have been counted, the new roads themselves are overcrowded. So a clamor arises to create other similar arteries and to provide more parking garages in the center of our metropolises; and the generous provision of these facilities expands the cycle of congestion, without any promise of relief until that terminal point when all the business and industry that originally gave rise to the congestion move out of the city, to escape strangulation, leaving a waste of expressways and garages behind them. This is pyramid building with a vengeance: a tomb of concrete roads and ramps covering the dead corpse of a city.

But before our cities reach this terminal point, they will suffer, as they now do, from a continued erosion of their social facilities: an erosion that might have been avoided if engineers had understood MacKaye's point that a motorway, properly planned, is another form of railroad for private use. Unfortunately, highway engineers, if one is to judge by their usual performance, lack both historic insight and social memory: accordingly, they have been repeating, with the audacity of confident ignorance, all the mistakes in urban planning committed by their predecessors who designed our railroads. The wide swathes of land devoted to cloverleaves and expressways, to parking lots and parking garages, in the very heart of the city, butcher up precious urban space in exactly the same way that freight yards and marshalling yards did when the railroads dumped their passengers and freight inside the city. These new arteries choke off the natural routes of circulation and limit the use of abutting properties, while at the points where they disgorge their traffic they create inevitable clots of congestion, which effectively cancel out such speed as they achieve in approaching these bottlenecks.

Today the highway engineers have no excuse for invading the city with their regional and transcontinental trunk systems: the change from the major artery to the local artery can now be achieved without breaking the bulk of goods or replacing the vehicle: that is precisely the advantage of the motorcar. Arterial roads, ideally speaking, should engirdle the metropolitan area and define where its greenbelt begins; and since American cities are still too impoverished and too improvident to acquire greenbelts, they should be planned to go through the zone where relatively high-density building gives way to low-density building. On this perimeter, through traffic will bypass the city, while cars that are headed for the center will drop off at the point closest to their destination.

Since I don't know a city whose highways have been planned on this basis, let me give as an exact parallel the new semicircular railroad line, with its suburban stations, that bypasses Amsterdam. That is good railroad planning, and it would be good highway planning, too, as the Dutch architect H. Th. Wijdeveld long ago pointed out. It is on relatively cheap land, on the edge of the city, that we should be building parking areas and garages: with free parking privileges, to tempt the commuter to leave his car and finish his daily journey on the public transportation system. The public officials who have been planning our highway system on just the opposite principle are likewise planning to make the central areas of our cities unworkable and uninhabitable.

Now, as noted before, the theory of the insulated, highspeed motorway, detached from local street and road systems, immune to the clutter of roadside 'developments,' was first worked out, not by highway engineers, but by Benton MacKaye, the regional planner who conceived the Appalachian Trail. He not merely put together its essential features, but identified its principal characteristic: the fact that to achieve speed it must bypass towns. He called it, in fact, the Townless Highway. (See 'The New Republic,' March 30, 1930.) Long before the highway engineers came through with Route 128, MacKaye pointed out the necessity for a motor bypass around the ring of suburbs that encircle Boston, in order to make every part of the metropolitan area accessible, and yet to provide a swift alternative route for through traffic.

MacKaye, not being a one-eyed specialist, visualized this circuit in all its potential dimensions and developments: he conceived, accordingly, a metropolitan recreation belt with a northbound motor road forming an arc on the inner flank and a southbound road on the outer flank – the two roads separated by a wide band of usable parkland, with footpaths and bicycle paths for recreation. In reducing MacKaye's conception to Route 128, without the greenbelt and without public control of the areas adjacent to the highway, the 'experts' shrank the multi-purpose Bay Circuit into the typical 'successful' expressway: so

successful in attracting industry and business from the center of the city that it already ceases to perform even its own limited functions of fast transportation, except during hours of the day when ordinary highways would serve almost as well. This, in contrast to MacKaye's scheme, is a classic example of how not to do it.

Just as highway engineers know too little about city planning to correct the mistakes made in introducing the early railroad systems into our cities, so, too, they have curiously forgotten our experience with the elevated railroad – and, unfortunately, most municipal authorities have been equally forgetful. In the middle of the nineteenth century the elevated seemed the most facile and up-to-date method of introducing a new kind of rapid transportation system into the city; and in America, New York led the way in creating four such lines on Manhattan Island alone. The noise of the trains and the overshadowing of the structure lowered the value of the abutting properties even for commercial purposes; and the supporting columns constituted a dangerous obstacle to surface transportation. So unsatisfactory was elevated transportation even in cities like Berlin, where the structures were, in contrast to New York, Philadelphia, and Chicago, rather handsome works of engineering, that by popular consent subway building replaced elevated railroad building in all big cities, even though no one could pretend that riding in a tunnel was nearly as pleasant to the rider as was travel in the open air. The destruction of the old elevated railroads in New York was, ironically, hailed as a triumph of progress precisely at the moment that a new series of elevated highways were being built, to repeat on a more colossal scale the same errors.

Like the railroad, again, the motorway has repeatedly taken possession of the most valuable recreation space the city possesses, not merely by thieving land once dedicated to park uses, but by cutting off easy access to the waterfront parks, and lowering their value for refreshment and repose by introducing the roar of traffic and the bad odor of exhausts, though both noise and gasoline exhaust are inimical to health. Witness the shocking spoilage of the Charles River Basin parks in Boston, the arterial blocking off of the Lake Front in Chicago (after the removal of the original usurpers, the railroads), the barbarous sacrifice of large areas of Fairmount Park in

Philadelphia, the insistent official efforts, despite public disapproval, to deface the San Francisco waterfront.[1]

One may match all these social crimes with a hundred other examples of barefaced highway robbery in every other metropolitan area. Even when the people who submit to these annexations and spoliations are dimly aware of what they are losing, they submit without more than a murmur of protest.

What they do not understand is that they are trading a permanent good for a very temporary advantage, since until we subordinate highway expansion to the more permanent requirements of regional planning, the flood of motor traffic will clog new channels. What they further fail to realize is that the vast sums of money that go into such enterprises drain necessary public monies from other functions of the city, and make it socially if not financially bankrupt.

Neither the highway engineer nor the urban planner can, beyond a certain point, plan his facilities to accommodate an expanding population. On the over-all problem of population pressure, regional and national policies must be developed for throwing open, within our country, new regions of settlement, if this pressure, which appeared so suddenly, does not in fact abate just as unexpectedly and just as suddenly. But there can be no sound planning anywhere until we understand the necessity for erecting norms, or ideal limits, for density of population. Most of our congested metropolises need a lower density of population, with more parks and open spaces, if they are to be attractive enough physically to retain even a portion of their population for day-and-night living; but most of our suburban and exurban communities must replan large areas at perhaps double their present densities in order to have the social, educational, recreational, and industrial facilities they need closer at hand. Both suburb and metropolis need a regional form of government, working in private organizations as well as public forms, to decentralize their strangled resources and facilities, so as to benefit the whole area.

To say this is to say that both metropolitan congestion and suburban scattering are obsolete.

This means that good planning must work to produce a radically new pattern for urban growth. On this matter, public policy in the United States is both contradictory and self-defeating. Instead of lowering central area densities, most urban renewal schemes, not least those aimed at housing the groups that must be subsidized, either maintain old levels of congestion, or create higher levels than existed in the slums they replaced. But the Home Loan agencies, on the other hand, have been subsidizing the wasteful, ill-planned, single-family house, on cheap land, ever remoter from the center of our cities; a policy that has done as much to promote the suburban drift as the ubiquitous motorcar.

In order to cement these errors in the most solid way possible, our highway policy maximizes congestion at the center and expands the area of suburban dispersion – what one might call the metropolitan 'fall-out.' The three public agencies concerned have no official connections with each other: but the total result of their efforts proves, once again, that chaos does not have to be planned.

Motorcar manufacturers look forward confidently to the time when every family will have two, if not three, cars. I would not deny them that hope, though I remember that it was first voiced in 1929, just before the fatal crash of our economic system, too enamored of high profits even to save itself by temporarily lowering prices. But if they don't want the motorcar to paralyze urban life, they must abandon their fantastic commitment to the indecently tumescent chariots they have been putting on the market. For long-distance travel, the big car, of course, has many advantages; but for town use, let us insist upon a car that fits the city's needs: it is absurd to make over the city to fit the swollen imaginations of Detroit. The Isetta and the Gogomobil have already pointed the way; but what we need is a less cramped but still compact vehicle, powered by electricity, delivered by a powerful storage cell, yet to be invented: the exact opposite of our insolent chariots.[2] Maneuverability and parkability are the prime urban virtues in cars; and the simplest way to achieve this is by designing smaller cars. These virtues are lacking in all but one of our current American models. But why should our cities be destroyed just so that Detroit's infantile fantasies should remain unchallenged and unchanged?

If we want to make the most of our national highway program, we must keep most of the proposed expressways in abeyance until we have done two other things. We must replan the inner city for pedestrian circulation, and we must rebuild and extend our public forms of mass transportation. In our entrancement with the motorcar, we have forgotten how much more efficient and how much more flexible the foot-goer is. Before there was any public transportation in London, something like fifty thousand people an hour used to pass over London Bridge on their way to work: a single artery. Mass public transportation can bring from forty to sixty thousand people per hour, along a single route, whereas our best expressways, using far more space, cannot move more than four to six thousand cars, and even if the average occupancy were more than one and a half passengers, as at present, this is obviously the most costly and inefficient means of handling the peak hours of traffic.

As for the pedestrian, one could move a hundred thousand people, by using all the existing streets, from, say, downtown Boston to the Common, in something like half an hour, and find plenty of room for them to stand. But how many weary hours would it take to move them in cars over these same streets? And what would one do with the cars after they had reached the Common? Or where, for that matter, could one assemble these cars in the first place? For open spaces, long distances, and low densities, the car is now essential; for urban space, short distances, and high densities, the pedestrian.

Every urban transportation plan should, accordingly, put the pedestrian at the center of all its proposals, if only to facilitate wheeled traffic. But to bring the pedestrian back into the picture, one must treat him with the respect and honor we now accord only to the automobile: we should provide him with pleasant walks, insulated from traffic, to take him to his destination, once he enters a business precinct or residential quarter. Every city should heed the example of Rotterdam in creating the Lijnbaan,

or of Coventry in creating its new shopping area. It is nonsense to say that this cannot be done in America, because no one wants to walk.

Where walking is exciting and visually stimulating, whether it is in a Detroit shopping center or along Fifth Avenue, Americans are perfectly ready to walk. The legs will come into their own again, as the ideal means of neighborhood transportation, once some provision is made for their exercise, as Philadelphia is now doing, both in its Independence Hall area and in Penn Center. But if we are to make walking attractive, we must not only provide trees and wide pavements and benches, beds of flowers and outdoor cafés, as they do in Zurich: we must also scrap the monotonous uniformities of American zoning practice, which turns vast areas, too spread out for pedestrian movement, into single-district zones, for commerce, industry, or residential purposes. (As a result, only the mixed zones are architecturally interesting today despite their often frowzy disorder.)

Why should anyone have to take a car or a taxi and drive miles to get domestic conveniences needed every day, as one must often do in a suburb? Why, on the other hand, should a growing minority of people not be able again to walk to work, by living in the interior of the city, or, for that matter, be able to walk home from the theater or the concert hall? Where urban facilities are compact, walking still delights the American: does he not travel many thousands of miles just to enjoy this privilege in the historic urban cores of Europe? And do not people now travel for miles, of an evening, from the outskirts of Pittsburgh, just for the pleasure of a stroll in Mellon Square? Nothing would do more to give life back to our blighted urban cores than to reinstate the pedestrian, in malls and pleasances designed to make circulation a delight. And what an opportunity for architecture!

While federal funds and subsidies pour without stint into highway improvements, the two most important modes of transportation for cities – the railroad for long distances and mass transportation, and the subway for shorter journeys – are permitted to languish and even to disappear. This is very much like what has happened to our postal system. While the time needed to deliver a letter across the continent has been reduced, the time needed for local delivery has been multiplied. What used to take two hours now sometimes takes two days. As a whole, our postal system has been degraded to a level that would have been regarded as intolerable even thirty years ago. In both cases, an efficient system has been sacrificed to a new industry, motorcars, telephones, airplanes; whereas, if the integrity of the system itself had been respected, each of these new inventions could have added enormously to the efficiency of the existing network.

If we could overcome the irrational drives that are now at work, promoting shortsighted decisions, the rational case for rebuilding the mass transportation system in our cities would be overwhelming. The current objection to mass transportation comes chiefly from the fact that it has been allowed to decay: this lapse itself reflects the general blight of the central areas. In order to maintain profits, or in many cases to reduce deficits, rates have been raised, services have decreased, and equipment has become obsolete, without being replaced and improved. Yet mass transportation, with far less acreage in roadbeds and rights of way, can deliver at least ten times more people per hour than the private motorcar. This means that if such means were allowed to lapse in our metropolitan centers – as the interurban electric trolley system, that beautiful and efficient network, was allowed to disappear in the nineteen-twenties – we should require probably five to ten times the existing number of arterial highways to bring the present number of commuters into the city, and at least ten times the existing parking space to accommodate them. In that tangled mass of highways, interchanges, and parking lots, the city would be nowhere: a mechanized nonentity ground under an endless procession of wheels. Witness Los Angeles, Detroit, Boston – indeed, every city whose municipal officials still stubbornly equate expressways, high-rise buildings, and parking facilities with urban progress.

That plain fact reduces a one-dimensional transportation system, by motorcar alone, to a calamitous absurdity, as far as urban development goes, even if the number of vehicles and the population count were not increasing year by year. Now it happens that the population of the core of our big cities has remained stable in recent years: in many cases, the decline which set in as early as 1910 in New York seems to have ceased. This means that it is now possible to set an upper limit for the daily inflow of workers, and to work out a permanent mass transportation system that will get them in and out again as pleasantly and efficiently as possible.

In time, if urban renewal projects become sufficient in number to permit the design of a system of minor urban throughways, at ground level, that will bypass the neighborhood, even circulation by motorcar may play a valuable part in the total scheme – provided, of course, that minuscule-size town cars take the place of the long-tailed dinosaurs that now lumber about our metropolitan swamps. But the notion that the private motorcar can be substituted for mass transportation should be put forward only by those who desire to see the city itself disappear, and with it the complex, many-sided civilization that the city makes possible.

There is no purely engineering solution to the problems of transportation in our age: nothing like a stable solution is possible without giving due weight to all the necessary elements in transportation – private motorcars, railroads, airplanes and helicopters, mass transportation services by trolley and bus, even ferryboats, and finally, not least, the pedestrian. To achieve the necessary over-all pattern, not merely must there be effective city and regional planning, before new routes or services are planned; we also need eventually – and the sooner the better – an adequate system of federated metropolitan government. Until these necessary tools of control have been created, most of our planning will be empirical and blundering; and the more we do, on our present premises, the more disastrous will be the results. What is needed is more thinking on the lines that Robert Mitchell, Edmund Bacon, and Colin Buchanan have been following, and less action, until this thinking has been embodied in a new conception of the needs and possibilities of contemporary urban life. We cannot have an efficient form for our transportation system until we can envisage a better permanent structure for our cities. And the first lesson we have to learn is that the city exists, not for the facile passage of motorcars, but for the care and culture of men.

Notes

1 The fact that the aroused citizens of San Francisco not only halted the engirdlement of the Bay but now also demand that the present structure be torn down may prove a turning point in the local community's relations with an entrenched and high-handed bureaucracy. An encouraging example to all other cities similarly threatened. In November 1967 the New York 'Times' headlined the happy news: "U.S. Road Plans Periled by Rising Urban Hostility."

2 Both the cell and the car are now, a decade later, on the way.

30

Designing Cities and Buildings as if They Were Ethical Choices

Jessica Woolliams

The way we design and live in our cities and build-ings has ethical implications. This article has two parts: the first argues that cities should be front and center in any environmental debate. The second argues that we as a society could design and live in our buildings in a way that would significantly lighten humanity's weight on the planet.

> The world's cities take up just 2 percent of the Earth's surface, yet account for roughly 78 per-cent of the carbon emissions from human activities, 76 percent of industrial wood use, and 60 percent of the water tapped for use by people.
> —*Molly O'Meara*, State of the World, *1999*

For over two and a half million years, human-ity lived as a hunting and gathering society. For less than ten thousand years, humanity has been farming. It has only been since the Industrial Revolution in Europe, beginning in the early nineteenth century, that humanity has lived in an industrial society. One of the characteristics of this industrialized society is urbanization. In 1950, the United Kingdom became the first nation to have over 50 percent of their population living in urban areas. It was soon followed by a host of European nations, and North America and

Japan were not far behind (O'Meara 1999). In the year 2000, for the first time in human history, over 50 percent of the human population lives in cities.

However, our collective understanding of humanity's impact on the planet has failed to catch up to our rapid urbanization. Humanity's understanding of their environmental impact is focused on the problems that we have created outside of cities. In 1993, 1,680 scientists from all over the world issued a "World Scientists' Warning to Humanity" concerning the environ-mental crisis. This comprehensive document outlined environmental problems in the areas of atmosphere, water resources, oceans, soil, forests, living species, and population. There was no talk of cities.

Recently, in what is being hailed as the most comprehensive study of global ecosystems to date, the World Bank, the UN Development Program, the UN Environment Program, and the World Resources Institute have come together to assess the health of the global ecosystem through the "Pilot Analysis of Global Ecosystems" (UNDP 2000). The study is organized into ecological assessments of the following areas: forests, fresh-water systems, coastal/marine habitats, grass-lands, and agricultural lands. Cities receive a few

From David Schmidtz and Elizabeth Willott, eds., *Environmental Ethics: What Really Matters, What Really Works* (Oxford: Oxford University Press, 2002), pp. 426–30. By permission of Oxford University Press, Inc.

sentences towards the end. It is as if humans, the animal with the largest impact on the planet, and cities, the locus of human activities, are not worth studying.[1]

This almost willful blindness toward the environmental impact of our cities is curious indeed. Cities have always been the loci of human consumption and waste, and now with over half of the world's people living in cities, cities are ever more the places humanity consumes the natural resources of the planet and gives back wastes. Cities in the wealthy world, of course, are responsible for an even greater throughput of energy and materials, and a greater production of waste. While consumption is not the central environmental problem in itself, the way resources are consumed in cities presently ensures that consumption is inexorably linked with the associated problems of global warming, habitat destruction leading to species loss, fishery destruction, and damage to water systems. As Rees (1997) puts it: "Half the people and three-quarters of the world's environmental problems reside in cities, and rich cities, mainly in the developed North, impose by far the greater load on the ecosphere and global commons." The way humanity chooses to design and live in its cities will decide whether it can sustain life on this planet, so there are strong ethical reasons to change the way cities are built.

Cities have the potential to be part of the solution rather than part of the problem. Certainly there is no going back to the land. Going back to low density rural living would be the surest way to cause accelerated environmental destruction and habitat loss (Lewis 1992). Gathering people together in cities allows for efficient use of resources, thereby minimizing pollution and environmental harm. A built environment that is designed to be dense, low-energy, low-consumption, and waste-assimilative can play a vital role in restoring global ecological health (Walker and Rees 1997; Roseland 1998).

The way we build our cities has a profound impact on the emission of greenhouse gases. A recent study by the David Suzuki Foundation found that in Canada emissions could be reduced by as much as 50 percent using standard technology. The largest portion of this reduction would come from retrofitting 80 percent of existing buildings, and all new ones, with better

insulation and windows. The other major portion of the reduction would come from Canadians switching to hybrid gasoline-electric cars that use three times less gas than conventional cars. In most industrial nations, buildings and transportation have the biggest impact on the environment and on global climate change. Because transportation patterns are in many ways influenced by building patterns, the focus here shall be on the way we as a society design buildings.

Buildings

> What would it be like if developments produced more energy than they consumed? What if they increased habitat and biodiversity, produced food and clean water?
> —*Amory Lovins*, Green Development, *1998*

The construction, renovation and operation of buildings worldwide devours more of the planet's resources than any other economic sector (ASMI 1999). Every year the buildings that we build and inhabit use as much as 40 percent of all of the raw materials and energy used on the planet (ASMI 1999). This means millions of tons of liquid and solid waste, toxic air pollution, and greenhouse gases. But what this means is that no other sector of the world economy has the potential to make such a large reduction in its impact on the environment. As architect Richard Rogers (1999) notes:

> The principal objective currently facing humanity is to allow a continued growth in living standards world-wide within diminishing resources. Architects have an important part to play, as they influence up to 75% of total energy use (50% in buildings, 25% in transport).

At a time when the World Bank is warning that the wars of the next century will be waged over access to fresh water supply, North America and most "developed" nations are literally flushing their resources down the toilet. Canada has the unenviable position of having the second largest domestic water consumption in the world, second only to the U.S.A. This waste of water reflects our values about our natural world and the creatures with which we share it.

Our consumption of energy also reflects our environmental values. The use of energy is arguably more significant because it dictates the rate at which other resources are consumed, as well as having an impact on greenhouse gases. Energy use per capita in Canada is roughly 500 percent more than world average. Clearly one way to reduce Canada's morally and logically offensive overuse of resources and energy and its incredible production of waste and pollution is to green our built environment. Green building guidelines have a role to play.

A green building or "high performance" building has a lighter impact on the environment, and it also encourages its users to have a lighter impact on the environment. It usually addresses at least some of the following areas: energy, water, landscape, materials, waste, construction management, and indoor environmental quality.

A green building addresses some of the challenges of global climate change by using less carbon-based energy than a standard building. Large reductions can be made to conventional energy use by using the tried and true techniques of increased insulation, better windows, passive solar heating, day-lighting, and natural cooling. Further reductions to carbon-based energy can be made by using more benign sources of energy, including solar water pre-heating, photovoltaic panels, wind power, geothermal heat exchange, local microhydro, or fuel cells. Encouraging walking, biking, and the use of public transit can make still further reductions. This can be done through locating the building close to pedestrian, bike, and transit routes and providing showers and secure lockers for cyclists.

A green building uses resources like water and materials with more care and creates less waste. Reductions in water use can start with installing water conserving fixtures, but can also include using biological waste water treatment systems for gray and blackwater; using waterless toilets or urinals; using composting toilets; or capturing on-site rainwater. It can also mean planting a landscape that is native and does not require watering. A green building should filter water pollution before it leaves site, recharge groundwater, preserve and encourage biodiversity, and use integrated pest management techniques. It should use materials that are not only durable but salvaged and salvageable, recycled and recyclable.

It should make it easy for occupants to recycle and compost, reuse construction and demolition waste, and avoid air pollutants.

The goal with green buildings is to use only as much energy and resources and create only as much waste as can be sustained by the environment. But can we really build these buildings? Yes, there are many being built. The C. K. Choi Institute of Asian Research at the University of British Columbia, Canada, was awarded one of the "2000 Earth Day Top Ten" by the American Institute of Architects. It is over 50 percent below the standard energy consumption level, uses extensive daylighting and natural ventilation, and uses extensive reused and recycled materials. Its nine composting toilets and three urinals require no water. Gray water and rainwater are used for irrigation. The various water-saving devices save about 1,500 gallons of potable water every day (Cole 1996). This is enough water to serve the domestic needs of 1,500 Haitians, 9 Americans, or 12 Canadians for one day. Are these buildings affordable? Yes, they require the same or lower capital cost to build and they should incur significantly lower operating and maintenance costs (Hawken, Lovins, and Lovins 1999).

In the last ten years a growing number of guidelines and rating systems have arisen to assist in the building and rating of green buildings. The city of Austin, Texas, developed the first green building rating system in North America in 1991. The purpose of Austin's Green Building Program is to encourage building professionals (architects, builders, and developers) to use sustainable practices in their buildings and to encourage consumers to value sustainable buildings. The program involves training the industry, marketing green ideas to consumers, and certifying "Green Buildings" with a rating of between one and five stars. Points can be received for following various building practices and ensuring a certain level of environmental performance.

Austin beat a trail for others to follow. Today there are many guideline programs. These include the Green Builder Program of Colorado, developed in 1995, which was the first to be developed on a state-wide basis; Santa Monica's Green Building Design and Construction Guidelines; New York City's High Performance Building Guidelines; and the Commonwealth of Pennsylvania's Guidelines for Creating High-Performance Buildings, the latter three all developed in 1999.

Perhaps the most far-reaching "Green Building" rating system is the U.S. Green Building Council's Leadership in Energy and Environmental Design (LEED) Green Building Rating System, developed in 1997 and further developed in 2000. Seattle has adopted LEED as a standard for all of its city buildings. So has the U.S. Navy, whose policy requires all its buildings to be built to these criteria. The Navy's policy alone ensures that roughly $5 billion worth of construction – roughly one percent of all US construction – is built using the LEED standards. Many other cities and jurisdictions are also considering adopting the LEED Rating System.

The LEED system is designed for new and existing institutional, commercial, and high-rise residential buildings. Certification is given to buildings that meet certain non-negotiable pre-requisites (like a minimum level of energy efficiency). The building designers may also choose among different options to receive a minimum number of credits. Like many other Green Building guideline systems, it was designed to be voluntary and driven by citizens' buying habits. There is much more work to be done on LEED to incorporate the overall urban form. There should also be a parallel standard for other building forms, like medium- and low-density residential, which are presently left out of the system. Like any standard, LEED will grow and change. However, it now stands as the first green building standard for the industry and consumers. Having an industrywide standard should enable the power of the market to be used to make more ethical decisions about our buildings and, ultimately, our cities. It is time for us as a society to start designing and living in our cities and buildings as if they were ethical choices, because that is what they are.

Notes

1   This exclusion of cities is also seen in the behavior of environmental activists, who have traditionally

been concerned with the environment that exists outside of cities. When Robert Hunter and Paul Watson established Greenpeace in Vancouver in 1971, the first issue on the agenda was nuclear testing in the arctic. Soon after, there were demonstrations against sealing and whaling, then deforestation and destruction of land from oil drilling. Today, urban environmental activists are beginning to lobby for bike lanes and tougher clean air standards in cities. Greenpeace, for example, has taken a crucial role in guiding the development of what is now the world's largest solar community, containing 665 solar-powered homes at the Sydney 2000 Olympics.

References

Athena Sustainable Materials Institute (ASMI), 1999, Web site http://www.athenasmi.ca/.

Raymond J. Cole, "Green Buildings: In Transit to a Sustainable World," *Canadian Architect* 41, no. 7 (1996): 13–19.

Paul Hawken, Amory Lovins, and L. Hunter Lovins, *Natural Capitalism: Creating the Next Industrial Revolution* (New York: Little, Brown and Company, 1999).

Martin Lewis, "Introduction," in *Green Delusions – An Environmentalist Critique of Radical Environmentalism* (New York: Duke University Press, 1992).

Molly O'Meara, "Exploring a New Vision for Cities," in *State of the World* (New York: Worldwatch Institute, 1999).

William Rees, "Is 'Sustainable City' an Oxymoron?" *Local Environment* 2 (1997): 303–10.

Richard Rogers Partnership, "The Environment," Richard Rogers Partnership Web page, http://www.richardrogers.co.uk/, January 20, 1999.

Mark Roseland, *Towards Sustainable Communities: Resources for Citizens and Their Governments* (Gabriola Island B.C.: New Society Publishers, 1998).

United Nations Development Programme (UNDP); United Nations Environment Programme (UNEP); World Bank (WB); World Resources Institute (WRI), *A Guide to World Resources 2000–2001* (Washington, D.C.: World Resources Institute, 2000).

Lyle Walker and William Rees, "Urban Density and Ecological Footprints – An Analysis of Canadian Households," in *Eco-City Dimensions: Healthy Communities, Healthy Planet*, ed. Mark Roseland. (Gabriola Island, B.C.: New Society Publishers, 1997)

31

The Local History of Space

Steven Moore

If space is produced, if there is a productive process, then we are dealing with history; . . . Since, ex hypothesi, *each mode of production has its own particular space, the shift from one to another must entail the production of a new space.*

Henri Lefebvre, *The Production of Space*, p. 46

. . . in the case of technological networks, we have no difficulty in reconciling their local aspect and their global dimension. They are composed of particular places, aligned by a series of branchings that cross other places and require other branchings in order to spread.

Bruno Latour, *Science in Action*, p. 131

The politics inscribed in the objects and social spaces of Blueprint Farm did not originate with Jim Hightower's plan to revolutionize Texas agriculture. Nor did they begin with Pliny Fisk's notion of "flexible farming." Nor with Israeli kibbutzim. To attempt an understanding of Blueprint Farm within the limits of recent events would be to suppress the cumulative history of space. The producers of Blueprint Farm did not invent their situation. Rather, they contributed to a continuum of events that have piled up in la Frontera Chica, or "the little border," as the inhabitants of Zapata, Starr, and Webb Counties refer to their region. To reconstruct how locals understand the historical continuum of their own practices will require some excavation.

This chapter is in three parts: First, it is necessary to define what I mean by the terms *place* and *technology* and to generally relate these constructs to each other. Second, these definitions provide a theoretical context for constructing what Henri Lefebvre would call a "local history of space." In Lefebvre's view, "A space is not a thing, but a set of relations between things."[1] Or, put even more directly," (Social) space is a (social) product."[2] The second goal of this chapter, then, is to reconstruct the production of la Frontera Chica as a social morphology. This will be derived from interviews with locals and from local histories. Third, I interpret that history of space through the abstract definitions of place and technology constructed in part one.

From Steven Moore, *Technology and Place: Sustainable Architecture and the Blueprint Farm* (Austin, TX: The University of Texas Press, 2001), pp. 45–85.

The chapter then concludes with an empirically derived hypothesis that relates the concepts of place and technology.

Place, Technology, and Technological Networks

My definitions of place and technology make use of arguments developed in the disciplines of cultural geography and sociology. From cultural geography my analysis relies most directly upon the works of John Agnew and Henri Lefebvre because they have been particularly influential in resurrecting the concept of place as a subject of serious study. I should be quick to point out that Lefebvre's discourse concerns the social construction of "space," not "place." These concepts are historically distinct, and occasionally allergic to each other. I will use them interchangeably because Lefebvre's project is itself a mixing of modern and postmodern, Marxist and Heideggerian, ecological and political assumptions. Neil Leach, for example, categorizes Lefebvre as a phenomenologist where others categorize him as a Marxist.[3] Such confusion about his assumptions makes Lefebvre the ideal precursor to a nonmodern thesis. From sociology my analysis relies upon texts by Donald MacKenzie, Judith Wajcman, and Bruno Latour.[4] Latour, in particular, has contributed terms to the critique of modern technology that are spatial, and thus related to the concerns of architecture.[5]

As a cultural geographer John Agnew argues that places cannot be understood within the limits of architecture or physical geography. There is a growing body of literature, originating with Heidegger, which meditates upon the question of boundaries.[6] These authors ask, where do places begin and end, and with what senses do we find them? Agnew's answer is that the qualities of place are complex, quantitatively and qualitatively. He offers three qualities through which we might understand the phenomenon of place: *location, locale,* and *sense of place*.[7] I will use these concepts as lenses through which to study la Frontera Chica.

By location, Agnew intends that a place can be understood as "the geographic area encompassing the settings for social interaction as defined by social and economic processes." This quality of place includes the objective structures of politics and economy that link one place to another: the EC (European Community) and the Monroe Doctrine are examples derived from formal political alliances and economic structures. The alliances of corporate economies also construct locations. Houston, for example, is effectively closer to the east coast of Scotland than to Arkansas because these oil-producing landscapes are managed by the same corporate structures. It is these structural conditions of political economy that most concern Marxist scholars.

If location is defined as a set of objective structures, Agnew argues for the existence of a sense of place as a set of intersubjective phenomena. By this term he means the local "structure of feeling" that pervades Being in a particular place. This quality of place includes the intersubjective realities that give a place what conventional language would describe as *character* or *quality of life*. For example, the reverence that the citizens of Austin, Texas, reserve for a swim in Barton Springs or the stylish ambition of street life that New Yorkers enjoy are ontological, rather than physical, qualities of place. It is in this mode that the complex human poetics of place are experienced. It is also in this mode that constructivist scholars study the intersubjective construction of reality.

Between the objective quality of location and the subjective quality of sense of place, Agnew establishes a middle ground, or locale. This quality of place is the setting in which social relations are constituted. Locale includes the institutional scale of living to which architecture contributes so much: the public square, the block, and the neighborhood. I have chosen this topos, or philosophical place, from which to observe Laredo in the third section of this chapter.[8] From this quasi-objective, quasi-subjective place I attempt to align the structures of location that hover above us with the sense of place that we experience on the ground. My intent in this operation is to avoid the overdetermination that derives from a Marxist preoccupation with the conditions of political economy and at the same time avoid the underdetermination that derives from a constructivist preoccupation with conditions of atomized reality.[9]

I should stress here that Agnew's distinction among location, locale, and sense of place is not simply a matter of macro-, meso-, and micro-scaled analysis. Rather, it is the "elastic" interaction among all three qualities that constitutes place in Agnew's terms. "It is the paths and projects of everyday life, to use the language of time-geography, that provide the practical glue for place in these three senses."[10]

If the concept of place requires such a multifaceted definition, what about technology? Just as conventional thought understands place as only physical in quality, technology is commonly understood as physical hardware. Such a physicalist definition tends to consider the social construction of automobiles or refrigerators, for example, as outside the competing interests of society.[11] In the tradition of positivism, technology is understood as the asocial application of scientific truths. In contrast, the literature of science and technology studies has demonstrated that technology, far from being constructed outside society, is a system that is inextricably part of society. Technology, like place, is a field where the struggle among competing interests *takes place*. MacKenzie and Wajcman have argued that the concept of technology, like place, includes three qualities. In their construction, technology includes *human knowledge, patterns of human activities*, and *sets of physical objects*.[12] I will also employ these concepts as a second set of lenses through which to study la Frontera Chica.

In MacKenzie and Wajcman's definition, knowledge – the first characteristic of technology – is required, not only to build the artifact, but also to relate the natural conditions upon which the artifact works, and to use the artifact. The second characteristic of technology, "patterns of human activity," or what I would prefer to call *human practices*, refers to the institutionalization, or routinization, of problem-solving that inevitably occurs in society. The practices of architecture, carpentry, or farming are examples. The third quality of technology, "sets of objects," is, of course, the most obvious – these are the things themselves. The point is, however, that computers, hammers, or tractors are useless without the human knowledge and practices that engage them.

What I want to argue here is that the definition of place offered by Agnew, and the definition of technology offered by MacKenzie and Wajcman, are related by a tripartite structure that is not accidental. Figure 31.2 will help to make this point clear.

The limited point of the diagram is threefold: First, that places and technologies are both spatial concepts with related structures. Second, that these qualities are dialogically related. And third, that modern forms of knowledge, like the economics of location, tend toward the abstract and overdetermined, while our understanding

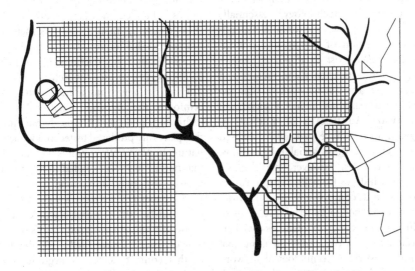

Figure 31.1 Map of downtown Laredo and Nuevo Laredo. The site of Blueprint Farm was on the grounds of Laredo Junior College at the left of the map between the international railroad and the bend in the river. Redrawn from Rand-McNally, Map of Laredo.

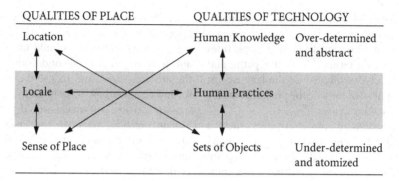

Figure 31.2   The dialogic qualities of place and technology.

of objects and sense of place tends toward the underdetermined and the atomized. These points serve only to magnify the centrality of locale and practices as the glue that holds the discourse of places and technologies together.

To argue that place is a spatial concept is a tautology and requires no further backing. However, to argue that technology is a spatial concept requires some explanation. Bruno Latour has argued that "Technological networks, as the name indicates, are nets thrown over spaces."[13] By "technological network," Latour refers, not just to "sets of objects," but to the social networks that construct a relationship among human knowledge, human practices, and nonhuman resources – the stuff from which the objects themselves are made. His point is that technology is essentially a spatial concept because its operation depends upon the mobilization of human and nonhuman resources *that exist in different places*.[14] For example, architects, clients, contractors, and bankers comprise a social network of building producers. Their relationship has a social and spatial quality to it. Advances in communications technology, however, have radically collapsed the spatial reality of these social relationships. When one recognizes, however, that lumber from Oregon, windows from Pittsburgh, carpet from Mobile, and compressors from Taiwan are required to realize the material intentions of the producers, the concrete qualities of their purely *social* network are materialized as a global *technological* network. A technological network produces spatial links that tie the social network of producers to those nonhuman resources required for construction. This is a central

argument of this study that has, as we shall see shortly, important implications for the social construction of la Frontera Chica as a place.

To follow Latour's argument and the relationship of technology and place constructed in Figure 31.2 leads to a central argument of this chapter. This proposes that technology is best understood not through history, but through geography. History tends to interpret reality as *human events in time*. Through temporal interpretation we might better understand the causal sequence in which humans construct artifacts. In contrast, geography tends to interpret reality as *human events in space*. Through spatial interpretation we are more likely to understand how technological networks operate to dominate the places inhabited by humans and nonhumans. It is geography, then, that offers methods more relevant to this inquiry. It is in the local history of space that the relationship of places and technologies becomes concrete.

Henri Lefebvre has argued two points that reinforce the dynamic relationship between technology and place that is claimed here. First, that places are produced by technology acting upon nature. Implicit in this point is the claim that original nature, if it ever existed at all, has long ago been incorporated into *second nature*, which is a work of society.[15] Lefebvre's second point is that each society – or, as Marxists would have it, each mode of production – produces its own peculiar type of space.[16] I have cited this point above, but in this context Lefebvre's argument serves to explain that the differing qualities of places are more a matter of technological practices than aesthetic choices.

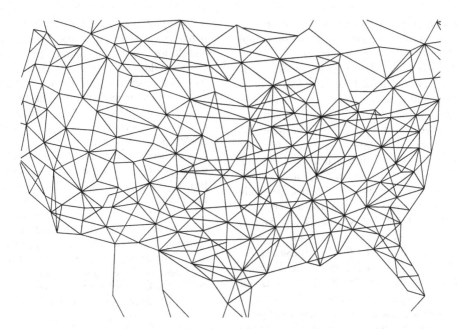

Figure 31.3 Highway linkages in the United States. Highways are, as Bruno Latour suggests, a literal net thrown over space. Redrawn from AAA road map of the United States.

In constructing this dialogic relationship between place and technology, I should make clear that I am not building a case for environmental determinism – places do not *cause* technologies. Given different cultural conditions, the sets of objects that dominate a place like Laredo might be different. Given constant environmental conditions, the interpretive flexibility of culture is entirely contingent. I will argue that environments do shape technologies, but are in turn shaped by them.[17] As a corollary, I am not building a case for technological determinism – technologies do not *cause* places. The same logic holds that technologies *do* shape places, but are also shaped by them.[18] [. . .] The point here is that the relationship of place and technology is both spatial and discursive. It is a dialogue of cause and effect, means and ends. They are inseparable, but contingent, concepts that lead inhabitants of a place to a dialogic narrowing of cultural horizons. This abstract proposition, however, requires concrete evidence to become relevant to our discussion.

[. . .]

The contingent condition of the economy in la Frontera Chica is magnified by the contingent condition of natural resources. Although the landscape of Tamaulipas has been softened by the importation of palm trees and buffle grass, and familiarized by generations of European settlement, it is an unbountiful place – it leaves recent immigrants like Rafael Bernadini "gasping." The Rio Grande, by the time it reaches Laredo, is one of the world's most polluted rivers. Life in such a physical environment is difficult to sustain, culturally and biologically. As in much of the United States, the values of Laredoans reflect, as Vera Sassoon put it, the "throw-away society" in which they live. Daily practices of nature preservation are not exactly commonplace in the lives of locals. Immigrants from Southern Mexico or Southeast Asia have not come to Laredo to preserve this austere natural world, but to exploit a cultural situation that incidentally includes the land itself. William de Buys argues that "the people of pioneer and subsistence cultures everywhere, have constantly underestimated their capacity for injuring the land."[19] Moderns romantically imagine that the rural peoples of any historical period live in closer harmony with the land than do those who live in metropolitan centers. There is, however, much evidence against such a generalization. The subsistence farmers of eighteenth-century Laredo, like their contemporary counterparts, are no exception.

Nor have the old ranching families of la Frontera Chica identified their interests with the natural world that surrounds them. In the eyes of locals like Vera Sassoon, Laredo remains ten to fifteen years behind these times of increasing environmental awareness. Unlike the citizens of other Southwestern cities – Austin and Phoenix come to mind – that depend upon a delicate ecology, Laredoans are, she claims, generally unconcerned with ecological issues.[20] Perhaps ironically, NAFTA has served as an external catalyst to stimulate local concern for degraded ecological conditions. In direct response to pressure from the U.S. government, Nuevo Laredo has improved the capacity of sewage treatment facilities to better match effluent load. Other Mexican cities still dump raw sewage directly into the river because they lack the authority to tax for infrastructure improvements at a level that reflects actual population growth. As a result, most Laredoans still have a decisively negative view of the Rio Grande. There is virtually no public access to the river because it is perceived as a dangerous place – the hangout of drug smugglers, pollution, and disease. Tom Vaughan, a founder of the Rio Grande International Study Center, which now occupies the site of Blueprint Farm, holds that few people on either side of the river are even aware that the river is their only source of drinking water.

Sissy Farenthold, a Democratic colleague of Jim Hightower's and the first woman ever to be nominated as a vice presidential candidate, sees the politics that link the *maquiladoras* and the environment in a particularly sobering light:

> In its 1991 study "Border Trouble: River in Peril," the National Toxics Campaign Fund reported on canals originating at U.S.-owned *maquiladoras* in Matamoros that are filled with toxins that pose a daily threat through ingestion, absorption, or contact with the polluted water, or through respiration of deadly chemicals as they evaporate. In Matamoros and other border cities, raw sewage flows in open ditches. In 1991, the American Medical Association declared the border region "a virtual cesspool and breeding ground for infectious diseases."

These problems are not limited to the disposal of human and toxic wastes into industrial canals and thence into the Rio Grande. The problem extends to the transportation of hazardous waste into, and sometimes from, Mexico. Disturbing and even more immediate is the effect of pesticides and contaminated water on the growing U.S.-bound agricultural goods. In a process labeled the "circle of poison" by Senator Patrick Leahy, these products are tainted with U.S.-made pesticides, long since banned in the United States, that are now sent back to the U.S. to be sold.

The 1983 La Paz Agreement, signed by President Reagan and President de la Madrid, was to initiate a new period of respect for and attention to environmental border issues. To date, this agreement has been ignored. It does not have the force of law, and the State of Texas has been particularly disrespectful of its terms.[21]

The increased public consciousness of such pollution of the watershed, and the extended drought of the 1990s, have forced people to acknowledge the need for water conservation measures. The local water master, whose job it is to enforce international water agreements related to the Falcon Dam, is now required to regularly monitor the meter at the intake at each irrigation system to prevent illegal takings. For those on the American side, the daily rinse of the family car and the watering of lawns have become a source of political debate, if not action. Across the river in Mexico, however, the rediscovered water shortage has stimulated a more severe crisis. The government of Tamaulipas has formally asked to renegotiate water rights to the Rio Grande with the U.S. government. Although Tamaulipas is not a desert on either side of the river, modern technology has failed to shield its residents from the desertlike conditions of scarcity. It is not a place to which visitors come expecting to find a health spa.

It is into this local history of space that the personalities and artifacts of Blueprint Farm were thrown by political circumstance. Their arrival was noticed only by those with a vested interest. When the Texas-Israel Exchange (TIE) was given authority to proceed by the Texas Legislature, Jim Hightower's political needs demanded the selection of a site that would yield instant political income. Members of the Laredo Jewish community were enlisted to locate the site, which turned out to be the old Slaughter farm. This undistinguished but easily accessible piece of property fit Hightower's ideological profile of a parcel of land no longer economically competitive because it was too small to justify the capital-intensive methods of agribusiness. Not only did this property have a history of produce production, but also the owners were in tax trouble and thus in jeopardy of losing their water rights. Out of these politically conducive conditions, a deal was hastily struck to rent the house, and the owners agreed to donate use of the land for experimental crop production. Within four months of occupancy in 1987, the first Israeli farm manager had produced a crop.

The promise of the jump-started relationship between the Slaughter family and TDA began to exhibit strain, however, as soon as a major grant from the Meadows Foundation made it possible to consider a capital investment in the rented property. When Pliny Fisk began to design the technological systems to be employed at the farm, he assumed that these would be constructed at the Slaughter property. As Alvaro Lacayo characterized the situation," In lieu of a long-term formal agreement with the Slaughters . . . the relationship ended in a fiasco."[22] The quickly negotiated lease agreement proved to be a small obstacle to the savvy Mr. Slaughter. Not only did the hastily conceived project have to look for a new home, but the initial investment in the Israeli drip irrigation system was lost in the process – that equipment stayed on the Slaughter farm, where it has remained in productive use.

Almost as quickly as the first site was selected, the project was moved to the campus of Laredo Junior College. The college occupies 200 acres on the river that was once the site of Fort McIntosh, the late-nineteenth-century U.S. military post that enforced American sovereignty over the disputed territory that lies between the Rio Grande and the Nueces River to the north.[23] Because the college had no long-range plan, Blueprint Farm was easily accommodated – some have said "dumped" – there. The site of Fort McIntosh and Laredo Junior College is thoroughly isolated from the town by a bend in the river and the railroad tracks that initially brought prosperity to Laredo. A few have concluded that this site was perfect for experimentation because of its proximity to the river. Pliny Fisk also thought it was an ideal site because, on state-owned property, the experimental technologies proposed would be exempt from local building code compliance. However, such pragmatic criteria also exempted the "demonstration" from a principal responsibility – the ability to be witnessed.

Narrowing Horizons of Spatial Discourse

The history of space produced by the inhabitants of la Frontera Chica can now be interpreted through those lenses donned in the first section of this chapter. As in Lefebvre's discourse on the production of space, I want to now argue that the history of local space "begins . . . with the spatio-temporal rhythms of nature transformed by social practice."[24] Lefebvre's emphasis upon "practices," rather than objects, supports the organization of Figure 31.2. If the reader will recall, in that figure I position "human practices" as central to the social production of both places and technologies. Based upon the structure of that figure, the conclusion of this chapter will consider the technological and spatial practices of Laredoans in relation to John Agnew's characterizations of place: *location, locale,* and *sense of place.*

The historical *location* of Laredo – Agnew's first characteristic of place – begins with the Spanish colonial Orders for Discovery and Settlement of 1573. It is essential to recognize that this ordinance was conceived as an instrument of sixteenth-century global political economy. In Lefebvre's analysis of that ordinance, he argues that

The very building of the [Spanish colonial] towns thus embodied a plan which would determine the mode of occupation of the territory and define how it was to be organized under the administrative and political authority of urban power.[25]

In Lefebvre's terms, the Spanish ordinance is best understood as an instrument of production – a technological code that served as a means to impose an emerging proto-capitalist order upon a premodern space. It is in this sense that the colonial ordinance was a "representation of space"

– what Lefebvre refers to as a kind of mental or conceptual map that directed the emplacement of the emerging European mode of production on top of that practiced by indigenous Americans.

A geography of technology concerned with location, or the structural conditions of political economy, is naturally concerned with competing territorial claims of European (and American) colonial powers. The mapping and remapping of Laredo by the four nations that have claimed the city exemplify these concerns.[26] Such political instability documents the stresses of political economy to which the region has been historically subject. I have been tempted to periodize the location of Laredo in the global economy as pre-1881 and post-1881, that being the year when the railroad reached the city. The early period – before 1881 – was generally dominated by the successive colonial economies to the south, and the later period – after 1881 – has been generally dominated by the market economy to the north. Such a rigid periodization, however, would tend to obscure the continuum of spatial development in the region. It would be better to argue that la Frontera Chica has been a transitional place in the evolving modern political economy – alternately claimed by competing national interests. Its current status as the "Gateway to Mexico" – the largest land-based port of entry into the United States – is thus not a newly minted condition.[27] My argument here is that the location of Laredo has always been a function of the globalized economy. What is changed is not the local presence of economic powers acting at a distance (what Latour describes as "centers of calculation"), but the relative location or strength of those powers. Rather than being tied to the court of Spanish monarchs, the French Empire, or Mexican revolutionaries, Laredo is now linked most strongly to the stock exchanges of New York, Frankfurt, and Tokyo.

A geography of technology concerned with locale, Agnew's second characteristic of place, is one concerned with the "setting for social relations." In the case of Laredo, the original eighty-nine Spanish land grants, or porciones, granted to settlers by Spain in 1767, created what might be described as a binary landscape. Each porción surveyed by colonial authorities included two parcels: a residential plot within the colonial grid of the town and a deeply linear, agricultural plot

with narrow river frontage that included water rights.[28] The patterns of property rights and the daily practices of citizens were thus ordered by a distinct separation between two kinds of space: the warped linear space of agriculture and the checkered space of society and trade.

As John Stilgoe has documented, the premodern European Landschaft also observed a distinction between agricultural and social space in that villages were densely clustered settlements surrounded by agricultural lands.[29] In that landscape, however, both agricultural and social spaces radiated organically from a central locus experienced by the community as a common spiritual center. In the colonial Spanish ordinance, however, spaces allocated for agriculture and society were conceived quite differently. Although social space was still surrounded by agricultural space, the mode of ordering these practices was geometrically and geographically distinct. The social space of the village was centered upon the abstract space of political, rather than spiritual, authority, and agricultural space was mathematically surveyed in relation to whatever local resources determined economic potential. In other words, agriculture in New Spain had already become conceptually, and thus spatially, disengaged from those more abstract political and economic practices assumed to dominate public life. Where premodern European space was perceived and lived as a monist, or singular, geography, the modern space of the New World was conceived as binary, or Cartesian.

In the period after the coming of the railroad to Laredo in 1881, interrelated networks of railways and highways have increasingly dominated the conditions of daily life and commerce. Those agricultural practices that once engaged the river to produce use-value have gradually been replaced by commercial practices that move people across the river to produce exchange-value. Laredoans – other than the elites who own the land – have generally rejected the stasis of farming and rootedness along the river as a way of life. What Laredoans say they want – at least the poor Hispanic families who eke out a living by working in the maquiladoras, tenant farming, or migrant laboring – is to escape the dark social history of their bondage to the land. As a locale, Laredo embodies the tension between stasis and movement, between farming and flight. The periodization of Laredo as pre- and post-1881 may be best portrayed as identifying a watershed in a continuum of change. It is reasonable to hold that in the early period the river was the conceptual center of the community. In that period the river was the net that bound the community to the natural cycles of agricultural production. In the later period, however, the river has become both a conceptual and real boundary – a marker that delineates changed social and economic conditions. The network of rails and roads, in the period after 1881, rather than centering the community upon cycles of agricultural production, has centered the community instead upon unequal rates of exchange that are determined by others at a great distance.

The ever-advancing commercial and suburban residential development that emanates from the Cartesian geography of the colonial city has gradually subsumed the agricultural geography of the river. The original binary geography that characterized the setting for social relations in the town has been transformed over time to a new monist physical geography – one based upon trading rather than agriculture as a way of life. My point is that the binary space established by the Spanish colonial ordinance in the sixteenth century was proto-modern. I mean by this term that the eighty-nine porciones surveyed in 1767 were a conceptual departure from the premodern pastoral monism of European space, but were not yet identical to modern capitalist monism. The space surveyed in the New Spain of 1767 was somewhere in between these singular visions of place. By monistic space I do not mean homogeneous space. The premodern space of the European Landschaft, like modern urban space in twentieth-century North America, was distinct and varied. By the term monism I mean only that the production of space was driven by a dominant idea, or set of concepts. In the case of Laredo, it was the Spanish who imported the seed of capitalistic monism that now dominates space on both sides of the Rio Grande.

I want to stress that the shift to a modern version of spatial monism in Laredo was not inevitable. Like the Spanish pobladores of the Las Trampas Land Grant in New Mexico, the inhabitants of la Frontera Chica might have made different technological choices. In New Mexico the colonial grid of densely settled urban, Cartesian space was generally abandoned in the early nineteenth century as soon as the threat of Indian attack had abated. Abandoning the colonial grid allowed these pastoralists to scatter their houses "along the edges of irrigation fields so that each family might better guard its crops from livestock and theft."[30] An organic, noncapitalist monistic space has been the result. Although the settlers of la Frontera Chica and Las Trampas began with the same binary geography as a conceptual map from which to build, they have over time produced very different geographies indeed. Where Laredo has become relentlessly competitive and capitalist, Las Trampas is characterized by vergüenza – a sense of "self-effacing probity" that restrains members of the community from advancing their own interests over those of the community.[31] Although the conditions of nature and economic forces acting at a distance certainly limited the choices made, each community might have made other choices. The Laws of the Indies did not predetermine the practices that sprung up in those places.

Finally, a geography of technology concerned with the sense of place, Agnew's third characteristic of place, is concerned with the "structure of feeling" of that place. Before its international partition in 1846, Laredo was a centered pastoral town on the north bank of the river. After partition, those who wished to retain their Mexican citizenship settled in Nuevo Laredo on the south bank, which was soon thereafter chartered as an autonomous Mexican municipality. Although Laredo is unique among U.S.-Mexican border towns in that both towns share a political structure and cultural identity dominated by Hispanics, the unequal economic forces at work in Los Dos Laredos create the sense of a split center. In this sense, the international city is still a binary construct, but one very different than the divided land-use patterns imported by the Spanish. On the north bank of the river, commodities are acquired and exchanged; one can almost hear the meters ticking. On the south bank of the river, an entertainment market has been created by the disparity between Mexican and American attitudes toward social regulation. Here the meters have a mariachi tempo. Where the colonial bifurcation of Laredo's locale structured how one moved through space, the more recent bifurcation of Laredo's sense of place structures how one thinks and feels *in* space – it is like the right and left side of one's brain.

Lefebvre might summarize this chapter by arguing that, in the "long history of space," the forces of political economy increasingly abstract the anthropological origins of daily life.[32] The trajectory of history, in Laredo and elsewhere, has generally been from the "absolute space" lived by archaic peoples to the "abstract space" surveyed by moderns. Because Laredo is a very old city, at least by the standards of the American West, we moderns tend to romanticize it as a place somehow different, or other, than those suburban landscapes now constructed by flexible capital. In Lefebvre's terms, however, Laredo was never a romantic assembly of "absolute spaces" linked to the spiritual practices of premodern life. Nor was it ever a mission intended to pacify souls and plow the earth so much as it has been a geographic opportunity created by displaced humans crossing the river to produce exchange value. In this sense, the colonial grid imposed upon Laredo in 1767 can be best understood as the technological apparatus through which the premodern space of Laredo was re-*located* by the Spanish masters.

If "every space has a history, one invariably grounded in nature," as Lefebvre argues, then the ecological conditions of Laredo, and the cultural conditions of la Frontera Chica, have not prefigured an agricultural future.[33] Nor have they precluded such choices. The ecological conditions and history of technological choices made in the Las Trampas Land Grant of New Mexico have been quite different. Similarly, where other towns in la Frontera Chica, like Roma, have become increasingly detached from centers of calculation through the weakness of their technological links, or through the strength of those who have made difficult technological choices, Laredo has reinforced its colonial potential as a trading town. It has done so by being better positioned to receive stronger technological links – the railroad, bridges, and Pan American Highway are only the most obvious examples. This argument does not erase the real, but marginal, agricultural history of la Frontera Chica. Rather, it acknowledges the principal point of this chapter: that most people have immigrated to Laredo, pre- and post-1881, to take their chances at trading – in the broadest meaning of that word – as a way of life. The space of Laredo has been increasingly abstracted by market forces acting at a distance and by those migrants who are passing through on their way to prosperity.

This argument has, as we will see, significant implications for the construction of an experimental farm. I want to be careful, however, not to leave the reader with the sense that Laredo has been transformed into an undifferentiated space indistinguishable from, say, suburban Columbus, Ohio. This is obviously not the case. If the location of Laredo is increasingly abstract, locals with deep roots in the city increasingly articulate its sense of place as "mestiza" – meaning that it is a place resistant to appropriation by the mechanisms of flexible capital. This unique sense of place, I will later argue, is a lost opportunity for the construction of an experimental farm.

In sum, it is worth repeating that Spanish colonials conceived Laredo as a binary geography that spatially alienated the practices of agriculture and trade. In the two centuries or so that have followed the deployment of that conceptual diagram, the binary quality of space has been gradually modified by the technological choices made by those immigrants who have been attracted to the political economy of the borderland. In the long public discourse between those who would farm and those who would trade for a living, the voices of traders have been alternately louder and better supported by distant networks. As a result, they have dominated the long history of local space in la Frontera Chica. In other locales, such as Las Trampas in New Mexico, farmers have prevailed, if not prospered. What those differentiated spaces share, however, is the dialogic mode in which technological choices shaped space.

To conclude this chapter I want to generalize the dialogic relationship of technologies and places that emerges from this particular story. The first step is to state that the qualities of place and technology are not interchangeable. They characterize different things, but the *location, locale,* and *sense of place* that describe Laredo as a place are largely congruent with the *human knowledge, human practices,* and *sets of physical objects* that describe the competing technologies that operate there. In the end, it is difficult to distinguish the boundaries of technologies and those of places. On the ground, these concepts are less distinct than assumed by the conventions of modern thought. This argument leads me to a general hypothesis:

Places and technologies are different things, but the processes of their social construction are dialogically related.

To clarify the distinction that I want to make between "dialectic" and "dialogic" relations it will be helpful to offer the reader two formal definitions. The definition of the dialectic is taken from J. K. Gibson-Graham's citation of Louis Althusser, who was in turn citing Marx:

[the dialectic] includes in the positive comprehension of the existing state of things at the same time also the comprehension of the negation of that state, of its inevitable breaking up; because it regards every developed form as in fluid movement and thus takes into account its transient nature, lets nothing impose upon it, and is in its essence critical and revolutionary.[34]

My operating definition of the "dialogic," so as to amplify similarities and differences to the Marxian dialectic, is simply a modification of the above:

The dialogic includes the positive comprehension of the contested meaning of things, and at the same time, it also includes the will to move that state toward a cultural horizon of meaning. Because the dialogic regards every developed form as a fluid movement, and thus takes into account its transient nature, it invites all to contribute to its development, and is in its essence life-enhancing and revolutionary.

[. . .]

Notes

1 Henri Lefebvre, *The Production of Space*, p. 83.
2 Ibid., p. 26.
3 See Neil Leach, *Rethinking Architecture: A Reader in Cultural Theory* (New York: Routledge, 1997), pp. 139–48.
4 Within social science, three traditions have evolved within science and technology studies: the network theory, pioneered by Bruno Latour and Michel Calion; the systems theory, developed by

Thomas Hughes; and the social constructivist theory, argued by Bijker, Law, MacKenzie, Wajcman, and others. Each of these positions has important contributions to make to this discussion: it is only the lack of space that prohibits their inclusion. These recent theories of technology take advantage of the work of Thomas Kuhn in the 1950s and 1960s, but depend most directly upon the pioneering research of the Frankfurt School beginning in the 1930s.

5 Bruno Latour, *We Have Never Been Modern*, p. 117.
6 See Martin Heidegger, *Being and Time*. Those who have expanded upon Heidegger include Edward Casey, *Getting Back into Place: Toward a Renewed Understanding of the Place-world* (Bloomington, IN: Indiana University Press, 1993) and David Seamon, A *Geography of the Lifeworld: Movement, Rest, and Encounter* (New York: St. Martin's Press, 1979).
7 John Agnew, *Place and Politics: The Geographical Mediation of State and Society*, p. 28. The definition of these terms is further amplified in his essay "Representing Space: Space, Scale, and Culture in Social Science."
8 The middle scale of inquiry to which Agnew refers, halfway between the macroscale of science and the microscale of poetry, is supported by Thomas Misa in "Retrieving Sociotechnical Change from Technological Determinism" in Smith, M. R. and Marx, L. *Does Technology Drive History?* (Cambridge, MA: MIT Press, 1995). p. 116. Misa argues that those who employ macro strategies of inquiry tend to arrive at technologically determined arguments, and those who employ microstrategies tend to arrive at socially determined arguments. A "middle-level" analysis tends to avoid the errors of either overdetermination or underdetermination.
9 See Thomas Misa, "Retrieving Sociotechnical Change from Technological Determinism."
10 Agnew, "Representing Space," p. 263.
11 Reductive, physicalist definitions of technology tend to be less sophisticated in their understanding of the social construction of artifacts. However, Bruce Bimber, in "Three Faces of Technological Determinism," in Smith, M. R. and Marx, L. *Does Technology Drive History?* (Cambridge, MA: MIT Press, 1995), develops a very scholarly, yet reductive, definition of technology as limited to apparatus. Bimber's project, however, leads to other ontological problems beyond the scope of this study.
12 See Donald MacKenzie and Judith Wajcman, *The Social Shaping of Technology*, pp. 3–4.
13 Latour, *We Have Never Been Modern*, p. 117.

14 See Bruno Latour's discussion of "immutable mobiles" in "Visualization and Cognition: Thinking with Eyes and Hands," in *Knowledge and Society: Studies in the Sociology of Culture Past and Present*, Vol. 3 (Greenwich, Conn: JAI Press, 1986), p. 9; see also *We Have Never Been Modern*, pp. 117–18.

15 Lefebvre, *The Production of Space*, p. 190.

16 Ibid., p. 31.

17 Ibid., p. 31.

18 Philip Brey has examined how "space-shaping technologies have disembedded the contemporary phenomenon of place." Where Brey's study has focused upon the role of "connectivity development" in transforming the experience of place, my own emphasis has been on what Brey terms "local development." See Philip Brey, "Space-Shaping Technologies and the Disembedding of Place: p. 242.

19 W. E. De Buys, *Enchantment and Exploitation*, (Albuquerque: University of New Mexico Press, 1985), p. 297.

20 "Vera Sassoon," author interview, June 21, 1995.

21 Sissy Farenthold, "Commerce without Conscience," *CITE: The Architecture and Design Review of Houston* 30 (Spring–Summer 1993): 12–14.

22 Alvaro author interview, June 23, 1995.

23 Fort Mcintosh, named for an Anglo officer killed in the war with Mexico, was one of a series of forts built along the disputed Mexican American frontier immediately following the Treaty of Guadalupe Hidalgo, signed in 1848. Jerry Thompson, *Sabers on the Rio Grande* (Austin, TX: Presidial Press, 1974), p. 166.

24 Lefebvre, *The Production of Space*, p. 117.

25 Ibid., p. 151.

26 These are Spain, Mexico, the Confederate States of America, and the United States of America. To make matters more complex, three additional governments influenced conditions in Laredo. France, although it never claimed Laredo in the era of Maximilian, certainly exerted influence there. The same holds true for the Republic of Texas, which never claimed territory south of the Nueces River. Finally, the Republic of the Rio Grande, which existed only for one year, claimed Laredo as its capital, but was never officially recognized by other governments.

27 See Hermalinda Aguirre Murillo, *History of Webb County* (Master's thesis, Southwest State Teachers College, 1941), p. 6. The title "Gateway to Mexico" was given to Laredo upon the connection of the Texas-Mexican Railroad from Corpus Christi with the National Railway of Mexico at Laredo in 1881. That title will be magnified by the pending appropriation of federal funds to construct the NAFTA Highway from Laredo to Canada. See Dan Feldstein, "House Approval of NAFTA Highway", *Houston Chronicle*, August 30, 1995.

28 N. Walsh, *The Founding of Laredo and St. Augustine Church*, Master's Thesis, The University of Texas at Austin, 1935. In Smith, M. R. and Marx, L. *Does Technology Drive History?* (Cambridge, MA: MIT Press, 1995), pp. 76–91. *Porciones* measuring 1,000 *veras* along the river and 30,000 *veras* in depth were apportioned by lottery.

29 J. Stilgoe, *Common Landscapes of America* (New Haven, Conn: Yale University Press, 1982), pp. 12–21.

30 De Buys, *Enchantment and Exploitation*, p. 175. De Buys and John Stilgoe are in apparent disagreement on this point. Stilgoe argues that the settlers of Chimayo clung to the nucleated colonial grid and de Buys argues the opposite, that settlers dispersed after the threat of Indian attack disappeared. I will assume that both are correct. In other words, the settlers of Chimayo documented by Stilgoe may have retained the colonial pattern, but other settlers in the Sangre de Cristo Mountains documented by de Buys have dispersed. My own informal observations of that region support this resolution.

31 De Buys, *Enchantment and Exploitation*, p. 195.

32 See Lefebvre, *The Production of Space*, p. 116.

33 Ibid., p. 110.

34 This passage is cited in J. K. Gibson-Graham, *The End of Capitalism (As We Knew It): A Feminist Critique of Political Economy* (Malden, MA: Blackwell, 1996), p. 26. It is attributed to L. Althusser, citing Marx, from *Politics and History* (London: New Left Books, 1972), p. 175.

32

Community

Joseph Grange

What makes intelligence-in-action possible on a widely regularized basis is the persistent presence of community as an integral part of the individual city dweller's life. Of course, by intelligence I mean much more than the capacity to be drawn by the iconic power of firstness or to follow efficiently the indexical pointers of secondness. All such modes of behavior, while intelligent in their own right, fail to promote the continuity, full meaning, and lived participation that are the marks of effective thirdness. To speak then of community is to raise once again the question of the possibility of intelligence-in-action in city life. But such a theme is made more difficult by the absence of real community as a widespread contemporary human experience, as well as by the tendency to invoke the term as a cure for any social problem. It is only recently that philosophers have begun to reach back to the beginnings of the American Republic for examples of living "civic community."[1] It appears that we know little of the philosophical history of the term and even less about how to bring it about.

This chapter begins by recalling the origins of the idea of community in the works of Plato. Next, it singles out the process that lies at the heart of any effective community – the making of connections. This activity finds its most concrete expression in "the desire of recognition by the other," which plays so profound a role in Hegel's systematic thought on the origins of human consciousness and social existence. This experience of desiring and desirability is then reinterpreted through the experience of sharing symbols. Finally, this fundamental act of urban cooperation is seen for what it really is: an invitation to make sharing in all its dimensions the ground of a concrete and equitable participatory democracy. In sum, the philosophical history of the idea of community is traced from its imagistic origins in the *Symposium* through its dialectical presence in Hegel's *Phenomenology* to its place as the very pivot of intelligence, freedom, and responsibility in the tradition of American naturalism and pragmatism. Urban praxis therefore follows Peirce by first painting an iconic likeness of community. Second, it opens up the workings of city life by identifying the activity lying at the base of urban community. Third, it argues for the central importance of democracy as the political form best suited for inculcating consistently good habits of sharing intelligently felt meanings and values. It concludes by presenting democracy as the epitome of the good community.

In Joseph Grange, *The City: An Urban Cosmology* (Albany, Suny Press: 1999), pp. 175–92.

Something In Between

The experience of community has played a decisive role in both the survival and successful evolution of the human race. Without it, knowledge understood as habits of successful reaction to all sorts of environmental stress would not have been possible. What to do when the lion strikes or where to go when the water holes dry up would never have become part of the intelligent traditions of human groups. Somehow or other humans became acutely aware of the need to double their chances of survival by sharing information, habits of response, and intelligence-in-action. What community fundamentally means is caught in this picture of human beings at their most desperate. Community is the recognition of "something in between" that bonds them into a whole greater than the sum of their individual existences.

It is precisely this understanding of community as something in between that characterizes Plato's discussion of eros in the *Symposium*. The theme, as is well known, is the nature of love: "Is Love a god or a mortal?" And while this particular form of the question appears, disappears, and then reappears as the various participants express their understanding of love, it is Socrates at the end of the dialogue who resurrects the question in its original form by recalling the words of Diotima concerning the nature of love. And it is from her iconic presentation of love as neither a god nor a mortal that the picture of *Koinonia* takes on such an engaging character:

In that case, Diotima, who *are* the people who love
wisdom, if they are neither wise nor ignorant?
That's obvious, she said. A child could tell you. Those
who love wisdom fall in between those two extremes. And
Love is one of them, because he is in love with what is
beautiful, and wisdom is extremely beautiful. It follows
that Love *must* be a lover of wisdom and, as such, is in
between being wise and being ignorant.[2]

Now, earlier in this dialogue Plato defines what it means to be in between:

[N]ow do you see? You don't believe Love is a god either!
Certainly not.
Then what is he?
He's like we mentioned before, she said. He is in between mortal and immortal.
What do you mean, Diotima?
He's a great spirit, Socrates. Everything spiritual, you see,
is in between god and mortal.
What is their function, I asked.
They are messengers who shuttle back and forth between
the two, conveying prayer and sacrifice from men to gods,
while to men they bring commands from the gods and gifts
in return for sacrifices. Being in the middle of the two, they
round out the whole and bind fast the all to all.
. . .
He who is wise in any of these ways is a man of spirit but he who is wise in any other way in a profession or any manual work, is merely a mechanic.[3]

What is being translated here as that which is in the middle and that which conveys meanings back and forth and as that which is between extremes is the experience of community, or as the Greek puts it, *Koinonia*. Three important dimensions are involved in this experience of *Koinonia*: philosophy, sharing meanings, and spiritual meeting. Each is an indispensable component of community as a human experience.

What makes possible each of these three aspects of community is the effective presence of "internal relations." (By this very abstract term I mean the very concrete experience of encountering within one's own being the lived values and real feelings of another human being.) I use this abstract term for now because I wish to raise in as general a way as possible the question of just how we get to know the feelings and values of other humans. Furthermore, an adequate answer to this question lies in forging a relation between the abstract disciplines of speculative philosophy and semiotics. Parts One and Two of this work dealt respectively with speculative philosophy and semiotics. In the final analysis, their ultimate reconciliation is made possible through the active presence of internal relations as a real factor in human experience. And that is the heart and soul of an intelligently effective urban praxis.

It is through internal relations that one feeler feels the feelings of another so as to make those feelings part of her own internal constitution. I insist on the abstractness of this formulation so as to avoid the smarmy smugness of such expressions as "I feel your pain." (Despite the power of the American presidency, the capacity to feel another's pain is not as easy as such a glib phrase might suggest.) Internal relations come about through the very hard work of bringing into one's sphere of experience the feelings and values of another. This can only be done by repeating back to another what they have expressed to you. This is what grounds Mead's "conversation of gestures." Unless I can reproduce in myself a simulacrum of what another has experienced, I cannot enter the conversation of gestures that makes up the domain of the socialized self. This simulacrum is the icon by which the other knows that I know what is going on in a particular situation. As instances of firstness, all such internal relations are about quality or value. The community begins with a gesture toward the value of others. That gesture is first of all a gesture of welcome. It take seriously the experiences of others:

> [T]he self to which we have been referring arises when the conversation of gestures is taken over into the individual form. When this conversation of gestures can be taken over into the individual's conduct so that the attitude of other forms can affect the organism, and the organism can reply with its corresponding gesture and thus arouse the attitude of the other in its own process, then a self arises. . . . Each individual has to take also the attitude of the community, the generalized attitude.[4]

From this perspective it can be seen just why the experience of community necessarily entails the active presence of philosophy, shared meanings, and spiritual meeting. Philosophy is the love of wisdom where that love is expressed through the effort to achieve the right balance between fact and value. This balance is brought about by reason of the adoption of an appropriate normative measure for balancing the roles of facts and values within particular concrete situations. To be a lover of wisdom is to know how factual

expressions (the gestures of the human body) display the real presence of values (the meaning of the gestures). Thus, the love that is an essential part of philosophy is a knowing grounded in the capacity to feel the real presence of another as a formative part of myself. Likewise, the sharing of meaning is essential for community since without it, no measure for balancing facts and values can be had. And lastly, spiritual meeting occurs when I encounter the other through his or her own experienced values. As Plato puts it: "Love shuttles back and forth" and "round[s] out the whole and binds fast all to all," for this is what happens in all communal "being in the middle."

All genuine spiritual meeting takes place within some sort of community, because a community is founded on the love of wisdom (which is the traditional goal of speculative philosophy) and shared meaning (whose achievement is the goal of semiotics). The something in between that is the very essence of community is a fusion of the love of wisdom and the sharing of meaning. This iconic presentation of community must now encounter the realm of secondness where the collision of values is worked out to the advantage of an enlarged sense of just what that "something in between" is really all about.

Making Connections

Just what is it that drives humans to such exertions on behalf of wisdom, love, and spiritual experience? What could possibly account for the expenditure of such energy and effort? Only some fierce and primal form of desire could explain such a prodigious drive for community. It is Hegel who provides the most compelling account of this desire and its place in human culture and self-understanding. In *Phenomenology of Spirit*, he tells us of the life and death struggle of the master and the slave as each seeks to win the recognition of the other as the way to their own self-awareness of who they are.[5] Recognition by the other of the worth of one's own being is the indispensable causal factor in the rise to concrete self-awareness on the part of individual human beings.

Now, in telling the story of the master and the slave and their dynamic failure to acknowledge each other in a free manner, Hegel points the way toward a true understanding of the connections that make or break community life. Without a free and full recognition of my being by another, I cease to be. (Just ask Captain Boycott what happened when the Irish refused to acknowledge his existence.) In the story of the master and the slave, the irony of course rests most heavily on the master since he must ask the slave to do what no slave can do: Grant full and free recognition to him as master. It is the very power of the master that thwarts the growth of his self-awareness. On the other hand, the irony deepens because it is the slave who can work (no master can work and still be a master) and thereby establish a realm of potential recognition within which his consciousness can in the future be recognized by another.

Thus, in terms of this study the primary work of the city is the establishment of modes of recognition that can stimulate full and free self-awareness. Those who decide to remain in the castle and/or any other privileged locus of isolation can create no history. They make no contribution to the manifold ways in which human self-consciousness gains recognition of its effective and real presence in culture. It is eros dialectically transformed (*aufhebung*) into the desire for recognition by another that sets free the human self toward its polymorphic modes of self-realization. When Mead speaks of the conversation of gestures wherein I repeat back to the other the form of his desire, he is reconstituting within the tradition of American pragmatism the essentials of the master/slave dialectic sketched in Hegel's *Phenomenology*. And insofar as the city remains the preeminent place for the establishment of human community, instituting, tending, and repairing connections generative of recognition become the paramount human activity.

But this is not the end of the story, for Hegel also reminds us of the potential for terror and willful blindness lurking in every corner of this drive for mutual recognition. For it is not beyond humankind to enslave the very persons whose freedom it needs so as to ensure for itself a false and empty form of self-consciousness. All forms of imperialism from Asia to the Middle East to the British Empire to American colonialism vitiate the conditions of full human growth in self-awareness. It really does not matter whether the deed of enforced recognition is carried out by the Bengal

lancers or loan officers of the World Bank. The result is the same: A vicious form of secondness freezes the possibilities of life into a bloody collision of forces bent on recognition through destruction.

When the need for recognition is stifled by futile collisions of cultural forms, then no matter who gains the upper hand, the ultimate victor is always some manifestation of the fallacy of simple location. As I argued in *Nature*, such a fallacy is bad enough when it infects our view of the value of the environment. Its viciousness is exponentially increased when it is used to found and structure false forms of community. Healthy and vibrant internal relations cannot survive the threat of rampant, violent acts of self-assertion. The very possibility of community as the love of wisdom, the sharing of values, and the sharing of spiritual experience is negated. What is demanded is capitulation, not dialogue. In and for and of themselves, genuine human connections are branded as contraband. Simply located human relations rest upon the logic of domination. Every attempt to seek another way is dismissed as foolish and irrational.

When forms of domination are the only permissible paths for social relations, the art of making connections is seriously compromised for then, forms of violence are the only media allowed entrance "into the middle of things." And if the middle is permeated by violence (however well disguised), then what hope for rationality exists? It is therefore no wonder that in the time of advanced late capitalism celebrations of the irrational and wholesale adoptions of *outré* lifestyles tend to proliferate among those who would rebel. But we ought not endorse rebellious postures as the only possible reaction to the tyranny of secondness disguising itself as thirdness. There is also the experience of work as the way out of the condition of bondage.

The work required to bring about the condition of free and full self-consciousness is the work of building a community. For it is within this realm that acts of social recognition can take place and thereby ground human awareness in the kinds of structures needed for its growth. Urban work involves making the kinds of connections that allow for the creation of meaning through shared experiences. It is here that eros joins hands with desire so as to form a base of rationality suffused with symbolic import. When philosophy is actively practiced as the love of wisdom, then the relation between facts and values assumes paramount importance. The love of wisdom demands the experience of shared meanings. This in turn requires the real presence of spiritual meeting on a widespread communal basis. And such spiritual meeting, Plato tells us, always involves the establishment of something in between as the ground and source of an ever-evolving rationality. Making connections in the city involves active participation in the symbolic realms that convey meaning throughout the various domains of city life. And such participation is rightfully called work because it demands philosophic effort, the communication of values, and the recognition of the deeply spiritual dimensions of human life.

Connections that embody reason are the most difficult to create for they must be grounded in intelligence-in-action. It is relatively easy to force connections (one can always get a bigger hammer). The true art of reason consists in locating those forms of praxis that persuade "by reason" of their effectiveness in providing solutions for seemingly intractable problems. And even if such a truly difficult feat is pulled off, there remains the task of persuading others to accept and practice the solution. This can only be done by invitation. Forced participation violates the intent of communal forms of reason. The connections that count as constitutive of community must be the outcome of a freely engaged praxis. This means that the community's internal relations must be forged through a variety of ritualized symbolic interactions. To the degree that such connections express depth and width, to that same degree they also propose powerful lures for community participation. In brief, ritual makes a real difference in the life of a community.

Connections are the concrete manifestation of the reality of internal relations. All such relations are signs of eros at work for they manifest the real presence of something in between. What is between is the desire for recognition that promotes healthy forms of self-consciousness. It is the task of philosophy to create such realms of community understanding. For the love of wisdom promotes the sharing of meanings through spiritual encounters. These encounters are occasioned by the effective presence of richly evocative symbols

in the life of the community. Working at the establishment of such symbols is what creates patches of intelligence in an otherwise brutal world. Working through such symbols is what transforms those patches into a widespread and effective range of intelligence-in-action. It is here that the symbol reveals itself most clearly as the royal road to shared experience.

Sharing Symbols

Speculative philosophy and semiotics are indispensable disciplines for understanding just how vital symbols are for the proper functioning of community life. I have already discussed the significance of symbolic reference for human perception. Also, the category of transmission was seen to function best when it was transferring symbols around the various domains of urban experience. Here, I wish to bring all these dimensions under a single theoretical umbrella whereby the symbol will be seen as the media of meaning itself. In so doing I am putting to work the theory of transformative symbolization elaborated by Susanne Langer and discussed in the previous chapter. Human beings survive and prosper by reason of successful efforts to transform the environment into a stable place for their growth and development. The prime mover of such transformations is the symbol.

The act of symbolic transformation takes place whenever a sign, object, and interpretant are brought together in order to render effective a particular approach to a particular environmental problem. All symbolic acts from puberty rites to learning traffic signals to funeral processions are similar in this sense: They attempt to transform situations into problems and thereby render them open to the beneficial presence of intelligence-in-action. But more is demanded, since finding ways to creatively embed these symbolic activities in the lived social body of the community is also necessary. This is what I mean by the somewhat awkward term *sharing symbols*. By it, I wish to suggest the supreme value of wedding reason to action through concretely felt aesthetic responses to environmental pressures. It also involves recognizing the power resident in the dynamical object to bring about qualitatively different symbolic responses to semiotic stimuli.

The communal act of sharing symbols must have some "give" built into its structure of learning so that the community's responses to problems are simultaneously general and specific. If this does not occur, then habit will quickly degenerate into instinctual response and the force of what Peirce called "the living idea" of all "mental phenomena" will dissipate and be replaced by blind instinct and frozen custom. The power of an effective urban semiotics depends entirely on its ability to develop a praxis that will consistently generate the abductive creativity demanded by the problems of existence.[6]

Put in metaphysical language, sharing symbols means learning how to carry out the symbolic transformation of reality in such a way as to keep alive the ultimate ontological fact that the world is always one and always many at the same time. Thus, when Sandra Rosenthal quotes Peirce to the effect that "the general idea is the mark of the habit,"[7] she is noting the fact that Peirce is saying many things at the same time. First, he is insisting on the importance of readily recognizable sensory cues that, secondly, can stimulate intelligent activities that, thirdly, bind themselves into some kind of effective structural whole that is appropriately normative for that time, that place, and that situation. Thus, habit is a process capable of adjusting itself to the special respects of special circumstances. And this is precisely what Robert Neville means when he suggests that Peirce could have added a fourthness to his triadic scheme – where this fourth would be the deftness of intuition to pick out just that "respect" with which the event in question expresses its own special *haeccitas*.[8] It is the objectively vague character of symbols that promotes the kind of flexibility needed to do justice to the uniqueness of each and every urban situation. And it is the specific application of the symbolic act that tests its adequacy.

Now, all this is to say that the place of aesthetics as the lead discipline for creating successful forms of urban praxis remains in force. For without a sense for the feel of a situation, there can be little hope for a sympathetic reading of the "respects of interpretation" that can fill out the form of the good that is normative for this special situation. As a practical matter, it therefore makes great pragmatic sense to make available to the community lessons in aesthetic training.

For institutions fail when they do not enlist the felt participation of the people affected. Recall my earlier analysis of the communities of Saint Jerome Church and the South Bronx Churches. It is precisely the exclusionary practices of the political culture that rendered them powerless. And by way of redress, these institutions must respond deftly and sensitively through forms of community praxis in order to regain political power and authority. Liturgy and ritual are not simply pretty "add-ons" to political action. They are precisely the ground and the source of the something in between that demands recognition by the political powers of City Hall as well as the power brokers in the Chancery Office of the Archdiocese of New York. What animates these urban experiences is eros transformed into the desire for recognition and a willingness to work for such moments of recognition.[9]

Also imperative for a fruitful communal sharing of symbols is the opportunity to allow for an active critique of a community's symbolic code. Another word for this community responsibility is participation – a concept that has a long and distinguished philosophical history. To participate in a good way entails the possession of appropriate normative measures for such acts of sharing. This brings the discussion back to the theme of thirdness, where reason grows by reason of its capacity to generate fitting responses to changing life situations. Thirdness is the realm of habit, rule, and order but it also involves a kind of reasoning that is at the same time a kind of feeling. In the course of this study it has been called various names. Sometimes it has been termed normative thinking, sometimes it has been called abduction, and sometimes it has simply been called felt intelligence. Whatever it is called and however it is identified, it always concerns the aim at experiencing the difference that "the truth of discovery" makes when it is at work in a particular situation.

The transformative power of this truth of discovery is grounded in what Professor Ralph Sleeper fortuitously called "the experienceable difference."[10] If the sharing of symbols does not make a difference in the lives of the community, then all such symbols fail the pragmatic test. For it is through the supremely practical act of participating in the symbolic life of the community that levels of value spread their influence through types of urban environments. What unites sharing and participation is the felt recognition of particular values within particular factual contexts. Now, this act of normative understanding demands standards that the community can use to ultimately identify such relations between facts and values. I say "ultimately" because it is here that Peirce's insistence on the "long-run" community of inquirers dedicated to the discovery of truth is most relevant. For such a requirement is both a sign of the possibility of normative truth and an admission of the steady presence of fallibilism in all our attempts to share symbols within communities of inquiry. It is precisely here that logic and ethics depend upon the active presence of an aesthetically sensitized community of inquirers who can feel the rightness or wrongness of a particular response to a situation. Thus, Peirce's insistence upon the importance of the convergence of interpretations in an idealized long run as the ultimate guarantor of truth is basically the same as the pragmatic maxim's assertion that the meaning of an idea is proportionate to its possible consequences.

Now, both the doctrine of the experienceable difference and that of the community of inquirers, as well as the pragmatic maxim itself, receive speculative and systematic enhancement in Robert Neville's hypotheses concerning network and content meaning. For what is felt intelligently in the long run is both the content meaning and the network meaning of a sign, its object, and its interpretants. In terms of intentional human activity, content meaning names the significance derived from the triadic use of object, sign, and interpretant as directed toward existent external objects. My categoreal scheme sees content meaning as the inscape of each event expressing its special value within an evolving urban environment. On the other hand, network meaning names the level of significance achieved when all the relevant semiotic connections are taken into consideration. In terms of my categoreal scheme, network meaning is akin to the category of transmission when that environmental function is working in "the respects" of a certain tradition of interpretation. Network meaning is therefore the functional equivalent of the postmodern emphasis on the importance of textuality, but with this difference: Genuinely pragmatic networks point toward really existent values and are not to be understood as merely self-referential structural skeins of words, images, and indices. Whether the case be that of content meaning or network meaning, the interest of the community always lies along the lines of the truth value of the semiotic experience. And since truth here means the "carryover of value" brought about by "true" representations, the next section of this chapter constitutes a final cumulative argument demonstrating the pragmatic effectiveness of this coordination of speculative philosophy, urban semiotics, and urban praxis.[11]

Community Is Democracy

Urban praxis finds its most direct expression in the ways in which symbolic codes enable participation in the wealth of experience generated through the ontological creativity and sign activity of a community. Therefore, this urban cosmology ought to be applicable to the value experience of city dwellers and also speak directly to their major experiential concerns. That experience has three dimensions: physical, biological, and cultural.[12] By physical I mean the important spacetime qualities of urban life. Most especially, I mean access to the richness of place as an indispensable physical element for the growth and development of human beings. For without place the possibility of community disappears as a real option. The placelessness infecting so much of contemporary urban environments (both rich and poor) is among the cruelest punishments visited on human beings in this age of advanced late capitalism. It does to city dwellers what the British did to the Irish when they forbade them the use of their language and religion. It deprives human community of the very physical foundations needed to help them grow even as it sets them adrift in a spaceless, timeless vacuum.

The biological level of urban participation is marked by the need for sound housing, healthy air and water, and the kind of food humans need for their growth and development. Foul air wastes human lungs, a filthy water supply poisons their bodies, and lack of nourishing food stifles all growth. All these facts of good urban biological participation are commonplaces. The real wonder is that arguments have to be continually put forward in order to win them again and again for the urban populace. But when the city is seen mostly as a place to earn a living through participation in "an increasingly competitive global marketplace," these self-evident biological levels of participation require continual justification.

But it is the cultural level of participation that witnesses the worst crimes against city dwellers. How cruel that even as they preen themselves on their cultural resources, the great cities deprive the mass of their citizens of the means to enjoy such wealth of experience. If the previous arguments on the importance of sharing symbols mean anything at all, then the viciousness of such inequality should be self-evident. What condemns the poor to their poverty is not simply the absence of money. In and of itself that could be remedied; rather, it is when the experiences that make life valuable are sealed off from human beings that the most awful kind of deprivation sets in. Depression, loss of insight and foresight, the urge for the quick fix – all these symptoms of urban sickness are for the most part directly related to the loss of significant participation on the cultural level.

What cultural participation is really all about is learning the symbolic code whereby the wealth of human experience can be transferred among the people of a city. Take away these vectors of meaning and you have killed the soul of the people. Deprive them of these vectors of truth and you have rubbed out the real and effective presence of value in their lives. Erase these vectors of felt intelligence and you have eliminated what is most distinctive about human beings – their capacity to feel the good and communicate it to others.

I quote again from John Sherrif's brief but powerful essay on Peirce's philosophy, for he spells out quite clearly what is at stake when the cultural level of participation is blocked: "Aesthetic experience is the awareness of the possibility of meaning; that is, the awareness of recognizable feelings."[13] If this study of the city has attempted anything bold, it is its effort to dispel forever the notion that culture is some kind of "nicety" humans can do without when other matters are more pressing. Culture is not about pretty things or even "beautiful" things. It is about those things that make life human: meaning,

value, and importances. Recalling Peirce's triadic ontology: What would happen if the field of quality experienced in firstness were forever sealed off from large segments of the city population? Would not a kind of numbness that dulls the eyes even as it threatens the soul steal over all those who lurk in our streets? Consider the clash of values characteristic of urban secondness: Would it not resemble Hobbes's "state of nature" – a war of all against all that takes place in full view of the Empire Skyline? And when we left that battlefield of secondness and returned to the neighborhood to seek out the gift of community which is thirdness, would not Eliot's words echo in our minds: "I did not think that death had undone so many"?[14] In sum, cultural deprivation constitutes the most permanent and damaging offense possible against the city dweller. It makes interaction impossible because it destroys the social bonds needed to forge a living connection between abductive intelligence and collective ideals.[15] It prevents confirmation of our values through concerted community action and thereby doubles the sense of helplessness felt at this level of urban cultural poverty.

In my earlier study of nature I used the philosophy of Spinoza to articulate certain important attitudes that should inform our character.[16] Here also, I wish to enlist Spinoza's wisdom but in connection with what he understands to be the great advantages of social life (and by extension, city experience). For Spinoza the decision on how we shall live with our fellow citizens comes down to a stark choice: Shall we love or shall we hate?[17]

To begin, it should be recalled that for Spinoza the most powerful form of knowing is not a knowing that knows a thing or object. Rather, Spinoza judges knowledge by the degree to which it enhances or diminishes our experience of union with what is to be known. Therefore, when Spinoza speaks of knowing God or Nature, he is indicating the degree of unity felt between the knower and the known rather than an abstract knowledge of something external to us. In other words, Spinoza's alleged rationalism is through and through imbued with felt intelligence-in-action. Thus, community is also to be judged by the standards it upholds concerning the importance of our relations with each other. In the *Ethics* Part IV, propositions 35–45, there is a stark contrast drawn between a society founded on hate and one that stresses the ultimate significance of internal relations. With characteristic directness and with the power brought about by a vast systematically normative deduction (*viz.*, the rest of the *Ethics*), Spinoza simply says: "Hatred can never be good. (Q.E.D)"

If hatred is never good, then it is one of the very few things in Spinoza's universe that does not take on its character from the context in which it is experienced. The reason for this absolute negative judgment of hatred is because it destroys all community and is therefore proof of the most supreme form of ignorance possible in the universe. This ultimate foolishness, this sickness of the mind that above all else requires healing[18] cancels out the something-in-between that is the very "substance" of Spinoza's universe of unity.

The result of hatred is a feeling of fear that destroys the possibility of the healthy presence of reason in our dealings with one another. A destructive mood invades our natural desire for recognition and overcomes any sense of sharing with our fellow humans. Hatred, of whatever sort, is the worst habit that can take over a human person or a social order. Until its invasive presence is expelled, there is no genuine possibility for community.

John Dewey's conception of democracy as the process best suited for gaining each individual what they require in terms of freedom, power, and respect mirrors this Spinozistic judgment:

> All deliberate action of mind is in a way an experiment with the world to see what it will stand for, what it will promote and frustrate. The world is tolerant and fairly hospitable. It permits and even encourages all sorts of experiments. But in the long run some are more welcomed than others. Here there can be no difference save one of depth and scope between the question of the relation of the world to a scheme of conduct in the form of church government or a form of art and that of its relation to democracy. If there be a difference, it is only because democracy is a form of desire and endeavor which reaches further and condenses itself into more uses.[19]

Besides noting the fact that these words (with their use of "desire" and "endeavor") could have been written by Spinoza himself, I wish to stress Dewey's insistence on the significance of "depth and scope" as the normative standards by which excellence in democractic community is to be judged. Now scope and depth are aesthetic terms with far-reaching implications. I have used them to express the importance, on the one hand, of perspective and on the other the ideal measure by which the value of an event's *haeccitas* and transmissive urban power can be estimated. Now, both the scope of an urban event as well as its depth is also directly related to its functional excellence as a symbolic form. Symbols, it will be recalled, do two things at the same time. First, they transform reality so that humans can handle it. This is a symbol's scope and it registers the "respects" with which any interpretation can or will proceed. Second, a symbol's success in terms of scope also touches upon the "dynamical power" resident in its object. This is its "depth." Both together, scope and depth, mark the boundaries of excellence within a democracy's attempt to create a healthy community. Such is the power of symbolic breadth.

The disastrous effects of failed institutions should now be evident. Due to a loss of contrast that springs from a failure to permit real otherness into its presence, institutions can suffer an absence of width, balance, depth, and integrity. Needless to say, as any one who has had to deal with a bureaucracy knows all too well, these institutions also suffer from a deadening absence of intensity. Taken together, these aesthetic failures account for precisely that loss of aesthetic sensibility that Peirce saw as the vital heart of his guess at the riddle of existence. Institutions that do not register feelings cannot possibly communicate fitting forms of the good to their clients. This loss of the ideal measures of depth, integrity, and wholeness is made far worse by the substitution of a bad form of intensity: competition. For when advanced late capitalism is granted free rein, the only standard of achievement it can offer is "more." What vanishes in this competition for more is the common good, for it is not within the proprietary interests of the great multinational corporation to pay attention to that which is outside the standard of profit. Any possibility of a "great community" is dashed on the rocks of secondness taken as the ultimate form of life. And likewise, any possibility of concrete egalitarian existence vanishes under the pressure of greed as a form of public and private virtue.

My reference to Dewey is deliberate, for more than any other American philosopher it is he who identifies community with democracy. It is democracy with its insistence on the interaction of individuals within a social setting that points the way toward a resolution of the conflict between individuals and the social order. From the perspective of building community there is for Dewey no irreconcilable struggle between these two levels of existence. The normative measure to be used in judging the rights and duties of both dimensions is how well they assist each other, not simply the degree of conflict they engender. Once again, it is the category of contrast that comes into play to push social situations toward moments of achieved thirdness. The community is built up out of the relations of individuals. It therefore requires strong individuality for its wellbeing. So also the individual requires a nurturing set of relations to foster growth. Thus, the questions are: "How numerous and varied are the interests which are consciously shared?" and "How full and free is the interplay with other forms of association."[20] Mutual support and mutual interest is the definition of community. It is also what marks democracy as the form of government best suited for realizing community in everyday life:

> [R]egarded as an ideal, democracy is not an alternative to other principles of associated life. It is the idea of community itself.
> [W]herever there is conjoint activity whose consequences are appreciated as good by all singular persons who take part in it, and where the realization of the good is such as to effect an energetic desire and effort to sustain it in being just because it is a good shared by all, there is in so far a community. The clear consciousness of a communal life, in all its implications, constitutes the idea of democracy.[21]

Now, this conjunction of democracy and community lays certain demands on the idea of the city. It says that the city must be a place where meaning is shared. Also, it demands that urban knowledge take on the form of an intelligence-in-action that can be felt by each member of the community. As Peter Manicas

puts it: "such knowledge is *shared*, . . . (it) funds experience with *common* meanings, transforms needs and wants into *mutually understood* goals and . . . thereby *consciously* directs *conjoint* activity."[22] These demands, understood in the context of the urban praxis developed here, are best met through the unification of urban cosmology and semiotics proposed by this study. The cosmology supplies the broad categoreal dimensions needed in order to maintain a good understanding of the city, while the semiotics offers a sign system keyed to the sharing of meanings and the solicitation of fitting acts of felt intelligence. Once again, it is the underlying aesthetic achievement of intense contrast felt and transmitted throughout the community that makes democracy the ideal form of city life. Speculative philosophy provides the necessary enlargement of awareness needed to welcome difference, and urban semiotics prepares for the task of providing appropriate symbolic inclusion of that increase of difference. Such is the heart of the democratic ideal.

This is not the end of the story. Dewey also insisted on the importance of keeping community localized within neighborhood settings. In fact, he saw local community as essential for the realization of the good life:

> In its deepest and richest sense a community must always remain a matter of face-to-face intercourse. . . . The Great Community, in the sense of free and full communication is conceivable. But it can never possess all the qualities which mark a local community. It will do its work in ordering the relations and enriching the experience of local associations.
> . . . Whatever the future may have in store, one thing is certain. Unless local community life can be restored the public cannot adequately resolve its most urgent problem: to find and identify itself.[23]

The tension between the local and the regional, the national and the global intensifies on a daily basis as civic values are driven more and more by market forces. So much so that, as the next chapter will show, the question of city justice can no longer be separated from economic policy. This final section could have been called: "Sharing the Wealth of Experience." The choice of the term *wealth* is deliberate, for experience as understood within this cosmology and semiotics is a form of value that brings wealth to community members. But this wealth is not in the form of monetary gain. Rather, it is, as should always be the case when talking about human beings, in the form of participating in the lived values of a social order bonded by internal relations and educated through dynamic symbols of thirdness. Furthermore, when the urban public is brought into the discussion, this wealth is always also communal, for community is the human meaning of thirdness. And when forms of political order are introduced into the discussion, then the question of democracy is also raised. But democracy also raises the question of justice, which in turn brings economics into the forefront of the analysis. For the great community is unthinkable without an examination of the relationship between community, democracy, and economics in the contemporary city.

Notes

1 See the work of Michael Sandel and William Sullivan. Both are forced back to the very beginnings of our political history as an independent state in order to find actual historical models for community in action. Michael Sandel, *Democracy's Discontent* (Cambridge: Harvard University Press, 1996) and William Sullivan, *Reconstructing Public Philosophy* (Berkeley: University of California Press, 1986).

2 *Symposium*, translated by Nehamas and Woodruff (Indianapolis: Hackett, 1989), 2204B. I remind the reader of Plato's earlier praise of the child as the model for the philosopher.

3 Ibid., 202D–203B.

4 George Herbert Mead, *Mind, Self, and Society*, ed. Charles Morris (Chicago: University of Chicago Press, 1967), p. 167.

5 G. W. F. Hegel, *The Phenomenology of Spirit*, trans. A. V. Miller (Oxford: Oxford University Press 1977), §§178–96.

6 See Sandra Rosenthal, *Charles S. Peirce's Pragmatic Pluralism* (Albany: State University of New York Press, 1994), p. 31.

7 Ibid.

8 See Robert Neville, *The Truth of Broken Symbols*, (Albany: SUNY Press, 1996), pp. 43–6 and *passim*. It is important to note how Neville places his thought within the great tradition of American

speculative philosophy by deploying the insights of Justus Buchler as well as his own to enlarge the understanding of Peirce's semiotics. This is using the history of philosophy in the most fruitful sense because it sees philosophy's history as that which is organic to its development. The arrogance behind the ignorant rejection of philosophy's history by varieties of postmodern thinking stands in stark contrast to this kind of intellectual responsibility and generosity. Neville's highly original article arguing for our capacity to intuitively recognize harmonic unities and judge their normative excellence is a fine example of this respectful use and at the same time departure from the American philosophical tradition. See Robert Neville, "Intuition," *The International Philosophical Quarterly* VII, No. 4 (December 1963): pp. 556–99.

9 The South Bronx Churches have adopted the well-known "Iron Rule" of Saul Alinsky's Industrial Areas Foundation: "Never do for another what they can do for themselves." Hegel would approve since it is only by this kind of work that authentic self-recognition through authentic community life comes into existence. Again, it is a matter of the deft recognition and use of the "respects of interpretation" that lay at the base of every community's system of symbols.

10 See Ralph Sleeper, "Pragmatism, Religion, and Experienceable Difference," in *American Philosophy and The Future*, ed. Michael Novak (New York: Scribner's, 1968), pp. 270–323.

11 See Robert Neville, *Recovery of The Measure*, (Albany: SUNY Press, 1989), for a discussion of network and content meanings.

12 I develop these three levels of environmental participation in *Nature: An Environmental Cosmology*, (Albany: The State University of New York Press, 1997), Chapter 2. They are also the ground of my contention throughout this work of a continuity as well as significant differences between the city and other forms of social dwelling.

13 John Sherrif, Charles S. Peirce's Guess at the Riddle (Indianapolis: Indiana University Press, 1994), p. 74.

14 T. S. Eliot, *The Waste Land* (New York: Harcourt, Brace and World, 1958), "The Burial of the Dead" ll. 60–70.

15 See the brilliant analysis of the problem of action in Hans Joas, *Pragmatism and Social Theory* (Chicago: The University of Chicago Press, 1993), pp. 245–59.

16 See *Nature: An Environmental Cosmology, op. cit.*, pp. 230–4.

17 See *Ethics*, Part IV, props. and *A Political Treatise*, trans. R. H. M. Elwes (New York: DoverBooks, 1951), Vol. I., c. 3, pp. 301–8.

18 See *On the Improvement of the Understanding*, which I argue is much better translated as "The Healing of The Mind." B. Spinoza, 'The Treatise on the Improvement of the Understanding' in *The Chief Works of Benedict de Spinoza: On the Improvement of the Understanding; The Ethics; Correspondence*. Trans. R. H. M. Elwes (New York: Dover Publications, 1995).

19 John Dewey, "Philosophy and Democracy," in *The Political Writings*, ed. Morris and Shapiro (Indianapolis: Hackett Publishing Company, 1993), p. 43.

20 John Dewey, *Democracy and Education* (New York: Free Press, 1966), p. 83.

21 John Dewey, *The Public and Its Problems* (Chicago: Swallow Press, 1954), pp. 148–9.

22 Peter Manicas, "John Dewey: Anarchism and the Political State," *Transactions of the Charles S. Peirce Society*, XVIII, No. 2 (Spring 1982): p. 144.

23 John Dewey, *The Public and Its Problems*, op. cit., pp. 211, 216.

33

Urban Ecological Citizenship

Andrew Light

There are many ways to describe cities. As a physical environment, more so than many other environments, they are at least an extension of our present intentions. But cities are not confined to the moment. Built spaces are also in conversation with the past and oriented toward the future as physical manifestations of our values and priorities. But even with all of the ways we have to describe cities we do not normally think of them as in any way akin to the "natural" environment. City and country, nature and culture, are opposed. We move through cities differently, with a different set of values, whether articulated or not. Consider just one small example: Even the most jaded urban dweller may hesitate to sully the environment around him when he perceives it to be something other than a product of the human community. Though we have all seen trash in a national park, we suspect (or at least hope) that there is a kind of hesitancy that occurs with a person wondering what to do with her candy wrappers on a trail in Yosemite. On the other hand, a visitor to my home, Greenwich Village, will usually not think twice about tossing his cigarette butts on the ground as he walks toward his next destination.

Especially on the weekends I notice this. The Village is a disaster area on Saturday and Sunday mornings. The streets are littered with trash from the previous night's revelry. Plastic cups, remnants of advertisements for free music venues, paper plates from a late night slice of pizza, and various other refuse fill the sidewalk garbage bins to the brim and spill out into the street (if we are lucky – much of it is tossed anywhere regardless of the proximity to a trash can). Visitors do not seem to care about these streets as an environment as they might think of Yosemite as an environment. We don't worry about it though. Those who are visiting move on to their homes in the suburbs or elsewhere; those who live here are confident that it will all be set right by Monday morning.

What would it take to make people think twice before they litter my neighborhood in the same way that I imagine them hesitating before they litter a path on the way to see a thermal spring? To begin to answer this question we must step back from the particular example of garbage on the streets. The more general question to be answered is what it would take to make people consider all of the city as an environment worth respecting, or if not respecting in and of itself, an environment worth treating with respect at least because one's fellows were trying to make a home of it? My intuition is that environmental ethicists could help answer this question by offering a vision of cities as more continuous

From *Journal of Social Philosophy*, 34 (1) (Spring 2003): 44–63. Reprinted by permission of Blackwell Publishing.

with other environments rather than separated from them.

In what follows I will first make a case for why the environment of the city is an environment worthy of attention to environmentalists. Second, I will consider a model of urban citizenship which includes a component entailing a range of obligations toward the urban environment, and third, I will illustrate this model with an example of how citizens can exercise their obligations toward the urban environment in a way that produces positive outcomes. I will throughout attempt to distill my conclusions into a set of "urban environmental theses" (UETs). These theses should be thought of as only starting places for a longer conversation – experimental hypotheses, if you will, for later debate among those who find this topic worthy of further consideration.

1. Urban Environments and the Ecological Future

Just prior to the turn of the year 2000 we were awash, as could be expected, in predictions. Not least were the warnings of impending environmental catastrophes. Often included as evidence of such problems was an appeal to the demographic trend toward increased global urbanization. It has been repeated often enough. In the 1990s we began to hear the dire warning that shortly into the next century more than half of the world's population would reside in large metropolitan areas.

The details of this prediction are without doubt a bit unsettling. The next quarter century, according to the United Nations, will see an "urban population explosion" of staggering proportions. The urban population of the developing world will double to four billion people, or increase by eighty million people per year. This is equivalent to the world's adding an additional Germany, Vietnam, or Colombia plus a South Africa every year, a Los Angeles every month, a Pittsburgh or Hanoi every week. Requirements to adequately meet this growth will entail the completion of fifty or more housing units every minute for twenty-five years in the developing world. By 2015 already, there will be 400 cities of over one million people in developing countries,

twice as many as in the 1990s, packing in twice the proportion of the third world's population. The urban population of India alone will grow to approximate the total population of the United States, Russia, and Japan combined, while China, now two-thirds rural, will become predominantly urban by 2025. Large city growth will be sustained, continuing from 1990 to 2015, at unprecedented increments (Lagos, 16.9 million; Bombay, 14 million; Dhaka, 13.2 million; and Karachi, 11.4 million), each city equivalent to the New York metropolitan area.[1]

What will all of this mean? Depending on whom we listen to, many terrible things. But most important for my purposes here, it has been largely assumed that such statistics do not bode well for the environment. No doubt unchecked population growth in and of itself is a problem. But what exactly would happen if over half of the world's population lived in cities? The claim seems to be that we would see an extrapolation of the worst environmental consequences which are assumed to be the result of urban dwelling: increased production of pollutants, increased population pressures, increased fuel costs for transportation of food into cities, and loss of public land. Cities separate us from nature, do they not? And more people living in cities spells an environmental disaster because of this separation. Just go to a large city and you can see it: People in urban areas do not understand, let alone care about, their relationship to the land, right?

William Rees is often cited as one convinced of the validity of such concerns. Rees, a Canadian geographer, is best known for his development of "ecological footprint" analysis. Ecological footprints are extrapolated maps of the environmental impacts of urban areas created with the help of geographical information systems (GISs). These maps help to track the full ecological burden of cities on the environment by spatially representing as a "footprint" the resources required from surrounding land to sustain an urban population and the direct environmental harms produced by cities in the form of waste generated. The point of such maps is to demonstrate that the environmental stress caused by cities is far greater than the actual physical borders of the city.[2] Evocative figures are derived from this analysis, such as the claim that the ecological footprint of London is larger than the entire island of Britain.

Here is a point in our analysis, however, where we can begin to retake some of the ground against cities in an environmental context. One thing that is critical to remember is that ecological footprint analysis gives us not simply a representation of the impact of cities on the environment per se, but a representation of the impact of cities on the environment as a function of the average consumption rates of a concentrated population. The ecological footprints of cities in developing countries are always smaller than those of first-world cities with a comparable population. The environmental problem that is revealed here is more precisely the unsustainable consumption patterns of people in the industrialized North. And importantly, it is not really the unsustainable consumption patterns only of urban dwellers that is implicated in this analysis, but the unsustainable consumption patterns of Northerners in general. (I hasten to add that while there is a persuasive historical account that one can give of the necessary importance of cities in the development of nineteenth-century production technologies, it is no longer the case that cities themselves uniquely generate advancements in Western technology. So even if there is a critical role that cities played in the genesis of unsustainable consumption patterns of the North, they do not currently play as important a role.) Footprint analysis provides a good indicator of unsustainable consumption patterns and a very helpful spatial metaphor to assist urbanites in understanding the larger ecological consequences of their consumption patterns. But perhaps more importantly, it also helps to demonstrate the unjust distribution of resources between first- and third-world countries. Ecological footprint analysis is therefore arguably a critique of unsustainable populations rather than cities themselves, absent evidence to the contrary.

Nonetheless, Rees and other advocates of this technique often use footprint analysis as the linchpin of a general urban critique which in recent years has grown louder as global population has shifted toward cities. In a short editorial in the *Chronicle of Higher Education*, Rees suggested that increased urbanization predicted into this century is evidence of humanity's "technological hubris." "Separating billions of people from the land that sustains them," he says, "is a giddy leap of faith with serious implications for ecological security."[3] Urbanization, we are told, removes people spatially and psychologically from the land. Cities create more intensive use of croplands and forests to sustain urban populations, and more hectares of productive land are then relied upon to sustain the populations of rich countries (presumably especially in cities). Most alarmingly, says Rees, in a world of "rapid change," cities are unsustainable because they are susceptible to adverse environmental changes and political instability in rural areas.[4] Political turmoil in a countryside can cut off urban populations from their sources of food, fuel, and other natural resources. Wouldn't it be better, then, for people to live in places not separated from their means of subsistence?

But as it turns out, the evidence here is at best mixed. There is actually no reason to believe that cities themselves are making it easier for populations to grow even though most cities do contain a demographic bulge at the critical childbearing ages of fifteen to twenty-nine. In developing countries the population is more likely to grow at a slower rate in cities because there is less need to replace children who die young in order to maintain family labor strength (for example, in order to work a farm). So, Rees' worries about urban populations, except for the strategic concern he mentions at the end, are all claims which apply generically to any population growth in the developed world, not just growth in cities as such. The convention of using the ecological footprint to critique cities amounts to little more than the old wine of population worries in the new bottles of GIS data.[5]

So, what then about the strategic worries that Rees raises? Is it the case that urbanization is unwise for ecological reasons because surrounding lands may become politically unstable? Perhaps. Thomas Homer-Dixon has argued that environmental problems are more likely to create political violence in rural than in urban areas.[6] At the same time, Homer-Dixon reports that, empirically, urban violence (at least political violence) does not necessarily increase with rural-urban migration. The extent to which urban advocates should worry about this consideration is at best unclear.

How then are we to evaluate the environmental worries of Rees and others about the demographic shift toward cities? Setting aside for the moment Rees' vague charge of "technological hubris," does the move toward cities necessarily herald a new stage of environmental disaster? I don't think so. In fact, I would go so far as to argue that we should be very thankful that the demographic shift is toward urbanization rather than away from it. Without that shift I would be very skeptical that we had a hopeful environmental future at all. I will call my reason for defending this intuition urban environmental thesis 1:

UET1: Densely populated human communities are inherently more environmentally sustainable than non-densely populated human communities, all other things being equal. Environmentalists should focus on defending and promoting urban density as a component of sustainability.

Such a suggestion flies in the face of the work of someone like Rees, as well as common perceptions of cities espoused by many environmentalists. But the empirical evidence on such a point is fairly strong. Densely populated cities, even without a concentrated attempt at conservation efforts, consume less energy than rural areas or regions which sustain human populations alongside wilderness areas.

For example, studies of the conservation gains made during the late 1970s and early 1980s in the United States show that more urbanized states tend to consume less energy per capita than less urbanized states. The lowest consumption rate went to New York state (215 BTUs per capita, on average) because so many residents of the state live in New York City apartments (sharing walls and hence sharing heat) and do not own cars (or do not regularly use them, if they own them). The highest consumption rate went to Alaska, with 1,139 BTUs on average, five times as much energy consumed as a New Yorker. While surely climactic differences in part explain this gap, other states with the highest individual consumption rates included those with more comparable weather. The lowest consumption rates included those of Rhode Island and Massachusetts, other highly urbanized states containing relatively substantial proportions of shared-wall housing stock.[7]

While this study reports only the outcome of one variable of measuring energy costs and savings, we should note how important a variable it is. Transportation costs for food and other goods are remarkably similar in most parts of the country. Most Americans now consume hothouse tomatoes from California (or some other equivalent by region); the amount of energy expended to bring food to Montana or Manhattan is, per capita, comparable. Each consumer dollar spent in the United States also releases some quantity of petroleum, but consumption is no greater per capita in large urban areas than in suburbs, small towns, or most rural areas (anecdotally, it may even be less given the relatively smaller living space of urbanites in a city like New York compared with their suburban and rural neighbors). The same is true of most production. The relative energy savings of cities should show how important my ceteris paribus clause was in UET1. The energy savings from effective public transportation and dense housing stock are arguably the most important structural differences between cities like New York and other residential areas in the United States, so long as everything else remains equal. If less densely populated areas were to undertake more substantial energy conservation measures then perhaps these comparisons would change. Unfortunately, however, they are not doing this, and, at least in the United States, the federal government has failed to pass any incentive measures to make energy consumption more sustainable across the board. Assuming that population increases and consumption rates remain constant, moving away from density becomes one of the single biggest hurdles for achieving environmental sustainability. The same would be true in other countries as well. It is the suburbanization of cities, and the flights of populations to the countryside, that must be dissuaded.

No matter what other problems they have, densely populated cities get around the lack of environmental leadership on energy conservation by creating and encouraging an infrastructure in which residents do not actively have to decide to change their lifestyles or priorities in order to live sustainably. The infrastructure itself makes it irrational to live otherwise. It is simply use of personal resources for a middle-income family in New York City to buy a sport utility vehicle given the ease and availability of public transportation and the high costs of keeping such an automobile in the city. So, if living lightly on Earth is indicative of environmental responsibility, what I will term "ecological citizenship" (or "civic environmentalism") below, then surely Manhattanites are doing more for the planet than their counterparts in Colorado. If some sort of environmental consciousness raising is required for environmental responsibility (as some of my colleagues in environmetal circles have argued), then one should be fearful that such a change in consciousness will not happen soon enough, or on a great enough scale, to forestall the worst consequences of the cumulative impact of humans on the planet. We should therefore nurture those forms of life that are structurally more sustainable. Of course, development of an explicit environmental consciousness would be better for urbanites too (again, the ceteris paribus clause of UET1 comes into play). But as things stand now, it is encouraging that all New Yorkers are in a weak sense "environmentalists" without having to do anything other than live as they live now. Such a consideration does not offset all other considerations that one may have about life in densely populated cities. Still the environmental case for cities is stronger than many may think.

2. Ecological Citizenship

It is for reasons such as these that I believe that environmental philosophers in particular, and environmental professionals in general, must turn in earnest to discussions of urban environments and environmental problems.[8] If cities are assumed to be part of the larger environmental problem (if not *the* problem, as Rees seems to think), rather than part of the solution, then the move toward environmental enlightenment may ironically follow the irrational move away from the inherent sustainability of cities. Rather than trashing cities, as is the wont of too many environmentalists, we must celebrate them as part of a complete environmentalism. Or rather, we must celebrate cities toward the end of promoting them as a better alternative for sustainable living than that available to most of us (how many of us will be able to choose to live in hay bale

houses with composting toilets "off the grid" anyway?).

At the same time, we must come to tackle environmental problems in cities head on. We must make them a priority on the environmental agenda. We have not been doing this. Mark Dowie, in his excellent survey of the recent history of the environmental movement, points out that the image of the environmentalist as backpacker and tree hugger has persisted throughout the history of environmentalism in America. While many see this focus as more a tendency of the so-called first wave of environmentalists at the turn of the century (for example, John Muir and his followers), even the second wave of environmentalism, which got off the ground in the 1960s and 1970s, embraces this wilderness focus: "Environmentalism means wildlife protection and wilderness conservation, while the environmental movement is identified with the Sierra Club and similar organizations."[9] David Schlosberg confirms Dowie's findings and argues that the recent rise of the environmental justice movement (which has certainly focused more on urban issues) has been in direct relation not only to the perceived lack of minority representation on the boards of the major environmental groups, but also to the focus of these groups. The "more telling complaint centered on the movement's focus on natural resources, wilderness, endangered species and the like, rather than toxics, public health, and the unjust distribution of environmental risks," exactly those issues that are of interest to low-income communities and committees of color, largely in urban areas.[10]

Finally, in addition to these priorities, even while we can show that part of the solution to ecological sustainability must come through dense urban living, we know that many of the world's cities – especially in the North where consumption is already most unsustainable – do not follow the pattern of the inherent environmental savings engendered in a place like Manhattan. The fastest-growing cities in North America have expanded on a suburban model that chafes against density in favor of single-family houses without shared walls, accessible to goods and services available only by car. We know them all by heart (Los Angeles, Houston, and Atlanta are the big three): pleasant as they can be at times, all three cities are a huge burden to the global possibility of sustainability in promoting a vision of the good life in which we all are thought to have a divine right to our own patch of Kentucky bluegrass. On the horizon are the fastest-growing cities in America, such as Las Vegas, which is in this context at the front lines of environmental struggles in the United States. The rate of growth in Las Vegas is staggering: a 62% increase between April 1990 and July 1999 to a total population of almost 1,400,000. Most of this growth has been in suburbs which cannot lay claim to the instrumental advantages of urban density. Two Vegas suburbs in the metro area, Henderson and North Las Vegas, were the fastest-growing cities of at least 100,000 in population in the United States in this period. Henderson grew 155.6% and North Las Vegas grew 112.4%. The accompanying drain on water, power, municipal services, and the like has been severe, to put it mildly. And even with some leanings toward "new urbanism" in the city government, the model of expansion is typical of the suburban ideal: expansion by automobile.

Part of the solution to reversing such trends must come from stronger zoning laws and tighter restrictions on automobile use such as progressive taxes on consumption directed to support the construction of workable mass transportation systems (not simply by providing funds for such systems, but by restricting growth so that such systems are workable). Following the path of more sustainable urban dwelling is often seen only as a last-ditch effort by a dying inner city to revive itself or as mandated from above by an unusually enlightened city council (it wasn't too surprising then when I learned that Amherst, Massachusetts, had enacted stricter zoning laws that prohibited extension of city services to new developments unless the new houses in those developments shared walls with their neighbors). We must therefore promote the building of suburbs more like densely populated cities, rather than building more cities like suburbs. My arguments here should not be interpreted as a claim that everyone should live in large cities like New York, but rather that towns and cities across all scales can become more sustainable through density.

Again, part of the success of such new directions in planning could be assured if cities are recognized as an ecological site worthy of commitment for environmental reasons. There would be many ways to try to achieve this goal, and as a moral pluralist I would claim that in principle there is not one single lens through which we may discern the natural values of cities and the connection of this range of values to the normative reasons which we may have for protecting or conserving the larger natural world. But from the way that I have set up the range of issues here, a model of citizenship as both urban and ecological is one good candidate. The reason, however, is not just out of my own philosophical predilections. It stems from a description of what a city is. A city is not just a place, it is an environment in which humans live in proximity to each other, in either a thin or thick community (or, if you prefer more technical language, in either an "instrumental" or a "constitutive" community).[11] The classical citizenship model of civic obligation may be uniquely well suited to contextualizing urban environmental concerns because cities are places where humans must be confronted with the advantages and disadvantages of living together.

Citizenship, conceived along classical republican lines, identifies a role for residents of a place by articulating a range of minimal obligations they have to each other for the sake of the larger community in which they live. On this account, citizenship is not satisfied merely by voting, or even less robustly, as only a legal category which one is either born into or becomes naturalized to. It is instead an "ethical citizenship," or a concept of "citizenship as vocation," where citizenship is a virtue met by active participation at some level of public affairs. Following a discussion of this issue by Richard Dagger, we need not catalogue all of the possible activities of a responsible citizen here so much as realize that "what matters is that these activities set him or her apart from those who regard politics as a nuisance to be avoided or a spectacle to be witnessed."[12]

Dagger and other republican theorists are quick to admit that this sense of citizenship is steadily in decline, if not dead on arrival in some sectors. Yet it does remain a powerful place to orient such discussions since its resuscitation is not beyond hope, has much promise, and is a language of moral responsibility that still resonates widely in many cultures. It is, as Dagger notes, something that those outside the academy do care to have concern and disagreements about. It is a ground that many still see as relevant to their lives and fortunes. Adding an environmental component to a classical republican model of citizenship becomes then the conceptual basis for a claim that the "larger community," to which the ethical citizen has obligations, is inclusive of the city as space, place, and environment, as well as people.

While one can imagine (though at times I find it difficult) an environmentalism at home not only in the wild but also in solitude – where the identity of environmentalists is sustained through and through by their separation from other human beings and their lonely quest for communion with the "more-than-human" world – such an environmentalism would be impossible in cities. I don't wish to drive too deep a wedge here. I think the idea of the wilderness advocate in solitude is fairly ludicrous as a model for an environmentalist who can get much done on any issue. Even though many of the icons of contemporary environmentalism are often mistakenly thought to have been disconnected from the larger human community (while admittedly there were quite a few misanthropes among them: Ed Abbey in particular comes to mind), very little by way of effective environmentalism will come through single actions by anyone on any particular cause. The larger human community must be persuaded, either at the ballot box, or in a court of law, or through other passions of religion or rhetoric, to go along with changes personal or public in the service of bettering the environment. Nonetheless, urban environmentalism, by its very nature, could not maintain such an isolated frame of reference, even as an ideal, for very long.

This intuition gives me:

UET2: An urban environmentalism ought to be a communal environmentalism, and if possible, a civic environmentalism.

While it would be possible to conceive of such an environmentalism along the lines of the predominant strain of nonanthroprocentric environmental ethics, it would be very difficult, if not impossible, for such an environmentalism to argue that the primary goal of environmental ethics is the articulation of the nonanthropocentric value of the nonhuman natural world.[13]

While there are many problems for such a view, one that can be highlighted here is that it does not always prioritize the importance of focusing environmental ethics or politics on the task of creating human bonds, which include nature as an object of concern, or as a common object of respect. Individuals, by themselves, can come to an understanding of the nonanthropocentric value of nature and then decide that the only way to respect those values is by leaving nature alone. But whatever natural values exist in cities, they cannot simply be left alone as a way of respecting them. The city is also the environment in which humans live. Any resource demands placed on the environment of cities is a demand that cannot be isolated from the needs of one's neighbors. So an urban environmentalism must focus at least as much on encouraging bonds of care or empathy with one's fellow urbanites (at least inclusive of their needs for clean air, water, open space, etc.) as it must focus on finding the normative source of bonds of care for nature or natural processes themselves. For now, I will try to remain agnostic on the question of whether an urban environmentalism must prioritize the search for values in nature itself.[14] However such an environmentalism comes to conceive of such pursuits, though, it cannot focus so exclusively on the search for natural value that it loses sight of the priority of building an environmentalism which provides part of the glue of building moral bonds between people as well.

If our goal then is to come up with a conception of ecological citizenship and then expand that to an urban ecological citizenship, or a civic environmentalism, how do we begin? First, by articulating a notion of citizenship which is, as much as possible, intrinsically ecological. Such a conception of citizenship should not be too far off for us, especially if, as just mentioned, we stick to a classical republican idea of citizenship, in particular one considered as an urban citizenship. Dagger has already given us sound reasons why an urban citizenship would need to be concerned with environmental issues, forming a civic environmentalism. Using the example of urban sprawl in his contribution to this symposium, Dagger argues that ethical citizens would need to fight sprawl because it threatens both the environmental and the civic fabric of a city: A sprawled city will only exacerbate the demise of civic associations which are necessary for the development of an urban citizenship.[15] Not to repeat Dagger's argument, I wish to point out here instead that it is also the case that a republican urban citizenship needs to be a form of civic environmentalism in order to save itself from its own dilemmas of scale. Just as sprawl thins out the ability of citizens to meaningfully engage with one another, the global city must be structurally decentered in order to make it amenable to a more robust conception of citizenship. But this decentering can become the fodder for notions of politics as reducible to personal identity. Conjoining environmentalism to urban citizenship by making environmental concerns one of, in Dagger's terms, the "virtues of citizenship" may help to ameliorate this problem. I will now try to more carefully spell out this argument.

Let us begin by backing up a bit and discussing the urban context of ethical citizenship. Contemporary republican theorists such as Dagger and Philip Pettit have already written much that helps us to conceive of citizenship as first urban. The classical conception of citizenship has historically been articulated in an urban context, back certainly to Aristotle, and hence back to a model connected to the polis and to urban life. While much has changed over the years since the birth of this model of citizenship, and its elaboration and extension in the work of other political theorists with strong ties to urban concerns, it is also still quite appropriate to life in today's cities. And, as Dagger notes, given global trends toward urbanization, it is necessary to connect republican thought to contemporary urban life if republicanism is to remain viable at all.[16]

But the question is begged concerning what unit of day-to-day life, which scale of experience with our fellows, we wish to find morally significant as a ground for urban citizenship. The polis of the past is gone as a meaningful and indeed realistic object of our project of building a thick conception of urban citizenship. The megacities of today, of the sort that I suggested earlier constitute our urban future, seem impermeable by those practices most conducive to bringing about a meaningful citizenship based on constitutive obligations among residents. Unfortunately, the environmental literature is of very little help here. Murray Bookchin, the founder of the school

of political ecology known as social ecology, has been one of the few environmental political theorists to discuss in any detail the role of cities and citizenship in a longterm ecological project.[17] But Bookchin's answers on the scale question ring hollow even though the reasoning behind them makes perfect sense in light of the larger literature on citizenship – Bookchin suggests a form of municipal confederalism which would advocate no rule larger than the small town. Real urban citizenship is possible only by reversing the current trends of scale. But bringing about such a transformation of the urban future seems unlikely, to say the least, no matter what environmental arguments we could weigh in to bolster it.

Discussions of citizenship cannot occur within a vacuum of understanding civic space. It goes without saying, I expect, that New York, and most other global cities, stand little chance of becoming a new agora at the level of their formal municipal boundaries. Even the early-twentieth-century language among urbanists of center and periphery does not capture the spatial dimension of the modern metropolis, where yesterday's suburb has become an identifiable and independent unit of potentially significant civic engagement today. It is important then to aim our understanding and expectations of urban citizenship at a continually updated picture of the scale of city life which seems most conducive to a more robust conception of citizenship.

Recognizing this problem, some urbanists focus on grand symbolic spaces, such as the development of the Potsdamer Platz in Berlin, as a locus for what it will mean, in some substantive sense, to be a Berliner. But even such heralded imagined urban centers are not much help to the project of grounding an urban citizenship which will have anything other than mere symbolic meaning. In New York, for example, spaces of meaningful democratic interaction cannot be encompassed only in symbolic spaces such as, perish the thought, the now Disneyfied Times Square. These are not the spaces of democratic interaction where responsibilities to others as part of the whole are acknowledged. They are sites of play, leisure, and increasingly, tourism – a home more to the visitor than the citizen. A ground for citizenship must be found at a smaller scale where citizens can know and interact with each other. One reason is that it is only at such a scale, as Dewey put it so well, that engagement and access to authority is possible by citizens in relation to each other. Reading Dewey's *The Public and Its Problems*, and bemoaning the descent into particularism in urban studies, Thomas Bender puts the point this way:

> A larger public, what [Dewey] called "the Great Community," is built upon the habits and accomplishments of *local publics*. Urban critics tend to focus on the diversity of public space and publics. Diversity is important; but so too are access to political institutions and the opportunity to give voice to public concerns. . . . I am proposing a public that is a public not because of mere propinquity, however important that is, but *because they propose to do something together*. The essential quality of what I call the "local" public is not, as we are inclined to think, sameness; it is accessibility to networks of informal power and to institutions of formal politics.[18]

I would add to this only that local public spaces also provide access to others as having shared lives, rather than serving only as symbolic representations of bonds of community. This access is necessary to ground a morally motivating empathy between citizens which is arguably necessary to make civic obligations coherent to individuals who might not otherwise see such obligations as in their best interests.

While it would take much more space to fully prosecute this argument than I have room to here, it seems at least plausible to me that a thicker form of citizenship in today's global metropolis must at least in part be dependent on the ability to engender civic obligation at a smaller scale within the metropolis. Dagger agrees, proposing that a small, stable and well-defined city, in civic republican terms, is in the range of 100,000 to 250,000 people. In turn, the goal of his republican-liberal citizenship when it comes to larger metropolitan areas is to create decentralized political structures, to create, in Bender's terms, "local publics." The number of these structures, according to Dagger, would vary with the population of the city, but we might imagine that a city of 1,000,000 would be divided into 20 districts of roughly 50,000 each, with each district subdivided into 10 wards. The wards and

districts would have their own councils, elected by and from their residents."[19] Whether Dagger is right about such figures will be largely an empirical issue. Nonetheless, he is not the first to argue that some reduced scale of civic association, if not governance itself, is a necessary condition for thickening citizenship. Though there is still much to discuss on this point I think it sufficient for now to warrant proposing:

> UET3: Urban citizenship must be grounded in institutions and practices across scales, including smaller scales at the level of the "local public."

Still, looking down the road toward a description of civic environmentalism, we can worry that the requirement of grounding citizenship in smaller scales might in the end be debilitating to solving larger environmental problems, which respect civic republican borders about as well as they respect national boundaries.[20] I want to ease up to this problem by first discussing what may sound like a very different worry concerning the tension between the citizen and identity models of public life. The resurgence of work on citizenship, it is important to remember, has been in part spurred by the search for a response to identity politics as the dominant framework of progressive political theory throughout the 1990s. The distinction between the two, as Bender notes, has often been lost: "Many scholars today do not even make a distinction between identity and citizenship, not realizing how much is being given up."[21] Will not a requirement of citizenship at a smaller scale only continue this confusion, as well as ultimately be too constraining for resolving environmental problems that cannot always be settled at the scale of the local public?

The worries here are pressing, given the competing demands of identity recognition on new theories of citizenship. Robert Beauregard and Anna Bounds, in their analysis of the obligations of urban citizenship, cluster a set of related concepts into five themes of rights and responsibilities for citizens: safety, tolerance, political engagement, recognition, and freedom.[22] For the most part these five themes are a good start. But when it comes to these authors' discussion of moral recognition, we can see Bender's worry about confusing identity with citizenship. For

Beauregard and Bounds "recognition" encompasses the moral responsibilities that citizens have toward each other in acknowledging different conceptions of the good. Accepting the common claim that one of the dominant forms of individual expression in contemporary societies is that of "identity," they suggest, following Iris Young, that "under an urban citizenship, citizens have the right to express their identity and to expect that those expressions will be recognized." Recognition of identify difference will then "enrich" the public realm, whereas suppression of identity, following Nancy Fraser, "weakens democracy by confining political behavior only to interests."[23]

We ought to be concerned, however, with embracing too narrow a conception of recognition as only concerning matters of identity. Wouldn't this require, for example, that when we break down metropolitan areas into smaller local publics we must first consider the cohesion of identity groups as best marking the boundaries of each local public, where applicable? If this is true, however, such a priority would be intuitively at odds with the ethical goals of urban citizenship. Recognition of identity should not be a prerequisite of citizenship. Rather, fulfillment of minimal citizen responsibilities should be a prerequisite for the recognition of identity. If this point is not palatable as a general claim, it is given further credence if we now return to the issue of including environmental concerns in our conception of citizenship. For if expression of identity was prior in a normative account of public life, then how could a meaningful environmentalism find a place among the rights and responsibilities of urban citizens? Suppose, for example, that my identity is ground in a cultural identity which demands conspicuous consumption of a particularly pernicious sort. Must this identity be respected even at the level of the local public, or must it capitulate to the larger demands of the broader public?

The answer to this question links us to our prior discussion concerning the appropriate scale of citizenship. I have already suggested that the local publics of a metropolitan area ought to be prescribed for a robust citizenship at a smaller scale. The priority of identity would most likely result if the boundaries of our smaller local publics were set with an eye to first respecting group

difference. There has traditionally been a strong tendency to identify neighborhoods – one version of a local public in larger metropolitan areas – on exactly this model as evolved ethnic enclaves. Are there alternatives? One alternative might be to identify local publics not as a function of recognition of identity groups, but instead at the level of the city conceived as a set of separable but integrated ecological processes, that is, at exactly the borders which I have used to bring out this tension between identity and citizen models of public engagement. As such we might juridically define local publics in order to create opportunities for a citizenship as vocation around watersheds, subsections of watersheds, or even constructed environments such as parks – something like a "subbioregionalism." Such orientations for neighborhood boundaries also have ample precedent even though they seem to be easily superseded by lines drawn around race, class, or ethnic identity.

But defining the scale of local publics in this way would be a decision we would have to make – it will not happen automatically, but only if we have reason to expand the obligations of citizens to include environmental obligations. But why do this? One reason is that a focus on smaller localities within cities, dominated by something like the identity affiliation of citizens, might encourage at least an ignorance of the extent to which most if not all environmental problems are not confined to a particular area, and at most degenerate into a NIMBYism where one neighborhood is pitted against another to ward off various environmental burdens. This is not to say that a local public ought not to be allowed to set its own priorities for environmental projects. One benefit of an urban environmentalism is that urban environmental issues are often not directed at historical or ecosystemic priorities – for example, what a particular neighborhood decides to do with a community garden or park.

If this alternative for defining local publics is infeasible, then at a minimum the incorporation of environmental issues into discussions of citizenship would serve as a reason to bring local publics back together rather than give them reasons to ignore each other's needs. For, after all, one advantage of ecosystemic issues is that they actually are important across the bounded communities of local publics, no matter how they are drawn, and they can serve to bring together differing publics in collaborative ventures for mutual benefit. This may not get us anything like a definitive argument for citizenship over identity. But it does point to an interesting way in which the active task of conjoining civic environmentalism to urban notions of citizenship can avoid the descent of urban citizenship into a fractured plurality of locales pitted against each other. So:

UET4: One way to temper a properly scaled urban citizenship is through the extension of citizen responsibilities to include urban environmental concerns which can either define the boundaries of local publics, or at minimum, cross them to create common civic priorities.

And to round out the picture here:

UET5: The practice of civic environmentalism is the fulfillment of ecological aims in a city concerned with both caring for ecosystems and building better civic communities.

Examples of civic environmentalism would include then both the resistance to forms of urban growth that degrade civic involvement (such as Dagger's proposed opposition to sprawl) and also common projects of preservation and restoration of vital natural areas in cities that cross a variety of communities.

One word of caution though: to argue that civic association should be made possible at a local scale is not to say that an argument for urban ecological citizenship means that all power for deciding the distribution or outcome of local public resources should be turned over solely to the local public. Such a suggestion would be absurd. Encouraging people to engage in the virtuous activity of participating in their local parent-teacher associations does not entail the claim that local PTAs should make all decisions concerning the fate of local schools. Clearly, we would want a variety of regulatory actors to step in to ensure that a consistent range of quality across districts is maintained for teacher training, classroom resources, etc. These sorts of considerations of equality cannot be ensured by divesting all power to local publics in the name of increasing citizen participation virtues. The argument then

is more simply that all levels of regulation are better when they are mediated through a robust form of local participation, be it in decisions over schools or over other public amenities such as environmental regulation. I will close then with an appeal for a specific parameter on local environmental projects, namely, that they maximize opportunities for citizen participation in the local environment as a way of creating opportunities for expressing an urban ecological citizenship.

3. The New Problem of Dirty Hands

If urban ecological citizenship can be captured in part by UET5, then how is it that we can encourage such an ethos in our citizens? Earlier I suggested that simply living in densely populated cities produces a profound good for the environment that I believe has been underappreciated in the recent history of the environmental movement. But a civic environmentalism cannot be content with simply encouraging a lifestyle that is unreflective of the environmental goods which it engenders. Certainly it helps if one's fellow citizens are, by their very lifestyle, part of the solution rather than part of the problem. Any environmentalism must in part aim to encourage more sustainable personal choices in the ways in which people live their lives. But such personal changes are themselves insufficient to achieve a meaningful level of environmental sustainability.

Recall again my ceteris paribus clause in UET1: The environmental gains of cities are perhaps most palpably felt when population and consumption patterns across a range of human communities remain constant. Consumption levels in the United States are unsustainable in their present form. Urbanites do not get a free pass when they live in densely populated environments. A civic environmentalism must also work on the problem of sustainable consumption. But the achievement of sustainable consumption levels is not a peculiar problem of urban life. More unique is the requirement of a civic environmentalism to make the city an object of civic environmental concern. Perhaps another way of thinking about this is that we must come to see cities as an object of stewardship in the same way that we often think of rural and wild environments as an appropriate object of stewardship. What makes this project in cities a more difficult task is the conceptual hurdle that must be overcome so that we in fact do see the city as an object of stewardship or, as I prefer, ecological citizenship.

If such a sense of citizenship involves environmental responsibility for one's own actions, the welfare of one's fellow citizens, and the welfare of one's environment, then how is that citizenship to be encouraged? Taking responsibility for one's political community is familiar enough to us, but how do we take responsibility for our environment as part of that community? My claim is that the first goal of the development of an urban ecological citizenship involves the stimulation of public participation in the maintenance of natural processes in cities.

At least part of the reason for this claim is that it is arguably true that a direct participatory relationship between local human communities and the nature they inhabit or are adjacent to, including urban natural areas, is a necessary condition for encouraging people to protect natural systems and landscapes around them rather than trade off these environments for short-term monetary gains from development. If someone is in a normative and participatory relationship with the land around, then she is less likely to allow it to be harmed further. But this does not mean that a relationship with nature is participatory only if it leads participants to make sacrifices for nature. It only means that such participation is a necessary condition for protecting nature as a foundation of ecological citizenship. Why? One reason is that environmental protection, as is the case with other laws governing common resources, often admits to free-rider problems, as well as outright violations of the law. If all environmental legislation were mandated from above and local populations had no reason to take an interest in environmental protection, then little would motivate citizens to respect laws other than threats of punitive consequences which are often difficult to enforce.

The relevance of this problem has been proven over and over in the history of environmental legislation. This point has been demonstrated through several well-publicized examples in the developing world where the export of the Western idea of a national park has done little to ensure environmental protection of the site

when local people do not feel engaged with the decision to grant protected status to the area. Deane Curtin provides a thorough account of some of these cases, including the failing attempt by the Nepali government to create the Chitwan National Park over the needs of local communities to collect firewood.[24] Local communities that have not participated in the process of setting aside land for preservation have little in the way of motivation to respect the boundaries of preserved land when they have the opportunity to circumvent the protected status of such areas. Curtin recalls a long trek through Nepal in search of the Chitwan park. Walking for several kilometers on a long dusty road, he is unable to find anything that appears to be a park. All the while, though, he continues to see women walking past him with bundles of firewood balanced on their heads. Finally Curtin realizes that he has indeed found the park: It is on the heads of the women walking past him in the opposite direction!

For reasons such as these I am tempted to gauge the relative importance of different environmental practices in terms of their ability to engender a more participatory relationship between humans and the nature around them. So:

UET6: A priority of a civic environmentalism should be to provide opportunities for citizens to engage in participatory practices, if possible in their neighborhoods, with the natural processes of the city.

One brief example of such a project is New York City's Bronx River Alliance, a project of the City of New York Parks and Recreation Department and the nonprofit City Parks Foundation. The alliance is organized by paid city employees who have brought together and coordinate sixty voluntary community groups, schools, and businesses in participatory projects along the twenty-three miles of the Bronx River. The focus is on not only the environmental priorities of the area, but also the opportunities afforded by it to create concrete links between the communities along the river by giving them a common project on which to focus their civic priorities. In the words of the alliance, the project is to "restore the Bronx River to a Healthy Community, Ecological, Economic and Recreational Resource." The activities of the alliance are thus jointly civic and environmental, and the scale of the environmental problem, crossing several distinct communities, helps to create a common interest between them. The environment becomes the civic glue between various local publics. Note, however, that the participation here is not abstract – entailing only participation on a citizen advisory panel – but is actually hands-on. People participate in the activities of the Bronx River Alliance primarily by actually getting their hands dirty.

The Bronx River Alliance did not emerge fully formed out of the good will of citizenry groups but was shaped by the New York City Parks and Recreation Department, attempting to follow other successful models such as the Central Park Conservancy, which has dramatically improved the ecological viability of Central Park as well as increased the level of citizen involvement in the maintenance of the park. Where such successful projects point the way, the citizenship model should try to formalize and institutionalize these models through legislation and policy. But what is happening in New York City is somewhat haphazard. While the encouragement of the alliance is favored by the current Parks and Recreation Department leadership (and was in part a response to financial exigencies which made it impossible for the department to maintain these parks solely through public expenditure), we can imagine that a different leadership or a different set of conditions would have produced a different outcome. If public participation in actual environmental projects encourages the development of an ecological citizenship, then it is appropriate to pursue legislation which would, where possible, encourage more developments such as the Bronx River Alliance.

This sort of analysis sets a clear goal: We want to encourage laws which would mandate local participation in environmental projects which are publicly funded (something like a right of first refusal for local communities and neighborhoods) because of the potential benefits to encouraging citizenship and grounding connections between local citizens and their local environments. If the state has an interest in citizenship in general, it should provide opportunities for such participation. If the state does not encourage such activities, then environmentalists must take an activist role in encouraging such participation. I have argued elsewhere that the criterion of

percentage of hands-on public participation in any given urban ecological project is just as important as scientific or technological proficiency and could even outweigh the value of such proficiency in some instances. The reason again is the claim that in the long term we will have environmental sustainability only when we all have vested interests in our local environments.[25]

Finally, the importance of public participation in urban environmental projects as a component not only of urban ecological citizenship, but also of a broader notion of ecological citizenship which is not necessarily tied to cities, must be emphasized. The reason is that such projects provide us with an opportunity to begin the task of shifting North American environmental consciousness, for lack of a better term, away from its destructive idealization of the rural and directing it back toward the city. Bill Jordan, founding editor of the journal *Ecological Restoration* (formerly *Restoration and Management Notes*), provides a nice summary of this point in a recent editorial for that journal. There, Jordan bemoans the growing tendency of Americans to flee the city, not this time in "white flight," but instead to try to connect with the natural world by choosing to live in a more rural area. This new phase of suburbanization, Jordan claims, is laid partly at the feet of environmentalists who have promoted the examples of figures like Thoreau, Muir, and Aldo Leopold, who also "fled the city and built their houses – or, in Leopold's case, their weekend retreats – in the country, acting out the idea that this is the way to achieve an intimate, harmonious relationship with nature."[26] The problem, of course, as the growth of my hometown of Atlanta exemplifies, is that the movement of thousands out from the city is a disaster for both the rural and the wild landscape. Jordan reports, for example, that the widening of U.S. Highway 12 in Wisconsin (according to him, "the very road Leopold drove to reach his Sand County retreat") to accommodate the increase in traffic in and out of the city has resulted in the extended destruction of roadside vegetation that "Leopold himself mourned in his elegy for the prairie in *A Sand County Almanac*." We need instead to find a way, as Jordan says, of "'doing' nature downtown," so as to make the city "more natural, making it a place where one can find and commune with nature without, in Leopold's own despairing phrase, loving it to death."[27] I couldn't agree more.

But the example of projects like the Bronx River Alliance as a possible practice embodying civic environmentalism does not go far enough. If environmentalists are to fully embrace the urban, then we must describe the "brown" space of the city to be as important a locus of normative consideration as the green space. An environmentalism expanded to include the city as an object of its concern needs to be founded in a specific set of moral and political ends that are justified through a broader conception of civic environmentalism. The city is not a landscape merely of human invention, spotted here and there with bits of green worthy of our expanded civic interaction, it is an embedded ecosystem within larger ecosystems and environmental processes. While there are more humans in this ecosystem than in nonurban areas, and while these systems can at times be more or less insulated from environmental impacts, a true civic environmentalism must not be confined only to particular places.

I imagine urban environmentalists hesitating to throw their trash on the ground in the city streets where I live, not because they fear a fine, but because they understand their larger civic role in caring for the city as an environment of hope – hope because of the space it gives us to live as citizens in a manner which is lighter on the land.

Notes

1 Figures derived from Department of Economic and Social Affairs, Population Division, United Nations, *World Urbanization Prospects* (New York: United Nations, 1998). My thanks to Michael Cohen at New School University for calling my attention to these numbers in his preliminary draft of a new report of the panel on Urban Population Dynamics for the National Academy of Sciences Committee on Population.

2 See William E. Rees and Mathis Wackernagel, *Our Ecological Footprint* (New York: New Society, 1995).

3 William E. Rees, "Life in the Lap of Luxury as Ecosystems Collapse," *Chronicle of Higher Education*, July 30, 1999, p. 1.

4 Ibid., 2.

5 While I have never carried such an experiment out, since first reading the literature on ecological footprints I have always wanted to run a GIS experiment in which the population of New York City is emptied out more equitably into the surrounding regions. It seems doubtful that the astounding forest regeneration that Bill McKibben and others have celebrated in upstate New York would be sustainable without the existence of a population concentration device as effective as the city.

6 Thomas F. Homer-Dixon, *Environment, Scarcity, and Violence* (Princeton: Princeton University Press, 1999), 155ff.

7 Allen R. Myerson, "Energy Addicted in America," *New York Times*, November 1, 1998, Week in Review, p. 5.

8 For a sustained argument for why environmental ethicists, at least, have not heretofore focused much on cities in their work, see my "The Urban Blind Spot in Environmental Ethics," *Environmental Politics* 10 (2001): 7–35.

9 Mark Dowie, *Losing Ground: American Environmentalism at the Close of the Twentieth Century* (Cambridge: MIT Press, 1996), 6. For a nice summary of the common supposition that American environmentalism contains three waves, from turn-of-the-century reformers to today's interest groups (often accused of being coopted by the business community), see David Schlosberg, *Environmental Justice and the New Pluralism* (Oxford: Oxford University Press, 1999).

10 Schlosberg, *Environmental Justice and the New Pluralism*, 9.

11 See Richard Dagger, *Civic Virtues: Rights, Citizenship, and Republican Liberalism* (Oxford: Oxford University Press, 1997), 49.

12 Richard Dagger, "Metropolis, Memory and Citizenship," in *Democracy, Citizenship and the Global City*, ed. Engin F. Isin (London: Routledge, 2000), 28.

13 Such is the environmentalism of objectivists with respect to "intrinsic value" in nature, such as Holmes Rolston III. For the implications of this view for an attention to urban environments see Light, "The Urban Blind Spot." While subjectivists with respect to intrinsic value do not fall prey to the same bind when it comes to assessments of natural value, leading figures in this wing of environmental ethics still define the field in these terms. J. Baird Callicott maintains that the central theoretical question of environmental ethics is the issue of whether nature has intrinsic value, arguing that "if nature lacks intrinsic value, then nonanthropocentric environmental ethics is ruled out." Callicott, *Beyond the Land Ethic:*

More Essays in Environmental Philosophy (Albany: SUNY Press, 1999), 241. Still, to his credit, Callicott has extended some of his views toward a positive understanding of cities.

14 My own view, for other reasons, is that a "weak" or "broad" anthropocentrism is the preferred way of thinking about the justification for environmental values. See my "The Case for a Practical Pluralism," in *Environmental Ethics: An Anthology*, ed. Andrew Light and Holmes Rolston III (Malden, Mass.: Blackwell Publishers, 2002), 229–47.

15 See Richard Dagger, "Stopping Sprawl for the Good of All: The Case for Civic Environmentalism," this issue, 28–43.

16 Dagger, *Civic Virtues*, 154.

17 See Murray Bookchin, *The Limits of the City* (New York: Harper Collins, 1974), and *The Rise of Urbanization and the Decline of Citizenship* (San Francisco: Sierra Club, 1987).

18 Thomas Bender, "The New Metropolitanism and a Pluralized Public," *Harvard Design Magazine* (Winter/Spring 2001): 73, emphasis added.

19 Dagger, *Civic Virtues*, 167–68.

20 Avner de-Shalit has made a compelling argument against such localism in *The Environment between Theory and Practice* (Oxford: Oxford University Press, 2000).

21 Thomas Bender, "Describing the World at the End of the Millennium," *Harvard Design Magazine* (Winter/Spring 2000): 70.

22 Robert A. Beauregard and Anna Bounds, "Urban Citizenship," in *Democracy, Citizenship and the Global City*, ed. Engin F. Isin (London: Routledge, 2000), 249–52.

23 Ibid., 251.

24 Deane Curtin, *Chinnagounder's Challenge: The Question of Ecological Citizenship* (Bloomington and Indianapolis: Indiana University Press, 1999).

25 See Andrew Light, "Restoring Ecological Citizenship," in *Democracy and the Claims of Nature*, ed. Ben Minteer and Bob Peperman Taylor (Lanham, Md.: Rowman and Littlefield, 2002), 153–72.

26 William R. Jordan III, "Ten Thousand Thoreaus," *Ecological Restoration* 18 (2000): 215.

27 Ibid.

Part IX
Urban Values

Introduction

The history of modern technology has also been a history of urbanization: the rapid growth of cities, the decline in agricultural labor, urban growth and decline, suburbanization, and stunning new architectural forms. However, while the city has received serious consideration in architecture and planning literature, with the exception of some aesthetic considerations, philosophers have been largely silent about the importance of built space. These essays will help us begin to evaluate the relations between changing city and architectural forms and values.

Lewis Mumford, 1895–1990, was one of the most important thinkers on subjects of urbanism, architecture, and technology in the twentieth century. In this essay from 1958 Mumford argues that we must aim to make cities that exist for the "care and culture" of humans and not for the easy movement of automobiles. His views predate and presage the work of Jane Jacobs in *The Life and Death of Great American Cities* where she argues that dense, low-height, richly integrated neighborhoods make for healthier and more humane cities, as well as contemporary movements such as "smart growth" and "New Urbanism." As we now anticipate ever-greater strains on supplies of fossil fuels, and try, even if fitfully, to address global warming, Mumford's critique of automobiled cities remains relevant. Addressing many of Mumford's concerns, Jessica Woolliams, Founding Director Light House Sustainable Building Centre and BC Director for the Cascadia Region Green Building Council, argues that cities should be at the center of any environmental debate, and that we can and ought to design buildings and cities that are significantly more environmentally sensitive.

In *Technology and Place: Sustainable Architecture and the Blueprint Farm*, Steven Moore, Bartlett Cocke Regents Professor of Architecture and Planning, and Director, Sustainable Design Program, School of Architecture, the University of Texas at Austin, demonstrates how the various stakeholders' competing definitions of "sustainability," "technology," and "place" ultimately enable and limit ecological and architectural practice. His argument focuses on the case of the Blueprint Farm, an experimental agricultural project intended to benefit displaced agricultural workers in the Rio Grande valley of Texas. In this chapter, he argues that we should balance our fascination with the world of appearances, and the traditional architectural emphasis on beauty, with the recognition that architecture is also an ecological, technological, and political practice.

In his book *The City: An Urban Cosmology*, Joseph Grange, Professor of Philosophy at the University of Maine, argues for the centrality of aesthetics to understanding the city and urban experience, including the possibilities of building more just cities. This is in part because of three factors: (1) the importance of embodiment, (2) (echoing Latour and Dewey) the fact that at bottom there is no metaphysically ultimate distinction to be made between the built and the natural, the organic and the artificial, and (3) the fact that life needs community (in the most simple case the community of a living body) in order to survive and prosper. But life remains at heart a wild moment creative of opportunity through its anarchic force. Healthy community thus needs order and room for spontaneity, and this, he argues, can be nurtured by constructing symbolic orders that tie people together in difference. These symbolic orders would be both in language as well as in architecture and urban form.

Andrew Light, Associate Professor of Philosophy and Public Affairs at the University of Washington, opens with a rumination on the tendency of visitors to his former Greenwich Village neighborhood to leave trash around as if the neighborhood were a dump. He then asks what we can do to encourage people to think of the entire city as an environment worth respecting. He develops a model of ecological citizenship appropriate to urban settings and offers a set of "urban environmental theses." He concludes with an examination of how civic and environmental values might be brought together by considering the Bronx River Alliance. Light's work stands clearly at the intersection of our concerns with the city and concerns with environmental issues.

34

Why Mow?

Michael Pollan

No lawn is an island, at least in America. Starting at my front stoop, this scruffy green carpet tumbles down a hill and leaps across a one-lane road into my neighbor's yard. From there it skips over some wooded patches and stone walls before finding its way across a dozen other unfenced properties that lead down into the Housatonic Valley, there to begin its march south toward the metropolitan area. Once below Danbury, the lawn – now purged of weeds and meticulously coiffed – races up and down the suburban lanes, heedless of property lines. It then heads west, crossing the New York border; moving now at a more stately pace, it strolls beneath the maples of Larchmont, unfurls across a dozen golf courses, and wraps itself around the pale blue pools of Scarsdale before pressing on toward the Hudson. New Jersey next is covered, an emerald postage stamp laid down front and back of ten thousand split-levels, before the broadening green river divides in two. One tributary pushes south, striding across the receptive hills of Virginia and Kentucky but refusing to pause until it has colonized the thin, sandy soils of Florida. The other branch dilates and spreads west, easily overtaking the Midwest's vast grid before running up against the inhospitable western states. But neither obdurate soil nor climate will impede the lawn's march to the Pacific:

it vaults the Rockies and, abetted by a monumental irrigation network, proceeds to green great stretches of western desert.

Nowhere in the world are lawns as prized as in America. In little more than a century, we've rolled a green mantle of it across the continent, with scant thought to the local conditions or expense. America has some 50,000 square *miles* of lawn under cultivation, on which we spend an estimated $30 billion a year – this according to the Lawn Institute, a Pleasant Hill, Tennessee, outfit devoted to publicizing the benefits of turf to Americans (surely a case of preaching to the converted). Like the interstate highway system, like fast-food chains, like television, the lawn has served to unify the American landscape; it is what makes the suburbs of Cleveland and Tucson, the streets of Eugene and Tampa, look more alike than not. According to Ann Leighton, the late historian of gardens, America has made essentially one important contribution to world garden design: the custom of "uniting the front lawns of however many houses there may be on both sides of a street to present an untroubled aspect of expansive green to the passerby." France has its formal, geometric gardens, England its picturesque parks, and America this unbounded democratic river of manicured lawn along which we array our houses.

From Michael Pollan, *Second Nature: A Gardener's Education* (New York: Dell Publishing, 1991), pp. 65–78.

To stand in the way of such a powerful current is not easily done. Since we have traditionally eschewed fences and hedges in America, the suburban vista can be marred by the negligence – or dissent – of a single property owner. This is why lawn care is regarded as such an important civic responsibility in the suburbs, and why, as I learned as a child, the majority will not tolerate the laggard or dissident. My father's experience with his neighbors in Farmingdale was not unique. Every few years a controversy erupts in some suburban community over the failure of a homeowner to mow his lawn. Not long ago, a couple that had moved to a $440,000 home in Potomac, Maryland, got behind in their lawn care and promptly found themselves pariahs in their new community. A note from a neighbor, anonymous and scrawled vigilante-style, appeared in their mailbox: *"Please, cut your lawn.* It is a disgrace to the entire neighborhood."* That subtle yet unmistakable frontier, where the crew-cut lawn rubs up against the shaggy one, is enough to disturb the peace of an entire neighborhood; it is a scar on the face of suburbia, an intolerable hint of trouble in paradise.

That same scar shows up in *The Great Gatsby*, when Nick Carraway rents the house next to Gatsby's and fails to maintain his lawn according to West Egg standards. The rift between the two lawns so troubles Gatsby that he dispatches his gardener to mow Nick's grass and thereby erase it. The neighbors in Potomac displayed somewhat less savoir faire. Some offered to lend the couple a lawn mower. Others complained to county authorities, until the offenders were hauled into court for violating a local ordinance under which any weed more than twelve inches tall is presumed to be "a menace to public health." Evidently, dubious laws of this kind are on the books in hundreds of American municipalities. In a suburb of Buffalo, New York, there lives a Thoreau scholar who has spent the last several years in court defending his right to grow a wildflower meadow in his front yard. After neighbors took it upon themselves to mow down the offending meadow, he erected a sign that said: "This yard is not an example of sloth. It is a natural yard, growing the way God intended." Citing an ordinance prohibiting "noxious weeds," a local judge ordered the Buffalo man to cut his lawn or face a fine of $50 a day. The Thoreau scholar defied the court order and, when last heard from, his act of suburban civil disobedience had cost him more than $25,000 in fines.

I wasn't prepared to take such a hard line on my own new lawn, at least not right off. So I bought a lawn mower, a Toro, and started mowing. Four hours every Saturday. At first I tried for a kind of Zen approach, clearing my mind of everything but the task at hand, immersing myself in the lawn-mowing here and now. I liked the idea that my weekly sessions with the grass would acquaint me with the minutest details of my yard. I soon knew by heart the precise location of every stump and stone, the tunnel route of each resident mole, the exact address of every anthill. I noticed that where rain collected white clover flourished, that it was on the drier rises that crabgrass thrived. After a few weekends I had in my head a map of the lawn that was as precise and comprehensive as the mental map one has to the back of his hand.

The finished product pleased me too, the fine scent and the sense of order restored that a new-cut lawn exhales. My house abuts woods on two sides, and mowing the lawn is, in both a real and a metaphorical sense, how I keep the forest at bay and preserve my place in this landscape. Much as we've come to distrust it, dominating nature is a deep human urge and lawn mowing answers to it. I thought of the lawn mower as civilization's knife and my lawn as the hospitable plane it carved out of the wilderness. My lawn was a part of nature made fit for human habitation.

So perhaps the allure of the lawn is in the genes. The sociobiologists think so: they've gone so far as to propose a "Savanna Syndrome" to explain our fondness for grass. Encoded in our DNA is a preference for an open grassy landscape resembling the shortgrass savannas of Africa on which we evolved and spent our first few thousand years. A grassy plain dotted with trees provides safety from predators and a suitable environment for grazing animals; this is said to explain why we have remade the wooded landscapes of Europe and North America in the image of East Africa. Thorstein Veblen, too, thought the popularity of lawns might be a throwback to our pastoral roots. "The close-cropped lawn," he wrote in *The Theory of the Leisure Class*, "is beautiful in the eyes of a people whose inherited bent it is to readily find pleasure in contemplating a well-preserved pasture or grazing land."

These theories go some way toward explaining the widespread appeal of grass, but they don't fully account for the American Lawn. They don't, for instance, account for the keen interest Jay Gatsby takes in Nick Carraway's lawn, or the scandal my father's unmowed lawn sparked in Farmingdale. Or the fact that, in America, we have taken down our fences and hedges in order to combine our lawns. And they don't account for the unmistakable odor of virtue that hovers in this country over a scrupulously maintained lawn.

To understand this you need to know something about the history of lawns in America. It turns out that the American lawn is a fairly recent invention, a product of the years following the Civil War, when the country's first suburban communities were laid out. If any individual can be said to have invented the American lawn, it is Frederick Law Olmsted. In 1868, he received a commission to design Riverside, outside of Chicago, one of the first planned suburban communities in America. Olmsted's design stipulated that each house be set back thirty feet from the road, and it prohibited walls. He was reacting against the "high dead walls" of England, which he felt made a row of homes there seem like "a series of private madhouses." In Riverside each owner would maintain one or two trees and a lawn that would flow seamlessly into his neighbors', creating the impression that all lived together in a single park.

Olmsted was part of a generation of American landscape designer/reformers – along with Andrew Jackson Downing, Calvert Vaux, and Frank J. Scott – who set out at mid-century to beautify the American landscape. That it needed beautification may seem surprising to us today, assuming as we do that the history of the landscape is a story of decline, but few at the time thought otherwise. William Cobbett, visiting from England, was struck at the "out-of-door slovenliness" of American homesteads. Each farmer, he wrote, was content with his "shell of boards, while all around him is as barren as the sea beach . . . though there is no English shrub, or flower, which will not grow and flourish here." The land looked like it had been shaped and cleared in a great hurry (as indeed it had): the landscape largely denuded of trees, makeshift fences outlining badly plowed fields, and tree stumps everywhere one looked. As soon as a plot of land was exhausted, farmers would simply clear a new one, leaving the first to languish. As Cobbett and many other nineteenth-century visitors noted, hardly anyone practiced ornamental gardening; the typical yard was "landscaped" in the style southerners would come to call white trash – a few chickens, some busted farm equipment, mud and weeds, an unkempt patch of vegetables.

This might do for farmers, but for the growing number of middle-class city people moving to the "borderland" in the years following the Civil War, something more respectable was called for. In 1870, Frank J. Scott, seeking to make Olmsted's and Downing's design ideas accessible to the middle class, published the first volume ever devoted to "suburban home embellishment": *The Art of Beautifying Suburban Home Grounds*, a book that probably did more than any other to determine the look of the suburban landscape in America. Like so many reformers of that time, Scott was nothing if not sure of himself: "A smooth, closely-shaven surface of grass is by far the most essential element of beauty on the grounds of a suburban house."

Americans like Olmsted and Scott did not invent the lawn – lawns had been popular in England since Tudor times. But in England lawns were usually found only on estates; the Americans democratized them, cutting the vast manorial greenswards into quarter-acre slices everyone could afford (especially after 1830, when Edwin Budding, a carpet manufacturer, patented the first practical lawn mower). Also, the English never considered the lawn an end in itself: it served as a setting for lawn games and as a backdrop for flower beds and trees. Scott subordinated all other elements of the landscape to the lawn; flowers were permissible, but only on the periphery of the grass: "Let your lawn be your home's velvet robe, and your flowers its not too promiscuous decoration."

But Scott's most radical departure from Old World practice was to dwell on the individual's responsibility to his neighbors. "It is unchristian," he declared, "to hedge from the sight of others the beauties of nature which it has been our good fortune to create or secure." One's lawn, Scott held, should contribute to the collective landscape. "The beauty obtained by throwing front grounds open together, is of that excellent quality which enriches all who take part in the exchange, and makes no man poorer." Scott, like Olmsted before him, sought to elevate an unassuming patch of turfgrass into an institution of democracy; those who would dissent from their plans were branded as "selfish," "unneighborly," "unchristian," and "undemocratic."

With our open-faced front lawns we declare our like-mindedness to our neighbors – and our distance from the English, who surround their yards with "inhospitable brick walls, topped with broken bottles" to thwart the envious gaze of the lower orders. The American lawn is an egalitarian conceit, implying that there is no reason to hide behind hedge or fence since we all occupy the same middle class. We are all property owners here, the lawn announces, and that suggests its other purpose: to provide a suitably grand stage for the proud display of one's own house. Noting that our yards were organized "to capture the admiration of the street" one landscape architect in 1921 attributed the popularity of open lawns to "our infantile instinct to cry 'hello!' to the passerby, [and] lift up our possessions to his gaze."

Of course the democratic front yard has its darker, more coercive side, as my family learned in Farmingdale. In commending the "plain style" of an unembellished lawn for American front yards, the mid-century designer/reformers were, like Puritan ministers, laying down rigid conventions governing our relationship to the land, our observance of which would henceforth be taken as an index to our character. And just as the Puritans would not tolerate any individual who sought to establish his or her own back-channel relationship with the divinity, the members of the suburban utopia do not tolerate the home-owner who establishes a relationship with the land that is not mediated by the group's conventions. The parallel is not as farfetched as it might sound, when you recall that nature in America has often been regarded as divine. Think of nature as Spirit, the collective suburban lawn as the Church, and lawn mowing as a kind of sacrament. You begin

to see why ornamental gardening would take so long to catch on in America, and why my father might seem an antinomian in the eyes of his neighbors. Like Hester Prynne, he claimed not to need their consecration for his actions; think of his initials in the front lawn as a kind of Emerald Letter.

Perhaps because it is this common land, rather than race or tribe, that makes us all American, we have developed a deep-seated distrust of individualistic approaches to the landscape. The land is too important to our identity as Americans to simply allow everybody to have their own way with it. And having decided that the land should serve as a vehicle of consensus, rather than as an arena for self-expression, the American lawn – collective, national, ritualized, and plain – presented the ideal solution. The lawn has come to express our attitudes toward the land as eloquently as Le Nôtre's confident geometries expressed the humanism of Renaissance France, or Capability Brown's picturesque parks expressed the stirrings of romanticism in England.

After my first season of lawn mowing, the Zen approach began to wear thin. I had by then taken up flower and vegetable gardening, and soon came to resent the four hours that my lawn demanded of me each week. I tired of the endless circuit, pushing the howling mower back and forth across the vast page of my yard, recopying the same green sentence over and over: "I am a conscientious homeowner. I share your middle-class values." Lawn care was gardening aimed at capturing "the admiration of the street," a ritual of consensus I did not have my heart in. I began to entertain idle fantasies of rebellion: Why couldn't I plant a hedge along the road, remove my property from the national stream of greensward, and do something else with it?

The third spring I planted fruit trees in the front lawn, apple, peach, cherry, and plum, hoping these would relieve the monotony and at least begin to make the lawn productive. In back I put in a perennial border. I built three raised beds out of old chestnut barn boards and planted two dozen different vegetable varieties. Hard work though it was, removing the grass from the site of my new beds proved a keen pleasure. First I outlined the beds with string. Then I made an incision in the lawn with the sharp edge of a spade. Starting at one end, I pried the sod from the soil and slowly rolled it up like a carpet. The grass made a tearing sound as I broke its grip on the earth. I felt a little like a pioneer subduing the forest with his ax; I day-dreamed of scalping the entire yard. But I didn't do it, didn't have the nerve – I continued to observe front-yard convention, mowing assiduously and locating all my new garden beds in the backyard.

The more serious about gardening I became, the more dubious lawns seemed. The problem for me was not, as it was for my father, with the relation to my neighbors that a lawn implied; it was with the lawn's relationship to nature. For however democratic a lawn may be with respect to one's neighbors, with respect to nature it is authoritarian. Under the Toro's brutal indiscriminate rotor, the landscape is subdued, homogenized, dominated utterly. I became convinced that lawn care had about as much to do with gardening as floor waxing, or road paving. Gardening was a subtle process of give-and-take with the landscape, a search for some middle ground between culture and nature. A lawn was nature under culture's boot.

Mowing the lawn, I felt like I was battling the earth rather than working it; each week it sent forth a green army and each week I beat it back with my infernal machine. Unlike every other plant in my garden, the grasses were anonymous, massified, deprived of any change or development whatsoever, not to mention any semblance of self-determination. I ruled a totalitarian landscape.

Hot monotonous hours behind the mower gave rise to existential speculations. I spent part of one afternoon trying to decide who, in the absurdist drama of lawn mowing, was Sisyphus. Me? The case could certainly be made. Or was it the grass, pushing up through the soil every week, one layer of cells at a time, only to be cut down and then, perversely, encouraged (with lime, fertilizer, etc.) to start the whole doomed process over again? Another day it occurred to me that time as we know it doesn't exist in the lawn, since grass never dies or is allowed to flower and set seed. Lawns are nature purged of sex or death. No wonder Americans like them so much.

And just where *was* my lawn, anyway? The answer's not as obvious as it seems. Gardening, I had by now come to appreciate, is a painstaking exploration of place; everything that happens in my garden – the thriving and dying of particular plants, the maraudings of various insects and other pests – teaches me to know this patch of land more intimately, its geology and microclimate, the particular ecology of its local weeds and animals and insects. My garden prospers to the extent I grasp these particularities and adapt to them. Lawns work on the opposite principle. They depend for their success on the *overcoming* of local conditions. Like Jefferson superimposing his great grid over the infinitely various topography of the Northwest Territory, we superimpose our lawns on the land. And since the geography and climate of much of this country is poorly suited to turfgrasses (none of which are native), this can't be accomplished without the tools of twentieth-century industrial civilization: its chemical fertilizers, pesticides, herbicides, machinery, and, often, computerized irrigation systems. For we won't settle for the lawn that will grow here; we want the one that grows *there*, that dense springy supergreen and weed-free carpet, that platonic ideal of a lawn featured in the Chemlawn commercials and magazine spreads, the kitschy sitcom yards, the sublime links and pristine diamonds. Our lawns exist less here than there; they drink from the national stream of images, lift our gaze from the real places we live, and fix it on unreal places elsewhere. Lawns are a form of television.

Need I point out that such an approach to "nature" is not likely to be environmentally sound? Lately we have begun to recognize that we are poisoning ourselves with our lawns, which receive, on average, more pesticide and herbicide per acre than any crop grown in this country. Suits fly against the national lawn-care companies, and lately interest has been kindled in more "organic" methods of lawn care. But the problem is larger than this. Lawns, I am convinced, are a symptom of, and a metaphor for, our skewed relationship to the land. They teach us that, with the help of petrochemicals and technology, we can bend nature to our will. Lawns stoke our hubris with regard to the land.

What is the alternative? To turn them into gardens. I'm not suggesting that there is no place for lawns *in* these gardens or that gardens by themselves will right our relationship to the land, but the habits of thought they foster can take us some way in that direction. Gardening, as compared to lawn care, tutors us in nature's ways, fostering an ethic of give-and-take with respect to the land. Gardens instruct us in the particularities of place. They lessen our dependence on distant sources of energy, technology, food, and, for that matter, interest. For if lawn mowing feels like copying the same sentence over and over, gardening is like writing out new ones, an infinitely variable process of invention and discovery. Gardens also teach the necessary if un-American lesson that nature and culture can be compromised, that there might be some middle ground between the lawn and the forest – between those who would complete the conquest of the planet in the name of progress, and those who believe it's time we abdicated our rule and left the earth in the care of its more innocent species. The garden suggests there might be a place where we can meet nature halfway.

Probably you will want to know if I have begun to practice what I'm preaching. Well, I have not ripped out my lawn entirely. But each spring larger and larger tracts of it give way to garden. Last year I took a half acre and planted a meadow of black-eyed Susans and ox-eye daisies. In return for a single annual scything, I am rewarded with a field of flowers from May until frost.

The lawn is shrinking, and I've hired a neighborhood kid to mow what's left of it. Any Saturday that Bon Jovi, Judas Priest, or Kiss isn't playing the Hartford Coliseum, this large blond teenage being is apt to show up with a 36-inch John Deere mower that sheers the lawn in less than an hour. It's $30 a week, and I don't particularly like having this kid around – his discourse consists principally of grunts, and he eyes my wife like he's waiting for a *Penthouse* letter to unfold – but he's freed me from my dark musings about the lawn and so given me more time in the garden.

Out in front, along the road where my lawn overlooks my neighbors', and in turn the rest of the country's, I have made my most radical move. I built a split-rail fence and have begun to plant a hedge along it – a rough one made up of forsythia, lilac, bittersweet, and bridal wreath. As soon as this hedge grows tall and thick, my secession from the national lawn will be complete. Anything then is possible. I *could* let it all go to meadow, or even forest, except that I'm not sure I go for that sort of self-effacement. I could put in a pumpkin patch, a lily pond, or maybe an apple orchard. And I could even leave an area of grass. But if I did choose to do that, this would be a very different lawn from the one I have now. For one thing, it would have a frame, which means it could accommodate plants more subtle and various than the screaming marigolds, fierce red salvias, and muscle-bound rhododendrons that people usually throw into the ring against a big unfenced lawn. Walled off from the neighbors, no longer a tributary of the national stream, my lawn would now form a distinct and private place – become part of a garden, rather than a substitute for one. Yes, there might well be a place for a small lawn in my new garden. But I think I'll wait until the hedge fills in before I make my decision.

35

Technology

Lori Gruen

Environmentalists tend to view technology with suspicion. In the wake of the Exxon Valdez oil spill in 1989 and the catastrophes at Bhopal (1984) and Chernobyl (1986), such suspicion appears warranted. And opponents of technology are not only concerned about such massive environmental disasters, even though this is often the time that environmentalist criticisms of technology reach the general public. Critics of technology see such disasters as an inevitable result of measuring progress in terms of our ability to manipulate the environment through technology. And thus, while critical of the environmental damage caused by the use of various technologies, critics have also raised concerns about the social and political consequences of technological use and development. In the first part of this chapter, I will discuss these criticisms. Even when the consequences of technology are not particularly problematic, some critics nonetheless believe that technology is bad. In the second part of this chapter, I will assess the normative arguments mounted by critics against technology, and explore the value assumptions upon which the criticisms of technology depend.

The Consequences of Technology

Many of the objectionable environmental consequences of technology are obvious. In addition to the disasters just mentioned, one need only pick up a daily newspaper to see that human reliance on technology is a large contributing factor in the deterioration of the environment. To take a recent example, consider global warming, which is a predicted outcome of increased emissions of greenhouse gases that trap heat in the atmosphere. Emissions of one of the main greenhouse gases, carbon dioxide, has increased exponentially since the beginning of the Industrial Revolution. Everincreasing human reliance on technologies that require fossil fuels is directly connected to the impending threat of global warming. And if some of the gloomier scientific predictions are right, such technologies will be implicated in significant damage to human and non-human health and life as well as the destruction of entire ecosystems.

Less obvious, but equally environmentally damaging, is the production of so-called "clean technologies" – computers and other high-tech equipment. The manufacturing process for

From Dale Jamieson, ed., *A Companion to Environmental Philosophy* (Malden, MA: Blackwell Publishing, 2003), pp. 339–448. Reprinted by permission of Blackwell Publishing.

microchips involves the use of many highly toxic chemicals, such as arsine, acetone, ethylene glycol, and xylene. These and other chemicals used by the high-tech industry have caused massive ground water contamination in the last thirty years. When a spill or leak of toxic chemicals occurs at the site of a high-tech company the pollution often spreads many miles from its origin and can affect vital sources of drinking water and precious wetlands. In Silicon Valley, the center of high-tech production, there are now more polluted sites prioritized for clean-up by the federal government than in any other county in the United States. So while the use of computers and other high-tech electronic equipment may not have particularly damaging environmental consequences, their production has proved destructive to humans, non-humans, and their environments.

It might be suggested that while the production and use of some technology is certainly responsible for a great deal of environmental damage, there are technologies that are not environmentally harmful. It might be argued that not all technology leads to bad consequences. With the development of environmentally friendly technologies we may be able to maintain current levels of human productivity and the high quality of life that technologies allow in those nations that can afford them. So, the environmentalist criticisms might only apply to environmentally destructive technologies.

While many environmentalists believe that technology can be used in environmentally sensitive and responsible ways, and may even be central in efforts to protect the natural world, others point out that there are additional negative consequences of its development and use that are often overlooked. These critics point to the economic and political organization of technology and the social consequences of technological "regimes." Neo-Luddites, ranging from journalist and author Kirkpatrick Sale (1995) and former advertising executive Jerry Mander (1991), to the "unabomber" Theodore Kaczinsky (widely available on the web), have highlighted the dangerous, undemocratic organization of virtually all technological advancements. They urge that the power relations that are implicit in technological development and use be examined. Critics point to the fact that technological

development requires both highly organized, hierarchical decision structures and the concentration of capital in the hands of a few. Nuclear power is a commonly cited example that illustrates this point. The development of nuclear power required massive capital, centralized control, and the establishment of governmental bureaucracies to monitor safety, distribution, and waste disposal. Communities that live in the areas that were to become sites for nuclear power facilities had little, if any, input into the decision process and the workers employed by the nuclear power industry often were not informed about the risks of their work.

The development of technology, the neo-Luddites also point out, plays a central role in the global economy with its fiercely competitive market expansion strategies. Not only does this contribute to environmental degradation, but it also leads to the destruction of small, localized, and environmentally sound ways of living. Vandana Shiva, Director of the Research Foundation for Science, Technology, and Natural Resource Policy in India, provides an interesting example of how first world technology exports have contributed to social and political turmoil in Punjab. During the 1960s, technologically enhanced agricultural strategies were being touted as the way for struggling, famine-afflicted areas to attain abundance and peace. The "green revolution," as the new strategy was called, involved a shift from locally controlled peasant farming to a reliance on imported and expensive agrochemical and miracle seeds. As Shiva writes:

> The Green Revolution technology requires heavy investments in fertilizers, pesticides, seed, water, and energy. Intensive agriculture generated severe ecological destruction, and created new kinds of scarcity and vulnerability, and new levels of inefficiency in resource use ... the Green Revolution created major changes in natural ecosystems and agrarian structures. New relationships between science and agriculture defined new links between the state and cultivators, between international interests and local communities. (1991, pp. 46–7)

According to Shiva, the green revolution created monocrops, new pests, new diseases, water shortages, economic debt, and dependence, and increased rather than decreased social instability. While the agricultural policies and political dynamics in Punjab, and throughout the third world, are complex, this example illustrates the way that even technologies that are meant to enhance well-being can permanently alter and damage traditional, sustainable ways of life.

The life-altering effects of technology extend beyond the agricultural practices and political life of various communities. Biotechnology also has the potential to physically change human and non-human beings. In the late 1980s, scientists began a multibillion-dollar, long-term project which attempts to identify and map all of the human genes. The ostensible purpose of the human genome project is to gain greater insight into human diseases. The basic goal is to identify genetic links to disease so that, eventually, individuals can be genetically tested for diseases and undergo treatment to prevent the diseases from manifesting. Two types of treatment, or therapy, are commonly discussed: somatic cell therapy and germ line therapy. Somatic cell therapy modifies or augments our non-reproductive cells; gene line therapy modifies or augments our reproductive cells. The former alters a particular individual and the therapy only affects that individual; the latter alters the genetic make-up of future generations. Critics of technology suggest that just as in the case of the green revolution, where the manipulation of seeds led to immediate, short-term gains in crop yield and productivity at the cost of far-reaching, and in some cases unanticipated, ecological and social consequences, so too biotechnological manipulation of human cells can have frightening consequences.

Critics of biotechnology do not have to rely on dystopian fantasies about the imagined consequences of genetic manipulation for their concerns to be understood, however, as the actual results of genetic engineering are already being felt ecologically and socially. Genetically engineered organisms are being created in laboratories and released into the environment, creating a whole new category of pollution – "biological pollution." Unlike chemical pollution, which in many cases can be contained and cleaned up, biological pollution does not eventually dissipate, but rather reproduces and can overrun indigenous plant and animal life. In the social realm, human growth hormone, which was originally engineered

from human pituitary hormones in order to help those suffering from dwarfism, has been marketed as a way to enhance the height of short children. Thousands of parents with the consent of their physicians are injecting human growth hormone into their children, usually the boys. Given that height enhancing technology is now available, parents who care about the social acceptance and success of their children (and who can afford it) now have a new option available to help their children get a good start in life. But what is the social cost? Parents who choose not to inject their children with growth hormones, either out of fear of unforeseen side-effects or because they do not want to condone what they perceive as groundless stereotypes, may be thought to be bad parents. Those who can afford the treatment, but forego it, may be thought to be selfish. If being a tall young man contributes to success in the world, those who can afford to provide their sons with the treatment put their children in a better social and economic position. What of the parents who cannot afford the treatment to help their shorter sons? The economic disadvantage their youngster faces will be multiplied by his height disadvantage.

The existence of human growth hormone raises interesting issues about the effects technology can have on what is considered normal. Though the consumption of technology appears to be optional, often the use of technology becomes very close to a social requirement. In most of the United States, for example, it is next to impossible to make a living without a car for transportation. In our high-tech society, individuals who do not have answering machines, faxes, pagers, or computers are excluded from certain forms of social and professional activities. And even if the individual without such devices is not excluded from these activities, she is often viewed as outside of the norm. The social pressure of failing to use certain technologies, even if not materially damaging, can nonetheless be intense (as anyone who doesn't have an answering machine or use email can testify). And once an individual is on the technological path, it is very difficult to step off. Many of us who use computers can hardly imagine what it would be like to go back to pen and paper. This technological mindset, as critics call it, is just one more damaging consequence of technology. Individuals in technological societies become dependent on technology and their ability to think and act outside of technology becomes limited.

All of the examples mentioned here raise complex questions which go beyond the scope of this brief discussion. The point these examples are meant to illustrate is that technology, even when it does not directly, immediately, or obviously affect the natural world, is nonetheless implicated in problematic social, political, psychological, and ecological consequences. But granting that the negative consequences outlined here exist, does that mean that all technology should be condemned? Or is technology that is organized in ways that avoid the undemocratic, capital intensive, socially, psychologically, and ecologically damaging consequences acceptable? In order to answer these questions, the value assumptions on both sides of the argument about technology must be explored.

Some of those who have analyzed technology, including those environmentalists who advocate "appropriate" technologies, have suggested that technology, in itself, is neutral, and they restrict their assessment of technology to the consequences that it brings about. They argue that recent technological advances, such as video cameras and the internet, can provide us with information about and glimpses of nature that will motivate people to work to protect non-humans and the environment. Their view is that when technology can aid in protecting the natural world and can be developed and implemented in a non-coercive and participatory way, then it is morally acceptable. When, however, technology is damaging to the environment or is undemocratically developed and forced on people it is open to criticism on moral grounds. The reasons for the moral condemnation are external to the technology itself. Other critics, however, have argued that technological neutrality is a myth and that there is something intrinsic to technology that makes it morally problematic. I will now look at two types of argument that might support such a view.

Nature Versus Culture

Technology is ubiquitous in human culture. It is so pervasive that technologies, those tools and devices used to shape and change the environment,

have become the mark of culture. It is often thought that a culture can be defined and assessed by how successful that culture is at controlling nature through technology. Culture is thus opposed to nature, and technology is the means by which this opposition is maintained. When culture is understood in opposition to nature, technology itself can be viewed as unnatural. For many environmentalists, no matter how well intentioned human cultural activities and technological innovations may be, no matter what the consequences of such activities and innovations, they are always artificial, and thus have a value that is fundamentally different from that of natural processes. The cultural is understood as a disruption of the natural and thus antithetical to it.

According to this perspective, artificial things, artifacts of human invention, are less valuable than natural things. Many environmentalists believe that nature has value because it is natural, that is, because it is not the product of human construction or creation. Such products have value insofar as they are useful to humans; they serve human ends. Automobiles, nuclear power facilities, biotechnologically engineered seeds, and computers are all created to serve human needs. Nature, on the other hand, exists for itself and thus is an altogether different kind of thing. Ecological and evolutionary processes occur quite independently of human activities. Nature existed long before humans did, and will continue long after humans are gone (assuming some human-generated technological disaster does not destroy everything). Some argue that this independent genesis and existence is what gives nature its value, and while nature can be and often is taken over by human cultural activity to serve human needs, its value goes beyond its use-value to humans.

Technology, on this view, is the mechanism by which nature is disrupted and destroyed. It is not simply an artifact of human creation, and thus less valuable than the natural world, but it is also that which diminishes natural value. Technology is the means by which valuable nature is transformed into products for human use and consumption. Even when technology is used for protecting and preserving nature, such as when various technologies are employed to identify and protect endangered species, it is still thought to be problematic by some environmentalists. At best, technology is a device humans can use to attempt to repair what their use of other technologies has destroyed. But technology cannot restore the natural value of the environment, nor can humans improve it. Once that value is interfered with by humans, it is gone forever.

There are a number of problems with this criticism of technology that is based on a dichotomy between nature and culture. First, viewing nature as a category that is distinct and independent from culture fails to take into account the various ways in which we interpret "nature" rather than merely report on it. How we understand natural processes, species, and the differences between ecosystems and their states of "health," for example, all require interpretation by cultural beings. The very division between culture and nature, one could say, is a cultural artifact. And technology often helps us to gain information to better interpret the natural world and thus can aid us in acquiring knowledge about what is valuable in nature.

A second problem with this view is that it presupposes that things can be neatly divided into either the cultural or the natural and this doesn't seem to be the case. At the extremes, the division may be clear – a plastic necklace is clearly cultural and artificial, a thunderstorm is clearly natural. But what about second-growth forests that have existed for over one hundred years or non-indigenous plants that have lived in and adapted to their non-native environments over several centuries? What about El Niño storms? While the forest, the plants, and the climate have in some ways been interfered with by the use of technologies, they have since flourished, evolved, and reacted independently of humans. Do they belong in the realm of the cultural or the natural or both?

A third problem with the conceptual framework that divides culture from nature is that it poses a dilemma for environmentalists. Either they must believe that humans are part of nature and thus what they do is natural, or that humans are separate from nature and thus human activity is unnatural. If they believe the former, then they must view the destruction of nature through the use of various technologies as natural, just as predation is natural. But then it would be difficult to say why the destruction of nature is

morally objectionable if what is natural is thought to be good. If they believe that humans are unnatural, then they would be committed to accepting a radical discontinuity between humans and other naturally evolved things (a view that has been convincingly challenged at least since the nineteenth-century British biologist Charles Darwin). This contradicts the notion that many environmentalists want to promote, namely that humans are a part of nature and that in order to solve our environmental problems we must recognize our connection to rather than separation from the natural world. This dilemma could be overcome by rejecting the normative implications of the culture/nature divide, that is, by denying the simple theory of value which maintains that what is natural is good and what is cultural is bad, and drawing on a more refined theory about what is valuable. Let me turn now to one such theory.

Technology Versus Authenticity

While many people view technology as a means to free us from the drudgery and tedium of thankless tasks, others argue that technology removes us from the workings of the natural world and in so doing mediates our experiences of it. What is wrong with technology, then, is not that it is a product of culture, but that it creates experiences that are unnatural or inauthentic. A thought experiment will help bring out the insight of this type of criticism of technology.

Imagine learning that your favorite place in nature is not what it seemed. The designated wilderness area you've been visiting every summer for the last ten years is not wilderness at all. The trees and mountains are synthetic, the lake is man-made, the "indigenous" wildlife has been farm-raised and then imported, and the fern, moss, and other plant life were manufactured in a factory outside of Gary, Indiana. The environmental engineers who created this place did a fine job. Their employers removed the natural resources in the area in order to make a profit, thereby destroying the original environment, and have replaced what was natural with a very good fake. For a decade you have enjoyed your time in what you thought was a pristine natural area. Only now you have learned that the wilderness was a technologically created copy.

What is wrong with this scenario? It might be argued that you will no longer enjoy your so-called wilderness experience, and since pleasure is what is ultimately valued, what is wrong here is that pleasure will no longer be generated from this experience. Assuming that no sentient beings suffered during the process of creating this place (an improbable assumption in a number of ways, but for the sake of argument let's assume that the native wildlife species of the area were removed and have now been released) and that human experiences of this area were not unpleasant, it looks like the only thing wrong with this scenario is that you and others experience disappointment upon learning that this natural area was not what it seemed. But is this really all that is wrong with faked nature?

Critics of technology can draw on the view that there is more wrong with this artificial environment than the mere disappointment you and others may experience. The value of the fake, they might suggest, is less than the value of real nature. This is, for example, how many environmentalists have responded to ecological restoration projects (see Elliot 1997). Environmental theorists have argued that when we destroy a natural area and then recreate or restore it through acts of technological wizardry, something of value is permanently lost. Some environmental theorists maintain that the value accompanying authentic wilderness cannot be restored, even when the area still provides positive "wilderness" experiences, because its value is not reducible to the experiences of valuers. The origin of a natural area contributes to its value and thus, because the faked wilderness area has a different origin, it does not have the same value as authentic or genuine nature. What is wrong with the above scenario, critics argue, is not that the subjective quality of our wilderness experience may have changed, but that the value of the actual objects of such experiences is diminished.

But is this view right? Consider the above scenario modified. Imagine that you never learned that the wilderness area you have so much enjoyed over the years is in fact fake. You continue to visit the area and experience much pleasure. Not only do you continue to maintain the false belief that you are experiencing authentic wilderness, but

other visitors share this belief. Imagine further that rather than being employed by a for-profit company, the environmental engineers were employed by an environmental organization that believed that it was important for as many people as possible to experience what they thought was nature. They believed, quite plausibly, that if everyone did experience real nature they would undoubtedly destroy it, so the environmental engineers were hired to create an excellent and durable substitute.

According to those who hold that what is ultimately valuable is pleasurable experiences, it looks as though the modified scenario is not only acceptable, but may even be morally required (assuming that a world in which many people had pleasurable experiences of what they think is real nature is a better world). But, critics might argue, it is not any pleasant experience that is valuable, but the quality of the experience. An experience that is based on a false belief, while pleasurable and thus of some value, is not as valuable as an experience that is based on a genuine, unmediated encounter with the world. Another way of putting it is that what is valuable is not just the belief that we have satisfied our desires, but their actual satisfaction. A person who desires to make the world more just, for example, by engaging in activities to resist racial or gender discrimination, would presumably not be satisfied if they were to learn that though they thought they were succeeding in changing discriminatory institutions, their success was illusory. Similarly, if one desired to experience authentic nature, and then learned that what they thought was authentic was artificial, their desire would not actually be satisfied.

Part of what contributes to a meaningful life, and thus a pleasurable one, is making plans and setting goals that one hopes will be accomplished. Part of what is pleasurable is the making of such plans, and seeing a plan through to fruition generates tremendous satisfaction. If people knew that there was a possibility that they could be deceived into thinking that they had actually succeeded in achieving a goal, when in fact they were just strapped to electrodes in a tank or were being lied to by everyone they knew, there is a good chance that they might not set out to make a plan or try to accomplish a goal. Given that the making of plans and the setting of goals is part of what contributes to overall pleasure, deception will negatively affect that pleasure. If one were asked whether their lives on the whole would be better if they were deceived, but nonetheless happy, I suspect many would answer negatively. In the modified case of faked nature, it can be argued that the very possibility of there being faked nature will undermine individuals' enjoyment of the real thing. The satisfaction that individuals would have experiencing nature could be compromised if they believed that what they thought they were experiencing might be fake. Further, the possibility that the deception might exist may cause individuals to not plan to visit natural areas and thus lose out on what could be valuable experiences. Most of us value authentic experiences and would be disturbed to learn that we have been deceived.

Drawing on the insights of this thought experiment, critics of technology might suggest that our experiences of the natural world, because they are mediated by technology are inauthentic, that our understanding of ourselves and the natural world is based on deception. Even though the natural world has not been recreated by environmental engineers (and the possibility of creating a realistic ecosystem is very remote indeed given that scientists working on artificial life have yet to create a realistic model of moths or cockroaches), our modern lives are so saturated with technology that our day-to-day experiences are not unlike those we would have in a faked environment. This is one of the reasons why people seek out natural areas, to escape to a more genuine reality. What is wrong with technology, critics might suggest, is that it creates a false, unnatural world.

While this criticism of technology is in many ways quite compelling in that it allows us to understand why technologically mediated experiences are of questionable value, it nonetheless presupposes an absolute rejection of technology that is difficult to defend. Perhaps this can best be seen by considering the extreme, but not unrealistic, situation in which an individual has no chance at all of experiencing genuine nature. Perhaps she is poor and lives in an overpopulated urban center. Most of her time is spent trying to survive and keep her children alive. On rare occasions, she and others like her are invited to view a nature video or experience nature in a

"virtual" way. This experience provides her with genuine pleasure, even though she knows her experience is mediated through technology and, we can assume further, such an experience allows her to recognize the value of nature. One of her hopes is that her children, or their children, might be able to experience genuine nature someday. While she has little time and no resources to devote to protecting the natural world, she nonetheless is glad that others can, so that her children may experience it. Critics of technology, though rightly critical of the conditions that lead to this woman's impoverished situation, would have to tell a richer story about what is morally wrong with this mediated experience of nature, in order for their sweeping moral condemnation of technology to hold.

Conclusion

Technology is surely implicated in much environmental damage, as well as the destruction of politically and psychologically sustainable ways of life. In the fast-paced world of video games, cell phones, laser-guided missiles, and the like, it is indeed tempting to think that technology, in itself, is objectionable. Such an argument, however, is hard to sustain. If nature is valuable, as many have argued, then technology can be used to inform, educate, and assist in promoting its value. Critics have done a tremendous service in helping us to think more critically about the use and development of various technologies. They are right to have us examine most of our current technological trends. But ultimately, the strongest arguments for or against technology will be based on the consequences of its use for people, for animals, and for the natural world.

References

Elliot, R. (1997) *Faking Nature* (London: Routledge). [An important book that discusses both the meta-ethical and normative issues of natural value as well as carefully exploring problems with technological intervention in the natural world.]

Mander, J. (1991) *In the Absence of the Sacred* (San Francisco: Sierra Club Books). [The former advertising executive and author of *Four Arguments for the Elimination of Television* explores the dangerous consequences of technology on native cultures and society in general.]

Sale, K. (1995) *Rebels Against the Future: The Luddites and Their War on the Industrial Revolution* (Reading, MA: Addison-Wesley). [A history of the nineteenth-century opposition to technology and the lessons we can learn from it.]

Shiva, V. (1991) *The Violence of the Green Revolution* (London: Zed Books Ltd). [A carefully documented examination of the dangers of technological intervention in India.]

Further Reading

Ellul, J. (1964) *The Technological Society* (New York: Vintage Books). [An early and thorough critique of technology.]

Ferré, F., ed. (1992) *Research in Philosophy and Technology: Technology and the Environment* (Greenwich, CN: JAI Press). [A wide-ranging book of essays that addresses both the theoretical and practical issues that technology raises.]

Mills, S., ed. (1997) *Turning Away from Technology* (San Francisco: Sierra Club Books). [Based on transcripts from the 1993 and 1994 international conferences on megatechnology, this volume contains insights from some of the leading critics of technology today.]

Sclove, R. (1995) *Democracy and Technology* (New York: Guilford Press). [Argues for the need for participatory technologies.]

Zerzan, J. and Carnes, A., eds. (1988) *Questioning Technology* (London: Freedom Press). [A collection of sometimes fictional, always disturbing, reflections on technology.]

36

Environment, Technology, and Ethics

Rajni Kothari

Twenty, even ten, years ago one had still to establish the 'case' for the environment. To this end, beginning with the Stockholm conference, a major intellectual and political effort was mounted, an effort that has proved successful. Unfortunately, this very success has been co-opted by the status quo, with the result that while everyone talks of the environment, the destruction of nature goes on apace, indeed at an increasing pace. The environment is proving to be a classic case of 'doublespeak', a lot of sophistry, and not a little deliberate duplicity and cunning.

A decade ago, there was reluctance on the part of national governments and international agencies to include the environmental dimension in their strategies of development. This reluctance has given way to acceptance, and 'sustainable development' has become a universal slogan. Yet nothing much seems to have changed in mainstream development policy. There is no genuine striving towards an alternative perspective on development, no ethical shift that makes sustainable development a reality. 'Sustainability' has been adopted as rhetoric, not as an ethical principle which restructures our relationship with the Earth and its creatures in the realm of knowledge and in arenas of action.

In the absence of an ethical imperative, environmentalism has been reduced to a technological fix, and as with all technological fixes, solutions are seen to lie once more in the hands of manager technocrats. Economic growth, propelled by intensive technology and fuelled by an excessive exploitation of nature, was once viewed as a major factor in environmental degradation; it has suddenly been given the central role in solving the environmental crisis. The market economy is given an even more significant role in organizing nature and society. The environmentalist label and the sustainability slogan have become deceptive jargons that are used as convenient covers for conducting business as usual. This is particularly the case with the world's privileged groups, whose privileges are tied to the status quo, and who will therefore hold on to those privileges as long as they can.

But there are other voices which give a different meaning to sustainability, one which is rooted in ethics, not in monetary policy, and which goes hand in hand with the striving towards an alternative mode of development. Without such striving, 'sustainability' is an empty term, because the current model of development destroys nature's wealth and hence is non-sustainable. And it is ecologically destructive *because* it is

From J. R. Engel and J. G. Engel, eds., *Ethics of Environment and Development: Global Challenge, International Response* (London: Belhaven Press, 1990), pp. 228–37.

ethically vacuous – not impelled by basic values, and not anchored in concepts of rights and responsibilities. Thinking and acting ecologically is basically a matter of ethics, of respecting the rights of other beings, both human and non-human.

[First I] will address these two opposing meanings of 'sustainability' and their respective development paradigms. [I] will differentiate between sustainability as a narrow economic ideal and sustainability as an ethical ideal, between sustainability of privileges and sustainability of life on Earth. Once the conflict was between 'environment' and 'development'. I now see a conflict between the two meanings of 'sustainable development', because sustainability has become everyone's catchword, even though it means entirely different things to different people.

Later I shall lay out the profile of an alternative design for development, one which is environmentally and ethically sound, and at the same time economically, socially, and politically just. But before I do, it is necessary to provide an analysis of the reasons why the present mode of development which once held out such promise and gave rise to the vision of 'continuous progress for all' has come to grief. In what follows, I shall provide such an analysis in the context of the fast-changing processes of history, their philosophical underpinning, and their consequences for the politics of development. We shall see that it is more from the striving of ordinary folk as they face the modern trauma that new possibilities might emerge than from the doings of counter-elites spawned by social movements, though the catalytic role of the latter should not be underestimated. The issue is less whose efforts should succeed than which interventions are likely to endure because they are ethically grounded.

The Crisis in World Order

We may begin with some fundamentals. The most fundamental point to grasp is that we live in a period of profound transformation which is engulfing and interlocking diverse regions, cultures, and ecosystems into a common enterprise, and in the process giving rise to new conflicts waged on a scale unheard of in earlier times. Whereas thinkers from time immemorial have defined the human predicament as the need to overcome conflict through some kind of a social order, most of them thought in terms of a single society, or of conflicts between two or more societies. *For the first time we are realizing that the human predicament is on a world scale.* And all actors in it, and perhaps most of all the weakest and the most deprived among them, need to think in terms that cover all persons and societies. The end of colonialism, the unprecedented increase in population, the urgency of the economic problem, the sudden sense of the bounties of nature drying up, and a feeling of scarcity of basic resources in place of a feeling of continuous progress – all point, on the one hand, to a scenario of growing conflict that will become worldwide in scope, and on the other hand, to the need to work out new solutions based on a new structure of human cooperation.

It is only by thinking in terms of a new concept of ordering the world as a whole that there can be any salvation for a humanity that has lost its moorings. This will require new ways of attending to human problems; but it is not impossible to do this once the problems become clear, and we are able to move out of the old grooves in which we habitually think.

It is necessary to grasp this point. For it is only at times of deep crisis that major changes become possible, for better or for worse, and human beings are capable of both. It is a time when we can either seize the opportunity by deciding to control our future and usher in a new era, or we may miss the opportunity and be pushed into a downward course by forces beyond our control, after which it may be difficult to retrieve lost ground. That we are caught in such a historic moment should be clear to anyone who reflects on the concrete realities of the world we live in and the developments taking place in different parts of this world – in economics, in politics, in the availability and distribution of resources, in the relationship between food and population, in patterns of trade and control of technology, and in the strategic and power relations in which the different nations and regions find themselves confronting each other.

The Causes of World Crisis

Now while it is recognized that the contemporary human condition is one of a deepening crisis, perception as to the nature of the crisis and its causes has changed over the last two decades. For a long time – and this view still persists – the crisis was perceived in terms of an ideological struggle between different ways of life and systems of belief, not infrequently associated with a struggle for power between rival blocs of countries. A very large part of human energy and world resources was devoted to this conflict, which is by no means over and which in no small way accounts for the terrible arms race that enveloped the world and still persists. Later, attention focused on something more immediate and very pressing, but which had somehow escaped human sensitivity for so long: the great economic schism that is dividing the world into extremes of affluence and deprivation, with concentrations of poverty, scarcity, and unemployment in one vast section, and over-abundance, over-production, and over-consumption in another and much smaller section. Furthermore, both these are in a relationship in which resources from the poorer regions have for long been drained, and continue to be drained, through new instruments of appropriation. The last few years have witnessed an increasing concern with this single problem of poverty and inequity on a global scale, though it must be admitted that very little has been done systematically to solve it; indeed, it has been getting worse.

All these perceptions of the nature of the human crisis are still relevant. But perhaps one needs to think beyond single dimensions and look to more fundamental causes. After all, the fact that a century of unprecedented material progress has also been one of sprawling misery and increasing domination of the world by just a few powers suggests that there is something basically wrong with our world and the global structures that have permeated it. Indeed, there *is* something basically wrong with the way modern humanity has gone about constructing its world.

Industrialization was supposed to be an end to the condition of scarcity for humankind as a whole; in fact, it has made even ordinary decent existence more scarce and inaccessible for an increasing number of human beings. Modern education was supposed to lead to continuous progress and enlightenment for all, and with that a greater equality among men and women; in fact, it has produced a world dominated by experts, bureaucrats, and technocrats, one in which the ordinary human being feels increasingly powerless. Similarly, modern communication and transportation were supposed to produce a 'small world' in which the fruits of knowledge and development in one part of the globe could become available to all the others; in fact, modern communication and transportation have produced a world in which a few metropolitan centres are sucking in a large part of world resources and depriving the other regions of whatever comforts, skills, and local resources they once used to enjoy. Surely then there is something more deeply wrong with the structure of this world than the mere production of nuclear weapons or the economic handicap of the poorer countries. The world in which we live is indeed very badly divided, but the divisions are more fundamental than those of ideology, or of military or economic power. Perhaps there is something wrong with the *basic model of life* humankind has created in the modern age.

Colonizing the Future

That this might indeed be the case is indicated by the rupture that has occurred for the first time in world history between the present and the future, the future including both the very young among us and the yet-unborn generations. While rational anticipation and prudence in preparing oneself against the future were inherent in all earlier thought, the future consequences of present action were never as morally relevant and urgent as they are today.

This is a result of the basic way of life we have created in the modern age, especially our creation of modern technology. Technology has a powerful impact on beings that have no voice in decisions regarding how technology is to be used. As the growing economic, energy, and environmental crises are now showing us, decisions taken at one point in time have the power to affect future generations in ways that are by and large irreversible. The consequences of what our parents and the older generation among us did – the ravaging of nature, the depletion of resources, the pattern of investment, the stock-piling of armaments, the building of highly cen-tralized economic and political structures that are difficult to change (except by long struggle and violence) – are being felt by the younger genera-tion of today. How is one to assure that the interests of the younger generation and the yet-to-be-born generations of the future are somehow represented in the present? They have no voice in the decision-making processes of modern society, least of all in representative systems of government of which only the old (whom we prefer to call 'adults') have a monopoly. It was once an assumption of planning and of prudence generally that one must sacrifice or postpone gratification in the present so that the future generations can live a better life. In fact, modern civilization does just the opposite. We are so involved in our own gratification in the present, stimulated by the mass media and advertisement agencies, that we are sacrificing the life chances of future generations.

Thus, just as decisions made in the metropo-litan centres of the world and their ever-rising consumption of finite resources are adversely affecting millions of people in far-off places, decisions made by the present generation are affecting and will continue to affect the future of the young and the yet-unborn generations. These are serious questions to which the present models of politics and economics provide no answer. They call for a different kind of con-sciousness, one which takes a total view of exis-tence; empathizes with the weak, the distant, the unborn, and the inarticulate; and intervenes in legislative and administrative processes at various levels of the world without, however, degenerat-ing into some kind of brahminical class that arrogates to itself all knowledge and wisdom. As yet such consciousness (which no doubt exists here and there) is still very dim and, at any rate, not very influential in the decision-making processes of business and government. But the need for someone to represent the future – the 'last child', the 'seventh generation' – in the decisions made in the present cannot be overemphasized.[1]

Sustainability cannot be real if the future itself is colonized. Sustainability therefore cannot be realized by those who have only learned how to act in the short term. For real lessons in sus-tainability we need to turn to peoples and cultures that have acted on behalf of future generations. Women, particularly Third World women, who produce sustenance for their children are inti-mately in touch with the future through their nurturance. It is little wonder that ecology movements, like Chipko in India, spring from these cultural pockets which have conserved the qualities of caring – caring for the 'last man' as Gandhi asked his countrymen to do, or the 'last child' as I would like to put it.[2] The Native Americans, who also have a special commitment to the future as part of their understanding of nature's ways, have conceptualized it in an even more telling way: to use and protect nature's creation 'so that seven generations from this day our children will enjoy the same things we have now'. Oren Lyons, spokesman for the traditional circles of elders, has been carrying this message into the contemporary world, pleading:

> Take care how you place your moccasins upon the earth, step with care, for the faces of the future generations are looking up from earth waiting their turn for life.
>
> Today belongs to us, tomorrow we'll give it to the children, but today is ours. You have the mandate, you have the responsibility. Take care of your people – not yourselves, your people.[3]

Colonizing Nature

That our basic model of life is wrong is also indicated by what we have done to other species and forms of life as well as to inanimate nature. We increasingly destroy other animal species, vegetation, the chemical sources of life, and the seabeds and rocky lands whose bounty has been the source of so much imagination, wonder, joy, and creativity. Springing from the unending acquisitiveness of our technological way of life and a concomitant decline in our sensitivity to other humans, we have been on a rampage that threat-ens our common organic bond with the whole of creation – and thus both our own survival and that of other species.

Modern humanity, and in particular Western technological humanity, has accumulated wealth by denying the rights of others to share in nature's bounty. These 'others' include marginal commun-ities (tribes and small villages), future generations, and other species. Inequality, non-sustainability, and ecological instability all arise from the selfish and arrogant notion that nature's gifts are for private exploitation, not for sharing. In contrast to this rapaciousness, many cultures of the world have based their relationship with nature on the assumption that human beings are members of the Earth family, and must respect the rights of other members of the family. In traditional India, human beings were believed to be part of the cosmic family – *Vasudhevkutamkam*. The belief that trees and plants, rivers and mountains have intrinsic value created ethical constraints on human use of the environment. Hunters and gatherers have always apologized to nature before killing plants and animals. Rural women in India offer leaves to the tree goddess, *Patnadevi*, before collecting fodder from the forest.

The living Earth has a right to life, and that right is the primary moral argument for sustainable life. As Aldo Leopold has pointed out, ethics is the recognition of constraints put on an individual as a member of a community, and the ecological ethic simply enlarges the boundaries of the community to include soils, waters, plants, and animals, or collectively, the land.[4] The modern West is slowly rediscovering the ethics of nature's rights as the basis for conservation and ecological recovery. The Earth is no longer just a bundle of resources. As James Lovelock has suggested, she is *Gaia*, a living being.[5] Animals are not just resources and game for human consumption. Peter Singer has argued for animal liberation as a component of human liberation.[6] And Christopher Stone has raised the issue of whether trees have rights.[7] The women of Gharwal, the backbone of the Chipko movement, who risked their lives to save their trees, clearly believe that trees have rights, and that the rights of trees are of a higher order than those of human beings because trees provide the *conditions* for life on Earth.

Sustainability ultimately rests on the democracy of all life, on the recognition that human beings are not masters but members of the Earth family.

Human Capacities for a New World Order

None of these issues – the rampage of technology, the divisions of global society, the sacrifice of the life chances of future generations, or the destruction of other species and other sources of life and sustenance – were adequately raised in earlier philosophical discussions about the human predicament. The predicament that faces humankind today includes all these issues; and the salvation that we must work out for ourselves and for the whole of nature must address itself to all these issues. In this sense, the crisis that we face is far more total than ever before.

And yet, human beings throughout history have shown an almost infinite capacity for identifying their own immediate purposes with larger purposes. We have come a long way from the primeval stage when we identified with just a few of our kind and cared little for others. Today we are able to identify not only with our own national or regional collectivity but with the whole of the human species, and even with non-human species. Our capacity to symbolize and identify with abstract entities enables us to think in cosmic terms and embrace entities and identities that range from the ephemeral to the eternal.

Nor is this entirely new. At many times in history humanity has shown a striking empathy for the whole of creation. The intellectual and religious movements that led to a deep sense of regard for life in all forms and an abhorrence of violence in all forms, including violence to other forms of life, had their mainsprings in this innate human power to symbolize and identify with creation and to revolt against human excesses. This is what powerful movements like Buddhism and Jainism represented in my own land. Similar movements took place in other regions. It is true that often, as in India, this kind of feeling for life produced a rather quiescent attitude toward life's purposes and even a metaphysic that undermined humanity's self-confidence. It will be necessary to guard against this kind of defeatist religiosity. But such an attitude is by no means inherent in developing a larger identity with creation.

The conclusion I draw is that if there is to be a moral imperative for sustainable development, there needs to be a sense of sanctity about the Earth. Concern for the environment has to emanate from the basic human capacity to experience the sacred, the capacity to wonder at the blessings that are still with us, to seek after the mysteries of the cosmic order, and a corresponding modesty of the self and its claims on that order. Respect for life has to be a fundamentally spiritual notion, based on faith in the inalienable rights of all living beings. The basic sanctions behind them are not contractual but transcendental. They are not primarily claims bestowed by law but are inherent in the very nature of life.

Humanity, then, does have the capacity to create a new world order. Indeed, of all the species only humanity has the capacity to transform its history on a global scale. Human beings are the abstracting animals, the historical animals, the aesthetic animals, the animal that through language, memory, empathy, and will – including the will to transcend the temptations of the moment – can integrate sense perceptions with intricate systems of knowledge, awareness, and morality, as well as with the as yet unknown and unravelled realms of mystery and wonder. Indeed, it is out of these unique capacities that our ultimate salvation must emerge. The predicaments that we face, however, are immediate; we need to move quickly beyond all the structures we have created – political, socioeconomic, and technological – and evolve new criteria for human effort and cooperation.

It is not that humanity must sacrifice all its activities, knowledge, and institutional structures, or surrender all its achievements and start all over again in a clean, new state. Evolution does not take such form. It is rather that our view of which values and purposes should inform our actions and institutions must be consciously re-viewed; and, wherever choices are called for, these should be exercised. We have the capacity to exercise such choices. Maybe some small technological 'breakthrough' in one field or another will again lull us to sleep. But we now know that all such breakthroughs are temporary and cannot take the place of a fundamental restructuring of society. Gadgets may temporarily overpower the mind, but ultimately the mind must come into its own and address itself to the moral questions of life.

The point is that every few hundred years a new challenge presents itself. And each time, it calls for new understanding and a new paradigm of action. Ours is one such moment in the history of humankind and the universe.

The Ethics of Sustainable Development

Contemporary concern for sustainable development is an authentic moral concern to the degree that it poses an alternative to the dominant model of modern development. Its moral significance lies not in the specialized concerns of experts and counter-experts (whether they be professional ethicists, scientists, or technicians), but in a vision of a new way of life that is at once comprehensible and accessible to all human beings.

One can identify four primary criteria for sustainable development when it is conceived as an ethical ideal: a holistic view of development; equity based on the autonomy and self-reliance of diverse entities instead of on a structure of dependence founded on aid and transfer of technology with a view to 'catching up'; an emphasis on participation; and an accent on the importance of local conditions and the value of diversity. To these we must add two still more basic concerns, or rather two broad considerations that should inform all our concerns. One is a fundamentally normative perspective on the future, particularly from the viewpoint of the coming generations for whom we are responsible. The other is a cosmic view of life as sacred.

The report of the World Commission on Environment and Development, *Our Common Future*,[8] is misleading if it suggests that the so-called 'underdeveloped' countries can experience the life they see on Western television programmes without further degradation of the global environment. 'Our common future' cannot lie in an affluence that is ecologically suicidal, and socially and economically exclusive. It can, and must, lie in a curtailment of wants, as Gandhi constantly reminded his countrymen and others.[9] We have more than enough empirical evidence that the destruction of the biosphere lies first and foremost in the wasteful lifestyles of the world's privileged groups, and that the problem of poverty emanates from this same source. Consumption, as an end in itself, excludes the rights of others, both because it makes heavy demands on resources, but also because, in self-gratification, it is blind to others' needs.

The moral approach to development suggested here also involves a certain understanding of how we can most authentically know and relate to the rest of nature. The presumption that the role of science and technology was to develop nature in the service of humankind has turned out to be an illusion. It was based on a view of science itself as an instrument of human power over nature, other men and women, other forms of life, and all the qualities of being that constitute the cosmic order. This must give place to the original purpose of science, namely, seeking to understand the mysteries of nature with a deep sense of humility and wonder. True science is practised by persons with a fundamental philosophical scepticism about the scope and limits of human knowledge, who never for a moment assume that all is knowable and that secular knowledge provides the key to "mastering" the universe. Such a moral vision will make for a partnership between science and nature, and – equally vital – between scientists and all peoples whose lives are rooted in the wisdom of their ancestors. There is a vast area of research and development that lies ahead in this field. The scientist will have to take on a more modest role as a participant in a total system of relationships. As was stressed earlier, one of the basic postulates of an alternative philosophy of development is to treat life as a whole and not in fragments. This calls for a perspective on science that is oriental rather than occidental, feminist rather than *macho*, rural rather than urban, one that draws on the accumulated wisdom of centuries (each succeeding century and generation refining the inherited pool) rather than one that rejects all that is past and traditional.

The shift to sustainable development is primarily an ethical shift. It is not a technological fix, nor a matter of new financial investment. It is a shift in values such that nature is valued in itself and for its life support functions, not merely for how it can be converted into resources and commodities to feed the engine of economic growth. Respect for nature's diversity, and the

responsibility to conserve that diversity, define sustainable development as an ethical ideal. Out of an ethics of respect for nature's diversity flows a respect for the diversity of cultures and livelihoods, the basis not only of sustainability, but also of justice and equity. The ecological crisis is in large part a matter of treating nature's diversity as dispensable, a process that has gone hand in hand with the view that a large portion of the human species is dispensable as well. To reverse the ecological decline we require an ethical shift that treats all life as indispensable.

Notes

1   R. Kothari, *Transformation and Survival: In Search of a Humane World Order* (New Delhi: Ajanta, 1988).

2   S. Bahuguna, ed., *Chipko Message* (Chipko Information Centre, Parvatiya Navjivan Mandal, P.O. Silyara, Tehri Gharwal, 1984); Core Group of the United Nations University's Major Project on Peace and Global Transformation, *The Last Child* (Delhi: Lokvani, 1990).

3   O. Lyons, "An Iroquois Perspective," in *American Indian Environments: Ecological Issues in Native American History*, ed. C. Vecsey and R. W. Venables (Syracuse, N.Y.: Syracuse University Press, 1980).

4   A. Leopold, *A Sand County Almanac, and Sketches Here and There* (New York: Oxford University Press, 1949).

5   J. Lovelock, *Gaia: A New Look at Life on Earth* (Oxford: Oxford University Press, 1979).

6   P. Singer, *Animal Liberation*, 2nd ed. (New York: Random House, 1990).

7   C. D. Stone, *Should Trees Have Standing? Toward Legal Rights for Natural Objects* (Los Altos, Calif.: William Kaufmann, 1974).

8   World Commission on Environment and Development, *Our Common Future* (Oxford: Oxford University Press, 1987).

9   J. D. Sethi, *Gandhi Today* (New Delhi: Vikas, 1978).

37

The Conceptual Foundations
of the Land Ethic

J. Baird Callicott

The two great cultural advances of the past century were the Darwinian theory and the development of geology . . . Just as important, however, as the origin of plants, animals, and soil is the question of how they operate as a community. That task has fallen to the new science of ecology, which is daily uncovering a web of interdependencies so intricate as to amaze – were he here – even Darwin himself, who, of all men, should have least cause to tremble before the veil.

(Aldo Leopold, fragment 6B16, no. 36, Leopold Papers,
University of Wisconsin – Madison Archives)

As Wallace Stegner observes, *A Sand County Almanac* is considered "almost a holy book in conservation circles," and Aldo Leopold a prophet, "an American Isaiah." And as Curt Meine points out, "The Land Ethic" is the climactic essay of *Sand County*, "the upshot of 'The Upshot.'"[1] One might, therefore, fairly say that the recommendation and justification of moral obligations on the part of people to nature is what the prophetic *A Sand County Almanac* is all about.

But, with few exceptions, "The Land Ethic" has not been favorably received by contemporary academic philosophers. Most have ignored it. Of those who have not, most have been either nonplussed or hostile. Distinguished Australian philosopher John Passmore dismissed it out of hand, in the first book-length academic discussion of the new philosophical subdiscipline called "environmental ethics."[2] In a more recent and more deliberate discussion, the equally distinguished Australian philosopher H. J. McCloskey patronized Aldo Leopold and saddled "The Land Ethic" with various far-fetched "interpretations." He concludes that "there is a real problem in attributing a coherent meaning to Leopold's statements, one that exhibits his land ethic as representing a major advance in ethics rather than a retrogression to a morality of a kind held by various primitive peoples."[3] Echoing McCloskey, English philosopher Robin Attfield went out of his way to impugn the philosophical respectability of "The Land Ethic." And Canadian philosopher L. W. Sumner has called it "dangerous nonsense."[4] Among those philosophers more favorably disposed, "The Land Ethic" has usually been simply quoted, as if it were little more

From J. Baird Callicott, *Companion to a Sand County Almanac* (Madison: University of Wisconsin Press, 1987), pp. 186–217.

than a noble, but naive, moral plea, altogether lacking a supporting theoretical framework – i.e., foundational principles and premises which lead, by compelling argument, to ethical precepts.

The professional neglect, confusion, and (in some cases) contempt for "The Land Ethic" may, in my judgment, be attributed to three things: (1) Leopold's extremely condensed prose style in which an entire conceptual complex may be conveyed in a few sentences, or even in a phrase or two; (2) his departure from the assumptions and paradigms of contemporary philosophical ethics; and (3) the unsettling practical implications to which a land ethic appears to lead. "The Land Ethic," in short, is, from a philosophical point of view, abbreviated, unfamiliar, and radical.

Here I first examine and elaborate the compactly expressed abstract elements of the land ethic and expose the "logic" which binds them into a proper, but revolutionary, moral theory. I then discuss the controversial features of the land ethic and defend them against actual and potential criticism. I hope to show that the land ethic cannot be ignored as merely the groundless emotive exhortations of a moonstruck conservationist or dismissed as entailing wildly untoward practical consequences. It poses, rather, a serious intellectual challenge to business-as-usual moral philosophy.

"The Land Ethic" opens with a charming and poetic evocation of Homer's Greece, the point of which is to suggest that today land is just as routinely and remorsely enslaved as human beings then were. A panoramic glance backward to our most distant cultural origins, Leopold suggests, reveals a slow but steady moral development over three millennia. More of our relationships and activities ("fields of conduct") have fallen under the aegis of moral principles ("ethical criteria") as civilization has grown and matured. If moral growth and development continue, as not only a synoptic review of history, but recent past experience suggest that it will, future generations will censure today's casual and universal environmental bondage as today we censure the casual and universal human bondage of three thousand years ago.

A cynically inclined critic might scoff at Leopold's sanguine portrayal of human history. Slavery survived as an institution in the "civilized" West, more particularly in the morally self-congratulatory United States, until a mere generation before Leopold's own birth. And Western history from imperial Athens and Rome to the Spanish Inquisition and the Third Reich has been a disgraceful series of wars, persecutions, tyrannies, pogroms, and other atrocities.

The history of moral practice, however, is not identical with the history of moral consciousness. Morality is not descriptive; it is prescriptive or normative. In light of this distinction, it is clear that today, despite rising rates of violent crime in the United States and institutional abuses of human rights in Iran, Chile, Ethiopia, Guatemala, South Africa, and many other places, and despite persistent organized social injustice and oppression in still others, moral consciousness is expanding more rapidly now than ever before. Civil rights, human rights, women's liberation, children's liberation, animal liberation, etc., all indicate, as expressions of newly emergent moral ideals, that ethical consciousness (as distinct from practice) has if anything recently accelerated – thus confirming Leopold's historical observation.

Leopold next points out that "this extension of ethics, so far studied only by philosophers" – and therefore, the implication is clear, not very satisfactorily studied – "is actually a process in ecological evolution" (202). What Leopold is saying here, simply, is that we may understand the history of ethics, fancifully alluded to by means of the Odysseus vignette, in biological as well as philosophical terms. From a biological point of view, an ethic is "a limitation on freedom of action in the struggle for existence" (202).

I had this passage in mind when I remarked that Leopold manages to convey a whole network of ideas in a couple of phrases. The phrase "struggle for existence" unmistakably calls to mind Darwinian evolution as the conceptual context in which a biological account of the origin and development of ethics must ultimately be located. And at once it points up a paradox: Given the unremitting competitive "struggle for existence" how could "limitations on freedom of action" ever have been conserved and spread through a population of *Homo sapiens* or their evolutionary progenitors?

[. . .]

Let me put the problem in perspective. How, we are asking, did ethics originate and, once in existence, grow in scope and complexity?

The oldest answer in living human memory is theological. God (or the gods) imposes morality on people. And God (or the gods) sanctions it. A most vivid and graphic example of this kind of account occurs in the Bible when Moses goes up on Mount Sinai to receive the Ten Commandments directly from God. That text also clearly illustrates the divine sanctions (plagues, pestilences, droughts, military defeats, etc.) for moral disobedience. Ongoing revelation of the divine will, of course, as handily and as simply explains subsequent moral growth and development.

Western philosophy, on the other hand, is almost unanimous in the opinion that the origin of ethics in human experience has somehow to do with human reason. Reason figures centrally and pivotally in the "social contract theory" of the origin and nature of morals in all its ancient, modern, and contemporary expressions from Protagoras, to Hobbes, to Rawls. Reason is the wellspring of virtue, according to both Plato and Aristotle, and of categorical imperatives, according to Kant. In short, the weight of Western philosophy inclines to the view that we are moral beings because we are rational beings. The ongoing sophistication of reason and the progressive illumination it sheds upon the good and the right explain "the ethical sequence," the historical growth and development of morality, noticed by Leopold.

An evolutionary natural historian, however, cannot be satisfied with either of these general accounts of the origin and development of ethics. The idea that God gave morals to man is ruled out in principle – as any supernatural explanation of a natural phenomenon is ruled out in principle in natural science. And while morality might *in principle* be a function of human reason (as, say, mathematical calculation clearly is), to suppose that it is so *in fact* would be to put the cart before the horse. Reason appears to be a delicate, variable, and recently emerged faculty. It cannot, under any circumstances, be

supposed to have evolved in the absence of complex linguistic capabilities which depend, in turn, for their evolution upon a highly developed social matrix. But we cannot have become social beings unless we assumed limitations on freedom of action in the struggle for existence. Hence we must have become ethical before we became rational.

Darwin, probably in consequence of reflections somewhat like these, turned to a minority tradition of modern philosophy for a moral psychology consistent with and useful to a general evolutionary account of ethical phenomena. A century earlier, Scottish philosophers David Hume and Adam Smith had argued that ethics rest upon feelings or "sentiments" – which, to be sure, may be both amplified and informed by reason.[5] And since in the animal kingdom feelings or sentiments are arguably far more common or widespread than reason, they would be a far more likely starting point for an evolutionary account of the origin and growth of ethics.

Darwin's account, to which Leopold unmistakably (if elliptically) alludes in "The Land Ethic," begins with the parental and filial affections common, perhaps, to all mammals.[6] Bonds of affection and sympathy between parents and offspring permitted the formation of small, closely kin social groups, Darwin argued. Should the parental and filial affections bonding family members chance to extend to less closely related individuals, that would permit an enlargement of the family group. And should the newly extended community more successfully defend itself and/or more efficiently provision itself, the inclusive fitness of its members severally would be increased, Darwin reasoned. Thus, the more diffuse familial affections, which Darwin (echoing Hume and Smith) calls the "social sentiments," would be spread throughout a population.[7]

Morality, properly speaking – i.e., morality as opposed to mere altruistic instinct – requires, in Darwin's terms, "intellectual powers" sufficient to recall the past and imagine the future, "the power of language" sufficient to express "common opinion," and "habituation" to patterns of behavior deemed, by common opinion, to be socially acceptable and beneficial.[8] Even so, ethics proper, in Darwin's account, remains firmly rooted in moral feelings or social sentiments which were – no less than physical faculties, he expressly avers – naturally selected, by the advantages for survival and especially for successful reproduction, afforded by society.[9]

The protosociobiological perspective on ethical phenomena, to which Leopold as a natural historian was heir, leads him to a generalization which is remarkably explicit in his condensed and often merely resonant rendering of Darwin's more deliberate and extended paradigm: Since "the thing [ethics] has its origin in the tendency of interdependent individuals or groups to evolve modes of co-operation, . . . all ethics so far evolved rest upon a single premise: that the individual is a member of a community of interdependent parts" (202–3).

Hence, we may expect to find that the scope and specific content of ethics will reflect both the perceived boundaries and actual structure or organization of a cooperative community or society. *Ethics and society or community are correlative.* This single, simple principle constitutes a powerful tool for the analysis of moral natural history, for the anticipation of future moral development (including, ultimately, the land ethic), and for systematically deriving the specific precepts, the prescriptions and proscriptions, of an emergent and culturally unprecedented ethic like a land or environmental ethic.

Anthropological studies of ethics reveal that in fact the boundaries of the moral community are generally coextensive with the perceived boundaries of society.[10] And the peculiar (and, from the urbane point of view, sometimes inverted) representation of virtue and vice in tribal society – the virtue, for example, of sharing to the point of personal destitution and the vice of privacy and private property – reflects and fosters the life way of tribal peoples.[11] Darwin, in his leisurely, anecdotal discussion, paints a vivid picture of the intensity, peculiarity, and sharp circumscription of "savage" mores: "A savage will risk his life to save that of a member of the same community, but will be wholly indifferent about a stranger."[12] As Darwin portrays them, tribespeople are at once paragons of virtue "within the limits of the same tribe" and enthusiastic thieves, manslaughterers, and torturers without.[13]

For purposes of more effective defense against common enemies, or because of increased population density, or in response to innovations in subsistence methods and technologies, or for some mix of these or other forces, human societies have grown in extent or scope and changed in form or structure. Nations – like the Iroquois nation or the Sioux nation – came into being upon the merger of previously separate and mutually hostile tribes. Animals and plants were domesticated and erstwhile hunter-gatherers became herders and farmers. Permanent habitations were established. Trade, craft, and (later) industry flourished. With each change in society came corresponding and correlative changes in ethics. The moral community expanded to become coextensive with the newly drawn boundaries of societies and the representation of virtue and vice, right and wrong, good and evil, changed to accommodate, foster, and preserve the economic and institutional organization of emergent social orders.

Today we are witnessing the painful birth of a human super-community, global in scope. Modern transportation and communication technologies, international economic interdependencies, international economic entities, and nuclear arms have brought into being a "global village." It has not yet become fully formed and it is at tension – a very dangerous tension – with its predecessor, the nation-state. Its eventual institutional structure, a global federalism or whatever it may turn out to be, is, at this point, completely unpredictable. Interestingly, however, a corresponding global human ethic – the "human rights" ethic, as it is popularly called – has been more definitely articulated.

Most educated people today pay lip service at least to the ethical precept that all members of the human species, regardless of race, creed, or national origin, are endowed with certain fundamental rights which it is wrong not to respect. According to the evolutionary scenario set out by Darwin, the contemporary moral ideal of human rights is a response to a perception – however vague and indefinite – that mankind worldwide is united into one society, one community – however indeterminate or yet institutionally unorganized. As Darwin presciently wrote:

As man advances in civilization, and small tribes are united into larger communities, the simplest reason would tell each individual that he ought to extend his social instincts and sympathies to all the members of the same nation, though personally unknown to him. This point being once reached, there is only an artificial barrier to prevent his sympathies extending to the men of all nations and races. If, indeed, such men are separated from him by great differences of appearance or habits, experience unfortunately shows us how long it is, before we look at them as our fellow-creatures.[14]

According to Leopold, the next step in this sequence beyond the still incomplete ethic of universal humanity, a step that is clearly discernible on the horizon, is the land ethic. The "community concept" has, so far, propelled the development of ethics from the savage clan to the family of man. "The land ethic simply enlarges the boundary of the community to include soils, waters, plants, and animals, or collectively: the land" (204).

As the foreword to *Sand County* makes plain, the overarching thematic principle of the book is the inculcation of the idea – through narrative description, discursive exposition, abstractive generalization, and occasional preachment – "that land is a community" (viii). The community concept is "the basic concept of ecology" (viii). Once land is popularly perceived as a biotic community – as it is professionally perceived in ecology – a correlative land ethic will emerge in the collective cultural consciousness.

Although anticipated as far back as the mid-eighteenth century – in the notion of an "economy of nature" – the concept of the biotic community was more fully and deliberately developed as a working model or paradigm for ecology by Charles Elton in the 1920s.[15] The natural world is organized as an intricate corporate society in which plants and animals occupy "niches," or as Elton alternatively called them, "roles" or "professions," in the economy of nature.[16] As in a feudal community, little or no socio-economic mobility (upward or otherwise) exists in the biotic community. One is born to one's trade.

Human society, Leopold argues, is founded, in large part, upon mutual security and economic interdependency and preserved only by limitations on freedom of action in the struggle for existence – that is, by ethical constraints. Since the biotic community exhibits, as modern ecology reveals, an analogous structure, it too can be preserved, given the newly amplified impact of "mechanized man," only by analogous limitations on freedom of action – that is, by a land ethic (viii). A land ethic, furthermore, is not only "an ecological necessity," but an "evolutionary possibility" because a moral response to the natural environment – Darwin's social sympathies, sentiments, and instincts translated and codified into a body of principles and precepts – would be automatically triggered in human beings by ecology's social representation of nature (203).

Therefore, the key to the emergence of a land ethic is, simply, universal ecological literacy.

The land ethic rests upon three scientific cornerstones: (1) evolutionary and (2) ecological biology set in a background of (3) Copernican astronomy. Evolutionary theory provides the conceptual link between ethics and social organization and development. It provides a sense of "kinship with fellow-creatures" as well, "fellow-voyagers" with us in the "odyssey of evolution" (109). It establishes a diachronic link between people and nonhuman nature.

Ecological theory provides a synchronic link – the community concept – a sense of social integration of human and nonhuman nature. Human beings, plants, animals, soils, and waters are "all interlocked in one humming community of cooperations and competitions, one biota."[17] The simplest reason, to paraphrase Darwin, should, therefore, tell each individual that he or she ought to extend his or her social instincts and sympathies to all the members of the biotic community though different from him or her in appearance or habits.

And although Leopold never directly mentions it in *A Sand County Almanac*, the Copernican perspective, the perception of the Earth as "a small planet" in an immense and utterly hostile universe beyond, contributes, perhaps subconsciously, but nevertheless very powerfully, to our sense of kinship, community, and interdependence with fellow denizens of the Earth household. It scales the Earth down to something like a cozy island paradise in a desert ocean.

Here in outline, then, are the conceptual and logical foundations of the land ethic: Its conceptual elements are a Copernican cosmology, a Darwinian protosociobiological natural history of ethics, Darwinian ties of kinship among all forms of life on Earth, and an Eltonian model of the structure of biocenoses all overlaid on a Humean-Smithian moral psychology. Its logic is that natural selection has endowed human beings with an affective moral response to perceived bonds of kinship and community membership and identity; that today the natural environment, the land, is represented as a community, the biotic community; and that, therefore, an environmental or land ethic is both possible – the biopsychological and cognitive conditions are in place – and necessary, since human beings collectively have acquired the power to destroy the integrity, diversity, and stability of the environing and supporting economy of nature. In the remainder of this essay I discuss special features and problems of the land ethic germane to moral philosophy.

The most salient feature of Leopold's land ethic is its provision of what Kenneth Goodpaster has carefully called "moral considerability" for the biotic community per se, not just for fellow members of the biotic community:[18]

> In short, a land ethic changes the role of *Homo sapiens* from conqueror of the land-community to plain member and citizen of it. It implies respect for his fellow-members, *and also respect for the community as such.* (204, emphasis added)

The land ethic, thus, has a holistic as well as an individualistic cast.

Indeed, as "The Land Ethic" develops, the focus of moral concern shifts gradually away from plants, animals, soils, and waters severally to the biotic community collectively. Toward the middle, in the subsection called Substitutes for a Land Ethic, Leopold invokes the "biotic rights" of *species* – as the context indicates – of wildflowers, songbirds, and predators. In The

Outlook, the climactic section of "The Land Ethic," nonhuman natural entities, first appearing as fellow members, then considered in profile as species, are not so much as mentioned in what might be called the "summary moral maxim" of the land ethic: "A thing is right when it tends to preserve the integrity, stability, and beauty of the biotic community. It is wrong when it tends otherwise" (224–25).

By this measure of right and wrong, not only would it be wrong for a farmer, in the interest of higher profits, to clear the woods off a 75 percent slope, turn his cows into the clearing, and dump its rainfall, rocks, and soil into the community creek, it would also be wrong for the federal fish and wildlife agency, in the interest of individual animal welfare, to permit populations of deer, rabbits, feral burros, or whatever to increase unchecked and thus to threaten the integrity, stability, and beauty of the biotic communities of which they are members. The land ethic not only provides moral considerability for the biotic community per se, but ethical consideration of its individual members is preempted by concern for the preservation of the integrity, stability, and beauty of the biotic community. The land ethic, thus, not only has a holistic aspect; it is holistic with a vengeance.

The holism of the land ethic, more than any other feature, sets it apart from the predominant paradigm of modern moral philosophy. It is, therefore, the feature of the land ethic which requires the most patient theoretical analysis and the most sensitive practical interpretation.

As Kenneth Goodpaster pointed out, mainstream modern ethical philosophy has taken egoism as its point of departure and reached a wider circle of moral entitlement by a process of generalization:[19] I am sure that *I*, the enveloped ego, am intrinsically or inherently valuable and thus that *my* interests ought to be considered, taken into account, by "others" when their actions may substantively affect *me*. My own claim to moral consideration, according to the conventional wisdom, ultimately rests upon a psychological capacity – rationality or sentiency were the classical candidates of Kant and Bentham, respectively – which is arguably valuable in itself and which thus qualifies *me* for moral standing.[20] However, then I am forced grudgingly to grant the same moral consideration I demand from others, on this basis, to those others who can also claim to possess the same general psychological characteristic.

A *criterion* of moral value and consideration is thus identified. Goodpaster convincingly argues that mainstream modern moral theory is based, when all the learned dust has settled, on this simple paradigm of ethical justification and logic exemplified by the Benthamic and Kantian prototypes.[21] If the criterion of moral values and consideration is pitched low enough – as it is in Bentham's criterion of sentiency – a wide variety of animals are admitted to moral entitlement.[22] If the criterion of moral value and consideration is pushed lower still – as it is in Albert Schweitzer's reverence-for-life ethic – all minimally conative things (plants as well as animals) would be extended moral considerability.[23] The contemporary animal liberation/rights, and reverence-for-life/life-principle ethics are, at bottom, simply direct applications of the modern classical paradigm of moral argument. But this standard modern model of ethical theory provides no possibility whatever for the moral consideration of wholes – of threatened *populations* of animals and plants, or of endemic, rare, or endangered *species*, or of biotic *communities*, or most expansively, of the *biosphere* in its totality – since wholes per se have no psychological experience of any kind.[24] Because mainstream modern moral theory has been "psychocentric," it has been radically and intractably individualistic or "atomistic" in its fundamental theoretical orientation.

Hume, Smith, and Darwin diverged from the prevailing theoretical model by recognizing that altruism is as fundamental and autochthonous in human nature as is egoism. According to their analysis, moral value is not identified with a natural quality objectively present in morally considerable beings – as reason and/or sentiency is objectively present in people and/or animals – it is, as it were, projected by valuing subjects.[25]

Hume and Darwin, furthermore, recognize inborn moral sentiments which have society as such as their natural object. Hume insists that "we must renounce the theory which accounts for every moral sentiment by the principle of self-love.

We must adopt a more *public affection* and allow that the *interests of society* are not, *even on their own account*, entirely indifferent to us."[26] And Darwin, somewhat ironically (since "Darwinian evolution" very often means natural selection operating exclusively with respect to individuals), sometimes writes as if morality had no other object than the commonweal, the welfare of the community as a corporate entity:

> We have now seen that actions are regarded by savages, and were probably so regarded by primeval man, as good or bad, solely as they obviously affect the welfare of the tribe, – not that of the species, nor that of the individual member of the tribe. This conclusion agrees well with the belief that the so-called moral sense is aboriginally derived from social instincts, for both relate at first exclusively to the community.[27]

Theoretically then, the biotic community owns what Leopold, in the lead paragraph of The Outlook, calls "value in the philosophical sense" – i.e., direct moral considerability – because it is a newly discovered proper object of a specially evolved "public affection" or "moral sense" which all psychologically normal human beings have inherited from a long line of ancestral social primates (223).[28]

In the land ethic, as in all earlier stages of social-ethical evolution, there exists a tension between the good of the community as a whole and the "rights" of its individual members considered severally. While The Ethical Sequence section of "The Land Ethic" clearly evokes Darwin's classical biosocial account of the origin and extension of morals, Leopold is actually more explicitly concerned, in that section, with the interplay between the holistic and individualistic moral sentiments – between sympathy and fellow-feeling on the one hand, and public affection for the commonweal on the other:

> The first ethics dealt with the relation between individuals; the Mosaic Decalogue is an example. Later accretions dealt with the relation between the individual and society. The Golden Rule tries to integrate the individual to society;

democracy to integrate social organization to the individual. (202–3)

Actually, it is doubtful that the first ethics dealt with the relation between individuals and not at all with the relation between the individual and society. (This, along with the remark that ethics replaced an "original free-for-all competition," suggests that Leopold's Darwinian line of thought has been uncritically tainted with Hobbesean elements. [202]. Of course, Hobbes's "state of nature," in which there prevailed a war of each against all, is absurd from an evolutionary point of view.) A century of ethnographic studies seems to confirm, rather, Darwin's conjecture that the relative weight of the holistic component is greater in tribal ethics – the tribal ethic of the Hebrews recorded in the Old Testament constitutes a vivid case in point – than in more recent accretions. The Golden Rule, on the other hand, does not mention, in any of its formulations, society per se. Rather, its primary concern seems to be "others," i.e., other human individuals. Democracy, with its stress on individual liberties and rights, seems to further rather than countervail the individualistic thrust of the Golden Rule.

In any case, the conceptual foundations of the land ethic provide a well-formed, self-consistent theoretical basis for including both fellow members of the biotic community and the biotic community itself (considered as a corporate entity) within the purview of morals. The preemptive emphasis, however, on the welfare of the community as a whole, in Leopold's articulation of the land ethic, while certainly *consistent* with its Humean-Darwinian theoretical foundations, is not *determined* by them alone. The overriding holism of the land ethic results, rather, more from the way our moral sensibilities are informed by ecology.

Ecological thought, historically, has tended to be holistic in outlook.[29] Ecology is the study of the *relationships* of organisms to one another and to the elemental environment. These relationships bind the *relata* – plants, animals, soils, and waters – into a seamless fabric. The ontological primacy of objects and the ontological subordination of relationships, characteristic of classical Western science, is, in fact, reversed in ecology.[30] Ecological relationships determine the nature of organisms rather than the other way around. A species is what it is because it has adapted to a niche in the ecosystem. The whole, the system itself, thus, literally and quite straightforwardly shapes and forms its component parts.

Antedating Charles Elton's community model of ecology was F. E. Clements' and S. A. Forbes' organism model.[31] Plants and animals, soils and waters, according to this paradigm, are integrated into one superorganism. Species are, as it were, its organs; specimens its cells. Although Elton's community paradigm (later modified, as we shall see, by Arthur Tansley's ecosystem idea) is the principal and morally fertile ecological concept of "The Land Ethic," the more radically holistic superorganism paradigm of Clements and Forbes resonates in "The Land Ethic" as an audible overtone. In the peroration of Land Health and the A-B Cleavage, for example, which immediately precedes The Outlook, Leopold insists that

in all of these cleavages, we see repeated the same basic paradoxes: man the conqueror *versus* man the biotic citizen; science the sharpener of his sword *versus* science the searchlight on his universe; land the slave and servant *versus* land the collective organism. (223)

And on more than one occasion Leopold, in the latter quarter of "The Land Ethic," talks about the "health" and "disease" of the land – terms which are at once descriptive and normative and which, taken literally, characterize only organisms proper.

In an early essay, "Some Fundamentals of Conservation in the Southwest," Leopold speculatively flirted with the intensely holistic superorganism model of the environment as a paradigm pregnant with moral implications:

It is at least not impossible to regard the earth's parts – soil, mountains, rivers, atmosphere, etc. – as organs or parts of organs, of *a coordinated whole*, each part with a definite function. And if we could see *this whole, as a whole*, through a great period of time, we might perceive not only organs with coordinated functions, but possibly also that process of consumption and replacement which in biology we call metabolism, or growth. In such a case we would have all the visible attributes of a living thing, which we do not realize to be such because it is too big, and its life processes too slow. And there would also follow that invisible attribute – a soul or consciousness – which . . . many philosophers of all ages ascribe to all living things and aggregates thereof, including the "dead" earth.

Possibly in our intuitive perceptions, which may be truer than our science and less impeded by words than our philosophies, we realize the indivisibility of the earth – its soil, mountains, rivers, forests, climate, plants, and animals – and *respect it collectively* not only as a useful servant but as a living being, vastly less alive than ourselves, but vastly greater than ourselves in time and space . . . Philosophy, then, suggests one reason why we cannot destroy the earth with moral impunity; namely, that the "dead" earth is an organism possessing a certain kind and degree of life, which we intuitively respect as such.[32]

Had Leopold retained this overall theoretical approach in "The Land Ethic," the land ethic would doubtless have enjoyed more critical attention from philosophers. The moral foundations of a land or, as he might then have called it, "earth" ethic, would rest upon the hypothesis that the Earth is alive and ensouled – possessing inherent psychological characteristics, logically parallel to reason and sentiency. This notion of a conative whole Earth could plausibly have served as a general criterion of intrinsic worth and moral considerability, in the familiar format of mainstream moral thought.

Part of the reason, therefore, that "The Land Ethic" emphasizes more and more the integrity, stability, and beauty of the environment as a whole, and less and less the "biotic right" of individual plants and animals to life, liberty, and the pursuit of happiness, is that the superorganism ecological paradigm invites one, much more than does the community paradigm, to hypostatize, to reify the whole, and to subordinate its individual members.

In any case, as we see, rereading "The Land Ethic" in light of "Some Fundamentals," the whole Earth organism image of nature is vestigially present in Leopold's later thinking. Leopold may have abandoned the "earth ethic" because ecology had abandoned the organism analogy,

in favor of the community analogy, as a working theoretical paradigm. And the community model was more suitably given moral implications by the social/sentimental ethical natural history of Hume and Darwin.

Meanwhile, the biotic community ecological paradigm itself had acquired, by the late thirties and forties, a more holistic cast of its own. In 1935 British ecologist Arthur Tansley pointed out that from the perspective of physics the "currency" of the "economy of nature" is energy.[33] Tansley suggested that Elton's qualitative and descriptive food chains, food webs, trophic niches, and biosocial professions could be quantitatively expressed by means of a thermodynamic flow model. It is Tansley's state-of-the-art thermodynamic paradigm of the environment that Leopold explicitly sets out as a "mental image of land" in relation to which "we can be ethical" (214). And it is the ecosystemic model of land which informs the cardinal practical precepts of the land ethic.

The Land Pyramid is the pivotal section of "The Land Ethic" – the section which effects a complete transition from concern for "fellowmembers" to the "community as such." It is also its longest and most technical section. A description of the "ecosystem" (Tansley's deliberately nonmetaphorical term) begins with the sun. Solar energy "flows through a circuit called the biota" (215). It enters the biota through the leaves of green plants and courses through plant-eating animals, and then on to omnivores and carnivores. At last the tiny fraction of solar energy converted to biomass by green plants remaining in the corpse of a predator, animal feces, plant detritus, or other dead organic material is garnered by decomposers – worms, fungi, and bacteria. They recycle the participating elements and degrade into entropic equilibrium any remaining energy. According to this paradigm

> land, then, is not merely soil; it is a fountain of energy flowing through a circuit of soils, plants, and animals. Food chains are the living channels which conduct energy upward; death and decay return it to the soil. The circuit is not closed; . . . but it is a sustained circuit, like a slowly augmented revolving fund of life. (216)

In this exceedingly abstract (albeit poetically expressed) model of nature, process precedes substance and energy is more fundamental than matter. Individual plants and animals become less autonomous beings than ephemeral structures in a patterned flux of energy. According to Yale biophysicist Harold Morowitz,

> viewed from the point of view of modern [ecology], each living thing... is a dissipative structure, that is it does not endure in and of itself but only as a result of the continual flow of energy in the system. An example might be instructive. Consider a vortex in a stream of flowing water. The vortex is a structure made of an ever-changing group of water molecules. It does not exist as an entity in the classical Western sense; it exists only because of the flow of water through the stream. In the same sense, the structures out of which biological entities are made are transient, unstable entities with constantly changing molecules, dependent on a constant flow of energy from food in order to maintain form and structure.... From this point of view the reality of individuals is problematic because they do not exist per se but only as local perturbations in this universal energy flow.[34]

Though less bluntly stated and made more palatable by the unfailing charm of his prose, Leopold's proffered mental image of land is just as expansive, systemic, and distanced as Morowitz'. The maintenance of "the complex structure of the land and its smooth functioning as an energy unit" emerges in The Land Pyramid as the *summum bonum* of the land ethic (216).

From this good Leopold derives several practical principles slightly less general, and therefore more substantive, than the summary moral maxim of the land ethic distilled in The Outlook. "The trend of evolution [not its "goal," since evolution is ateleological] is to elaborate and diversify the biota" (216). Hence, among our cardinal duties is the duty to preserve what species we can, especially those at the apex of the pyramid – the top carnivores. "In the beginning, the pyramid of life was low and squat; the food chains short and simple. Evolution has added layer after layer, link after link" (215–16). Human activities today, especially those, like systematic deforestation in the tropics, resulting in abrupt massive extinctions

of species, are in effect "devolutionary"; they flatten the biotic pyramid; they choke off some of the channels and gorge others (those which terminate in our own species).[35]

The land ethic does not enshrine the ecological status quo and devalue the dynamic dimension of nature. Leopold explains that "evolution is a long series of self-induced changes, the net result of which has been to elaborate the flow mechanism and to lengthen the circuit. Evolutionary changes, however, are usually slow and local. Man's invention of tools has enabled him to make changes of unprecedented violence, rapidity, and scope" (216–17). "Natural" species extinction, i.e., species extinction in the normal course of evolution, occurs when a species is replaced by competitive exclusion or evolves into another form.[36] Normally speciation outpaces extinction. Mankind inherited a richer, more diverse world than had ever existed before in the 3.5 billion-year odyssey of life on Earth.[37] What is wrong with anthropogenic species extirpation and extinction is the *rate* at which it is occurring and the *result*: biological impoverishment instead of enrichment.

Leopold goes on here to condemn, in terms of its impact on the ecosystem, "the worldwide pooling of faunas and floras," i.e., the indiscriminate introduction of exotic and domestic species and the dislocation of native and endemic species; mining the soil for its stored biotic energy, leading ultimately to diminished fertility and to erosion; and polluting and damming water courses (217).

According to the land ethic, therefore: Thou shalt not extirpate or render species extinct; thou shalt exercise great caution in introducing exotic and domestic species into local ecosystems, in extracting energy from the soil and releasing it into the biota, and in damming or polluting water courses; and thou shalt be especially solicitous of predatory birds and mammals. Here in brief are the express moral precepts of the land ethic. They are all explicitly informed – not to say derived – from the energy circuit model of the environment.

The living channels – "food chains" – through which energy courses are composed of individual plants and animals. A central, stark fact lies at the heart of ecological processes: Energy, the currency of the economy nature, passes from one organism to another, not from hand to hand, like coined money, but, so to speak, from stomach to stomach. Eating *and being eaten*, living *and dying* are what make the biotic community hum.

The precepts of the land ethic, like those of all previous accretions, reflect and reinforce the structure of the community to which it is correlative. Trophic asymmetries constitute the kernel of the biotic community. It seems unjust, unfair. But that is how the economy of nature is organized (and has been for thousands of millions of years). The land ethic, thus, affirms as good, and strives to preserve, the very inequities in nature whose social counterparts in human communities are condemned as bad and would be eradicated by familiar social ethics, especially by the more recent Christian and secular egalitarian exemplars. A "right to life" for individual members is not consistent with the structure of the biotic community and hence is not mandated by the land ethic. This disparity between the land ethic and its more familiar social precedents contributes to the apparent devaluation of individual *members* of the biotic community and augments and reinforces the tendency of the land ethic, driven by the systemic vision of ecology, toward a more holistic or community-per-se orientation.

Of the few moral philosophers who have given the land ethic a moment's serious thought, most have regarded it with horror because of its emphasis on the good of the community and its deemphasis on the welfare of individual members of the community. Not only are other sentient creatures members of the biotic community and subordinate to its integrity, beauty, and stability; so are *we*. Thus, if it is not only morally permissible, from the point of view of the land ethic, but morally required, that members of certain species be abandoned to predation and other vicissitudes of wild life or even deliberately culled (as in the case of alert and sentient whitetail deer) for the sake of the integrity, stability, and beauty of the biotic community, how can we consistently exempt ourselves from a similar draconian regime? We too are only "plain members and citizens" of the biotic community. And our global population is growing unchecked.

According to William Aiken, from the point of view of the land ethic, therefore, "massive human diebacks would be good. It is our duty to cause them. It is our species' duty, relative to the whole, to eliminate 90 percent of our numbers." Thus, according to Tom Regan, the land ethic is a clear case of "environmental fascism."[38]

Of course Leopold never intended the land ethic to have either inhumane or antihumanitarian implications or consequences. But whether he intended them or not, a logically consistent deduction from the theoretical premises of the land ethic might force such untoward conclusions. And given their magnitude and monstrosity, these derivations would constitute a *reductio ad absurdum* of the whole land ethic enterprise and entrench and reinforce our current human chauvinism and moral alienation from nature. If this is what membership in the biotic community entails, then all but the most radical misanthropes would surely want to opt out.

The land ethic, happily, implies neither inhumane nor inhuman consequences. That some philosophers think it must follows more from their own theoretical presuppositions than from the theoretical elements of the land ethic itself. Conventional modern ethical theory rests moral entitlement, as I earlier pointed out, on a criterion or qualification. If a candidate meets the criterion – rationality or sentiency are the most commonly posited – he, she, or it is entitled to equal moral standing with others who possess the same qualification in equal degree. Hence, reasoning in this philosophically orthodox way, and forcing Leopold's theory to conform: if human beings are, with other animals, plants, soils, and waters, equally members of the biotic community, and if community membership is the criterion of equal moral consideration, then not only do animals, plants, soils, and waters have equal (highly attenuated) "rights," but human beings are equally subject to the same subordination of individual welfare and rights in respect to the good of the community as a whole.

But the land ethic, as I have been at pains to point out, is heir to a line of moral analysis different from that institutionalized in contemporary moral philosophy. From the biosocial evolutionary analysis of ethics upon which Leopold builds the land ethic, it (the land ethic) neither replaces nor overrides previous accretions. Prior moral sensibilities and obligations attendant upon and correlative to prior strata of social involvement remain operative and preemptive.

Being citizens of the United States, or the United Kingdom, or the Soviet Union, or Venezuela, or some other nation-state, and therefore having national obligations and patriotic duties, does not mean that we are not also members of smaller communities or social groups – cities or townships, neighborhoods, and families – or that we are relieved of the peculiar moral responsibilities attendant upon and correlative to these memberships as well. Similarly, our recognition of the biotic community and our immersion in it does not imply that we do not also remain members of the human community – the "family of man" or "global village" – or that we are relieved of the attendant and correlative moral responsibilities of that membership, among them to respect universal human rights and uphold the principles of individual human worth and dignity. The biosocial development of morality does not grow in extent like an expanding balloon, leaving no trace of its previous boundaries, so much as like the circumference of a tree.[39] Each emergent, and larger, social unit is layered over the more primitive, and intimate, ones.

Moreover, as a general rule, the duties correlative to the inner social circles to which we belong eclipse those correlative to the rings farther from the heartwood when conflicts arise. Consider our moral revulsion when zealous ideological nationalists encourage children to turn their parents in to the authorities if their parents should dissent from the political or economic doctrines of the ruling party. A zealous environmentalist who advocated visiting war, famine, or pestilence on human populations (those existing somewhere else, of course) in the name of the integrity, beauty, and stability of the biotic community would be similarly perverse. Family obligations in general come before nationalistic duties and humanitarian obligations in general come before environmental duties. The land ethic, therefore, is not draconian or fascist. It does not cancel human morality. The land ethic

may, however, as with any new accretion, demand choices which affect, in turn, the demands of the more interior social-ethical circles. Taxes and the military draft may conflict with family-level obligations. While the land ethic, certainly, does not cancel human morality, neither does it leave it unaffected.

Nor is the land ethic inhumane. Nonhuman fellow members of the biotic community have no "human rights," because they are not, by definition, members of the human community. As fellow members of the biotic community, however, they deserve respect.

How exactly to express or manifest respect, while at the same time abandoning our fellow members of the biotic community to their several fates or even actively consuming them for our own needs (and wants), or deliberately making them casualties of wildlife management for ecological integrity, is a difficult and delicate question.

Fortunately, American Indian and other traditional patterns of human-nature interaction provide rich and detailed models. Algonkian woodland peoples, for instance, represented animals, plants, birds, waters, and minerals as other-than-human persons engaged in reciprocal, mutually beneficial socioeconomic intercourse with human beings.[40] Tokens of payment, together with expressions of apology, were routinely offered to the beings whom it was necessary for these Indians to exploit. Care not to waste the usable parts, and care in the disposal of unusable animal and plant remains, were also an aspect of the respectful, albeit necessarily consumptive, Algonkian relationship with fellow members of the land community. As I have more fully argued elsewhere, the Algonkian portrayal of human-nature relationships is, indeed, although certainly different in specifics, identical in abstract form to that recommended by Leopold in the land ethic.[41]

[. . .]

This brings us face to face with the paradox posed by Peter Fritzell:[42] Either we are plain members and citizens of the biotic community, on a par with other creatures, or we are not. If we are, then we have no moral obligations to our fellow members or to the community per se because, as understood from a modern scientific perspective, nature and natural phenomena are amoral. Wolves and alligators do no wrong in killing and eating deer and dogs (respectively). Elephants cannot be blamed for bulldozing acacia trees and generally wreaking havoc in their natural habitats. If human beings are natural beings, then human behavior, however destructive, is natural behavior and is as blameless, from a natural point of view, as any other behavioral phenomenon exhibited by other natural beings. On the other hand, we are moral beings, the implication seems clear, precisely to the extent that we are civilized, that we have removed ourselves from nature. We are more than natural beings; we are metanatural – not to say, "supernatural" – beings. But then our moral community is limited to only those beings who share our transcendence of nature, i.e., to human beings (and perhaps to pets who have joined our civilized community as surrogate persons) and to the human community. Hence, have it either way – we are members of the biotic community or we are not – a land or environmental ethic is aborted by either choice.

But nature is *not* amoral. The tacit assumption that we are deliberating, choice-making ethical beings only to the extent that we are metanatural, civilized beings, generates this dilemma. The biosocial analysis of human moral behavior, in which the land ethic is grounded, is designed precisely to show that in fact intelligent moral behavior *is* natural behavior. Hence, we are moral beings not in spite of, but in accordance with, nature. To the extent that nature has produced at least one ethical species, *Homo sapiens*, nature is not amoral.

Alligators, wolves, and elephants are not subject to reciprocal interspecies duties or land ethical obligations themselves because they are incapable of conceiving and/or assuming them. Alligators, as mostly solitary, entrepreneurial reptiles, have no apparent moral sentiments or social instincts whatever. And while wolves and elephants certainly do have social instincts and at least protomoral sentiments, as their social behavior amply indicates, their conception or imagination of community appears to be less culturally plastic than ours and less amenable to cognitive information. Thus, while we might regard them as ethical beings, they are not able, as we are, to form the concept of a universal biotic community, and hence conceive an all-inclusive, holistic land ethic.

The paradox of the land ethic, elaborately noticed by Fritzell, may be cast more generally

still in more conventional philosophical terms: Is the land ethic prudential or deontological? Is the land ethic, in other words, a matter of enlightened (collective, human) self-interest, or does it genuinely admit nonhuman natural entities and nature as a whole to true moral standing?

The conceptual foundations of the land ethic, as I have here set them out, and much of Leopold's hortatory rhetoric, would certainly indicate that the land ethic is deontological (or duty oriented) rather than prudential. In the section significantly titled The Ecological Conscience, Leopold complains that the then-current conservation philosophy is inadequate because "it defines no right or wrong, assigns no obligation, calls for no sacrifice, implies no change in the current philosophy of values. In respect of land-use, it urges *only* enlightened self-interest" (207–8, emphasis added). Clearly, Leopold himself thinks that the land ethic goes beyond prudence. In this section he disparages mere "self-interest" two more times, and concludes that "obligations have no meaning without conscience, and the problem we face is the extension of the social conscience from people to land" (209).

In the next section, Substitutes for a Land Ethic, he mentions rights twice – the "biotic right" of birds to continuance and the absence of a right on the part of human special interest to exterminate predators.

Finally, the first sentences of The Outlook read: "It is inconceivable to me that an ethical relation to land can exist without love, respect, and admiration for land, and a high regard for its value. By value, I of course mean something far broader than mere economic value; I mean value in the philosophical sense" (223). By "value in the philosophical sense," Leopold can only mean what philosophers more technically call "intrinsic value" or "inherent worth."[43] Something that has intrinsic value or inherent worth is valuable in and of itself, not because of what it can do for us. "Obligation," "sacrifice," "conscience," "respect," the ascription of rights, and intrinsic value – all of these are consistently opposed to self-interest and seem to indicate decisively that the land ethic is of the deontological type.

Some philosophers, however, have seen it differently. Scott Lehmann, for example, writes,

Although Leopold claims for communities of plants and animals a "right to continued existence," his argument is homocentric, appealing to the human stake in preservation. Basically it is an argument from enlightened self-interest, where the self in question is not an individual human being but humanity – present and future – as a whole . . .[44]

Lehmann's claim has some merits, even though it flies in the face of Leopold's express commitments. Leopold does frequently lapse into the language of (collective, long-range, human) self-interest. Early on, for example, he remarks, "in human history, we have learned (I hope) that the conqueror role is eventually *self*-defeating" (204, emphasis added). And later, of the 95 percent of Wisconsin's species which cannot be "sold, fed, eaten, or otherwise put to economic use," Leopold reminds us that "these creatures are members of the biotic community, and if (as I believe) its stability depends on its integrity, they are entitled to continuance" (210). The implication is clear: the economic 5 percent cannot survive if a significant portion of the uneconomic 95 percent are extirpated; nor may *we*, it goes without saying, survive without these "resources."

Leopold, in fact, seems to be consciously aware of this moral paradox. Consistent with the biosocial foundations of his theory, he expresses it in sociobiological terms:

An ethic may be regarded as a mode of guidance for meeting ecological situations so new or intricate, or involving such deferred reactions, that the path of social expediency is not discernible to the average individual. Animal instincts are modes of guidance for the individual in meeting such situations. Ethics are possibly a kind of community instinct in-the-making. (203)

From an objective, descriptive sociobiological point of view, ethics evolve because they contribute to the inclusive fitness of their carriers (or, more reductively still, to the multiplication of their carriers' genes); they are expedient. However, the path to self-interest (or to the self-interest of the selfish gene) is not discernible to the participating individuals (nor, certainly, to their genes). Hence, ethics are grounded in instinctive feeling – love, sympathy, respect – not in self-conscious

calculating intelligence. Somewhat like the paradox of hedonism – the notion that one cannot achieve happiness if one directly pursues happiness per se and not other things – one can only secure self-interest by putting the interests of others on a par with one's own (in this case long-range collective human self-interest and the interest of other forms of life and of the biotic community per se).

So, is the land ethic deontological or prudential, after all? It is both – self-consistently both – depending upon point of view. From the inside, from the lived, felt point of view of the community member with evolved moral sensibilities, it is deontological. It involves an affective-cognitive posture of genuine love, respect, admiration, obligation, self-sacrifice, conscience, duty, and the ascription of intrinsic value and biotic rights. From the outside, from the objective and analytic scientific point of view, it is prudential. "There is no other way for land to survive the impact of mechanized man," nor, therefore, for mechanized man to survive his own impact upon the land (viii).

Notes

1   Wallace Stegner, "The Legacy of Aldo Leopold"; Curt Meine, "Building 'The Land Ethic'"; both in *Companion to a Sand County Almanac: Interpretive and Critical Essays* (Madison, WI: University of Wisconsin Press, 1987). The oft-repeated characterization of Leopold as a prophet appears traceable to Roberts Mann, "Aldo Leopold: Priest and Prophet," *American Forests* 60, no. 8 (August 1954): 23, 42–43; it was picked up, apparently, by Ernest Swift, "Aldo Leopold: Wisconsin's Conservationist Prophet," *Wisconsin Tales and Trails* 2, no. 2 (September 1961): 2–5; Roderick Nash institutionalized it in his chapter, "Aldo Leopold: Prophet," in *Wilderness and the American Mind* (New Haven: Yale University Press, 1967; revised edition, 1982).

2   John Passmore, *Man's Responsibility for* [significantly not "*to*"] *Nature: Ecological Problems and Western Traditions* (New York: Charles Scribner's Sons, 1974).

3   H. J. McCloskey, *Ecological Ethics and Politics* (Totowa, N. J.: Rowman and Littlefield, 1983), 56.

4   Robin Attfield, in "Value in the Wilderness," *Metaphilosophy* 15 (1984), writes, "Leopold the philosopher is something of a disaster, and I

dread the thought of the student whose concept of philosophy is modeled principally on these extracts. (Can value 'in the philosophical sense' be contrasted with instrumental value? If concepts of right and wrong did not apply to slaves in Homeric Greece, how could Odysseus suspect the slavegirls of 'misbehavior'? If all ethics rest on interdependence how are obligations to infants and small children possible? And how can 'obligations have no meaning without conscience,' granted that the notion of conscience is conceptually dependent on that of obligation?)" (294). L. W. Sumner, "Review of Robin Attfield, *The Ethics of Environmental Concern*," *Environmental Ethics* 8 (1986): 77.

5   See Adam Smith, *Theory of the Moral Sentiments* (London and Edinburgh: A Millar, A. Kinkaid, and J. Bell, 1759) and David Hume, *An Enquiry Concerning the Principles of Morals* (Oxford: The Clarendon Press, 1777, first published in 1751). Darwin cites both works in the key fourth chapter of *Descent* (pp. 106 and 109, respectively).

6   Darwin, *Descent*, 98ff.

7   Ibid., 105f.

8   Ibid., 113ff.

9   Ibid., 105.

10 See, for example, Elman R. Service, *Primitive Social Organization: An Evolutionary Perspective* (New York: Random House, 1962).

11 See Marshall Sahlins, *Stone Age Economics* (Chicago: Aldine Atherton, 1972).

12 Darwin, *Descent*, III.

13 Ibid., 117ff. The quoted phrase occurs on p. 118.

14 Ibid., 124.

15 See Donald Worster, *Nature's Economy: The Roots of Ecology* (San Francisco: Sierra Club Books, 1977).

16 Charles Elton, *Animal Ecology* (New York: Macmillan, 1927).

17 Aldo Leopold, *Round River* (New York: Oxford University Press, 1953), 148.

18 Kenneth Goodpaster, "On Being Morally Considerable," *Journal of Philosophy* 22 (1978): 308–25. Goodpaster wisely avoids the term *rights*, defined so strictly albeit so variously by philosophers, and used so loosely by nonphilosophers.

19 Kenneth Goodpaster, "From Egoism to Environmentalism" in *Ethics and Problems of the 21st Century*, ed. K. E. Goodpaster and K. M. Sayre (Notre Dame, Ind.: University of Notre Dame Press, 1979), 21–35.

20 See Immanuel Kant, *Foundations of the Metaphysics of Morals* (New York: Bobbs-Merrill, 1959; first published in 1785); and Jeremy Bentham, *An Introduction to the Principles of Morals and Legislation*, new edition (Oxford: The Clarendon Press, 1823).

21 Goodpaster, "Egoism to Environmentalism." Actually Goodpaster regards *Hume* and Kant as the cofountainheads of this sort of moral philosophy. But Hume does not reason in this way. For Hume, the other-oriented sentiments are as primitive as self-love.

22 See Peter Singer, *Animal Liberation: A New Ethics for Our Treatment of Animals* (New York: Avon Books, 1975) for animal liberation; and see Tom Regan, *All That Dwell Therein: Animal Rights and Environmental Ethics* (Berkeley: University of California Press, 1982) for animal rights.

23 See Albert Schweitzer, *Philosophy of Civilization: Civilization and Ethics*, trans. John Naish (London: A. & C. Black, 1923). For a fuller discussion see J. Baird Callicott, "On the Intrinsic Value of Non-human Species," in *The Preservation of Species*, ed. Bryan Norton (Princeton: Princeton University Press, 1986), 138–72.

24 Peter Singer and Tom Regan are both proud of this circumstance and consider it a virtue. See Peter Singer, "Not for Humans Only: The Place of Nonhumans in Environmental Issues" in *Ethics and Problems of the 21st Century*, 191–206; and Tom Regan, "Ethical Vegetarianism and Commercial Animal Farming" in *Contemporary Moral Problems*, ed. James E. White (St. Paul, Minn.: West Publishing Co., 1985), 279–94.

25 See J. Baird Callicott, "Hume's Is/Ought Dichotomy and the Relation of Ecology to Leopold's Land Ethic," *Environmental Ethics* 4 (1982): 163–74, and "Non-anthropocentric Value Theory and Environmental Ethics," *American Philosophical Quarterly* 21 (1984): 299–309, for an elaboration.

26 Hume, *Enquiry*, 219.

27 Darwin, *Descent*, 120.

28 I have elsewhere argued that "value in the philosophical sense" means "intrinsic" or "inherent" value. See J. Baird Callicott, "The Philosophical Value of Wildlife," in *Valuing Wildlife: Economic and Social Values of Wildlife*, ed. Daniel J. Decker and Gary Goff (Boulder, Col.: Westview Press, 1986), 214–221.

29 See Worster, *Nature's Economy*.

30 See J. Baird Callicott, "The Metaphysical Implications of Ecology," *Environmental Ethics* 8 (1986): 300–315, for an elaboration of this point.

31 Robert P. McIntosh, *The Background of Ecology: Concept and Theory* (Cambridge: Cambridge University Press, 1985).

32 Aldo Leopold, "Some Fundamentals of Conservation in the Southwest," *Environmental Ethics* 1 (1979): 139–40, emphasis added.

33 Arthur Tansley, "The Use and Abuse of Vegetational Concepts and Terms," *Ecology* 16 (1935): 292–303.

34 Harold J. Morowitz, "Biology as a Cosmological Science," *Main Currents in Modern Thought* 28 (1972): 156.

35 I borrow the term "devolution" from Austin Meredith, "Devolution," *Journal of Theoretical Biology* 96 (1982): 49–65.

36 Holmes Rolston, III, "Duties to Endangered Species," *Bioscience* 35 (1985): 718–26. See also Geerat Vermeij, "The Biology of Human-Caused Extinction," in Norton, *Preservation of Species*, 28–49.

37 See D. M. Raup and J. J. Sepkoski, Jr., "Mass Extinctions in the Marine Fossil Record," *Science* 215 (1982): 1501–3.

38 William Aiken, "Ethical Issues in Agriculture," in *Earthbound: New Introductory Essays in Environmental Ethics*, ed. Tom Regan (New York: Random House, 1984), 269. Tom Regan, *The Case for Animal Rights* (Berkeley: University of California Press, 1983) 262, and "Ethical Vegetarianism," 291. See also Eliott Sober, "Philosophical Problems for Environmentalism," in Norton, *Preservation of Species*, 173–94.

39 I owe the tree-ring analogy to Richard and Val Routley (now Sylvan and Plumwood, respectively), "Human Chauvinism and Environmental Ethics," in *Environmental Philosophy*, ed. D. Mannison, M. McRobbie, and R. Routley (Canberra: Department of Philosophy, Research School of the Social Sciences, Australian National University, 1980), 96–189. A good illustration of the balloon analogy may be found in Peter Singer, *The Expanding Circle: Ethics and Sociobiology* (New York: Farrar, Straus and Giroux, 1983).

40 For an elaboration see Thomas W. Overholt and J. Baird Callicott, *Clothed-in-Fur and Other Tales: An Introduction to an Ojibwa World View* (Washington, D. C.: University Press of America, 1982).

41 J. Baird Callicott, "Traditional American Indian and Western European Attitudes Toward Nature: An Overview," *Environmental Ethics* 4 (1982): 163–74.

42 Peter Fritzell, "The Conflicts of Ecological Conscience," in this volume.

43 See Worster, *Nature's Economy*.

44 Scott Lehmann, "Do Wildernesses Have Rights?" *Environmental Ethics* 3 (1981): 131.

38

Deep Ecology

Bill Devall and George Sessions

The term *deep ecology* was coined by Arne Naess in his 1973 article, "The Shallow and the Deep, Long-Range Ecology Movements."[1] Naess was attempting to describe the deeper, more spiritual approach to Nature exemplified in the writings of Aldo Leopold and Rachel Carson. He thought that this deeper approach resulted from a more sensitive openness to ourselves and nonhuman life around us. The essence of deep ecology is to keep asking more searching questions about human life, society, and Nature as in the Western philosophical tradition of Socrates. As examples of this deep questioning, Naess points out "that we ask why and how, where others do not. For instance, ecology as a science does not ask what kind of a society would be the best for maintaining a particular ecosystem – that is considered a question for value theory, for politics, for ethics." Thus deep ecology goes beyond the so-called factual scientific level to the level of self and Earth wisdom.

Deep ecology goes beyond a limited piecemeal shallow approach to environmental problems and attempts to articulate a comprehensive religious and philosophical worldview. The foundations of deep ecology are the basic intuitions and experiencing of ourselves and Nature which comprise ecological consciousness. Certain outlooks on politics and public policy flow naturally from this consciousness. And in the context of this book, we discuss the minority tradition as the type of community most conducive both to cultivating ecological consciousness and to asking the basic questions of values and ethics addressed in these pages.

Many of these questions are perennial philosophical and religious questions faced by humans in all cultures over the ages. What does it mean to be a unique human individual? How can the individual self maintain and increase its uniqueness while also being an inseparable aspect of the whole system wherein there are no sharp breaks between self and the *other*? An ecological perspective, in this deeper sense, results in what Theodore Roszak calls "an awakening of wholes greater than the sum of their parts. In spirit, the discipline is contemplative and therapeutic."[2]

Ecological consciousness and deep ecology are in sharp contrast with the dominant worldview of technocratic-industrial societies which regards humans as isolated and fundamentally separate from the rest of Nature, as superior to, and in charge of, the rest of creation. But the view of humans as separate and superior to the rest of Nature is only part of larger cultural patterns. For thousands of years, Western culture has become increasingly obsessed with the idea of *dominance*: with dominance of humans over nonhuman Nature, masculine over the feminine, wealthy and powerful over the poor, with the dominance

From Bill Devall and George Sessions, *Deep Ecology* (Salt Lake, UT: Peregrine Smith, 1985), pp. 65–77.

of the West over non-Western cultures. Deep ecological consciousness allows us to see through these erroneous and dangerous illusions.

For deep ecology, the study of our place in the Earth household includes the study of ourselves as part of the organic whole. Going beyond a narrowly materialist scientific understanding of reality, the spiritual and the material aspects of reality fuse together. While the leading intellectuals of the dominant worldview have tended to view religion as "just superstition," and have looked upon ancient spiritual practice and enlightenment, such as found in Zen Buddhism, as essentially subjective, the search for deep ecological consciousness is the search for a more objective consciousness and state of being through an active deep questioning and meditative process and way of life.

Many people have asked these deeper questions and cultivated ecological consciousness within the context of different spiritual traditions – Christianity, Taoism, Buddhism, and Native American rituals, for example. While differing greatly in other regards, many in these traditions agree with the basic principles of deep ecology.

Warwick Fox, an Australian philosopher, has succinctly expressed the central intuition of deep ecology: "It is the idea that we can make no firm ontological divide in the field of existence: That there is no bifurcation in reality between the human and the non-human realms . . . to the extent that we perceive boundaries, we fall short of deep ecological consciousness."[3]

From this most basic insight or characteristic of deep ecological consciousness, Arne Naess has developed two *ultimate norms* or intuitions which are themselves not derivable from other principles or intuitions. They are arrived at by the deep questioning process and reveal the importance of moving to the philosophical and religious level of wisdom. They cannot be validated, of course, by the methodology of modern science based on its usual mechanistic assumptions and its very narrow definition of data. These ultimate norms are *self-realization* and *bio-centric equality*.

Self-Realization

In keeping with the spiritual traditions of many of the world's religions, the deep ecology norm of self-realization goes beyond the modern Western *self* which is defined as an isolated ego striving primarily for hedonistic gratification or for a narrow sense of individual salvation in this life or the next. This socially programmed sense of the narrow self or social self dislocates us, and leaves us prey to whatever fad or fashion is prevalent in our society or social reference group. We are thus robbed of beginning the search for our unique spiritual/biological personhood. Spiritual growth, or unfolding, begins when we cease to understand or see ourselves as isolated and narrow competing egos and begin to identify with other humans from our family and friends to, eventually, our species. But the deep ecology sense of self requires a further maturity and growth, an identification which goes beyond humanity to include the nonhuman world. We must see beyond our narrow contemporary cultural assumptions and values, and the conventional wisdom of our time and place, and this is best achieved by the meditative deep questioning process. Only in this way can we hope to attain full mature personhood and uniqueness.

A nurturing nondominating society can help in the "real work" of becoming a whole person. The "real work" can be summarized symbolically as the realization of "self-in-Self" where "Self" stands for organic wholeness. This process of the full unfolding of the self can also be summarized by the phrase, "No one is saved until we are all saved," where the phrase "one" includes not only me, an individual human, but all humans, whales, grizzly bears, whole rain forest ecosystems, mountains and rivers, the tiniest microbes in the soil, and so on.

Biocentric Equality

The intuition of biocentric equality is that all things in the biosphere have an equal right to live and blossom and to reach their own individual forms of unfolding and self-realization within the larger Self-realization. This basic intuition is that all organisms and entities in the ecosphere, as parts of the interrelated whole, are equal in intrinsic worth. Naess suggests that biocentric equality as an intuition is true in principle, although in the process of living, all species use each other as food, shelter, etc. Mutual predation is a biological fact of life, and many of the world's religions have struggled with the spiritual implications of this. Some animal liberationists who attempt to side-step this problem by advocating vegetarianism are forced to say that the entire plant kingdom including rain forests have no right to their own existence. This evasion flies in the face of the basic intuition of equality.[4] Aldo Leopold expressed this intuition when he said humans are "plain citizens" of the biotic community, not lord and master over all other species.

Biocentric equality is intimately related to the all-inclusive Self-realization in the sense that if we harm the rest of Nature then we are harming ourselves. There are no boundaries and everything is interrelated. But insofar as we perceive things as individual organisms or entities, the insight draws us to respect all human and nonhuman individuals in their own right as parts of the whole without feeling the need to set up hierarchies of species with humans at the top.

The practical implications of this intuition or norm suggest that we should live with minimum rather than maximum impact on other species and on the Earth in general. Thus we see another as our guiding principle: "simple in means, rich in ends." . . .

A fuller discussion of the biocentric norm as it unfolds itself in practice begins with the realization that we, as individual humans, and as communities of humans, have vital needs which go beyond such basics as food, water, and shelter to include love, play, creative expression, intimate relationships with a particular landscape (or Nature taken in its entirety) as well as intimate relationships with other humans, and the vital need for spiritual growth, for becoming a mature human being.

Our vital material needs are probably more simple than many realize. In technocratic-industrial societies there is overwhelming propaganda and advertising which encourages false needs and destructive desires designed to foster increased production and consumption of goods. Most of this actually diverts us from facing reality in an objective way and from beginning the "real work" of spiritual growth and maturity.

Many people who do not see themselves as supporters of deep ecology nevertheless recognize an overriding vital human need for a healthy and high-quality natural environment for humans, if not for all life, with minimum intrusion of toxic waste, nuclear radiation from human enterprises, minimum acid rain and smog, and enough free flowing wilderness so humans can get in touch with their sources, the natural rhythms and the flow of time and place.

Drawing from the minority tradition and from the wisdom of many who have offered the insight of interconnectedness, we recognize that deep ecologists can offer suggestions for gaining maturity and encouraging the processes of harmony with Nature, but that there is no grand solution which is guaranteed to save us from ourselves.

The ultimate norms of deep ecology suggest a view of the nature of reality and our place as an individual (many in the one) in the larger scheme of things. They cannot be fully grasped intellectually but are ultimately experiential. . . .

Basic Principles of Deep Ecology

In April 1984, during the advent of spring and John Muir's birthday, George Sessions and Arne Naess summarized fifteen years of thinking on the principles of deep ecology while camping in Death Valley, California. In this great and special place, they articulated these principles in a literal, somewhat neutral way, hoping that they would be understood and accepted by persons coming from different philosophical and religious positions.

Readers are encouraged to elaborate their own versions of deep ecology, clarify key concepts and think through the consequences of acting from these principles.

Basic principles

1   The well-being and flourishing of human and nonhuman Life on Earth have value in themselves (synonyms: intrinsic value, inherent value). These values are independent of the usefulness of the nonhuman world for human purposes.

2   Richness and diversity of life forms contribute to the realization of these values and are also values in themselves.

3 Humans have no right to reduce this richness and diversity except to satisfy *vital* needs.

4 The flourishing of human life and cultures is compatible with a substantial decrease of the human population. The flourishing of nonhuman life requires such a decrease.

5 Present human interference with the nonhuman world is excessive, and the situation is rapidly worsening.

6 Policies must therefore be changed. These policies affect basic economic, technological, and ideological structures. The resulting state of affairs will be deeply different from the present.

7 The ideological change is mainly that of appreciating *life quality* (dwelling in situations of inherent value) rather than adhering to an increasingly higher standard of living. There will be a profound awareness of the difference between big and great.

8 Those who subscribe to the foregoing points have an obligation directly or indirectly to try to implement the necessary changes.

Naess and Sessions provide comments on the basic principles

Re (1)

This formulation refers to the biosphere, or more accurately, to the ecosphere as a whole. This includes individuals, species, populations, habitat, as well as human and nonhuman cultures. From our current knowledge of all-pervasive intimate relationships, this implies a fundamental deep concern and respect. Ecological processes of the planet should, on the whole, remain intact. "The world environment should remain 'natural'" (Gary Snyder).

The term "life" is used here in a more comprehensive nontechnical way to refer also to what biologists classify as "nonliving"; rivers (watersheds), landscapes, ecosystems. For supporters of deep ecology, slogans such as "Let the river live" illustrate this broader usage so common in most cultures.

Inherent value as used in (1) is common in deep ecology literature ("The presence of inherent value in a natural object is independent of any awareness, interest, or appreciation of it by a conscious being").[5]

Re (2)

More technically, this is a formulation concerning diversity and complexity. From an ecological standpoint, complexity and symbiosis are conditions for maximizing diversity. So-called simple, lower, or primitive species of plants and animals contribute essentially to the richness and diversity of life. They have value in themselves and are not merely steps toward the so-called higher or rational life forms. The second principle presupposes that life itself, as a process over evolutionary time, implies an increase of diversity and richness. The refusal to acknowledge that some life forms have greater or lesser intrinsic value than others (see points 1 and 2) runs counter to the formulations of some ecological philosophers and New Age writers.

Complexity, as referred to here, is different from complication. Urban life may be more complicated than life in a natural setting without being more complex in the sense of multifaceted quality.

Re (3)

The term "vital need" is left deliberately vague to allow for considerable latitude in judgment. Differences in climate and related factors, together with differences in the structures of societies as they now exist, need to be considered (for some Eskimos, snowmobiles are necessary today to satisfy vital needs).

People in the materially richest countries cannot be expected to reduce their excessive interference with the nonhuman world to a moderate level overnight. The stabilization and reduction of the human population will take time. Interim strategies need to be developed. But this in no way excuses the present complacency – the extreme seriousness of our current situation must first be realized. But the longer we wait the more drastic will be the measures needed. Until deep changes are made substantial decreases in richness and diversity are liable to occur: the rate of extinction of species will be ten to one hundred times greater than any other period of earth history.

Re (4)

The United Nations Fund for Population Activities in their State of World Population Report (1984) said that high human population growth rates (over 2.0 percent per annum) in many developing countries "were diminishing the quality of life for many millions of people." During the decade 1974–1984, the world population grew by nearly 800 million – more than the size of India. "And we will be adding about one Bangladesh (population 93 million) per annum between now and the year 2000."

The report noted that "The growth rate of the human population has declined for the first time in human history. But at the same time, the number of people being added to the human population is bigger than at any time in history because the population base is larger."

Most of the nations in the developing world (including India and China) have as their official government policy the goal of reducing the rate of human population increase, but there are debates over the types of measures to take (contraception, abortion, etc.) consistent with human rights and feasibility.

The report concludes that if all governments set specific population targets as public policy to help alleviate poverty and advance the quality of life, the current situation could be improved.

As many ecologists have pointed out, it is also absolutely crucial to curb population growth in the so-called developed (i.e., overdeveloped) industrial societies. Given the tremendous rate of consumption and waste production of individuals in these societies, they represent a much greater threat and impact on the biosphere per capita than individuals in Second and Third World countries.

Re (5)

This formulation is mild. For a realistic assessment of the situation, see the unabbreviated version of the I.U.C.N.'s *World Conservation Strategy*. There are other works to be highly recommended, such as Gerald Barney's *Global 2000 Report to the President of the United States*.

The slogan of "noninterference" does not imply that humans should not modify some ecosystems as do other species. Humans have modified the earth and will probably continue to do so. At issue is the nature and extent of such interference.

The fight to preserve and extend areas of wilderness or near wilderness should continue and should focus on the general ecological functions of these areas (one such function: large wilder-

ness areas are required in the biosphere to allow for continued evolutionary speciation of animals and plants). Most present designated wilderness areas and game preserves are not large enough to allow for such speciation.

Re (6)

Economic growth as conceived and implemented today by the industrial states is incompatible with (1)–(5). There is only a faint resemblance between ideal sustainable forms of economic growth and present policies of the industrial societies. And "sustainable" still means "sustainable in relation to humans."

Present ideology tends to value things because they are scarce and because they have a commodity value. There is prestige in vast consumption and waste (to mention only several relevant factors).

Whereas "self-determination," "local community," and "think globally, act locally," will remain key terms in the ecology of human societies, nevertheless the implementation of deep changes requires increasingly global action – action across borders.

Governments in Third World countries (with the exception of Costa Rica and a few others) are uninterested in deep ecological issues. When the governments of industrial societies try to promote ecological measures through Third World governments, practically nothing is accomplished (e.g., with problems of desertification). Given this situation, support for global action through non-governmental international organizations becomes increasingly important. Many of these organizations are able to act globally "from grassroots to grassroots," thus avoiding negative governmental interference.

Cultural diversity today requires advanced technology, that is, techniques that advance the basic goals of each culture. So-called soft, intermediate, and alternative technologies are steps in this direction.

Re (7)

Some economists criticize the term "quality of life" because it is supposed to be vague. But on closer inspection, what they consider to be vague is actually the nonquantitative nature of the term. One cannot quantify adequately what is important for the quality of life as discussed here, and there is no need to do so.

RE (8)

There is ample room for different opinions about priorities: what should be done first, what next? What is most urgent? What is clearly necessary as opposed to what is highly desirable but not absolutely pressing?

Notes

1 Arne Naess, "The Shallow and The Deep, Long-Range Ecology Movements: A Summary," *Inquiry* 16 (Oslo, 1973), pp. 95–100.

2 Theodore Roszak, *Where the Wasteland Ends* (New York: Anchor, 1972).

3 Warwick Fox, "The Intuition of Deep Ecology" (Paper presented at the Ecology and Philosophy Conference, Australian National University, September, 1983). To appear in *The Ecologist* (England, Fall 1984).

4 Tom Regan, *The Case for Animal Rights* (New York: Random House, 1983). For excellent critiques of the animal rights movement, see John Rodman, "The Liberation of Nature?" *Inquiry* 20 (Oslo, 1977). J. Baird Callicott, "Animal Liberation," *Environmental Ethics* 2, 4, (1980); see also John Rodman, "Four Forms of Ecological Consciousness Reconsidered" in T. Attig and D. Scherer, eds., *Ethics and the Environment* (Englewood Cliffs, N.J.: Prentice-Hall, 1983).

5 Tom Regan, "The Nature and Possibility of an Environmental Ethic," *Environmental Ethics* 3 (1981), pp. 19–34.

39

Radical American Environmentalism and Wilderness Preservation: A Third World Critique

Ramachandra Guha

I present a Third World critique of the trend in American environmentalism known as deep ecology, analyzing each of deep ecology's central tenets: the distinction between anthropocentrism and biocentrism, the focus on wilderness preservation, the invocation of Eastern traditions, and the belief that it represents the most radical trend within environmentalism. I argue that the anthropocentrism/biocentrism distinction is of little use in understanding the dynamics of environmental degradation, that the implementation of the wilderness agenda is causing serious deprivation in the Third World, that the deep ecologist's interpretation of Eastern tradition is highly selective, and that in other cultural contexts (e.g., West Germany and India) radical environmentalism manifests itself quite differently, with a far greater emphasis on equity and the integration of ecological concerns with livelihood and work. I conclude that despite its claims to universality, deep ecology is firmly rooted in American environmental and cultural history and is inappropriate when applied to the Third World.

> Even God dare not appear to the poor man except in the form of bread.
> —*Mahatma Gandhi*

Introduction

[...]

In this article I develop a critique of deep ecology from the perspective of a sympathetic outsider. I critique deep ecology not as a general (or even a foot soldier) in the continuing struggle between the ghosts of Gifford Pinchot and John Muir over control of the U.S. environmental movement, but as an outsider to these battles. I speak admittedly as a partisan, but of the environmental movement in India, a country with an ecological diversity comparable to the U.S., but with a radically dissimilar cultural and social history.

[...] Specifically, I examine the cultural rootedness of a philosophy that likes to present itself in universalistic terms. I make two main arguments: first, that deep ecology is uniquely American, and despite superficial similarity in rhetorical style, the social and political goals of radical environmentalism in other cultural contexts (e.g., West Germany and India) are quite different; second, that the social consequences of putting deep ecology into practice on a worldwide basis (what its practitioners are aiming for) are very grave indeed.

From *Environmental Ethics* 11 (1989): 71–83. Reprinted by permission of Ramachandra Guha.

The Tenets of Deep Ecology

While I am aware that the term *deep ecology* was coined by the Norwegian philosopher Arne Naess, this article refers specifically to the American variant.[1] Adherents of the deep ecological perspective in this country, while arguing intensely among themselves over its political and philosophical implications, share some fundamental premises about human-nature interactions. As I see it, the defining characteristics of deep ecology are fourfold:

First, deep ecology argues that the environmental movement must shift from an "anthropocentric" to a "biocentric" perspective. In many respects, an acceptance of the primacy of this distinction constitutes the litmus test of deep ecology. A considerable effort is expended by deep ecologists in showing that the dominant motif in Western philosophy has been anthropocentric – i.e., the belief that man and his works are the center of the universe – and conversely, in identifying those lonely thinkers (Leopold, Thoreau, Muir, Aldous Huxley, Santayana, etc.) who, in assigning man a more humble place in the natural order, anticipated deep ecological thinking. In the political realm, meanwhile, establishment environmentalism (shallow ecology) is chided for casting its arguments in human-centered terms. Preserving nature, the deep ecologists say, has an intrinsic worth quite apart from any benefits preservation may convey to future human generations. The anthropocentric-biocentric distinction is accepted as axiomatic by deep ecologists, it structures their discourse, and much of the present discussion remains mired within it.

The second characteristic of deep ecology is its focus on the preservation of unspoilt wilderness – and the restoration of degraded areas to a more pristine condition – to the relative (and sometimes absolute) neglect of other issues on the environmental agenda. I later identify the cultural roots and portentous consequences of this obsession with wilderness. For the moment, let me indicate three distinct sources from which it springs. Historically, it represents a playing out of the preservationist (read *radical*) and utilitarian (read *reformist*) dichotomy that has plagued American environmentalism since the turn of the century. Morally, it is an imperative that

follows from the biocentric perspective; other species of plants and animals, and nature itself, have an intrinsic right to exist. And finally, the preservation of wilderness also turns on a scientific argument – viz., the value of biological diversity in stabilizing ecological regimes and in retaining a gene pool for future generations. Truly radical policy proposals have been put forward by deep ecologists on the basis of these arguments. The influential poet Gary Snyder, for example, would like to see a 90 percent reduction in human populations to allow a restoration of pristine environments, while others have argued forcefully that a large portion of the globe must be immediately cordoned off from human beings.[2]

Third, there is a widespread invocation of Eastern spiritual traditions as forerunners of deep ecology. Deep ecology, it is suggested, was practiced both by major religious traditions and at a more popular level by "primal" peoples in non-Western settings. This complements the search for an authentic lineage in Western thought. At one level, the task is to recover those dissenting voices within the Judeo-Christian tradition; at another, to suggest that religious traditions in other cultures are, in contrast, dominantly if not exclusively "biocentric" in their orientation. This coupling of (ancient) Eastern and (modern) ecological wisdom seemingly helps consolidate the claim that deep ecology is a philosophy of universal significance.

Fourth, deep ecologists, whatever their internal differences, share the belief that they are the "leading edge" of the environmental movement. As the polarity of the shallow/deep and anthropocentric/biocentric distinctions makes clear, they see themselves as the spiritual, philosophical, and political vanguard of American and world environmentalism.

Toward a Critique

Although I analyze each of these tenets independently, it is important to recognize, as deep ecologists are fond of remarking in reference to nature, the interconnectedness and unity of these individual themes.

(1) Insofar as it has begun to act as a check on man's arrogance and ecological hubris, the transition from an anthropocentric (human-centered)

to a biocentric (humans as only one element in the ecosystem) view in both religious and scientific traditions is only to be welcomed.[3] What is unacceptable are the radical conclusions drawn by deep ecology, in particular, that intervention in nature should be guided primarily by the need to preserve biotic integrity rather than by the needs of humans. The latter for deep ecologists is anthropocentric, the former biocentric. This dichotomy is, however, of very little use in understanding the dynamics of environmental degradation. The two fundamental ecological problems facing the globe are (i) overconsumption by the industrialized world and by urban elites in the Third World and (ii) growing militarization, both in a short-term sense (i.e., ongoing regional wars) and in a long-term sense (i.e., the arms race and the prospect of nuclear annihilation). Neither of these problems has any tangible connection to the anthropocentric-biocentric distinction. Indeed, the agents of these processes would barely comprehend this philosophical dichotomy. The proximate causes of the ecologically wasteful characteristics of industrial society and of militarization are far more mundane: at an aggregate level, the dialectic of economic and political structures, and at a micro-level, the life style choices of individuals. These causes cannot be reduced, whatever the level of analysis, to a deeper anthropocentric attitude toward nature; on the contrary, by constituting a grave threat to human survival, the ecological degradation they cause does not even serve the best interests of human beings! If my identification of the major dangers to the integrity of the natural world is correct, invoking the bogy of anthropocentrism is at best irrelevant and at worst a dangerous obfuscation.

(2) If the above dichotomy is irrelevant, the emphasis on wilderness is positively harmful when applied to the Third World. If in the U.S. the preservationist/utilitarian division is seen as mirroring the conflict between "people" and "interests," in countries such as India the situation is very nearly the reverse. Because India is a long settled and densely populated country in which agrarian populations have a finely balanced relationship with nature, the setting aside of wilderness areas has resulted in a direct transfer of resources from the poor to the rich. Thus, Project Tiger, a network of parks hailed by the international conservation community as an outstanding success, sharply posits the interests of the tiger against those of poor peasants living in and around the reserve. The designation of tiger reserves was made possible only by the physical displacement of existing villages and their inhabitants; their management requires the continuing exclusion of peasants and livestock. The initial impetus for setting up parks for the tiger and other large mammals such as the rhinoceros and elephant came from two social groups, first, a class of ex-hunters turned conservationists belonging mostly to the declining Indian feudal elite and second, representatives of international agencies, such as the World Wildlife Fund (WWF) and the International Union for the Conservation of Nature and Natural Resources (IUCN), seeking to transplant the American system of national parks onto Indian soil. In no case have the needs of the local population been taken into account, and as in many parts of Africa, the designated wildlands are managed primarily for the benefit of rich tourists. Until very recently, wildlands preservation has been identified with environmentalism by the state and the conservation elite; in consequence, environmental problems that impinge far more directly on the lives of the poor – e.g., fuel, fodder, water shortages, soil erosion, and air and water pollution – have not been adequately addressed.

Deep ecology provides, perhaps unwittingly, a justification for the continuation of such narrow and inequitable conservation practices under a newly acquired radical guise. Increasingly, the international conservation elite is using the philosophical, moral, and scientific arguments used by deep ecologists in advancing their wilderness crusade. A striking but by no means atypical example is the recent plea by a prominent American biologist for the takeover of large portions of the globe by the author and his scientific colleagues. Writing in a prestigious scientific forum, the *Annual Review of Ecology and Systematics*, Daniel Janzen argues that only biologists have the competence to decide how the tropical landscape should be used. As "the representatives of the natural world," biologists are "in charge of the future of tropical ecology," and only they have the expertise and mandate to "determine whether the tropical agroscape is to be populated only by humans, their mutualists, commensals, and parasites, or whether it will also contain some islands of the greater nature – the nature that spawned humans, yet has been vanquished by them." Janzen exhorts his colleagues to advance their territorial claims on the tropical world more forcefully, warning that the very existence of these areas is at stake: "if biologists want a tropics in which to biologize, they are going to have to buy it with care, energy, effort, strategy, tactics, time, and cash."[4]

This frankly imperialist manifesto highlights the multiple dangers of the preoccupation with wilderness preservation that is characteristic of deep ecology. As I have suggested, it seriously compounds the neglect by the American movement of far more pressing environmental problems within the Third World. But perhaps more importantly, and in a more insidious fashion, it also provides an impetus to the imperialist yearning of Western biologists and their financial sponsors, organizations such as the WWF and the IUCN. The wholesale transfer of a movement culturally rooted in American conservation history can only result in the social uprooting of human populations in other parts of the globe.

(3) I come now to the persistent invocation of Eastern philosophies as antecedent in point of time but convergent in their structure with deep ecology. Complex and internally differentiated religious traditions – Hinduism, Buddhism, and Taoism – are lumped together as holding a view of nature believed to be quintessentially biocentric. Individual philosophers such as the Taoist Lao Tzu are identified as being forerunners of deep ecology. Even an intensely political, pragmatic, and Christian influenced thinker such as Gandhi has been accorded a wholly undeserved place in the deep ecological pantheon. Thus the Zen teacher Robert Aitken Roshi makes the strange claim that Gandhi's thought was not human-centered and that he practiced an embryonic form of deep ecology which is "traditionally Eastern and is found with differing emphasis in Hinduism, Taoism and in Theravada and Mahayana Buddhism."[5] Moving away from the realm of high philosophy and scriptural religion, deep ecologists make the further claim that at the level of material and spiritual practice "primal" peoples subordinated themselves to the integrity of the biotic universe they inhabited.

I have indicated that this appropriation of Eastern traditions is in part dictated by the need to construct an authentic lineage and in part a desire to present deep ecology as a universalistic philosophy. Indeed, in his substantial and quixotic biography of John Muir, Michael Cohen goes so far as to suggest that Muir was the "Taoist of the [American] West."[6] This reading of Eastern traditions is selective and does not bother to differentiate between alternate (and changing) religious and cultural traditions; as it stands, it does considerable violence to the historical record. Throughout most recorded history the characteristic form of human activity in the "East" has been a finely tuned but nonetheless conscious and dynamic manipulation of nature. Although mystics such as Lao Tzu did reflect on the spiritual essence of human relations with nature, it must be recognized that such ascetics and their reflections were supported by a society of cultivators whose relationship with nature was a far more *active* one. Many agricultural communities do have a sophisticated knowledge of the natural environment that may equal (and sometimes surpass) codified "scientific" knowledge; yet, the elaboration of such traditional ecological knowledge (in both material and spiritual contexts) can hardly be said to rest on a mystical affinity with nature of a deep ecological kind. Nor is such knowledge infallible; as the archaeological record powerfully suggests, modern Western man has no monopoly on ecological disasters.

In a brilliant article, the Chicago historian Ronald Inden points out that this romantic and essentially positive view of the East is a mirror image of the scientific and essentially pejorative view normally upheld by Western scholars of the Orient. In either case, the East constitutes the Other, a body wholly separate and alien from the West; it is defined by a uniquely spiritual and nonrational "essence", even if this essence is valorized quite differently by the two schools. Eastern man exhibits a spiritual dependence with respect to nature – on the one hand, this is symptomatic of his prescientific and backward self, on the other, of his ecological wisdom and deep ecological consciousness. Both views are monolithic, simplistic, and have the characteristic effect – intended in one case, perhaps unintended in the other – of denying agency and reason to the East and making it the privileged orbit of Western thinkers.

The two apparently opposed perspectives have then a common underlying structure of discourse in which the East merely serves as a vehicle for Western projections. Varying images of the East are raw material for political and cultural battles being played out in the West; they tell us far more about the Western commentator and his desires than about the "East." Inden's remarks apply not merely to Western scholarship on India, but to Orientalist constructions of China and Japan as well:

Although these two views appear to be strongly opposed, they often combine together. Both have a similar interest in sustaining the Otherness of India. The holders of the dominant view, best exemplified in the past in imperial administrative discourse (and today probably by that of "development economics"), would place a traditional, superstition-ridden India in a position of perpetual tutelage to a modern, rational West. The adherents of the romantic view, best exemplified academically in the discourses of Christian liberalism and analytic psychology, concede the realm of the public and impersonal to the positivist. Taking their succour not from governments and big business, but from a plethora of religious foundations and self-help institutes, and from allies in the "consciousness industry," not to mention the important industry of tourism, the romantics insist that India embodies a private realm of the imagination and the religious which modern, western man lacks but needs. They, therefore, like the positivists, but for just the opposite reason, have a vested interest in seeing that the Orientalist view of India as "spiritual," "mysterious," and "exotic" is perpetuated.[7]

(4) How radical, finally, are the deep ecologists? Notwithstanding their self-image and strident rhetoric (in which the label "shallow ecology" has an opprobrium similar to that reserved for "social democratic" by Marxist-Leninists), even within the American context their radicalism is limited and it manifests itself quite differently elsewhere.

To my mind deep ecology is best viewed as a radical trend within the wilderness preservation movement. Although advancing philosophical

rather than aesthetic arguments and encouraging political militancy rather than negotiation, its practical emphasis – viz., preservation of unspoilt nature – is virtually identical. For the mainstream movement, the function of wilderness is to provide a temporary antidote to modern civilization. As a special institution within an industrialized society, the national park "provides an opportunity for respite, contrast, contemplation, and affirmation of values for those who live most of their lives in the workaday world."[8] Indeed, the rapid increase in visitations to the national parks in postwar America is a direct consequence of economic expansion. The emergence of a popular interest in wilderness sites, the historian Samuel Hays points out, was "not a throwback to the primitive, but an integral part of the modern standard of living as people sought to add new 'amenity' and 'aesthetic' goals and desires to their earlier preoccupation with necessities and conveniences."[9]

Here, the enjoyment of nature is an integral part of the consumer society. The private automobile (and the life style it has spawned) is in many respects the ultimate ecological villain, and an untouched wilderness the prototype of ecological harmony; yet, for most Americans it is perfectly consistent to drive a thousand miles to spend a holiday in a national park. They possess a vast, beautiful, and sparsely populated continent and are also able to draw upon the natural resources of large portions of the globe by virtue of their economic and political dominance. In consequence, America can simultaneously enjoy the material benefits of an expanding economy and the aesthetic benefits of unspoilt natures. The two poles of "wilderness" and "civilization" mutually coexist in an internally coherent whole, and philosophers of both poles are assigned a prominent place in this culture. Paradoxically as it may seem, it is no accident that Star Wars technology and deep ecology both find their fullest expression in that leading sector of Western civilization, California.

Deep ecology runs parallel to the consumer society without seriously questioning its ecological and socio-political basis. In its celebration of American wilderness, it also displays an uncomfortable convergence with the prevailing climate of nationalism in the American wilderness movement. For spokesmen such as the historian

Roderick Nash, the national park system is America's distinctive cultural contribution to the world, reflective not merely of its economic but of its philosophical and ecological maturity as well. In what Walter Lippman called the American century, the "American invention of national parks" must be exported worldwide. Betraying an economic determinism that would make even a Marxist shudder, Nash believes that environmental preservation is a "full stomach" phenomenon that is confined to the rich, urban, and sophisticated. Nonetheless, he hopes that "the less developed nations may eventually evolve economically and intellectually to the point where nature preservation is more than a business."[10]

The error which Nash makes (and which deep ecology in some respects encourages) is to equate environmental protection with the protection of the wilderness. This is a distinctively American notion, borne out of a unique social and environmental history. The archetypal concerns of radical environmentalists in other cultural contexts are in fact quite different. The German Greens, for example, have elaborated a devastating critique of industrial society which turns on the acceptance of environmental limits to growth. Pointing to the intimate links between industrialization, militarization, and conquest, the Greens argue that economic growth in the West has historically rested on the economic and ecological exploitation of the Third World. Rudolf Bahro is characteristically blunt:

> The working class here [in the West] is the richest lower class in the world. And if I look at the problem from the point of view of the whole of humanity, not just from that of Europe, then I must say that the metropolitan working class is the worst exploiting class in history.... What made poverty bearable in eighteenth or nineteenth-century Europe was the prospect of escaping it through exploitation of the periphery. But this is no longer a possibility, and continued industrialism in the Third World will mean poverty for whole generations and hunger for millions.[11]

Here the roots of global ecological problems lie in the disproportionate share of resources consumed by the industrialized countries as a whole and the urban elite within the Third World. Since it is impossible to reproduce an industrial monoculture worldwide, the ecological movement in the West must begin by cleaning up its own act. The Greens advocate the creation of a "no growth" economy, to be achieved by scaling down current (and clearly unsustainable) consumption levels. This radical shift in consumption and production patterns requires the creation of alternate economic and political structures – smaller in scale and more amenable to social participation – but it rests equally on a shift in cultural values. The expansionist character of modern Western man will have to give way to an ethic of renunciation and self-limitation, in which spiritual and communal values play an increasing role in sustaining social life. This revolution in cultural values, however, has as its point of departure an understanding of environmental processes quite different from deep ecology.

Many elements of the Green program find a strong resonance in countries such as India, where a history of Western colonialism and industrial development has benefited only a tiny elite while exacting tremendous social and environmental costs. The ecological battles presently being fought in India have as their epicenter the conflict over nature between the subsistence and largely rural sector and the vastly more powerful commercial-industrial sector. Perhaps the most celebrated of these battles concerns the Chipko (Hug the Tree) movement, a peasant movement against deforestation in the Himalayan foothills. Chipko is only one of several movements that have sharply questioned the non-sustainable demand being placed on the land and vegetative base by urban centers and industry. These include opposition to large dams by displaced peasants, the conflict between small artisan fishing and large-scale trawler fishing for export, the countrywide movements against commercial forest operations, and opposition to industrial pollution among downstream agricultural and fishing communities.

Two features distinguish these environmental movements from their Western counterparts. First, for the sections of society most critically affected by environmental degradation – poor and landless peasants, women, and tribals – it is a question of sheer survival, not of enhancing the quality of life. Second, and as a consequence, the environmental solutions they articulate deeply involve questions of equity as well as economic and political redistribution. Highlighting these differences, a leading Indian environmentalist stresses that "environmental protection per se is of least concern to most of these groups. Their main concern is about the use of the environment and who should benefit from it."[12] They seek to wrest control of nature away from the state and the industrial sector and place it in the hands of rural communities who live within that environment but are increasingly denied access to it. These communities have far more basic needs, their demands on the environment are far less intense, and they can draw upon a reservoir of cooperative social institutions and local ecological knowledge in managing the "commons" – forest, grasslands, and the waters – on a sustainable basis. If colonial and capitalist expansion has both accentuated social inequalities and signaled a precipitous fall in ecological wisdom, an alternate ecology must rest on an alternate society and polity as well.

This brief overview of German and Indian environmentalism has some major implications for deep ecology. Both German and Indian environmental traditions allow for a greater integration of ecological concerns with livelihood and work. They also place a greater emphasis on equity and social justice (both within individual countries and on a global scale) on the grounds that in the absence of social regeneration environmental regeneration has very little chance of succeeding. Finally, and perhaps most significantly, they have escaped the preoccupation with wilderness preservation so characteristic of American cultural and environmental history.

A Homily

In 1958, the economist J. K. Galbraith referred to overconsumption as the unasked question of the American conservation movement. There is a marked selectivity, he wrote, "in the conservationist's approach to materials consumption. If we are concerned about our great appetite for materials, it is plausible to seek to increase the supply, to decrease waste, to make better use of the stocks available, and to develop substitutes.

But what of the appetite itself? Surely this is the ultimate source of the problem. If it continues its geometric course, will it not one day have to be restrained? Yet in the literature of the resource problem this is the forbidden question. Over it hangs a nearly total silence."[13]

[. . .]

In their widely noticed book, Bill Devall and George Sessions make no mention of militarization or the movements for peace, while activists whose practical focus is on developing ecologically responsible life styles (e.g., Wendell Berry) are derided as "falling short of deep ecological awareness."[14] A truly radical ecology in the American context ought to work toward a synthesis of the appropriate technology, alternate life style, and peace movements. By making the (largely spurious) anthropocentric-biocentric distinction central to the debate, deep ecologists may have appropriated the moral high ground, but they are at the same time doing a serious disservice to American and global environmentalism.[15]

Notes

1 One of the major criticisms I make in this essay concerns deep ecology's lack of concern with inequalities *within* human society. In the article in which he coined the term *deep ecology*, Naess himself expresses concerns about inequalities between and within nations. However, his concern with social cleavages and their impact on resource utilization patterns and ecological destruction is not very visible in the later writings of deep ecologists. See Arne Naess, "The Shallow and the Deep, Long-Range Ecology Movement: A Summary, "*Inquiry* 16 (1973): 96 (I am grateful to Tom Birch for this reference).

2 Gary Snyder, quoted in Sale, "The Forest for the Trees: Can Today's Environmentalists Tell the Difference", *Mother Jones* 11, no. 8 (November 1986): 32.

3 See, for example, Donald Worster, *Nature's Economy: The Roots of Ecology* (San Francisco: Sierra Club Books, 1977).

4 Daniel Janzen, "The Future of Tropical Ecology," *Annual Review of Ecology and Systematics* 17 (1986): 305–306; emphasis added.

5 Robert Aitken Roshi, "Gandhi, Dogen, and Deep Ecology," reprinted as appendix C in Bill Devall and George Sessions, *Deep Ecology: Living as if*

Nature Mattered (Salt Lake City: Peregrine Smith Books, 1985).

6   Michael Cohen, *The Pathless Way* (Madison: University of Wisconsin Press, 1984), p. 120.

7   Ronald Inden, "Orientalist Constructions of India," *Modern Asian Studies* 20 (1986): 442. Inden draws inspiration from Edward Said's forceful polemic, *Orientalism* (New York: Basic Books, 1980). It must be noted, however, that there is a salient difference between Western perceptions of Middle Eastern and Far Eastern cultures respectively. Due perhaps to the long history of Christian conflict with Islam, Middle Eastern cultures (as Said documents) are consistently presented in pejorative terms. The juxtaposition of hostile and worshiping attitudes that Inden talks of applies only to Western attitudes toward Buddhist and Hindu societies.

8   Joseph Sax, *Mountains Without Handrails: Reflections on the National Parks* (Ann Arbor: University of Michigan Press, 1980), p. 42.

9   Samuel Hays, "From Conservation to Environment: Environmental Politics in the United States since World War Two," *Environmental Review* 6 (1982): 21. See also the same author's book entitled *Beauty, Health and Permanence: Environmental Politics in the United States, 1955–1985* (New York: Cambridge University Press, 1987).

10 Roderick Nash, *Wilderness and the American Mind*, 3rd ed. (New Haven: Yale University Press, 1982).

11 Rudolf Bahro, *From Red to Green* (London: Verso Books, 1984).

12 Anil Agarwal, "Human-Nature Interactions in a Third World Country," *The Environmentalist* 6, no. 3 (1986): 167.

13 John Kenneth Galbraith, "How Much Should a Country Consume?" in *Perspectives on Conservation*, ed. Henry Jarrett (Baltimore: Johns Hopkins Press, 1958), pp. 91–92.

14 Devall and Sessions, *Deep Ecology*, p. 122. For Wendell Berry's own assessment of deep ecology, see his "Amplications: Preserving Wildness," *Wilderness* 50 (Spring 1987): 39–40, 50–54.

15 In this sense, my critique of deep ecology, although that of an outsider, may facilitate the reassertion of those elements in the American environmental tradition for which there is a profound sympathy in other parts of the globe. A global perspective may also lead to a critical reassessment of figures such as Aldo Leopold and John Muir, the two patron saints of deep ecology. As Donald Worster has pointed out, the message of Muir (and, I would argue, of Leopold as well) makes sense only in an American context; he has very little to say to other cultures. See Worster's review of Stephen Fox's *John Muir and His Legacy*, in *Environmental Ethics* 5 (1983): 277–81.

40

Just Garbage

Peter S. Wenz

Environmental racism is evident in practices that expose racial minorities in the United States, and people of color around the world, to disproportionate shares of environmental hazards. These include toxic chemicals in factories, toxic herbicides and pesticides in agriculture, radiation from uranium mining, lead from paint on older buildings, toxic wastes illegally dumped, and toxic wastes legally stored. In this chapter, which concentrates on issues of toxic waste, both illegally dumped and legally stored, I will examine the justness of current practices as well as the arguments commonly given in their defense. I will then propose an alternative practice that is consistent with prevailing principles of justice.

A Defense of Current Practices

Defenders often claim that because economic, not racial, considerations account for disproportionate impacts on nonwhites, current practices are neither racist nor morally objectionable. Their reasoning recalls the Doctrine of Double Effect. According to that doctrine, an effect whose production is usually blameworthy becomes blameless when it is incidental to, although predictably conjoined with, the production of another effect whose production is morally justified. The classic case concerns a pregnant woman with uterine cancer. A common, acceptable treatment for uterine cancer is hysterectomy. This will predictably end the pregnancy, as would an abortion. However, Roman Catholic scholars who usually consider abortion blameworthy consider it blameless in this context because it is merely incidental to hysterectomy, which is morally justified to treat uterine cancer. The hysterectomy would be performed in the absence of pregnancy, so the abortion effect is produced neither as an end-in-itself, nor as a means to reach the desired end, which is the cure of cancer.

Defenders of practices that disproportionately disadvantage non-whites seem to claim, in keeping with the Doctrine of Double Effect, that racial effects are blameless because they are sought neither as ends-in-themselves nor as means to reach a desired goal. They are merely predictable side effects of economic and political practices that disproportionately expose poor people to toxic substances. The argument is that burial of toxic wastes, and other locally undesirable land uses (LULUs), lower property values. People who can afford to move elsewhere do so. They are replaced by buyers (or renters) who are

From Laura Westra and Bill E. Lawson, *Faces of Environmental Racism* (Lanham, MD: Rowman and Littlefield, 2001), pp. 57–71. Reprinted by permission of Rowman and Littlefield Publishers, Inc.

predominantly poor and cannot afford housing in more desirable areas. Law professor Vicki Been puts it this way: "As long as the market allows the existing distribution of wealth to allocate goods and services, it would be surprising indeed if, over the long run, LULUs did not impose a disproportionate burden upon the poor." People of color are disproportionately burdened due primarily to poverty, not racism.[1] This defense against charges of racism is important in the American context because racial discrimination is illegal in the United States in circumstances where economic discrimination is permitted.[2] Thus, legal remedies to disproportionate exposure of nonwhites to toxic wastes are available if racism is the cause, but not if people of color are exposed merely because they are poor.

There is strong evidence against claims of racial neutrality. Professor Been acknowledges that even if there is no racism in the process of siting LULUs, racism plays at least some part in the disproportionate exposure of African Americans to them. She cites evidence that "racial discrimination in the sale and rental of housing relegates people of color (especially African Americans) to the least desirable neighborhoods, regardless of their income level."[3]

Without acknowledging for a moment, then, that racism plays no part in the disproportionate exposure of nonwhites to toxic waste, I will ignore this issue to display a weakness in the argument that justice is served when economic discrimination alone is influential. I claim that even if the only discrimination is economic, justice requires redress and significant alteration of current practices. Recourse to the Doctrine of Double Effect presupposes that the primary effect, with which a second effect is incidentally conjoined, is morally justifiable. In the classic case, abortion is justified only because hysterectomy is justified as treatment for uterine cancer. I argue that disproportionate impacts on poor people violate principles of distributive justice, and so are not morally justifiable in the first place. Thus, current practices disproportionately exposing nonwhites to toxic substances are not justifiable even if incidental to the exposure of poor people.

Alternate practices that comply with acceptable principles of distributive justice are suggested below. They would largely solve problems of environmental racism (disproportionate impacts on nonwhites) while ameliorating the injustice of disproportionately exposing poor people to toxic hazards. They would also discourage production of toxic substances, thereby reducing humanity's negative impact on the environment.

The Principle of Commensurate Burdens and Benefit

We usually assume that, other things being equal, those who derive benefits should sustain commensurate burdens. We typically associate the burden of work with the benefit of receiving money, and the burdens of monetary payment and tort liability with the benefits of ownership.

There are many exceptions. For example, people can inherit money without working, and be given ownership without purchase. Another exception, which dissociates the benefit of ownership from the burden of tort liability, is the use of tax money to protect the public from hazards associated with private property, as in Superfund legislation. Again, the benefit of money is dissociated from the burden of work when governments support people who are unemployed.

The fact that these exceptions require justification, however, indicates an abiding assumption that people who derive benefits should shoulder commensurate burdens. The ability to inherit without work is justified as a benefit owed to those who wish to bequeath their wealth (which someone in the line of inheritance is assumed to have shouldered burdens to acquire). The same reasoning applies to gifts.

Using tax money (public money) to protect the public from dangerous private property is justified as encouraging private industry and commerce, which are supposed to increase public wealth. The system also protects victims in case private owners become bankrupt as, for example, in Times Beach, Missouri, where the government bought homes made worthless due to dioxin pollution. The company responsible for the pollution was bankrupt.

Tax money is used to help people who are out of work to help them find a job, improve their credentials, or feed their children. This promotes economic growth and equal opportunity. These exceptions prove the rule by the fact that justification for any deviation from the commensuration of benefits and burdens is considered necessary.

Further indication of an abiding belief that benefits and burdens should be commensurate is grumbling that, for example, many professional athletes and corporate executives are overpaid. Although the athletes and executives shoulder the burden of work, the complaint is that their benefits are disproportionate to their burdens. People on welfare are sometimes criticized for receiving even modest amounts of taxpayer money without shouldering the burdens of work, hence recurrent calls for "welfare reform." Even though these calls are often justified as means to reducing government budget deficits, the moral issue is more basic than the economic. Welfare expenditures are minor compared to other programs, and alternatives that require poor people to work are often more expensive than welfare as we know it.

The principle of commensuration between benefits and burdens is not the only moral principle governing distributive justice, and may not be the most important, but it is basic. Practices can be justified by showing them to conform, all things considered, to this principle. Thus, there is no move to "reform" the receipt of moderate pay for ordinary work, because it exemplifies the principle. On the other hand, practices that do not conform are liable to attack and require alternate justification, as we have seen in the cases of inheritance, gifts, Superfund legislation, and welfare.

Applying the principle of commensuration between burdens and benefits to the issue at hand yields the following: In the absence of countervailing considerations, the burdens of ill health associated with toxic hazards should be related to benefits derived from processes and products that create these hazards.

Toxic Hazards and Consumerism

In order to assess, in light of the principle of commensuration between benefits and burdens, the justice of current distributions of toxic hazards, the benefits of their generation must be considered. Toxic wastes result from many manufacturing processes, including those for a host of common items and materials, such as paint, solvents, plastics, and most petrochemical-based materials. These materials surround us in the paint on our houses, in our refrigerator containers, in our clothing, in our plumbing, in our garbage pails, and elsewhere.

Toxins are released into the environment in greater quantities now than ever before because we now have a consumer-oriented society where the acquisition, use, and disposal of individually owned items is greatly desired. We associate the numerical dollar value of the items at our disposal with our "standard of living," and assume that a higher standard is conducive to, if not identical with, a better life. So toxic wastes needing disposal are produced as by-products of the general pursuit of what our society defines as valuable, that is, the consumption of material goods.

Our economy requires increasing consumer demand to keep people working (to produce what is demanded). This is why there is concern each Christmas season, for example, that shoppers may not buy enough. If demand is insufficient, people may be put out of work. Demand must increase, not merely hold steady, because commercial competition improves labor efficiency in manufacture (and now in the service sector as well), so fewer workers can produce desired items. More items must be desired to forestall labor efficiency-induced unemployment, which is grave in a society where people depend primarily on wages to secure life's necessities.

Demand is kept high largely by convincing people that their lives require improvement, which consumer purchases will effect. When improvements are seen as needed, not merely desired, people purchase more readily. So our culture encourages economic expansion by blurring the distinction between wants and needs.

One way the distinction is blurred is through promotion of worry. If one feels insecure without the desired item or service, and so worries about life without it, then its provision is easily seen as a need. Commercials, and other shapers of social expectations, keep people worried by adjusting downward toward the trivial what people are expected to worry about. People worry about the provision of food, clothing, and housing without much inducement. When these basic needs are satisfied, however, attention shifts to indoor plumbing, for example, then to stylish indoor plumbing. The process continues with needs for a second or third bathroom, a kitchen disposal, and a refrigerator attached to the plumbing so that ice is made automatically in the freezer, and cold water can be obtained without even opening the refrigerator door. The same kind of progression results in cars with CD players, cellular phones, and automatic readouts of average fuel consumption per mile.

Abraham Maslow was not accurately describing people in our society when he claimed that after physiological, safety, love, and (self-) esteem needs are met, people work toward self-actualization, becoming increasingly their own unique selves by fully developing their talents. Maslow's Hierarchy of Needs describes people in our society less than Wenz's Lowerarchy of Worry. When one source of worry is put to rest by an appropriate purchase, some matter less inherently or obviously worrisome takes its place as the focus of concern. Such worry-substitution must be amenable to indefinite repetition in order to motivate purchases needed to keep the economy growing without inherent limit. If commercial society is supported by consumer demand, it is worry all the way down. Toxic wastes are produced in this context.

People tend to worry about ill health and early death without much inducement. These concerns are heightened in a society dependent upon the production of worry, so expenditure on health care consumes an increasing percentage of the gross domestic product. As knowledge of health impairment due to toxic substances increases, people are decreasingly tolerant of risks associated with their proximity. Thus, the same mindset of worry that elicits production that generates toxic wastes, exacerbates reaction to their proximity. The result is a desire for their placement elsewhere, hence the NIMBY syndrome – Not In My Back Yard. On this account, NIMBYism is not aberrantly selfish behavior, but integral to the cultural value system required for great volumes of toxic waste to be generated in the first place.

Combined with the principle of Commensurate Burdens and Benefits, that value system indicates who should suffer the burden of proximity to toxic wastes. Other things being equal, those who benefit most from the production of waste should shoulder the greatest share of burdens associated with its disposal. In our society, consumption of goods is valued highly and constitutes the principal benefit associated with the generation of toxic wastes. Such consumption is generally correlated with income and wealth. So other things being equal, justice requires that people's proximity to toxic wastes be related positively to their income and wealth. This is exactly opposite to the predominant tendency in our society, where poor people are more proximate to toxic wastes dumped illegally and stored legally.

Rejected Theories of Justice

Proponents of some theories of distributive justice may claim that current practices are justified. In this section I will explore such claims.

A widely held view of justice is that all people deserve to have their interests given equal weight. John Rawls's popular thought experiment in which people choose principles of justice while ignorant of their personal identities dramatizes the importance of equal consideration of interests. Even selfish people behind the "veil of ignorance" in Rawls's "original position" would choose to accord equal consideration to everyone's interests because, they reason, they may themselves be the victims of any inequality. Equal consideration is a basic moral premise lacking serious challenge in our culture, so it is presupposed in what follows. Disagreement centers on application of the principle.

Libertarianism

Libertarians claim that each individual has an equal right to be free of interference from other people. All burdens imposed by other people are unjustified unless part of, or consequent upon, agreement by the party being burdened. So no individual who has not consented should be burdened by burial of toxic wastes (or the emission of air pollutants, or the use of agricultural pesticides, etc.) that may increase risks of disease, disablement, or death. Discussing the effects of air pollution, libertarian Murray Rothbard writes, "The remedy is simply to enjoin anyone from injecting pollutants into the air, and thereby invading the rights of persons and property. Period."[4] Libertarians John Hospers and Tibor R. Machan seem to endorse Rothbard's position.[5]

The problem is that implementation of this theory is impractical and unjust in the context of our civilization. Industrial life as we know it inevitably includes production of pollutants and toxic substances that threaten human life and health. It is impractical to secure the agreement of every individual to the placement, whether on land, in the air, or in water, of every chemical that may adversely affect the life or health of the individuals in question. After being duly informed of the hazard, someone potentially affected is bound to object, making the placement illegitimate by libertarian criteria.

In effect, libertarians give veto power to each individual over the continuation of industrial society. This seems a poor way to accord equal consideration to everyone's interests because the interest in physical safety of any one individual is allowed to override all other interests of all other individuals in the continuation of modern life. Whether or not such life is worth pursuing, it seems unjust to put the decision for everyone in the hands of any one person.

Utilitarianism

Utilitarians consider the interests of all individuals equally, and advocate pursuing courses of action that promise to produce results containing the greatest (net) sum of good. However, irrespective of how "good" is defined, problems with utilitarian accounts of justice are many and notorious.

Utilitarianism suffers in part because its direct interest is exclusively in the sum total of good, and in the future. Since the sum of good is all that counts in utilitarianism, there is no guarantee that the good of some will not be sacrificed for the greater good of others. Famous people could receive (justifiably according to utilitarians) particularly harsh sentences for criminal activity to effect general deterrence. Even when fame results from honest pursuits, a famous felon's sentence is likely to attract more attention than sentences in other cases of similar criminal activity. Because potential criminals are more likely to respond to sentences in such cases,

harsh punishment is justified for utilitarian reasons on grounds that are unrelated to the crime.

Utilitarianism suffers in cases like this not only from its exclusive attention to the sum total of good, but also from its exclusive preoccupation with future consequences, which makes the relevance of past conduct indirect. This affects not only retribution, but also reciprocity and gratitude, which utilitarians endorse only to produce the greatest sum of future benefits. The direct relevance of past agreements and benefits, which common sense assumes, disappears in utilitarianism. So does direct application of the principle of Commensurate Burdens and Benefits.

The merits of the utilitarian rejection of common sense morality need not be assessed, however, because utilitarianism seems impossible to put into practice. Utilitarian support for any particular conclusion is undermined by the inability of anyone actually to perform the kinds of calculations that utilitarians profess to use. Whether the good is identified with happiness or preference-satisfaction, the two leading contenders at the moment, utilitarians announce the conclusions of their calculations without ever being able to show the calculation itself.

When I was in school, math teachers suspected that students who could never show their work were copying answers from other students. I suspect similarly that utilitarians, whose "calculations" often support conclusions that others reach by recourse to principles of gratitude, retributive justice, commensuration between burdens and benefits, and so forth, reach conclusions on grounds of intuitions influenced predominantly by these very principles.

Utilitarians may claim that, contrary to superficial appearances, these principles are themselves supported by utilitarian calculations. But, again, no one has produced a relevant calculation. Some principles seem *prima facie* opposed to utilitarianism, such as the one prescribing special solicitude of parents for their own children. It would seem that in cold climates more good would be produced if people bought winter coats for needy children, instead of special dress coats and ski attire for their own children. But utilitarians defend the principle of special parental concern. They declare this principle consistent with utilitarianism by appeal to entirely untested, unsubstantiated assumptions about counterfactuals. It is a kind of "Just So" story that explains how good is maximized by adherence to current standards. There is no calculation at all.

Another indication that utilitarians cannot perform the calculations they profess to rely upon concerns principles whose worth is in genuine dispute. Utilitarians offer no calculations that help to settle the matter. For example, many people wonder today whether or not patriotism is a worthy moral principle. Detailed utilitarian calculations play no part in the discussion.

These are some of the reasons why utilitarianism provides no help to those deciding whether or not disproportionate exposure of poor people to toxic wastes is just.

Free market approach

Toxic wastes, a burden, could be placed where residents accept them in return for monetary payment, a benefit. Since market transactions often satisfactorily commensurate burdens and benefits, this approach may seem to honor the principle of commensuration between burdens and benefits.

Unlike many market transactions, however, whole communities, acting as corporate bodies, would have to contract with those seeking to bury wastes. Otherwise, any single individual in the community could veto the transaction, resulting in the impasse attending libertarian approaches.[6] Communities could receive money to improve such public facilities as schools, parks, and hospitals, in addition to obtaining tax revenues and jobs that result ordinarily from business expansion.

The major problem with this free market approach is that it fails to accord equal consideration to everyone's interests. Where basic or vital goods and services are at issue, we usually think equal consideration of interests requires ameliorating inequalities of distribution that markets tend to produce. For example, one reason, although not the only reason, for public education is to provide every child with the basic intellectual tools necessary for success in our society. A purely free market approach, by contrast, would result in excellent education for children of wealthy parents and little or no

education for children of the nation's poorest residents. Opportunities for children of poor parents would be so inferior that we would say the children's interests had not been given equal consideration.

The reasoning is similar where vital goods are concerned. The United States has the Medicaid program for poor people to supplement market transactions in health care precisely because equal consideration of interests requires that everyone be given access to health care. The 1994 health care debate in the United States was, ostensibly, about how to achieve universal coverage, not about whether or not justice required such coverage. With the exception of South Africa, every other industrialized country already has universal coverage for health care. Where vital needs are concerned, markets are supplemented or avoided in order to give equal consideration to everyone's interests.

Another example concerns military service in time of war. The United States employed conscription during the Civil War, both world wars, the Korean War, and the war in Vietnam. When the national interest requires placing many people in mortal danger, it is considered just that exposure be largely unrelated to income and market transactions.

The United States does not currently provide genuine equality in education or health care, nor did universal conscription (of males) put all men at equal risk in time of war. In all three areas, advantage accrues to those with greater income and wealth. (During the Civil War, paying for a substitute was legal in many cases.) Imperfection in practice, however, should not obscure general agreement in theory that justice requires equal consideration of interests, and that such equal consideration requires rejecting purely free market approaches where basic or vital needs are concerned.

Toxic substances affect basic and vital interests. Lead, arsenic, and cadmium in the vicinity of children's homes can result in mental retardation of the children.[7] Navaho teens exposed to radiation from uranium mine tailings have seventeen times the national average of reproductive organ cancer.[8] Environmental Protection Agency (EPA) officials estimate that toxic air pollution in areas of South Chicago increase cancer risks one hundred to one thousand times.[9] Pollution from

Otis Air Force base in Massachusetts is associated with alarming increases in cancer rates.[10] Non-Hodgkin's Lymphoma is related to living near stone, clay, and glass industry facilities, and leukemia is related to living near chemical and petroleum plants.[11] In general, cancer rates are higher in the United States near industries that use toxic substances and discard them nearby.[12]

In sum, the placement of toxic wastes affects basic and vital interests just as do education, health care, and wartime military service. Exemption from market decisions is required to avoid unjust impositions on the poor, and to respect people's interests equally. A child dying of cancer receives little benefit from the community's new swimming pool.

Cost-benefit analysis (CBA)

CBA is an economist's version of utilitarianism, where the sum to be maximized is society's wealth, as measured in monetary units, instead of happiness or preference satisfaction. Society's wealth is computed by noting (and estimating where necessary) what people are willing to pay for goods and services. The more people are willing to pay for what exists in society, the better off society is, according to CBA.

CBA will characteristically require placement of toxic wastes near poor people. Such placement usually lowers land values (what people are willing to pay for property). Land that is already cheap, where poor people live, will not lose as much value as land that is currently expensive, where wealthier people live, so a smaller loss of social wealth attends placement of toxic wastes near poor people. This is just the opposite of what the Principle of Commensurate Burdens and Benefits requires.

The use of CBA also violates equal consideration of interests, operating much like free market approaches. Where a vital concern is at issue, equal consideration of interests requires that people be considered irrespective of income. The placement of toxic wastes affects vital interests. Yet CBA would have poor people exposed disproportionately to such wastes.[13]

In sum, libertarianism, utilitarianism, free market distribution, and cost-benefit analysis are inadequate principles and methodologies to guide the just distribution of toxic wastes.

LULU Points

An approach that avoids these difficulties assigns points to different types of locally undesirable land uses (LULUs) and requires that all communities earn LULU points.[14] In keeping with the Principle of Commensurate Benefits and Burdens, wealthy communities would be required to earn more LULU points than poorer ones. Communities would be identified by currently existing political divisions, such as villages, towns, city wards, cities, and counties.

Toxic waste dumps are only one kind of LULU. Others include prisons, half-way houses, municipal waste sites, low-income housing, and power plants, whether nuclear or coal fired. A large deposit of extremely toxic waste, for example, may be assigned twenty points when properly buried but fifty points when illegally dumped. A much smaller deposit of properly buried toxic waste may be assigned only ten points, as may a coal-fired power plant. A nuclear power plant may be assigned twenty-five points, while municipal waste sites are only five points, and one hundred units of low-income housing are eight points.

These numbers are only speculations. Points would be assigned by considering probable effects of different LULUs on basic needs, and responses to questionnaires investigating people's levels of discomfort with LULUs of various sorts. Once numbers are assigned, the total number of LULU points to be distributed in a given time period could be calculated by considering planned development and needs for prisons, power plants, low-income housing, and so on. One could also calculate points for a community's already existing LULUs. Communities could then be required to host LULUs in proportion to their income or wealth, with new allocation of LULUs (and associated points) correcting for currently existing deviations from the rule of proportionality.

Wherever significant differences of wealth or income exist between two areas, these areas should be considered part of different communities if there is any political division between them. Thus, a county with rich and poor areas would not be considered a single community for purposes of locating LULUs. Instead, villages or towns may be so considered. A city with rich and poor areas may similarly be reduced to its wards.

The purpose of segregating areas of different income or wealth from one another is to permit the imposition of greater LULU burdens on wealthier communities. When wealthy and poor areas are considered as one larger community, there is the danger that the community will earn its LULU points by placing hazardous waste near its poorer members. This possibility is reduced when only relatively wealthy people live in a smaller community that must earn LULU points.

Practical Implications

Political strategy is beyond the scope of this chapter, so I will refrain from commenting on problems and prospects for securing passage and implementation of the foregoing proposal. I maintain that the proposal is just. In a society where injustice is common, it is no surprise that proposals for rectification meet stiff resistance.

Were the LULU points proposal implemented, environmental racism would be reduced enormously. To the extent that poor people exposed to environmental hazards are members of racial minorities, relieving the poor of disproportionate exposure would also relieve people of color.

This is not to say that environmental racism would be ended completely. Implementation of the proposal requires judgment in particular cases. Until racism is itself ended, such judgment will predictably be exercised at times to the disadvantage of minority populations. However, because most people of color currently burdened by environmental racism are relatively poor, implementing the proposal would remove 80 to 90 percent of the effects of environmental racism. While efforts to end racism at all levels should continue, reducing the burdens of racism is generally advantageous to people of color. Such reductions are especially worthy when integral to policies that improve distributive justice generally.

Besides improving distributive justice and reducing the burdens of environmental racism, implementing the LULU points proposal would benefit life on earth generally by reducing the generation of toxic hazards. When people of wealth, who exercise control of manufacturing processes, marketing campaigns, and media coverage, are themselves threatened disproportionately by toxic hazards, the culture will evolve quickly to find their production largely unnecessary. It will be discovered, for example, that many plastic items can be made of wood, just as it was discovered in the late 1980s that the production of many ozone-destroying chemicals is unnecessary. Similarly, necessity being the mother of invention, it was discovered during World War II that many women could work in factories. When certain interests are threatened, the impossible does not even take longer.

The above approach to environmental injustice should, of course, be applied internationally and intranationally within all countries. The same considerations of justice condemn universally, all other things being equal, exposing poor people to vital dangers whose generation predominantly benefits the rich. This implies that rich countries should not ship their toxic wastes to poor countries. Since many poorer countries, such as those in Africa, are inhabited primarily by non-whites, prohibiting shipments of toxic wastes to them would reduce significantly worldwide environmental racism. A prohibition on such shipments would also discourage production of dangerous wastes, as it would require people in rich countries to live with whatever dangers they create. If the principle of LULU points were applied in all countries, including poor ones, elites in those countries would lose interest in earning foreign currency credits through importation of waste, as they would be disproportionately exposed to imported toxins.

In sum, we could reduce environmental injustice considerably through a general program of distributive justice concerning environmental hazards. Pollution would not thereby be eliminated, since to live is to pollute. But such a program would motivate significant reduction in the generation of toxic wastes, and help the poor, especially people of color, as well as the environment.

Notes

1   Vicki Been, "Market Forces, Not Racist Practices, May Affect the Siting of Locally Undesirable Land Uses," in *At Issue: Environmental Justice*, ed. by Jonathan Petrikin (San Diego, Calif.: Greenhaven Press, 1995), 41.

2   See *San Antonio Independent School District v. Rodriguez*, 411 R.S. 1 (1973) and *Village of*

Arlington Heights v. Metropolitan Housing Development Corporation, 429 U.S. 252 (1977).

3   Been, 41.

4   Murray Rothbard, "The Great Ecology Issue," *The Individualist* 21, no. 2 (February 1970): 5.

5   See Peter S. Wenz, *Environmental Justice* (Albany, N.Y.: State University of New York Press, 1988), 65–67 and associated endnotes.

6   Christopher Boerner and Thomas Lambert, "Environmental Justice Can Be Achieved Through Negotiated Compensation," in *At Issue: Environmental Justice.*

7   F. Diaz-Barriga et al., "Arsenic and Cadmium Exposure in Children Living Near to Both Zinc and Copper Smelters," summarized in *Archives of Environmental Health* 46, no. 2 (March/April 1991): 119.

8   Dick Russell, "Environmental Racism," *Amicus Journal* (Spring 1989): 22–32, 24.

9   Marianne Lavelle, "The Minorities Equation," *National Law Journal* 21 (September 1992): 3.

10   Christopher Hallowell, "Water Crisis on the Cape," *Audubon* (July/August 1991): 65–74, especially 66 and 70.

11   Athena Linos et al., "Leukemia and Non-Hodgkin's Lymphoma and Residential Proximity to Industrial Plants," *Archives of Environmental Health* 46, no. 2 (March/April 1991): 70–74.

12   L. W. Pickle et al., *Atlas of Cancer Mortality among Whites: 1950–1980,* HHS publication # (NIH) 87-2900 (Washington, D.C.: U.S. Department of Health and Human Services, Government Printing Office: 1987).

13   Wenz, 216–18.

14   The idea of LULU points comes to me from Frank J. Popper, "LULUs and Their Blockage," in *Confronting Regional Challenges: Approaches to LULUs, Growth, and Other Vexing Governance Problems,* ed. by Joseph DiMento and Le Roy Graymer (Los Angeles, Calif.: Lincoln Institute of Land Policy, 1991), 13–27, especially 24.

Part XI

Immediate Challenges: Information Technologies, Technological Systems and the Future of Human Values

Introduction

Next to biotechnologies, the area of techno-logical development that has received the most recent attention has been the revolution in information technologies over the past 30 years. Information technologies are given credit for the economic boom of the 1990s and criticized for undermining local cultures and individual privacy. They are considered instrumental in the process of globalization. These essays examine relations between information technologies, technological systems, and values.

Carl Mitcham is Professor and Director of the Hennenbach Program in the Humanities at the Colorado School of Mines. This section opens with his essay exploring some of the metaphysical issues that underlie value questions in informa-tion technology. Sven Birkerts, director of the Bennington College Writing Seminars, is con-cerned with the fate of reading, literacy, and the printed word as we increasingly rely on electronic information technologies. He warns of three troubling developments: language erosion, a flattening of historical perspective, and a waning of the private self. Each of these is a tendency in existing culture; witness the increased use of the condensed languages of email or mobile-device texting, history as fodder for electronic information while libraries increasingly give away printed collections, and the rise of social-networking sites on which many live most of their lives in full transparent view (with diaries and journals available for general viewing and

comments). He notes that these are trends, and perhaps are not inevitable. Nonetheless, his posi-tion comes close to the dystopian pessimism of Ellul.

Wendell Berry, conservationist, farmer, essay-ist, novelist, former professor of English at the University of Kentucky, and poet, argues that selecting for a particular technology is a complex decision that implies many changes that are often not acknowledged. In this case, opting for a computer would mean changes not only in his immediate work habits, but also in the life of the household. His essay thus ties together technology assessment and an examination of technological systems with reflections on the relations between technology and everyday life.

Shoshana Zuboff, retired Charles Edward Wilson Professor of Business Administration at Harvard University, presents an optimistic account of information technologies. She argues that computers bring the potential to revolu-tionize the workplace by distributing authority, allowing better information flow and therefore a better understanding of the entire process by all involved.

Paul Duguid, a cultural studies specialist at UC-Berkeley and a professorial research fellow at Queen Mary, University of London, and John Seely Brown, former chief research scientist and director of the Palo Alto Research Center (PARC) at Xerox, argue that the many revolu-tions promised by new information technologies have not, and will not, come to pass. Echoing

Feenberg's work, they argue that information requires context, and the existing research programs and products in information technologies tend to ignore the social context that makes information meaningful.

Founding director of the Human-Computer Interface Lab and Professor of Computer Science at the University of Maryland, Ben Shneiderman argues that far too much attention has been paid to what the machine can do, and we should turn our attention instead to what users can do with computers. On this second point, his analysis recapitulates some of the concerns of Feenberg. Although his focus on the capacities of computers echoes the fears of Ellul that technology will become an autonomous and determining system, Shneiderman believes that thoughtful people can direct technological development in ways that will unleash previously unrealized human potentials.

41

Philosophy of Information Technology

Carl Mitcham

Philosophy of information technology may be seen as a special case of the philosophy of technology. Philosophical reflection on technology aims in general to comprehend the nature and meaning of the making and using, especially of things made and used. Such reflection nevertheless exhibits a tension between two major traditions: one arising within engineering, another in the humanities (Mitcham 1994). For the former or expansionist view, technology is deeply and comprehensively human, and thus properly extended into all areas of life; according to the latter or limitationist perspective, technology is a restricted and properly circumscribed dimension of the human. This distinction and corresponding tensions may also be seen at play in the philosophy of information technology (IT), between those who would critically celebrate and extend IT and those who would cautiously subordinate and delimit it. Diverse metaphysical, epistemological, and ethical arguments are marshaled to defend one position over the other, as well as to build bridges between these two philosophical poles.

Philosophies of *x* commonly begin with attempts to define *x*. Philosophy of science, for instance, logically opens with the demarcation problem, by considering various proposals for distinguishing science from other forms of knowledge or human activity. The philosophy of information technology, like the philosophy of computer science, is properly initiated by the effort to define that on which it seeks to reflect. Once preliminary definitions are negotiated, philosophies of *x*, often against a historico-philosophical background, recapitulate in differentially weighted forms the main branches of philosophy *tout court* – metaphysics, epistemology, and ethics – with particular emphases reflecting both the unique philosophical challenges of *x* and the context of presentation. In the present case, for example, although ethical and political issues play a prominent role in the philosophy of information technology, they are treated lightly here. Here the stress is on theoretical issues concerning especially metaphysical assessments of information technology.

What is Information Technology?

Information technology – or such closely related terms as "information systems" and "media technology" – is commonly described as that technology constituted by the merging of data-processing and telecommunications (with diverse input devices, processing programs, communications systems, storage formats, and output displays). It arose from earlier forms of electronic

From Luciano Floridi, ed., *The Blackwell Guide to Philosophy of Computing and Information Technology* (Oxford: Blackwell Publishing, 2003), pp. 327–36. Reprinted by permission of Blackwell Publishing.

communications technology (telegraph, telephone, phonograph, radio, motion pictures, television) by way of computers and cybernetics, an earlier term that still casts its shadow over IT, as in such coinages as "cyberspace" and other cognates. It may nevertheless be useful to begin by attempting to rethink what is perhaps too facile in such a description.

The terms "information" and "technology" are both subject to narrow and broad, not to say engineering and humanities, definitions. Developed by Claude Shannon (Shannon & Weaver 1949), the technical concept of information is defined as the probability of a signal being transmitted from device A to device B, which can be mathematically quantified. The theory of information opened up by this distinct conceptual analysis has become the basis for constructing and analyzing digital computational devices and a whole range of information (also called communication) technologies, from telephones to televisions and the internet.

In contrast to information (and information technologies) in the technical sense is the concept of information in a broader or semantic sense. Semantic information is not a two-term relation – that is, a signal being transmitted from A to B – but a three-term relation: a signal transmitted from one device to another, which is then understood as saying something to a person C. Although information technologies in the technical sense readily become information technologies in the semantic sense, there is no precise relation between technical and semantic information. Independent of its probability as a signal, some particular transmission may possess any number of different semantic meanings. A signal in the form of two short clicks or light flashes (Morse Code for the letter "i"), could be

a self-referential pronoun, part of the word "in," a notation in Latin numerals of the number one – or any number of other possibilities. Absent the context, a signal is not a message. Kenneth Sayre (1976) and Fred I. Dretske (1981) are nevertheless two important attempts to develop semantic theories of information grounded in the technical concept of information.

"Technology" too is a term with narrow and broad definitions. In the narrow or engineering sense, technology is constituted by the systematic study and practice of the making and using of artifacts (cf. the curricula of technological universities), and to some extent by the physical artifacts themselves (from hammers to cars and computers). Indeed, a distinction is often drawn between premodern *techne* or technics and modern technology. For thousands of years human making and using proceeded by intuitive, trial-and-error methods, remained mostly small-scale and dwarfed by natural phenomena. With the rise of modern methods for making and using, these activities became systematically pursued (often on the basis of scientific theories) and their products began to rival natural phenomena in scale and scope. In a broader humanities parlance, technology covers both intuitive, small-scale and scientific, large-scale making and using in all its modes – as knowledge, as artifact, as activity, and even as volition.

Given these narrow and broad definitions for each element in the compound term, one may postulate a two by two matrix and imagine four different information technology exemplars (Table 41.1). In what follows, a significant sample from among these possible information technologies will be analyzed in order to illustrate diverse facets of a potentially comprehensive philosophy of information technology.

Table 41.1   Information technology exemplars

	Premodern technology	*Scientific technology*
Technical sense of information	Alphabetic writing	Electronic and source code signal transmission
Semantic sense of information	Books and related texts	Works of high representational electronic communications media (movies, TV programs, hypertexts, etc.)

Information Technology in Historico-philosophical Perspective

Philosophy is not coeval with human thought, but emerges from and against prephilosophical reflection that it nevertheless continues or mirrors. Prior to the rise of philosophy, mythological and poetic narratives often expressed the ambivalence of the human experience of tool making and using. Stories of the conflict between Cain (builder of cities) and Abel (pastoral shepherd), of Prometheus (who stole fire for humans from the gods), of Hephaestus (the deformed god of the forge), and of Icarus (the inventor who went too far) all attest to the problematic character of human engagement with what has come to be called technology. The story of the Tower of Babel (Genesis 11) even suggests the destructive linguistic repercussions of an excessive pursuit of technological prowess.

By contrast, when the prophet Ezekiel learns in the desert to infuse dry bones (alphabetic consonants) with the breath of the spirit (unwritten vowels), it is as if God were speaking directly through him (Ezekiel 37). Indeed, God himself creates through speech or *logos* (Genesis 1), and writes the law both in stone and in the hearts of a people. Thus, information technologies in their earliest forms – speech and writing – manifest at least two fundamental experiences of the human condition: sin or hubris and transcendence, the demonic and the divine.

Greek philosophical reflection on *techne* likewise noted the two-fold tendency of human skill in the making and using of artifacts to be pursued in isolation from the good and to participate in the divine. This is as true of information *technai*, such as oratory and writing, as it is of the mechanical and military arts. In Plato's *Gorgias*, for instance, Socrates challenges the sophist to reintegrate the techniques of rhetoric with the pursuit of truth, to eschew the tricks of gaining power divorced from knowledge of the good. In the *Phaedrus*, Socrates tells the story of how King Thamus rejected the Egyptian god Theuth's invention of writing on the grounds that it would replace real with merely virtual memory (*Phaedrus* 274d ff.). Socrates himself comments on the silence of written words, and Plato famously remarked on the limitations of writing even in his own works (*Letter* VII, 341b–e and

344c–d). The *Politicus* (300c ff.), however, concludes with a modest defense of written laws, and the *Ion* presents the poet as one inspired by the gods.

Aristotle, in an analysis that echoes Plato's assessment in *The Republic* of artifice and poetry as thrice-removed from being itself, notes the inability of *techne* to effect a substantial unity of form and matter. "If a bed were to sprout," says Aristotle, "not a bed would come up but a tree" (*Physics* II, i, 193a12–16). In a parallel analysis of the relation between experiences, spoken words, and writing at the beginning of *On Interpretation*, Aristotle places the written word at two removes from experience and three removes from the things experienced, thus implying a dilution of contact with reality as one moves from the information technology of speech to that of writing. Spoken words refer to experience; written words to spoken words.

In contrast to Aristotle's characterization of words in strictly human terms, Christianity reaffirms the divine character of the transcendent word incarnate (John 1) and of the transmission of the gospel through that preaching which represents the word (Romans 10:17). Indeed, according to Augustine, Christian preaching unites truth and language with an efficacy that the Platonists could not imagine (*De vera religione* i, 1 ff.). This is an argument that has been revived in Catherine Pickstock's theological interpretation of that information technology known as liturgy (Pickstock 1998). At the same time, the meaning of the words of revelation in Scripture is not always obvious, thus requiring the development of principles of interpretation (see Augustine's *De doctrina christiana*). Faith in the Scriptures as the word of God solves, as it were, the technical question concerning the extent to which the signal has been accurately transmitted from A to B (God to humans), but not the semantic question of what this signal means (to whom it speaks and about what). The meaning of revelation requires a science of interpretation or hermeneutics if its information (from the Latin *informare*, to give form) is truly to convert those who receive it.

This dedication to the development of techniques of interpretation led to a unique medieval flowering of logical, rhetorical, and hermeneutic prowess. Reflecting the effulgence of poetic exegesis of sacred texts, Thomas Aquinas defends the metaphorical "hiding of truth in figures" as fitting to the word of God, and argues the power of Scripture to signify by way of multiple references: historical or literal, allegorical, tropological or moral, and anagogical or eschatological (*Summa theologiae* I, q.1, art.9–10). What is equally remarkable is that – no doubt stimulated by the literal and spiritual interpretations of revelation as granting the world a certain autonomy and calling upon human beings to exercise positive mastery over it – the flowering of semantic studies was paralleled by an equally unprecedented blossoming of physical technologies. Examples include the waterwheel and windmill, the moldboard plow, the horse collar, the lateen sail, and the mechanical clock.

The modern world opens, paradoxically, by pitting the second form of technological progress (physical inventions) against the first (poetic creativity). Metaphorical words are to be rejected in the pursuit of real things and ever more powerful technologies (see especially the arguments of Francis Bacon and René Descartes). The historical result was to turn exegesis into criticism and semantic analysis into a drive for conceptual clarity, in a reform of the techniques of communication that became most manifest in the new rhetoric of modern natural science – as well as in the invention of a whole new information technology known as moveable type.

The invention of the printing press and the consequent democratization of reading can be associated with a manifold of social transformations: religious, political, economic, and cultural. The philosophical influences of such changes have been legion. To cite but one example, as the world was increasingly filled with texts, and texts themselves were severed from stable lifeworlds of interpreters, philosophy became increasingly linguistic philosophy, in two forms. In continental Europe, hermeneutics was redefined by Friedrich Schleiermacher as the interpretation of all (not just sacred) texts, by Wilhelm Dilthey as the foundation of the *Geistwissenschaften* or humanistic sciences, and by Martin Heidegger as the essence of *Dasein* or human being. In this same milieu, Ferdinand de Saussure invented the science of linguistics, focusing neither on efficient signal transmission nor on multiple levels of external reference but on language as a system of words that mutually define one another through

their internal relations. In the Anglo-American world, especially under the influence of Ludwig Wittgenstein, philosophy became linguistic philosophy, which takes the meaning of words to be constituted by their uses, thus calling attention to multiple contexts of use, what Wittgenstein called ways of life. Indeed, in some forms the resultant philosophy of language turns into a kind of behaviorism or is able to make common cause with pragmatism.

In another instance, theories were posited about the relation between changes in information technologies and cultural orders. The contrast between orality and literacy has been elaborated by a series of scholars – from Albert Lord and Milman Parry to Marshall McLuhan, Walter Ong, and Ivan Illich – who have posited complementary theories about relations between information technology transformations and cultures. With McLuhan, for example, there is a turn not just from technical signal to semantic message, but an attempt to look at the whole new electronic signal transmitting and receiving technology (never mind any specific semantic content) as itself a message. In his own condensed formulation: the medium (or particular form of information technology) is the message (McLuhan 1964).

Stimulated especially by McLuhan, reviews of the historical influences between philosophy and IT begin to mesh into a philosophy of history that privileges IT experience the way G. W. F. Hegel privileged politics and ideas. Here Paul Levinson's "natural history of information technology" (1997) is a worthy illustration.

Information Technology and Metaphysics

Although the historico-philosophical background points to an emergence, in conjunction with information technology, of new cultural constellations in human affairs, pointing alone is insufficient to constitute philosophy. Popular attempts to think the new IT lifeworld have emphasized economics and politics, in which issues are decided about e-banking and e-commerce on the basis of market forces and political power. The ethics of information technology, as an initiation into philosophical reflection – that is, into thought in which issues are assessed on the basis of argument and insight rather than money and votes – has highlighted issues of privacy, equity, and accountability. Yet given that the fundamental question for ethics concerns how to act in accord with what really is, there are reasons to inquire into the kind of reality disclosed by IT – that is, to raise metaphysical (beyond the physical or empirical) and ontological (from *ontos*, the Greek word for "being") questions.

What are the fundamental structures of the IT phenomenon? What is real and what is appearance with regard to IT? Richard Coyne (1995), for instance, argues that it is illusory to view IT as simply a novel instrument available for the effective realization of traditional projects for conserving and manipulating data. Albert Borgmann (1999) insightfully distinguishes between information about reality (science), information for reality (engineering design), and information as reality (the high-definition representations and creations emerging from IT) – and further the increasing prominence, glamor, and malleability of information as reality is having the effect of diminishing human engagement with more fundamental realities. With regard to the kinds of metaphysical issues raised by Coyne, Borgmann, and others, it is useful to distinguish again expansionist and limitationist approaches to the nature and meaning for information technology.

The expansionist approach has its roots in technical thinking about IT, first in terms of physical entities. At least since Norbert Wiener (1948) effectively posited that, along with matter and energy, information is a fundamental constituent of reality, questions have been raised about the metaphysical status of information. Building on Wiener's own analysis, distinctions may be drawn between three fundamentally different kinds of technology: those which transform matter (hammers and assembly lines), those which produce and transform energy (power plants and motors), and those which transform information (communication systems and computers).

A related phenomenology of human engagement would observe how the being of IT differs from tools and machines. Unlike tools (which do not function without human energy input and guidance) or machines (which derive energy from nonhuman sources but still require human

guidance), information technologies are in distinctive ways independent of the human with regard to energy and immediate guidance; they are self-regulating (cybernetic). In this sense, steam engines with mechanical governors on them or thermostatically controlled heating systems are examples of information machines. Insofar as the operation of more electronically advanced IT is subject to human guidance, guidance ceases to be direct or mechanical and is mediated by humanly constructed programs (electronically coded plans). What is the ontological status of programs? What are their relations to intentions? Indeed, in IT, operation and use appear to have become distinguishable. IT is a new species of artifact, a hybrid that is part machine running on its own and part utility structure like a road waiting to be driven on – hence the term "new media" (as both means and environment). The static availability of such structures is contingent on their semi-autonomous dynamic functioning.

Second, in terms of the cognitive capabilities of IT, transempirical questions arise about the extent to which computers (as pervasive elements in IT) imitate human cognitive processes. Do computers think? What kind of intelligence is artificial intelligence (AI)? Are the different kinds of AI – algorithmic, heuristic, connectionist, embodied, etc. – different forms of intelligence? Such ontological questions now blend into others, concerning the extent to which high-tech artifacts are different from living organisms. Biotechnology has breached Aristotle's distinction between natural tree and artificial bed, growth and construction, the born and the made. Soon computer programs may also be able not just to mimic patterns of growth on the screen, but autonomous, artificial agents that are able to reproduce themselves. At the nano-scale, robotic design will hardly be distinct from genetic engineering. Will any differences in being remain?

From the technical perspective, information is ubiquitous in both the organic and the artificial worlds. The wall between the two is vanishing, although, insofar as the technical concept of information becomes a category of explanation in biology, it has also been argued to have distinct ideological roots (see Kay 2000 on this point). The cyborg (cybernetic organism) is a living machine, not a goddess (Haraway 1991). Within such a reality, the ethical imperative becomes experimenting with ourselves, what Coyne (1995) calls a pragmatic interaction with advancing IT. This is an attitude widely present among leading IT designers such as Mark Weiser at the famous Xerox Palo Alto Research Center (PARC), the ethos of which is commonly celebrated in *Wired* magazine. It has also been given philosophical articulation by media philosopher Wolfgang Schirmacher. For Schirmacher (1994), IT is a kind of artificial nature, a post-technology in which we are free (and obligated, if we would act in harmony with the new way of being in the world) to live without predeterminations, playfully and aesthetically.

The limitationist approach originates in a different, more skeptical stance. Issues are no doubt oversimplified by characterizing one approach as pro-IT and another as con-IT – although such a contrast captures some measure of real difference (but see Gordon Graham, 1999, for a down-to-earth philosophical utilization of this contrast using the terms technophiles and neoluddites). Perhaps a better contrast would be that of Hegel versus Socrates: the comprehensive critical affirmation as opposed to the argumentative gadfly. From the Hegelian perspective there is something both adolescent and irresponsible about an ongoing Socratic negativity that refuses to take responsibility for world creation. Indeed, Socratic negativity easily becomes a philosophically clichéd substitute for true thinking. From the Socratic perspective, however, the expansionist approach comes on the scene as a court philosophy, especially insofar as it flatters the king and counsels expanding an already popular and widely affirmed domain of influence. In a state already dominated by information technology, the Socratic tradition thus finds expression in repeatedly questioning the nature and meaning of IT – a questioning that must ultimately go metaphysical.

At a first level, however, the questioning of IT will be, as already suggested, ethical. For instance, does IT not threaten privacy? Even more profoundly, does the IT mediation of human action in complex software programs, which are created by multiple technicians and are not even in principle able to be fully tested (Zimmerli 1986), not challenge the very notion of moral accountability? At a second level are political questions:

Is the internet structured so as to promote social justice through equity of access? Is it compatible with democracy? Furthermore, IT exists on the back of a substantial industrial base, whose environmental sustainability is at least debatable. Insofar as IT depends on an unsustainable base, might not its own justice and goodness be compromised? At still a third level are psychological questions, blending into epistemological ones. Does the exponential growth of information availability not challenge the human ability to make sense of it? Information overload or information anxiety (see Wurman 2001) is one of the most widely cited paradoxes of IT life. Finally, at a fourth level are psychological-anthropological questions about the social implications of the new "mode of information" (Poster 1990), what it means to live a "virtual life" (Brook & Boal 1995) and "life on the screen" (Turkle 1995).

The third and fourth dimensions of limitationist, Socratic questioning – that is, the epistemological and anthropological levels – hint at the metaphysical. Information technology may hide reality from us in a much more fundamental way than simply by means of information overload. It may deform our being at deeper levels than the psychological. To develop this possibility it is useful to refer at some length to Martin Heidegger, the most influential exponent of this position.

According to Heidegger's highly influential argument in "The Question Concerning Technology" (1977 [1954]), technology is constituted not so much by machines or even instrumental means in general as by its disclosure of reality, its unhiding, its truth. Premodern technology in the form of *poiesis* functioned as a bringing or leading forth that worked with nature, and as such revealed Being as alive with its own bringing forth, the way a seed blossoms into a flower or an acorn grows into an oak tree. Modern technology, by contrast, is not so much a bringing forth as a challenging-forth that reveals the world as *Bestand* or manipulatable resource.

In reading Heidegger it is crucial to recognize that he felt it necessary to couch his insights in a special vocabulary ("bringing forth," "challenging forth," "*Bestand*"), because of the way ordinary concepts are sedimented with assumptions that themselves help conceal the dimensions of reality to which he invites attention. In Heidegger's own words:

> The revealing that rules throughout modern technology has the character of a setting-upon, in the sense of a challenging-forth. That challenging happens in that the energy concealed in nature is unlocked, what is unlocked is transformed, what is transformed is stored up, what is stored up is, in turn, distributed, and what is distributed is switched about ever anew. Unlocking, transforming, storing, distributing, and switching about are ways of revealing. (Heidegger 1977 [1954]: 297–8)

To this distinctive way of revealing Heidegger also gives a special name: *Gestell* or enframing.

Although Heidegger seems to be thinking here of electric power generation, the same description would in many ways be applicable to information technology. There is a challenging that happens when digitally concealed information is unlocked (from, say, a computer disk), transformed (by some software program), stored up (on a hard drive), distributed (by internet), and switched about (forwarded, reprocessed, data mined, etc.). Indeed, in another text Heidegger makes the reference to IT explicit, although under the name of cybernetics. "Cybernetics," he writes, "transforms language into an exchange of news. The arts become regulated-regulating instruments of information" (Heidegger 1977 [1966]: 376). Modern information technology thus does to language what modern non-information technology does to the material world: turns it into *Bestand*, that is, a resource for human manipulation.

What is wrong with this? The basic answer is that modern technology, including modern information technology, conceals as well as it reveals. Insofar as we persist in emphasizing the revealing and ignore the concealing, concealing will actually dominate. We will not be fully aware of what is going on. To develop this point requires a brief elaboration of Heidegger's theory of hermeneutics. In his version of hermeneutics, which argues interpretation (more than rationality) as the defining characteristic of the human, Heidegger makes two basic claims.

The first is that no revealing (the acquisition of information in the semantic sense) is ever simple; it always involves the process of interpretation. Interpretation itself proceeds in texts, in perception, in thinking, and in life by means of a dialectic between part and whole, what is called the hermeneutic circle. The part is only revealed in terms of the whole, and the whole in terms of the parts. As a result, Heideggerian hermeneutics postulates a pregivenness in all revealing or, as he also likes to say, unconcealing. Our minds and our lives open not as with a *tabula rasa*, but with an immanent reality (both part and whole) waiting to be brought forth into the light of appearances. Understanding proceeds by means of a process of moving from part to whole and *vice versa*, repeatedly to make the implicit explicit, to reveal the concealed, analogous especially to the ways that premodern technology also worked to till the fields and to fashion handcrafted artifacts. The upshot is that not only is all information subject to interpretation, but that all information technology is part of a larger lifeworld and cannot be understood apart from such an implicit whole. To think otherwise is a metaphysical mistake.

The second claim is that any unconcealing is at one and the very same time a concealing. This second claim has even more profound implications for information technology, which through its expanding realms makes information more and more omnipresent. Information technology appears to reveal with a vengeance. According to Heidegger, however, this is ultimately an illusion – and dangerous to what it means to be human. The problem is not just one of sensory or information overload, but of information as a concealing of Being itself, the fundamental nature of reality, of the distinctly human relation to such reality.

For Heidegger the rise of modern technology, and its culmination in cybernetics or information technology, is the culmination of a historico-philosophical trajectory of thinking that began with the Greeks. With Plato and Aristotle, Being was first revealed, however tentatively and minimally, as a presence that could be re-presented in thought or rationalized. Over the course of its 2,500-year history, philosophy has successively spun off the various scientific disciplines as specialized ways to re-present the world: in mathematics, in logic, in astronomy, in physics, in chemistry, in biology, in cosmology, and now in the interdisciplinary fields of molecular biology, cognitive science, and more. This continuing development is the end of philosophy in two senses: its perfection and its termination. The very success of scientific revealing grew out of a specialization of thinking as philosophy that entailed leaving behind or concealing thinking in a more fundamental sense, something that Heidegger refers to as *Lichtung*, translated variously as "lighting" or "opening." "Perhaps there is a thinking," Heidegger writes, "which is more sober-minded than the incessant frenzy of rationalization and the intoxicating quality of cybernetics" (Heidegger 1977 [1966]: 391).

In another text, Heidegger describes this "new task of thinking" at "the end of philosophy" by means of a comparison between what he calls calculative and meditative thinking. "Calculative thinking never stops, never collects itself. Calculative thinking is not meditative thinking, not thinking which contemplates the meaning which reigns in everything that is" (Heidegger 1966 [1955]L 46). Meditative thinking, premodern and even preclassical Greek philosophical thinking, which was once in touch with the root of human existence, and out of which by means of a narrowing and intensified calculative thinking has emerged, has been replaced by calculative thinking in the form of "all that with which modern techniques of communication stimulate, assail, and drive human beings" (Heidegger 1966 [1955]: 48). Technology, especially information technology, conceals this meditative thinking, which Heidegger terms *Gelassenheit*, releasement or detachment. "Releasement toward things and openness to the mystery . . . promise us a new ground and foundation upon which we can stand and endure the world of technology without being imperiled by it" (Heidegger 1966 [1955]: 55). The fundamental threat in information technology is thus a threat to the human being's "essential nature" and the "issue of keeping meditative thinking alive" (ibid.: 56).

Current Research and Open Issues

What is most remarkable is the fact that Heidegger's radical critique of technology in general and information technology in particular has been subject to significant practical appropriations by IT users and designers, thus building bridges between the engineering and humanities, the expansionist and limitationist, traditions in the philosophy of information technology. Raphael Capurro (1986), for instance, brings Heidegger to bear on the field of library and information science. Hubert Dreyfus (2001) examines the Internet from a philosophical perspective indebted to Heidegger. With slightly more expansion, one may also reference two other leading examples: Terry Winograd and Fernado Flores, and Richard Coyne. At the same time serious challenges have been raised by Mark Poster to the adequacy of a Heideggerian approach to IT.

In the mid-1980s, computer scientists Winograd and Flores argued at length that Heideggerian analyses could disclose the reasons behind the failures of information technologies to function as well in the office as computer scientists predicted. In Winograd and Flores (1987) they argue that Heideggerian insights can thus be a stimulus for redesigning computer systems.

A decade later architectural theorist Coyne (1995) goes even further, arguing that not just Heidegger but the post-Heideggerian thought of Jacques Derrida provides a philosophical account of what is going on among leading-edge information technology designers. Building on Heidegger's notion that all revealing involves a simultaneous concealing, Derrida proposes to deconstruct specific concepts, methods, and disciplinary formations precisely to bring to light their hidden aspects, that on which they depend without knowing or acknowledging it. For Coyne this opens the way for and justifies the turn from a commitment to rational method in information technology design to the renewed reliance on metaphor.

Heidegger and Derrida thus revalidate the creative significance of metaphor – of thinking of a computer operating system as "windows," of a screen "desktop" with "icons," even of the mind as a computer. It is precisely a play with such "irrational" connections that facilitates advances in information technology design. With Aquinas, Coyne seeks to defend the metaphorical "hiding of truth in figures" as functional not just in theology but also in technology. Whether either Aquinas or Heidegger would counsel such appropriation of their philosophies of information technology is, of course, seriously in doubt.

As if to reinforce such doubt about such creative appropriations, Poster argues at length that Heidegger "captures the revealing of modern technology only, not postmodern technology." Indeed, "some information technologies, in their complex assemblages, partake not only of [*Gestell*] but also of forms of revealing that do not conceal but solicit participants to a relation to Being as freedom" (Poster 2001: 32–3). For Poster a more adequate approach to the philosophical understanding of IT is through Felix Guattari's image of the rhizome and a phenomenology of the enunciative properties of specific technologies. A potentially comprehensive philosophy of IT thus remains, not unlike all philosophy, suspended in and energized by its fundamental alternatives.

References

Borgmann, A. 1999. *Holding On To Reality: The Nature of Information at the Turn of the Millennium*. Chicago: University of Chicago Press. [Distinguishes natural information (about reality), cultural information (for constructing reality), and technological information (information becoming a reality in its own right). Seeks to establish guidelines for assessing and limiting information as reality.]

Brook, J. and Boal, I. A. 1995. *Resisting the Virtual Life: The Culture and Politics of Information*. San Francisco: City Lights. [Twenty-one critical essays on IT inequities, impacts on the body, the degrading of the workplace, and cultural deformations.]

Capurro, R. 1986. *Hermeneutik der Fachinformation*. Freiburg: Alber. [For a short English paper that reviews the thesis of this book, see R. Capurro, "Hermeneutics and the phenomenon of information," *Research in Philosophy and Technology* 19: 79–85.]

Coyne, R. 1995. *Designing Information Technology in the Postmodern Age: From Method to Metaphor*. Cambridge, MA: MIT Press. [Advances in information technology are examined from the diverse philosophical perspectives of analytic philosophy, pragmatism, phenomenology, critical theory, and hermeneutics, in order to reveal their different implications for the design and development of new electronic communications media.]

Dretske, F. 1983. *Knowledge and the Flow of Information*. Cambridge, MA: MIT Press. [The most well-developed theory of perception and empirical knowledge based on mathematical information theory.]

Dreyfus, H. L. 2001. *On the Internet*. New York: Routledge. [A phenomenologically influenced but interdisciplinary critique.]

Graham, G. 1999. *The Internet: A Philosophical Inquiry*. New York: Routledge. [Questions and criticizes both neoluddite and technophile claims about the dangers and implications of the internet.]

Haraway, D. 1991. *Simians, Cyborgs, and Women: The Reinvention of Nature*. New York: Routledge. [See especially the "Manifesto for Cyborgs" included in this book.]

Heidegger, M. 1966 [1955]. *Discourse on Thinking*, tr. J. M. Anderson and E. H. Freund. New York: Harper and Row. [This includes translation of Heidegger's essay, "Gelassenheit."]

Heidegger M. 1977 [1954]. "The question concerning technology," tr. W. Lovitt. In M. Heidegger, *Basic Writings*. New York: Harper and Row, pp. 287–317. [Heidegger's most important critique of technology.]

Heidegger M. 1977 [1966]. "The end of philosophy and the task of thinking," tr. J. Stambaugh. In M. Heidegger, *Basic Writings*. New York: Harper and Row, pp. 373–92. [A brief statement of Heidegger's philosophy of the history of philosophy.]

Kay, L. E. 2000. *Who Wrote the Book of Life: A History of the Genetic Code*. Stanford: Stanford University Press. [A critical assessment of information as a metaphor in biology.]

Levinson, P. 1997. *The Soft Edge: A Natural History and Future of the Information Revolution*. New York: Routledge. [A new information medium (such as computers) does not so much replace an old medium (such as the telephone) as complement it.]

McLuhan, M. 1964. *Understanding Media: The Extensions of Man*. New York: McGraw-Hill. [It is not the information content of a medium (such as speech or television) that is most influential on a culture, but the character or structure of the medium itself. Electronic media are structurally distinct from, say, books. "The medium is the message."]

Mitcham, C. 1994. *Thinking through Technology: The Path between Engineering and Philosophy*. Chicago: University of Chicago Press. [A general introduction to the philosophy of technology that distinguishes two major traditions: engineering and humanities philosophy of technology. The former argues for the expansion, the latter for the delimitation of technology as object, knowledge, activity, and volition.]

Pickstock, C. 1998. *After Writing: On the Liturgical Consummation of Philosophy*. Oxford, UK: Blackwell. [A critique of Derrida and defense of information as subordinate to the context created by linguistic and bodily performance in a historical tradition.]

Poster, Mark. 1990. *The Mode of Information: Poststructuralism and Social Context.* Chicago: University of Chicago Press. [Argues that four new modes of information – TV ads, data bases, electronic writing, and computer science – create a world in which humans are socially constituted differently than in pre-electronic IT history.]

Poster, Mark. 2001. *What's the Matter with the Internet.* Minneapolis, MI: University of Minnesota Press. [A critique of applying Heidegger's analysis of technology to information technology, with special reference to postmodern thinkers such as Felix Guattari.]

Sayre, K. M. 1976. *Cybernetics and the Philosophy of Mind.* London: Routledge & Kegan Paul. [Argues a naturalist theory of mind based on mathematical information theory.]

Schirmacher, W. 1994. "Media and postmodern technology." In G. Bender and T. Druckrey, eds., *Culture on the Brink: Ideologies of Technologies.* Seattle: Bay Press. [The other contributions to this book are useful as well.]

Shannon, C. and Weaver, W. 1949. *The Mathematical Theory of Communication.* Urbana: University of Illinois Press. [Contains two classic papers: Shannon's, from the *Bell System Technical Journal* (1948); and Weaver's, from *Scientific American* (1949).]

Turkle, Sherry. 1995. *Life on the Screen: Identity in the Age of the Internet.* New York: Simon & Schuster. [A psychologist's analysis of emerging forms of self-definition unique to the internet experience.]

Wiener, N. 1948. *Cybernetics: Or, Control and Communication in the Animal and the Machine.* Cambridge, MA: MIT Press. [The classic statement of the engineering theory of cybernetics. In other works Wiener also examined the social and ethical implications of his theories.]

Winograd, T. and Flores, F. 1987. *Understanding Computers and Cognition: A New Foundation for Design.* Reading, MA: Addison-Wesley. [A Heideggerian analysis by two computer scientists.]

Wurman, R. S. 2001. *Information Anxiety 2.* Indianapolis, IN: Que. [This is the second edition of a widely cited critique of information overload by a well-known architect and student of the work of Louis Kahn.]

Zimmerli, W. 1986. "Who is to blame for data pollution?" In C. Mitcham and A. Huning, eds., *Philosophy and Technology II: Information Technology and Computers in Theory and Practice.* Boston: Reidel. [One of 20 original papers from a conference, introduced by an overview of "Information technology and computers as themes in the philosophy of technology," and followed by an annotated bibliography on philosophical studies of information technology and computers.]

42

Into the Electronic Millennium

Sven Birkerts

Some years ago, a friend and I comanaged a used and rare book shop in Ann Arbor, Michigan. We were often asked to appraise and purchase libraries – by retiring academics, widows, and disgruntled graduate students. One day we took a call from a professor of English at one of the community colleges outside Detroit. When he answered the buzzer I did a double take – he looked to be only a year or two older than we were. "I'm selling everything," he said, leading the way through a large apartment. As he opened the door of his study I felt a nudge from my partner. The room was wall-to-wall books and as neat as a chapel.

The professor had a remarkable collection. It reflected not only the needs of his vocation – he taught nineteenth- and twentieth-century literature – but a book lover's sensibility as well. The shelves were strictly arranged, and the books themselves were in superb condition. When he left the room we set to work inspecting, counting, and estimating. This is always a delicate procedure, for the buyer is at once anxious to avoid insult to the seller and eager to get the goods for the best price. We adopted our usual strategy, working out a lower offer and a more generous fallback price. But there was no need to worry. The professor took our first offer without batting an eye.

As we boxed up the books, we chatted. My partner asked the man if he was moving. "No," he said, "but I am getting out." We both looked up. "Out of the teaching business, I mean. Out of books." He then said that he wanted to show us something. And indeed, as soon as the books were packed and loaded, he led us back through the apartment and down a set of stairs. When we reached the basement, he flicked on the light. There, on a long table, displayed like an exhibit in the Space Museum, was a computer. I didn't know what kind it was then, nor could I tell you now, fifteen years later. But the professor was keen to explain and demonstrate.

While he and my partner hunched over the terminal, I roamed to and fro, inspecting the shelves. It was purely a reflex gesture, for they held nothing but thick binders and paperbound manuals. "I'm changing my life," the ex-professor was saying. "This is definitely where it's all going to happen." He told us that he already had several good job offers. And the books? I asked. Why was he selling them all? He paused for a few beats. "The whole profession represents a lot of pain to me," he said. "I don't want to see any of these books again."

The scene has stuck with me. It is now a kind of marker in my mental life. That afternoon I got

From Sven Birkerts, *The Gutenberg Elegies: The Fate of Reading in an Electronic Age* (New York: Fawcett Columbine: 1994), pp. 117–33.

my first serious inkling that all was not well in the world of print and letters. All sorts of corroborations followed. Our professor was by no means an isolated case. Over a period of two years we met with several others like him. New men and new women who had glimpsed the future and had decided to get out while the getting was good. The selling off of books was sometimes done for financial reasons, but the need to burn bridges was usually there as well. It was as if heading to the future also required the destruction of tokens from the past.

A change is upon us – nothing could be clearer. The printed word is part of a vestigial order that we are moving away from – by choice and by societal compulsion. I'm not just talking about disaffected academics, either. This shift is happening throughout our culture, away from the patterns and habits of the printed page and toward a new world distinguished by its reliance on electronic communications.

This is not, of course, the first such shift in our long history. In Greece, in the time of Socrates, several centuries after Homer, the dominant oral culture was overtaken by the writing technology. And in Europe another epochal transition was effected in the late fifteenth century after Gutenberg invented movable type. In both cases the long-term societal effects were overwhelming, as they will be for us in the years to come.

The evidence of the change is all around us, though possibly in the manner of the forest that we cannot see for the trees. The electronic media, while conspicuous in gadgetry, are very nearly invisible in their functioning. They have slipped deeply and irrevocably into our midst, creating sluices and circulating through them. I'm not referring to any one product or function in isolation, such as television or fax machines or the networks that make them possible. I mean the interdependent totality that has arisen from the conjoining of parts – the disk drives hooked to modems, transmissions linked to technologies of reception, recording, duplication, and storage. Numbers and codes and frequencies. Buttons and signals. And this is no longer "the future," except for the poor or the self-consciously atavistic – it is now. Next to the new technologies, the scheme of things represented by print and the snail-paced linearity of the reading act looks stodgy and dull. Many educators say that our students are less and less able to read, or analyze, or write with clarity and purpose. Who can blame the students? Everything they meet with in the world around them gives the signal: That was then, and electronic communications are now.

Do I exaggerate? If all this is the case, why haven't we heard more about it? Why hasn't somebody stepped forward with a bow tie and a pointer stick to explain what is going on? Valid questions, but they also beg the question. They assume that we are all plugged into a total system – where else would that "somebody" appear if not on the screen at the communal hearth?

Media theorist Mark Crispin Miller has given one explanation for our situation in his discussions of television in *Boxed In: The Culture of TV*. The medium, he proposes, has long since diffused itself throughout the entire system. Through sheer omnipresence it has vanquished the possibility of comparative perspectives. We cannot see the role that television (or, for our purposes, all electronic communications) has assumed in our lives because there is no independent ledge where we might secure our footing. The medium has absorbed and eradicated the idea of a pretelevision past; in place of what used to be we get an ever-new and ever-renewable present. The only way we can hope to understand what is happening, or what has already happened, is by way of a severe and unnatural dissociation of sensibility.

To get a sense of the enormity of the change, you must force yourself to imagine – deeply and in nontelevisual terms – what the world was like a hundred, even fifty, years ago. If the feat is too difficult, spend some time with a novel from the period. Read between the lines and reconstruct. Move through the sequence of a character's day and then juxtapose the images and sensations you find with those in the life of the average urban or suburban dweller today.

Inevitably, one of the first realizations is that a communications net, a soft and pliable mesh woven from invisible threads, has fallen over everything. The so-called natural world, the place we used to live, which served us so long as the yardstick for all measurements, can now only be perceived through a scrim. Nature was then; this is now. Trees and rocks have receded. And the great geographical Other, the faraway rest of the world, has been transformed by the pure possibility of access. The numbers of distance and time no longer mean what they used to. Every place, once unique, itself, is strangely shot through with radiations from every other place. "There" was then; "here" is now.

Think of it. Fifty to a hundred million people (maybe a conservative estimate) form their ideas about what is going on in America and in the world from the same basic package of edited images – to the extent that the image itself has lost much of its once-fearsome power. Daily newspapers, with their long columns of print, struggle against declining sales. Fewer and fewer people under the age of fifty read them; computers will soon make packaged information a custom product. But if the printed sheet is heading for obsolescence, people are tuning in to the signals. The screen is where the information and entertainment wars will be fought. The communications conglomerates are waging bitter takeover battles in their zeal to establish global empires. As Jonathan Crary has written in "The Eclipse of the Spectacle," "Telecommunications is the new arterial network, analogous in part to what railroads were for capitalism in the nineteenth century. And it is this electronic substitute for geography that corporate and national entities are now carving up." Maybe one reason why the news of the change is not part of the common currency is that such news can only sensibly be communicated through the more analytic sequences of print.

To underscore my point, I have been making it sound as if we were all abruptly walking out of one room and into another, leaving our books to the moths while we settle ourselves in front of our state-of-the-art terminals. The truth is that we are living through a period of overlap; one way of being is pushed athwart another. Antonio Gramsci's often-cited sentence comes inevitably to mind: "The crisis consists precisely in the fact that the old is dying and the new cannot be born; in this interregnum a great variety of morbid symptoms appears." The old surely is dying, but I'm not so sure that the new is having any great difficulty being born. As for the morbid symptoms, these we have in abundance.

The overlap in communications modes, and the ways of living that they are associated with, invites comparison with the transitional epoch in ancient Greek society, certainly in terms of the relative degree of disturbance. Historian Eric Havelock designated that period as one of "proto-literacy," of which his fellow scholar Oswyn Murray has written:

> To him [Havelock] the basic shift from oral to literate culture was a slow process; for centuries, despite the existence of writing, Greece remained essentially an oral culture. This culture was one which depended heavily on the encoding of information in poetic texts, to be learned by rote and to provide a cultural encyclopedia of conduct. It was not until the age of Plato in the fourth century that the dominance of poetry in an oral culture was challenged in the final triumph of literacy.

That challenge came in the form of philosophy, among other things, and poetry has never recovered its cultural primacy. What oral poetry was for the Greeks, printed books in general are for us. But our historical moment, which we might call "proto-electronic," will not require a transition period of two centuries. The very essence of electronic transmissions is to surmount impedances and to hasten transitions. Fifty years, I'm sure, will suffice. As for what the conversion will bring – and *mean* – to us, we might glean a few clues by looking to some of the "morbid symptoms" of the change. But to understand what these portend, we need to remark a few of the more obvious ways in which our various technologies condition our senses and sensibilities.

I won't tire my reader with an extended rehash of the differences between the print orientation and that of electronic systems. Media theorists from Marshall McLuhan to Walter Ong to Neil Postman have discoursed upon these at length. What's more, they are reasonably commonsensical. I therefore will abbreviate.

The order of print is linear, and is bound to logic by the imperatives of syntax. Syntax is the substructure of discourse, a mapping of the ways that the mind makes sense through language. Print communication requires the active engagement of the reader's attention, for reading is fundamentally an act of translation. Symbols are turned into their verbal referents and these are in turn interpreted. The print engagement is essentially private. While it does represent an act of communication, the contents pass from the privacy of the sender to the privacy of the receiver. Print also posits a time axis; the turning of pages, not to mention the vertical descent down the page, is a forward-moving succession, with earlier contents at every point serving as a ground for what follows. Moreover, the printed material is static – it is the reader, not the book, that moves forward. The physical arrangements of print are in accord with our traditional sense of history. Materials are layered; they lend themselves to rereading and to sustained attention. The pace of reading is variable, with progress determined by the reader's focus and comprehension.

The electronic order is in most ways opposite. Information and contents do not simply move from one private space to another, but they travel along a network. Engagement is intrinsically public, taking place within a circuit of larger connectedness. The vast resources of the network are always there, potential, even if they do not impinge on the immediate communication. Electronic communication can be passive, as with television watching, or interactive, as with computers. Contents, unless they are printed out (at which point they become part of the static order of print) are felt to be evanescent. They can be changed or deleted with the stroke of a key. With visual media (television, projected graphs, highlighted "bullets") impression and image take precedence over logic and concept, and detail and linear sequentiality are sacrificed. The pace is rapid, driven by jump-cut increments, and the basic movement is laterally associative rather than vertically cumulative. The presentation structures the reception and, in time, the expectation about how information is organized.

Further, the visual and nonvisual technology in every way encourages in the user a heightened and ever-changing awareness of the present. It works against historical perception, which must depend on the inimical notions of logic and sequential succession. If the print medium exalts the word, fixing it into permanence, the electronic counterpart reduces it to a signal, a means to an end.

Transitions like the one from print to electronic media do not take place without rippling or, more likely, *reweaving* the entire social and cultural web. The tendencies outlined above are already at work. We don't need to look far to find their effects. We can begin with the newspaper headlines and the millennial lamentations sounded in the op-ed pages: that our educational systems are in decline; that our students are less and less able to read and comprehend their required texts, and that their aptitude scores have leveled off well below those of previous generations. Tag-line communication, called "bite-speak" by some, is destroying the last remnants of political discourse; spin doctors and media consultants are our new shamans. As communications empires fight for control of all information outlets, including publishers, the latter have succumbed to the tyranny of the bottom line; they are less and less willing to publish work, however worthy, that will not make a tidy profit. And, on every front, funding for the arts is being cut while the arts themselves appear to be suffering a deep crisis of relevance. And so on.

Every one of these developments is, of course, overdetermined, but there can be no doubt that they are connected, perhaps profoundly, to the transition that is underway.

Certain other trends bear watching. One could argue, for instance, that the entire movement of postmodernism in the arts is a consequence of this same macroscopic shift. For what is postmodernism at root but an aesthetic that rebukes the idea of an historical time line, as well as previously uncontested assumptions of cultural hierarchy. The postmodern artifact manipulates its stylistic signatures like Lego blocks and makes free with combinations from the formerly sequestered spheres of high and popular art. Its combinatory momentum and relentless referencing of the surrounding culture mirror perfectly the associative dynamics of electronic media.

One might argue likewise, that the virulent debate within academia over the canon and multiculturalism may not be a simple struggle between the entrenched ideologies of white male elites and the forces of formerly disenfranchised gender, racial, and cultural groups. Many of those who would revise the canon (or end it altogether) are trying to outflank the assumption of historical tradition itself. The underlying question, avoided by many, may be not only whether the tradition is relevant, but whether it might not be too taxing a system for students to comprehend. Both the traditionalists and the progressives have valid arguments, and we must certainly have sympathy for those who would try to expose and eradicate the hidden assumptions of bias in the Western tradition. But it also seems clear that this debate could only have taken the form it has in a society that has begun to come loose from its textual moorings. To challenge repression is salutary. To challenge history itself, proclaiming it to be simply an archive of repressions and justifications, is idiotic.[1]

Then there are the more specific sorts of developments. Consider the multibillion-dollar initiative by Whittle Communications to bring commercially sponsored education packages into the classroom. The underlying premise is staggeringly simple: If electronic media are the one thing that the young are at ease with, why not exploit the fact? Why not stop bucking television and use it instead, with corporate America picking up the tab in exchange for a few minutes of valuable airtime for commercials? As the *Boston Globe* reports:

Here's how it would work:

Participating schools would receive, free of charge, $50,000 worth of electronic paraphernalia, including a satellite dish and classroom video monitors. In return, the schools would agree to air the show.

The show would resemble a network news program, but with 18- to 24-year-old anchors.

A prototype includes a report on a United Nations Security Council meeting on terrorism, a space shuttle update, a U2 music video tribute to Martin Luther King, a feature on the environment, a "fast fact" ('Arachibutyrophobia is the fear of peanut butter sticking to the roof of your mouth') and two minutes of commercial advertising.

"You have to remember that the children of today have grown up with the visual media," said Robert Calabrese [Billerica School Superintendent]. "They know no other way and we're simply capitalizing on that to enhance learning."

Calabrese's observation on the preconditioning of a whole generation of students raises troubling questions: Should we suppose that American education will begin to tailor itself to the aptitudes of its students, presenting more and more of its materials in newly packaged forms? And what will happen when educators find that not very many

of the old materials will "play" – that is, capture student enthusiasm? Is the *what* of learning to be determined by the *how*? And at what point do vicious cycles begin to reveal their viciousness?

A collective change of sensibility may already be upon us. We need to take seriously the possibility that the young truly "know no other way," that they are not made of the same stuff that their elders are. In her *Harper's* magazine debate with Neil Postman, Camille Paglia observed:

Some people have more developed sensoriums than others. I've found that most people born before World War II are turned off by the modern media. They can't understand how we who were born after the war can read and watch TV at the same time. But we *can*. When I wrote my book, I had earphones on, blasting rock music or Puccini and Brahms. The soap operas – with the sound turned down – flickered on my TV. I'd be talking on the phone at the same time. Baby boomers have a multilayered, multitrack ability to deal with the world.

I don't know whether to be impressed or depressed by Paglia's ability to disperse her focus in so many directions. Nor can I say, not having read her book, in what ways her multitrack sensibility has informed her prose. But I'm baffled by what she means when she talks about an ability to "deal with the world." From the context, "dealing" sounds more like a matter of incessantly repositioning the self within a barrage of onrushing stimuli.

Paglia's is hardly the only testimony in this matter. A *New York Times* article on the cult success of Mark Leyner (author of *I Smell Esther Williams* and *My Cousin, My Gastroenterologist*) reports suggestively:

His fans say, variously, that his writing is like MTV, or rap music, or rock music, or simply like everything in the world put together: fast and furious and intense, full of illusion and allusion and fantasy and science and excrement.

Larry McCaffery, a professor of literature at San Diego State University and co-editor of Fiction International, a literary journal, said his students get excited about Mr. Leyner's writing, which he considers important and unique: "It speaks to them, somehow, about this weird milieu they're swimming through. It's this dissolving, discontinuous world." While older people might find Mr. Leyner's world bizarre or unreal, Professor McCaffery said, it doesn't seem so to people who grew up with Walkmen and computers and VCR's, with so many choices, so much bombardment, that they have never experienced a sensation singly.

The article continues:

There is no traditional narrative, although the book is called a novel. And there is much use of facts, though it is called fiction. Seldom does the end of a sentence have any obvious relation to the beginning. "You don't know where you're going, but you don't mind taking the leap," said R. J. Cutler, the producer of "Heat," who invited Mr. Leyner to be on the show after he picked up the galleys of his book and found it mesmerizing. "He taps into a specific cultural perspective where thoughtful literary world view meets pop culture and the TV generation."

My final exhibit – I don't know if it qualifies as a morbid symptom as such – is drawn from a *Washington Post Magazine* essay on the future of the Library of Congress, our national shrine to the printed word. One of the individuals interviewed in the piece is Robert Zich, so-called "special projects czar" of the institution. Zich, too, has seen the future, and he is surprisingly candid with his interlocutor. Before long, Zich maintains, people will be able to get what information they want directly off their terminals. The function of the Library of Congress (and perhaps libraries in general) will change. He envisions his library becoming more like a museum: "Just as you go to the National Gallery to see its Leonardo or go to the Smithsonian to see the Spirit of St. Louis and so on, you will want to go to libraries to see the Gutenberg or the original printing of Shakespeare's plays or to see Lincoln's hand-written version of the Gettysburg Address."

Zich is outspoken, voicing what other administrators must be thinking privately. The big research libraries, he says, "and the great national libraries and their buildings will go the way of the railroad stations and the movie palaces of an earlier era which were really vital institutions in their time . . . Somehow folks moved away from that when the technology changed."

And books? Zich expresses excitement about Sony's hand-held electronic book, and a miniature encyclopedia coming from Franklin Electronic Publishers. "Slip it in your pocket," he says. "Little keyboard, punch in your words and it will do the full text searching and all the rest of it. Its limitation, of course, is that it's devoted just to that one book." Zich is likewise interested in the possibility of memory cards. What he likes about the Sony product is the portability: one machine, a screen that will display the contents of whatever electronic card you feed it.

I cite Zich's views at some length here because he is not some Silicon Valley research and development visionary, but a highly placed executive at what might be called, in a very literal sense, our most conservative public institution. When men like Zich embrace the electronic future, we can be sure it's well on its way.

Others might argue that the technologies cited by Zich merely represent a modification in the "form" of reading, and that reading itself will be unaffected, as there is little difference between following words on a pocket screen or a printed page. Here I have to hold my line. The context cannot but condition the process. Screen and book may exhibit the same string of words, but the assumptions that underlie their significance are entirely different depending on whether we are staring at a book or a circuit-generated text. As the nature of looking – at the natural world, at paintings – changed with the arrival of photography and mechanical reproduction, so will the collective relation to language alter as new modes of dissemination prevail.

Whether all of this sounds dire or merely "different" will depend upon the reader's own values and priorities. I find these portents of change depressing, but also exhilarating – at least to speculate about. On the one hand, I have a great feeling of loss and a fear about what habitations will exist for self and soul in the future. But there is also a quickening, a sense that important things are on the line. As Heraclitus once observed, "The mixture that is not shaken soon stagnates." Well, the mixture is being shaken, no doubt about it. And here are some of the kinds of developments we might watch for as our "proto-electronic" era yields to an all-electronic future:

1. *Language erosion.* There is no question but that the transition from the culture of the book

to the culture of electronic communication will radically alter the ways in which we use language on every societal level. The complexity and distinctiveness of spoken and written expression, which are deeply bound to traditions of print literacy, will gradually be replaced by a more telegraphic sort of "plainspeak." Syntactic masonry is already a dying art. Neil Postman and others have already suggested what losses have been incurred by the advent of telegraphy and television – how the complex discourse patterns of the nineteenth century were flattened by the requirements of communication over distances. That tendency runs riot as the layers of mediation thicken. Simple linguistic prefab is now the norm, while ambiguity, paradox, irony, subtlety, and wit are fast disappearing. In their place, the simple "vision thing" and myriad other "things." Verbal intelligence, which has long been viewed as suspect as the act of reading, will come to seem positively conspiratorial. The greater part of any articulate person's energy will be deployed in dumbing-down her discourse.

Language will grow increasingly impoverished through a series of vicious cycles. For, of course, the usages of literature and scholarship are connected in fundamental ways to the general speech of the tribe. We can expect that curricula will be further streamlined, and difficult texts in the humanities will be pruned and glossed. One need only compare a college textbook from twenty years ago to its contemporary version. A poem by Milton, a play by Shakespeare – one can hardly find the text among the explanatory notes nowadays. Fewer and fewer people will be able to contend with the so-called masterworks of literature or ideas. Joyce, Woolf, Soyinka, not to mention the masters who preceded them, will go unread, and the civilizing energies of their prose will circulate aimlessly between closed covers.

2. *Flattening of historical perspectives.* As the circuit supplants the printed page, and as more and more of our communications involve us in network processes – which of their nature plant us in a perpetual present – our perception of history will inevitably alter. Changes in information storage and access are bound to impinge on our historical memory. The depth of field that is our sense of the past is not only a linguistic construct, but is in some essential way represented by the book and the physical accumulation of books in library spaces. In the contemplation of the single volume, or mass of volumes, we form a picture of time past as a growing deposit of sediment; we capture a sense of its depth and dimensionality. Moreover, we meet the past as much in the presentation of words in books of specific vintage as we do in any isolated fact or statistic. The database, useful as it is, expunges this context, this sense of chronology, and admits us to a weightless order in which all information is equally accessible.

If we take the etymological tack, history (cognate with "story") is affiliated in complex ways with its texts. Once the materials of the past are unhoused from their pages, they will surely *mean* differently. The printed page is itself a link, at least along the imaginative continuum, and when that link is broken, the past can only start to recede. At the same time it will become a body of disjunct data available for retrieval and, in the hands of our canny dream merchants, a mythology. The more we grow rooted in the consciousness of the now, the more it will seem utterly extraordinary that things were ever any different. The idea of a farmer plowing a field – an historical constant for millennia – will be something for a theme park. For, naturally, the entertainment industry, which reads the collective unconscious unerringly, will seize the advantage. The past that has slipped away will be rendered ever more glorious, ever more a fantasy play with heroes, villains, and quaint settings and props. Small-town American life returns as "Andy of Mayberry" – at first enjoyed with recognition, later accepted as a faithful portrait of how things used to be.

3. *The waning of the private self.* We may even now be in the first stages of a process of social collectivization that will over time all but vanquish the ideal of the isolated individual. For some decades now we have been edging away from the perception of private life as something opaque, closed off to the world; we increasingly accept the transparency of a life lived within a set of systems, electronic or otherwise. Our technologies are not bound by season or light – it's always the same time in the circuit. And so long as time is money and money matters, those circuits will keep humming. The doors and walls of our habitations matter less and less – the world sweeps through

the wires as it needs to, or as we need it to. The monitor light is always blinking; we are always potentially on-line.

I am not suggesting that we are all about to become mindless, soulless robots, or that personality will disappear altogether into an oceanic homogeneity. But certainly the idea of what it means to be a person living a life will be much changed. The figure-ground model, which has always featured a solitary self before a background that is the society of other selves, is romantic in the extreme. It is ever less tenable in the world as it is becoming. There are no more wildernesses, no more lonely homesteads, and, outside of cinema, no more emblems of the exalted individual.

The self must change as the nature of subjective space changes. And one of the many incremental transformations of our age has been the slow but steady destruction of subjective space. The physical and psychological distance between individuals has been shrinking for at least a century. In the process, the figure-ground image has begun to blur its boundary distinctions. One day we will conduct our public and private lives within networks so dense, among so many channels of instantaneous information, that it will make almost no sense to speak of the differentiations of subjective individualism.

We are already captive in our webs. Our slight solitudes are transected by codes, wires, and pulsations. We punch a number to check in with the answering machine, another to tape a show that we are too busy to watch. The strands of the web grow finer and finer – this is obvious. What is no less obvious is the fact that they will continue to proliferate, gaining in sophistication, merging functions so that one can bank by phone, shop via television, and so on. The natural tendency is toward streamlining: The smart dollar keeps finding ways to shorten the path, double-up the function. We might think in terms of a circuit-board model, picturing ourselves as the contact points. The expansion of electronic options is always at the cost of contractions in the private sphere. We will soon be navigating with ease among cataracts of organized pulsations, putting out and taking in signals. We will bring our terminals, our modems, and menus further and further into our former privacies; we will implicate ourselves by degrees in the unitary life,

and there may come a day when we no longer remember that there was any other life.

While I was brewing these somewhat melancholy thoughts, I chanced to read in an old *New Republic* the text of Joseph Brodsky's 1987 Nobel Prize acceptance speech. I felt as though I had opened a door leading to the great vault of the nineteenth century. The poet's passionate plea on behalf of the book at once corroborated and countered everything I had been thinking. What he upheld in faith were the very ideals I was saying good-bye to. I greeted his words with an agitated skepticism, fashioning from them something more like a valediction. Here are four passages:

If art teaches anything . . . it is the privateness of the human condition. Being the most ancient as well as the most literal form of private enterprise, it fosters in a man, knowingly or unwittingly, a sense of his uniqueness, of individuality, of separateness – thus turning him from a social animal into an autonomous "I."

The great Baratynsky, speaking of his Muse, characterized her as possessing an "uncommon visage." It's in acquiring this "uncommon visage" that the meaning of human existence seems to lie, since for this uncommonness we are, as it were, prepared genetically.

Aesthetic choice is a highly individual matter, and aesthetic experience is always a private one. Every new aesthetic reality makes one's experience even more private; and this kind of privacy, assuming at times the guise of literary (or some other) taste, can in itself turn out to be, if not a guarantee, then a form of defense, against enslavement.

In the history of our species, in the history of Homo sapiens, the book is an anthropological development, similar essentially to the invention of the wheel. Having emerged in order to give us some idea not so much of our origins as of what that sapiens is capable of, a book constitutes a means of transportation through the space of experience, at the speed of a turning page. This movement, like every movement, becomes flight from the common denominator . . . This flight is the flight in the direction of "uncommon visage," in the direction of the numerator, in the direction of autonomy, in the direction of privacy.

Brodsky is addressing the relation between art and totalitarianism, and within that context his words make passionate sense. But I was reading from a different vantage. What I had in mind was not a vision of political totalitarianism, but rather of something that might be called "societal totalism" – that movement toward deindividuation, or electronic collectivization, that I discussed above. And from that perspective our era appears to be in a headlong flight *from* the "uncommon visage" named by the poet.

Trafficking with tendencies – extrapolating and projecting as I have been doing – must finally remain a kind of gambling. One bets high on the validity of a notion and low on the human capacity for resistance and for unpredictable initiatives. No one can really predict how we will adapt to the transformations taking place all around us. We may discover, too, that language is a hardier thing than I have allowed. It may flourish among the beep and the click and the monitor as readily as it ever did on the printed page. I hope so, for language is the soul's ozone layer and we thin it at our peril.

Note

1 The outcry against the modification of the canon can be seen as a plea for old reflexes and routines. And the cry for multicultural representation may be a last-ditch bid for connection to the fading legacy of print. The logic is simple. When a resource is threatened – made scarce – people fight over it. In this case the struggle is over textual power in an increasingly nontextual age. The future of books and reading is what is at stake, and a dim intuition of this drives the contending factions.

As Katha Pollitt argued so shrewdly in her much-cited article in *The Nation*: If we were a nation of readers, there would be no issue. No one would be arguing about whether to put Toni Morrison on the syllabus because her work would be a staple of the reader's regular diet anyway. These lists are suddenly so important because they represent, very often, the only serious works that the student is ever likely to be exposed to. Whoever controls the lists comes out ahead in the struggle for the hearts and minds of the young.

43

Why I Am not Going To Buy A Computer

Wendell Berry

Like almost everybody else, I am hooked to the energy corporations, which I do not admire. I hope to become less hooked to them. In my work, I try to be as little hooked to them as possible. As a farmer, I do almost all of my work with horses. As a writer, I work with a pencil or a pen and a piece of paper.

My wife types my work on a Royal standard typewriter bought new in 1956 and as good now as it was then. As she types, she sees things that are wrong and marks them with small checks in the margins. She is my best critic because she is the one most familiar with my habitual errors and weaknesses. She also understands, sometimes better than I do, what *ought* to be said. We have, I think, a literary cottage industry that works well and pleasantly. I do not see anything wrong with it.

A number of people, by now, have told me that I could greatly improve things by buying a computer. My answer is that I am not going to do it. I have several reasons, and they are good ones.

The first is the one I mentioned at the beginning. I would hate to think that my work as a writer could not be done without a direct dependence on strip-mined coal. How could I write conscientiously against the rape of nature if I were, in the act of writing, implicated in the rape? For the same reason, it matters to me that my writing is done in the daytime, without electric light.

I do not admire the computer manufacturers a great deal more than I admire the energy industries. I have seen their advertisements, attempting to seduce struggling or failing farmers into the belief that they can solve their problems by buying yet another piece of expensive equipment. I am familiar with their propaganda campaigns that have put computers into public schools in need of books. That computers are expected to become as common as TV sets in "the future" does not impress me or matter to me. I do not own a TV set. I do not see that computers are bringing us one step nearer to anything that does matter to me: peace, economic justice, ecological health, political honesty, family and community stability, good work.

What would a computer cost me? More money, for one thing, than I can afford, and more than I wish to pay to people whom I do not admire. But the cost would not be just monetary. It is well understood that technological innovation always requires the discarding of the "old model" – the "old model" in this case being not just our old Royal standard, but my wife, my critic, my closest reader, my fellow worker. Thus (and I think this is typical of

From Wendell Berry, *What are People For?* (New York: North Point Press, 2000), pp. 171–7.

present-day technological innovation), what would be superseded would be not only something, but somebody. In order to be technologically up-to-date as a writer, I would have to sacrifice an association that I am dependent upon and that I treasure.

My final and perhaps my best reason for not owning a computer is that I do not wish to fool myself. I disbelieve, and therefore strongly resent, the assertion that I or anybody else could write better or more easily with a computer than with a pencil. I do not see why I should not be as scientific about this as the next fellow: when somebody has used a computer to write work that is demonstrably better than Dante's, and when this better is demonstrably attributable to the use of a computer, then I will speak of computers with a more respectful tone of voice, though I still will not buy one.

To make myself as plain as I can, I should give my standards for technological innovation in my own work. They are as follows:

1 The new tool should be cheaper than the one it replaces.
2 It should be at least as small in scale as the one it replaces.
3 It should do work that is clearly and demonstrably better than the one it replaces.
4 It should use less energy than the one it replaces.
5 If possible, it should use some form of solar energy, such as that of the body.
6 It should be repairable by a person of ordinary intelligence, provided that he or she has the necessary tools.
7 It should be purchasable and repairable as near to home as possible.
8 It should come from a small, privately owned shop or store that will take it back for maintenance and repair.
9 It should not replace or disrupt anything good that already exists, and this includes family and community relationships.

1987

After the foregoing essay, first published in the *New England Review and Bread Loaf Quarterly*, was reprinted in *Harper's*, the *Harper's* editors published the following letters in response and permitted me a reply.   W. B.

Letters

Wendell Berry provides writers enslaved by the computer with a handy alternative: Wife – a low-tech energy-saving device. Drop a pile of handwritten notes on Wife and you get back a finished manuscript, edited while it was typed. What computer can do that? Wife meets all of Berry's uncompromising standards for technological innovation: she's cheap, repairable near home, and good for the family structure. Best of all, Wife is politically correct because she breaks a writer's "direct dependence on strip-mined coal."

History teaches us that Wife can also be used to beat rugs and wash clothes by hand, thus eliminating the need for the vacuum cleaner and washing machine, two more nasty machines that threaten the act of writing.

Gordon Inkeles
Miranda, Calif.

I have no quarrel with Berry because he prefers to write with pencil and paper; that is his choice. But he implies that I and others are somehow impure because we choose to write on a computer. I do not admire the energy corporations, either. Their shortcoming is not that they produce electricity but how they go about it. They are poorly managed because they are blind to long-term consequences. To solve this problem, wouldn't it make more sense to correct the precise error they are making rather than simply ignore their product? I would be happy to join Berry in a protest against strip mining, but I intend to keep plugging this computer into the wall with a clear conscience.

James Rhoads
Battle Creek, Mich.

I enjoyed reading Berry's declaration of intent never to buy a personal computer in the same way that I enjoy reading about the belief systems of unfamiliar tribal cultures. I tried to imagine a tool that would meet Berry's criteria for superiority to his old manual typewriter. The clear winner is the quill pen. It is cheaper, smaller, more energy-efficient, human-powered, easily repaired, and non-disruptive of existing relationships.

Berry also requires that this tool must be "clearly and demonstrably better" than the one it replaces. But surely we all recognize by now that "better" is in the mind of the beholder. To the quill pen aficionado, the benefits obtained from elegant calligraphy might well outweigh all others.

I have no particular desire to see Berry use a word processor; if he doesn't like computers, that's fine with me. However, I do object to his portrayal of this reluctance as a moral virtue. Many of us have found that computers can be an invaluable tool in the fight to protect our environment. In addition to helping me write, my personal computer gives me access to up-to-the-minute reports on the workings of the EPA and the nuclear industry. I participate in electronic bulletin boards on which environmental activists discuss strategy and warn each other about urgent legislative issues. Perhaps Berry feels that the Sierra Club should eschew modern printing technology, which is highly wasteful of energy, in favor of having its members hand-copy the club's magazines and other mailings each month?

Nathaniel S. Borenstein
Pittsburgh, Pa.

The value of a computer to a writer is that it is a tool not for generating ideas but for typing and editing words. It is cheaper than a secretary (or a wife!) and arguably more fuel-efficient. And it enables spouses who are not inclined to provide free labor more time to concentrate on *their* own work.

We should support alternatives both to coal-generated electricity and to IBM-style technocracy. But I am reluctant to entertain alternatives that presuppose the traditional subservience of one class to another. Let the PCs come and the wives and servants go seek more meaningful work.

Toby Koosman
Knoxville, Tenn.

Berry asks how he could write conscientiously against the rape of nature if in the act of writing on a computer he was implicated in the rape. I find it ironic that a writer who sees the under-lying connectedness of things would allow his diatribe against computers to be published in a magazine that carries ads for the National Rural Electric Cooperative Association, Marlboro, Phillips Petroleum, McDonnell Douglas, and yes,

even Smith-Corona. If Berry rests comfortably at night, he must be using sleeping pills.

Bradley C. Johnson
Grand Forks, N.D.

Wendell Berry Replies

The foregoing letters surprised me with the intensity of the feelings they expressed. According to the writers' testimony, there is nothing wrong with their computers; they are utterly satisfied with them and all that they stand for. My correspondents are certain that I am wrong and that I am, moreover, on the losing side, a side already relegated to the dustbin of history. And yet they grow huffy and condescending over my tiny dissent. What are they so anxious about?

I can only conclude that I have scratched the skin of a technological fundamentalism that, like other fundamentalisms, wishes to monopolize a whole society and, therefore, cannot tolerate the smallest difference of opinion. At the slightest hint of a threat to their complacency, they repeat, like a chorus of toads, the notes sounded by their leaders in industry. The past was gloomy, drudgery-ridden, servile, meaningless, and slow. The present, thanks only to purchasable products, is meaningful, bright, lively, centralized, and fast. The future, thanks only to more purchasable products, is going to be even better. Thus consumers become salesmen, and the world is made safer for corporations.

I am also surprised by the meanness with which two of these writers refer to my wife. In order to imply that I am a tyrant, they suggest by both direct statement and innuendo that she is subservient, characterless, and stupid – a mere "device" easily forced to provide meaningless "free labor." I understand that it is impossible to make an adequate public defense of one's private life, and so I will only point out that there are a number of kinder possibilities that my critics have disdained to imagine: that my wife may do this work because she wants to and likes to; that she may find some use and some meaning in it; that she may not work for nothing. These gentlemen obviously think themselves feminists of the most correct and principled sort, and yet they do not hesitate to stereotype and insult, on the basis of one fact, a woman they do not know. They are audacious and irresponsible gossips.

In his letter, Bradley C. Johnson rushes past the possibility of sense in what I said in my essay by implying that I am or ought to be a fanatic. That I am a person of this century and am implicated in many practices that I regret is fully acknowledged at the beginning of my essay. I did not say that I proposed to end forthwith all my involvement in harmful technology, for I do not know how to do that. I said merely that I want to limit such involvement, and to a certain extent I do know how to do that. If some technology does damage to the world – as two of the above letters seem to agree that it does – then why is it not reasonable, and indeed moral, to try to limit one's use of that technology? *Of course*, I think that I am right to do this.

I would not think so, obviously, if I agreed with Nathaniel S. Borenstein that "'better' is in the mind of the beholder." But if he truly believes this, I do not see why he bothers with his personal computer's "up-to-the-minute reports on the workings of the EPA and the nuclear industry" or why he wishes to be warned about "urgent legislative issues." According to his system, the "better" in a bureaucratic, industrial, or legislative mind is as good as the "better" in his. His mind apparently is being subverted by an objective standard of some sort, and he had better look out.

Borenstein does not say what he does after his computer has drummed him awake. I assume from his letter that he must send donations to conservation organizations and letters to officials. Like James Rhoads, at any rate, he has a clear conscience. But this is what is wrong with the conservation movement. It has a clear conscience. The guilty are always other people, and the wrong is always somewhere else. That is why Borenstein finds his "electronic bulletin board" so handy. To the conservation movement, it is only production that causes environmental degradation;

the consumption that supports the production is rarely acknowledged to be at fault. The ideal of the run-of-the-mill conservationist is to impose restraints upon production without limiting consumption or burdening the consciences of consumers.

But virtually all of our consumption now is extravagant, and virtually all of it consumes the world. It is not beside the point that most electrical power comes from strip-mined coal. The history of the exploitation of the Appalachian coal fields is long, and it is available to readers. I do not see how anyone can read it and plug in any appliance with a clear conscience. If Rhoads can do so, that does not mean that his conscience is clear; it means that his conscience is not working.

To the extent that we consume, in our present circumstances, we are guilty. To the extent that we guilty consumers are conservationists, we are absurd. But what can we do? Must we go on writing letters to politicians and donating to conservation organizations until the majority of our fellow citizens agree with us? Or can we do something directly to solve our share of the problem?

I am a conservationist. I believe wholeheartedly in putting pressure on the politicians and in maintaining the conservation organizations. But I wrote my little essay partly in distrust of centralization. I don't think that the government and the conservation organizations alone will ever make us a conserving society. "Why do I need a centralized computer system to alert me to environmental crises? That I live every hour of every day in an environmental crisis I know from all my senses. Why then is not my first duty to reduce, so far as I can, my own consumption?

Finally, it seems to me that none of my correspondents recognizes the innovativeness of my essay. If the use of a computer is a new idea, then a newer idea is not to use one.

44

In the Age of the Smart Machine

Shoshana Zuboff

The history of technology is that of human history in all its diversity. That is why specialist historians of technology hardly ever manage to grasp it entirely in their hands.

— FERNAND BRAUDEL
The Structures of Everyday Life

We don't know what will be happening to us in the future. Modern technology is taking over. What will be our place?

— A Piney Wood worker

Piney Wood, one of the nation's largest pulp mills, was in the throes of a massive modernization effort that would place every aspect of the production process under computer control. Six workers were crowded around a table in the snack area outside what they called the Star Trek Suite, one of the first control rooms to have been completely converted to microprocessor-based instrumentation. It looked enough like a NASA control room to have earned its name.

It was almost midnight, but despite the late hour and the approach of the shift change, each of the six workers was at once animated and thoughtful. "Knowledge and technology are changing so fast," they said, "what will happen to us?" Their visions of the future foresaw wrenching change. They feared that today's working assumptions could not be relied upon to carry them through, that the future would not resemble the past or the present. More frightening still was the sense of a future moving out of reach so rapidly that there was little opportunity to plan or make choices. The speed of dissolution and renovation seemed to leave no time for assurances that we were not heading toward calamity – and it would be all the more regrettable for having been something of an accident.

The discussion around the table betrayed a grudging admiration for the new technology – its power, its intelligence, and the aura of progress surrounding it. That admiration, however, bore a sense of grief. Each expression of gee-whiz-Buck-Rogers breathless wonder brought with it an aching dread conveyed in images of a future that rendered their authors obsolete. In what ways would computer technology transform

From Shoshana Zuboff, *In the Age of the Smart Machine* (New York: Basic Books, 1989), pp. 3–12. Reprinted by permission of Basic Books, a member of Perseus Books Group.

their work lives? Did it promise the Big Rock Candy Mountain or a silent graveyard?

> In fifteen years there will be nothing for the worker to do. The technology will be so good it will operate itself. You will just sit there behind a desk running two or three areas of the mill yourself and get bored.

The group concluded that the worker of the future would need "an extremely flexible personality" so that he or she would not be "mentally affected" by the velocity of change. They anticipated that workers would need a great deal of education and training in order to "breed flexibility." "We find it all to be a great stress," they said, "but it won't be that way for the new flexible people." Nor did they perceive any real choice, for most agreed that without an investment in the new technology, the company could not remain competitive. They also knew that without their additional flexibility, the technology would not fly right. "We are in a bind," one man groaned, "and there is no way out." The most they could do, it was agreed, was to avoid thinking too hard about the loss of overtime pay, the diminished probability of jobs for their sons and daughters, the fears of seeming incompetent in a strange new milieu, or the possibility that the company might welsh on its promise not to lay off workers.

During the conversation, a woman in stained overalls had remained silent with her head bowed, apparently lost in thought. Suddenly, she raised her face to us. It was lined with decades of hard work, her brow drawn together. Her hands lay quietly on the table. They were calloused and swollen, but her deep brown eyes were luminous, youthful, and kind. She seemed frozen, chilled by her own insight, as she solemnly delivered her conclusion:

> I think the country has a problem. The managers want everything to be run by computers. But if no one has a job, no one will know how to do anything anymore. Who will pay the taxes? What kind of society will it be when people have lost their knowledge and depend on computers for everything?

Her voice trailed off as the men stared at her in dazzled silence. They slowly turned their heads to look at one another and nodded in agreement. The forecast seemed true enough. Yes, there was a problem. They looked as though they had just run a hard race, only to stop short at the edge of a cliff. As their heels skidded in the dirt, they could see nothing ahead but a steep drop downward.

Must it be so? Should the advent of the smart machine be taken as an invitation to relax the demands upon human comprehension and critical judgment? Does the massive diffusion of computer technology throughout our workplaces necessarily entail an equally dramatic loss of meaningful employment opportunities? Must the new electronic milieu engender a world in which individuals have lost control over their daily work lives? Do these visions of the future represent the price of economic success or might they signal an industrial legacy that must be overcome if intelligent technology is to yield its full value? Will the new information technology represent an opportunity for the rejuvenation of competitiveness, productive vitality, and organizational ingenuity? Which aspects of the future of working life can we predict, and which will depend upon the choices we make today?

The workers outside the Star Trek Suite knew that the so-called technological choices we face are really much more than that. Their consternation puts us on alert. There is a world to be lost and a world to be gained. Choices that appear to be merely technical will redefine our lives together at work. This means more than simply contemplating the implications or consequences of a new technology. It means that a powerful new technology, such as that represented by the computer, fundamentally reorganizes the infrastructure of our material world. It eliminates former alternatives. It creates new possibilities. It necessitates fresh choices.

The choices that we face concern the conception and distribution of knowledge in the workplace. Imagine the following scenario: Intelligence is lodged in the smart machine at the expense of the human capacity for critical judgment. Organizational members become ever more dependent, docile, and secretly cynical. As more tasks must be accomplished through the medium of information technology (I call this "computer-mediated work"), the sentient body loses its salience as a source of knowledge, resulting in profound disorientation and loss of meaning. People intensify their search for avenues of escape through drugs, apathy, or adversarial conflict, as the majority of jobs in our offices and factories become increasingly isolated, remote, routine, and perfunctory. Alternatively, imagine this scenario: Organizational leaders recognize the new forms of skill and knowledge needed to truly exploit the potential of an intelligent technology. They direct their resources toward creating a work force that can exercise critical judgment as it manages the surrounding machine systems. Work becomes more abstract as it depends upon understanding and manipulating information. This marks the beginning of new forms of mastery and provides an opportunity to imbue jobs with more comprehensive meaning. A new array of work tasks offer unprecedented opportunities for a wide range of employees to add value to products and services.

The choices that we make will shape relations of authority in the workplace. Once more, imagine: Managers struggle to retain their traditional sources of authority, which have depended in an important way upon their exclusive control of the organization's knowledge base. They use the new technology to structure organizational experience in ways that help reproduce the legitimacy of their traditional roles. Managers insist on the prerogatives of command and seek methods that protect the hierarchical distance that distinguishes them from their subordinates. Employees barred from the new forms of mastery relinquish their sense of responsibility for the organization's work and use obedience to authority as a means of expressing their resentment. Imagine an alternative: This technological transformation engenders a new approach to organizational behavior, one in which relationships are more intricate, collaborative, and bound by the mutual responsibilities of colleagues. As the new technology integrates information across time and space, managers and workers each overcome their narrow functional perspectives and create new roles that are better suited to enhancing value-adding activities in a data-rich environment. As the quality of skills at each organizational level becomes similar, hierarchical distinctions begin to blur. Authority comes to depend more upon an appropriate fit between knowledge and responsibility than upon the ranking rules of the traditional organizational pyramid.

The choices that we make will determine the techniques of administration that color the psychological ambience and shape communicative behavior in the emerging workplace. Imagine this scenario: The new technology becomes the source of surveillance techniques that are used to ensnare organizational members or to subtly bully them into conformity. Managers employ the technology to circumvent the demanding work of face-to-face engagement, substituting instead techniques of remote management and automated administration. The new technological infrastructure becomes a battlefield of techniques, with managers inventing novel ways to enhance certainty and control while employees discover new methods of self-protection and even sabotage. Imagine the alternative: The new technological milieu becomes a resource from which are fashioned innovative methods of information sharing and social exchange. These methods in turn produce a deepened sense of collective responsibility and joint ownership, as access to ever-broader domains of information lend new objectivity to data and preempt the dictates of hierarchical authority.

This book is about these alternative futures. Computer-based technologies are not neutral; they embody essential characteristics that are bound to alter the nature of work within our factories and offices, and among workers, professionals, and managers. New choices are laid open by these technologies, and these choices are being confronted in the daily lives of men and women across the landscape of modern organizations. This book is an effort to understand the deep structure of these choices – the historical, psychological, and organizational forces that imbue our conduct and sensibility. It is also a vision of a fruitful future, a call for action that can lead us beyond the stale reproduction of the past into an era that offers a historic opportunity to more fully develop the economic and human potential of our work organizations.

The Two Faces of Intelligent Technology

The past twenty years have seen their share of soothsayers ready to predict with conviction one extreme or another of the alternative futures I have presented. From the unmanned factory to the automated cockpit, visions of the future hail information technology as the final answer to "the labor question," the ultimate opportunity to rid ourselves of the thorny problems associated with training and managing a competent and committed work force. These very same technologies have been applauded as the hallmark of a second industrial revolution, in which the classic conflicts of knowledge and power associated with an earlier age will be synthesized in an array of organizational innovations and new procedures for the production of goods and services, all characterized by an unprecedented degree of labor harmony and widespread participation in management process. Why the paradox? How can the very same technologies be interpreted in these different ways? Is this evidence that the technology is indeed neutral, a blank screen upon which managers project their biases and encounter only their own limitations? Alternatively, might it tell us something else about the interior structure of information technology?

Throughout history, humans have designed mechanisms to reproduce and extend the capacity of the human body as an instrument of work. The industrial age has carried this principle to a dramatic new level of sophistication with machines that can substitute for and amplify the abilities of the human body. Because machines are mute, and because they are precise and repetitive, they can be controlled according to a set of rational principles in a way that human bodies cannot.

There is no doubt that information technology can provide substitutes for the human body that reach an even greater degree of certainty and precision. When a task is automated by a computer, it must first be broken down to its smallest components. Whether the activity involves spraying paint on an automobile or performing a clerical transaction, it is the information contained in this analysis that translates human agency into a computer program. The resulting software can be used to automatically guide equipment, as in the case of a robot, or to execute an information transaction, as in the case of an automated teller machine.

A computer program makes it possible to rationalize activities more comprehensively than if they had been undertaken by a human being. Programmability means, for example, that a robot will respond with unwavering precision because the instructions that guide it are themselves unvarying, or that office transactions will be uniform because the instructions that guide them have been standardized. Events and processes can be rationalized to the extent that human agency can be analyzed and translated into a computer program.

What is it, then, that distinguishes information technology from earlier generations of machine technology? As information technology is used to reproduce, extend, and improve upon the process of substituting machines for human agency, it simultaneously accomplishes something quite different. The devices that automate by translating information into action also register data about those automated activities, thus generating new streams of information. For example, computer-based, numerically controlled machine tools or microprocessor-based sensing devices not only apply programmed instructions to equipment but also convert the current state of equipment, product, or process into data. Scanner devices in supermarkets automate the checkout process and simultaneously generate data that can be used for inventory control, warehousing, scheduling of deliveries, and market analysis. The same systems that make it possible to automate office transactions also create a vast overview of an organization's operations, with many levels of data coordinated and accessible for a variety of analytical efforts.

Thus, information technology, even when it is applied to automatically reproduce a finite activity, is not mute. It not only imposes information (in the form of programmed instructions) but also produces information. It both accomplishes tasks and translates them into information. The action of a machine is entirely invested in its object, the product. Information technology, on the other hand, introduces an additional dimension of reflexivity: it makes its contribution to the product, but it also reflects back on its activities and on the system of activities to which it is related. Information technology not only produces action but also produces a voice that symbolically renders events, objects, and processes so that they become visible, knowable, and shareable in a new way.

Viewed from this interior perspective, information technology is characterized by a fundamental duality that has not yet been fully appreciated. On the one hand, the technology can be applied to automating operations according to a logic that hardly differs from that of the nineteenth-century machine system – replace the human body with a technology that enables the same processes to be performed with more continuity and control. On the other, the same technology simultaneously generates information about the underlying productive and administrative processes through which an organization accomplishes its work. It provides a deeper level of transparency to activities that had been either partially or completely opaque. In this way information technology supersedes the traditional logic of automation. The word that I have coined to describe this unique capacity is *informate*. Activities, events, and objects are translated into and made visible by information when a technology *informates* as well as *automates*.

The informating power of intelligent technology can be seen in the manufacturing environment when microprocessor-based devices such as robots, programmable logic controllers, or sensors are used to translate the three-dimensional production process into digitized data. These data are then made available within a two-dimensional space, typically on the screen of a video display terminal or on a computer printout, in the form of electronic symbols, numbers, letters, and graphics. These data constitute a quality of information that did not exist before. The programmable controller not only tells the machine what to do – imposing information that guides operating equipment – but also tells what the machine has done – translating the production process and making it visible.

In the office environment, the combination of on-line transaction systems, information systems, and communications systems creates a vast information presence that now includes data formerly stored in people's heads, in face-to-face conversations, in metal file drawers, and on widely dispersed pieces of paper. The same

technology that processes documents more rapidly, and with less intervention, than a mechanical typewriter or pen and ink can be used to display those documents in a communications network. As more of the underlying transactional and communicative processes of an organization become automated, they too become available as items in a growing organizational data base.

In its capacity as an automating technology, information technology has a vast potential to displace the human presence. Its implications as an informating technology, on the other hand, are not well understood. The distinction between *automate* and *informate* provides one way to understand how this technology represents both continuities and discontinuities with the traditions of industrial history. As long as the technology is treated narrowly in its automating function, it perpetuates the logic of the industrial machine that, over the course of this century, has made it possible to rationalize work while decreasing the dependence on human skills. However, when the technology also informates the processes to which it is applied, it increases the explicit information content of tasks and sets into motion a series of dynamics that will ultimately reconfigure the nature of work and the social relationships that organize productive activity.

Because this duality of intelligent technology has not been clearly recognized, the consequences of the technology's informating capacity are often regarded as unintended. Its effects are not planned, and the potential that it lays open remains relatively unexploited. Because the informating process is poorly defined, it often evades the conventional categories of description that are used to gauge the effects of industrial technology.

These dual capacities of information technology are not opposites; they are hierarchically integrated. Informating derives from and builds upon automation. Automation is a necessary but not sufficient condition for informating. It is quite possible to proceed with automation without reference to how it will contribute to the technology's informating potential. When this occurs, informating is experienced as an unintended consequence of automation. This is one point at which choices are laid open. Managers can choose to exploit the emergent informating capacity and explore the organizational innovations required to sustain and develop it. Alternatively, they can choose to ignore or suppress the informating process. In contrast, it is possible to consider informating objectives at the start of an automation process. When this occurs, the choices that are made with respect to how and what to automate are guided by criteria that reflect developmental goals associated with using the technology's unique informating power.

Information technology is frequently hailed as "revolutionary." What are the implications of this term? *Revolution* means a pervasive, marked, radical change, but *revolution* also refers to a movement around a fixed course that returns to the starting point. Each sense of the word has relevance for the central problem of this book. The informating capacity of the new computer-based technologies brings about radical change as it alters the intrinsic character of work – the way millions of people experience daily life on the job. It also poses fundamentally new choices for our organizational futures, and the ways in which labor and management respond to these new choices will finally determine whether our era becomes a time for radical change or a return to the familiar patterns and pitfalls of the traditional workplace. An emphasis on the informating capacity of intelligent technology can provide a point of origin for new conceptions of work and power. A more restricted emphasis on its automating capacity can provide the occasion for that second kind of revolution – a return to the familiar grounds of industrial society with divergent interests battling for control, augmented by an array of new material resources with which to attack and defend.

The questions that we face today are finally about leadership. Will there be leaders who are able to recognize the historical moment and the choices it presents? Will they find ways to create the organizational conditions in which new visions, new concepts, and a new language of workplace relations can emerge? Will they be able to create organizational innovations that can exploit the unique capacities of the new technology and thus mobilize their organization's productive potential to meet the heightened rigors of global competition? Will there be leaders who understand the crucial role that human beings from each organizational stratum can play in adding value to the production of goods and services? If not, we will be stranded in a new world with old solutions. We will suffer through the unintended consequences of change, because we have failed to understand this technology and how it differs from what came before. By neglecting the unique informating capacity of advanced computer-based technology and ignoring the need for a new vision of work and organization, we will have forfeited the dramatic business benefits it can provide. Instead, we will find ways to absorb the dysfunctions, putting out brush fires and patching wounds in a slow-burning bewilderment.

[...]

45

The Social Life of Information

John Seely Brown and Paul Duguid

On an average weekday the New York Times *contains more information than any contemporary of Shakespeare's would have acquired in a lifetime.*

– Anonymous (and ubiquitous)

Every year, better methods are being devised to quantify information and distill it into quadrillions of atomistic packets of data.

– Bill Gates

By 2047 . . . all information about physical objects, including humans, buildings, processes and organizations, will be online. This is both desirable and inevitable.

– Gordon Bell and Jim Gray

This is the datafication of shared knowledge.

– Tom Phillips, Deja News[1]

It now seems a curiously innocent time, though not that long ago, when the lack of information appeared to be one of society's fundamental problems. Theorists talked about humanity's "bounded rationality" and the difficulty of making decisions in conditions of limited or imperfect information. Chronic information shortages threatened work, education, research, innovation, and economic decision making – whether at the level of government policy, business strategy, or household shopping. The one thing we all apparently needed was more information.

So it's not surprising that infoenthusiasts exult in the simple volume of information that technology now makes available. They count the bits, bytes, and packets enthusiastically. They cheer the disaggregation of knowledge into data (and provide a new word – *datafication* – to describe it). As the lumps break down and the bits pile up, words like *quadrillion, terabyte,* and *megaflop* have become the measure of value.

Despite the cheers, however, for many people famine has quickly turned to glut. Concern about access to information has given way to concern

From John Seely Brown and Paul Duguid, *The Social Life of Information* (Boston, MA: Harvard Business School Press, 2000), pp. 11–33. © President and Fellows of Harvard College. Reprinted by permission of Harvard Business School Press.

about coping with the amounts to which we do have access. The Internet is rightly championed as a major information resource. Yet a little time in the nether regions of the Web can make you feel like the SETI researchers at the University of California, Berkeley, searching through an unstoppable flood of meaningless information from outer space for signs of intelligent life.[2]

With the information spigot barely turned on – the effect has seemed more like breaching a dam than turning a tap – controlling the flow has quickly become the critical issue. Where once there seemed too little to swim in, now it's hard to stay afloat. The "third wave" has rapidly grown into a tsunami.[3] Faced by cheery enthusiasts, many less optimistic people resemble the poor swimmer in Stevie Smith's poem, lamenting that

I was much too far out all my life
And not waving, but drowning.

Yet still raw information by the quadrillion seems to fascinate.

Could Less Be More?

Of course, it's easy to get foolishly romantic about the pleasures of the "simpler" times. Few people really want to abandon information technology. Hours spent in a bank line, when the ATM in the supermarket can do the job in seconds, have little charm. Lose your papers in a less-developed country and trudge, as locals must do all the time, from line to line, from form to form, from office to office and you quickly realize that life without information technology, like life without modern sanitation, may seem simpler and even more "authentic," but for those who have to live it, it is not necessarily easier or more pleasant.

Even those people who continue to resist computers, faxes, e-mail, personal digital assistants, let alone the Internet and the World Wide Web, can hardly avoid taking advantage of the embedded microchips and invisible processors that make phones easier to use, cars safer to drive, appliances more reliable, utilities more predictable, toys and games more enjoyable, and the trains run on time. Though any of these technologies can undoubtedly be infuriating,

most people who complain want improvements, not to go back to life without them.[4]

Nonetheless, there is little reason for complacency. Information technology has been wonderfully successful in many ways. But those successes have extended its ambition without necessarily broadening its outlook. Information is still the tool for all tasks. Consequently, living and working in the midst of information resources like the Internet and the World Wide Web can resemble watching a firefighter attempt to extinguish a fire with napalm. If your Web page is hard to understand, link to another. If a "help" system gets overburdened, add a "help on using help." If your answer isn't here, then click on through another 1,000 pages. Problems with information? Add more.

Life at Xerox has made us sensitive to this sort of trap. As the old flip cards that provided instructions on copiers became increasingly difficult to navigate, it was once suggested that a second set be added to explain the first set. No doubt, had this happened, there would have been a third a few years later, then a fourth, and soon a whole laundry line of cards explaining other cards.

The power and speed of information technology can make this trap both hard to see and hard to escape. When information burdens start to loom, many of the standard responses fall into a category we call "Moore's Law" solutions. The law, an important one, is named after Gordon Moore, one of the founders of the chip maker Intel. He predicted that the computer power available on a chip would approximately double every eighteen months. This law has held up for the past decade and looks like it will continue to do so for the next.[5] (It's this law that can make it hard to buy a computer. Whenever you buy, you always know that within eighteen months the same capabilities will be available at half the price.)

But while the law is insightful, Moore's Law solutions are usually less so. They take it on faith that more power will somehow solve the very problems that they have helped to create. Time alone, such solutions seem to say, with the inevitable cycles of the Law, will solve the problem. More information, better processing, improved data mining, faster connections, wider bandwidth, stronger cryptography – these are the answers. Instead of thinking hard, we are encouraged simply to "embrace dumb power."[6]

More power may be helpful. To the same degree, it is likely to be more problematic, too. So as information technology tunnels deeper and deeper into everyday life, it's time to think not simply in terms of the next quadrillion packets or the next megaflop of processing power, but to look instead to things that lie beyond information.

Drowning and Didn't Know It

If, as one of our opening quotations suggests, "all information about physical objects, including humans, buildings, processes and organizations, will be online," it's sometimes hard to fathom what there is beyond information to talk about.

Let us begin by taking a cue from MIT's Nicholas Negroponte. His handbook for the information age, *Being Digital*, encouraged every-one to think about the differences between atoms, a fundamental unit of matter, and bits, the fundamental unit of information.[7] Here was a provocative and useful thought experiment in contrasts. Moreover, it can be useful to consider possible similarities between the two as well.

Consider, for example, the industrial revolution, the information revolution's role model. It was a period in which society learned how to process, sort, rearrange, recombine, and transport atoms in unprecedented fashion. Yet people didn't complain that they were drown-ing in atoms. They didn't worry about "atom overload." Because, of course, while the world may be composed of atoms, people don't per-ceive it that way. They perceive it as buses and books and tables and chairs, buildings and cof-fee mugs, laptops and cell phones, and so forth. Similarly, while information may come to us in quadrillions of bits, we don't consider it that way. The information reflected in bits comes to us, for example, as stories, documents, diagrams, pictures, or narratives, as knowledge and mean-ing, and in communities, organizations, and institutions.[8]

The difficulty of looking to these various forms through which information has conven-tionally come to us, however, is that infocentric visions tend to dismiss them as irrelevant.

Infoenthusiasts insist, for example, not only that information technology will see the end of documents, break narratives into hypertext, and reduce knowledge to data, but that such things as organizations and institutions are little more than relics of a discredited old regime.

Indeed, the rise of the information age has brought about a good deal of "endism." New technology is widely predicted to bring about, among other things,

the end of the press, television, and mass media
the end of brokers and other intermediaries
the end of firms, bureaucracies, and similar
  organizations
the end of universities
the end of politics
the end of government
the end of cities and regions
the end of the nation–state

There's no doubt that in all these categories par-ticular institutions and particular organizations are under pressure and many will not survive long. There's nothing sacred here. But it's one thing to argue that many "second wave" tools, institutions, and organizations will not survive the onset of the "third wave." It's another to argue that in the "third wave" there is no need for social institu-tions and organizations at all.

The strong claim seems to be that in the new world individuals can hack it alone with only information by their side. Everyone will return to frontier life, living in the undifferentiated global village.[9] Here such things as organizations and institutions are only in the way. Consequently, where we see solutions to information's burdens, others see only burdens on information.

Origin Myths

From all the talk about electronic frontiers, global villages, and such things as electronic cottages, it's clear that the romanticism about the past we talked about earlier is not limited to technophobes.[10] Villages and cottages, after all, are curious survivors from the old world applied to the conditions of the new. They remind us that the information age, highly rationalist though it seems, is easily trapped by its own myths. One of the most interesting may be its origin myth, which is a myth of separation.

Historians frequently trace the beginnings of the information age not to the Internet, the computer, or even the telephone, but to the telegraph. With the telegraph, the speed of information essentially separated itself from the speed of human travel. People traveled at the speed of the train. Information began to travel at the speed of light. In some versions of this origin story (which tends to forget that fire and smoke had long been used to convey messages over a distance at the speed of light), information takes on not only a speed of its own, but a life of its own. (It is even capable, in some formulations, of "wanting" to be free.)[11] And some scholars contend that with the computer, this decisive separation entered a second phase. Information technologies became capable not simply of transmitting and storing information, but of producing information independent of human intervention.[12]

No one doubts the importance of Samuel Morse's invention. But with the all-but-death of the telegraph and the final laying to rest in 1999 of Morse code, it might be time to celebrate less speed and separation and more the ways information and society intertwine. Similarly, it's important not to overlook the significance of information's power to breed upon itself. But it might be time to retreat from exuberance (or depression) at the volume of information and to consider its value more carefully.[13] The ends of information, after all, are human ends. The logic of information must ultimately be the logic of humanity. For all information's independence and extent, it is people, in their communities, organizations, and institutions, who ultimately decide what it all means and why it matters.

Yet it can be easy for a logic of information to push aside the more practical logic of humanity. For example, by focusing on a logic of information, it was easy for *Business Week* in 1975 to predict that the "paperless office" was close. Five years later, one futurist was firmly insisting that "making paper copies of anything" was "primitive."[14] Yet printers and copiers were running faster and faster for longer and longer periods over the following decade. Moreover, in the middle of the decade, the fax rose to become an essential paper-based piece of office equipment. Inevitably, this too was seen as a breach of good

taste. Another analyst snorted that the merely useful fax "is a serious blemish on the information landscape, a step backward, whose ramifications will be felt for a long time."[15]

But the fax holds on. Rather like the pencil – whose departure was predicted in 1938 by the *New York Times* in the face of ever more sophisticated typewriters – the fax, the copier, and paper documents refuse to be dismissed.[16] People find them useful. Paper has wonderful properties – properties that lie beyond information, helping people work, communicate, and think together.

If only a logic of information, rather than the logic of humanity, is taken into account, then all these other aspects remain invisible. And futurists, while raging against the illogic of humankind and the primitive preferences that lead it astray, will continue to tell us where we ought to go. By taking more account of people and a little less of information, they might instead tell us where we are going, which would be more difficult but also more helpful.

Hammering Information

Caught in the headlights of infologic, it occasionally feels as though we have met the man with the proverbial hammer to whom everything looks like a nail. If you have a problem, define it in terms of information and you have the answer. This redefining is a critical strategy not only for futurology, but also for design. In particular, it allows people to slip quickly from questions to answers.

If indeed Morse did launch the information age, he at least had the modesty to do it with a famously open-ended question. "What," he piously asked in his first message, "hath God wrought?" Now, "we have answers," or "solutions" or "all the answers you need" (11,000 according to Oracle's Web site). Similarly, IBM claims that a single computer can contain "answers to all the questions you ever had."[17] So if Morse were to ask his question again today, he would no doubt be offered an answer beginning "http://www....."

True, Microsoft advertises itself with a question: "Where do you want to go today?" But that is itself a revealing question. It suggests that Microsoft has the answers. Further, Microsoft's pictures of people sitting eagerly at computers also suggest that whatever the question, the answer lies in digital, computer-ready information. For though it asks where you want to go, Microsoft isn't offering to take you anywhere. (The question, after all, would be quite different if Microsoft's Washington neighbor Boeing had asked it.) Atoms are not expected to move, only bits. No doubt to the regret of the airlines, the ad curiously redefines "go" as "stay." Stay where you are, it suggests, and technology will bring virtually anything you want to you in the comfort of your own home. (Bill Gates himself intriguingly refers to the computer as a "passport.")[18] Information offers to satisfy your wanderlust without the need to wander from the keyboard.[19]

Refining, or Merely Redefining?

In the end, Microsoft's view of your wants is plausible so long as whatever you do and whatever you want translates into information – and whatever gets left behind doesn't matter. From this viewpoint, value lies in information, which technology can refine away from the raw and uninteresting husk of the physical world.

Thus you don't need to look far these days to find much that is familiar in the world redefined as information. Books are portrayed as information containers, libraries as information warehouses, universities as information providers, and learning as information absorption. Organizations are depicted as information coordinators, meetings as information consolidators, talk as information exchange, markets as information-driven stimulus and response.

This desire to see things in information's light no doubt drives what we think of as "infoprefixation." *Info* gives new life to a lot of old words in compounds such as *infotainment, infomatics, infomating,* and *infomediary.* It also gives new promise to a lot of new companies, from InfoAmerica to InfoUSA, hoping to indicate that their business is information. Adding *info* or something similar to your name doesn't simply add to but multiplies your market value.

Undoubtedly, information is critical to every part of life. Nevertheless, some of the attempts to squeeze everything into an information perspective recall the work of the Greek mythological bandit Procrustes. He stretched travelers who were too short and cut off the legs of those who were too long until all fitted his bed. And we suspect that the stretching and cutting done to meet the requirements of the infobed distorts much that is critically human. Can it really be useful, after all, to address people as information processors or to redefine complex human issues such as trust as "simply information?"[20]

6-D Vision

Overreliance on information leads to what we think of as "6-D vision." Unfortunately, this is not necessarily twice as good as the ordinary 3-D kind. Indeed, in many cases it is not as good, relying as it does on a one-dimensional, infocentric view.

The *D* in our 6-D notion stands for the *de-* or *dis-* in such futurist-favored words as

demassification
decentralization
denationalization
despacialization
disintermediation
disaggregation[21]

These are said to represent forces that, unleashed by information technology, will break society down into its fundamental constituents, principally individuals and information. (As we scan the Ds, it sometimes feels as though the only things that will hold up against this irresistible decomposition are the futurists' increasingly long words.)

We should say at once that none of these D-visions is inherently mistaken or uninteresting. Each provides a powerful lens on an increasingly complicated world. They help expose and explain important trends and pressures in society. Nonetheless, the Ds too easily suggest a linear direction to society – parallel movements from complex to simple, from group to individual, from personal knowledge to ubiquitous information, or more generally from composite to unit.

Yet it does not feel that modern life is moving in one direction, particularly in the direction from complex to simple. To most of us, society seems capable of moving in almost any direction, and often in the direction of chaos rather than simplicity. Indeed, many shifts that the 6-Ds reveal are not the first step in an unresisting downward spiral from complex to simple. Rather, they are parts of profound and often dramatic shifts in society's dynamic equilibrium, taking society from one kind of complex arrangement to another, as a quick review of a few Ds will suggest.

Dimensions of the Ds

Much talk about disaggregation and demassification readily assumes that the new economy will be a place of ever-smaller firms, light, agile, and unencumbered. It was once commonplace, for example, to compare the old Goliath, GM, against the new David, Microsoft. As Microsoft's market capitalization passed GM's, the latter had some 600,000 employees and the former barely 25,000. The difference is stark. Not, though, stark enough to step from here to what the business writers Larry Downes and Chunka Mui call the "Law of Diminishing Firms." After all, it's GM that's shrinking. Microsoft continues to grow while other high-tech start-ups compete for the title of "fastest growing ever."[22]

Downes and Mui draw on the theory of the firm proposed by the Nobel Prize-winning economist Ronald Coase. Coase developed the notion of *transaction costs*. These are the costs of using the marketplace, of searching, evaluating, contracting, and enforcing. When it is cheaper to do these as an organization than as an individual, organizations will form. Conversely, as transaction costs fall, this glue dissolves and firms and organizations break apart. Ultimately, the theory suggests, if transaction costs become low enough, there will be no formal organizations, but only individuals in market relations. And, Downes and Mui argue, information technology is relentlessly driving down these costs.

Though he produced elegant economic theory, Coase had strong empirical leanings. He developed his theory of transaction costs in the 1930s to bridge the gap between theoretical accounts of the marketplace and what he saw in the actual marketplace – particularly when he traveled in the United States. There, business was dominated by huge and still-growing firms. These defied the purity and simplicity of the theoretical predictions, which envisaged markets comprising primarily individual entrepreneurs.[23]

In honor of Coase's empiricism, it's important to look around now. When we began work on this book, Justice Department lawyers opened their case against Microsoft, accusing it of monopolistic practices. David now resembles Goliath. At the same time, other Justice Department lawyers were testifying that 1998 would be the first two-trillion-dollar year for mergers. Seven of the ten largest mergers in history had occurred in the first six months alone. We began keeping a list of firms involved. These included Amoco, AT&T, Bankers Trust, BMW, British Petroleum, Chrysler, Citibank, Deutsche Bank, Exxon, Ford, IBM, MCI, Mercedes, Mobil, Travelers, and many more.

Nor were these large firms buying up minnows. They were buying up each other. Ninety years after the era of trust busting, oil, banking, and tobacco, the initial targets, were all consolidating again.[24] As the *Economist* put it, after Exxon's merger with Mobil followed British Petroleum's purchase of Amoco: "Big Oil is Dead. Long Live Enormous Oil."[25]

Whatever else was apparent, we soon realized that whenever the book came out, any list of ours would be profoundly out of date. The only successful strategy in such conditions would be to imitate the great comic novelist of the eighteenth century, Laurence Sterne, who faced with an impossible description inserted a blank page into his manuscript and told the readers to take up their own pen and do it for themselves. As we were revising the manuscript, the two behemoths of the information age, AT&T and Microsoft, began their own extraordinary mating dance. That we found well beyond the reach of our pens.

Undoubtedly, several of the mergers we mentioned may represent the last struggles of dinosaurs to protect their ecological niche before other forces destroy it. Investment and even retail banking, for example, may have particularly precarious futures.

But massification is not occurring in dying "second wave" sectors alone. Many mergers have involved firms based in the "third wave" information sectors. Here mergers often involve not so much dinosaurs as phoenixes rising from the ashes of old business models. These might include AT&T's absorption of TCI and Time-Warner's of Turner Broadcasting. They surely do include Internet-driven combinations such as MCI's merger with WorldCom, IBM's takeover of Lotus, and AT&T's purchase of IBM's Global Network. Meanwhile, firms wholly within the new economy, such as AOL, Microsoft, Amazon, and eBay, go on regular shopping sprees for other companies.

Elsewhere in the information sector, Sir John Daniel, vice-chancellor of Britain's Open University, points to the rise of the "mega-university." Daniel presides over some 160,000 students, but his school hardly qualifies as "mega" in a field in which the largest – China's TV University System – has 580,000 students in degree programs. According to Daniel's figures, two universities break the half-million mark, one exceeds one-third of a million, and three are approaching a quarter million.[26] These are all "distance" universities, using a variety of information technologies to reach their students. So no simple demassification here either. Similarly, the concentration of the media in recent years challenges any simple idea of media demassification.[27]

It doesn't feel then as if firms are shrinking under an iron law. Rather, it feels more as if, as the economist Paul Krugman puts it, "We've gone from an economy where most people worked in manufacturing – in fairly large companies that were producing manufactured goods and engaged in things like transportation – to an economy where most people work for fairly large companies producing services."[28]

The resilience of the large organization is not all that surprising. Given that information technologies are particularly good at taking advantage of large networks, the information economy in certain circumstances actually favors the aggregated, massified firm.[29] These are firms that can or have knit diverse networks together, as AOL hopes to do with its purchase of Netscape or as Microsoft hopes to do with the insertion of Windows into television set-top boxes. Consequently, the small, agile firm with big ideas and little money is less likely to be the viable start-up of legend. (As a recent article in *Red Herring* put it, referring to the famous garage-based start-ups of Silicon Valley, the "garage door slams shut.")[30] And any that do start up in the traditional way are likely to be snatched up by the giants of the industry.

So, while stories abound about the new "niche" markets exploited through the Internet, the examples often come not from niche firms, but from large ones with well-established networks. The paradoxical phrase "mass customizing" suggests that fortune favors the latter. It is possible, for example, to have jeans cut to your personal dimensions. But it is quite probably Levi's that will do it for you. Here the strategy for customized goods relies on a large firm with a large market and a highly standardized product. So the demassification of production relies on the massification of markets and consumption. The Henry Ford of the new economy would tell us that we can all have jeans made to measure, so long as they are Levi's.

Finally, firms are not merely taking power from one another. They are accumulating power that once lay elsewhere. The political scientist Saskia Sassen traces the decline of the nation–state not to the sweeping effects of demassification and disaggregation, but to the rise of powerful, concentrated transnational corporations. The new economic citizen of the world, in her view, is not the individual in the global village but the transnational corporation, often so formidable that it has "power over individual governments."[31] The state and the firm, then, are not falling together along a single trajectory. At least in some areas, one is rising at the other's expense.

In sum, as people try to plot the effects of technology, it's important to understand that information technologies represent powerful forces at work in society. These forces are also remarkably complex. Consequently, while some sectors show disaggregation and demassification, others show the opposite. On the evidence of the 6-Ds, attempts to explain outcomes in terms of information alone miss the way these forces combine and conflict.

So while it might seem reasonable to propose a law of increasing, not diminishing, firms, that too would be a mistake. It would merely replace one linear argument with another. It's not so much the actual direction that worries us about infocentrism and the 6-Ds as the assumption of a single direction. The landscape is more complex. Infocentricity represents it as disarmingly simple. The direction of organizational change is especially hard to discern. The 6-Ds present it as a foregone conclusion.

More dimensions

Similarly, despite talk of disintermediation and decentralization, the forces involved are less predictable and unidirectional than a quick glance might suggest.[32] First, the evidence for disintermediation is far from clear. Organizations, as we shall see, are not necessarily becoming flatter. And second, where it does occur, disintermediation doesn't necessarily do away with intermediaries. Often it merely puts intermediation into fewer hands with a larger grasp. The struggle to be one of those few explains several of the takeovers that we mentioned above. It also explains the "browser wars" between Netscape and Microsoft, the courtship of AT&T and Microsoft, and the continuing struggle for dominance between Internet Service Providers (ISPs). Each of these examples points not to the dwindling significance but to the continuing importance of mediation on the 'Net (as does the new term *infomediary*, another case of infoprefixation). Moreover this kind of limited disintermediation often leads to a *centralization* of control. These two Ds, then, are often pulling not together, but against one another.

Not flatter Francis Fukuyama and Abram Shulsky conducted a RAND study in 1997 into the relationship between disintermediation, flat organizations, and centralization on behalf of the army. They began by studying the private sector. Here they give little hope for any direct link between information technology and flatter organizations. Indeed, like us, they believe that the conventional argument that information technology (IT) will lead to flatter organizations is an infocentric one

> [that] focuses on a single, if very important, function of middle management: the aggregation, filtering, and transmission of information. It is of course precisely with respect to this function that the advances in IT suggest that flattening is desirable, since IT facilitates the automation of much of this work. On the other hand, middle management serves other functions as well.[33]

If managers are primarily information processors, then information-processing equipment might replace them, and organizations will be flatter. If, on the other hand, there is more to management

than information processing, then linear predictions about disintermediation within firms are too simple.

Empirical evidence suggests such predictions are indeed over-simplified. Despite the talk of increasingly flatter and leaner organizations, Paul Attewell, a workplace sociologist, argues that "administrative overhead, far from being curtailed by the introduction of office automation and subsequent information technologies, has increased steadily across a broad range of industries."[34] Attewell's data from the U.S. Bureau of Labor Statistics show that the growth of nonproduction employees in manufacturing and the growth of managerial employment as a percentage of the nation's workforce has risen steadily as the workplace has been infomated.

Nor more egalitarian Fukuyama and Shulsky also argue that in instances where information technology has led to disintermediation, this has not necessarily produced decentralization. "Despite talk about modern computer technology being necessarily democratizing," they argue, "a number of important productivity-enhancing applications of information technology over the past decade or two have involved highly centralized data systems that are successful because all their parts conform to a single architecture dictated from the top."[35] Among the successful examples they give are Wal-Mart and FedEx, both of which have famously centralized decision making.

These two are merely recent examples of a clear historical trend whereby information technology centralizes authority. Harold Innis, an early communications theorist, noted how the international telegraph and telephone lines linking European capitals to their overseas colonies radically reduced the independence of overseas administrators. Previously, messages took so long to travel that most decisions had to be made locally. With rapid communication, they could be centralized. Similarly, histories of transnational firms suggest that with the appearance of the telegraph, overseas partners, once both financially and executively autonomous, were quickly absorbed by the "home" office.[36]

Less innocent than infoenthusiasts, commanders in the U.S. Navy understood the potential of information technology to disempower when they resisted the introduction of Marconi's ship-to-shore radio.[37] They realized that, once orders could be sent to them on-board ship, they would lose their independence of action. (Their resistance recalls a story of the famous British admiral Lord Nelson, who "turned a blind eye" to his telescope at the Battle of Copenhagen to avoid seeing his commander's signal to disengage.)[38]

In contemplating assumptions about the decentralizing role of information technology, Shoshona Zuboff, a professor at Harvard Business School, confessed to becoming much more pessimistic in the decade since she wrote her pathbreaking book on the infomated workplace, *In the Age of the Smart Machine*: "The paradise of shared knowledge and a more egalitarian working environment," she notes, "just isn't happening. Knowledge isn't really shared because management doesn't want to share authority and power."[39]

Of course this need not be the outcome. As Zuboff argues, it's a problem of management, not technology.[40] Smaller organizations, less management, greater individual freedom, less centralization, more autonomy, better organization, and similar desirable goals – these arguments suggest – will not emerge spontaneously from information's abundance and the relentless power of the 6-Ds. Rather, that abundance is presenting us with new and complex problems that another few cycles of Moore's Law or "a few strokes of the keyboard" will not magically overcome.[41] The tight focus on information, with the implicit assumption that if we look after information everything else will fall into place, is ultimately a sort of social and moral blindness.

The Myth of Information

6-D vision, while giving a clear and compelling view of the influence of the 'Net and its effects on everything from the firm to the nation, achieves its clarity by oversimplifying the forces at work. First, it isolates information and the informational aspects of life and discounts all else. This makes it blind to other forces at work in society. Second, as our colleague Geoffrey Nunberg has argued, such predictions tend to take the most rapid point of change and to extrapolate from there into the future, without noticing other forces that may be regrouping.[42]

This sort of reductive focus is a common feature of futurology. It accounts, for example, for all those confident predictions of the 1950s that by the turn of the century local and even domestic nuclear power stations would provide all the electricity needed at no cost. Not only did such predictions overlook some of the technological problems ahead, they also overlooked the social forces that confronted nuclear power with the rise of environmentalism. (Fifties futurism also managed to miss the significance of feminism, civil rights, and student protest while continually pointing to the imminence of the videophone and the jet pack.)

We began this chapter with a brief look back to the industrial revolution. In many ways the train epitomized that earlier revolution. Its development was an extraordinary phenomenon, spreading from a 12-mile line in the 1830s to a network of nearly 25,000 miles in little more than a decade.[43] The railway captured the imagination not only of Britain, where it began, but of the world. Every society that could afford a railway, and some that couldn't, quickly built one. Standards were developed to allow for interconnections. Information brokers emerged to deal with the multiple systems involved.[44] The train also sparked an extraordinarily speculative bubble, with experienced and first-time investors putting millions of pounds and dollars into companies with literally no track record, no income, and little sign of profitability. Unsurprisingly, in popular imagination, both at the time and since, the train has presented itself as a driving force of social and economic revolution.

Economic and social historians have long argued, however, that the story of the industrial revolution cannot be told by looking at the train alone. Historians might as well whistle for all the effect they have had. The myth of the train is far more powerful.

Today, it's the myth of information that is overpowering richer explanations. To say this is not to belittle information and its technologies. These are making critical and unprecedented contributions to the changes society is experiencing. But it is clear that the causes of those changes include much more than information itself. So the myth significantly blinds society to the character of and forces behind those changes.

In particular, the myth tends to wage a continual war against aspects of society that play a critical role in shaping not only society, but information itself, making information useful and giving it value and meaning. It's hard to see what there is other than information when identity is reduced to "life on the screen," community thought of as the users of eBay.com, organization envisaged only as self-organization, and institutions merely demonized as "second wave."

We do not believe that society is relentlessly demassifying and disaggregating. Though we admit it would be much easier to understand if it were. The social forces that resist these decompositions, like them or not, are both robust and resourceful. They shaped the development of the railroad, determining where it ran, how it ran, and who ran it. And they will continue to shape the development of information networks. As we hope to show in the course of this book, to participate in that shaping and not merely to be shaped requires understanding such social organization, not just counting on (or counting up) information.[45]

Notes

1 Gates, 1995, p. 21; Gordon Bell and James N. Gray, quoted in Nardi and O'Day, 1999; Phillips, quoted in Napoli, 1999.

2 SETI stands for Search for Extra Terrestrial Intelligence. For more information, see http://seti.ssl.berkeley.edu, retrieved 21 July 1999.

3 The notion of the "third wave" comes from the futurist Alvin Toffler's (1980) book of the same name.

4 Indeed, the degree to which we complain about such technologies when they do go wrong indicates the extent to which we have all become dependent on them to go right. The apprehension about the Y2K bug, which some predict may allow tiny embedded chips to bring huge social institutions to a halt, is itself a sign of how deeply enmeshed these things are in our lives.

5 Sometime around 2012, it has been predicted, Moore's Law will come up against the physical limitations of current microchip components, though by then solid-state components may well have been replaced.

6 Kelly, 1997.

7 Negroponte, 1995. John Tukey coined the term 'bit'. It stands for "binary digit."

8 It's worth remembering that formal information theory, while it holds the bit as a central concern, is indifferent to meaning. Claude Shannon, who with Warren Weaver laid the foundations of modern information theory, is quite clear about this: "the semantic aspects of communication are irrelevant to the engineering aspects" (Shannon and Weaver, 1964, p. 8).

9 The pervasive image of the new open frontier only works if we forget the presence on the old frontier of the US Army, the Church of Latter-Day Saints, and myriad other organizations and institutions large and small (let alone its native inhabitants) that shaped the frontier.

10 For "electronic frontier," see Rheingold, 1993; for "global village," see McLuhan, 1962; and for "electronic cottages," see Toffler, 1980.

11 The phrase "information wants to be free" is usually attributed to the author Bruce Sterling.

12 Castells, 1996.

13 The SETI project (see note 2), after all, acknowledges that the universe has long been capable of producing raw information. Finding intelligence in its midst is a different matter.

14 *Business Week*, 30 June 1975, p. 48.; Toffler, 1980, p. 205.

15 Negroponte, 1995, p. 187.

16 For the dismissal of the pencil, see Petrosky, 1990, p. 331.

17 See the IBM advertisement in the *New York Times*, 30 June 1999, sec. C, pp. 13–20.

18 Gates, 1995, p. 5.

19 The idea is certainly a tempting one to the weary business traveler and is echoed in the curious enigma of the laptop. Ads for these suggest that laptops can be so "loaded" with communications software that you can travel anywhere and remain a virtual presence in your own office. Yet in suggesting this possibility, they make you wonder why you need to travel at all.

20 Reagle, 1996.

21 Readers of Toffler's (1980) *Third Wave* will recognize the first three terms here, particularly the first, *demassification*, to which Toffler adds three subtypes: demassification of media, production, and society. Notions of *disintermediation* and *decentralization* are features, for example, in the work of George Gilder or Kevin Kelly's (1997) writing on the "new economy." There are more "Ds" that could be added, such as Kevin Kelly's *displacement* and *devolution*.

22 Downes and Mui, 1998.

23 Coase, 1937. Coase's theory should be seen not so much as an attack on neoclassical individualism as an attempt to save it from itself.

24 Among the targets of early, landmark trust cases were Northern Securities (1911), Standard Oil (1911), and American Tobacco (1911). In November 1998, Philip Morris acquired several brands from the Ligget corporation.

25 *The Economist*, 13 December 1997.

26 Daniel, 1996, table 3.2.

27 The concentration of media ownership was the subject of a special edition of *The Nation* (March 13, 1997). The picture drawn there recalled Frank Norris's famous image of the railroads as an "octopus," but, as with any list we might produce, the *Nation's* octopus is already well out of date.

28 Paul Krugman, quoted in Kelly, 1998.

29 The appearance of the fax provides a conventional example of network effects at work. At first, for most people, it was hardly worth owning a fax machine because so few others did. But with each new buyer, owning a fax became more useful and so more valuable. Each addition to the network meant that everyone had more people to fax to and receive faxes from. By the late 1980s, the network had grown to such a size that owning one became almost a necessity.

The fax spread on the back of nonproprietary standards that allowed many different firms to compete in the fax market, as they still do. But if someone had owned those standards, the results would have been quite different. Here the obvious example is not the fax, but the video recorder and the battle between VHS and Betamax. This battle was in effect a race between the two to establish network effects. In such races, a small lead can produce a "tipping" effect, where whoever gets ahead can quickly end up taking it all. In such conditions, firms will grow, not shrink.

30 Raik-Allen, 1998.

31 Sassen, 1996, p. 38.

32 With the rise of the Internet, many people argue that information can travel straight from its producer to its consumer, without need for any intermediaries. Kelly (1997), for example, envisages music going from musicians to listeners and then leaps to gourmet coffee from producers to drinkers. Undoubtedly disintermediation along these lines is happening; however, there's a good deal of wishful endism here, too. Specific examples always seem to include unpopular professions that people want to see ended, with banking and real estate falling in behind travel agents and car salespeople. One curious example of a job relentlessly pushed off the stage yet refusing to sing

its own swan song is that of the humble meter reader. Most recently, Jeremy Rifkin (1995) sang a requiem for this job in 1995. Curiously, so did articles on the effects of computers in 1970 in *National Geographic*, and as far back as 1952 in *Fortune*.

33 Fukuyama and Shulsky, 1997, p. 63.

34 Attewell, 1994, p. 36.

35 Fukuyama and Shulsky, 1997, p. 11.

36 Innis, 1991. For a case study of previously decentralized business organizations that were later centralized, see Duguid and Silva Lopes, 1999.

37 Fidler, 1997.

38 The same long arm of technology has led to the direct intervention of politics into battlefield planning; for example, though President Bush said he would leave decisions in the Gulf War to local commanders, the White House began to exercise veto power and control over the conduct of the war, after a smart bomb went astray and drew bad press.

39 Shoshona Zuboff, quoted in Lohr, 1996.

40 Meanwhile, Royal Dutch/Shell, one of the most well known and widely applauded decentralizers of the past decade, has announced that it will recentralize its "treasury." Decentralization had been too costly and inefficient. More generally, the *Economist* magazine's "Intelligence Unit" has noted a trend toward "shared services" in large corporations - and what they describe reads very much like recentralization.

41 Negroponte (1995) suggests that "a few strokes of the keyboard" will close the generation gap (p. 203).

42 Nunberg, 1996. Coffman and Odlyzko (1998) show how many projections about the current and future growth of the 'Net extrapolate from a particularly rapid point in its development, failing to note that both before and after this point, the growth curve, though impressive, was far less steep.

43 Hobsbawm, 1977, pp. 61–2. The spread of the telephone was more dramatic, growing from 14 miles in 1845 in the United States to 670,000 in 1886, but the infrastructure of the telephone was far easier to build than railway lines. The spread of the radio was more impressive yet.

44 See Campbell-Kelly and Aspray, 1996.

45 Wellman (1988) provides one of the few worthwhile studies of the effects of information technologies on social communities and networks.

References

Paul Attewell, 1994. "Information Technology and the Productivity Paradox," In: D. Harris (editor).

Organizational Linkages: Understanding the Productivity Paradox. Washington, DC: National Academy Press, pp. 13–53.

Martin Campbell-Kelly and William Aspray, 1996. *Computer: A History of the Information Machine.* New York: Basic Books.

Manuel Castells, 1996. *The Rise of the Networked Society.* Volume 1 of *The Information Age: Economy, Society, Culture.* Oxford, Eng.: Blackwell.

Ronald H. Coase, 1937. "The Nature of the Firm," *Economica* NS, volume 4, number 16, pp. 386–405.

K. G. Coffman and Andrew M. Odlyzko, 1998. "The Size and Growth Rate of the Internet," *First Monday* volume 3, number 10 (October), at http://firstmonday.org/issues/issue3_10/coffman/, retrieved 21 July 1999.

John S. Daniel, 1996. *Mega-Universities.* London: Kogan Page.

Larry Downes and Chunka Mui, 1998. *Unleashing the Killer Ap: Digital Strategies for Market Dominance.* Boston: Harvard Business School Press.

Paul Duguid and Teresa Silva Lopes, 1999. "Ambiguous Company: Institutions and Organizations in the Port Wine Trade, 1814–1834," *Scandinavian Journal Economic History Review,* volume 47, number 1, pp. 84–102.

Roger Fidler, 1997. *Mediamorphosis: Understanding New Media.* Thousand Oaks, Calif.: Pine Forge Press.

Francis Fukyama and Abram Shulsky, 1997. *The "Virtual Corporation" and Army Organization.* Santa Monica, Calif.: Rand.

Bill Gates, 1995. *The Road Ahead.* New York: Viking.

Eric J. Hobsbawm, 1977. *The Age of Revolution, 1789–1848.* London: Abacus.

Harold Innis, 1991. *The Bias of Communication.* Toronto: University of Toronto Press.

Kevin Kelly, 1998. "New Economy? What New Economy?" *Wired* volume 6, number 5 (May), pp. 146–7, and at www.wired.com/wired/archive/6.05/krugman.html, retrieved 21 July 1999.

Kevin Kelly, 1997. "New Rules for the New Economy," *Wired* volume 5, number 9 (September), pp. 140–4, 186, 188, 190, 192, 194, 196–7, and at www.wired.com/wired/archive/5.09/newrules.html, retrieved 21 July 1999.

Steve Lohr, 1996. "The Network Computer as the PC's Evil Twin," *New York Times* (4 November), section D, p. 1.

Marshall McLuhan, 1962. *The Gutenberg Galaxy: The Making of Typographic Man.* Toronto: University of Toronto Press.

Lisa Napoli, 1999. "Turning 'Sticky traffic' into Advertising Dollars," *New York Times* (10 May), section C, p. 4.

Bonnie A. Nardi and Vicki L. O'Day, 1999. *Information Ecologies: Using Technology with Heart.* Cambridge, Mass.: MIT Press.

Nicholas Negroponte, 1995. *Being Digital.* New York: Basic Books.

Geoffrey Nunberg, 1996. "Farewell to the Information Age," In: Geoffrey Nunberg (editor). *The Future of the Book.* Berkeley, Calif.: University of California Press, pp. 103–38. Also available at www.parc.xerox.com/istl/members/nunberg/Farewell.html, retrieved 21 July 1999.

Henry Petrosky, 1990. *The Pencil: A History of Design and Circumstance.* New York: Knopf.

Georgie Raik-Allen, 1998. "Garage Door Slams Shut," *Red Herring Online* (3 November), at www.redherring.com/insider/1998/1103/garagedoor.html, retrieved 21 July 1999.

Joseph M. Reagle, Jr., 1996. "Trust in Electronic Markets: The Convergence of Cryptographers and Economists," *First Monday,* volume 1, number 2 (August), http://firstmonday.org/issues/issue2/markets/, retrieved 21 July 1999.

Howard Rheingold, 1993. *The Virtual Community: Homesteading on the Electronic Frontier.* Reading, Mass.: Addison-Wesley.

Jeremy Rifkin, 1995. *The End of Work: The Decline of the Global Work Force and the Dawn of the Post-market Era.* New York: Putnam.

Saskia Sassen, 1996. *Losing Control: Sovereignty in an Age of Globalization.* New York: Columbia University Press.

Alvin Toffler, 1980. *The Third Wave.* New York: William Morrow.

Barry Wellman, 1988. "The Community Question Reevaluated," In: Michael Peter Smith (editor). *Power, Community, and the City.* New Brunswick, N.J.: Transaction Books, pp. 81–107.

46

The Quest for Universal Usability

Ben Shneiderman

I feel ... an ardent desire to see knowledge so disseminated through the mass of mankind that it may ... reach even the extremes of society: beggars and kings.
– Thomas Jefferson, Reply to American Philosophical Society, 1808

Defining Universal Usability

An important step toward the new computing is to promote the compelling goal of universal access to information and communication services. Enthusiastic networking innovators, business leaders, and government policymakers see opportunities and benefits from widespread usage. But even if they succeed in bringing low costs through economies of scale, new computing professionals will still have much work to do. They will have to deal with the difficult question, How can information and communication technologies be made usable for every user?

Designing for experienced frequent users is difficult enough, but designing for a broad audience of unskilled users is a far greater challenge. Scaling up from a listserv for 100 software engineers to 100,000 schoolteachers to 100,000,000 registered voters will take inspiration and perspiration.

Users of information and communication technologies also have a vital role to play in pushing for what they want and need. Customer-oriented pressure will accelerate efforts to make new computing technologies usable and useful. Older technologies such as postal services, telephones, and television are universally usable, but computing technology is still too hard for too many people to use. Low-cost hardware, software, and networking will benefit many users, but interface and information design improvements are necessary to achieve higher levels of success.

We can define universal usability as more than 90 percent of all households being successful users of information and communication technologies at least once a week. A 2000 survey of U.S. households shows that 51 percent have computers and 42 percent use Internet-based e-mail or other services (NTIA 2000), but the percentages decline in poorer and less educated areas. Mario Morino's leadership in promoting technology access and social development in low-income communities provides a realistic vision based on ten premises. His 2001 report, "From Access to Outcomes: Raising the

From Ben Shneiderman, *Leonardo's Laptop: Human Needs and the New Computing Technologies* (Cambridge, MA: MIT Press, 2003), pp. 35–49. © 2002 Massachusetts Institute of Technology. By permission of MIT Press.

Aspirations for Technology Initiatives in Low-Income Communities," encourages working through trusted community leaders to build community capacity with affordable housing, health clinics, public transportation, and other services. He recognizes the catalytic power of e-mail lists, the importance of skills training, and the value of improved software design.

Internationally, meeting the challenge of universal usability is still harder. Internet dissemination is much lower in other countries around the world than in the United States. Many European countries have Internet usage rates approaching 50%, but in South America, the leading Internet-using country is Brazil, with only 3% usage. In many African and Asian countries there is only one Internet Service Provider, and usage is well below 1% of the population.[1] The United Nations Development Programme and the United Nations Information Technology Service seek to apply information technology for community building at an international level.[2] They coordinate activities with groups such as British Partnerships Online, which is devoted to developing information community centers, and Volunteers in Technical Assistance (VITA), which promotes appropriate communication, agricultural, and manufacturing technologies in Benin, Mali, Guinea, and other developing countries. Cost is a central issue for many, but hardware limitations, perceived difficulty, and lack of utility discourage others. It is hard to overstate the importance of addressing international digital divides because of the potential for accelerating economic development that benefits all nations and the opportunity to promote social initiatives that support constructive, rather than violent, movements. If countries are to meet the goal of universal usability, then researchers and technology developers need to aggressively improve current products, tune them to the realities of local needs, and increase the relevance of Web services.

Leonardo would probably be a promoter of universal usability. He was described as having "a spiritual kinship for the underprivileged" (Frere 1995, 59) and sought to please and serve the needs of rich and poor. Many of his mechanical inventions, public artworks, and urban plans were meant to benefit the full spectrum of Florentine and Milanese citizens. His weapons and defensive fortifications supported urban protection from invaders. His theatrical constructions and clever toys are other examples of Leonardo's desire to please a wide range of people. Leonardo was not a scholar isolated in a university or a scientist dealing with esoteric theories. He may have worked for the nobility, but he was very much a man of the people, wandering through the Florentine markets and sketching ordinary citizens as well as painting portraits for the nobility.

Our modern understanding of universal access is usually linked to the U.S. Communications Act of 1934 covering telephone, telegraph, and radio services. It sought to ensure "adequate facilities at reasonable charges," especially in rural areas, and prevent "discrimination on the basis of race, color, religion, national origin, or sex." The term *universal access* has been applied to computing services, but the greater complexity of computing services means that access is not sufficient to ensure successful usage.

Therefore *universal usability* has emerged as an important issue. The complexity of information and communication technologies stems, in part, from the high degree of interactivity that is necessary for information exploration, commercial applications, and creative activities. The Internet is compelling because of its support for interpersonal communications and decentralized initiatives: entrepreneurs can open businesses, journalists can start publications, and citizens can organize political movements.

The increased pressure for universal usability is a happy by-product of the growth of the Internet. Since communication, e-business services (shopping, finance, and travel), e-learning, and e-healthcare are expanding and users are becoming dependent on them, there is a strong push to ensure that the widest possible audience can participate. A particularly strong argument for universal usability is tied to e-government applications such as access to national digital libraries and the movement towards citizen services at federal, state, and local levels. These services include tax rules and filing, Social Security benefits, foreign travel information, commercial licensing, recreation and parks, and police and fire departments. Another circle of support for universal usability includes daily life needs provided by employment agencies, training centers, parent-teacher associations, public interest groups, community services, and charitable organizations.

Critics worry about the creation of an information-poor minority, or worse, Internet apartheid. Although the digital divide (Campaine 2001) in Internet usage has been shrinking between men and women, and between old and young, the digital divide is growing between rich and poor (NTIA 2001). (See figure 46.1.) Well-off households are three times as likely to have Internet access as poor households. Similarly, education levels produce a digital divide between well and poorly educated households. (See figure 46.2.) Less well documented is the continuing

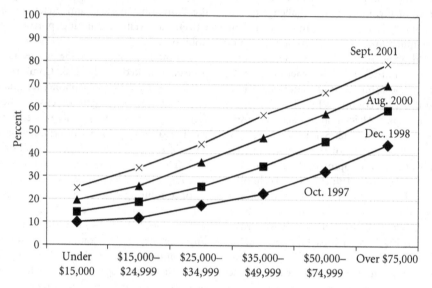

Figure 46.1   Percent of US households with Internet access, by income ($000s), 1998 and 2000.
Source: NTIA 2001.

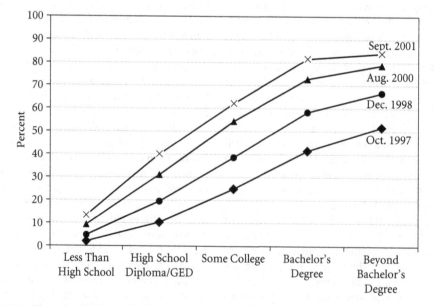

Figure 46.2   Percent of US households with Internet access, by education, 1998 and 2000.
Source: NTIA 2001.

separation between cultural and racial groups, and the low rates of usage by disadvantaged users whose unemployment, homelessness, poor health, or cognitive limitations raise further barriers. Even more challenging is the situation of international users or refugees whose literacy skills may be poor and whose languages may not be well represented on the World Wide Web. Improved design and multilingual capabilities can do much to reduce the digital divide, but training at community centers and schools will be important contributors to reducing this divide.[3]

There are other criticisms of information and communication systems. These include concerns about breakdown of community social systems, alienation of individuals that leads to crime and violence, and expansion of bureaucracies. Further threats come from loss of privacy, inadequate attention to communications or power breakdowns, and exposure to malicious viruses or hostile attacks. Open public discussion of these issues by way of participatory design strategies and town meeting forums can help to improve proposals and win public support. I have proposed that significant plans for new technology applications by government agencies should be accompanied by publication of social impact statements (Shneiderman and Rose 1996). These documents, modeled after environmental impact statements, are written to make them accessible to the public and might elicit widespread awareness and diverse proposals that reduce negative and unanticipated side effects.

Technology enthusiasts can be proud of what they have accomplished and of the number of successful Internet users, but deeper insights will come from understanding the problems of frustrated users and of those who have stayed away. Each step to broaden participation and reach these forgotten users by providing useful and usable services will be welcomed.

Universal usability is sometimes tied to meeting the needs of users who are disabled or work in disabling conditions. This important direction is likely to benefit all users. The adaptability needed for users with diverse physical, visual, auditory, or cognitive disabilities is likely to benefit users with differing preferences, tasks, skills, hardware, and so on. The recent growth of interest in disability access is tied to the standards for "comparable access" by disabled

users, as specified under section 508 of the U.S. Rehabilitation Act as amended by Congress in 1998. (Access Board 2000). The beneficial effects of changes to Web sites and other technologies may spill over to bring positive changes for all users.

Advocates who promote accommodation of disabled users often describe the curbcut – a scooped-out piece of sidewalk to allow wheelchair users to cross streets. Adding curbcuts after the curbs have been built is expensive, but building them in advance reduces costs because less material is needed. The benefits extend to baby carriage pushers, delivery service workers, bicyclists, and travelers with roller bags. Computer-related accommodations that benefit many users are power switches in the front of computers, adjustable keyboards, and user control over audio volume, screen brightness, and monitor position.

Automobile designers have long understood the benefits of accommodating a wide range of users. They feature adjustable seats, steering wheels, mirrors, and lighting levels as standard equipment and offer optional equipment for those who need additional flexibility.

Reaching a broad audience is more than a democratic ideal; it makes good business sense. The case for *network externalities*, the concept that all users benefit from expanded participation, has been made repeatedly. Facilitating access and improving usability expands markets and increases participation of diverse users whose contributions may be valuable to many. Broadening participation is not only an issue of reducing costs for new equipment. As the number of users grows, the capacity to rapidly replace a majority of equipment declines, so strategies that accommodate a wide range of equipment will become even more in demand.

Coping with Technology Variety

Supporting a broad range of hardware, software, and networks is not an easy task. The job is even more challenging when one considers the need to accommodate old hardware and software as well as new features and environments.

My friend's 93-year-old grandmother is a successful computer user, but she struggles in isolation because of her inability to keep up. She has a 1985 computer with a ten-megabyte hard disk, a character-based green screen, and an ancient word processor. She types and prints letters, but getting connected to e-mail would require a lot of changes. Adequate help is hard to get, and no company supports her technologies. Her grandson comes by every so often and gives her a hand, which does create a nice bond between them, but moving up to newer technology seems too difficult.

Every user of computers has to decide about keeping up with change. The new features can be attractive, but the fear of upgrades has become a national source of anxiety. Most users have stories of how their last upgrade caused unexpected failures or how it took them weeks to convert files. The stabilizing forces of standard hardware, operating systems, network protocols, file formats, and user interfaces are undermined by the rapid pace of technological change. Technology developers delight in novelty and improved features. They see competitive advantage in advanced designs, but these changes disrupt efforts to broaden audiences and markets. Limiting progress is one solution, but a more appealing strategy is to pressure developers to make information content, Web services, and user interfaces more malleable and adaptable to change.

The range of processor speeds in use varies by a factor of 1,000 or more. Moore's law, which states that processor speeds double every eighteen months, means that after ten years the newest processors are one hundred times faster than the old ones. Designers who wish to take advantage of new technologies risk excluding users with older machines. Similar changes for random access memory (RAM) and hard disk space also inhibit current designers who wish to reach a wide audience. Other hardware improvements such as increased screen size and improved input devices also threaten to limit access. Accommodating varying processor speed, RAM, hard disk, screen size, and input devices could help cope with this challenge. Shouldn't software be designed so that users could run the same calendar program on a palm-sized device, a laptop, and a wall-sized display?

Software changes are also a concern. As applications programs mature and operating systems evolve, software becomes obsolete because newer versions fail to preserve file format compatibility. Some changes are necessary to support new features, but modular designs are needed that promote evolution while ensuring compatibility and bidirectional file conversion. The Java movement is a step in the right direction, since it proposes to support platform independence, but its struggles indicate the difficulty of the problems.

Another concern is the variety of network connection speeds. Some users will continue to be limited to slower telephone dial-up modems while others will use much faster cable or DSL modems. This hundredfold differential creates an enormous chasm between user communities. Since many Web pages contain large graphics, user control of byte counts would be a huge benefit. Most browsers allow users to inhibit graphics, but more flexible strategies are needed. You should be able to select information-bearing graphics only or reduced byte count graphics, and invoke procedures on the server to compress the image from 300K to 80K or to 20K.

Another development that is needed is software to convert interfaces and information across media or devices. If you want your Web page contents read to you over the telephone, as many blind users do, there are already some services.[4] However, improvements are needed to speed delivery and extract the content appropriately. A more advanced idea, a generalization of the Universal Serial Bus, is a total access system that would allow you to attach a wider array of input or output devices to a computer.[5] This would enable users with disabilities or special needs to connect their specialized equipment to any computer, just as they bring along their own eyeglasses or hearing aids.

Accommodating Diverse Users

Users vary with respect to computer skills, knowledge, age, gender, disabilities, disabling conditions (mobility, sunlight, noise), literacy, culture, and income.

Since skill levels with computing vary greatly, search engines usually provide a basic and an advanced dialog box for query formulation. As a novice you can proceed without too many impediments, whereas experts can fine-tune their search strategy. Since knowledge levels in an application domain vary greatly, some sites provide two or more versions of the content. For example, the National Cancer Institute provides introductory cancer information for patients and in-depth details for physicians.[6] Since children differ from adults in their needs, NASA provides a children's section on its space mission pages.[7] Universities often segment their sites for applicants, current students, or alumni but then provide links to shared content of mutual interest.

Similar segmenting strategies could be applied to accommodate users with poor reading skills or users who require other natural languages. While there are some services to automatically convert Web pages to multiple languages,[8] the quality of human translations is still higher. If an e-commerce site maintains multiple-language versions of a product catalog, then it should make simultaneous changes to product price (possibly in different currencies), name (possibly in different character sets), or description (possibly tuned to regional variations).

As a consumer you should expect Web sites to accommodate your needs depending on your interests, income, cultural background, or religion. You should be able to find music, food, or clothing catalogs attuned to your needs so that you can easily find the products you want and not be offended by things you are not interested in. You'll probably find yourself revisiting the e-commerce sites that follow these strategies. If you are looking for Mozart symphonies, you shouldn't have to page through long lists of B. B. King songs and vice versa.

For disabled users, the needs are even more critical – if the Web site designer has not accommodated their needs, they are unhappy visitors or lost customers. Many systems allow partially sighted users, especially elderly users, to increase the font size or contrast in documents. This is good news, but a complete solution includes allowing users to improve readability in control panels, help messages, or dialog boxes. Blind users will be more active users of information and communication services if they can receive documents by speech generation or in Braille, and provide input by voice or through their customized interface devices. Physically disabled users will eagerly use services if they can connect their customized interfaces to standard graphical user interfaces, even though they may work at a much slower pace. Cognitively impaired users with mild learning disabilities, dyslexia, poor memory, and other special needs could also be accommodated with modest changes to improve layouts, control vocabulary, and limit short-term memory demands.

Expert and frequent users also have special needs. Enabling customization that speeds high-volume users, shortcuts to support repeated operations, and special-purpose devices could improve interfaces for all users. Such expert or professional customizations may represent an important business opportunity.

Finally, appropriate services for a broader range of users need to be developed, tested, and refined. Corporate knowledge workers are the primary audience for many contemporary software projects, so the interface and information needs of the unemployed, homemakers, the disabled, or migrant workers usually get less attention. This has been an appropriate business decision till now, but as the market broadens and key societal services are provided electronically, the forgotten users must be accommodated. For example, Microsoft Word provides templates for marketing plans and corporate reports, but "every-citizen" interfaces might help with job applications, babysitting cooperatives, or templates for letters to City Hall. And what about first-aid, 911 emergency assistance, crime reporting, or poison control on the Web? We should expect these services. In disaster and crisis situations, Internet services may be more reliable than telephones, but little attention has been devoted to such needs.

The growth of online support communities, medical first-aid guides, neighborhood-improvement councils, and parent-teacher associations will be accelerated as improved interface and information designs appear. Community-oriented plans for preventing drug or alcohol abuse, domestic violence, or crime could also benefit from improved interface and information designs. Such improvements are especially important for government Web sites, since they are moving toward providing basic services such as driver registration, business licenses, municipal services, tax filing, and eventually voting. Respect for the differing needs of users will do much to attract them to advanced technologies.[9] As citizens we should demand good service for all.

Bridging the Gap Between What Users Know and What They Need to Know

Every user of computers must learn how to use interfaces to accomplish his or her task. Whether you are trying to manage your retirement funds or find an apartment in a new city, there are new concepts to learn and new information to acquire. What is a margin account? How can I set a stop limit order? How can I get a map to show me how far it is from Lincoln Park to Hyde Park?

The challenge for technology developers is to bridge the gap between what users know and what they need to know. Many users don't know how to begin, what to choose in dialog boxes, how to handle crashes, or what to do about viruses. You should look for software that applies fadeable scaffolding (instructional aids that can be removed as your skills improve), training wheels (limited features to prevent errors for beginners), and just-in-time training (instructions accessible whenever the user runs into trouble).

There are many competing theories about how to train users, but too little study of what really works. One popular theory is the minimal manual approach that suggests reducing up-front instruction and getting users to be active quickly, even if they make some mistakes. This theory is applied in the short "Getting Started" guides for new hardware or software. Two other theories are the constructivist (gets the user to carry out practical projects soon) and social construction (engages pairs or larger groups of users in working together to learn).

Users approach new software tools with diverse skills and multiple intelligences. Some users need only a few minutes of orientation to understand the novelties and begin to use new tools successfully. Others need more time to acquire knowledge about the objects and actions in the applications domain and the user interface. Current interfaces could be improved with more lucid instructions, better error prevention, regular use of graphical overviews, and more effective tutorials for novices. Intermittent users would benefit from more well-designed online help, and experts need compact presentations of guidance materials and shortcuts for frequent tasks. Other helpful aids include easily reversible actions and detailed history keeping for review and consultation with peers and mentors.

A fundamental interface improvement would be support for evolutionary learning and a level-structured approach to design (Baecker et al. 2000). Why can't you begin with an interface that contains only basic features (say 5 percent of the full system) and become expert at this level within a few minutes? Game designers have created clever introductions that gracefully present new features as users acquire skill at the first level of complexity. Could similar techniques apply to the numerous features in modern word processors, e-mail handlers, and Web browsers? A good beginning has been made in some advanced systems with features such as training wheels and "Getting Started" guides, but scaling up and broader application are happening slowly. (Carroll and Carrithers 1984). A good level-structured design in the interface must be accompanied by levels in the tutorials, online help, and the error messages.

Finally, many users might be assisted through online help by way of e-mail, telephone, video conferencing, and shared screens. There is no single best way – you should be able to get help in the way that you find most comfortable at the moment. Some users like to read stories of how other users have solved their problems or used new technologies. For these users, Web sites with case studies, best practices, common problems, and frequently asked question (FAQ) lists are helpful. Many users like to talk about their problems and ask for help in highly social ways involving peers rather than experts. If you are one of these users, you might try chat rooms, news groups, or online communities.

The Skeptic's Corner

This chapter focuses on three universal usability challenges: technology variety, user diversity, and gaps in user knowledge. Skeptics caution that accommodating lower end technology and lower ability users, or users with fewer skills will result in a lowest-common-denominator system that will be less useful to most users. This scenario, called "dumbing down," is a reasonable fear, but my experience supports a brighter outcome. I believe that accommodating a broader spectrum of usage situations forces technology developers to consider a wider range of designs and often leads to innovations that benefit all users. For example, Web browsers, unlike word processors, reformat text to match the width of the window. This accommodates users with small displays (narrower than 640 pixels), and provides a nice benefit for users with larger displays (wider than 1,024 pixels), who can view more of a Web page with less scrolling. Accommodating narrower (less than 400 pixels) or wider (more than 1,200 pixels) displays presents just the kind of challenge that may push designers to develop new ideas. For example, they could consider reducing font and image sizes for small displays, moving to a multicolumn format for large displays, exploring paging strategies (instead of scrolling), and developing overviews.

A second skeptics' caution, called innovation restriction, is that attempts to accommodate the low end (technology, ability, or skill) will constrain innovations for the high end. This is again a reasonable caution, but if designers are aware of this concern, the dangers are easily avoidable. A basic Web page accommodates low-end users, and sophisticated user interfaces using Java applets or Flash plug-ins can be added for users with advanced hardware and software and fast network connections. New technologies can often be provided as an add-on or plug-in rather than as a replacement. As new technologies become perfected and widely accepted, they become the new standard. Layered approaches have been successful in the past, and they are compelling for accommodating a wide range of users. They are easy to implement when planned in advance but often difficult to retrofit.

Notes

1 From CyberAtlas, Geographies, http://cyberatlas. internet.com.

2 United Nations Development Programme, www.undp.org/; United Nations Information Technology Service, www.unites.org/; Partnerships Online: Creating Online Communities for Neighbourhoods and Networks, www.partnerships. org.uk/c-; Volunteers in Technical Assistance, www.vita.org/.

3 Digital Divide Network, www.digitaldividenetwork. org/, developed by the Benton Foundation, www.benton.org/.

4 See www.conversa.com/.

5 See Neil Scott's Archimedes Project at Stanford University, http://archimedes.stanford.edu/.

6 The US National Cancer Institute is at www.nci.nih.gov/, and the cancer information is at www.nci.nih.gov/cancerinfo/index.hrrnl.

7 NASA's regular page is at www.nasa.gov/, and the page for kids is at www.nasa.gov/kids.htm!.

8 Altavista, http://world.altavista.com/, provides the Systran services, www.systransofc.corn/. The Seattle Community Network offers a wonderful guide to resources for translating its content to many languages, www.scn.org/spanish.html.

9 There are several resources for universal design. *Professional groups*; The ACM SIGCHI (Special Interest Group on Computer-Human Interaction), www. acm.org/sigchi/, focuses on design of useful, usable, and universal user interfaces. SIGCHI promotes diversity with its outreach efforts to seniors, kids, teachers, and international groups, and it sponsors the Conferences on Universal Usability, www1. acm.org/sigs/sigchi/cuu/. The ACM's SIGCAPH (Special Interest Group on Computers and the Physically Handicapped), www.acm.org/sigcaph/, has long promoted accessibility for disabled users, and its ASSETS series of conference proceedings, www1.acm.org/sigs/sigcaph/assers/, provides useful guidance. The European conferences on User Interfaces for All, <http://ui4all.ics.forth.gr/index. html>, also deal with interface design strategies. The Web Accessibility Initiative, www.w3.org/WAll, of the World Wide Web Consortium has a guidelines document with fourteen thoughtful content design items to support disabled users. Corporate Web sites; Sun Microsystems, www.sun.corn/access/o, offers Java-specific recommendations. Thoughtful Web sites from IBM, www.ibm.com/easy/, and Microsoft, www.microsoft.com/enable/, describe processes and designs for supporting diverse users. University Web sites; North Carolina State University's Center for Universal Design, www.design.ncsu.edu/cud/c-, lists seven key principles, and the University of Wisconsin's TRACE Center, http://trace.wisc.edu/world/, offers links to many resources. Another source, http://universalusability.org/, has a taxonomy of topics and links, plus information on the Universal Usability Policy template for inclusion on Web sites. Students at the University of Maryland have created a Universal Usability in Practice Web site, www.otal.umd.eduluupractice/, with design guidelines.

References

Access Board. *Electronic and Information Technology* 2000. www.access-board.gov/sec508/status.htm

Baecker, R., K. Booth, S. Jovicic, J. McGrenere, and G. Moore. "Reducing the Gap Between What Users Know and What They Need to Know." In *Proceedings of User Interface Software and Technology Symposium 2001* (New York: ACM Press, 2001).

Carroll, J. and C. Carrithers, "Training Wheels in a User Interface," *Communication of the ACM* 27 1984 (8): 800–6.

Compaigne, D. (ed.) *The Digital Divide: Facing Crisis or Creating a Myth?* (Cambridge, MA: MIT Press, 2001).

Frere, J. C. *Leonardo: Painter, Inventor, Visionary, Mathematician, Philosopher, Engineer* (Paris: Terrail, 1995).

NTIA (National Telecommunications and Information Administration) US Department of Commerce. 2000. *Falling through the Net: Toward Digital Inclusion* www.ntia.doc.gov/ntiahome/fttn00/contents00.html.

NTIA (National Telecommunications and Information Administration) US Department of Commerce. 2001 www.ntia.doc.gov/reports/anol/index.html

Shneiderman, B. and A. Rose "Social Impact Statement: Engaging Public Participation in Information Technology Design," in *Proceedings of CQL 96, ACM SIGCAS, Symposium on Computers and the Quality of Life, 90–96.* Also in *Human Values and the Design of Computer Technology,* ed. B. Friedman, (New York: Cambridge University Press, 1997) 117–33.

Bibliography

Achterhuis, H. *American Philosophy of Technology.* Bloomington: Indiana University Press, 2001.

Adas, M. *Machines as the Measure of Men: Science, Technology, and the Ideologies of Western Dominance.* Ithaca, NY: Cornell University Press, 1989.

Adorno, T. *The Culture Industry.* London: Routledge, 2001.

Adorno, T. and M. Horkheimer. *The Dialectic of Enlightenment.* New York: Continuum, 1976.

Agassi, J. *Technology: Philosophical and Social Aspects.* Dordrecht: Reidel, 1985.

Aibar, E. 1996. "The Evaluative Relevance of Social Studies of Technology." *Society for Philosophy and Technology,* no. 1–2.

Alford, C. F. *Science and the Revenge of Nature: Marcuse and Habermas.* Gainesville, FL: University Presses of Florida, 1985.

Appadurai, A. *The Social Life of Things.* Cambridge: Cambridge University Press, 1986.

Aronowitz, S. and M. Mesner (eds.), *Technoscience and Cyberculture.* New York: Routledge, 1996.

Bachelard, G. *The Poetics of Space.* Boston: Beacon Press, 1994.

Baker, N. *Double Fold: Libraries and the Assault on Paper.* New York: Vintage 2002.

Basalla, G. *The Evolution of Technology.* Cambridge: Cambridge University Press, 1989.

Baudrillard, J. *The Mirror of Production.* New York: Telos Press, 1975.

Baudrillard, J. *Simulacra and Simulation.* Ann Arbor: University of Michigan Press, 1994.

Baudrillard, J. *The Gulf War Did Not Take Place,* trans. P. Patton. Bloomington, IN: Indiana University Press, 1995.

Baudrillard, J. and J. Nouvel. *The Singular Objects of Architecture.* Minneapolis: University of Minnesota Press, 2002.

Bauer, M. (ed.) *Resistance to New Technology: Nuclear Power, Information Technology, Biotechnology.* Cambridge: Cambridge University Press, 1995.

Bell, D. *The End of Ideology.* New York: Free Press, 1962.

Bell, D. *The Coming of Post-Industrial Society: A Venture in Social Forecasting.* New York: Basic, 1973.

Bell, D. *The Cultural Contradictions of Capitalism.* New York: Basic, 1976.

Beniger, J. R. *The Control Revolution: Technological and Economic Origins of the Information Society.* Cambridge, MA: Harvard University Press, 1986.

Benjamin, W. *Illuminations,* ed. Hannah Arendt, trans. Harry Zohn. Schocken, 1968/1936.

Bent, S., R. L. Schwaab, D. G. Conlin, and D. D. Jeffrey. *Intellectual Property Rights in Biotechnology Worldwide.* New York: Stockton Press, 1991.

Berger, John. *Ways of Seeing.* London: BBC and Penguin Books, 1972.

Berger, P. and T. Luckmann. *The Social Construction of Reality.* New York: Doubleday, 1966.

Berger, P., B. Berger, and H. Kellner. *The Homeless Mind: Modernization and Consciousness.* New York: Random House, 1973.

Berman, M. *All That is Solid Melts Into Air: The Experience of Modernity.* New York: Simon and Schuster, 1982.

Berry, W. *The Gift of Good Land: Further Essays Cultural & Agricultural.* Albany, CA: North Point Press, 1982.

Berry, W. *Home Economics.* Albany, CA: North Point Press, 1987.

Berry, W. *What Are People For?* New York: North Point Press, 1990.

Berry, W. *Sex, Economy, Freedom & Community.* NY: Pantheon, 1994.

Berry, W. *Life Is a Miracle: An Essay Against Modern Superstition.* Washington, DC: Counterpoint, 2001.

Berry, W. *The Unsettling of America: Culture & Agriculture.* San Francisco, CA: Sierra Club Books, 2004.

Best, S. and D. Kellner. *Postmodern Theory: Critical Interrogations.* London: Macmillan Press, 1991.

Best, S. and D. Kellner, *The Postmodern Adventure: Science, Technology, and Cultural Studies at the Third Millennium.* New York: Guilford Press, 2001.

Bijker, W. "Do not Despair: There is Life after Constructivism." *Science, Technology & Human Values* 18 (1993): 113–38.

Bijker, W. *Bikes, Bakelite, and Bulbs: Steps Toward a Theory of Socio-Technical Change.* Cambridge, MA: MIT Press, 1995.

Bijker, W. and J. Law (eds.). *Shaping Technology/ Building Society: Studies in Sociotechnical Change.* Cambridge, MA: MIT Press, 1992.

Bijker, W., T. Pinch, and T. Hughes (eds.). *The Social Construction of Technological Systems: New Directions in the Sociology and History of Technology.* Cambridge, MA: MIT Press, 1987.

Birkerts, S. *The Gutenberg Elegies: The Fate of Reading in an Electronic Age.* New York: Fawcett Columbine, 1994.

Bloor, D. *Knowledge and Social Imagery.* London: Routledge and Kegan Paul, 1976.

Borgmann, A. *Technology and the Character of Contemporary Life.* Chicago: University of Chicago Press, 1984.

Borgmann, A. *Crossing the Postmodern Divide.* Chicago: University of Chicago Press, 1992.

Borgmann, A. *Holding on to Reality: The Nature of Information at the Turn of the Millennium.* Chicago: University of Chicago Press, 1999.

Borlaug, N. "Ending World Hunger. The Promise of Biotechnology and the Threat of Antiscience Zealotry." *Plant Physiology* (2000) 124: 487–90.

Brey, P. "Social Constructivism for Philosophers of Technology: A Shopper's Guide." *Techné: Journal of the Society for Philosophy and Technology* 2 (1997): 3–4, 56–78.

Brey, P. A. E. "Sustainable Technology and the Limits of Ecological Modernization." *Ludus Vitalis. Revista De Filosofia de Las Ciencias de la Vida / Journal of Philosophy of Life Sciences* (1997): 17–30.

Brey, P. "Space-Shaping Technologies and the Geographical Disembedding of Place." In A. Light and B. Smith (eds.), *Philosophy & Geography vol. 3: Philosophies of Place.* New York and London: Rowman & Littlefield, 1998, pp. 239–63.

Brey, P. A. E. 'Artifacts as Social Agents.' In H. Harbers (ed.), *Inside the Politics of Technology. Agency and Normativity in the Production of Technology and Society,* Amsterdam: Amsterdam University Press, 2005.

Brey, P. A. E. "The Social Agency of Technological Artifacts. A Typologie." In P. P. C. C. Verbeek and A. Slob (eds.), *User Behavior and Technology Development.* Dordrecht: Springer, 2006.

Brey, P. A. E. "Computer Ethics in (Higher) Education." In G. Dodig-Crnkovic and S. Stuart (eds.), *Computation, Information Cognition: The Nexus and the Liminal.* Cambridge: Cambridge Scholars Press, 2007.

Brey, P. A. E. "Is Information Ethics Culture-Relative?" *International Journal of Technology and Human Interaction* 3:3 (2007): 12–24.

Brin, David. *The Transparent Society: Will Technology Force Us to Choose Between Privacy and Freedom?* New York: Addison-Wesley, 1998.

Brooke, Collin Gifford. "The Fate of Rhetoric in an Electronic Age." *Enculturation: Cultural Theories & Rhetorics* 1.1 (Spring 1997).

Brown, D. E., M. Fox and M. R. Pelletier (eds.). *Sustainable Architecture: White Papers.* New York: Earth Pledge Foundation, 2001.

Bruner, Jerome. "The Narrative Construction of Reality." *Critical Inquiry* 18 1.1 (Autumn 1991): 1–21.

Bruthiaux, Paul. "Missing in Action: Verbal Metaphor for Information Technology." *English Today* 17.3 (July 2001): 24–30.

Buliarello, G. and D. B. Doner (eds.). *History of Philosophy and Technology.* Urbana, IL: University of Illinois Press, 1979.

Bunge, M. *Treatise on Basic Philosophy;* vol. 7: Part II: "Life Science, Social Science and Technology." Dordrecht-Boston: Kluwer Academic Publishers, 1985.

Bunge, M. *Treatise on Basic Philosophy;* vol. 8: *Ethics.* Dordrecht-Boston: Kluwer Academic Publishers, 1989.

Bunge, M. *Social Science Under Debate.* Toronto: University of Toronto Press, 1998.

Bunge, M. *Philosophy of Science;* vol. 2: *From Explanation to Justification.* New Brunswick, NJ: Transaction Publishers, 1998.

Callahan, D. *False Hopes.* New York: Simon & Schuster, 1998.

Callahan, D. *The Research Imperative: What Price Better Health?* Berkeley, CA: University of California Press, 2003.

Callahan, D. *Medicine and the Market: Equity vs. Choice.* Baltimore: Johns Hopkins University Press, 2006.

Callicott, J. B. (ed.). *Companion to a Sand County Almanac: Interpretive and Critical Essays.* Madison, WI: University of Wisconsin Press, 1987.

Callicott, J. B. *Earth's Insights*. Berkeley: University of California Press, 1994.

Callicott, J. B. "After the Industrial Paradigm, What?" In C. Mitcham et al. (eds.), *Research in Philosophy and Technology*; vol. 18: *Philosophies of the Environment and Technology*. Stamford, CT: JAI Press, 1999, pp. 13–26.

Callicott, J. B. *Beyond the Land Ethic*. Albany: State University of New York Press, 1999.

Callicott, J. B. and M. P. Nelson. *American Indian Environmental Ethics: An Ojibwa Case Study*. Upper Saddle River, NJ: Prentice Hall, 2003.

Callon, M. and B. Latour, 'Don't Throw the Baby Out with the Bath School! A Reply to Collins and Yearley.' In A. Pickering (ed.), *Science as Practice and Culture*. Chicago: University of Chicago Press, 1992.

Carson, R. *Silent Spring*. Boston: Houghton-Mifflin, 1962.

Castells, Manuel. *High Technology, Space and Society*. Beverly Hills, CA: Sage Publications, 1985.

Castells, M., *The Information Age: Economy, Society and Culture (Vol. 1), The Rise of the Network Society*. Oxford: Blackwell, 1996.

Castells, M. *The Information Age: Economy, Society and Culture (Vol. 2), The Power of Identity*. Oxford: Blackwell, 1997.

Castells, M. *The Information Age: Economy, Society and Culture (Vol. 3), End of Millennium*. Oxford: Blackwell, 1998.

Christensen, S. H., M. Meganck, and B. Delahousse (eds.). *Philosophy in Engineering*. Aarhuus: Academica, 2007.

Cockburn, C. and R. F. Dilic (eds.). *Bringing Technology Home: Gender and Technology in Changing Europe*. Buckingham: Open University Press, 1994.

Cohen, Cynthia B. *Renewing the Stuff of Life: Stem Cells, Ethics, and Public Policy*. New York: Oxford University Press, 2007.

Collins, H. M. *Changing Order: Replication and Induction in Scientific Practice*. Beverly Hills, CA: Sage, 1985.

Collins, H. and T. Pinch. *The Golem at Large: What You Should Know About Technology*. Cambridge: Cambridge University Press, 1998.

Collins, H. and S. Yearley. "Epistemological Chicken." In A. Pickering (ed.), *Science as Practice and Culture*. Chicago: University of Chicago Press, 1992.

Comstock, G. *Vexing Nature: On the Ethical Case Against Agriculutural Biotechnology*. Boston/Dordrecht/London: Kluwer Academic Publishers, 2000.

Cook, D. "Time, Technology, and Writing: On the Far Side of the Digital Divide." *English Record* 52.3 (Spring–Summer 2002): 15–24.

Cowan, R. S. *More Work For Mother: The Ironies of Household Technology from the Open Hearth to the Microwave*. New York: Basic Books, 1983.

Cowan, R. S. *A Social History of American Technology*. New York: Oxford University Press, 1997.

Crease, R. and E. Selinger. *The Philosophy of Expertise*. New York: Columbia University Press, 2006.

Cutcliffe, S. H. *Ideas, Machines and Values*. Lanham, MD: Rowman & Littlefield, 2000.

Davis, D. S. *Genetic Dilemmas: Reproductive Techno-logy, Parental Choices, and Children's Futures*. New York: Routledge, 2001.

De Botton, A. *The Architecture of Happiness*. New York: Pantheon Books, 2006.

DeGrazia, D. "Human-Animal Chimeras: Human Dignity, Moral Status, and Species Prejudice." *Metaphilosophy* (2007) 38: 2–3, 309–29.

Deleuze, G. *Bergsonism*. Cambridge, MA: Zone Books, 1966.

Deleuze, G. and F. Guattari. *A Thousand Plateaus*. New York: Continuum, 2004.

Deleuze, G. and F. Guattari. *Anti-Oedipus: Capitalism and Schizophrenia*. Minneapolis: University of Minnesota Press, 1983.

de Melo-Martín, I. *Making Babies: Biomedical Technologies, Reproductive Ethics, and Public Policy*. Dordrecht: Kluwer Academic Publishers, 1998.

de Melo-Martín, I. "Ethics and Uncertainty: Biomedical Technologies and Risks to Women's Health." *Risk: Issues in Health, Safety, and Environment* 9:3 (1998): 201–27.

de Melo-Martín, I. "New Assisted-Conception Technologies: Social Dimensions and Democratic Participation." *Research in Philosophy and Techno-logy* 17 (1998): 219–38.

de Melo-Martín, I. "On Cloning Human Beings." *Bioethics* 16:3 (2002): 246–65.

de Melo-Martín I. "On our Obligation to Select the Best Children: a Reply to Savulescu." *Bioethics* 18:1 (2004): 72–83.

de Melo-Martín I., and I. N. Cholst, "Researching Human Oocyte Cryopreservation: Ethical Issues." *Fertility and Sterility* 89:3 (March 2008): 523–8.

de Melo-Martín, I. and C. Hanks. "Genetic Technologies and Women: The Importance of Context." *Bulletin of Science, Technology & Society* 21:5 (2001): 354–60.

de Melo-Martín, I. and Z. Meghani, "Beyond Risk. A More Realistic Risk-benefit Analysis of Agricultural Biotechnologies." *EMBO Reports* 9:4 (April 2008): 302–6.

de Melo-Martín I., Z. Rosenwaks, and J. J. Fins, "New Methods for Deriving Embryonic Stem Cell Lines: are the Ethical Problems Solved?" *Fertility and Sterility* 86:5 (Nov 2006): 1330–2.

Devall, B. and G. Sessions: *Deep Ecology: Living as if Nature Mattered*. Layton, Utah: Gibbs Smith, 2001.

de Vries, M. J. *Teaching About Technology: An Introduction to the Philosophy of Technology for Non-*

philosophers. Dordrecht, Netherlands: Springer, 2005.

Dewey, J. *The Collected Works of John Dewey, 1882–1953*, ed. J. A. Boydston. Carbondale, IL: Southern Illinois Press, 1969–1991 (including the full text of the Early Works, Middle Works, and Later Works).

Dewey, J. *The Collected Works of John Dewey, 1882–1953: The Electronic Edition*, ed. L. A. Hickman. Charlottesville, VA: InteLex Corporation, 1996 (including the full electronic text of the Early Works, Middle Works, and Later Works).

Dickson, D. *Alternative Technology and The Politics of Technological Change*. New York: University Books, 1974.

Douglas, M. and A. Wildavsky, *Risk and Culture: An Essay on the Selection of Technical and Environmental Dangers*. Berkeley: University of California Press, 1982.

Drengson, A. "Shifting Paradigms: From the Technocratic to the Person-Planetary." *Environmental Ethics* 3 (1980): 221–40.

Drengson, Alan, "Toward a Philosophy of Technology." *Philosophy Today* (Summer 1982): 103–17.

Drengson, A. and Y. Inoue (eds.). *The Deep Ecology Movement: An Introductory Anthology*. Berkeley: North Atlantic Books, 1995.

Dreyfus, H. *What Computers Can't Do: The Limits of Artificial Intelligence*. New York: Harper and Row, 1972.

Dreyfus, H. *What Computers Still Can't Do: A Critique of Artificial Reason*. Cambridge, MA: MIT Press, 1992.

Dreyfus, H. and S. E. Dreyfus. *Mind Over Machine: The Power of Human Intuition and Expertise in the Era of the Computer*. New York: Free Press, 1986.

Duany, A., E. Plater-Zyberk, and J. Speck. *Suburban Nation: The Rise of Sprawl and the Decline of the American Dream*. New York: North Point Press, 2001.

Duguid, P. and J. S. Brown. *The Social Life of Information*. Cambridge, MA: Harvard Business School Press, 2002.

Durbin, P. T. "Review of Albert Borgmann's Technology and the Character of Contemporary Life." *Man and World* 21 (1988): 231–5.

Durbin, P. T. *Social Responsibility in Science, Technology, and Medicine*. Bethlehem, PA: Lehigh University Press, 1992.

Durbin, P. T. "Philosophy of Science, Technology, and Medicine." In C. Mitcham and W. Williams (eds.), *The Best in Science, Technology, and Medicine*, vol. 5 of *The Reader's Adviser*, 14th edn., New York: Bowker, 1994, pp. 59–104.

Durbin, P. T. "Philosophy of Technology: Retrospective and Prospective Views." In E. Higgs, A. Light,

and D. Strong (eds.), *Technology and the Good Life*. Chicago: University of Chicago Press, 2000, pp. 38–49.

Dusek, V. *The Holistic Inspirations of Physics: An Underground History of Electromagnetic Theory*. New Brunswick, NJ: Rutgers University Press, 1999.

Dusek, V. *Philosophy of Technology: An Introduction*. Malden, MA: Blackwell, 2006.

Edgerton, D. *The Shock of the Old: Technology and Global History since 1900*. New York: Oxford University Press, 2006.

Edwards, A. R. *The Sustainability Revolution: Portrait of a Paradigm Shift*. Gabriola Island, BC, Canada: New Society Publishers, 2005.

Elam, M. "Anti Anticonstructivism or Laying the Fears of a Langdon Winner to Rest." *Science, Technology, & Human Values* 19 (1994): 101–6.

Ellul, J. *The Technological Society*. New York: Random House, 1964.

Ellul, J. *The Politics of God and the Politics of Man*. Grand Rapids, MI: Eerdmans, 1972.

Ellul, J. *The Ethics of Freedom*. Grand Rapids, MI: Eerdmans, 1976.

Ellul, J. *The Technological System*. New York: Continuum, 1980.

Ellul, J. *The Technological Bluff*. Grand Rapids, MI: Eerdmans, 1990.

Ellul, J. *The Meaning of the City*. Grand Rapids, MI: Eerdmans, 1993.

Elster, J. *Explaining Technical Change*. Cambridge: Cambridge University Press, 1993.

Engels, Friedrich. *Anti-Duhring*. Moscow: Foreign Languages Publishing House, 1962.

Ess, C. *Philosophical Perspectives on Computer-Mediated Communication*. Albany, NY: SUNY Press, 1996.

Federoff, N. V. and N. M. Brown. *Mendel in the Kitchen*. Washington, DC: Joseph Henry Press, 2004.

Feenberg, A. *Critical Theory of Technology*. New York: Oxford University Press, 1991.

Feenberg, A. *Alternative Modernity: The Technical Turn in Philosophy and Social Theory*. Berkeley: University of California Press, 1995.

Feenberg, A. *Questioning Technology*. London: Routledge, 1999.

Feenberg, A. *Transforming Technology: A Critical Theory Revisited*. New York: Oxford University Press, 2002.

Feenberg, A. and A. Hannay (eds.). *Technology and the Politics of Knowledge*. Bloomington: Indiana University Press, 1995.

Ferre, F. *The Philosophy of Technology*. Englewood Cliffs, NJ: Prentice Hall, 1988.

Fisher, E. "Lessons learned from the Ethical, Legal and Social Implications program (ELSI): Planning societal implications research for the National Nanotechnology Program." *Technology in Society* 27 (2005): 321–8.

Fledermann, C. *Engineering Ethics.* New York: Prentice Hall, 2003.

Florman, S. *Blaming Technology.* New York: St Martin's, 1982.

Florman, S. *The Civilized Engineer.* New York: St Martin's, 1988.

Florman, S. *The Existential Pleasures of Engineering.* New York: St Martin's, 1996.

Florman, S. *The Introspective Engineer.* New York: St Martin's, 1997.

Forrester, T. *High-Tech Society.* Oxford: Basil Blackwell, 1987.

Foucault, M. *Discipline and Punish: The Birth of the Prison,* trans. A. Sheridan. New York: Pantheon, 1977.

Foucault, M. *History of Sexuality, Volume, 1,* trans R. Hurley. New York: Pantheon, 1978.

Foucault, M. *The Archaeology of Knowledge & The Discourse on Language.* New York: Pantheon, 1982.

Foucault, M. *The Order of Things: An Archaeology of Human Sciences.* New York: Vintage, 1994.

Fox, R. (ed.). *Technological Change: Methods and Themes in the History of Technology.* Amsterdam: Harwood, 1999.

Fox, W. "The Deep Ecology-Ecofeminism Debate and Its Parallels." *Environmental Ethics* 11:1 (1989): 5–26.

Frankel, B. *The Post-Industrial Utopians.* Oxford: Basil Blackwell, 1987.

Franssen, M. "Technological Regime as a Key Concept in Explaining Technical Inertia and Change: a Critical Analysis." *International Journal of Technology, Policy and Management* 2 (2002) 455–70.

Franssen, M. "The Normativity of Artefacts." *Studies in History and Philosophy of Science* (2006) 37, 42–57.

Friedman, T. *The World Is Flat: A Brief History of the Twenty-First Century.* New York: Farrar, Straus and Giroux, 2006.

Fukuyama, F. *Our Posthuman Future: Consequences of the Biotechnology Revolution.* New York: Picador, 2003.

Galbraith, J. K. *The New Industrial State.* New York: New American Library, 1967.

Garrigou-Lagrange, M. *In Season, Out of Season: An Introduction to the Thought of Jacques Ellul.* San Francisco: Harper & Row, 1982.

Garson, B. *The Electronic Sweatshop: How Computers are Transforming the Office of the Future into the Factory of the Past.* New York: Penguin Books, 1988.

Gehlen, A. *Man in the Age of Technology,* trans. Patricia Lipscomb. New York: Columbia University Press, 1980.

Gendron, B. *Technology and the Human Condition.* New York: St Martin's, 1977.

Giedion, S. *Mechanization Takes Command.* New York: Oxford University Press, 1948.

Gill, R. 'Power, Social Transformation, and the New Determinism: A Comment on Grint and Woolgar.' *Science, Technology, & Human Values* 21 (1996): 347–53.

Gore, A. Jr. *Earth in the Balance: Ecology and the Human Spirit.* Boston: Houghton Mifflin, 1992.

Gottdeiner, Mark. *The Social Production of Urban Space.* Austin: University of Texas Press, 1985.

Grange, J. *Nature: An Environmental Cosmology.* Albany: The State University of New York Press, 1997.

Grange, J. *The City: An Urban Cosmology.* Albany: The State University of New York Press, 1999.

Grint, K. and R. Gill. *The Gender-Technology Relation: Contemporary Theory and Research.* London: Taylor and Francis, 1995.

Grint, K. and S. Woolgar, "On Some Failures of Nerve in Constructivist and Feminist Analyses of Technology." *Science, Technology, & Human Values* (1995) 20: 286–310.

Grint, K., and S. Woolgar, "A Further Decisive Refutation of the Assumption That Political Action Depends on the 'Truth' and a Suggestion That We Need to Go beyond This Level of Debate: A Reply to Rosalind Gill." *Science, Technology, & Human Values* 21 (1996): 354–7.

Gruen, L., with L. Grabel and P. Singer (eds.). *Stem Cell Research: The Ethical Issues.* Cambridge, MA: Wiley-Blackwell Publishing, 2007.

Guha, R. *The Unquiet Woods: Ecological Change and Peasant Resistance in the Himalaya.* Berkeley, CA: University of California, 1989.

Guha, R. *How Much Should a Person Consume? Thinking Through the Environment.* Berkeley, CA: University of California, 2006.

Guha, R. and M. Gadgil. *Ecology and Equity.* London: Penguin, 1995.

Guha, R. and T. N. Maden (eds.). *Social Ecology.* New York: Oxford University Press, 1994.

Habermas, J. *Toward a Rational Society.* Boston: Beacon Press, 1970.

Habermas, J. *Knowledge and Human Interests.* Boston: Beacon Press, 1971.

Habermas, J. *Legitimation Crisis,* trans. Thomas McCarthy. Boston: Beacon Press, 1975.

Habermas, J. *Communication and the Evolution of Society.* Boston: Beacon Press, 1979.

Habermas, J. *Theory of Communicative Action,* 2 vols., trans. Thomas McCarthy. Boston: Beacon Press, 1984–1987.

Habermas, J. *The Philosophical Discourse of Modernity: Twelve Lectures.* Cambridge: MIT Press, 1987.

Habermas, J. *Structural Transformation of the Public Sphere*, trans. Thomas Burger. Cambridge, MA: MIT Press, 1989.

Habermas, J. *The Past As Future*. Lincoln: University of Nebraska Press, 1994.

Habermas, J. *Between Facts and Norms*. Cambridge: MIT Press, 1996.

Habermas, J. *The Inclusion of the Other: Studies in Political Theory*. Cambridge, MA: MIT Press, 1999.

Handa, C. "Digital Literacy and Rhetoric." *Computers & Composition* 18:2 (2001): 195–202.

Hanks, J. M. *Jacques Ellul: A Comprehensive Bibliography*. Greenwich, CN: JAI Press, 1984.

Haraway, D. *Primate Visions: Gender, Race, and Nature in the World of Modern Science*. New York: Routledge, 1989.

Haraway, D. *Simians, Cyborgs, and Women*. London: Routledge, 1991.

Hård, M. and A. Jamison (eds.). *The Intellectual Appropriation of Technology: Discourses on Modernity, 1900–1939*. Cambridge, MA: MIT Press, 1998.

Harris, C. E., M. S. Pritchard, and M. J. Rabins. *Engineering Ethics: Concepts and Cases*. Belmont, CA, 2000.

Hayles, N. K. *How We Became Post-Human*. Chicago: University of Chicago Press, 1999.

Hayles, N. K. "Deeper into the Machine: Learning to Speak Digital." *Computers & Compostion* 19:4 (2002): 371–86.

Heidegger, M. *Being and Time*. Oxford: Blackwell, 1962.

Heidegger, M. *The Question Concerning Technology and Other Essays*. New York: Harper & Row, 1977.

Heim, M. *The Metaphysics of Virtual Reality*. New York: Oxford U P, 1994.

Heim, M. *Virtual Realism*. New York: Oxford U P, 2000.

Hettinger, N. "Patenting Life: Biotechnology, Intellectual Property, and Environmental Ethics." *Environmental Affairs* 22 (1995): 267–305.

Hickman, L. A. *John Dewey's Pragmatic Technology*. Bloomington: Indiana U P, 1990.

Hickman, L. A. *Philosophical Tools for Technological Culture: Putting Pragmatism to Work*. Bloomington: Indiana University Press, 2001.

Hickman, L. A. (ed.). *Technology as a Human Affair*. New York: McGraw-Hill, 1990.

Hickman, L. A. (ed.). *Reading Dewey*. Bloomington: Indiana U P, 1998.

Hopkins, P. D. (ed.). *Sex/Machine: Readings in Culture, Gender, and Technology*. Bloomington: Indiana University Press, 1998.

Houkes, W. N. and A. W. M. Meijers. "The Ontology of Artefacts: The Hard Problem." *Studies in History and Philosophy of Science* (2006) 118–31.

Hughes, R. *The Shock of the New*. New York: Alfred A. Knopf, 1982.

Hughes, T. P. *Rescuing Prometheus*. New York: Pantheon, 1998.

Hughes, T. P *Human-Built World: How to Think about Technology and Culture*. Chicago: University of Chicago Press, 2005.

Ihde, D. *Technics and Praxis*. Dordrecht: Reidel, 1979.

Ihde, D. *Existential Technics*. Albany: State University of New York Press, 1983.

Ihde, D. *Technology and the Lifeworld: From Garden to Earth*. Indianapolis: Indiana University Press, 1990.

Ihde, D. *Instrumental Realism*. Indianapolis: Indiana University Press, 1991.

Ihde, D. *Philosophy of Technology: An Introduction*. New York: Paragon, 1993.

Ihde, D. with E. Selinger. *Chasing Technoscience*. Indianapolis: Indiana University Press, 2003.

Ingram, D. *Critical Theory and Philosophy*. New York: Paragon House, 1990.

Innis, R. E. "Dewey's Aesthetic Theory and the Critique of Technology." *Phänomenologische Forschungen* (1983) 15, 7–42.

Innis, R. E. "Technics and the Bias of Perception." *Philosophy and Social Criticism* 11:1 (1984): 7–89.

Jacobs, J. *The Life and Death of Great American Cities*. New York: Modern Library, 1993.

Jameson, F. *Postmodernism, or, The Cultural Logic of Late Capitalism*. Durham, NC: Duke University Press, 1991.

Jameson, F. *Archaeologies of the Future*. New York: Verso, 2005.

Johnson, M. and G. Lakoff. *Metaphors We Live By*. Chicago: University of Chicago Press, 1983.

Jonas, H. "Toward a Philosophy of Technology." *Hastings Center Report* 9 (February 1979): 34–43.

Jonas, H. *The Imperative of Responsibility: In Search of an Ethics for the Technological Age*. Chicago: University of Chicago Press, 1984.

Kaplan, D. (ed.). *Readings in Philosophy of Technology*. Lanham, MD: Rowman and Littlefield, 2004.

Kass, L. *Human Cloning and Human Dignity: The Report of the President's Council on Bioethics*. New York: Public Affairs, 2002.

Kass, L. *Life, Liberty, and the Defense of Dignity: The Challenge for Bioethics*. San Francisco: AEI Press, 2004.

Kavoulakos, K. 'Nature, Science, and Technology in Habermas.' *Democracy and Nature* (1998) 112–45.

Kellert, S. R. and E. O. Wilson (eds.). *The Biophilia Hypothesis*. Washington, DC: Island Press, 1993.

Kellner, D. *Jean Baudrillard: From Marxism to Postmodernism and Beyond*. Cambridge: Polity Press, 1989.

Kellner, D. *Television and the Crisis of Democracy*. Boulder, CO: Westview, 1990.

Kellner, D. *Media Culture: Cultural Studies, Identity and Politics Between the Modern and the Postmodern*. New York: Routledge, 1995.

Kellner, D. *Media Spectacle*. New York: Routledge, 2003.

Kellner, D. (ed.), *Baudrillard: A Critical Reader*. Cambridge, MA: Blackwell, 1994.

Kidder, T. *The Soul of a New Machine*. New York: Atlantic-Little, Brown, 1981.

Kitcher, P. *The Lives to Come: The Genetic Revolution and Human Possibilities*. New York: Simon & Schuster, 1996.

Kline, R. and T. Pinch, "Users as Agents of Technological Change: The Social Construction of the Automobile in the Rural United States." *Technology and Culture* (1996) 37: 763–95.

Kling, R. "Audiences, Narratives, and Human Values in Social Studies of Technology." *Science, Technology & Human Values* 17 (1992): 349–65.

Koen, B. V. *Discussion of the Method: Conducting the Engineer's Approach to Problem Solving*. New York: Oxford University Press, 2003.

Kothari, R. "Environment, Technology, and Ethics." In J. R. Engel and J. G. Engel (eds.), *Ethics of Environment and Development: Global Challenge, International Response*. London: Belhaven Press, 1990, pp. 27–35.

Kraft, M. E. and N. J. Vig (eds.). *Technology and Politics*. Durham, NC: Duke University Press, 1988.

Krakauer, E. L. *The Disposition of the Subject: Readings Adorno's Philosophy of Technology*. Evansville, IL: Northwestern University Press, 1998.

Kramer, M. "The Ruination of the Tomato." *Atlantic Monthly* (January 1980).

Kranzberg, M. and C. W. Pursell, Jr. (eds.). *Technology in Western Civilization*, 2 vols. New York: Oxford University Press, 1967.

Kroes, P. A. *Technological Development and Science in the Industrial Age*. Dordrecht: Kluwer Acad. Publishers, 1992.

Kroes, P. A. "Coherence of Structural and Functional Descriptions of Technical Artefacts." *Studies in History and Philosophy of Science* (2006): 137–51.

Lacey, H. "Assessing The Value of Transgenic Crops." *Science and Engineering Ethics* 8:4 (2002): 497–511.

Landauer, T. K. *The Trouble with Computers: Usefulness, Usability, and Productivity*. Cambridge, MA: MIT Press, 1995.

Latour, B. *Science in Action*. Cambridge, MA: Harvard University Press, 1987.

Latour, B. *We Have Never Been Modern*. Cambridge, MA: Harvard University Press, 1993.

Latour, B. *Aramis, or the Love of Technology*. Cambridge, MA: Harvard University Press, 1996.

Latour, B. and S. Woolgar. *Laboratory Life: The Construction of Scientific Facts*. London: Sage, 1986.

Lavery, J. V., E. R. Wahl, C. Grady, and E. J. Emanuel. *Ethical Issues in International Biomedical Research: A Casebook*. New York: Oxford University Press, 2007.

Layton, E. *Technology as Knowledge*. Technology and Culture Vol. 15. 1974.

Le Corbusier. *Towards a New Architecture*. New York: Dover Publications, 1985.

Lefebvre, Henri. *The Production of Space*. Malden, MA: Wiley-Blackwell, 1992.

Levidow, W. and L. Robins. *Cyborg Worlds: The Military Information Society*. London: Free Association Books, 1989.

Lie, M. and K. Sørenson (eds.). *Making Technology our Own? Domesticating Technology into Everyday Life*. Oslo: Scandinavian University Press, 1996.

Light, A. and A. de-Shalit. *Moral and Political Reasoning in Environmental Practice*. Cambridge, MA: Harvard University Press, 2003.

Light, A. and E. Katz (eds.). *Environmental Pragmatism*. New York: Routledge, 1996.

Light, A. and E. McKenna (eds.). *Animal Pragmatism: Rethinking Human-Nonhuman Relationships*. Bloomington: Indiana University Press, 2004.

Light, A. and H. Rolston, III (eds.). *Environmental Ethics: An Anthology*. Cambridge, MA: Blackwell, 2003.

Light, A. and J. M. Smith. *Philosophies of Place*. Lanham, MD: Rowman and Littlefield, 1999.

Light, A. and J. M. Smith. *The Aesthetics of Everyday Life*. New York: Columbia University Press, 2005.

Light, A., E. Higgs, and D. Strong (eds.). *Technology and the Good Life?* Chicago: University of Chicago Press, 2000.

Light, A., E. Katz, and D. Rothenberg (eds.). *Beneath the Surface: Critical Essays on the Philosophy of Deep Ecology*. Cambridge, MA: MIT Press, 2000.

Light, A., E. Katz, and W. Thompson (eds.). *Controlling Technology*, 2nd edn. Amherst, NY: Prometheus Books, 2003.

Lovekin, David. *Technique, Discourse, and Consciousness: An Introduction to the Philosophy of Jacques Ellul*. Bethlehem, PA: Lehigh University Press, 1991.

Lovelock, J. *A New Look at Life on Earth*. New York: Oxford University Press, 1987.

Lowrance, W. W. *Of Acceptable Risk: Science and the Determination of Safety*. New York: William Kaufmann, Inc., 1976.

Lowrance, W. W. *Modern Science and Human Values*. New York: Oxford University Press, 1985.

Lyotard, J.-F. *The Postmodern Condition: A Report on Knowledge*. Minneapolis, MN: University of Minnesota Press, 1984.

MacKenzie, D. *Inventing Accuracy: A Historical Sociology of Nuclear Missile Guidance*. Cambridge, MA: MIT Press, 1990.

MacKenzie, D. and J. Wajcman (eds.). *The Social Shaping of Technology*. Milton Keynes: Open University Press, 1985.

MacKenzie, N. R. *Science and Technology Today: Readings for Writers*. New York: St Martin's Press, 1995.

Mahowald, M. *Genes, Women, Equality*. New York: Oxford University Press, 2000.

Mahowald, M., A. Silvers, and D. Wasserman. *Disability, Difference, Discrimination: Perspectives on Justice in Bioethics and Public Policy*. Lanham, MD: Rowman and Littlefield, 1998.

Mahowald, M., et al. (eds.). *Genetics in the Clinic: Clinical, Ethical, and Social Implications for Primary Care*. St. Louis: Mosby, 2001.

Mander, J. *Four Arguments for the Elimination of Television*. New York: Harper, 1978.

Mander, J. *In the Absence of the Sacred: The Failure of Technology and the Survival of the Indian Nations*. San Francisco, CA: Sierra Club, 1992.

Mannheim, K. *Ideology and Utopia*, trans. L. Wirth and E. Shils. London: Routledge & Kegan Paul, 1936.

Mannheim, K. *Freedom, Power, and Democratic Planning*. New York: Oxford University Press, 1950.

Marcuse, H. *One Dimensional Man: Studies in the Ideology of Advanced Industrial Society*. Boston: Beacon Press, 1968.

Marcuse, H. *Negations*. Boston: Beacon Press, 1968.

Marcuse, H. *Eros and Civilization*. Boston: Beacon Press, 1969.

Marcuse, H. *An Essay on Liberation*. Boston: Beacon Press, 1969.

Marcuse, H. *Counter-Revolution and Revolt*. Boston: Beacon Press, 1972.

Marcuse, H. *The Aesthetic Dimension*. Boston: Beacon Press, 1978.

Marcuse, H. "Some Implications of Modern Technology." In A. Arato and E. Gebhart (eds.). *The Essential Frankfurt School Reader*. New York: Continuum, 1990.

Martin, M. and R. Schinzinger. *Ethics in Engineering*. New York: McGraw-Hill, 1996.

Marx, K. *Economic and Philosophic Manuscripts of 1844*. New York: International Publishers, 1964.

Marx, K. *A Contribution to the Critique of Political Economy*. New York: International Publishers, 1970.

Marx, K. *Grundrisse*. New York: Vintage, 1973.

Marx, K. *Capital*, 3 vols. New York: Penguin, 1992–3.

McCarthy, J. and P. Wright, *Technology as Experience*. Cambridge, MA: MIT Press, 2007.

McClellen, J. E., and H. Dorn. *Science and Technology in World History: An Introduction*. Baltimore: The Johns Hopkins University Press, 2006.

McDermott, J. J. *The Culture of Experience: Philosophical Essays in the American Grain*. New York: New York University Press, 1976.

McDermott, J. J. *Streams of Experience: Reflections on the History and Philosophy of American Culture*. Amherst: University of Massachusetts Press, 1986.

Mcdonough, W. and M. Braungart. *Cradle to Cradle: Remaking the Way We Make Things*. New York: North Point Press, 2002.

McGee, T. and P. Ericsson. "The Politics of the Program: MS WORD as the Invisible Grammarian." *Computers & Composition* 19.4 (2002): 453–70.

McGinn, R. *Science, Technology and Society*. New York: Prentice Hall, 1990.

McKenna, E. *The Task of Utopia*. Lanham, MD: Rowman and Littlefield, 2002.

McKibben, B. *Deep Economy: The Wealth of Communities and the Durable Future*. New York: Holt Paperbacks, 2008.

McLaughlin, Andrew. *Regarding Nature: Industrialism and Deep Ecology*. New York: State University of New York Press, 1993.

McLennan, J. F. *The Philosophy of Sustainable Design*. Bainbridge Island, WA: Ecotone Publishing Company LLC, 2004.

McLuhan, M. *The Gutenberg Galaxy*. New York: New American Library, 1969.

McLuhan, M. *Understanding Media*. Cambridge, MA: MIT Press, 1994.

Mead, G. H. *Mind, Self, and Society*. Chicago: University of Chicago Press, 1934.

Meadows, D. L., D. H. Meadows, and J. Randers. *Beyond the Limits: Confronting Global Collapse: Envisioning a Sustainable Future*. Post Mills, VT: Chelsea Green, 1992.

Mesthene, E. *Technological Change: Its Impact on Man and Society*. Cambridge, MA: Harvard University Press, 1970.

Misa, T., P. Brey, and A. Feenberg. *Modernity and Technology*. Cambridge, MA: MIT Press, 2003.

Mitcham, C. *Philosophy and Technology*. New York: Free Press, 1983.

Mitcham, C. *Thinking Through Technology*. Chicago: University of Chicago Press, 1994.

Mitcham, C. *Thinking Ethics in Technology*. Golden, CO: Colorado School of the Mines Press, 1997.

Mitcham, C. *Engineer's Toolkit: Engineering Ethics*. New York: Prentice Hall, 1999.

Mitcham, C. *Technological Reflection on Campus*. Lanham, MD: Rowman and Littlefield, 2004.

Mitcham, C. (ed.). *Research in Philosophy and Technology*, vol. 15: *Social and Philosophical Constructions of Technology*. Greenwich, CT: JAI Press, 1995.

Mitcham, C. (ed.). *Encyclopedia of Science, Technology and Ethics*, New York: Macmillan Reference, 2005.

Mitcham, C. and J. Grote. *Theology and Technology*. Lanham, MD: Rowman and Littlefield, 1984.

Moore, S. *Technology and Place: Sustainable Architecture and the Blueprint Farm.* Austin, TX: University of Texas Press, 2001.

Moore, S. *Alternative Routes to the Sustainable City: Austin, Curitiba and Frankfurt.* Lanham, MD: Rowman & Littlefield/Lexington, 2007.

Moore, S. and S. Guy (eds.). *Sustainable Architectures: Natures and Cultures in Europe and North America.* London: Routledge, 2005.

Moser, M. A. with D. McLeod. *Immersed in Technology: Art and Virtual Environments.* Cambridge, MA: MIT Press, 1996.

Mumford, L. *Technics and Civilization.* New York: Harcourt Brace, 1934.

Mumford, L. *Technics and Civilization.* New York: Harcourt Brace Jovanovich, 1961.

Mumford, L. *The Myth of the Machine.* New York: Harcourt Brace, 1967, 1970.

Mumford, L. *Art and Technics.* New York: Columbia U P, 2000/1952.

Murphie, A. and J. Potts. *Culture and Technology.* New York: Palgrave Macmillan, 2003.

Naess, A. "The Shallow and the Deep, Long-Range Ecology Movements: A Summary." *Inquiry* 1:16 (1973).

Naess, A. *Ecology, Community and Lifestyle,* trans. and rev. David Rothenberg. New York: Cambridge University Press, 1989.

Naess, A. "The Third World, Wilderness, and Deep Ecology." in G. Sessions (ed.). *Deep Ecology for the 21st Century.* Boston: Shambala, 1995, 397–407.

Negroponte, N. *Being Digital.* London, Hodder & Stoughton, 1995.

Nelkin, D. *Controversy: Politics of Technological Change.* London: Sage, 1984.

Nelson, M. P. and J. B. Callicott (eds.). *The Wilderness Debate Rages On.* Athens, GA: University of Georgia Press, 2007.

Noble, D. *America by Design: Science, Technology, and the Rise of Corporate Capitalism.* New York: Knopf, 1977.

Noble, D. *Forces of Production: A Social History of Industrial Automation.* New York: Knopf, 1984.

Norman, D. *Things that Make Us Smart: Defending Human Attributes in the Age of the Machine.* New York: Addison-Wesley Publishing Company, 1993.

Norman, D. A. *The Design of Everyday Things.* New York: Basic Books, 2002.

Norman, D. A. *The Design of Future Things.* New York: Basic Books, 2007.

Nunberg, G. *The Future of the Book.* Berkeley, CA: University of California Press, 1996.

Nye, D. E. *Narratives and Spaces: Technology and the Construction of American Culture.* Exeter: University of Exeter Press, 1997.

Nye, D. E *Technology Matters: Questions to Live With.* Cambridge, MA: MIT Press, 2007.

Oldenziel, R. *Making Technology Masculine: Men, Women and Modern Machines in America, 1870–1945.* Amsterdam: Amsterdam University Press, 1999.

Olsen, J.-K.B. and E. Selinger. *Philosophy of Technology.* Automatic Press, 2007.

Olsen, J.-K.B., E. Selinger. and S. Riis. *New Waves in Philosophy of Technology.* New York: Palgrave Macmillan, 2008.

Ormiston, G. L. *From Artifact to Habitat.* Bethlehem: Lehigh University Press, 1990.

Ortega y Gasset, J. *History as a System and Other Essays Toward a History of Philosophy.* New York: W. W. Norton & Co., 1961.

Pacey, A. *Technology in World Civilization.* Cambridge, MA: MIT Press, 1990.

Pacey, A. *The Culture of Technology.* Cambridge, MA: MIT Press, 1983.

Pacey, A. *Meaning in Technology.* Cambridge, MA: MIT Press, 1999.

Parkin, A. "On the Practical Relevance of Habermas's Theory of Communicative Action." *Social Theory and Practice* 22:3 (Fall 1996): 417–42.

Pence, G. *Flesh of My Flesh: The Ethics of Human Cloning.* Lanham, MD: Rowman and Littlefield, 1998.

Pence, G. *Who's Afraid of Human Cloning?* Lanham, MD: Rowman and Littlefield, 1998.

Pence, G. *The Ethics of Food.* Lanham, MD: Rowman and Littlefield, 2002.

Pence, G. *Designer Food.* Lanham, MD: Rowman and Littlefield, 2002.

Pence, G. *Brave New Bioethics.* Lanham, MD: Rowman and Littlefield, 2003.

Petroski, H. *The Pencil: A History of Design and Circumstance.* New York: Alfred A, Knopf, 1989.

Petroski, H. *To Engineer is Human.* New York: Vintage, 1992.

Petroski, H. *The Evolution of Useful Things.* New York: Vintage, 1994.

Petroski, H. *Invention by Design.* Cambridge, MA: Harvard University Press, 1998.

Petroski, H. *Small Things Considered: Why There is No Perfect Design.* New York: Knopf, 2003.

Pirsig, R. M. *Zen and the Art of Motorcycle Maintenance: An Inquiry into Values.* New York: Bantam, 1975.

Pitt, J. "On The Philosophy of Technology, Past and Future." *Techne* 1:1–2 (Fall 1995).

Pitt, J. *Thinking About Technology: Foundations of the Philosophy of Technology.* New York: Seven Bridges Press, 1999.

Plumwood, V. *Gender and Ecology: Feminism and the Mastery of Nature.* London: Routledge, 1992.

Pollan, M. *Second Nature: A Gardener's Education*. New York: Atlantic Monthly Press, 1991.

Pollan, M. *Place of My Own: The Education of an Amateur Builder*. New York: Random House, 1997.

Pollan, M. *The Botany of Desire: A Plant's-Eye View of the World*. New York: Random House, 2001.

Pollan, M. *The Omnivore's Dilemma: A Natural History of Four Meals*. New York: Penguin Press, 2006.

Pollan, M. *In Defense of Food: An Eater's Manifesto*. New York: Penguin Press, 2008.

Postman, N. *Technopoly: The Surrender of Culture to Technology*. New York: Vintage/Random House, 1992.

Purdy, L. *Reproducing Persons: Issues in Feminist Bioethics* Ithaca, New York: Cornell University Press, 1996.

Pursell, C. "The Rise and Fall of the Appropriate Technology Movement in the United States, 1961–1985." *Technology and Culture* 34 (1993): 629–37.

Radder, H. "Normative Reflexions on Constructivist Approaches to Science and Technology." *Social Studies of Science* 22 (1992): 141–73.

Radder, H. *In and About the World: Philosophical Studies of Science and Technology*. Albany: SUNY Press, 1996.

Rapp, F. (ed.). *Contributions to a Philosophy of Technology*. Dordrecht: Kluwer, 1974.

Reingold, H. *The Virtual Community*. New York: Harper, 1994.

Rip, A., T. J. Misa, and J. Schot (eds.). *Managing Technology in Society: The Approach of Constructive Technology Assessment*. London: Pinter Publishers, 1995.

Robert, J. S. "The Science and Ethics of Making Part-Human Animals in Stem Cell Biology." *FASEB Journal* 20:7 (2006): 838–45.

Rochlin, G. I. *Trapped in the Net: The Unanticipated Consequences of Computerization*. Princeton: Princeton University Press, 1997.

Roeser, S. and L. Asveld (eds.). *The Ethics of Technological Risk*. London: Earthscan Publishers, 2008.

Rolston III, H. *Philosophy Gone Wild: Essays in Environmental Ethics*. Buffalo, NY: Prometheus, 1986.

Romanyshyn, R. D. *Technology as Symptom and Dream*. New York: Routledge, 1989.

Rosen, P. "The Social Construction of Mountain Bikes: Technology and Postmodernity in the Cycle Industry." *Social Studies of Science* 23 (1993): 479–513.

Ross, A. *Strange Weather: Culture, Science, and Technology in an Age of Limits*. London: Verso, 1991.

Rothman, B. K. *The Book of Life: A Personal and Ethical Guide to Race, Normality, and the Implications of the Human Genome Project*. Boston: Beacon Press, 2001.

Rothschild, J. (ed.). *Machina ex Dea: Feminist Perspectives on Technology*. Oxford: Pergamon Press, 1983.

Rowell, A. *Green Backlash: Global Subversion of the Environmental Movement*. New York: Routledge, 1996.

Rutsky, R. L. *High Techne: Art and Technology from the Machine Aesthetic to the Posthuman*. Minneapolis: University of Minnesota Press, 1999.

Ryan, M. L. *Cyberspace Textuality: Computer Technology and Literary Theory*. Bloomington, IN: Indiana U P, 1999.

Sandel, M. *The Case against Perfection: Ethics in the Age of Genetic Engineering*. Cambridge, MA: Harvard University Press, 2007.

Sassower, R. *Cultural Collisions: Postmodern Technoscience*. New York: Routledge, 1995.

Sassower, R. *Technoscientific Angst: Ethics and Responsibility*. Minneapolis: University of Minnesota Press, 1997.

Sassower, R. *Confronting Disaster: An Existential Approach to Technoscience*. Lanham, Boulder, New York and Oxford: Lexington Books, 2004.

Sassower, R. and M. A. Cutter. *Ethical Choices for Contemporary Medicine: Integrative Bioethics*. Montreal, CA: McGill-Queen's University Press, 2008.

Sassower, R. and G. L. Ormiston. *Narrative Experiments: The Discursive Authority of Science and Technology*. Minneapolis, MN: University of Minnesota Press, 1989.

Savulescu, J. "Procreative Beneficence: Why we Should Select the Best Children." *Bioethics* 15 (2001): 413–26.

Savulescu, J. "New Breeds of Humans: the Moral Obligation to Enhance." *Reproductive Biomedicine Online* 10:Suppl 1 (March 2005): 36–9.

Savulescu, J. "Justice, Fairness, and Enhancement." *Annals of the New York Academy of Science* 1093 (Dec. 2006): 321–38.

Savulescu, J. "In Defence of Procreative Beneficence." *Journal of Medical Ethics* 33:5 (May 2007): 284–8.

Scharff, R. S. and Dusek, V. (eds.). *Philosophy of Technology: The Technological Condition*. Oxford: Blackwell, 2003.

Sclove, R. *Democracy and Technology*. New York: Guilford Press, 1995.

Schumacher, F. *Small is Beautiful: A Study of Economics as if People Mattered*. London: Abacus, 1974.

Segal, H. P. *Technological Utopianism in American Culture*. Chicago: University of Chicago Press, 1985.

Sessions, G. (ed.). *Deep Ecology for the 21st Century: Readings on the Philosophy and Practice of the New Environmentalism*. Boston: Shambhala, 1995.

Sherlock, R. and J. D. Morrey (eds.). *Ethical Issues in Biotechnology*. Lanham, MD: Rowman and Littlefield, 2002.

Shneiderman, B. *Leonardo's Laptop: Human Needs and the New Computing Technologies*. Cambridge, MA: MIT Press, 2003.

Shrader-Frechette, K. *Environmental Ethics*. Pacific Grove, CA: Boxwood Press, 1991.

Shrader-Frechette, K. *Risk and Rationality*. Berkeley, CA: University of California Press, 1991.

Shrader-Frechette, K. *Environmental Justice: Creating Equality, Reclaiming Democracy*. New York: Oxford, 2002.

Shrader-Frechette, K. and E. D. Mccoy. *Method in Ecology*. Cambridge: Cambridge University Press, 1993.

Shrader-Frechette, K. and L. Westra (eds.). *Technology and Values*. Lanham, MD: Rowman and Littlefield, 1997.

Simpson, L. C. *Technology, Time, and the Conversations of Modernity*. New York: Routledge, 1995.

Smith, M. R. and L. Marx. *Does Technology Drive History? The Dilemma of Technological Determinism*. Cambridge, MA: MIT Press, 1994.

Snow, C. P. *The Two Cultures*. Cambridge: Cambridge University Press, 1993.

Soper, K. "Feminism and Ecology: Realism and Rhetoric in the Discourses of Nature." *Science, Technology & Human Values* 20 (1995): 311–31.

Stone, A. R. *The War of Desire and Technology at the Close of the Mechanical Age*. Cambridge, MA: MIT Press, 1995.

Suzuki, D. and P. Knudtson. *Genethics: The Clash Between the New Genetics and Human Values*. Cambridge, MA: Harvard University Press, 1989.

Tabachinick, D. and T. Koivukoski. *Globalization, Technology, and Philosophy*. Albany, NY: SUNY Press, 2004.

Tenner, E. *Why Things Bite Back: Technology and the Revenge of Unintended Consequences*. New York: Vintage, 1997.

Thompson, P. B. *The Ethics of Aid and Trade*. Cambridge: Cambridge University Press, 1992.

Thompson, P. B. *The Spirit of the Soil*. New York: Routledge, 1994.

Thompson, P. B. *Food Biotechnology in Ethical Perspective*. Dordrecht: Kluwer, 1997.

Thompson, P. B. *Agricultural Ethics: Research, Teaching and Public Policy*. Ames, IA: Iowa State University Press, 1998.

Thompson, P. B. and K. David. *What Can Nanotechnology Learn From Biotechnology? Social and Ethical Lessons for Nanoscience from the Debate over Agrifood Biotechnology and GMOs*. Burlington, MA: Academic Press, 2008.

Tichi, C. *Shifting Gears: Technology, Literature, Culture in Modernist America*. Chapel Hill: University of North Carolina Press, 1987.

Toffler, A. *Future Shock*. New York: Random House, 1970.

Toffler, A. *The Third Wave*. New York: Morrow, 1980.

Toulmin, S. *Cosmopolis: The Hidden Agenda of Modernity*. New York: Free Press, 1990.

Turkle, S. *The Second Self: Computers and the Human Spirit*. New York: Simon and Schuster, 1984.

Turkle, S. *Life On Screen: Identity in the Age of the Internet*. New York: Simon and Schuster, 1995.

Unger, S. *Controlling Technology*. New York: John Wiley and Sons, 1994.

van den Hoven, M. J. and J. Weckert (eds.). *Information Technology and Moral Philosophy*. Cambridge: Cambridge University Press, 2005.

van de Poel, I. R. "Ethics in Engineering Practice." In *Philosophy in Engineering*, edited by S. Hylgaard Christensen, M. Meganck and B. Delahousse. Aarhuus: Academica, 2007.

van de Poel, I. R., and A. C. van Gorp. "Degrees of Responsibility in Engineering Design: Type of Design and Design Hierarchy." *Science, Technology & Human Values* 31:3 (2006): 333–60.

van de Poel, I. R., and P. A. Kroes, "Technology and Normativity." *Techne* (2006): 1–6.

van de Poel, I. R., and P. P. C. C. Verbeek, "Ethics and Engineering Design." *Science, Technology and Human Values* 31:3 (2006): 223–36.

van Lente, H. *Promising Technologies*. Delft: Eburon, 1993.

van Wyck, P. C. *Primitives in the Wilderness: Deep Ecology and the Missing Human Subject*. Albany, NY: State University of New York Press, 1997.

Veblen, T. *The Theory of the Leisure Class*. New York: Mentor, 1953.

Veblen, T. *The Engineers and the Price System*. New Brunswick, NJ: Transaction Publishers, 1983.

Verbeek, P.-P. *What Things Do: Philosophical Reflections on Technology, Agency, and Design*, trans. Robert P. Crease. University Park, PA: Pennsylvania University Press, 2005.

Verbeek, P.-P. and A. Slob (eds.). *User Behavior and Technology Development*. Dordrecht: Springer, 2006.

Vermaas, P. E. "The Physical Connection: Engineering Function Ascriptions to Technical Artefacts and their Components." *Studies in History and Philosophy of Science* 37 (2006): 62–75.

Vermaas, P. E., P. Kroes, A. Light, and S. A. Moore (eds.). *Philosophy and Design: From Engineering to Architecture*. Dordrecht: Springer, 2008.

Vincenti, W. G. *What Engineers Know and How They Know It*. Baltimore: Johns Hopkins University Press, 1990.

Volti, R. *Society and Technological Change.* New York: St Martin's Press, 1992.

Wajcman, J. *Feminism Confronts Technology.* State College, PA: Penn State University Press, 1991.

Walker, S. *Sustainable by Design: Explorations in Theory and Practice.* London: Earthscan Publications Ltd, 2006.

Waring, S. *Taylorism Transformed: Scientific Management Theory Since 1945.* Chapel Hill, NC: University of North Carolina Press, 1991.

Weber, M. *General Economic History.* New York: Greenberg Press, 1927.

Weber, M. *Theory of Social and Economic Organization.* Glencoe, IL: Free Press, 1957.

Weber, M. *The Protestant Ethic and the Spirit of Capitalism.* New York: Scribner's, 1958.

Webster, F. *Theories of the Information Society.* London: Routledge, 1995.

Wenz, P. S. *Environmental Justice.* New York: SUNY Press, 1988.

Wenz, P. S. *Abortion Rights as Religious Freedom.* Philadelphia: Temple University Press, 1992.

Wenz, P. S. *Nature's Keeper.* Philadelphia: Temple University Press, 1996.

Wenz, P. S. *Environmental Ethics Today.* Oxford: Oxford University Press, 2001.

Wenz, P. "Just Garbage." In L. Westra and B. E. Lawson (eds.), *Faces of Environmental Racism.* Lanham, MD: Rowman and Littlefield, 2001, pp. 57–71.

Wenz, P. S. and L. Westra (eds.). *Faces of Environmental Racism.* Lanham MD: Rowman and Littlefield, 1995.

Whitbeck, C. *Ethics in Engineering Practice and Research.* Cambridge: Cambridge University Press, 1998.

White, L. Jr., *Medieval Technology and Social Change.* London: Oxford University Press, 1962.

Whitehead, A. N. *Science and the Modern World.* New York: Free Press, 1997.

Widdows, H. "Is Global Ethics Moral Neo-colonialism? An Investigation of the Issue in the Context of Bioethics." *Bioethics* 21:6 (July 2007).

Winner, L. *Autonomous Technology: Technics Out-of-Control as a Theme in Political Thought.* Cambridge, MA: MIT Press, 1977.

Winner, L. "Do Artifacts Have Politics?" *Daedalus* (1980) 109, 121–36.

Winner, L. *The Whale and the Reactor: A Search for Limits in an Age of High Technology.* Chicago: Univ. of Chicago Press, 1986.

Winner, L. "Upon Opening the Black Box and Finding it Empty: Social Constructivism and the Philosophy of Technology." In J. Pitt, and E. Lugo (eds.), *The Technology of Discovery and the Discovery of Technology.* Blacksburg, VA: Society for Philosophy and Technology, 1991.

Winner, L. "Reply to Mark Elam." *Science, Technology, & Human Values* 19 (1994): 107–9.

Winner, L. "Technology Today: Utopia or Dystopia?" *Social Research* 64:3 (1997): 989–1017.

Winner, L. (ed.). *Democracy in a Technological Society.* Dordrecht: Kluwer, 1992.

Woolgar, S. "The Turn to Technology in Social Studies of Science." *Science, Technology & Human Values* 16 (1991): 20–50.

Woolgar, S. "What's at Stake in the Sociology of Technology? A Reply to Pinch and Winner." *Science, Technology, & Human Values* 18 (1993): 523–9.

Woolliams, J. "Designing Cities and Buildings as if They Were Ethical Choices." In D. Schmidtz and E. Willott (eds.), *Environmental Ethics.* New York: Oxford University Press, 2000, pp. 426–30.

Wright, B. D. (ed.). *Women, Work, and Technology.* Ann Arbor, MI: University of Michigan Press, 1987.

Zimdahl, R. L. *Agriculture's Ethical Horizon.* Burlington, MA: Academic Press, 2006.

Zimmerman, M. E. *Heidegger's Confrontation with Modernity: Technology, Politics, Art.* Bloomington, IN: Indiana University Press, 1990.

Zuboff, S. *In The Age of the Smart Machine.* New York: Basic Books, 1989.

Zumthor, P. *Thinking Architecture,* 2nd edn. Basel, Switzerland: Birkhäuser Basel, 2006.